Advanced Energy Harvesting Technology—Volume I

Advanced Energy Harvesting Technology—Volume I

Topic Editors

Mengying Xie
Kean C. Aw
Junlei Wang
Hailing Fu
Wee Chee Gan

MDPI

Basel • Beijing • Wuhan • Barcelona • Belgrade • Novi Sad • Cluj • Manchester

Topic Editors

Mengying Xie
School of Precision Instruments and Opto-Electronics Engineering
Tianjin University
Tianjin
China

Kean C. Aw
Department of Mechanical Engineering
University of Auckland
Auckland
New Zealand

Junlei Wang
School of Mechanical and Power Engineering
Zhengzhou University
Zhengzhou
China

Hailing Fu
School of Automation
Beijing Institute of Technology
Beijing
China

Wee Chee Gan
School of Energy and Chemical Engineering
Xiamen University Malaysia
Bandar Sunsuria
Malaysia

Editorial Office
MDPI AG
Grosspeteranlage 5
4052 Basel, Switzerland

This is a reprint of the Topic, published open access by the journals *Applied Sciences* (ISSN 2076-3417), *Energies* (ISSN 1996-1073), *Materials* (ISSN 1996-1944), *Micromachines* (ISSN 2072-666X) and *Sensors* (ISSN 1424-8220), freely accessible at: https://www.mdpi.com/topics/energy_harvesting.

For citation purposes, cite each article independently as indicated on the article page online and as indicated below:

Lastname, A.A.; Lastname, B.B. Article Title. *Journal Name* **Year**, *Volume Number*, Page Range.

Volume I
ISBN 978-3-7258-5635-0 (Hbk)
ISBN 978-3-7258-5636-7 (PDF)
https://doi.org/10.3390/books978-3-7258-5636-7

Set
ISBN 978-3-7258-5541-4 (Hbk)
ISBN 978-3-7258-5542-1 (PDF)

Contents

About the Editors

Mengying Xie

Mengying Xie is currently an Associate Professor at School of Precision Instrument and Opto-electronics Engineering, Tianjin University, China. She received her bachelor's degree in applied physics from Tianjin University, China, in 2010, and received the Ph.D. degree in Mechanical Electronic Engineering from the University of Auckland, Auckland, New Zealand, in 2015. From 2016 to 2018, Dr. Xie worked as a Postdoctoral Researcher at the University of Bath, Bath, U.K., and from 2018 to 2019 as a Senior Researcher at the Ritsumeikan University, Kyoto, Japan. She is Vice Chair of the IEEE Industrial Electronic Society Technical Committee on MEMS. Her research interests include the flexible tactile sensors, flexible multifunctional sensors, energy harvesters, and their application in wearable devices and soft robotics.

Kean C. Aw

Kean C. Aw was born in Georgetown, Malaysia and is a Professor at the Department of Mechanical and Mechatronics Engineering at the University of Auckland. He joined the university in 2004. Prior to this academic position, he worked at Intel, Delta Prisma, Altera and Navman for a total of 15 years. He earned his undergraduate degree in Electrical and Electronics Engineering, M.Sc in Advanced Manufacturing Systems and PhD in Applied Physics (Semiconductor Physics). His main interests are in micro-systems and deployment of smart/functional materials and structures such as conducting polymers, metallic oxides, etc., which act as sensors and actuators in various applications, such as bio-sensors, medical/rehabilitation robots, micro-pumps, micro-manipulators, MEMS, energy harvesters, etc.

Junlei Wang

Junlei Wang is a Professor at the School of Mechanical and Power Engineering, Zhengzhou University. His research focuses on flow-induced vibration and piezoelectric energy harvesting, with particular emphasis on self-powered sensing in air, ground, and marine environments, as well as vibration suppression and high-efficiency energy harvesting mechanisms. He has been recognized as a Henan Provincial Outstanding Young Talent and a Science and Technology Innovation Talent by the Department of Education of Henan Province. He was also named among Elsevier's Highly Cited Chinese Researchers in 2023–2024 and included in Stanford University's lifetime list of the world's top 2% scientists. Professor Wang has led multiple national and provincial research projects, including three funded by the National Natural Science Foundation of China. He has published more than 70 papers in leading journals and IEEE/ASME Transactions as first or corresponding author, with nine ESI highly cited papers and two hot papers, accumulating 7,342 citations on Google Scholar. His contributions have been recognized with the First Prize of the Science and Technology Award (Basic Research) from the Chinese Society for Vibration Engineering, the Second Prize of the Science and Technology Progress Award from the China Petroleum and Chemical Industry Federation, and the Second Prize of the Henan Provincial Science and Technology Progress Award. His academic contributions have been recognized with several prestigious awards, including the First Prize of the Science and Technology Award (Basic Research) from the Chinese Society for Vibration Engineering, the Second Prize of the Science and Technology Progress Award from the China Petroleum and Chemical Industry Federation, and the Second Prize of the Henan Provincial Science and Technology Progress Award. With sustained contributions in theoretical research, experimental innovation, and practical applications.

Hailing Fu

Hailing Fu is a Professor of Autonomous and Intelligent Perception at Beijing Institute of Technology (BIT)since 2023. He gained his PhD at Imperial College London (ICL) in 2018 and then worked as a research associate in the Structural Integrity and Heath Monitoring Group at ICL, and a lecturer at Loughborough University from 2019 to 2022, before he joined BIT as a professor. He is a Senior Member of IEEE and Chair of the Industrial Electronic Society Technical Committee on MEMS and nanotechnology and serviced as the TPC member for PowerMEMS and Transducers conferences. His research interests include active acoustic sensing, wireless sensors, piezoelectric transducers, energy harvesters and self-powered sensing. Dr. Fu has acted as an Associate Editor for *IEEE Sensors Journal* and *Journal of Intelligent Manufacturing*, a member of the Advisory Editorial Board for *Hybrid Advances* (Elsevier) and *Next Energy* (Elsevier), and a Guest Editor for multiple top-tier journals including *Smart Materials and Structures*. He has published more than 70 publications and filed 10 China and UK patents with Outstanding Paper Finalist Awards in PowerMEMS 2019 and Best Paper Awards in VEH 2025.

Wee Chee Gan

Wee Chen Gan received his Ph.D. in Advanced Materials from the University of Malaya. He is currently working as an Associate Professor at the School of Energy and Chemical Engineering, Xiamen University Malaysia, with an adjunct appointment at the College of Chemistry and Chemical Engineering, Xiamen University. He is also the Head of Programme of New Energy Science and Engineering (Undergraduate Degree). His research focuses on developing self-powered, sustainable energy harvesters using piezoelectric and triboelectric nanogenerators, while investigating the electroactive properties of smart materials for advanced energy generation and sensing technologies.

micromachines

Article

An Investigation on Energy Harvesting Behavior of an Array Piezoelectric Coupled Disc Damper

Xiangdong Xie *, Xunnan Huang, Junjie Wang, Zijing Wang, Bowen Zhou and Jiankun Zhang

School of Urban Construction, Yangtze University, Jingzhou 434000, China; 15027174591@163.com (X.H.); wangjunjie4316@163.com (J.W.); 15071562521@163.com (Z.W.); z2443353863@163.com (B.Z.); zjk9709@163.com (J.Z.)

* Correspondence: xdxie@yangtzeu.edu.cn

Abstract: In order to make full use of the vibration energy in the process of attenuating vibration, an array piezoelectric coupled disc damper is developed, which works by converting part vibration energy into electrical energy. The piezoelectric damper is made of a pair of piezoelectric coupled discs built in a case cylinder. Its energy harvesting behavior is studied by a series of forced-vibration experiments and simulations. The influences of some factors, such as the excitation frequency, substrate thickness, the size of the piezoelectric patch, the paste form of the piezoelectric patch and the load resistance, on the energy harvesting behavior of the damper are analyzed and concluded. The experimental results show that the maximum peak-to-peak voltage and average power from one piezoelectric patch with an inner diameter of 35 mm, an outer diameter of 80 mm, and a thickness of 1 mm can reach up to 163 V and 161 mW, respectively. This research provides a practical piezoelectric damper attenuating harmful vibration by converting them into useful electric power, and the corresponding theoretical models are derived to predict its electrical output.

Keywords: piezoelectric damper; excitation frequency; energy harvesting behavior; peak-to-peak voltage; piezoelectric coupled disc

1. Introduction

Due to the action of energy dissipation and vibration absorption, dampers are widely used in mechanical manufacture, civil engineering, automotive industry, and other fields. The traditional dampers, usually including liquid dampers, friction dampers, and regenerative dampers, have been researched and applied widely in different fields for many years [1–4].

Liquid damper, for its low initial capital, easy installation, almost no maintenance, and easy to match the tuned frequency, has been studied and used in various devices, such as bridges [5], buildings [6–8], wind turbines [9] and so on. Wang et al. [10] proposed a damping system by embedding tuned liquid multiple column damper into reinforced concrete shear wall systems to improve the energy dissipation capabilities of traditional structural component. Rozas et al. [11] presented a bidirectional tuned liquid column damper to reduce the seismic response of buildings. Zhang et al. [12] evaluated the performance of a full-scale tuned liquid damper in mitigating lateral tower vibrations of multi-megawatt wind turbines. In addition, Sarkar and Chakraborty [13], Zhang and Høeg [14] and Chen et al. [15] developed different liquid dampers, which all displayed a good effect in controlling the vibration. However, the thing with liquid damper is, its dissipated damping is often affected by the environment temperature. In addition, in order to enhance the robustness, multiple viscous liquid dampers are needed at the same time, which causes the complicated structure and the higher cost. Therefore, researchers devoted their interest to friction dampers with a steady property and simple structure. Latour et al. [16] proposed a removable friction damper for steel beam-to-column joints to improve the moment resisting ability of the steel frames. Ghorbani and Rofooei [17] developed a novel double slip loads friction damper to control the seismic response of structures, which can effectively reduce the seismic response of the structures compared with the traditional single slip load friction damper. Furthermore, Pardo-Varela and Llera [18], and Chen et al. [19,20] also provided their respective friction damper to reduce the seismic response of the structures in different aspects. All dampers aforesaid can achieve a high level of vibration dissipation. However, they cannot transfer the vibration energy into electric power for recycling applications. In view of this, a series of regenerative dampers are developed, which have a good efficiency in dissipating vibration while achieving an efficient energy harvesting by electromagnetic technology. Li et al. [21] modeled two types of rotational electromagnetic regenerative dampers used for vehicle suspension. The research results show that both dampers can make vehicles more comfortable in the driving process. Meanwhile, the dampers can generate considerable electric power outputs for the reuse of vehicles. Li et al. [22], Gao et al. [23], Sung and Choi [24], respectively, developed their electromagnetic dampers and applied them to the suspension of the vehicle. In addition, Li et al. [25] and Jung et al. [26] fabricated their electromagnetic dampers suitable to the cable structures, the research results reveal that the dampers had a good ability to control vibration and harvest energy simultaneously. There are also many scholars, such as Li and Zhu [27], Hu et al. [28], Luo et al. [29], and Kopylov et al. [30] proposed their own dampers by different ways, which can effectively transfer the vibration into electricity through electromagnetic technology.

Although the aforementioned regenerative dampers can convert some of the vibration energy of devices into electric energy in the attenuation process, the conversion mechanism of these dampers is electromagnetic transaction which has a lower energy conversion density compared with the piezoelectric transaction [31,32]. Meanwhile, the piezoelectric materials can be widely coupled with different structures to achieve piezoelectric transaction [33–37]. Yoon et al. [38] proposed a PVDF–silicone rubber damper using to attenuate the mutual collision of the offshore structures and studied the deformation and energy harvesting behavior of the sandwiched and coiled sheet geometries of the device. The experiment results show that the output power density of the coiled-type damper can reach up to 0.982 mW/cm^3. Khan and Ali [39] presented a vibration isolator composed of a piezoelectric disc embedded in the silicone rubber. This isolator can convert the environmental vibration into electrical energy, and it can obtain an optimum power of 2.12 mW at a matching load of 340 kΩ and frequency (resonant) of 56 Hz. However, the devices above cannot efficiently harvest the vibration energy and just generate a small amount of electric energy.

In view of this, in this research, a practical piezoelectric damper attenuating harmful vibration by converting them into useful electric power, and the corresponding theoretical models are derived to predict its electrical output. The novel damper is made of a pair of linear arrays of piezoelectric coupled discs built-in a case cylinder, and the theoretical models are based on a series of experiments and simulations. According to the test results, the average power (power density) can reach up to 161.03 mW (39.6 mW/cm^3) from a piezoelectric patch. Correspondingly, the total output power of the piezoelectric damper consisting of 20 unimorph PCDs can reach up to 3.22 W, which can serve for many electric appliances such as sensors, actuators, signal transmitters.

2. Development of the Piezoelectric Damper

The structure design and working mechanism of the new type of piezoelectric damper are displayed in Figures 1–3. The PCD is an aluminum alloy substrate disc bonded with one/two concentric circular piezoelectric patch of PZT4 (Lead Zirconate Titanate) by epoxy resin (shown in Figure 1a). The normal section of unimorph PCD is displayed in Figure 1b, in which a PZT4 patch with an inner diameter of d_p, an outer diameter of D_p and a thickness of t_p is pasted on one surface of the concentric substrate disc. The normal section of bimorph PCD is revealed in Figure 1c, in which two PZT4 patches are pasted on the two surfaces of the concentric substrate disc with an inner diameter of d_s, an outer diameter of D_s and a thickness of t_s. In this experiment, the array of PCDs shown as Figure 2a has the actions of a damper and spring, and can absorb and transfer part of the vibration energy from the outer excitation into electric energy to attenuate the outer vibration. It is obvious that two arrays of PCDs shown in Figure 2a need to match two arrays of inner supporting rings shown in Figure 2b, two arrays of outer supporting rings shown in Figure 2c, and a load transfer unit shown in Figure 2d to form a piezoelectric damper shown in Figure 2e. The inner and outer support rings are used to isolate the PCDs and fix them in an array, which can be seen in Figure 2e. The load transfer unit is composed of two pilot bearings, two nuts, a spring, and two bolts. The pilot bearings, inlaid in the concentric holes of the top and bottom covers, can make the bolts just move in the axial direction and reduce the friction between bolts and covers. Two pairs of nuts on the bolts are used to make the array of PCDs move with the bolts. The installation of these components is displayed in Figure 2e. The outer support rings are in contact with the inner surface of the cylinder case shown in Figure 3a. The PCDs are isolated by inner and outer support rings and restricted by the cap (Figure 3b), the covers and the nuts. The outer excitation is exerted on the array of PCDs through the load transfer unit. The axial displacement of the bolts is restricted in a range by the spring (Figure 3c) connecting the two bolts, which can protect the PCDs to avoid broken. The inner support ring shown in Figure 3d circles around the bolt and make PCDs has a uniform deformation in the move of the bolts.

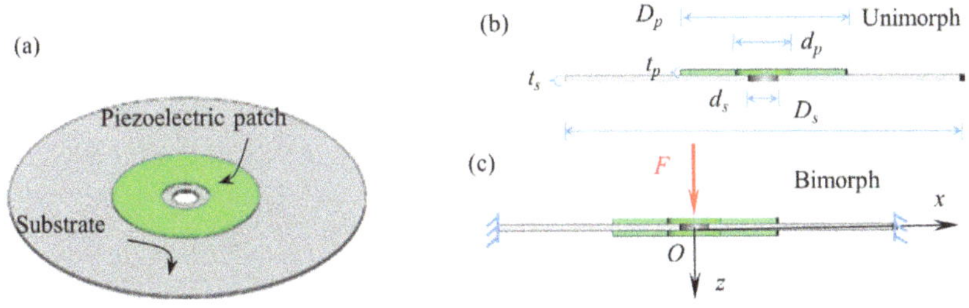

Figure 1. Set-up of a piezoelectric coupled disc (PCD). (**a**) 3D view of PCD; (**b**) normal section view of unimorph PCD; (**c**) normal section view of bimorph PCD.

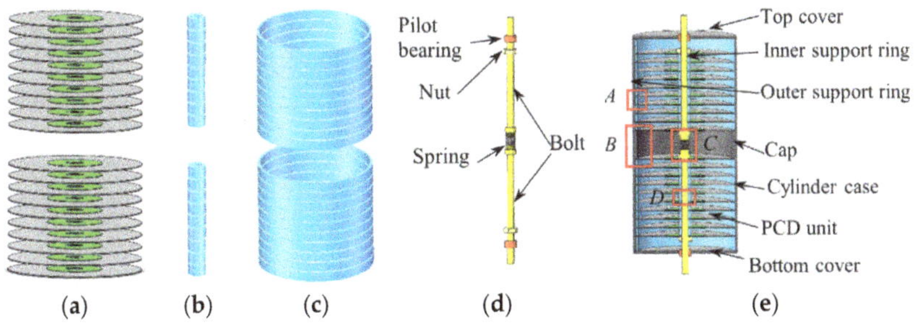

Figure 2. Set-up of a piezoelectric damper. (**a**) PCD array; (**b**) inner support ring array; (**c**) outer support ring array; (**d**) load transfer unit; (**e**) section view of piezoelectric damper.

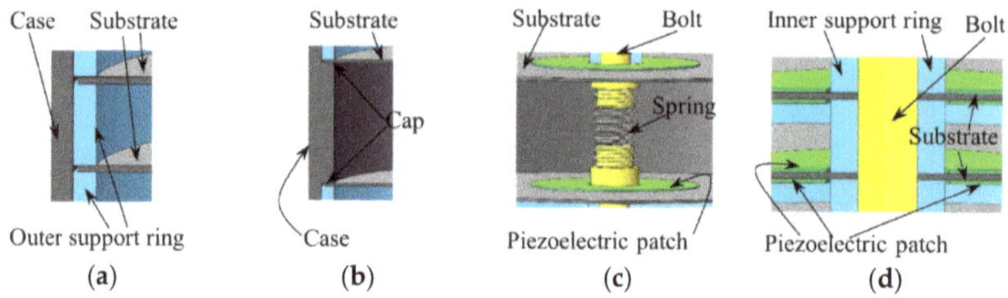

Figure 3. Local enlarged view of points *A*, *B*, *C*, and *D* in Figure 2. (**a**) Local view of point *A*; (**b**) local view of point *B*; (**c**) local view of point *C*; (**d**) local view of point *D*.

3. Experimental Study

The dimensions and physical parameters of substrates S1 and S2 in the experiments are revealed in Table 1, including inner diameter, outer diameter, thickness, density, and damping ratio.

Table 1. Dimensions and Physical parameters of substrates.

Substrates	Inner Diameter d_s (mm)	Outer Diameter D_s (mm)	Thickness t_s (mm)	Density (kg/m^3)	Damping Ratio ξ
S1	15	218	0.6	2650	0.0319
S2	15	218	1	2650	0.0478

The inner diameter, outer diameter, thickness, Young's modulus, shear modulus, passion's ratio, and capacitance of PZT4 patches of P1–P8 are specifically listed in Table 2.

Table 2. Dimensions and physical parameters of PZT4 patches.

PZT4 Patch	Inner Diameter d_p (mm)	Outer Diameter D_p (mm)	Thickness t_p (mm)	Young's Modulus E (Gpa)	Shear Modulus E_s (Gpa)	Poisson's Ratio μ	Capacitance C (nF)
P1	20	60	1	69.49	22.61	0.337	38.14
P2	20	65	1	69.49	22.61	0.337	42.42
P3	20	70	1	69.49	22.61	0.337	50.2
P4	20	75	1	69.49	22.61	0.337	55.2
P5	20	80	1	69.49	22.61	0.337	63.3
P6	25	80	1	69.49	22.61	0.337	62.7
P7	30	80	1	69.49	22.61	0.337	62.2
P8	35	80	1	69.49	22.61	0.337	58.4

The combination style, experimental specimen, the weight, and the stiffness of PCD units from 1# to 11# are displayed in Table 3.

Table 3. Material properties of piezoelectric harvester units.

Piezoelectric Harvester Unit	Combination Styles	Experimental Specimen	Weight (g)	Stiffness (N/m)
1# (P1 + S1)	Unimorph		80.8	13,823
2# (P2 + S1)	Unimorph		83.8	14,715
3# (P3 + S1)	Unimorph		88.5	19,834

Table 3. *Cont.*

Piezoelectric Harvester Unit	Combination Styles	Experimental Specimen	Weight (g)	Stiffness (N/m)
4# (P4 + S1)	Unimorph		91.9	23,532
5# (P5 + S1)	Unimorph		97.9	24,010
6# (P6 + S1)	Unimorph		97.1	19,007
7# (P7 + S1)	Unimorph		96.1	16,292
8# (P8 + S1)	Unimorph		94.9	13,417
9# (P8 + S2)	Unimorph		131.3	19,834
10# (P8 + S1)	Bimorph		125.2	23,252
11# (P8 + S2)	Bimorph		161.6	36,987

Specially, in order to study the influence of the outer diameter of the PZT4 patch on the energy harvesting ability of the PCD unit, PCDs of 1#–5# are fabricated by pasting PZT4 patches of P1–P5 on one surface of the substrate of S1, respectively. The PZT4 patches of P1–P5 have the same inner diameter of 20 mm, the thickness of 1 mm, and different outer diameters in turn of 60 mm, 65 mm, 70 mm, 75 mm, and 80 mm. Similarly, to study the effect of the inner diameter on the energy harvesting behavior of the piezoelectric unit, PCDs of 5#–8# are manufactured through pasting PZT4 patches of P5–P8 on one surface of the substrate S1, respectively. The PZT4 patches of P5–P8 have the same outer diameter of 80 mm, thickness of 1 mm, and different inner diameter in turn of 20 mm, 25 mm, 30 mm, and 35 mm. In order to research the effect of the substrate thickness on the energy harvesting behavior of the piezoelectric unit, a PZT4 patch with an inner/outer diameter of 35 mm/80 mm and a thickness of 1 mm is pasted on the surfaces of the substrates with a thickness of 0.6 mm and 1 mm to form the PCDs of 8# and 9#, respectively. To study energy harvesting behavior of the bimorph PCD, four PZT4 patches of P8 are pasted on two sides of the substrates of S1 and S2 to form PCD units of 10# and 11#, the output voltage and power from PCD units of 10# and 11# will be compared with those from PCD units of 8# and 9#, respectively. The outer surface of the PZT4 patch is connected with a wire as a positive electrode, and the inner surface between the PZT4 patch and the substrate is connected with a copper foil as a negative electrode, then the copper foil is connected with a wire. The experimental model of the piezoelectric damper is fixed on the fixed pedestal shown in Figure 4. In the experiment, the sine wave signal of the shaker (SA-JZ040) is generated by the vibration controller (VT-9008) and amplified by the power amplifier (SA-PA010). The piezoelectric signal collector (SC-HY-PZT-2.0, ±50 V, ±1%) is used to record the harvested electric signal, and the resistor box (ZX1M 0–1000 KΩ, ±0.1%) in the circuit aims for adjusting the load resistance of the access circuit to achieve the maximum output power. The outer vibration is transferred to the PCDs through the bolt to make the PZT4 patch strain and generate electricity in the vibration. Furthermore, the vibration controller (VT-9008) can control the acceleration, displacement and speed of the shaker at different frequencies. It is noted that in the following experiments the excitation acceleration is set to be 1.5 g and the excitation amplitude is set to be 5 mm.

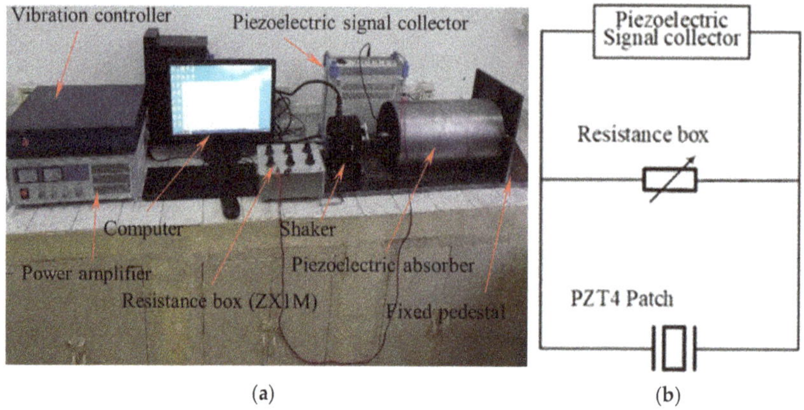

Figure 4. The experiment setup and electric circuit of the piezoelectric damper. (**a**) experiment setup; (**b**) electric circuit.

4. Results and Analysis

4.1. The Effect of the Outer Diameter of the PZT4 Patch

The peak-to-peak voltages (open-loop) of 1#–5# PCDs in different excitation frequencies are displayed in Figure 5a. It can be observed from Figure 5a that the peak-to-peak voltages and the 1st resonant frequencies of the unimorph PCDs all show a trend of increase with the increase of the outer diameter of the PZT4 patches. Specifically, when the outer diameter of the PZT4 patch is taken as 60 mm, 65 mm, 70 mm, 75 mm, and 80 mm, the maximum peak-to-peak voltage and the corresponding 1st resonant frequency of the PCD are (110 V, 30 Hz), (125 V, 35 Hz), (134 V, 44 Hz), (141 V, 48 Hz), and (143 V, 53 Hz), respectively. This phenomenon indicated that in the same excitation amplitude condition, the strain born by the PZT4 patch increases with the increase of its outer diameter, which leads to the increase of the voltage from the PZT4 patch. At the same time, with the increase of the outer diameter, the bending rigidity of the PCD increases, which causes the increase of the natural frequency of the PZT4 patch. In the 1st resonance, the changes of the average powers of 1#–5# PCDs along the increase of the load resistance are presented in Figure 5b. As displayed in Figure 5b, with the increase of the outer diameter of the PZT4 patch as 60 mm, 65 mm, 70 mm, 75 mm, and 80 mm, the maximum average powers from 1#–5# unimorph PCDs are 32 mW, 40 mW, 78 mW, 89 mW, and 125 mW corresponding to the optimal load resistances of 19.61 KΩ, 19.61 KΩ, 9.9 KΩ, 9.9 KΩ, and 8.9 KΩ, respectively. Obviously, the average output powers from 1#–5# PCDs all sharply increase up to their respective peaks and then sharply decrease till the load resistance increases to 100 KΩ, and then the increase of the load resistance hardly affects the average output powers from 1#–5# PCDs. In addition, in the 1st resonance, the peak-to-peak voltages (close-loop) of 1#–5# PCDs are measured in the different load resistance and the corresponding results are illustrated in Figure 5c, from which it can be observed that the voltages from 1#–5# PCDs all increase sharply with the increase of the load resistance (resistance box) from 0.9 KΩ to 19 KΩ, and then tend to a stable value with the increase of the load resistance from 19 KΩ to 500 KΩ. This phenomenon is mainly because with the increase of the load resistance from 0.9 KΩ to 19 KΩ, the proportion of the load resistance in the total resistance of the circuit rapidly increases, and after 19 KΩ, the inner resistance of the PZT4 patch is insignificant compared to the load resistance. Namely, inner resistance of the PZT4 patch has a negligible effect on the circuit current and the voltage exerting on the load resistance after the load resistance is greater than 19 KΩ. This process can also be explained by Equation (1) given below. Based on the discussion above, a conclusion can be found that the peak-to-peak voltage (open/close-loop) and the maximum average power, and the

1st natural frequency of the PCD all increase with the increase of the outer diameter of the PZT4 patch from 60 mm to 80 mm.

$$V_L = IR_L = \frac{R_L}{R_I + R_L}V, \tag{1}$$

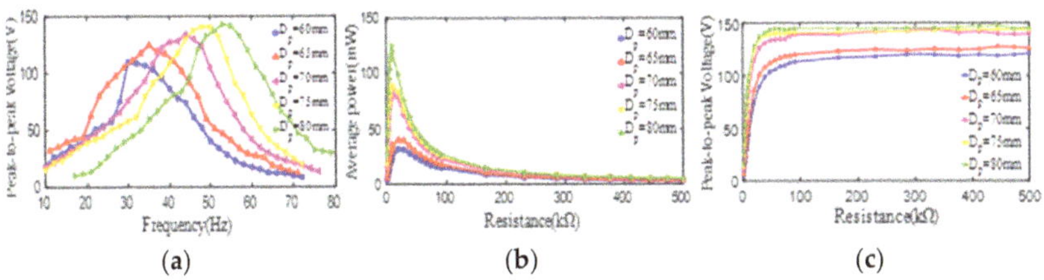

Figure 5. The effect of the outer diameter of the PZT4 patch on its peak-to-peak voltage and average power. (**a**) peak-to-peak voltage versus excitation frequency; (**b**) average power versus load resistance; (**c**) peak-to-peak voltage versus load resistance.

In which, V and I denote the total voltage and the current in the electric circuit shown as Figure 4b; R_L and V_L express the load resistance and its voltage; R_I is the inner resistance of the PZT4 patch.

4.2. The Effect of the Inner Diameter of the PZT4 Patch

Following the research on the effect of the outer diameter of the PZT4 patch, the effect of its inner diameter on its output electric parameters is studied. The peak-to-peak voltages (open-loop) from unimorph PCDs of 5#–8# at different excitation frequencies are measured and displayed in Figure 6a, in which the trend of the curves of 5#–8# PCDs are roughly the same, increasing firstly up to the peak and then decrease continually. As Figure 6a, the maximum peak-to-peak voltages and their corresponding 1st resonant frequencies are (143 V, 53 Hz), (151 V, 52 Hz), (158 V, 50 Hz), and (163 V, 47 Hz) corresponding to the inner diameter of 20 mm, 25 mm, 30 mm, and 35 mm, respectively. These data indicate that with the increase of the inner diameter of the PZT4 patch, the maximum voltage can be improved to some extent. This is because with the increase of the inner diameter (from 20 mm to 35 mm), the bending rigidity of the PCD decreases, which leads to the increase of the strain and output voltage of the PZT4 patch indicated as Equations (2)–(4). The average powers of 5#–8# PCDs all show a sharply linear increase up to the peak value along the increase of the load resistance and then a sharply nonlinear decrease followed by a gentle down, which can be seen in Figure 6b. Meanwhile, it can be found the maximum average power of the unimorph PCD increases with the increase of the inner diameter of the PZT4 patch. Specifically, for the inner diameter of 20 mm, 25 mm, 30 mm, and 35 mm, the maximum average powers and the corresponding optimal load resistances are (125 mW, 8.9 KΩ), (141 mW, 8.9 KΩ), (148 mW, 9.9 KΩ), and (160 mW, 9.9 KΩ), respectively. According to Figure 5c, it is obvious that the peak-to-peak voltages (close-loop) from unimorph PCDs of 5#–8# also show a sharply linear increase up to the maximum and then basically keep it with the increase of the load resistance. In addition, it can be observed that the output voltages of the unimorph PCDs increase with the increase of the inner diameter of the PZT4 patch when the load resistance changes from 9.9 KΩ to 500 KΩ.

In a word, from Figure 6, it can be observed that with the increase of the inner diameter of the PCD, its peak-to-peak voltage (open/close-loop) and maximum average power increase, but the 1st natural frequency of the PCD decreases.

$$V(t) = Q(t)/C = d_{31}\varepsilon(t)E_p t_p/\varepsilon_0\varepsilon_r, \tag{2}$$

$$Q(t) = d_{31}\varepsilon(t)E_p A_p, \tag{3}$$

$$C = \varepsilon_0 \varepsilon_r A_p / t_p, \tag{4}$$

where, $Q(t)$, $V(t)$ and $\varepsilon(t)$ are the generated electric charge, voltage for an open-loop circuit, and strain at time t from the PZT4 patch; C is the capacitance of the PZT4 patch; d_{31} is the piezoelectric constant; E_p, A_p, and t_p are the Young's modulus, surface area, and thickness of the PZT4 patch; ε_0 is the permittivity of air and ε_r is the dielectric constant of the PZT4 patch.

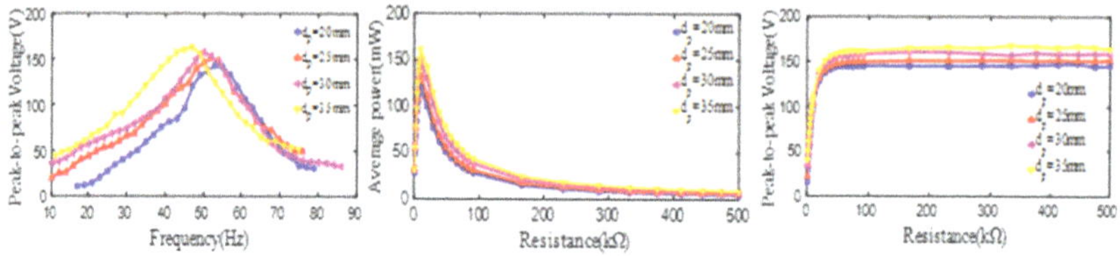

(a) (b) (c)

Figure 6. The effect of the inner diameter of the PZT4 patch on its peak-to-peak voltage and average power. (**a**) peak-to-peak voltage versus excitation frequency; (**b**) average power versus load resistance; (**c**) peak-to-peak voltage versus load resistance.

4.3. The Effect of the Thickness of the Substrate

In order to study the influence of the substrate thickness on the output electric parameters of unimorph PCD, the same PZT4 patches with an inner diameter of 35 mm, an outer diameter of 80 mm, and a thickness of 1 mm are bonded on the substrates with a thickness of 0.6 mm and 1.0 mm to form PCDs of 8# and 9#, respectively. Shown as Figure 7a, the maximum peak-to-peak voltages (open-loop) from unimorph PCDs of 8# and 9# can reach up to 163 V and 104 V when they work at the 1st resonances of 47 Hz and 78 Hz, respectively. To further investigate the influence of the load resistance on the average output power and the peak-to-peak voltage from PCDs, the parallel resistance with a measuring range of 0.9 kΩ to 500 kΩ is introduced to the experimental circuit. The maximum average powers from unimorph PCDs of 8# and 9# all sharply increase up to their respective peaks and then sharply decrease along the increase of the load resistance till 100 KΩ, and then show a gently reducing tendency, which is displayed in Figure 7b. Furthermore, it can be obtained that the maximum average power from unimorph PCDs of 8# and 9# and the corresponding optimal load resistance are (161 mW, 9.9 kΩ) and (42 mW, 9.9 kΩ), respectively. Shown as Figure 7c, the peak-to-peak voltages (close-loop) from unimorph PCDs of 8# and 9# sharply increase up to 159 V and 98 V firstly and then basically keep a constant with the increase of the load resistance.

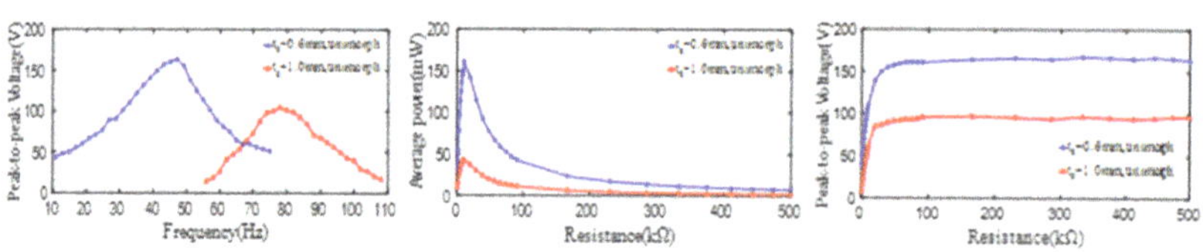

(a) (b) (c)

Figure 7. The effect of the substrate thickness on the peak-to-peak voltage and average power of one PZT4 patch bonded on the substrate. (**a**) peak-to-peak voltage versus excitation frequency; (**b**) average power versus load resistance; (**c**) peak-to-peak voltage versus load resistance.

In short, with the increase of the substrate thickness from 0.6 mm to 1.0 mm, the peak-to-peak voltages (close-loop) from unimorph PCDs and the average power decrease. However, the 1st natural frequency of the PCD increases.

To further study the effect of the substrate thickness on the output peak-to-peak voltage and average power from PZT4 patch of bimorph PCD, the experimental results of the voltages and powers from the bimorph PCD with a substrate thickness of 0.6 mm (10#) and 1 mm (11#) are taken out for analysis. The changing of the peak-to-peak voltages (open-loop) from one PZT4 patch along the increase of the excitation frequency is revealed in Figure 8a, in which the relationship curves all similar to a hill with the maximum peak-to-peak voltage of 121 V at 55 Hz for substrate thickness of 0.6 mm and 97 V at 89 Hz for substrate thickness of 1 mm. This phenomenon once again indicates that the PCD with the thinner substrate is better for generating a larger peak-to-peak voltage. In order to further investigate the influence of load resistance on the output power and peak-to-peak voltage from bimorph PCD of the piezoelectric damper. The maximum output powers of the PCDs are extracted and shown in Figure 8b, in which the powers all increase firstly and then decrease with the increase of the load resistance. Furthermore, it can be found that the output power and corresponding optimal resistance from the bimorph PCDs of 10# and 11# are (119 mW, 9.9 kΩ) and (34 mW, 9.9 kΩ), respectively. The peak-to-peak voltages (close-loop) from the bimorph PCDs of 10# and 11# in different load resistance are measured and displayed in Figure 8c, in which the curves show an increase sharply and then basically keep a stable value of 115 V for a substrate thickness of 0.6 mm and 87 V for substrate thickness of 1 mm.

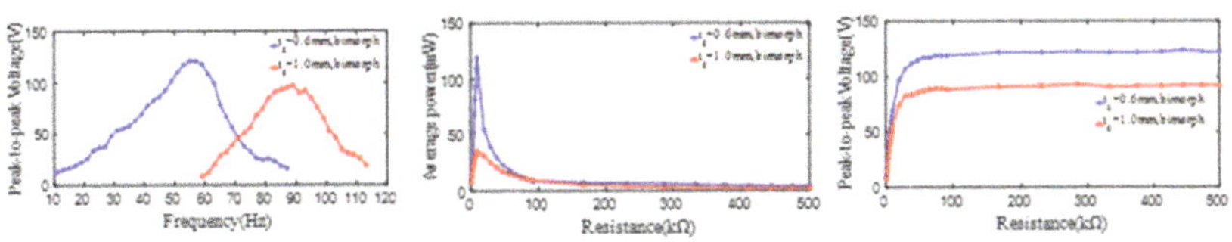

(a) (b) (c)

Figure 8. The effect of the substrate thickness on the peak-to-peak voltage and average power of two PZT4 patches bonded on the substrate. (**a**) peak-to-peak voltage versus excitation frequency; (**b**) average power versus load resistance; (**c**) peak-to-peak voltage versus load resistance.

Obviously, for bimorph PCD, its peak-to-peak voltages (open/close-loop), average power, and 1st natural frequency have the same changing rule as unimorph PCD with the increase of the substrate thickness.

4.4. The Effect of the Unimorph/Bimorph PCD

In order to investigate the effect of the unimorph PCD and bimorph PCD on their energy harvesting ability, the peak-to-peak voltages and average powers from the unimorph and bimorph PCDs (d_p = 35 mm, D_p = 80 mm, t_s = 0.6 mm) are extracted and revealed in Figure 9. Figure 9a demonstrated that the voltage (open-loop) changes similar to a gentle hill along the increase of the excitation frequency from 10 Hz to 90 Hz. Obviously, the voltage from unimorph PCD is larger than the corresponding one from bimorph PCD; meanwhile, the 1st resonant frequency of the unimorph PCD is less than that of the bimorph PCD. The maximum peak-to-peak voltages (open-loop) from one PZT4 patch of unimorph and bimorph PCDs are 163 V and 121 V at the 1st resonant frequencies of 47 Hz and 55 Hz, respectively. It is noted that the difference between the peak-to-peak voltages from unimorph PCD and bimorph PCD can reach up to 42 V. This phenomenon is because the unimorph PCD has a smaller bending rigidity and a bigger deformation than that of bimorph PCD. Figure 9b displayed the average powers of the unimorph and bimorph

PCDs increase similar to a steep mountain along the increase of the load resistance from 0.9 kΩ to 100 kΩ and then attenuate close to zero value. It can be found that the average output powers from one PZT4 patch of unimorph and bimorph PCDs can reach up to 161 mW and 119 mW at the same optimal resistance of 9.9 kΩ. Obviously, the difference of the maximum average powers is up to 41 mW. Of course, the total average output powers of the two PZT4 patches (whether in series or parallel) of bimorph PCD is far bigger than that of unimorph PCD, but it is at the expense of inputting bigger excitation force. As shown in Figure 9c, with the increase of the load resistance, the peak-to-peak voltages (close-loop) from one PZT4 patch show a steep rising up to the peak of 159 V and 115 V for unimorph and bimorph PCDs and then tend to a steady value.

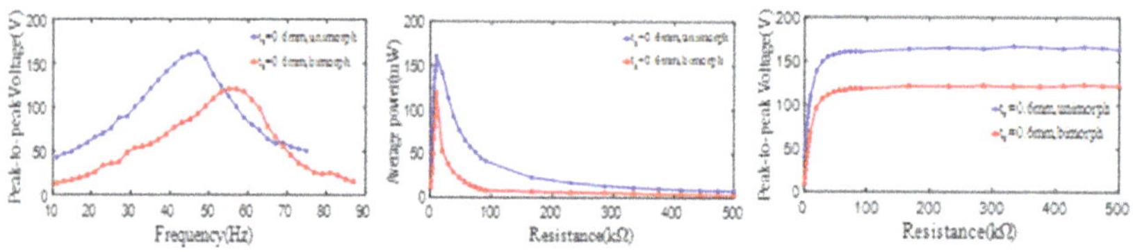

(**a**) (**b**) (**c**)

Figure 9. The effect of unimorph PCD and bimorph PCD on the peak-to-peak voltage and average power of one PZT4 patch (t_s = 0.6 mm). (**a**) peak-to-peak voltage versus excitation frequency; (**b**) average power versus load resistance; (**c**) peak-to-peak voltage versus load resistance.

To further explore the influence of unimorph and bimorph PCDs, the peak-to-peak voltages and average powers from unimorph PCD (9#) and bimorph PCD (11#) with a substrate thickness of 1 mm are also extracted and revealed in Figure 10. It can be obtained from Figure 10a that the peak-to-peak voltages (open-loop) from the unimorph and bimorph PCDs (d_p = 35 mm, D_p = 80 mm, t_s = 1 mm) can reach up to 104 V at its 1st resonant frequency of 78 Hz and 97 V at its 1st resonant frequency of 89 Hz. This phenomenon is also caused by the smaller bending rigidity of the unimorph PCD compared to that of bimorph PCD. In addition, the maximum average power from unimorph and bimorph PCDs can reach up to 42 mW and 34 mW, respectively, at the same optimal resistance of 9.9 kΩ, which can be seen in Figure 10b. It also can be found from Figure 10c that the peak-to-peak voltages (close-loop) from unimorph PCD (9#) and bimorph PCD (11#) sharply rise up to a stable value of 104 V for unimorph PCD and 97 V for bimorph PCD with the increase of the load resistance and then basically keep a constant. From Figure 10, it can be found that the maximum peak-to-peak voltages or the maximum average powers from the unimorph PCD are slightly larger than the corresponding ones from the bimorph PCD. Specially, the difference of the maximum peak-to-peak voltages is 7 V, and the difference between the maximum average powers is 8 mW, which is far less than the corresponding difference between the voltages and the average powers from PCDs with a substrate thickness of 0.6 mm. This phenomenon indicates that with the increase of the substrate thickness, the energy harvesting ability of one PZT4 patch from bimorph PCD is gradually close to that from unimorph PCD. That is to say, with the increase of the substrate thickness, the energy harvesting ability of bimorph PCD is basically two times that of unimorph PCD. At the same time, by comparing with the experimental results in the last paragraph, it is obvious that when the substrate thickness is 0.6 mm, unimorph/bimorph PCD has a bigger difference in energy harvesting shown in Figures 9c and 10c.

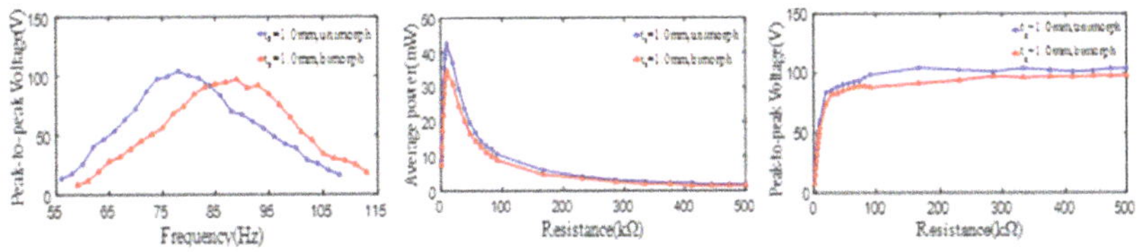

(a) (b) (c)

Figure 10. The effect of unimorph PCD and bimorph PCD on the output peak-to-peak voltage and average power of one PZT4 patch (t_s = 1 mm). (**a**) peak-to-peak voltage versus excitation frequency; (**b**) average power versus load resistance; (**c**) peak-to-peak voltage versus load resistance.

5. COMSOL Simulation

In order to comprehensively research the effects of the inner/outer diameter and the thickness of the PZT4 patch and the substrate thickness on the energy harvesting behavior of the PCDs, more PCD models with different dimension are built by COMSOL software. To visually display the stress distribution of the PCD under the axial force, the stress contours of the PCD are simulated by COMSOL software and revealed in Figure 11. It can be intuitively found that the deformation focuses on the area bonding PZT4 patch and its neighborhood. In addition, to verify the accuracy and the reliability of the COMSOL models, the peak-to-peak voltages of the experimental models of 2#–4# PCDs are simulated by corresponding COMSOL models. The results from experiments and simulations are extracted and shown in Figure 12. It can be found from Figure 12 that the curves of the peak-to-peak voltages from the simulation are roughly the same as those from the experiment, which can be seen in Figure 12a–c for 2#, 3#, and 4# PCD, respectively. These phenomena indicate that the simulation models of PCDs are reliable and the simulation results of the voltages of PCDs are accurate. Therefore, using COMSOL software to extend the research of PCDs with more dimensions is feasible. Just to be clear, in the following description, the size parameters of PCDs with star mean the simulation models. For example, D_p^* denotes the outer diameter of the PZT4 patch for COMSOL simulation models.

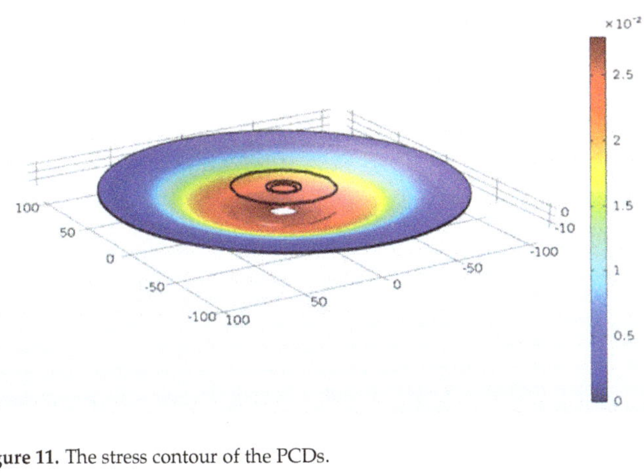

Figure 11. The stress contour of the PCDs.

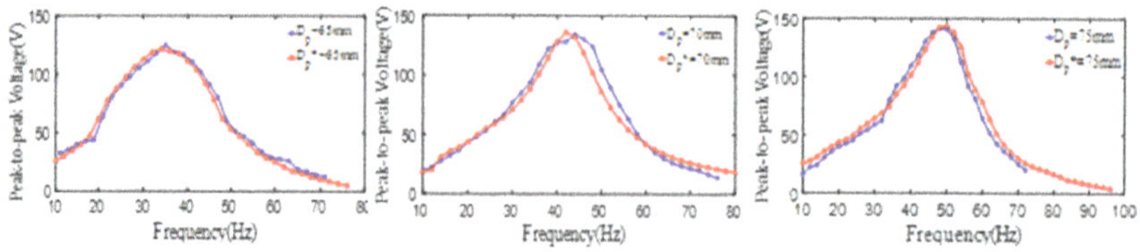

Figure 12. The relationship curves of the peak-to-peak voltage and the excitation frequency of the unimorph PCD from experimental results (blue line) and simulation results (red line). (**a**) peak-to-peak voltage of PCD with D_p = 65 mm; (**b**) peak-to-peak voltage of PCD with D_p = 70 mm; (**c**) peak-to-peak voltage of PCD with D_p = 75 mm.

In order to further explore the effect of the outer diameter of the PZT4 patch on the output voltage of the PCD, a series of PCD models are built by COMSOL software. In these models, the material properties are taken as those of the experimental specimens shown in Tables 1 and 2, the dimensions of the PCDs are set to be: d_p^* = 20 mm, t_p^* = 1 mm, t_s^* = 0.6 mm, d_s^* = 15 mm, and D_s^* = 218 mm, in addition, the outer diameters of the PZT4 patch include 75 mm, 105 mm, 135 mm, 165 mm, and 195 mm. According to the simulation, the output voltages of each PCD model all show an increase first and then decrease with the increase of the excitation frequency, which can be seen in Figure 13. The maximum peak-to-peak voltage and the corresponding 1st resonant frequency are (143 V, 50 Hz), (159 V, 62 Hz), (191 V, 84 Hz), (210 V, 100 Hz), and (144 V, 104 Hz) for the PZT4 patch with an outer diameter of 75 mm, 105 mm, 135 mm, 165 mm, and 195 mm, respectively. Obviously, the maximum peak-to-peak voltages increase with the increase of the outer diameter from 75 mm to 165 mm and then decrease when the outer diameter is more than 165 mm. Meanwhile, it can be found that the 1st resonant frequency of the PCD keeps a continual increase with the increase of the outer diameter of the PZT4 patch. This phenomenon indicates that there is an optimal outer diameter of the PZT4 patch for the PCD to output maximum voltage, rather than the bigger diameter of the PZT4 patch is always better for the PCD to output maximum voltage.

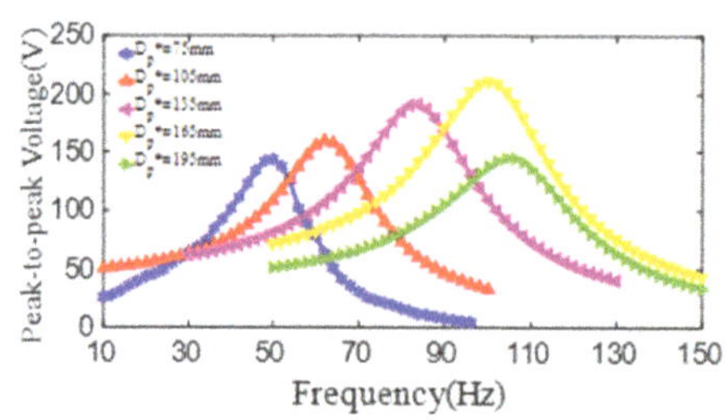

Figure 13. The relationship curves of the peak-to-peak voltage and the excitation frequency of PZT4 patch with different outer diameter (d_p^* = 20 mm, t_p^* = 1 mm, t_s^* = 0.6 mm).

Based on the test and simulation results, the formula for the peak-to-peak voltages (V_{pp}) of the PCD and the outer diameter of D_p can be fitted as below:

$$V_{pp} = 552 - 12D_p + 0.11D_p^2 + 3.12 * 10^{-4}D_p^3 (D_p = 75 \sim 195\,\text{mm}) \quad (5)$$

To explore the effect of the thickness of the PZT4 patch on the energy harvesting behavior of the PCD, the simulation models of the damper, in which the PZT4 patch with different thicknesses of t_p^*, are built. In these simulations, the material properties are displayed in Tables 1 and 2, and the dimensions are taken as d_p^* = 20 mm, D_p^* = 75 mm,

t_s^* = 0.6 mm, d_s^* = 15 mm, and D_s^* = 218 mm. As shown in Figure 14, corresponding to the thickness t_p^* of 0.5 mm, 1.0 mm, 1.5 mm, 2.0 mm, and 2.5 mm, the maximum peak-to-peak voltages and the corresponding 1st resonant frequency are, respectively (164 V, 46 Hz), (143 V, 49 Hz), (119 V, 50 Hz), (104 V, 50 Hz), and (91 V, 51 Hz). The curves of the peak-to-peak voltage and the excitation frequency all increase firstly to a maximum value and then decrease. It can be concluded that the peak-to-peak voltages decrease with the increase of the thickness of the PZT4 patch, the reason is that with the increase of the thickness of the PZT4 patch, the stiffness of the PCD increase, which leads to the notable decrease of the strain and output voltage of the PZT4 patch although its thickness increases (indicated by Equation (1)).

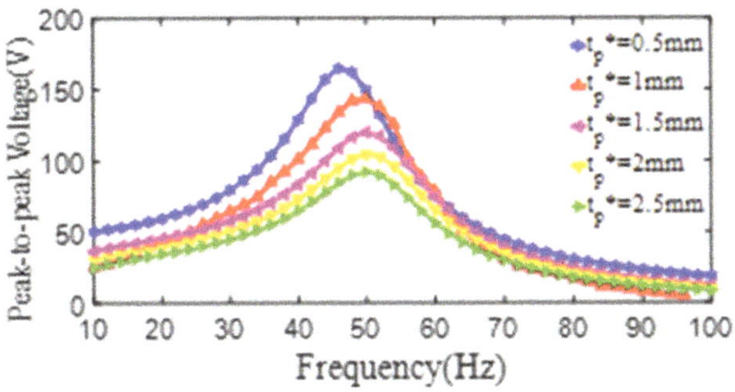

Figure 14. The relationship curves of the peak-to-peak voltage and the excitation frequency of PZT4 patch with different thickness (d_p^* = 20 mm, D_p^* = 75 mm, t_s^* = 0.6 mm).

For the convenience of evaluating the effect of the PZT4 patch thickness on the output voltage of the PCD, the peak-to-peak voltage (V_{pp}) of the PCD and the corresponding thickness of t_p are extracted and fitted as the formula as follows:

$$V_{pp} = 192.2 - 58.43t_p + 7.14t_p^2 (t_p = 0.5 \sim 2.5\,\text{mm}), \quad (6)$$

To expand the study on the relationship of the inner diameter of the PZT4 patch and the output voltage of the PCD, a series of simulation models of the damper are modeled, in which the substrate discs are bonded by PZT4 patches with different inner diameters. In these simulations, the dimensions of PZT4 patch and substrate are set to be D_p^* = 80 mm, t_p^* = 1 mm, t_s^* = 0.6 mm, d_s^* = 15 mm, and D_s^* = 218 mm. It can be obtained from Figure 15 that the maximum peak-to-peak voltages and the 1st resonant frequency are (163 V, 47 Hz), (181 V, 46 Hz), (221 V, 45 Hz), (288 V, 44 Hz), and (327 V, 42 Hz), when the inner diameter d_p^* of the PZT4 patch increases as 35 mm, 45 mm, 55 mm, 65 mm, and 75 mm, respectively. Obviously, with the increase of the inner diameter, the peak-to-peak voltage increases. This phenomenon is caused by the decrease of the bending rigidity of the PCD, which further causes the increases of the surface strain and the output voltage of the PZT4 patch. Another finding is that with the increase of the inner diameter, the 1st resonant frequency of the PCD keeps a continuous decrease. This is the same as the rule in experiments and for the same reason as experiments.

To theoretically value the effect of the inner diameter of the PZT4 patch on the output electric parameter of the PCD, the peak-to-peak voltage of the PCD corresponding to the inner diameter of d_p is extracted and their relationship is fitted as the formula as below:

$$V_{pp} = 614 - 9.77d_p + 0.049d_p^2 (d_p = 35 \sim 75\,\text{mm}), \quad (7)$$

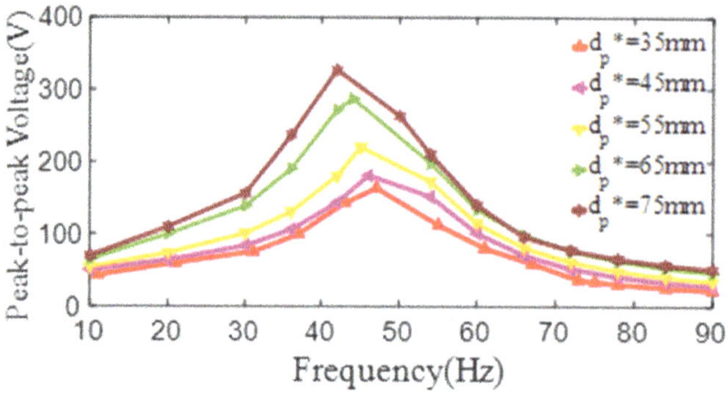

Figure 15. The relationship curves of the peak-to-peak voltage and the excitation frequency of PZT4 patch with different inner diameter ($D_p{}^*$ = 80 mm, $t_p{}^*$ = 1 mm, $t_s{}^*$ = 0.6 mm).

The effect of the substrate thickness on the output voltage of the PCD is also further researched by the simulation method. In this computation process, the parameters are taken as $d_p{}^*$ = 80 mm, $D_p{}^*$ = 135 mm, $t_p{}^*$ = 1 mm, $d_s{}^*$ = 15 mm, and $D_s{}^*$ = 218 mm; and the substrate thickness is changed from 0.6 mm to 2.5 mm. As displayed in Figure 16, when the substrate thickness of $t_s{}^*$ is rising to 0.6 mm, 1.0 mm, 1.5 mm, 2.0 mm and 2.5 mm, the maximum peak-to-peak voltage from the PZT4 patch is 201 V at 65 Hz, 163 V at 82 Hz, 121 V at 115 Hz, 91 V at 150 Hz and 71 V at 180 Hz, respectively. From the data above, it can be founded that as the thickness of the substrate increases, the 1st resonant frequency of the PCD increases markedly, but the peak-to-peak voltage at this frequency decreases. Obviously, with the increase of its substrate thickness, the bending stiffness and the natural frequency of the PCD increases. Another key reason is that the excitation amplitude of the shaker decreases with the increase of its excitation frequency, which also leads to the decreases of the deformation and the generated voltage (at 1st resonance) of the PZT4 patch of the PCD with a bigger substrate thickness.

On the basis of the experiment and simulation, the relationship curve of the peak-to-peak voltage of the PZT4 patch of the PCD and the substrate thickness of t_s can be fitted using the equation as below:

$$V = 253 - 107.54t_s + 13.71t_s^2 (t_s = 0.6 \sim 2.5\,\text{mm}), \quad (8)$$

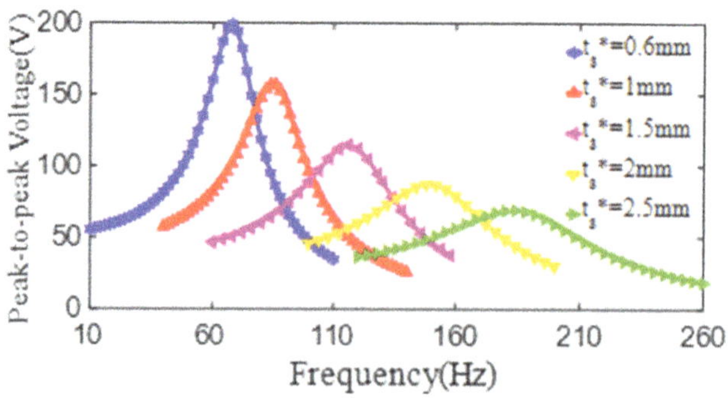

Figure 16. The relationship curves of the peak-to-peak voltage and the excitation frequency of PZT4 patch with different substrate thickness ($d_p{}^*$ = 80 mm, $D_p{}^*$ = 135 mm, $t_p{}^*$ = 1 mm).

6. Discussion

The piezoelectric damper proposed in this research is practical and ingenious. Firstly, the proposed damper can ingeniously attenuate the vibration by transferring part of the vibration energy into electric power for recycling utilization; Secondly, the restrained condition of the PCD and the force style make its strain uniform and focus on the piezoelectric coupling area shown as Figure 11; Thirdly, the design of the array PCDs make the damper have a great potential to attenuate the vibration and generate electric power. The key highlight for this damper is that it can bear a huge impact and output considerable electric energy for the huge bending rigidity of the array PCDs. The bearing capacity and electric generation capacity of the damper can be adjusted to suit any practical requirement by changing the number and the dimension of the PCD in the array. In addition, it is worth mentioning that the ingenious design of the damper makes the small PZT4 patch play a great role, which makes the piezoelectric damper can be used in many fields, including civil structure or mechanical structure, and generates appreciable electric energy. For example, the total output power of the piezoelectric damper including 20 of unimorph PCDs can reach up to 3.22 W, in which the average output power and power density of one PZT4 patch (D_p = 80 mm, d_p = 35 mm, and t_p = 1 mm) can reach up to 161 mW and 39.61 μW/mm^3. This density is larger than the maximum output power densities from L-shaped harvester (2.1 μW/mm^3) [33] and U-shaped harvester (28.1 μW/mm^3) [34]. Of course, with the increase of the number and/or the dimension of the PCD, the bearing capacity and electric generation capacity of the damper can be improved further to meet the vibration attenuation of large-scale structures.

However, the PCD of the proposed damper just can bear a small axial displacement, this restraint makes the damper just to be used in the small deformation or displacement situation. For example, the axial movement of the damper in this design should be on the millimeter scale. Otherwise, the PCD will be broken by the overrun strain/stress. In view of this, in our next research, a new experiment will be conducted by replacing the outer/inner supporting ring with an outer/inner supporting spring with the same stiffness to increase the axial movement range of the PCD in the damper. Meanwhile, the damping performance of the damper will also be studied.

7. Conclusions

A new type of piezoelectric damper, including two PCD arrays, two inner support ring arrays, two outer support ring arrays, and a load transfer unit, is developed. A series of forced-vibration experiments about the piezoelectric damper are conducted and comprehensively simulation research is carried out, in which the effects of the design dimensions of the damper, unimorph/bimorph PCD, and the load resistance on the peak-to-peak voltage and average power are acquired and analyzed. According to the research results, some important conclusions can be drawn as follows.

With the increase of the outer diameter of the PZT4 patch from 60 mm to 80 mm, the peak-to-peak voltage and average power from the unimorph PCD always increase, when the outer diameter of the piezoelectric patch increase from 80 mm to 195 mm, the peak-to-peak voltage increase up to the peak firstly and then decrease.

With the increase of the inner diameter of the PZT4 patch from 20 mm to 75 mm, the peak-to-peak voltage and average power from the unimorph PCD always increase.

In the condition of the same excitation and design parameters, the unimorph PCD is better than bimorph for generating larger peak-to-peak voltage and average output power.

In the condition of the same excitation, the PCD with a thinner substrate thickness is better to generate a bigger peak-to-peak voltage and average output power.

With the increase of the substrate thickness, the difference of the peak-to-peak voltages and the average powers from the PZT4 patch of unimorph PCD and bimorph PCD decrease.

A peak-to-peak voltage of 163 V and an average power of 161 mW can be achieved from a PZT4 patch of a unimorph PCD (D_p = 80 mm, d_p = 35 mm, and t_p = 1 mm), a total output power of 3.22 W can be obtained for a proposed damper including 20 of unimorph PCDs.

Author Contributions: Conceptualization: X.X. and X.H.; writing—original draft preparation, J.W., Z.W. and B.Z.; writing—review and editing, X.X., X.H. and J.Z.; supervision, X.X.; funding acquisition and project administration, X.X. and J.W. All authors have read and agreed to the published version of the manuscript.

Funding: This research received no external funding.

Acknowledgments: This work was partly supported by the College Students Innovative Entrepreneurial Training Plan Program (20191008122234846).

Conflicts of Interest: The authors declare no conflict of interest.

References

1. Li, Z.; Fang, Z.; Li, J.; Feng, Z. Static and Rotordynamic Characteristics for Two Types of Novel Mixed Liquid Damper Seals With Hole-Pattern/Pocket-Textured Stator and Helically Grooved Rotor. *J. Eng. Gas Turbines Power* **2021**, *143*, 041019. [CrossRef]
2. Li, Y.; Li, S.; Sun, B.; Liu, M.; Chen, Z. Effectiveness of a tiny tuned liquid damper on mitigating wind-induced responses of cylindrical solar tower based on elastic wind tunnel tests. *J. Wind Eng. Ind. Aerodyn.* **2020**, *208*, 104455. [CrossRef]
3. Sun, Y.; Yuan, J.; Pesaresi, L.; Denimal, E.; Salles, L. Parametric Study and Uncertainty Quantification of the Nonlinear Modal Properties of Frictional Dampers. *J. Vib. Acoust.* **2020**, *142*, 051102. [CrossRef]
4. Amini, A.; Ekici, Ö.; Yakut, K. Experimental Study of Regenerative Rotational Damper in Low Frequencies. *Int. J. Auto. Technol.* **2020**, *21*, 83–90. [CrossRef]
5. An, Y.; Wang, Z.; Ou, G.; Pan, S.; Ou, J. Vibration Mitigation of Suspension Bridge Suspender Cables Using a Ring-Shaped Tuned Liquid Damper. *J. Bridg. Eng.* **2019**, *24*, 04019020. [CrossRef]
6. Ross, A.S.; Damatty, A.E.; Ansary, A.M.E. Application of tuned liquid dampers in controlling the torsional vibration of high rise buildings. *Wind Struct.* **2015**, *21*, 537–564. [CrossRef]
7. Kamgar, R.; Gholami, F.; Sanayei, H.R.Z.; Heidarzadeh, H. Modified Tuned Liquid Dampers for Seismic Protection of Buildings Considering Soil–Structure Interaction Effects. *Iran. J. Sci. Technol. Trans. Civ. Eng.* **2019**, *44*, 339–354. [CrossRef]
8. Shin, J.-H.; Kwak, M.K.; Kim, S.-M.; Baek, K.-H. Vibration control of multi-story building structure by hybrid control using tuned liquid damper and active mass damper. *J. Mech. Sci. Technol.* **2020**, *34*, 5005–5015. [CrossRef]
9. Hemmati, A.; Oterkus, E.; Khorasanchi, M. Vibration suppression of offshore wind turbine foundations using tuned liquid column dampers and tuned mass dampers. *Ocean Eng.* **2019**, *172*, 286–295. [CrossRef]
10. Wang, Z.; Cao, L.; Ubertini, F.; Laflamme, S. Numerical Investigation and Design of Reinforced Concrete Shear Wall Equipped with Tuned Liquid Multiple Columns Dampers. *Shock Vib.* **2021**, *2021*, 6610811. [CrossRef]
11. Rozas, L.; Boroschek, R.L.; Tamburrino, A.; Rojas, M. A bidirectional tuned liquid column damper for reducing the seismic response of buildings. *Struct. Control Health Monit.* **2016**, *23*, 621–640. [CrossRef]
12. Zhang, Z.; Staino, A.; Basu, B.; Nielsen, S.R. Performance evaluation of full-scale tuned liquid dampers (TLDs) for vibration control of large wind turbines using real-time hybrid testing. *Eng. Struct.* **2016**, *126*, 417–431. [CrossRef]
13. Sarkar, S.; Chakraborty, A. Development of semi-active vibration control strategy for horizontal axis wind turbine tower using multiple magneto-rheological tuned liquid column dampers. *J. Sound Vib.* **2019**, *457*, 15–36. [CrossRef]
14. Zhang, Z.; Høeg, C. Dynamics and control of spar-type floating offshore wind turbines with tuned liquid column dampers. *Struct. Control Health Monit.* **2020**, *27*, e2532. [CrossRef]
15. Chen, J.; Zhan, G.; Zhao, Y. Application of spherical tuned liquid damper in vibration control of wind turbine due to earthquake excitations. *Struct. Des. Tall Spec.* **2016**, *25*, 431–443. [CrossRef]
16. Latour, M.; D'Aniello, M.; Zimbru, M.; Rizzano, G.; Piluso, V.; Landolfo, R. Removable friction dampers for low-damage steel beam-to-column joints. *Soil Dyn. Earthq. Eng.* **2018**, *115*, 66–81. [CrossRef]
17. Ghorbani, H.R.; Rofooei, F.R. A novel double slip loads friction damper to control the seismic response of structures. *Eng. Struct.* **2020**, *225*, 111273. [CrossRef]
18. Pardo-Varela, J.; de la Llera, J.C. A Semi-active piezoelectric friction damper. *Earthq. Eng. Struct. Dyn.* **2015**, *44*, 333–354. [CrossRef]
19. Chen, B.; Weng, S.; Zhi, L.; Li, D. Response control of a large transmission tower-line system under seismic excitations using friction dampers. *Adv. Struct. Eng.* **2017**, *20*, 1155–1173. [CrossRef]
20. Chen, B.; Yang, D.; Zheng, Y.; Feng, K.; Ouyang, Y. Response Control of a High-Rise Television Tower under Seismic Excitations by Friction Dampers. *Int. J. Struct. Stab. Dyn.* **2018**, *18*, 1850140. [CrossRef]
21. Li, P.; Zuo, L. Influences of the electromagnetic regenerative dampers on the vehicle suspension performance. *Proc. Inst. Mech. Eng. Part D J. Automob. Eng.* **2017**, *231*, 383–394. [CrossRef]
22. Li, S.; Xu, J.; Pu, X.; Tao, T.; Gao, H.; Mei, X. Energy-harvesting variable/constant damping suspension system with motor based electromagnetic damper. *Energy* **2019**, *189*, 116199. [CrossRef]
23. Gao, X.; Niu, J.; Jia, R.; Liu, Z. Influential characteristics of electromagnetic parameters on self-powered MR damper and its application in vehicle suspension system. *Proc. Inst. Mech. Eng. Part K J. Multi-Body Dyn.* **2020**, *234*, 38–49. [CrossRef]

24. Ataei, M.; Asadi, E.; Goodarzi, A.; Khajepour, A.; Khamesee, M.B. Multi-objective optimization of a hybrid electromagnetic suspension system for ride comfort, road holding and regenerated power. *J. Vib. Control* **2017**, *23*, 782–793. [CrossRef]
25. Li, Y.; Shen, W.; Zhu, H. Vibration mitigation of stay cables using electromagnetic inertial mass dampers: Full-scale experiment and analysis. *Eng. Struct.* **2019**, *200*, 109693. [CrossRef]
26. Jung, H.Y.; Kim, I.H.; Jung, H.J. Feasibility Study of the Electromagnetic Damper for Cable Structures Using Real-Time Hybrid Simulation. *Sensors* **2017**, *17*, 2499. [CrossRef] [PubMed]
27. Li, J.-Y.; Zhu, S. Tunable electromagnetic damper with synthetic impedance and self-powered functions. *Mech. Syst. Signal Process.* **2021**, *159*, 107822. [CrossRef]
28. Hu, G.; Yi, F.; Liu, H.; Zeng, L. Performance Analysis of a Novel Magnetorheological Damper with Displacement Self-Sensing and Energy Harvesting Capability. *J. Vib. Eng. Technol.* **2020**, *9*, 85–103. [CrossRef]
29. Luo, Y.; Sun, H.; Wang, X.; Chen, A.; Zuo, L. Parametric optimization of electromagnetic tuned inerter damper for structural vibration suppression. *Struct. Control Health Monit.* **2021**, *28*, e2711. [CrossRef]
30. Kopylov, S.; Chen, Z.; Abdelkareem, M.A. Back-iron design-based electromagnetic regenerative tuned mass damper. *Proc. Inst. Mech. Eng. Part K J. Multi-Body Dyn.* **2020**, *234*, 607–622. [CrossRef]
31. Wang, J.; Li, G.; Zhang, M.; Zhao, G.; Jin, Z.; Xu, K.; Zhang, Z. Energy harvesting from flow-induced vibration: A lumped parameter model. *Energy Sources Part A Recover. Util. Environ. Eff.* **2018**, *40*, 2903–2913. [CrossRef]
32. Wang, J.; Zhao, W.; Su, Z.; Zhang, G.; Li, P.; Yurchenko, D. Enhancing vortex-induced vibrations of a cylinder with rod attachments for hydrokinetic power generation. *Mech. Syst. Signal Process.* **2020**, *145*, 106912. [CrossRef]
33. Xie, X.; Wang, Z.; Liu, D.; Du, G.; Zhang, J. An experimental study on a novel cylinder harvester made of L-shaped piezoelectric coupled beams with a high efficiency. *Energy* **2020**, *212*, 118752. [CrossRef]
34. Xie, X.; Zhang, J.; Wang, Z.; Song, G. An experimental study on a high-efficient multifunctional U-shaped piezoelectric coupled beam. *Energy Convers. Manag.* **2020**, *224*, 113330. [CrossRef]
35. Xie, X.; Wang, Q.; Wu, N. Energy harvesting from transverse ocean waves by a piezoelectric plate. *Int. J. Eng. Sci.* **2014**, *81*, 41–48. [CrossRef]
36. Xie, X.; Wu, N.; Yuen, K.; Wang, Q. Energy harvesting from high-rise buildings by a piezoelectric coupled cantilever with a proof mass. *Int. J. Eng. Sci.* **2013**, *72*, 98–106. [CrossRef]
37. Hu, X.; Li, Y.; Xie, X. A study on a U-shaped piezoelectric coupled beam and its corresponding ingenious harvester. *Energy* **2019**, *185*, 938–950. [CrossRef]
38. Yoon, S.-J.; Arakawa, K.; Uchino, M. Development of an energy harvesting damper using PVDF film. *Int. J. Energy Res.* **2015**, *39*, 1545–1553. [CrossRef]
39. Khan, F.; Ali, T. A piezoelectric based energy harvester for simultaneous energy generation and vibration isolation. *Int. J. Energy Res.* **2019**, *43*, 5922–5931. [CrossRef]

energies

MDPI

Article

Design of a V-Twin with Crank-Slider Mechanism Wind Energy Harvester Using Faraday's Law of Electromagnetic Induction for Powering Small Scale Electronic Devices

Jamshid Farzidayeri * and Vishwas Bedekar

Department of Engineering Technology, Middle Tennessee State University, 1301 E Main St., Murfreesboro, TN 37132, USA
* Correspondence: jef4k@mtmail.mtsu.edu; Tel.: +1-316-807-0111

Abstract: The maintenance of wireless sensor networks involves challenges such as the periodic replacement of batteries or energy sources in remote locations that are often inaccessible. Therefore, onboard energy harvesting solutions can provide a viable alternative. Experimental energy harvesting from fluid flow, specifically from air flow, is typically restricted to a rotor and stator design or a model that strikes a piezoelectric. On the other hand, energy harvesting from mechanical vibrations routinely uses the linear motion of a magnet passing through a coil or vibrating piezoelectric elements. In this paper, we propose a novel V-twin harvester design that converts wind energy from a rotational input into the linear motion of a magnet inside a coil via a crank-slider mechanism. This design allows for high performance with a smoother voltage output when compared to a reference rotor/stator harvester design or piezoelectric method. At 0.5 Hz, a single crank-slider generated a voltage of 0.176 Vpp with an output power of 0.147 mW, whereas the reference harvester generated 0.14 mW at 1.0 Hz with a 0.432 Vpp. A single crank-slider operating at regulated frequencies of 0.5, 1, 2, and 3 Hz, with a stroke length of 50 mm and a generated continuous power of 0.147, 0.452, 2.00, and 4.48 mW, respectively. We found that under ambient wind speeds of 3.4 and 4.1 m/s the V-twin formation with the optimized configuration, in which the coils and loads were both connected in series, generated 27.0 and 42.2 mW, respectively.

Keywords: energy; electromagnetic; vibration; optimization; sustainability; crank; slider; wind

1. Introduction

The quantity of electronic devices used in our everyday interactions creates a significant demand for electrical power. However, that power comes at a cost, be it monetary, pollution, radiation, or resource depletion. Since 1950, the volume ofretail sales of US electricity has increased from 0.3 trillion kWh to 4.0 trillion kWh [1]. A major concern is whether civilization will be able to globally sustain this rate of consumption. Current estimates suggest that the world's oil reserves may last for only the next 50 years [2]. This limited supply of resources has pushed society to look towards alternative and innovative methods of developing usable energy. The most common large-scale energy alternatives used today are nuclear, wind, solar, hydro, thermo, and bio-fuels. Each of these alternatives has its own benefits and drawbacks. One of the major limitations of each of these is the scope and scale in which they are viable. These listed alternatives require a significant investment in order to be sustainable but their outputs can reach the gigawatt range.

Although power can and is harvested from the previously mentioned alternative sources in order to satisfy the needs of large-scale consumption, there are options available for small-scale devices to harvest their own energy. For example, self-winding watches use the motion of the human body to self-wind the mainspring in order to continually power the device. This energy-providingmotion is readily available, not limited by our depleting natural resources, and easily offsets the power consumption that would otherwise be used by the device. A small-scale energy harvesting device is, by definition, a mechanism that converts energy from an external source into useful electric energy that can provide on-board energy solutions to micro-robots, wrist watches, etc. [3]. Several of the options that are readily available to the average consumer have the following shortcomings:

- Wind/hydro/fluid flow: Requires adequate wind speed, orientation, and a very high contact area.
- Solar: Requires ideal conditions with no cloud cover and a proper angle. Additionally, its components have a high decay rate.
- Electromagnetic: Has size requirements and its moving components can experience failure;
- Piezoelectric: Outputs a high voltage with a low current and its moving components can experience failure.
- Vibration/linear: Must be in line with the device, operate at natural resonance for the maximum output, and its moving components can experience failure.
- Rotational: Produces significant forces when the masses rotate at a large radius, has moving components that can experience failure, and is generally associated with fluid flow, which has several additional restrictions (see above).

Electromagnetic energy harvesting is one such small-scale harvesting method that utilizes mechanical vibrations and oscillations coupled with piezoelectric, electrostatic, magnetostrictive, and electromagnetic devices to generate electricity [4–6]. For example, a piezoelectric tile used in an energy harvesting device placed in a person's shoe can generate up to 39 V [7]. Electrostatic devices are also capable of high power output, obtaining up to 549 μW using a 1 cm^2 chip with an acceleration of 0.6 g [8]. At the current state of development, small-scale devices can range in power consumption from 1 μW for micro-bots up to 10 mW for wireless sensors and hearing aids [9]. With optimization, there are now energy harvesters that generate over 0.1 mW/cm^3 [10]. Myers et al. designed a small-scale windmill, harvesting 5 mW of continuous power at an average wind speed of 10 miles per hour [6]. Dinulovic et al. developed a rotational electromagnetic harvesting transducer that generated 4 mJ at a load of 10 ohm [11]. Luong et al. used a magnetic force exciter to vibrate a piezocomposite generating element in a small-scale windmill [12]. It was able to charge a 40 mA battery in approximately 3 h using natural wind in an urban area. Wang et al. developed a wind energy harvester that generated high power at a high wind speed of 20.3 m/s [13]. Based on several of these experimental designs, an energy harvester the size of a shoe box would have the capacity to charge a cell phone which would use approximately 2 W when charging. Considering the fact that there are over 4 billion cell phones in the world, energy harvesters have the potential to dramatically offset the usage of limited resources. A barrel of oil represents approximately 1700 kWh of energy [14]. Assuming that each cell phone user only charged their device once per week for two hours, harvesters would save roughly half a million barrels of oil per year.

Even though energy harvesters do not fulfil the role of providing energy storage, due to improvements in electronics efficiencies, cost, demand, and flexibility [15], they present an interesting option for creating self-powered electronics, emergency energy assistance, supplemental energy, and so on. A long-term and resilient self-powered device that requires little or no maintenance would serve not only to offset resource consumption but would provide devices that fit a variety of unique situations. For example, medical implants and sensors that monitor or assist in human bodily functions would greatly benefit from a long-term alternative to recharging the device externally and periodically by utilizing piezoelectric or similar devices [16]. Additionally, with ongoing expansion in space exploration, there is a potential demand to develop self-powered electronics. Currently, many such devices have focused on solar power dependence [17–19], yet other sources remain widely unexplored for space applications.

Although the abovementioned mechanisms, such as the piezoelectric effect, the magnetoelectric effect, and magnetostriction, can utilize mechanical vibrations or mechanical energy and convert that into electrical energy [6,20,21], the actual power output and efficiency of these devices is much lower compared to the energy requirements of electronic devices (on the order of several milliwatts [10,22,23]). Hence, in order to harvest high power at lower frequencies, inductive energy harvesting becomes a more reliable solution. Of all the vibration-based electromagnetic harvesters, perhaps one of the most common and simplest is the permanent magnet and coil configuration, which is a linear harvester, as shown in Figure 1. In this arrangement, a coil experiences a changing magnetic field resulting from an oscillating permanent magnet. It is well known from Faraday's law that electromotive force ($\mathcal{E}$) is effected by the number of coils (N) in a changing magnetic flux (Φ) over a change in time (t), as seen in Formula (1). When keeping the volume of the coil space constant, $\mathcal{E}$ can be increased by decreasing the diameter of the wire gauge, which in turn increases the number of turns of the coil. Additionally, increasing the frequency of oscillation also increases $\mathcal{E}$. Finally, improving $\Delta\Phi$ will also increase the output of the harvester. An advantage of this type of harvester is that it can also be easily magnetically coupled or even have its dampening tuned to increase the overall output [23,24].

$$\mathcal{E} = -N\frac{\Delta\Phi}{\Delta t} \tag{1}$$

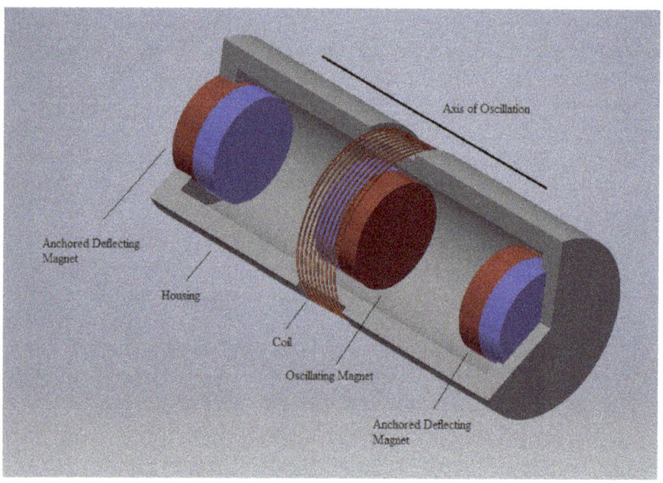

Figure 1. Example of a common small-scale electromagnetic energy harvester, which typically uses one degree of freedom along the axis of oscillation as a method of harvesting energy using vibrations.

Magnetism and magnetic materials have been studied for several centuries for their unique properties and applications. Using Faraday's law of induction, an electromotive force is generated when an electrical conductor is subjected to a changing magnetic field. Several studies have been carried out to investigate energy generation using electromagnetic induction from a variety of sources, such as mechanical vibrations, fluid flow, wind, and mechanical movement. In this study, we investigated the conversion of small-scale rotational energy into electricity using Faraday's law of electromagnetic induction by incorporating a a crank-slider design with a linear electromagnetic energy harvesting system. We believe that the energy generated by such a device can be used for powering small-scale electronic devices.

Researchers are continually exploring several additional styles of experimental harvesting devices. One such design, which uses the typical magnet and coil configuration, is the swing-magnet type, which is used in harvesting bicycle vibrations [25]. Another more common and established configuration of energy harvesting is the previously mentioned large-scale harvesting systems using turbine-style harvesters with wind or water. The most common and recognizable wind harvesters are the large two- and three-bladed versions of the horizontal-axis wind turbines (HAWT) [26,27] generally seen in large wind farms.

Similarly to how an airplane flies, these systems use aerodynamic lift to turn the blades to rotate a shaft that turns the generator [28]. With this type of system, the blades, shaft, and generator are all aligned on the same axis of rotation along the central hub holding the blades and through the generator [27,28].

One of the major uncertaintiesassociated with these HAWT-style systems, or any similar-style harvester, is where to locate the magnets and coils required for electromagnetic harvesting. A rotating object is subject to significant forces (F_c) that are directly proportional to the location of mass (m) and the square of the velocity (V) at the point of rotation, as shown in Formula (2). Furthermore, as shown in Formula (3), the inertia (I) is also greatly affected by the location of mass (m). Therefore, attaching a dense material such as a neodymium magnet or a copper coil with densities of approximately 7300 kg/m^3 and 8920 kg/m^3, respectively, as part of the spinning blades would be unfavorable compared to the commonly used carbon epoxy composite or graphite epoxy composite, which have respective densities of 1446.2 kg/m^3 and 1580 kg/m^3 [29].

$$F_c = m\frac{V^2}{r} \tag{2}$$

$$I = mr^2 \tag{3}$$

Some novel smaller designs have attempted to attach the magnets as part of the blades or as separately rotating components at a radius equivalent to the blades. In these designs, presented in Figures 2 and 3, the magnets move about the axis of rotation at some relatively significant radius. Although they are effective, these designs introduce problems with rotational forces and inertia, as presented in Formulas (2) and (3). As a method to further evaluate our proposed crank-slider design, in this paper we construct and analyze a rotor/stator device based on the designs presented in Figures 2 and 3 and refer to it as the reference harvester.

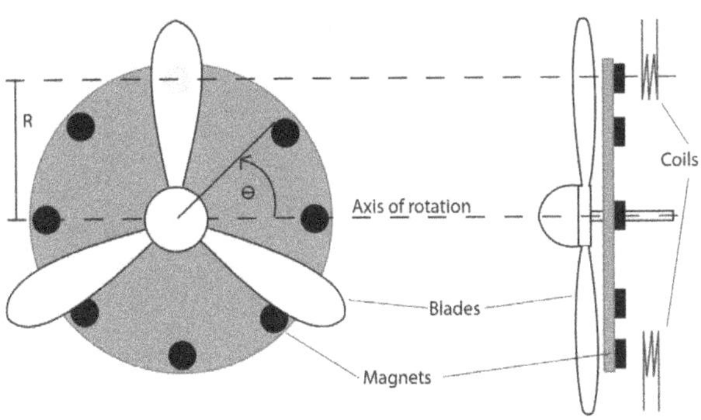

Figure 2. Example of a rotational energy harvester that has the magnets rotating at a radius similar to that of the fan blades [22,30]. Here, although the magnets are not attached directly to the blades, they still have a significantly large mass spinning at a relatively fast velocity and at a large radius from the axis of rotation.

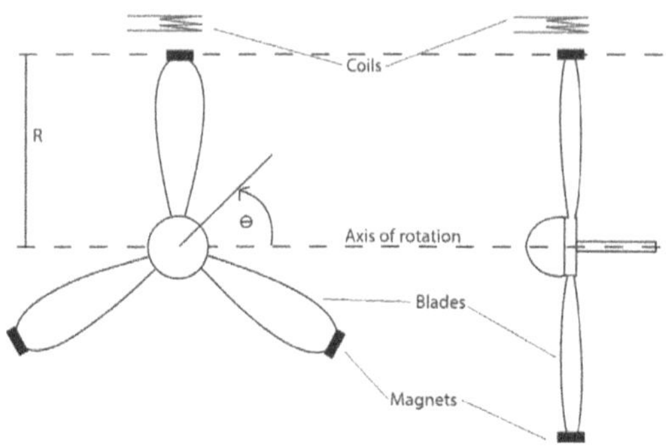

Figure 3. Example of a rotational energy harvester that has the magnets attached to the fan blades [31–33]. In this design, the mass of the magnets at such an extreme radius has a significant effect on the stability, inertia, and structural integrity of the system.

This paper is organized as follows. In the introduction we present a background on electromagnetic harvesters, the need for innovative designs, and the current state of small-scale energy harvesting. In the Section 2, we cover the proposed design, the kinematics of a crank-slider, and the design of the reference harvester. Additionally, in the Section 2, we discuss the formulas used for the evaluations and theoretical outputs of the design. In this section, the formula and kinematics presented are used in combination with COMSOL Multiphysics software for the simulation and modeling of magnetic fields as a method to generate computer-driven predictions for theoretical outputs for the system. For further theoretical evaluations, in the Section 2 we provide a measured prediction, obtained using a Teslameter to physically gather data from the magnetic field. These data were used as the inputs for the formulas to provide an additional set of theoretical outputs for the system. The Section 3 comprises two subsections which separate the single crank-slider harvester topic from the modified V-twin harvester. The first subsection covers the plots produced by the experimental device, including optimized values for power production for the single crank-slider harvester alone. This subsection also presents a comparison of the crank-slider design with the reference harvester, along with their strengths and weaknesses. The second subsection presents a V-twin harvester that utilizes two of the crank-slider harvesters. This subsection provides several output arrangements and compares them to obtain the optimal configuration. Finally, in the Section 4, we present the reader with a novel expansion of the experimental crank-slider harvester, obtained by modifying the design to include the addition of a second crank-slider in the form of a V-Twin-style harvesting device. The Conclusions include relevant data gathered from the V-Twin harvester, along with its possible applications. The final V-twin harvester configuration proposed in this study consists of the following:

- A wind harvester that has the capacity to function at low speeds (3.4 m/s);
- A novel way of converting rotational motion to linear motion via a crank-slider mechanism in a harvester;
- A harvester that is easily transported and small-scale;
- A harvester design that can be easily adapted for higher output by adding more crank-slider mechanisms to the system;
- A rotational design that experiences lower detrimental forces on its components compared to similar designs; and
- A configuration that has a comparable power output to other similar devices, without experiencing potentially damaging high peak voltages.

2. Materials and Methods

Figure 4 shows the crank-slider harvester prototype, along with a reference harvester similar to the ones presented in Figures 2 and 3. The crank-slider kinematics for the dimensions in Figure 5 are outlined in Formulas (4) through (10). The crank-slider harvester has a crank radius (r) of 25.3 mm, an arm length (l) of 222 mm, and a piston distance (d) of 13.9 mm. Formulas (11) through (14) express the rotational mechanics associated with the experimental reference harvester. The experimental reference harvester has a radius (R) of 192 mm. Both the crank-slider and the reference harvester each have two N52 axially magnetized disk magnets with a diameter of 25.4 mm and a thickness of 3.175 that are attached for a total thickness of 6.35 mm and a $B_r max$ of 1.48 T. Both coils (one for the crank-slider and one for the reference harvester) are composed of coated AWG 32 wire, with 150 turns, an inner coil diameter of 29.2 mm, a width of 7.5 mm, a resistance of 8.1 Ω, and a wire length of 14.03 m. In Figure 6, the components included in this device are included within the "Harvester Collection" portion. In order to operate the device, it can be placed on a flat surface and it will operate under wind speeds as low as 3.4 m/s. Furthermore, it can be incorporated with a "conversion and storage" system,which is not included as part of this study, as a method to power or recharge small devices and low-energy sensors and to provide supplemental clean energy.

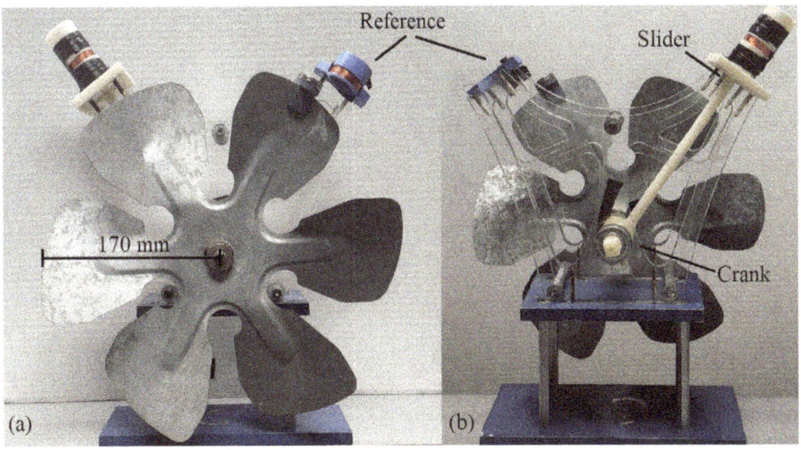

Figure 4. (**a**) Front section of the harvester which includes the crank-slider mechanism, along with the reference harvester as indicated in the style presented in Figures 2 and 3. (**b**) Back section showing the crank-slider mechanism.

The kinematic analysis for the magnet attached to the piston of the crank-slider along the axis of oscillation is as follows. The point x in Figure 5 represents where the crank-slider harvester's magnet is in relation to its distance from the axis of rotation from the driving wheel and is represented as x in Formula (4). This does not take into account the position of the coil, which could be theoretically placed anywhere along the travel path of x. It does include the additional mechanical components required to attach the magnet to the slider, which are represented as d, the distance from the center of the magnet along the axis of travel connecting at the point l.

$$x = rcos(\theta) + \sqrt{l^2 - r^2 sin^2(\theta)} + d \tag{4}$$

The velocity ($\dot{x}$) of the magnet at point x along the travel path in Figure 5 is represented by Formula (5). As shown in Figure 5, the relation of ϕ and θ is expressed in Formula (6) and rearranged into Formula (7). Using Formula (7), ϕ in Formula (5) is replaced, resulting in Formula (7).

$$\dot{x} = -r\omega\left(sin(\theta) + \frac{rsin(2\theta)}{2lcos(\phi)}\right) \tag{5}$$

$$rsin(\theta) = lsin(\phi) \tag{6}$$

$$cos(\phi) = \sqrt{1 - (\frac{rsin(\theta)}{l})^2} \tag{7}$$

$$\dot{x} = -r\omega \left(sin(\theta) + \frac{rsin(2\theta)}{2l\sqrt{1 - (\frac{rsin(\theta)}{l})^2}} \right) \tag{8}$$

In this paper, l will always be assumed to be at least four times greater than r; therefore, the acceleration ($\ddot{x}$) of the magnet at point x at along the travel path in Figure 5 can be approximated with Formula (9), where ϕ is no longer a component [34].

$$\ddot{x} = -r\alpha \left(sin(\theta) + \frac{rsin(2\theta)}{2l} \right) - r\omega^2 \left(cos(\theta) + \frac{rcos(2\theta)}{l} \right) \tag{9}$$

Furthermore, it is assumed that the propulsion of the fan is at a constant velocity in order to evaluate the system. With a constant velocity, the angular acceleration (α) is zero and the first component of Formula (9) can be eliminated, resulting in the reduced Formula (10) [34].

$$\ddot{x} = -r\omega^2 \left(cos(\theta) + \frac{rcos(2\theta)}{l} \right) \tag{10}$$

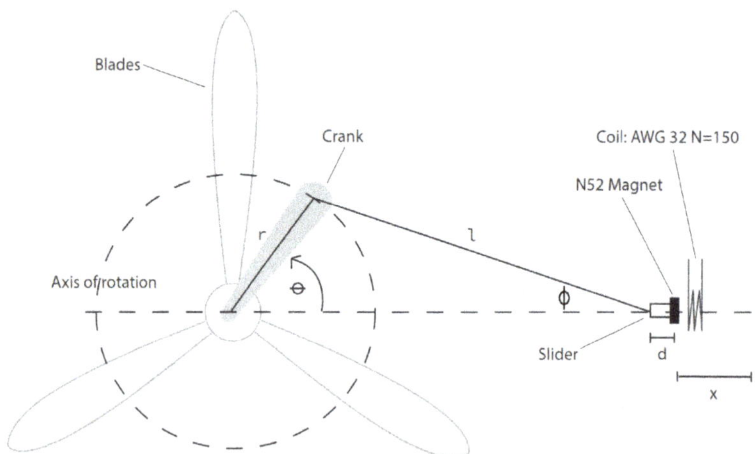

Figure 5. Crank slider mechanism in which the slider is aligned along the axis of rotation for the crank.

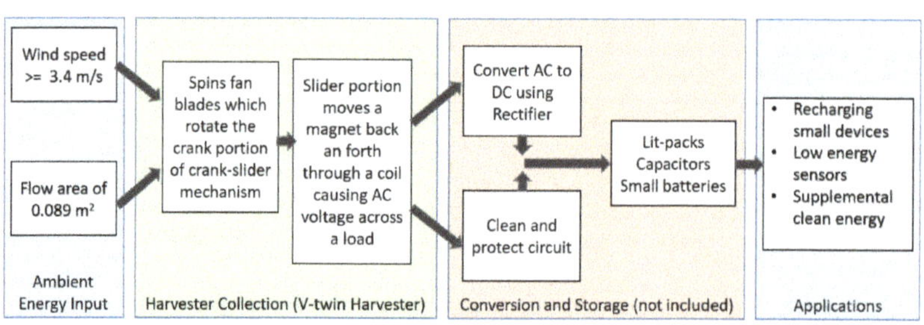

Figure 6. Schematic of wind harvesting process.

Figure 7 is a sample output of $\dot{x}$ from Formula (8) and $\ddot{x}$ from Formula (10) with the given values for ω, r, and l for this experiment. It is significant to note the critical components of motion for the magnet. At θ values equal to 0 or 2π and multiples thereof, the magnet is outside the coil, its $\dot{x}$ is zero, and the magnet is at its minimum and maximum positions relative to the crank. At θ of π, 3π, and their multiples, the magnet's $\dot{x}$ is zero and this is where the theoretical and experimental central position of the coil ismounted. Additionally, the maximum acceleration of the magnet portion of the crank-slider harvester occurs when θ is 0, where the mechanism is at its elongated maximum, and the magnet is outside the coil.

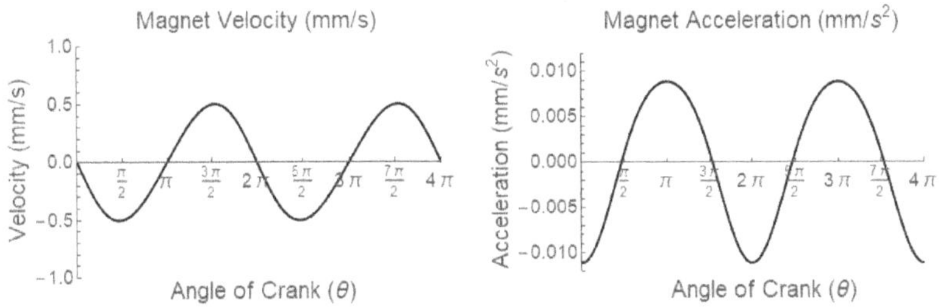

Figure 7. Sample velocity ($\dot{x}$) and acceleration ($\ddot{x}$) of the magnet portion of the crank-slider harvester with ω of 2 π, r = 2.5 mm, and l = 222 mm, using Formulas (8) and (10), respectively [35].

An evaluation of magnet's movement when rotating at the circumference of the edge of the fan blade, as shown in Figures 4 and 8, was performed using the following well-known formulas. The arc distance traveled (D) by the magnet from the initial position is given in Formula (11). The speed (V) of the magnet in Figure 8, given a uniform circular motion, is represented in Formula (12). The centripetal acceleration (a_c) and force (F_c) on the magnet due to a_c are given in Formulas (13) and (14), respectively.

$$D = R\Theta \quad (11)$$

$$V = \omega R \quad (12)$$

$$a_c = \frac{V^2}{R} = R\omega^2 \quad (13)$$

$$F_c = m_{magnet}R\omega^2 = m_{magnet}\frac{V^2}{R} \quad (14)$$

For a crank-slider harvester which utilizes the style presented in Figure 1, the induced voltage (U_v) from the motion of a conductor within a magnetic field (B) over a length of wire (L) at a velocity ($\dot{x}$) is shown in Formulas (15), (16), and reduced to (17) [36]. The values for B are assessed for theoretical analysis in two ways: using the COMSOL values seen in Figure 9 and using the physical readings from a WT10A Teslameter. Those values are plotted in Figure 10.

$$dU_V = (\vec{\dot{x}} \times \vec{B}) \cdot d\vec{L} \quad (15)$$

$$U_V = \oint (\vec{\dot{x}} \times \vec{B}) \cdot d\vec{L} \quad (16)$$

$$U_V = BL\dot{x} \quad (17)$$

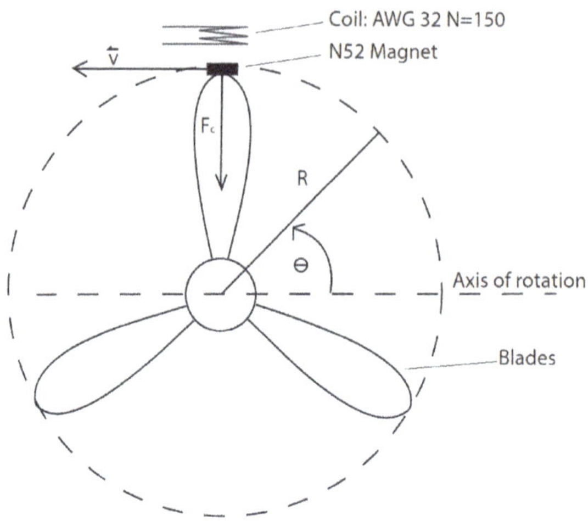

Figure 8. Reference turbine-style harvester with magnets attached to the tips (max radius) of the blades and coil attached to the stationary surface, similar to the designs presented by [22,30].

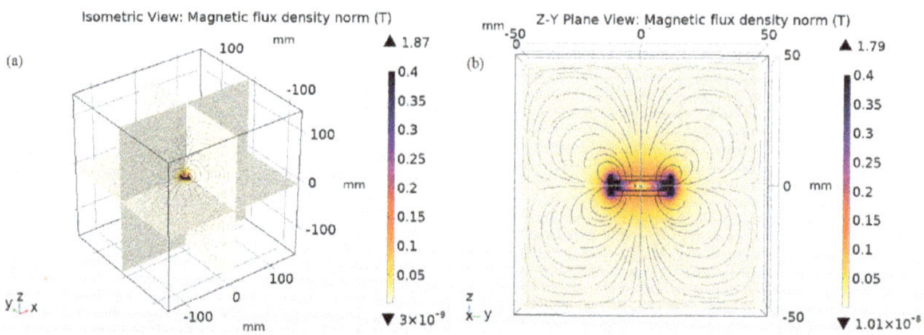

Figure 9. COMSOL-generated field based on two N52 axially magnetized disk magnets with a diameter of 25.4 mm and a thickness of 3.175 that are attached for a total thickness of 6.35 mm and a $B_r max$ of 1.48 T along the z-axis, as referenced in Figures 5 and 8. (**a**) Isometric view with the original boundary conditions of −150 mm to 150 mm used for the data simulation process and theoretical predictions. (**b**) Emphasis of the field at close proximity to the magnet surface along the ZY plane.

In Formula (17), BL is also referred to as the transduction coefficient (Φ); see Formula (18) [36]. The values for B from Figure 10 were used to generate Φ in Figure 11.

$$\Phi = BL \tag{18}$$

For the crank-slider harvester, the theoretical voltage is evaluated by replacing $\dot{x}$ in Formula (17) with the non-linear kinematic evaluation presented in Formula (8) and Figure 5, resulting in Formula (19). In Figure 12, the theoretical voltage is represented using Formula (19) and both measured and COMSOL values for Φ from Figure 11 in comparison to the actual voltage, measured using a RIGOL DS1054 oscilloscope.

$$U_V = \Phi r\omega \left(sin(\theta) + \frac{rsin(2\theta)}{2l\sqrt{1 - (\frac{rsin(\theta)}{l})^2}} \right) \tag{19}$$

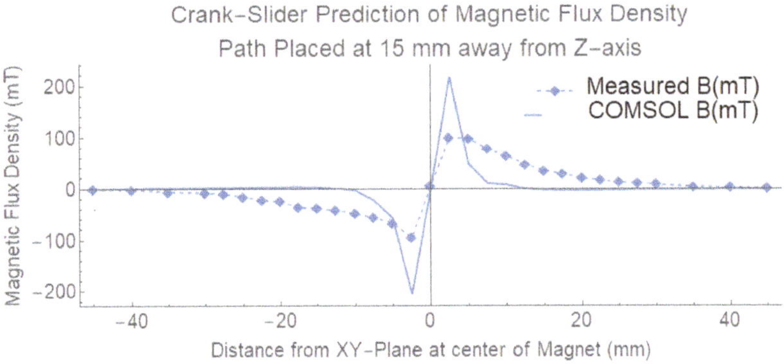

Figure 10. A sample of values for B through the coil along the axis of oscillation. The measured values were physically collected using a WT10 Teslameter at the indicated distances. The COMSOL values were extracted from Figure 9 at a distance of 15 mm from the z-axis (the center of the magnet) and along the axis of oscillation extending ±25 mm from the XY-plane.

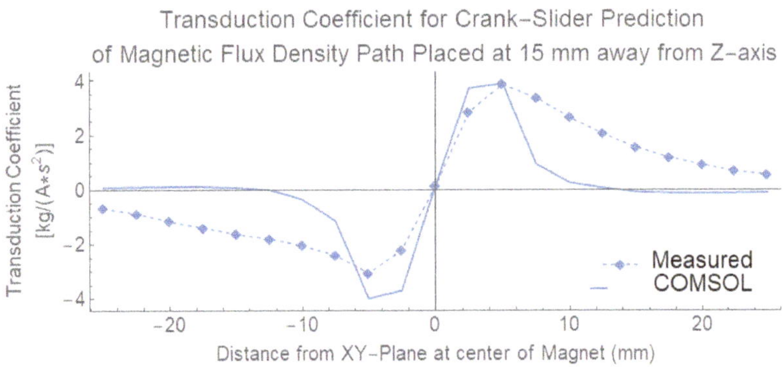

Figure 11. Representation of the transduction coefficient (Φ) based both the measured and COMSOL prediction values from Figure 10 with the l value for the crank-slider harvester. This image represents a one-half stroke from 0 to π of the crank, which would originate at −25 mm and extend to +25 mm from the center of the coil.

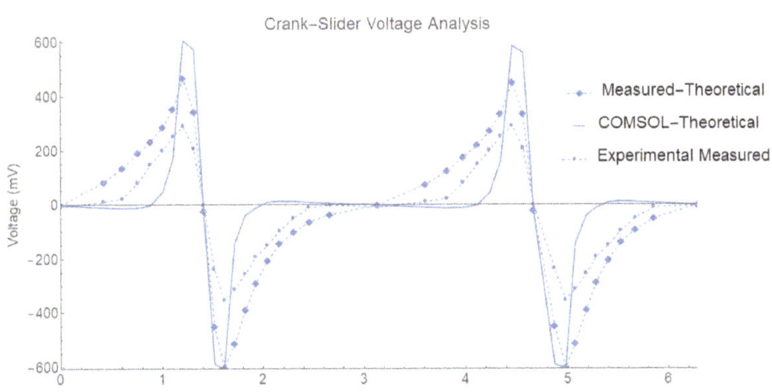

Figure 12. Sample of open circuit voltage output from the crank-slider harvester at 1 Hz. These data represent a full stroke of the crank starting at 0 and ending at 2π, which includes 2 passes through

the coil. The first pass occurs from 0 to π as the slider extends. The second pass through the coil occurs when the slider retracts as the crank moves from π to 2π. The measured-theoretical and COMSOL-theoretical values were both evaluated using the data generated from Figure 11. The experimental measured values are the physically measured values generated by a RIGOL oscilloscope for each of the data points. All measurements and predictions were evaluated at 1 Hz.

Although Formula (19) does not include the additional characteristics of the system, namely, the resistance, inductance, parasitic dampening (D_p), or number of turns (N), an electrical power system produces maximum power when coil impedance matches load impedance [36,37]. In Formula (20), the optimal load occurs when the proportionality constant (k) is less than the first term [37].

$$R_{optimal} = \left(\frac{1}{D_p}\left(\frac{d\phi}{dx}\right)^2 - k\right)N^2 \quad (20)$$

3. Results and Discussion

3.1. Single Crank-Slider Harvester

In Figure 12, the actual output of the crank-slider harvester is lower than the theoretical COMSOL and predicted measured values under the outlined conditions. This is typical of many experimental electromagnetic energy harvesters due to a variety of conditions, such as sub-optimal conditions, the exclusion of friction, and parasitic dampening [36–38]. The measured experimental voltage and power results for the crank-slider harvester at 0.5, 1, 2, and 3 Hz each with a stroke length of 50 mm are shown in Figure 13. In each of the figures, the optimal power can be observed around 8 Ω, which is close to the measured resistance of the coil used, and this is consistent with what was predicted [36,37]. The maximum power produced for the device at regulated frequencies of 0.5 Hz (30 rpm), 1 Hz (60 rpm), 2 Hz (120 rpm), and 3 Hz (180 rpm) were 0.147, 0.452, 2.00, and 4.48 mW, respectively.

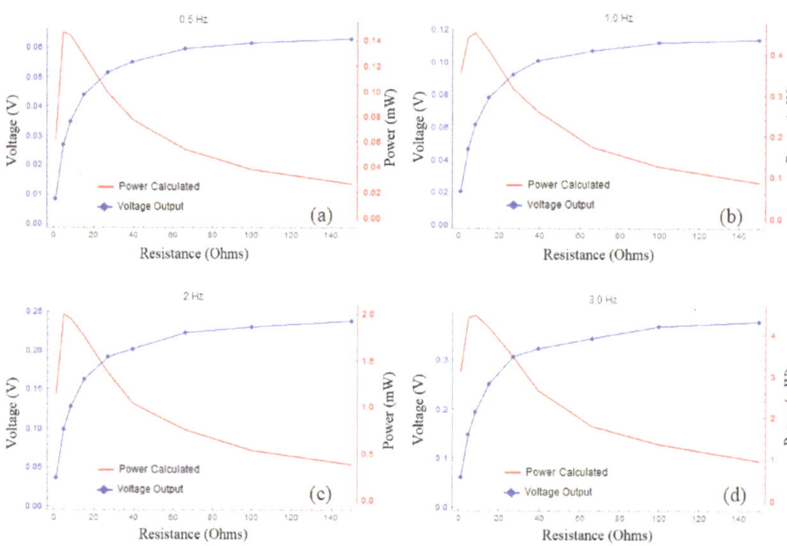

Figure 13. Voltage and power outputs for the crank-slider harvester at 0.5, 1.0, 2.0, and 3.0 Hz. The devices depicted in each figure (**a–d**) had a 50 mm stroke length with the coil located at the center of oscillation. The coil had an internal resistance of 8.1 Ω.

The reference harvester was measured under similar conditions in order to analyze how the crank-slider style of harvester compared to the existing experimental harvester techniques shown in Figures 2 and 3 [22,30–32]. The coil size, wire length, ω, load resistance, and magnet strength of the reference harvester were consistent with the crank-slider during each measurement. The distance R in Figure 8, representing the rotational radius of the rotor's magnet portion in that style of harvester, was measured at 190 mm. In Figure 14, a sample output of the crank-slider was compared to that of the reference harvester at 3 Hz across a load of 8 Ω to indicate several key factors. First, due to the mechanical nature of the crank-slider mechanism, during each rotation of the driving fan, the stroke cycle of the crank-slider's magnet passes through its coil twice for every single pass the reference harvester makes across its coil, respectively. Second, the peak voltage and peak-to-peak voltage (V_{pp}), and thus the peak current of the reference harvester were significantly higher than those of the crank-slider harvester, as shown in Figure 15. Finally, in Figure 16, while still under all the same conditions, the crank-slider produced more power than the reference harvester. It could be argued that the reference harvester was not optimized to compete against the crank-slider. Therefore, one could evaluate the data in Figure 16 at a point at which both systems had a similar power output. At 0.5 Hz, the crank-slider produced 0.147 mW, which matched the power output of 0.14 mW that the reference harvester generated at 1.0 Hz. At those chosen points, the crank-slider has 0.176 V_{pp} and the reference harvester has 0.432 V_{pp} at a coil resistance of 8.1 Ω, which equates to an instantaneous current of 21.5 mA and 53.3 mA, respectively. One way to improve the output of any harvester is to use a smaller gauge wire with more turns (N) for the coil. These significantly larger voltage and current spikes on the part of the reference harvester restrict the allowable wire gauges usable for the system. For example, AWG 38 wire has a max current rating of 22.8 mA, which is out of range for use in the reference harvester under these conditions, yet is still usable by the crank-slider design [39].

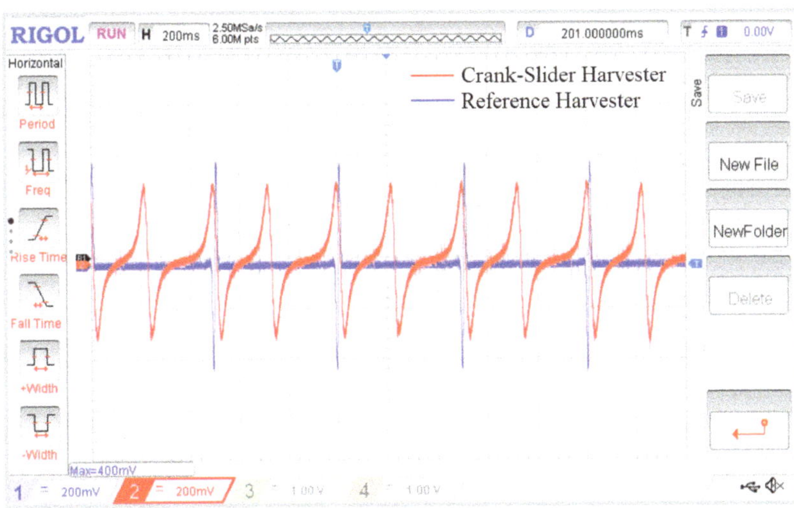

Figure 14. Output at 2 Hz across a 8.4 Ω resistor for the crank-slider harvester vs. the reference harvester, as measured using an RIGOL DS1054 oscilloscope. The crank-slider oscillates across its coil twice for every single pass of the reference harvester. The reference harvester also has a higher peak voltage and a smaller curve area.

Finally, due to the previously discussed mass disparity between the very dense neodymium (7000 kg/m^3) and much less dense connecting materials, such as the ABS plastic (1020 kg/m^3) used, we only compared the forces encountered only by the action of magnet mass measuredat 23.57 g. This excludes the additional forces of friction, gravity, and resistance, as well as excluding any masses of the material used to connect the magnets to the rest of the system. In the reference case, using Formula (14) at 1 Hz, the outward force was 0.177 N. Using Formula (21) and substituting $\ddot{x}$ from Formula (10) to obtain

Formula (22), we can evaluate the force of the magnet on the crank-slider. However, in Figure 7, it can be observed that in this case, where l is much greater than r, the max acceleration occurred when r and l were extended to their longest range, which was when θ equaled zero. The resulting crank-slider harvester maximum force at that moment was 2.35×10^{-3} N. Based on Formula (14) compared to Formula (22), these 1 Hz force calculations increased equivalently for both models at a value of ω^2. Therefore, in all cases of ω for this experiment, the reference harvester would have significantly larger forces acting upon it. Having dense materials located at critical points such as the tips of the blades on the reference harvester, similarly to those presented in Figures 2 and 3, would be problematic considering the fact that the majority of failures for wind turbines are due to mass imbalances [40].

$$F = ma \tag{21}$$

$$F = mr\omega^2\left(cos(\theta) + \frac{rcos(2\theta)}{l}\right) \tag{22}$$

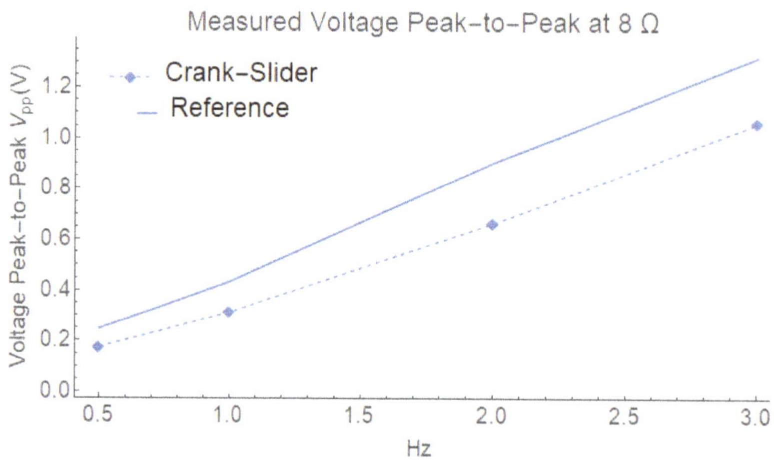

Figure 15. Measured peak-to-peak voltage demonstrating that the reference harvester had higher values than the crank-slider harvester at 0.5, 1, 2, and 3 Hz with a 8.4 Ω load, signifying that the crank-slider harvester would have a greater range of wire gauge available.

3.2. *V-Twin Harvester Utilizing Two Crank-Slider Mechanisms*

Figure 17 depicts a combination of two crank-slider harvesters in a V-twin formation similar to common designs used in motorcycle engines. Due to the magnets being positioned at a large distance relative the drop in magnetic field strength, it is assumed that neither magnet interferes with the other's coil when referring to $\Delta\Phi$ in Formula (1). This configuration allows for several methods of analyzing the output of the system by re-configuring the wiring of the two load resistors and two coils according to the wiring diagram column in Table 1.

This proposed device is simple in design, easily transported, and could be readily re-manufactured with most components being 3D-printed. In addition, the V-twin has the potential to be used in various remote, hazardous, or emergency locations or situations. This design has room to be improved by adding more crank-slider harvesters to the V-twin design, including a rectifier and an energy storage system, or improving the coil structure by using a smaller-gauge wire with more turns.

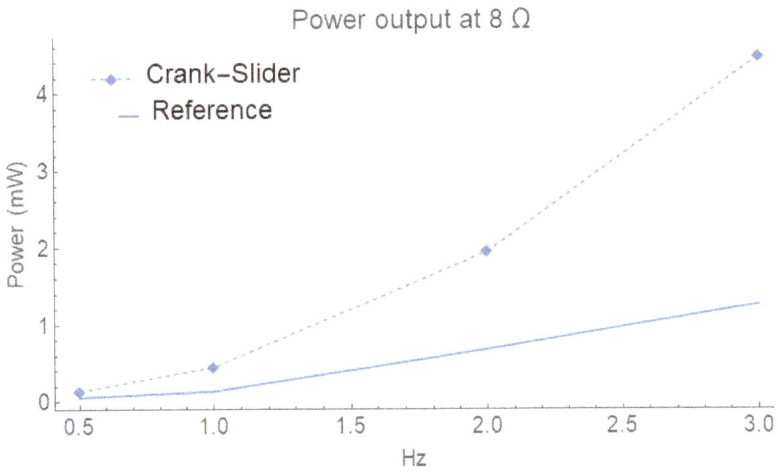

Figure 16. Power output demonstrating that the reference harvester had a lower power output than the crank-slider harvester at 0.5, 1, 2, and 3 Hz with a 8.4 Ω load.

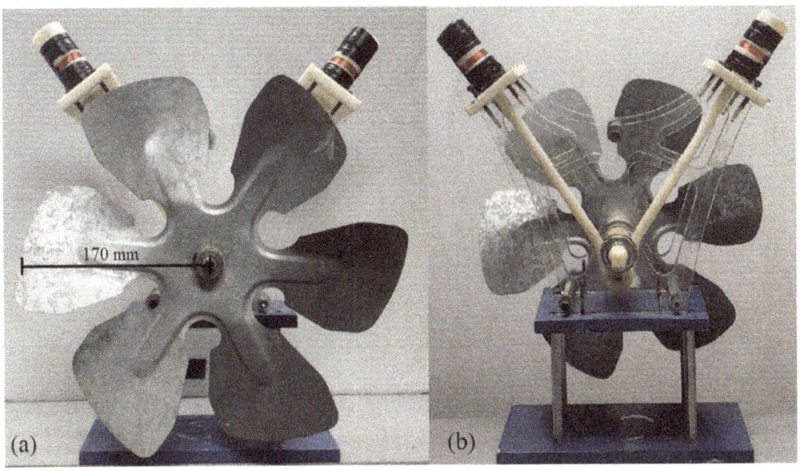

Figure 17. Front (**a**) and back (**b**) V-twin harvester formation.

In Table 1, each measurement was made at 2 Hz for all configurations. In each configuration, the main difference when swapping the leads was observed when using the in- and out-of-sync patterns for the output. For example, in the first configuration image (a) under the oscilloscope output column in Table 1, the output from the V-twin system was reminiscent of a typical sine wave output generated by the typical linear oscillating electromagnetic harvester akin to that in Figure 1. However, swapping the leads, as in image (b), resulted in an interesting output that showed improved performance across all configurations.

In the first configuration, which includes images (a) and (b) from Table 1, both cylinders and their loads were wired together in series with the other cylinder. In (b), the leads from the right cylinder were swapped with those from (a). In the second configuration, which includes images (c) and (d), each cylinder along with its load were wired in parallel with the other cylinder and load. These four configurations performed poorly, with the power output of both harvesters under-performing when having only a single harvester in operation. The highest power was generated in the third configuration, in which the system was wired with the coils in series with each other, and the loads in series but in parallel with the coils as shown in (e) and (f). Considering that a single crank-slider harvester under these conditions generated 2 mW, then it would be reasonable to assume that the standard for two would be 4 mW due to series stacking on phase output. However, as shown in image (f), with an output of 7.30 mW, this configuration produced 45.2% more power than what would be expected by the standard. Conversely, having the system in parallel reduced the overall output, with image (c) showing the poorest performance due to parallel stacking.

Table 1. Output of V-twin harvesters in both series and parallel configuration at 2 Hz. (a) each cylinder and its respective load resistance of 8.2 Ω were wired in series as units. The resulting total load resistance was 16.4 Ω. (b) the output leads of the right cylinder were interchanged from the system shown inimage (a). (c) each cylinder and its respective load resistance of 8.2 Ω in parallel as units. (d) the output leads of the right cylinder were switched from the system shown in image (c). In (e), each cylinder was wired in series and both loads were wired in series; however, both cylinders were kept in parallel with both loads. In (f), the output leads of the right cylinder were switched with those from the system shown in image (e).

Image	Wiring Diagram	Oscilloscope Output	V_{rms} (mV)	Power (mW)
(a)	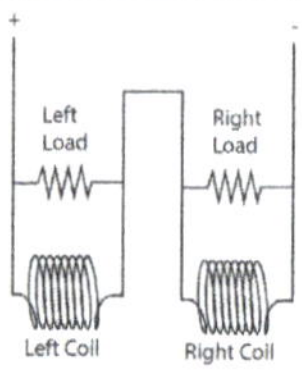	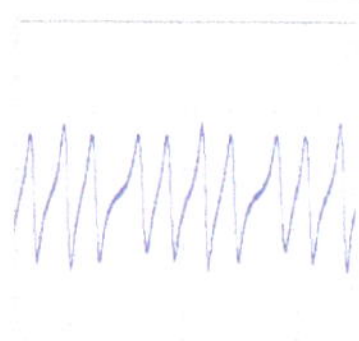	184	2.05
(b)	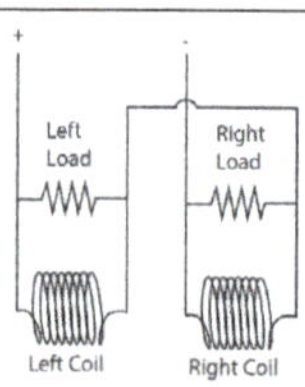	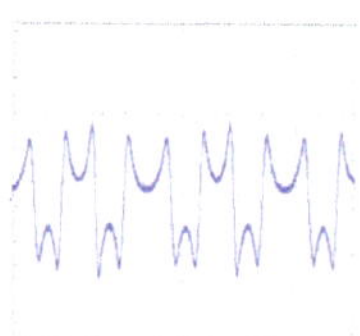	210	2.7
(c)	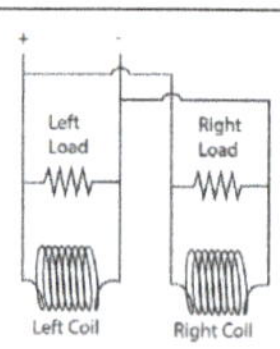	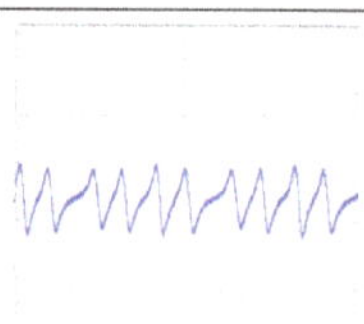	92.5	0.5
(d)	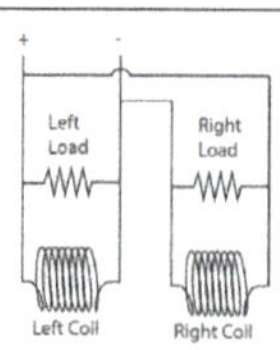	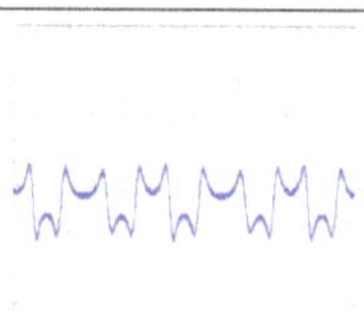	105	0.65
(e)	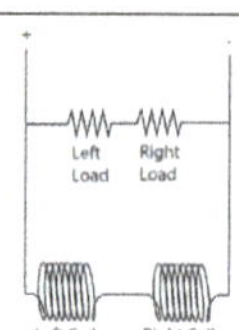	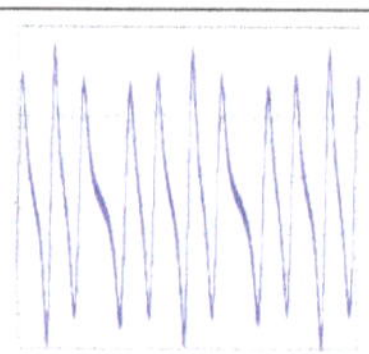	308	5.78

Table 1. *Cont.*

Image	Wiring Diagram	Oscilloscope Output	V_{rms} (mV)	Power (mW)
(f)	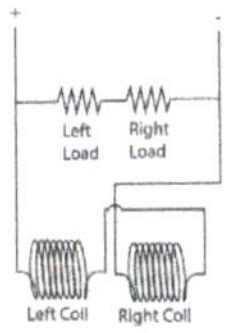	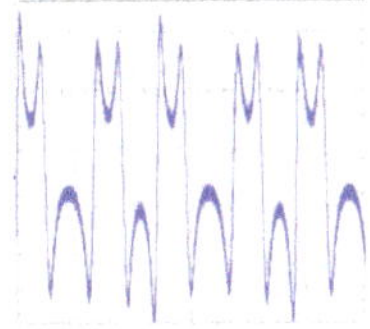	346	7.30

A residential fan was used to supply input wind energy to the device using three standard speeds—3.2, 3.4, and 4.1 m/s—measured at 65 mm from the surface using an XRCLIF-818 Anemometer. We found that the threshold speed fell between 3.2 and 3.4 m/s. At 3.2 m/s, the V-twin harvester would require a small force to overcome initial resisting forces. However, at 3.4 m/s the system could start harvesting with only the residential fan acting upon it.

Table 2, summarizes the performances of comparable style and size for the harvesting devices and indicates that the proposed V-twin harvesters displayed the highest wind speed to power conversion. As shown in Table 2, the V-twin system presented in this paper used the optimal set up shown in image (f) in Table 1 across a load of 16.3 Ω and had measurements taken at 3.4 and 4.1 m/s. The small-scale windmill design by Myers et al. [6] utilizes wind energy harvesting in combination with piezoelectric bimorph transducers. We compared the highest readings (from lab results) listed in a study on highway bridge vibrations by Peigney and Siegert [41].Both vehicle-mounted harvesters developed by Li et al. utilize wind harvesting to excite a piezoelectric device [42]. In their designs, tge difference between FPEH and EVEH was that of a 'fluttering' device versus a linear electromagnetic harvester. A triboelectric-hybrid study by Ye et al. ubvikved multiple devices and configurations; however, the highest performance was observed for the FB-EMG design that used a rotor and stator to harvest wind energy [43]. The outdoor IoT harvester designed by Fang et al. is similar to our reference harvester, utilizing rotating magnets at a large radius to harvest wind [30]. For the galloping wind harvester studied by Wang et al., the SHPTWEH had the highest performance, out of the several versions that were developed [44]. Cao et al. developed a Canyon Bridge system that used a combination of both piezoelectric and electromagnetic harvesters in their design [45].

Table 2. Harvester wind speed testing comparison.

Design	Reference Number	Winds Speed (m/s)	Harvester Frequency (Hz)	Power (mW)	Conversion Mechanism
V-Twin	(f)	3.4	6.38	27.0	Dynamo with rotation
V-Twin	(f)	4.1	7.72	42.2	Dynamo with rotation
small-scale Windmill	[6]	4.47	4.5	5	Piezoelectric with rotation
Highway Bridge Vibrations	[41]	N/A	4.1	1.8	Piezoelectric with vibration
Vehicle-Mounted Harvester (FPEH)	[42]	18	N/A	14.5	Piezoelectric with vibration
Vehicle-Mounted Harvester (EVEH)	[42]	18	N/A	31.8	Dynamo with vibration
Triboelectric-Hybrid (FB-EMG)	[43]	6.96	N/A	4.23	Rotor/stator with rotation
Outdoor IoT Harvester	[30]	12	4.45	62	Rotor/stator with rotation

Table 2. *Cont.*

Design	Reference Number	Winds Speed (m/s)	Harvester Frequency (Hz)	Power (mW)	Conversion Mechanism
Galloping Wind Harvester (SHPTWEH)	[44]	14	N/A	0.238	Piezoelectric—triboelectric with vibration
Canyon Bridge (WEHS)	[45]	6.5	40.56	19.24	Hybrid using both piezoelectric electromagnetic with rotation

4. Conclusions

Based on the analysis of the crank-slider design presented in the Section 3, in this study we designed and fabricated a novel V-twin-shaped energy harvester to convert rotational energy from wind into electrical energy using inductive energy harvesting. From the single magnet and coil crank-slider system operating at a regulated 2 Hz across an optimized load of 8.4 Ω, the average power generated was 2.0 mW. In order to improve the power output at low frequency, a V-twin harvester was designed. This improved design combined two of the crank-slider designs into a single harvester output and it was evaluated under several operating conditions. We experimentally observed that the V-twin system wired in a series configuration produced 7.30 mW at the same regulated frequency of 2 Hz. The results demonstrate an increase in performance over two crank-slider harvesters of 45.2%. We also found that the V-twin harvester generated 27.0 and 42.2 mW of power at wind speeds of 3.4 and 4.1 m/s for the optimized configuration, i.e., series-series electrical connections for coils and loads.

The results show that this is a promising solution for the harvesting of low-to-high-speed wind energy as a method of powering small-scale electronic devices such as cell phones or smart devices. We believe that the energy harvester design proposed here is much cleaner, with less interface, than those of commercially available wind energy harvesters used for generating electricity on a large scale.

Author Contributions: Writing—original draft, J.F. and V.B. All authors have read and agreed to the published version of the manuscript.

Funding: This research received no external funding.

Conflicts of Interest: The authors declare no conflict of interest.

References

1. U.S. EIA (U.S. Energy Information Administration). *Independent Statistics and Analysis;* U.S. EIA: Washington, DC, USA, 2022.
2. BP Plc. *bp Statistical Review of World Energy 2020;* Addison-Wesley: London, UK, 2020.
3. Kiziroglou, M.; Yeatman, E. 17—Materials and techniques for energy harvesting. In *Functional Materials for Sustainable Energy Applications;* Kilner, J.A., Skinner, S.J., Irvine, S.J., Edwards, P.P., Eds.; Woodhead Publishing Series in Energy; Woodhead Publishing: Cambridge, UK, 2012; pp. 541–572. [CrossRef]
4. Khan, F.U.; Ahmad, I. Review of Energy Harvesters Utilizing Bridge Vibrations. *Shock Vib.* **2016**, *2016*, 1340402. [CrossRef]
5. Roundy, S. On the Effectiveness of Vibration-based Energy Harvesting. *J. Intell. Mater. Syst. Struct.* **2005**, *16*, 809–823. [CrossRef]
6. Myers, R.; Vickers, M.; Kim, H.; Priya, S. Small scale windmill. *Appl. Phys. Lett.* **2007**, *90*, 054106. [CrossRef]
7. Dutta, M.; Shrimoyee, P. Footstep voltage generator using piezo-electric transducers. *Int. J. Sci. Eng. Res.* **2017**, *8*, 117–120.
8. Crovetto, A.; Wang, F.; Hansen, O. Modeling and Optimization of an Electrostatic Energy Harvesting Device. *J. Microelectromech. Syst.* **2014**, *23*, 1141–1155. [CrossRef]
9. Annapureddy, V.; Palneedi, H.; Hwang, G.T.; Peddigari, M.; Jeong, D.Y.; Yoon, W.H.; Kim, K.H.; Ryu, J. Magnetic energy harvesting with magnetoelectrics: An emerging technology for self-powered autonomous systems. *Sustain. Energy Fuels* **2017**, *1*, 2039–2052. [CrossRef]
10. Chen, W.T.; Gurdal, A.E.; Tuncdemir, S.; Gambal, J.; Chen, X.M.; Randall, C.A. Introducing an extremely high output power and high temperature piezoelectric bimorph energy harvester technology based on the ferroelectric system $Bi(Me)O_3$-$PbTiO_3$. *J. Appl. Phys.* **2020**, *128*, 144102. [CrossRef]
11. Dinulovic, D.; Brooks, M.; Haug, M.; Petrovic, T. Rotational Electromagnetic Energy Harvesting System. *Phys. Procedia* **2015**, *75*, 1244–1251. [CrossRef]

12. Luong, H.T.; Goo, N.S. Use of a magnetic force exciter to vibrate a piezocomposite generating element in a small-scale windmill. *Smart Mater. Struct.* **2012**, *21*, 025017. [CrossRef]
13. Wang, X.; Pan, C.L.; Liu, Y.B.; Feng, Z.H. Electromagnetic resonant cavity wind energy harvester with optimized reed design and effective magnetic loop. *Sens. Actuators A Phys.* **2014**, *205*, 63–71. [CrossRef]
14. Chen, J. Barrel of Oil Equivalent (BOE). 2020. Available online: https://www.investopedia.com/terms/b/barrelofoilequivalent.asp#:\protect\T1\textdollar\sim\protect\T1\textdollar:text=%20There%20are%2042%20gallons%20(approximately,have%20slightly%20different%20energy%20%20equivalents (accessed on 27 July 2022).
15. Cepnik, C.; Lausecker, R.; Wallrabe, U. Review on electrodynamic energy harvesters—A classification approach. *Micromachines* **2013**, *4*, 168–196. [CrossRef]
16. Shi, B.; Li, Z.; Fan, Y. Implantable energy-harvesting devices. *Adv. Mater.* **2018**, *30*, 1801511. [CrossRef] [PubMed]
17. Michaud, S.; Schneider, A.; Bertrand, R.; Lamon, P.; Siegwart, R.; Winnendael, M.v.; Schiele, A. SOLERO: Solar powered exploration rover. In Proceedings of the 7th ESA Workshop on Advanced Space Technologies for Robotics and Automation (ASTRA), Noordwijk, The Netherlands, 19–21 November 2002; ETH-Zürich: Zurich, Switzerland, 2002.
18. Lamarre, O.; Kelly, J. Overcoming the challenges of solar rover autonomy: Enabling long-duration planetary navigation. *arXiv* **2018**, arXiv:1805.05451.
19. Landis, G.A. Exploring mars with solar-powered rovers. In Proceedings of the Conference Record of the Thirty-First IEEE Photovoltaic Specialists Conference, Lake Buena Vista, FL, USA, 3–7 January 2005; IEEE: New York, NY, USA, 2005; pp. 858–861.
20. Zhao, L.C.; Zou, H.X.; Yan, G.; Liu, F.R.; Tan, T.; Wei, K.X.; Zhang, W.M. Magnetic coupling and flextensional amplification mechanisms for high-robustness ambient wind energy harvesting. *Energy Convers. Manag.* **2019**, *201*, 112166. [CrossRef]
21. Izadgoshasb, I.; Lim, Y.Y.; Vasquez Padilla, R.; Sedighi, M.; Novak, J.P. Performance enhancement of a multiresonant piezoelectric energy harvester for low frequency vibrations. *Energies* **2019**, *12*, 2770. [CrossRef]
22. Fan, K.; Qu, H.; Wu, Y.; Wen, T.; Wang, F. Design and development of a rotational energy harvester for ultralow frequency vibrations and irregular human motions. *Renew. Energy* **2020**, *156*, 1028–1039. [CrossRef]
23. Li, X.; Meng, J.; Yang, C.; Zhang, H.; Zhang, L.; Song, R. A magnetically coupled electromagnetic energy harvester with low operating frequency for human body kinetic energy. *Micromachines* **2021**, *12*, 1300. [CrossRef] [PubMed]
24. Tang, X.; Zuo, L. Simulation and experiment validation of simultaneous vibration control and energy harvesting from buildings using tuned mass dampers. In Proceedings of the 2011 American Control Conference, San Francisco, CA, USA, 29 June–1 July 2011; IEEE: New York, NY, USA, 2011; pp. 3134–3139.
25. Sato, M.; Takemura, T.; Mizuno, T. Voltage Improvement of a Swing-Magnet-Type Generator for Harvesting Bicycle Vibrations. *Energies* **2022**, *15*, 4630. [CrossRef]
26. National Geographic Society (NGS). *Wind Energy*; NGS: Washington, DC, USA, 2022.
27. Contreras Montoya, L.T.; Hayyani, M.Y.; Issa, M.; Ilinca, A.; Ibrahim, H.; Rezkallah, M. 8—Wind power plant planning and modeling. In *Hybrid Renewable Energy Systems and Microgrids*; Kabalci, E., Ed.; Academic Press: Cambridge, MA, USA, 2021; pp. 259–312. [CrossRef]
28. Hyams, M. Wind energy in the built environment. In *Metropolitan Sustainability*; Elsevier: Amsterdam, The Netherlands, 2012; pp. 457–499.
29. Appadurai, M.; Raj, E.F.I. Finite Element Analysis of Composite Wind Turbine Blades. In Proceedings of the 2021 7th International Conference on Electrical Energy Systems (ICEES), Virtual, 11–13 February 2021; pp. 585–589. [CrossRef]
30. Fang, Y.; Tang, T.; Li, Y.; Hou, C.; Wen, F.; Yang, Z.; Chen, T.; Sun, L.; Liu, H.; Lee, C. A high-performance triboelectric-electromagnetic hybrid wind energy harvester based on rotational tapered rollers aiming at outdoor IoT applications. *iScience* **2021**, *24*, 102300. [CrossRef]
31. Pozo, B.; Gárate, J.; Araujo, J.; Ferreiro, S. Energy Harvesting Technologies and Equivalent Electronic Structural Models—Review. *Electronics* **2019**, *8*, 486. [CrossRef]
32. Pozo, B.; Araujo, J.; Zessin, H.; Mateu, L.; Gárate, J.; Spies, P. Mini Wind Harvester and a Low Power Three-Phase AC/DC Converter to Power IoT Devices: Analysis, Simulation, Test and Design. *Appl. Sci.* **2020**, *10*, 6347. [CrossRef]
33. Narolia, T.; Gupta, V.; Parinov, I. Design and analysis of a shear mode piezoelectric energy harvester for rotational motion system. *J. Adv. Dielectr.* **2020**, *10*, 2050008. [CrossRef]
34. Shigley, J.E. *Series in Mechanical Engineering*; McGraw Hill Book Company Inc.: New York, NY, USA, 1961.
35. McCaslin, S.; Brown, F. Slider and Crank Mechanism. 2010. Available online: https://demonstrations.wolfram.com/SliderAndCrankMechanism/ (accessed on 27 July 2022).
36. Bedekar, V.; Oliver, J.; Priya, S. Pen harvester for powering a pulse rate sensor. *J. Phys. D Appl. Phys.* **2009**, *42*, 105105. [CrossRef]
37. Saha, C.R.; O'Donnell, T.; Loder, H.; Beeby, S.; Tudor, J. Optimization of an electromagnetic energy harvesting device. *IEEE Trans. Magn.* **2006**, *42*, 3509–3511. [CrossRef]
38. Xu, Z.; Shan, X.; Chen, D.; Xie, T. A novel tunable multi-frequency hybrid vibration energy harvester using piezoelectric and electromagnetic conversion mechanisms. *Appl. Sci.* **2016**, *6*, 10. [CrossRef]
39. Teja, R. Wire Gauge Chart | American Wire Gauge (AWG) Wire Size Chart. 2021. Available online: https://www.electronicshub.org/wire-gauge-chart/ (accessed on 27 July 2022).
40. Chen, J.; Hu, W.; Cao, D.; Zhang, B.; Huang, Q.; Chen, Z.; Blaabjerg, F. An imbalance fault detection algorithm for variable-speed wind turbines: A deep learning approach. *Energies* **2019**, *12*, 2764. [CrossRef]

41. Peigney, M.; Siegert, D. Low-frequency electromagnetic energy harvesting from highway bridge vibrations. *J. Bridge Eng.* **2020**, *25*, 04020056. [CrossRef]
42. Li, X.; Li, Z.; Bi, C.; Liu, B.; Su, Y. Study on wind energy harvesting effect of a vehicle-mounted piezo-electromagnetic hybrid energy harvester. *IEEE Access* **2020**, *8*, 167631–167646. [CrossRef]
43. Ye, C.; Dong, K.; An, J.; Yi, J.; Peng, X.; Ning, C.; Wang, Z.L. A triboelectric–electromagnetic hybrid nanogenerator with broadband working range for wind energy harvesting and a self-powered wind speed sensor. *ACS Energy Lett.* **2021**, *6*, 1443–1452. [CrossRef]
44. Wang, Q.; Zou, H.X.; Zhao, L.C.; Li, M.; Wei, K.X.; Huang, L.P.; Zhang, W.M. A synergetic hybrid mechanism of piezoelectric and triboelectric for galloping wind energy harvesting. *Appl. Phys. Lett.* **2020**, *117*, 043902. [CrossRef]
45. Cao, H.; Wu, X.; Wu, H.; Pan, Y.; Luo, D.; Azam, A.; Zhang, Z. A Hybrid Self-Powered System Based on Wind Energy Harvesting for Low-Power Sensors on Canyon Bridges. *Int. J. Precis. Eng. Manuf. -Green Technol.* **2022**. [CrossRef]

energies

MDPI

Article

Design Optimization of a Rotary Thermomagnetic Motor for More Efficient Heat Energy Harvesting

Jonathan Hey [1,*], Maheswar Repaka [2], Tao Li [1] and Jun Liang Tan [1]

[1] Singapore Institute of Manufacturing Technology, 2 Fusionopolis Way, Singapore 138634, Singapore
[2] Institute of Materials Research and Engineering, 2 Fusionopolis Way, Singapore 138634, Singapore
* Correspondence: jonathan-hey@simtech.a-star.edu.sg

Abstract: A rotary thermomagnetic motor that is designed for heat energy harvesting is presented in this paper. The power output, power density, and efficiency of the device is estimated using a mathematical model coupling the heat transfer, magnetic interactions, and rotor dynamics. The design analysis shows that the efficiency of the device is maximized, when there is a balance between the volume of thermomagnetic material used against the rate of heating and cooling of the material. On the other hand, the power output is determined largely by the size of the rotor, while the power density tends to peak at a particular aspect (length to diameter) ratio of the rotor. It is also observed that a higher rate of cooling leads to more output, especially when this is matched to a similar rate of heat supplied to the thermomagnetic motor. The result from the design optimization points to an 'optimal' design configuration and corresponding operating conditions that results in the largest power output, highest power density and best efficiency. After the optimization, it is estimated that the rotary thermomagnetic motor is able to produce up to 88 W of power with a power density of approximately 27 kW/m^3 of thermomagnetic material used, while a maximum thermal-to-mechanical energy conversion efficiency of 2.1% is achievable. The results obtained from this design analysis and optimization shows the potential for such a rotary thermomagnetic motor to be implemented at a larger scale for heat energy harvesting application.

Keywords: design optimization; thermomagnetic motor; energy harvesting

1. Introduction

Low-grade heat is potentially a source of energy as it is a by-product of many physical processes. The ability to harvest such low-grade heat would allow for some part of this otherwise wasted heat to be reused or recovered [1]. There are various types of thermo-responsive materials that have been used in heat energy harvesting devices, such as thermoelectric, pyroelectric, thermomagnetic, etc. [2]. Thermomagnetic material exhibits changes in its permeability to a magnetic field, when subjected to a small step change in temperature. This makes such materials suited for harvesting heat energy by converting it into other useful forms, such as electric potential. Thermomagnetic materials that have low Curie temperatures show high energy conversion efficiency [3]. There have been recent developments in the synthesis of new soft magnetic materials that have Curie temperatures that are even closer to room temperature [4,5]. These advancement in the materials make them even more suitable for low-grade heat harvesting (heat source with temperature of less than 80 °C).

There are mainly two categories of thermomagnetic devices that are developed for heat energy harvesting. Firstly, thermomagnetic generators makes use of the thermomagnetic material to turn 'on' and 'off' an external magnetic field to create a fluctuating magnetic field in a receiving coil which produces an electric potential that can be harnessed [6]. Secondly, thermomagnetic motors makes use of the difference in permeability in the material to directly produce a force or torque [7]. In both type of devices, the output is often limited by the change of the material's permeability against temperature and indirectly by its temperature response to the external heat source [8,9]. It is also a fact that not all the heat from the source can be fully utilized due to the limits of heat transfer at a lower temperature, while other parasitic heat losses tend to dominate when operating at higher temperatures [10]. There are also inherent losses in the material, such as the hysteresis that happens during each cycle of magnetization and demagnetization [11]. This is partly why even though the thermodynamic efficiency of the material is considerably high, the practical devices tend to have significantly lower energy conversion efficiency.

Typical power output from thermomagnetic devices tends to be in the range of up to several hundreds of mili Watts [12]. However, earlier development of a rotary disc type thermomagnetic motor shows that up to 6 W of continuous mechanical output can be achieved or an equivalent power density of 60 kW/m^3 per unit volume of thermomagnetic material used [13]. Rotary thermomagnetic devices have been shown to have higher power outputs due to the continuous nature in which the reluctance torque is harnessed to generate the output. Rotating thermomagnetic topology has added mechanical advantage due to the inertia of the rotating assembly, which works in favor of the power transfer after the initial breakaway. More recently, industrial scale rotary thermomagnetic motor prototype developed by Swiss Blue Energy AG is able to produce up to 1 kW of power [14]. Rotating topology would allow for the power density to scale beyond the current threshold, which is often limited by the frequency of heating and cooling cycle [15,16]. Yet, the output is limited due to the eddy current braking and viscous friction, especially at higher rotation speeds [13,17]. Although there have been similar rotating thermomagnetic motors reported in the prior art, there is still a lack of published work about the scalability of such rotary thermomagnetic devices [7,18] as well as the potential efficiency gains through further optimization [8,19]. Furthermore, there has been a wide range of power densities reported so far, but there is little insight into what actually limits the power density of such devices [12,20].

In this paper, a rotary thermomagnetic motor is presented with a mathematical model that couples the heat transfer, magnetic interactions, and rotor dynamics. This multi-physics model is used to make predictions of its power output, power density, and efficiency against the various design and operating parameters. For instance, the critical dimensions of thermomagnetic material and rotor as well as operation conditions, such as the temperature of the heat source, the rate of heating, and cooling, are considered in this design analysis. Furthermore, an optimization method is implemented to determine the optimal design configuration and the corresponding operating conditions that maximizes each of the output parameters. Critical design and operating parameters are identified through the analysis. In addition, recommendations are given for the suitable selection of parameters for the desired output characteristics.

2. Material Selection and Characterization

Thermomagnetic materials exhibit the magnetocaloric effect (MCE) whereby the temperature of the material changes in response to the magnetic field. The process is reversible in which a temperature change also brings about a change in magnetization in the material. Although naturally occurring ferromagnetic materials, such as iron, nickel, and cobalt, exhibit the MCE at their Curie temperature; this typically happens at a high temperature, which makes these materials less useful for low-grade heat energy harvesting applications. Alloys of such pure metals infused with other metals, such as chromium, manganese, and germanium, show more desirable characteristics and lower the Curie temperature [11]. For instance, manganese based thermomagnetic materials have garnered increasing interest due to the large magnetocaloric effect [21]. A suitable composition of the thermomagnetic material is selected to enable the thermomagnetic motor to harvest low-grade heat. This would require the thermomagnetic material to have a sufficiently low Curie transition temperature of significantly less than 80 °C. The selection criteria includes a high level of change in the magnetization level following the phase transformation that is induced by temperature changes or a large MCE. In addition, there should be minimal or little hysteresis that occurs following a transition.

The thermomagnetic material undergoes a series of changes in its state of magnetization and temperature in the thermomagnetic cycle illustrated in Figure 1a. The magnetic field is usually generated externally by permanent magnets. Assuming that the initial temperature state of the material is below the Curie temperature ($T_1 < T_C$) and it is unmagnetized (at point 1). At this point, the material is ferromagnetic, which means that it allows the magnetic field lines from the external source to pass through it relatively easily. It is then brought near to the magnetic source and becomes magnetized (at point 2). At the next step, the heat load is introduced to the thermomagnetic material, which causes its temperature to rise above the curie temperature, $T_2 > T_C$. At this point, the material becomes paramagnetic, which results in fewer magnetic field lines passing through it (at point 3). Thereafter, the external magnetic field is removed and the cooling process is started until it returns to its initial ferromagnetic state (at point 4), where the cycle can be repeated again.

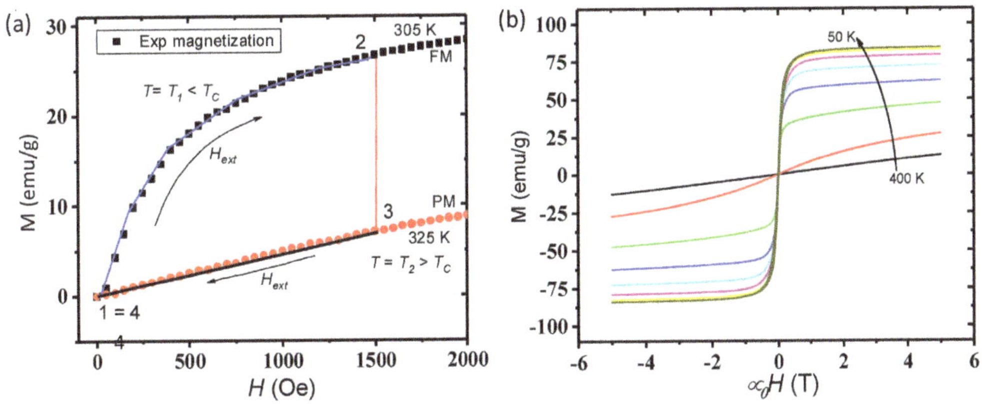

Figure 1. (**a**) Thermomagnetic cycle and (**b**) Magnetization and field strength (M–H) measurement data.

It is necessary to quantify the magnetic energy conversion and to evaluate the conversion efficiency in order to select the appropriate thermomagnetic material for heat harvesting application. The maximum magnetic energy conversion that happens in a material during magnetization and demagnetization is obtained from the area enclosed by the M–H curve (as illustrated in Figure 1a). The area can be expressed mathematically as:

$$W = \mu_0 \oint HdM \tag{1}$$

The magnetic energy converted in each cycle is equivalent to the net work done by the material. The total heat gained by the material is equivalent to the sensible heat gain due to temperature changes as well as any entropy generated by the expression in (2). The isothermal heat addition due to entropy is ignored in this case as it is much lower compared to the sensible heat addition in a heat harvesting application [3].

$$Q = \rho \int_{T_c}^{T_h} c_p(T)dT + T \int dS_m \approx \rho \int_{T_c}^{T_h} c_p(T)dT \tag{2}$$

The efficiency of the thermodynamic cycle is measured by the ratio of the work done to the total heat addition per unit volume of the material as defined in (3).

$$\eta = \frac{W}{Q} \approx \frac{\mu_0 \oint HdM}{\rho \int_{T_c}^{T_h} C_p(T)dT} \tag{3}$$

The thermomagnetic material is processed by the powder metallurgy method. The process starts with ball milling of the powder of the raw materials mixed according to the final composition of the thermomagnetic material that is desired. This is then followed by the compaction of the finely mixed powders under high pressure. Finally, the compacted part goes through a heat treatment via a sintering process. Thermomagnetic material with the following composition, Mn_2Fe_1Sn, has been synthesized and produced to the desired shape and size (Figure 2). The final part that is produced has a density of approximately $\rho = 7.2\ \mathrm{g/cm^3}$. Samples of the material are characterized using the physical property measurement system (PPMS) and vibrating sample magnetometer (VSM) instrument to determine its temperature dependent magnetic properties. Figure 1b shows the M–H curve for the thermomagnetic material at different temperatures obtained through the PPMS. From this set of measurement data, it is possible to work out the thermomagnetic material energy conversion efficiency as defined by (3).

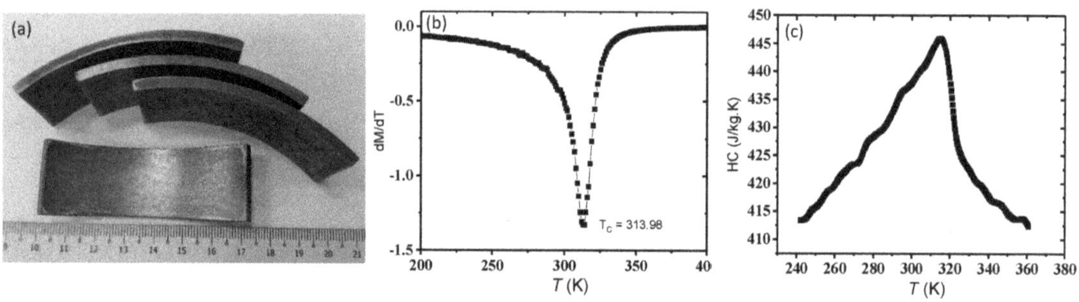

Figure 2. (**a**) Shape and size of the thermomagnetic material, (**b**) Curie temperature, and (**c**) heat capacity measurement data.

Firstly, it is necessary to identify the four operating points from this set of data that describes the thermodynamic cycle illustrated earlier in Figure 1a. The curie temperature of this composition of material is approximately $T_{cu} = 314$ K based on the measurement data shown in Figure 2b. The temperature difference between the hot and cool state is kept at $\Delta T = 20$ K. This means that $T_h = 324$ K, while $T_c = 304$ K to achieve the targeted temperature fluctuation around the Curie temperature. The applied magnetic field is estimated to be approximately 1500 Oe, which is the magnetic field strength at some distance away from the permanent magnet source. Based on the above operating conditions, the magnetic energy converted in each cycle is equivalent to approximately $W = 1.7\ \mathrm{J/cm^3}$. On the other hand, the sensible heat addition works out to be approximately $Q = 63.4\ \mathrm{J/cm^3}$ (from Figure 2c). This gives a cycle averaged thermal to magnetic energy conversion efficiency of approximately 2.6%. This would serve as the basis for comparison with the energy conversion efficiency that can be achieved by the rotary thermomagnetic motor, which will be discussed shortly.

3. Mathematical Modelling

3.1. Magnetic Analysis

The resultant magnetic fields from the permanent magnets and their interaction with the thermomagnetic material is estimated with Maxwell's equations of electromagnetics described here. Specifically, Gauss law states that the divergence of the magnetic field of enclosed volume is equal to zero or $\nabla \cdot B = 0$. The curl of the magnetic field is equal to current density or $J = \nabla \times H$ as defined by the Ampere–Maxwell law. In addition, when there is no external magnetic field present, the flux density is proportional to the field intensity ($B = \mu_0 H$), which leads to the following expression:

$$\nabla \times B = \mu_0 J \tag{4}$$

From the vector mathematics and Gauss law, the flux density B can be represented as curl of a vector potential A or $B = \nabla \times A$, which when substituted into (4) yields the following description of the field in terms of the vector potential:

$$\nabla \times (\nabla \times A) = \mu_0 J \tag{5}$$

When the coulomb gauge condition is met, then (5) simplifies further to yield the equation of magnetostatics given by (6) which is applicable when a steady current is the only magnetic source [22,23].

$$\nabla^2 A = -\mu_0 J \tag{6}$$

In the regions where there are permanent magnetic materials present, the constitutive relationship, $B = \mu_0(H + M)$, is true whereby M is the magnetization field vector contributed by the magnetic material, which can be expressed as an equivalent current density $(J_M = \nabla \times M)$. For linear and isotropic magnetic materials, the magnetization of the material is proportional to the magnetic field intensity such that $M = \chi H$ whereby χ is the magnetic susceptibility of the material. This gives rise to the following form of the constitutive relationship, $B = \mu_0(1 + \chi)H$, which is expressed in terms of relative permeability of the material, $\mu_r = (1 + \chi)$.

The magnetostatic equation defined in (6) can be further generalized into (7) including the effects of the material.

$$\begin{aligned} \nabla \times H &= J_M \\ \nabla \times \left(\tfrac{1}{\mu_0 \mu_r} B\right) &= J_M \end{aligned} \tag{7}$$

Additionally, with the earlier definition of the vector potential, the magnetostatic equation can be expressed in terms of the vector potential as:

$$\nabla \times \left(\frac{1}{\mu_0 \mu_r} \nabla \times A\right) = J_M \tag{8}$$

Equation (8) is valid where linear and isotropic permanent magnetic materials are present.

The magnetic force acting on the volume of thermomagnetic material to produce torque under an external magnetic field with flux density (B) is given by:

$$F = \frac{1}{\mu_0} \iiint dV\, (\nabla \times B) \times B \tag{9}$$

The volume integral can be expressed as a surface integral by taking the integral of the normal components of flux acting on the bounding surface S that encloses the whole volume. Therefore, the magnetic force can be represented in terms of the surface flux density using the stress tensor [24] as:

$$F = \oint_s n \cdot \sigma\, dS \tag{10}$$

where n is the vector normal to the surface S and the stress tensor (σ) is defined as:

$$\sigma = \frac{1}{\mu_0} \left[BB^T - \frac{1}{2} B \cdot B \right] \tag{11}$$

The torque that is generated on this volume is then evaluated by integrating the cross product of the force and moment arm along the whole surface (S).

$$\tau = \frac{1}{\mu_0} \oint_s (r - r_0) \times n \cdot \sigma\, dS = \frac{1}{\mu_0} \oint (r - r_0) \times n \cdot \left[BB^T - \frac{1}{2} B \cdot B \right] dS \tag{12}$$

where $(r - r_0)$ is the normal distance from the axis of rotation.

3.2. Thermal Analysis

The rotor is made up of equally divided segments. Each alternating segment is subjected to heating and cooling by jets of hot or cold fluid, as shown in Figure 5. A lumped parameter analysis approach is used to study the time dependent temperature response of the thermomagnetic material. This method assumes that the object has a uniform internal temperature, which in this analysis is reasonable, as the Biot number is sufficiently small for the given geometric shape of the thermomagnetic material [25]. Based on the conservation of energy, the rate of change of internal energy is equal to the net rate of heat transferred ($\dot{Q}$) and the rate of work done on it $\left(\dot{W}\right)$.

$$\frac{dU}{dt} = \dot{Q} + \dot{W} \tag{13}$$

For a solid object with constant density and specific heat capacity with no net work done on it, then the preceding formulation can be expressed as:

$$\rho c V \frac{dT}{dt} = K(T - T_{\infty}) \tag{14}$$

where ρ, V, and c is the density, volume, and specific heat capacity, respectively, while T is the average temperature of the object and K is the thermal conductance.

Although the lumped parameter method assumes that the elements of interest have a uniform internal temperature, but the neighboring elements may differ in temperature. This could be due to the presence of other materials or poor interfacial contact. As a result, a temperature gradient between the elements exists. The temperature gradient can be characterized by the thermal conductance (K) of the interface. A general definition of the thermal conductance is given here:

$$K = \frac{kA}{l} \text{ or } K = hA \tag{15}$$

where k and l are the thermal conductivity and thickness of the material, respectively. Should the element be exposed to a fluid, the thermal conductance at the external boundary is characterized by the heat transfer coefficient (h). The effective area of the interface or boundary where heat transfer occurs by either conduction and convection is denoted by A.

Even though the earlier description of the lumped parameter method applies to a single object, it can be adopted for the heat transfer analysis for a network of objects by making some generalization to the formula. As the elements in the network are physically connected, exchange of energy through the interfaces occurs. The following first order differential equation shows the energy balance of a solid object, with a thermal capacitance of ($C = \rho c V$) and a total of m heat transfer paths on the external surfaces.

$$C_i \frac{dT_i}{dt} = -K_i T_i + \sum_{j=1}^{m} \delta_{i,j} K_{i,j} T_{i,j} \text{ where } \delta_{i,j} = \begin{cases} -1, & \text{if } i \neq 1 \text{ and } i < j \\ 1, & \text{if } i \neq 1 \text{ and } i > j \text{ for } i = 1, 2, \ldots, m \\ 0, & \text{if } i = j \end{cases} \tag{16}$$

where T_i is the temperature of the ith element offset from a reference temperature (T_0), which is often taken as the ambient temperature. The sink can be the ambient surrounding in the case of natural cooling or the temperature of a fluid in the case of forced cooling. K_i refers to the thermal conductance of the direct path from the ith element to the sink or source. The notation can be further simplified by defining $T_{i,j} = T_j - T_i$ as the temperature difference between the ith and jth element. Similarly, $K_{i,j}$ refers to the thermal conductance of the heat transfer path between the ith and jth element such that $\delta_{i,j}$ is the signed permutation of the possible connections between two points in a network of m points.

The differential term in (16) expressed in discrete form by using a forward differencing scheme as shown here:

$$\frac{dT_i}{dt} = \frac{T_i(n+1) - T_i(n)}{\delta t} \tag{17}$$

where δt is the time step. The derivation is completed by introducing the term $\Delta T_i(n+1) = T_i(n+1) - T_i(n)$, which is defined as temperature rise and α_i, which is defined as $\alpha_i = \delta t / C_i$. The temperature rise of the ith element in a network of m elements is given by the following difference equation.

$$T_i(n+1) = T_i(n) - \alpha_i K_i T_i(n) + \alpha_i \sum_{j=1}^{m} \delta_{i,j} K_{i,j} T_{i,j}(n) \text{ for } i = 1, 2, \ldots, m \quad (18)$$

Further details of the method can be found in the reference [25]. The heat transfer coefficient that is first introduced in (15) is defined here as:

$$h = \frac{\dot{Q}}{A\left(T_s - T_{f,in}\right)} \quad (19)$$

where $\dot{Q}$ is the rate of heat flow and A is the area exposed to the fluid, while T_s is the temperature of the thermomagnetic material and $T_{f,in}$ is the temperature of the fluid at the source. The rate of heat transferred to or from the fluid is given by the following formulation:

$$\dot{Q} = \rho_f c_f \dot{V}\left(T_{f,out} - T_{f,in}\right) \quad (20)$$

where $T_{f,out}$ is the temperature of the fluid at the outlet. The volume flow rate of the fluid is denoted by $\dot{V}$, while ρ_f is the density and c_f is the heat capacity of the fluid. The heat transfer coefficient at the surfaces exposed to the hot and cool fluid is then determined from the empirical data using (19) and (20).

4. Experimental Measurements

A scaled rotor assembly is constructed to collect experimental data to evaluate the heat transfer parameters described in the previous section. This scaled rotor is constructed from steel with the dimensions shown in Figure 3. The dimensions of the scaled rotor are chosen to be between the lower and upper bounds of the design parameters that are used in the subsequent design analysis presented in Section 5. The scaled rotor assembly is subjected to impingement heating and cooling using hot water and cool air as the heat transfer medium. Hot water is supplied from a hot water bath through a pump via a network of pipe and nozzles which directs the water to the outer surface of the rotor. Cool air is supplied by a pressurized air source through a series of piping which directs the air to the inner surface of the rotor. The hot water and cool air is supplied sequentially by alternating the output from either source via a control valve. At the same time, the rotor is locked in position throughout the experiment so only one segment of the rotor assembly is heated and then cooled sequentially.

The temperature measurement taken at several points of interest is shown in Figure 3. The rotor surface temperature measurement is taken as reference temperature of the thermomagnetic material during both heating and cooling. The data shown in Figure 4 is a collection of several runs of the experiments that was conducted to obtain the average temperature. The line plot gives the mean value of the several data points collected, while the error bar shows variation of the data within one standard deviation. The different colored plots correspond to the temperature data collected at various flow rates of the hot water supplied. The cool air supply is kept constant at a regulated pressure of approximately 1.5 bar, which gives rise to an overall air flow rate of approximately 0.05 m^3/s or around 106 CFM. The equivalent heat transfer coefficient is evaluated using (19) and (20), with the results shown in Figure 4.

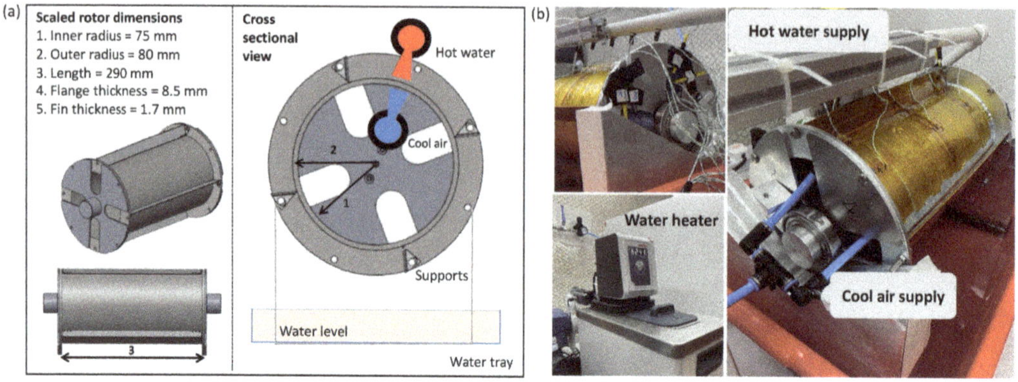

Figure 3. (**a**) Dimensions of the scaled rotor assembly and (**b**) experimental set-up.

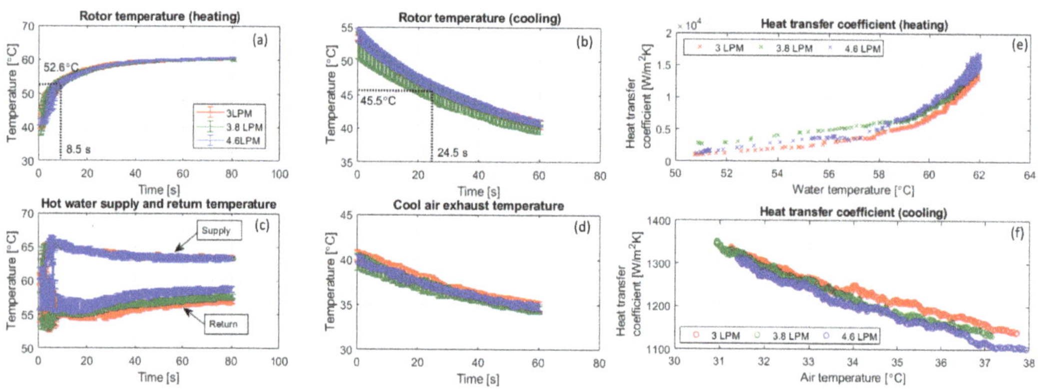

Figure 4. Measured rotor temperature response during (**a**) heating and (**b**) cooling as well as the (**c**) water supply and return temperature, (**d**) cool air exhaust temperature and the heat transfer coefficient evaluated during (**e**) heating and (**f**) cooling.

The plot shows that the heat transfer coefficient during heating varies between 1200 to 16,000 W/m^{2}K, while it ranges from 1100 to 1350 W/m^{2}K during cooling. It is expected that the heat transfer increases with the rate of hot water supplied as indicated by upward shifted plots in Figure 4. The heat transfer during cooling remains relatively constant as air supply is kept at the same level across the several experiments. The averaged heat transfer coefficient during heating is approximately $h_h = 11000$ W/m^{2}K, while it is approximately $h_c = 1200$ W/m^{2}K during cooling. The data presented here shows that there is variation in the heat transfer parameters with the operating conditions, such as the temperature of the object and fluid [26,27]. The data obtained from this experimental investigation provides an estimate of the upper and lower bounds of the heat transfer coefficient that will be considered in the subsequent design analysis.

5. Design Analysis and Optimization

The rotary thermomagnetic motor presented in this paper consists of a rotor and a stator, as shown in Figure 5a. In this design, the thermomagnets are introduced to the rotor, which is divided into an even number of segments, while the permanent magnets are housed in an internal stator assembly. These thermomagnets are separated physically and thermally insulated from each other by spacers found on the rotor body. Alternate segments of the thermomagnets are heated and cooled in the hot and cool zones, as shown in Figure 5b. In the hot zone, hot water is supplied to the outer rotor surface through a series of nozzles in the housing, and the water is then directed toward the outer periphery of the rotor by the spacers. In the cool zone, the cool air is directed towards the inner surface of the rotor through internal channels in the stator. As the rotor rotates, each thermomagnet segment will go through the hot and cool zones sequentially.

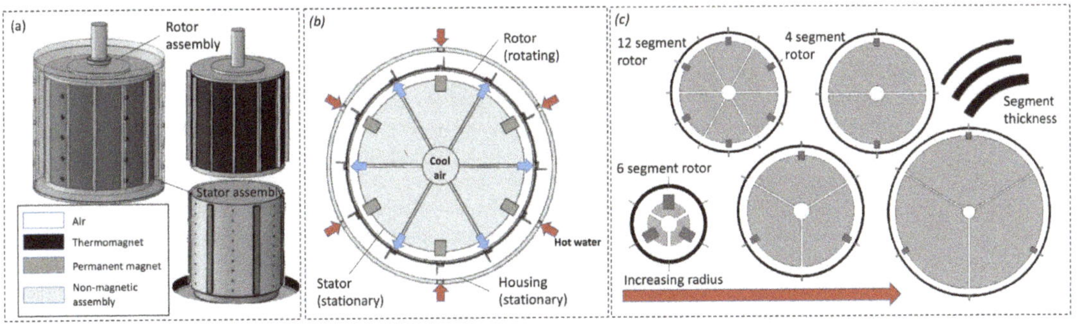

Figure 5. (**a**) Model of the thermomagnetic motor, (**b**) cross-sectional view of the heating and cooling zones, and (**c**) illustrations of the design parameters.

The dimensions of the rotor will be one of the key design considerations along with the number of rotor segments. In addition, operating conditions, such as the temperature of the heat source, the rate of heating, and cooling, will also be considered in the subsequent design analysis. The list of design variables are given in Table 1, with the associate range of values to provide some constraints over the design space. There is a periodicity that is observed in the rotor assembly, which is illustrated in the insert of Figure 5c. One period of rotation is equivalent to the arc angle covered by two segments of the rotor. For example, a 12-segment rotor would have a periodicity that is equivalent to 60°, while a 6-segment rotor would have a periodicity of 120° and so on. This makes it possible to analyze rotor configurations with a different number of segments. It should be pointed out that the alternating heating and cooling arrangement remains unchanged even though the number of segments may vary.

Table 1. List of design and operating parameters with its associated values considered in the design analysis.

Parameters	Initial Estimate (Nominal)	Lower Bound	Upper Bound
1. Rotor diameter (D) [m]	0.5	0.1	0.5
2. Rotor length (L) [m]	1	0.1	1
3. Number of rotor segments (n)	12	4	12
4. Rotor segment thickness (λ) [mm]	7.5	2.5	7.5
5. Temperature of heat source (T_h) [°C]	80	60	80
6. Rate of heating (h_h) [W/m^{2}K]	16,000	0	16,000
7. Rate of cooling (h_c) [W/m^{2}K]	1350	0	1350
8. Load torque (τ_{load}) [Nm]	1	0	2

The rotor is driven by the torque generated due to changes in reluctance of the thermomagnetic material, as described in Section 3.1. The reluctance torque that is generated on the rotor is described by (9), which is a function of the magnetic flux density on the rotor surface. The flux that is generated is in turn dependent on the permeability of the material as well as its position relative to the permanent magnet. It should be noted that the reluctance torque is position dependent. The rotor dynamics is described by the following equation:

$$\tau - b\omega = J\frac{d\omega}{dt} \tag{21}$$

where J is the inertia of the rotating mass and viscous friction is expressed as a function of the rotation speed (ω) and friction coefficient (b). The rotor speed is determined by solving (21) and the rotor position is evaluated through the time integration of the speed over time.

The position of the thermomagnet relative to the permanent magnet and the strength of the magnetic field as well the temperature all has a significant effect on the torque produced by the thermomagnetic motor. The design process can be divided into several steps using the various models presented, as illustrated in the flowchart in Figure 6. Firstly, the rate of heating or cooling of the material is determined by the position of the thermomagnet relative to the heating or cooling zone. Next, the temperature of the thermomagnet is then estimated using (18). Besides the temperature, the permeability of the material is also affected by the magnetic field intensity, which is also dependent on the position of the thermomagnet. The final magnetic field distribution in and around the thermomagnet is evaluated from the magnetostatic analysis described in Section 3.1. Finally, the magnetic force acting on the thermomagnet is then evaluated using (12).

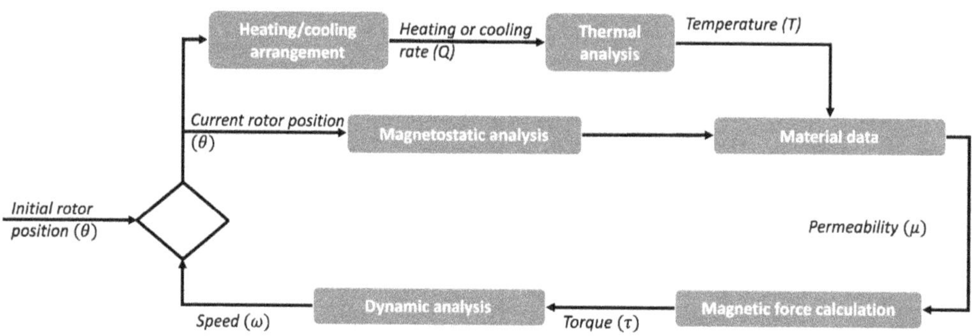

Figure 6. The design process flow and the various model considered in the analysis.

Figure 7a shows the magnetic field intensity produced by the permanent magnets at a different angular position at a radial distance of 10 mm away from the top surface of the magnet. The different colored plots correspond to the various configuration of the rotary magnetic assembly that is considered in this design analysis. Additionally, several measurement data points are plotted against data obtained from the mathematical model for comparison. The measurement is taken from an experimental set-up shown in Figure 7b using the Lakeshore 460 Gauss meter. The permanent magnet assembly shown goes with a six-segment rotor that has an effective radius of 0.07 m. There is good agreement between the measurement and the model predictions at the point of interest. It is expected that the magnetic field intensity decreases with distance from the permanent magnet. The rotor is positioned at an offset distance of approximately 10 mm away from the top surface of the magnet. This would ensure that the magnetic field intensity experienced by the thermomagnets will be within the range of values that is measured and shown here in Figure 7c.

The permeability of the thermomagnetic material is evaluated from the magnetization data shown here in Figure 8a, which is obtained from the material characterization procedure presented earlier in Section 2. The permeability data presented here is converted into a look-up-table with the temperature and magnetic field intensity as inputs. With the material properties fully defined, the torque that is produced by the device as well as the rotor speed can be estimated using (21). Finally, we can work out the position of the rotor in time through the time integration of the instantaneous rotor speed. This analysis is performed at each instance of time to obtain the time evolution of the rotor speed and torque generated. The interdependence between the various variables that links the different physical processes that happens in the device makes the design analysis non-trivial. The results obtained from this analysis will be presented in the following section.

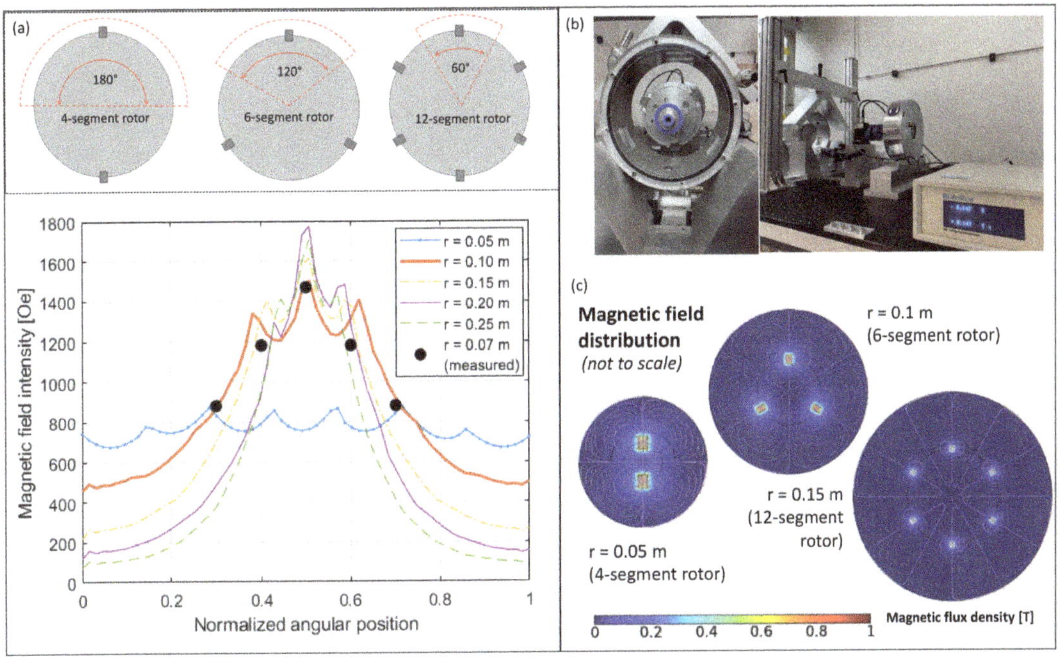

Figure 7. (**a**) Magnetic field generated by the permanent magnet stator assembly, (**b**) experimental set-up to take the magnetic measurements, and (**c**) simulated magnetic flux density.

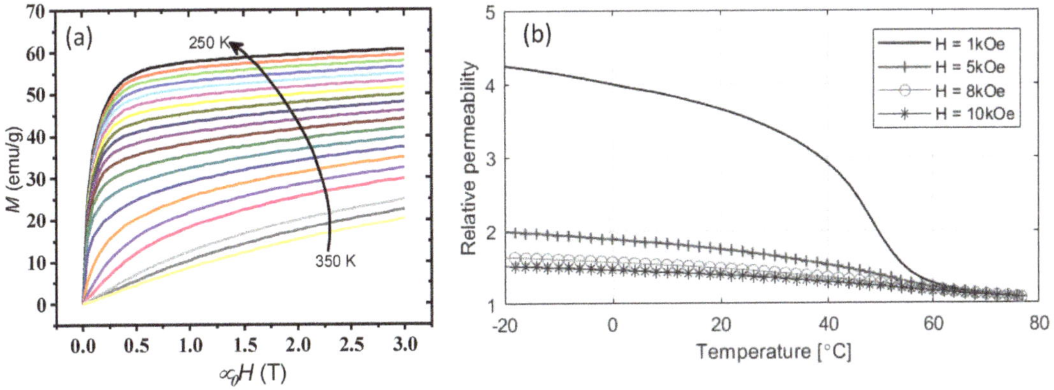

Figure 8. (**a**) Magnetization and field strength (M-H) measurement data, (**b**) Relative permeability of thermomagnetic material as a function of temperature and magnetic field intensity.

Figure 9 shows the plot of the torque that is generated as a function of the rotor position with rotor as well as the temperature state of the thermomagnet. It is clear from the data that there is a range of rotor positions that produces a positive torque (counter clockwise). The maximum positive torque occurs at a rotor position that corresponds to approximately 95% of one period of rotation. The output torque level diminishes after 25% of the rotation period is completed and turns negative reaching a minimum value when approximately 52% of the rotation period is attained. The output then turns positive and reaches the maximum value again when almost one period of rotation is completed. The simulation result shows that the torque output is increased when alternate segments are either hot or cool. The torque produced is maximized when the hot segment enters the magnetic zone (position 1). On the other hand, a large negative torque is produced when the hot segment leaves the magnetic zone (position 2).

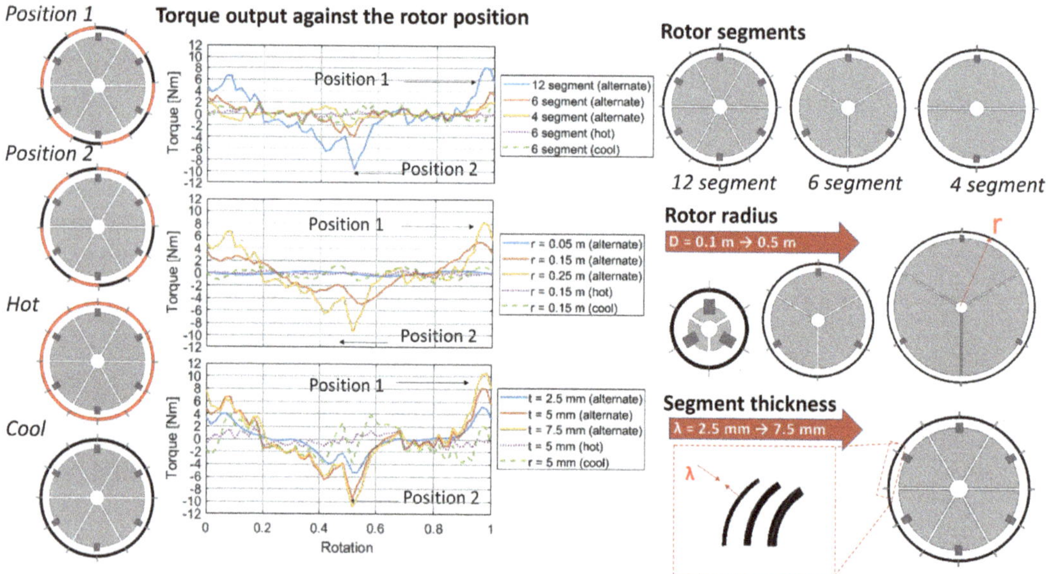

Figure 9. The reluctance torque generated against the rotor diameter as a function of the rotor position.

The simulation data shows that an alternating hot and cool segment is required to produce the positive torque. This leads to an alternating heating and cooling zone arrangement relative to the permanent magnets that are shown earlier in Figure 5. The simulation data also shows that the peak torque tends to increase with the number of rotor segments as shown by the increase peak value in Figure 9. This is also true when the rotor radius is increased as shown by the plots in Figure 9. Lastly, the peak torque is also amplified when the thickness of the segment is thickened. This data yields important insights into the key characteristics of the rotor design that leads to greater torque output. However, the eventual torque that is produced by the thermomagnetic motor still depends on the temperature that can be attained under dynamic conditions. It should be pointed out that the segments will be in motion once the torque is produced and this rotation does have an effect on the amount of heating or cooling experienced by each segment of thermomagnetic material. This coupled thermal-magnetic effect on the rotor dynamics will be presented in the Results section.

Design Optimization

An optimization problem is defined to determine the optimal design of the thermomagnetic motor that maximizes output parameters, such as its power output (P), power density (P'''), and efficiency (η). The optimization problem is typically expressed mathematically as an objective function $f(\mathbf{x})$ in the following form:

$$\mathbf{x}^* = \arg_{\mathbf{x}}\min f(\mathbf{x}) \tag{22}$$

whereby $\mathbf{x}$ is the set of design variables given by $\mathbf{x} = [x_1\ x_2\ \cdots\ x_n]^T$ and $\mathbf{x}^*$ is the solution of this minimization problem. Maximum value problems are converted to a minimization problem by substituting the objective function with its negative such that $f(\mathbf{x}) = -f(\mathbf{x})$. The list of design variables considered in this optimization problem are listed earlier in Table 1. There will be some restrictions to the physical design and these are expressed mathematically as constraints. The constraint usually applies on the design variables, such as a limit on the minimum or maximum value that a design variable can carry. Moreover, some design variables only take discrete values and additional constraints can be defined to cater for such discrete variables.

Gradient based optimization methods require the evaluation of the objective function and its derivative at various points in the solution space. If the necessary conditions are not met, then the solution is advanced to the next feasible solution by a step change in $\mathbf{x}$ given by $\Delta\mathbf{x} = \mathbf{x} - \mathbf{x}^*$. Linear or quadratic approximations of the function $f(\mathbf{x})$ is typically used in place of the actual objective function during the optimization process. The method progresses in a step wise fashion to search for the optimal solution. Firstly, a direction of search is determined, then the step size is selected with the aim of minimizing the objective function at each iteration. There are various methods used to decide on the search direction and step size. The following description provides details of the key steps involved in a Quadratic Programming (QP) method.

For a constrained optimization problem, the objective function is first replaced by a Lagrange function, which is defined as such:

$$L(\mathbf{x}, \mathbf{v}) = f(\mathbf{x}) + \sum_{i=1}^{p} v_i h_i(\mathbf{x}) \tag{23}$$

where $f(\mathbf{x})$ is the objective function and $h_i(\mathbf{x})$ is the ith constraint out of a total of p constraints while v_i is the associated Lagrange multiplier. The following quadratic approximation of the Lagrange function is used to determine the search direction, $\mathbf{d} = \Delta\mathbf{x}$

$$L = \mathbf{c}^T\mathbf{d} + 0.5\mathbf{d}^T\mathbf{H}\mathbf{d} + \sum_{i=1}^{p} v_i\left(h_i + \mathbf{n}_i^T\mathbf{d}\right) \tag{24}$$

where $\mathbf{c} = \nabla f(\mathbf{x})$ is the gradient of the objective function and $\mathbf{H} = \nabla^2 L(\mathbf{x})$ is the Hessian matrix. The search direction is determined by solving (25), which is derived from the necessary conditions for a minimum solution of (24)

$$\begin{gathered} \mathbf{c} + \mathbf{H}\mathbf{d} + \mathbf{N}\mathbf{v} = 0 \\ h_i + \mathbf{n}_i^T\mathbf{d} = 0, \quad \text{for } i = 1 \text{ to } p \end{gathered} \tag{25}$$

where $\mathbf{n}_i$ is the gradient of the ith constraint function such that $\mathbf{n}_i = \begin{bmatrix} \frac{\partial h_i}{\partial x_1} & \cdots & \frac{\partial h_i}{\partial x_n} \end{bmatrix}$ and $\mathbf{N} = \begin{bmatrix} \mathbf{n}_1 & \cdots & \mathbf{n}_p \end{bmatrix}$, while $\mathbf{v}$ is the Lagrange multiplier first defined in (23) expressed in vector form.

Iterative numerical methods are typically employed to solve such constrained optimization problems. Approximations of the objective and constraint functions (or Lagrange function) as well as its derivatives are evaluated at each iteration. The solution is then moved from one point of the solution space to the next based on the search direction that minimizes the objective function and its derivative. This continues until a termination criteria is met, which is often when the optimal conditions are met or when there is no further noticeable change in the objective function value. The calculations performed in each iteration of the algorithm are listed here:

1. Obtain initial estimate of the design variable $\mathbf{x}_{(0)}$ and set the Hessian matrix $\mathbf{H}_{(0)} = \mathbf{I}$. Initialize the penalty parameter $R_{(0)}$, permissible constraint violation ϵ_1 and convergence parameter ϵ_2;
2. Evaluate the constraint functions and determine the maximum constraint violation, which is defined as $V(k) = \max\{|h_1| \quad \cdots \quad |h_p|\}$;
3. Determine the search direction $\mathbf{d}_{(k)}$ and Lagrange multipliers $\mathbf{v}_{(k)}$ by solving (25);
4. Stop if the termination criteria $\left|\mathbf{d}_{(k)}\right| \leq \epsilon_2$ is met and the maximum constraint violation is $V_{(k)} \leq \epsilon_1$ otherwise continue;
5. Evaluate the sum of the Lagrange multiplier defined as $r_{(k)} = \sum_{i=1}^{p} |v_i|$ and set the penalty factor $R_{(k)} = max\left\{R_{(k)}, r_{(k)}\right\}$;
6. Update the design variable $x_{(k+1)} = x_{(k)} + \alpha_{(k)} d_{(k)}$ where $\alpha_{(k)}$ is defined as the step size determined from a line search method;
7. Update the penalty parameter $R_{(k+1)} = R_{(k)}$;
8. Update the Hessian matrix using (26)–(32) and increase the counter $k = k + 1$, then return to step 2.

The gradient of the Lagrange function is used to estimate the Hessian matrix in a constrained optimization problem based on the quasi-Newton Hessian approximation method proposed by Broyden–Fletcher–Goldfarb–Shanno (BFGS) [28]. Correction factors are introduced to the Lagrange function in the BFGS method to ensure that the Hessian matrix is definitively positive. Furthermore, additional variables are introduced in the intermediate steps before a final estimate of the Hessian matrix can be obtained. The approximation method first starts with the definition of the step change given here:

$$\mathbf{D}_{(k)} = \alpha_{(k)} \mathbf{d}_{(k)} \tag{26}$$

where the working variables $\mathbf{Z}_{(k)}$, $\mathbf{Y}_{(k)}$, and $\mathbf{Z}_{(k)}$ are defined by (24), (25), and (26), respectively.

$$\mathbf{Z}_{(k)} = \mathbf{H}_{(k)} \mathbf{D}_{(k)} \tag{27}$$

$$\mathbf{Y}_{(k)} = \nabla L_{(k+1)} - \nabla L_{(k)} \tag{28}$$

$$\mathbf{W}_{(k)} = \gamma \mathbf{Y}_{(k)} + (1-\gamma) \mathbf{Z}_{(k)} \tag{29}$$

γ is determined from the following cases

$$\gamma = \begin{cases} 1, & \text{if } \beta_1 > 0.2\beta_2 \\ 0.8\left(\frac{\beta_2}{\beta_2 - \beta_1}\right), & \text{otherwise} \end{cases} \tag{30}$$

where β_1 and β_2 are defined in (28)

$$\beta_1 = \mathbf{D}_{(k)} \cdot \mathbf{Y}_{(k)}, \ \beta_2 = \mathbf{D}_{(k)} \cdot \mathbf{Z}_{(k)} \tag{31}$$

Finally, the Hessian matrix is updated according to (32)

$$\mathbf{H}_{(k+1)} = \mathbf{H}_{(k)} + \frac{1}{\beta_3} \mathbf{W}_{(k)} \mathbf{W}_{(k)}^T - \frac{1}{\beta_3} \mathbf{Z}_{(k)} \mathbf{Z}_{(k)}^T \tag{32}$$

where $\beta_3 = \mathbf{D}_{(k)} \cdot \mathbf{W}_{(k)}$

The sequential quadratic programming (SQP) method described here is the preferred numerical method in many commercial software packages, such as Matlab, because it is stable, accurate, and efficient to compute. In the subsequent design optimization, the 'fmincon' solver available in the Optimization Toolbox of Matlab using the SQP algorithm is deployed during the optimization process.

6. Results and Discussion

Figure 10 shows the output characteristics of the thermomagnetic motor with nominal parameters. The torque produced by the motor climbs to a final maximum value of 0.9 Nm after approximately 40 s, which it maintains for the remaining duration of the simulation. The motor finally settles at a rotation speed of approximately 17 RPM which gives an averaged power output of approximately 1.7 W. While the torque and speed settle at a more or less steady value, the temperature is seen to be oscillating below and above 75 °C. This fluctuation is required for a continuous positive torque to be produced, as discussed earlier in Section 5. However, the temperature variation happens at an elevated temperature beyond the Curie temperature of the material. This shows that the device is not yet running optimally. Figure 10c shows the heat flow in and out of the material evaluated using (17). There is a huge flow of heat into the material at the start in order to heat up the material from the ambient temperature. Thereafter, the rate of heat flow reduces to an average value of 23 kW when only the inflow of heat is considered.

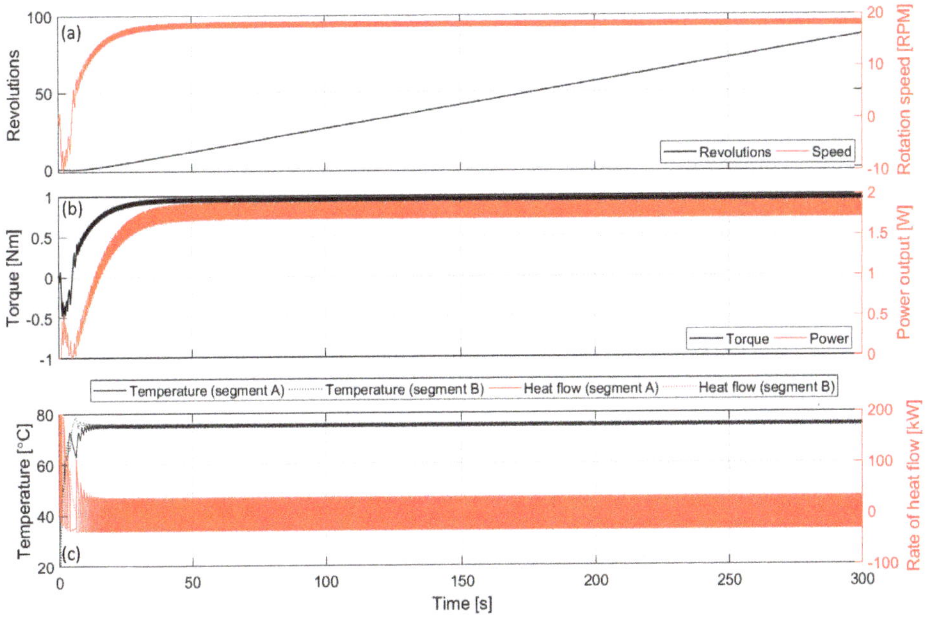

Figure 10. (**a**) Output speed and revolutions and (**b**) output torque and power generated by the thermomagnetic motor, (**c**) the thermomagnet temperature and heat flow into the material.

Based on the data generated from the model, the power output and efficiency of the thermomagnetic motor with the nominal parameters is fairly low and quite insignificant. This is because the nominal parameters are selected to maximize the inputs to the device as well as the amount of thermomagnetic material used. However, this does not necessarily yield the most output from the device since the operating conditions should be selected to match the design of the thermomagnetic motor. For instance, using more thermomagnetic material would mean more heat is required to power the device which could lead to a lower overall efficiency. On the other hand, using too little material would mean less reluctance torque generated leading to lower outputs. These interactions will be analyzed further in the subsequent design analysis which shows how the output characteristics of the device can be further improved through careful selection of the design and operating parameters.

The output torque generated by the thermomagnetic motor and the power output are critical measures of performance. In addition, the power density and efficiency of the thermomagnetic motor provides a useful measure of the output against the amount of material used as well as the amount of heat energy required. These measures form the basis of comparison with other similar heat energy harvesting devices. The instantaneous torque and speed are derived from (21), while the instantaneous power is given by the following expression:

$$P(t) = \tau(t) \cdot \omega(t) \tag{33}$$

The time averaged values of the various output parameters (collectively referred as y) are evaluated from the RMS value of the parameters using the expression given in (34).

$$\overline{Y} = \sqrt{\frac{1}{T}\int_0^T [y(t)]^2\, dt} \tag{34}$$

where T is the time duration of the analysis, which is chosen to be 300 s. The efficiency of the thermomagnetic motor is defined as the percentage ratio of the time averaged power output $(\overline{P})$ to the averaged heat supplied $(\overline{Q})$ such that:

$$\overline{\eta} = \frac{\overline{P}}{\overline{Q}} \times 100 \tag{35}$$

Finally, the power density (P''') is given by the expression in (36) where V is the volume of the thermomagnetic material used.

$$P''' = \frac{\overline{P}}{V} \tag{36}$$

The subsequent design analysis is performed in a step wise fashion in which only the parameters of interest are varied at each step of the analysis, while the other parameters are kept constant based on the nominal values given in Table 1. The average power produced by the thermomagnetic motor with nominal design parameters is plotted against the ratio of the heating to cooling rate as shown in Figure 11. The colored plots correspond to a higher or lower rate of cooling, while the different markers correspond to the various temperatures at which the hot water is supplied to the device. The sub-plots in Figure 11 show the variation in the power generated by devices with varying thickness of thermomagnetic material used. Generally, the power output is enhanced at a higher rate of cooling, i.e., when $h_c = 1200\ \mathrm{W/m^2k}$. In addition, the data shows that the power generated tends to peak when the rate of heating is equivalent to the rate of cooling (at points where x-axis is equal to 1). There are some instances where a lower heating rate is beneficial, such as when thicker thermomagnetic material is used. The temperature of the hot water supply does not appear to have a significant effect on the output of the device. On the other hand, the thickness of the material has a significant effect on the power output. There is almost an output increase of 10 times when the thickness of the material is increased from 2.5 mm to 5 mm. However, any further increase in the thermomagnetic material thickness does not yield any more gains in the output.

The thermal boundary conditions that are necessary for the thermomagnetic motor to generate a higher power output is shown (by the black arrow) in Figure 11. These conditions are used in the subsequent design analysis. The sub-plots in Figure 12 show the power (shaded) and torque (line) generated by the thermomagnetic motor with different thicknesses of the material used. This set of data is generated assuming that the rate of heating and cooling are closely matched according to the 'ideal' conditions described earlier. The data shows that there is an optimal speed at which power transfer and the corresponding output torque level are maximized. Therefore, it is important to carefully select the braking torque applied by the load onto the device to ensure that the thermomagnetic motor will operate at a speed that maximizes the power transfer. The point where power transfer is maximized is highlighted (by the black arrow) in Figure 12 and the torque applied onto the load is kept constant in the subsequent analysis.

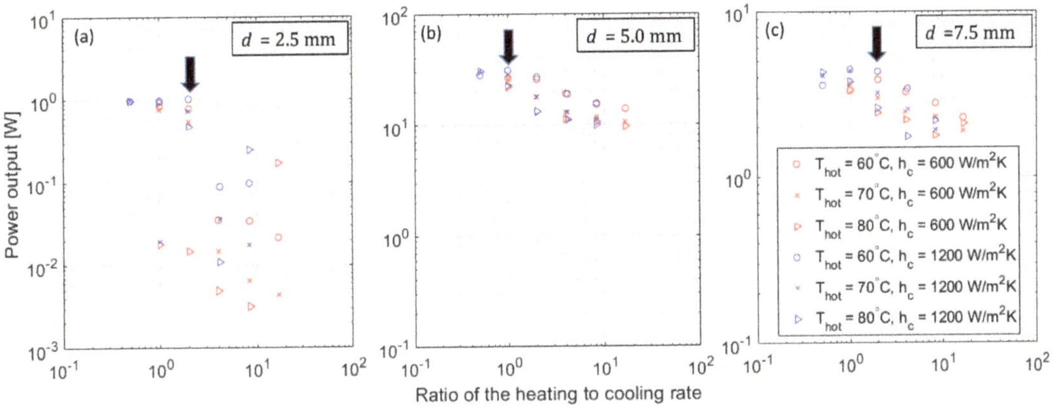

Figure 11. Power output from the thermomagnetic motor plotted against the ratio of heating to cooling rate when using (**a**) 2.5 mm, (**b**) 5 mm, and (**c**) 7.5 mm thick thermomagnetic material.

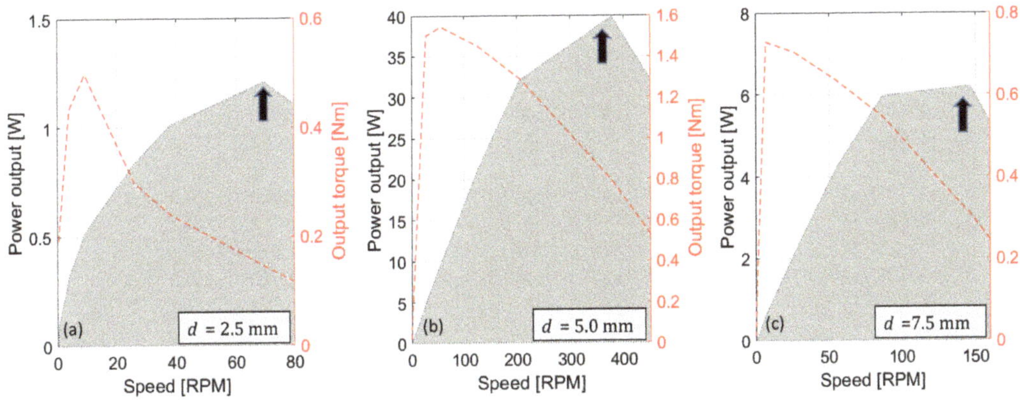

Figure 12. Power output and torque generated by the thermomagnetic motor against the rotation speed when using (**a**) 2.5 mm, (**b**) 5 mm, and (**c**) 7.5 mm thick thermomagnetic material.

The power density is first defined in (36) and it is plotted here in Figure 13 against the aspect ratio of the thermomagnetic motor. The aspect ratio refers to the ratio of the external length to the diameter of the rotor. The different colored plots refer to data generated from rotors with a varying number of segments, while markers are used to differentiate rotors of different diameters. The data generated from devices with varying thickness of thermomagnetic material is shown in the sub-plots in Figure 13. The data shows that the power density tends to peak at aspect ratios between 1–2. This trend is observed with rotors of different diameters, but it is most pronounced for rotors with larger diameters. From this set of data, it is observed that the higher power density can be generated from a 12-segment rotor. However, when thicker thermomagnetic material is used, higher power density is achieved with a smaller number of rotor segments instead. Nevertheless, the highest power density recorded is still from a 12-segment rotor with a diameter of 0.4 m and a length of 0.5 m.

The design analysis shows that the power density tends to increase with the number of segments as shown by the data provided in Figure 13. This is expected since having more segments increases the useful area of the rotor, where heat can be introduced. This also means that there is more effective use of the thermomagnetic material to produce the torque for a rotor of a particular radius. However, there is a practical limit of the number of segments of thermomagnetic material that can be fit into a rotor. In addition, increasing the number of segments would mean that there is less time for the material to be heated up and cooled down at a given rotational speed. With the limits of heat transfer to and from the fluid, the speed of rotation has to be reduced correspondingly when there are more segments of material used. Therefore, the net effect of increasing the number of segments on the power output is not always positive since the power output is a product of both the torque and the speed.

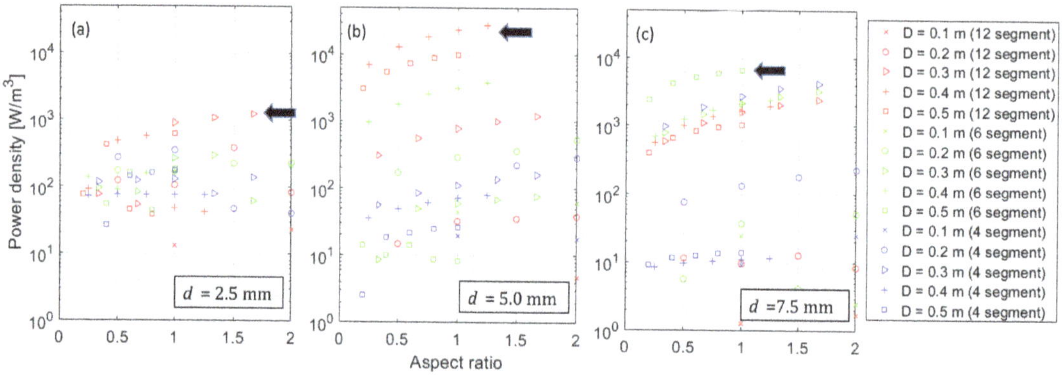

Figure 13. Power density of the thermomagnetic motor with various rotor diameter (D) plotted against the aspect ratio $\left(\frac{L}{D}\right)$ when using (**a**) 2.5 mm, (**b**) 5 mm, and (**c**) 7.5 mm thick thermomagnetic material.

The design analysis presented earlier shows that the thermomagnetic motor output is sensitive to the following parameters: (i) ratio of the heating and cooling rate, (ii) aspect ratio of the rotor, (iii) number of rotor segments, and (iv) thickness of thermomagnetic material. The preceding discussion shows how the output changes against selected parameters, while the other parameters are kept constant. It can be expected that there are further interactions between these parameters, and it is possible that there is a set of parameters that yields the 'optimal' output. A design optimization is performed based on the optimization method presented earlier in Section 5. The objective of this design optimization is to maximize each of the following output parameters: (a) power output, (b) power density, and (c) efficiency of the device. Three separate runs of the optimization routine were performed with the data generated in each run plotted in the sub-plots in Figure 14.

The design variables are normalized before they are plotted on the same scale in Figure 14. There is significant variation in the variables as the optimization progresses. The largest change is observed in the heating rate (h_h) as well as the thickness (λ) of the thermomagnetic material. This is indicative of the high sensitivity of the output to these variables. The data also shows an increasing objective function value as the optimization progress, which implies that the quadratic programming method implemented here in this study is able to move the solution toward the optimal. However, further fine tuning of optimization parameters or algorithm is not considered. The results obtained from the design optimization are shown in Table 2, with the corresponding set of parameters that maximizes each of the targeted output.

There is a slight variation in the operating parameters of the thermomagnetic motor depending on the desired output characteristics. For instance, the rate of cooling and heating is maintained lower in order to maximize the efficiency of the device. On the other hand, a higher rate of heating and cooling is required to maximize both the power and the power density. It is also not necessary to increase the heat source temperature beyond the 60 °C in all instances. The result from the design optimization also shows that there is an 'optimal' set of design parameters that give the most desirable output characteristics. A 12-segment rotor that is 0.5 m long with a diameter of 0.4 m using thermomagnetic material with a thickness of 5 mm is the design configuration that leads to the most favorable outcome for all 3 of the output measures considered.

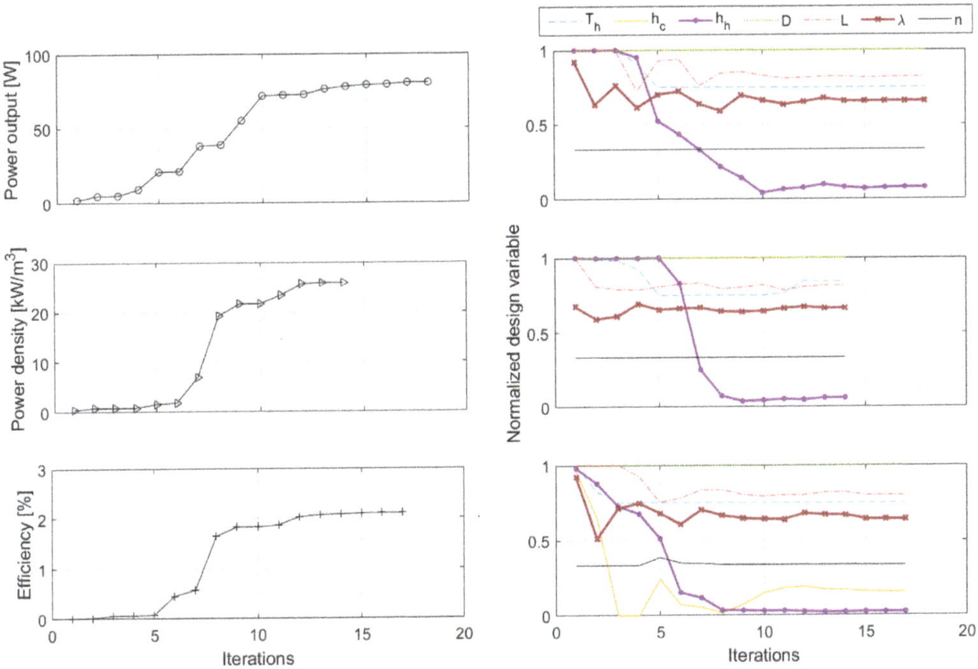

Figure 14. Design variables and objective function value variation during optimization.

Table 2. Summary of parameters obtained from the design optimization.

Parameters	Run 1: Maximize Power Output	Run 2: Maximize Power Density	Run 3: Maximize Efficiency
Temperature of heat source (T_h) [°C]	60	67	60
Rate of cooling (h_c) [W/m^{2}K]	1350	1350	210
Rate of heating (h_h) [W/m^{2}K]	1287	965	403
Rotor length (L) [m]	0.5	0.5	0.5
Rotor diameter (D) [m]	0.4	0.4	0.4
Rotor segment thickness (λ) [mm]	5	5	5
Number of rotor segment (n)	12	12	12
Power output [W]	88	87	56
Power density [kW/m^3]	27	27	17
Efficiency [%]	0.9	0.9	2.1

The design analysis and optimization show that the thickness of the thermomagnetic material has a significant effect on the output characteristics of the thermomagnetic motor. Thicker material tends to increase the production of the reluctance torque but it also requires more heat and time to bring about a change in temperature which indirectly reduces the output over time. These counteracting factors therefore lead to an 'optimal' thickness of the material that yields larger outputs. It is also observed that a larger rotor diameter and longer rotor yields higher power outputs. This is expected since the reluctance torque produced is proportional to the length and diameter based on (12). However, the maximum power output is obtained not from the largest diameter but at an intermediate diameter instead. This indicates again there are conflicting factors that lead to an 'optimal' diameter.

Additionally, increasing the length of the rotor with a particular diameter increases the power density but only up to a point where it starts to level off. This shows that the power output of the device will scale proportionately with the rotor dimensions but any further increase does not yield significant gains in output per unit of additional thermomagnetic material used.

7. Conclusions

A rotary thermomagnetic motor using a soft magnetic material (Mn_2FeSn) with a low Curie temperature ($T_{cu} = 41$ °C) that is suited for heat energy harvesting is introduced in this paper. A design analysis and optimization of this thermomagnetic motor is presented. A mathematical model coupling the heat transfer, magnetic interactions, and rotor dynamics in the thermomagnetic motor is developed in this work. Measurement data obtained from the material characterization, scaled rotor, and permanent magnet assembly provide reference input values to the model. The multi-physics model is used to make predictions of the power output, power density, and efficiency of the motor against the various design and operating parameters. From the design analysis, it is determined that the thickness of the thermomagnetic material has the most significant effect on the output characteristics. Thicker material tends to increase the torque production but it also requires more heat and time to bring about a change in temperature, which indirectly reduces the output over time.

Increasing the number of segments, length, and diameter of the rotor increases the power output. At the same time, the power density is also increased but only up to a point beyond which it does not yield any significant gains in output per unit volume of additional thermomagnetic material used. The result from the design optimization also shows that there is an 'optimal' design configuration that results in the largest power output, power density, and efficiency. It is also observed that a higher rate of cooling leads to more output, especially when this is matched to a similar rate of heat supplied to the thermomagnetic motor. The data shows that there is an optimal speed at which power transfer is maximized and so it is necessary to load the thermomagnetic motor appropriately in order to drive it at the desired speed. After the optimization, it is estimated that the rotary thermomagnetic motor is able to produce up to 88 W of power with a power density of approximately 27 kW/m^3 of thermomagnetic material used, while a maximum thermal-to-mechanical energy conversion efficiency of 2.1% is possible. With further optimization, it is possible to determine the suitable operating parameters that achieve the right balance between the power generated and the conversion efficiency of the device to meet the requirements of the heat energy harvesting application.

Author Contributions: Writing—original draft, J.H.; Writing—review & editing, M.R., T.L. and J.L.T. All authors have read and agreed to the published version of the manuscript.

Funding: This research was funded by Agency for Science, Technology, and Research (A*STAR) of Singapore with grant number (202D800030) and the APC was funded A*STAR.

Institutional Review Board Statement: Not applicable.

Informed Consent Statement: Not applicable.

Data Availability Statement: Not applicable.

Acknowledgments: The authors would like to acknowledge the Agency for Science, Technology, and Research of Singapore for supporting this research through the Career Development Award.

Conflicts of Interest: The authors declare no conflict of interest.

Nomenclature

A	Area [m^2]
b	Viscous damping coefficient [Nm/(rad/s)]
B	Magnetic flus density [T]
c	Specific heat capacity [J/kgK]
D	Rotor diameter [m]
F	Force [N]
h	Heat transfer coefficient [W/m^{2}K]
H	Magnetic field strength [A/m]
J	Current density [A/m^2]
J	Inertia [kgm^2]
K	Thermal conductance [W/K]
L	Rotor length [m]
M	Magnetization [A/m]
n	Number of rotor segments
$\dot{Q}$	Rate of heat flow [W]
T	Temperature [K]
V	Volume [m^3]
$\dot{W}$	Rate of work done [W]
Subscripts	
cu	Curie temperature
h	Hot temperature
c	Cool temperature
Greek Letters	
η	Efficiency
ρ	Density [kg/m^3]
μ_0	Permeability of free space [H/m]
μ_r	Relative permeability
λ	Rotor segment thickness [mm]
σ	Stress tensor [N/m^2]
τ	Torque [Nm]
χ	Susceptibility
ω	Angular speed [rad/s]
Mathematical Functions	
$\mathbf{x}$	Design variables
$f(\mathbf{x})$	Objective function
$L(\mathbf{x},\mathbf{v})$	Langrange function
$c(\mathbf{x})$	Gradient function
$H(\mathbf{x})$	Hessian function
$\mathbf{v}$	Lagrange variables

References

1. Forman, C.; Muritala, I.K.; Pardemann, R.; Meyer, B. Estimating the global waste heat potential. *Renew. Sustain. Energy Rev.* **2016**, *57*, 1568–1579. [CrossRef]
2. Zabek, D.; Morini, F. Solid state generators and energy harvesters for waste heat recovery and thermal energy harvesting. *Therm. Sci. Eng. Prog.* **2018**, *9*, 235–247. [CrossRef]
3. Sandoval, S.M.; Wetzlar, K.P.; Carman, G.P. Thermomagnetic conversion efficiencies for ferromagnetic materials. *J. Appl. Phys.* **2011**, *110*, 123923. [CrossRef]
4. Chen, H.; Ma, Z.; Liu, X.; Qiao, K.; Xie, L.; Li, Z.; Shen, J.; Dai, W.; Ou, Z.; Yibole, H.; et al. Evaluation of thermomagnetic generation performance of classic magnetocaloric materials for harvesting low-grade waste heat. *Appl. Energy* **2021**, *306*, 117999. [CrossRef]
5. Felez, M.R.; Monteiro, J.C.B.; De Florio, D.Z.; Gandra, F.C.; Gama, S. Effects of Fe substitution on Mn2Sn alloy on its structural, magnetic and magnetocaloric properties. *J. Magn. Magn. Mater.* **2020**, *505*, 166676. [CrossRef]
6. Waske, A.; Dzekan, D.; Sellschopp, K.; Berger, D.; Stork, A.; Nielsch, K.; Fähler, S. Energy harvesting near room temperature using a thermomagnetic generator with a pretzel-like magnetic flux topology. *Nat. Energy* **2018**, *4*, 68–74. [CrossRef]
7. Takahashi, Y.; Yamamoto, K.; Nishikawa, M. Fundamental performance of triple magnetic circuit type cylindrical thermomagnetic engine. *Electr. Eng. Jpn.* **2006**, *154*, 68–74. [CrossRef]

8. Jiang, C.; Zhu, S.; Yu, G.; Luo, E.; Li, K. Numerical and experimental investigations on a regenerative static thermomagnetic generator for low-grade thermal energy recovery. *Appl. Energy* **2022**, *311*, 118585. [CrossRef]
9. Kishore, R.A.; Davis, B.; Greathouse, J.; Hannon, A.; Kennedy, D.E.; Millar, A.; Mittel, D.; Nozariasbmarz, A.; Kang, M.G.; Kang, H.B.; et al. Energy scavenging from ultra-low temperature gradients. *Energy Environ. Sci.* **2019**, *12*, 1008–1018. [CrossRef]
10. Zeeshan; Panigrahi, B.K.; Ahmed, R.; Mehmood, M.U.; Park, J.C.; Kim, Y.; Chun, W. Operation of a low-temperature differential heat engine for power generation via hybrid nanogenerators. *Appl. Energy* **2021**, *285*, 116385. [CrossRef]
11. Deepak, K.; Ramanujan, R. Near room temperature giant magnetocaloric effect in (MnNiSi)1-x(Fe2Ge)x alloys. *J. Alloys Compd.* **2018**, *743*, 494–505. [CrossRef]
12. Kishore, R.A.; Priya, S. A review on design and performance of thermomagnetic devices. *Renew. Sustain. Energy Rev.* **2018**, *81*, 33–44. [CrossRef]
13. Takahashi, Y.; Matsuzawa, T.; Nishikawa, M. Fundamental performance of the disc-type thermomagnetic engine. *Electr. Eng. Jpn.* **2004**, *148*, 26–33. [CrossRef]
14. Coray, P.S.; Wymann, D.; Brunner, R.; Vida, N. Fully Operational Prototype of a 1 kW Thermo-Magnetic Motor for Generating Electricity from <80 °C Heat. In Proceedings of the Delft Days on Magneto Calorics (DDMC), Delft, The Netherlands, 2–3 November 2015.
15. Evaristo, E.; Colman, F.; Alves, C.; Trevizoli, P. Mathematical modelling and simulation results of a linear thermomagnetic motor with gravity return. *J. Magn. Magn. Mater.* **2021**, *544*, 168668. [CrossRef]
16. Deepak, K.; Varma, V.B.; Prasanna, G.; Ramanujan, R.V. Hybrid thermomagnetic oscillator for cooling and direct waste heat conversion to electricity. *Appl. Energy* **2019**, *233–234*, 312–320. [CrossRef]
17. Alves, C.; Colman, F.; Foleiss, G.; Szpak, W.; Vieira, G.; Bento, A. Simulation of solar Curie wheel using NiFe alloy and Gd. *Int. J. Refrig.* **2014**, *37*, 215–222. [CrossRef]
18. Joshi, K.B.; Priya, S. Multi-physics model of a thermo-magnetic energy harvester. *Smart Mater. Struct.* **2013**, *22*, 055005. [CrossRef]
19. Fajar, D.M.; Khotimah, S.N. Khairurrijal Thermodynamic and Mechanical Analysis of a Thermomagnetic Rotary Engine. *J. Physics Conf. Ser.* **2016**, *739*, 12028. [CrossRef]
20. Arnold, D.P. Review of Microscale Magnetic Power Generation. *IEEE Trans. Magn.* **2007**, *43*, 3940–3951. [CrossRef]
21. Felez, M.R.; Coelho, A.A.; Gama, S. Magnetic properties of Mn3-xFexSn compounds with tuneable Curie temperature by Fe content for thermomagnetic motors. *J. Magn. Magn. Mater.* **2017**, *444*, 280–283. [CrossRef]
22. Silvester, P.P.; Ferrari, R.L. Electromagnetics of finite elements. In *Finite Elements for Electrical Engineers*, 3rd ed.; Cambridge University Press: Cambridge, UK, 1996; pp. 68–125. [CrossRef]
23. Hey, J.; Liang, T.J.; Hao, T.Z. An evaluation of thermomagnetic motors for heat energy harvesting. In Proceedings of the 2022 IEEE/ASME International Conference on Advanced Intelligent Mechatronics (AIM), Hokkaido, Japan, 11–15 July 2022.
24. Salon, S.J. Calculation of Force and Torque. In *Finite Element Analysis of Electrical Machines*; Springer: New York, NY, USA, 1995; pp. 97–123. [CrossRef]
25. Hey, J.; Malloy, A.; Martinez-Botas, R.; Lamperth, M. Experimentally determined thermal parameters of an energy conversion device using a constraint least square parameter estimation method coupled with an analytical thermal model. In Proceedings of the 15th International Heat Transfer Conference (IHTC-15), Kyoto, Japan, 10–15 August 2014.
26. Jambunathan, K.; Lai, E.; Moss, M.; Button, B. A review of heat transfer data for single circular jet impingement. *Int. J. Heat Fluid Flow* **1992**, *13*, 106–115. [CrossRef]
27. Oliveira, A.V.S.; Maréchal, D.; Borean, J.-L.; Schick, V.; Teixeira, J.; Denis, S.; Gradeck, M. Experimental study of the heat transfer of single-jet impingement cooling onto a large heated plate near industrial conditions. *Int. J. Heat Mass Transf.* **2021**, *184*, 121998. [CrossRef]
28. Arora, J. *Introduction to Optimum Design*; Elsevier Science: Amsterdam, The Netherlands, 2011.

energies

MDPI

Review

Quasi-Zero Stiffness-Based Synchronous Vibration Isolation and Energy Harvesting: A Comprehensive Review

Zhiwen Chen [1], Zhongsheng Chen [1,2,*] and Yongxiang Wei [1]

1 College of Electrical & Information Engineering, Hunan University of Technology, Zhuzhou 417002, China
2 College of Automotive Engineering, Changzhou Institute of Technology, Changzhou 213032, China
* Correspondence: chenzs@czu.cn

Abstract: In recent years, the advantages of nonlinearity in vibration isolation and energy harvesting have become increasingly apparent. The quasi-zero stiffness (QZS) of the nonlinear term provided by the negative stiffness element can achieve vibration isolation under low-frequency environments while improving the efficiency of energy harvesting. The QZS provides a new research idea for simultaneous vibration isolation and energy harvesting. The main purpose of this paper is to review past research results, summarize possible problems, and discuss trends. After briefly analyzing the basic principle of QZS vibration isolation, the progress of QZS in vibration isolation and energy harvesting in recent years is reviewed. At the same time, main challenges of QZS in realizing synchronous vibration isolation and energy harvesting are also discussed. Finally, according to the existing QZS challenges, the future development trend of QZS is proposed. This paper would provide a quick guide for future newcomers to this field.

Keywords: nonlinearity; QZS; vibration isolation; energy harvesting; low-frequency vibrations

1. Introduction

As a common phenomenon in nature, vibration always exists in various forms, such as the swing of leaves, the surging of waves, the vibration of mechanical equipment, and so on [1,2]. Generally speaking, vibrations can have beneficial or destructive effects in practice [3–5]. On one hand, vibrations can help us to produce musical sounds, transmit signals and generate electricity [6–8]. However, on the other hand, once the vibration exceeds a certain threshold, it will bring adverse consequences [9–11]. For example, vibration may decrease the accuracy of mechanical equipment, cause mechanical damage, and shorten the useful life of mechanical equipment [12–16]. Therefore, accurate vibration isolation is very necessary for controlling such vibrations within a limited range on many occasions [17–19].

Vibration isolators are widely used to control the dynamic response of a structure, which are usually composed of dampers and elastic elements [20]. To date, different dampers have been rapidly developed, and the corresponding vibration isolation performance has also been improved greatly. Vibration isolation technologies can be divided into passive [21–24], semi-active [25–28], and active [29–31] methods according to the damping mechanism. Passive vibration isolators function without external power sources and they are driven by the vibration itself. In most cases, they have the simplest structure and require little or no maintenance. However, they still have some shortcomings in practical applications. The first one is that the stiffness and damping are not adjustable so that good vibration isolation performance cannot be achieved under various working conditions. The second is that the adaptability and stability cannot be balanced simultaneously. The third one is that they cannot work effectively under vibrations of high frequencies and low magnitudes. Active vibration isolators reduce vibrations by the additional actuator and are most effective under vibrations of low frequencies. However, they require power for operating the sensors, processing the sensors' signals, and driving the actuator. Meanwhile, they have a higher cost and possibility of failure. Semi-active vibration isolators are intermediate between active and passive ones, which have variable stiffness and damping according to the characteristics of vibrations. However, they also require external power.

As stated above, semi-active and active vibration isolators are promising in vibration isolations. However, the disadvantage is that both of them need external electric energy to achieve the damping effect. At the same time, vibration energy in structures is dissipated into the air. Therefore, there is a strong conflict from the viewpoint of energy conservation. In order to overcome this issue, a valuable way is to harvest vibration energy and convert it into electrical energy. By this way, simultaneous vibration isolation and energy harvesting (SVIEH) can be carried out. Moreover, the harvested electrical energy can be used for semi-active or active dampers. In recent years, SVIEH has been widely studied by many researchers. Relevant professionals have done some work on the above research, which is also known as vibration isolation and energy harvesting, vibration suppression and energy harvesting. According to the above keywords, the relevant literature in the past 20 years was obtained from the Web of Science document retrieval system. As shown in Figure 1, it can be seen from the figure that both China and the United States have carried out a lot of research in the field of vibration isolation and energy harvesting. Adhikari et al. [32] first proposed the idea of energy harvesting using dynamic vibration absorbers, which was carried out by placing a piezoelectric stack between the primary and secondary vibration isolators. Brennan et al. [33] investigated SVIEH under both random and harmonic excitations. The results demonstrated that enough energy was harvested under random excitation and the maximum vibration isolation ability was achieved at the same time. While under harmonic excitations, the excitation frequency should be equal to the natural frequency of the system in order to achieve both optimal goals. Bardaweel et al. [34] developed a new type of low-frequency vibration isolation and energy harvesting device, which is composed of a negative stiffness honeycomb spring and a positive stiffness magnetic spring. Nitin et al. [35] proposed an energy-harvesting shock absorber for vehicle suspension, which consisted of a linear generator and a vibration amplification unit. Shen et al. [36] designed an electromagnetic damper for SVIEH in bridge diagonal cables, which could harvest vibration energy and then provide sufficient damping for the cable.

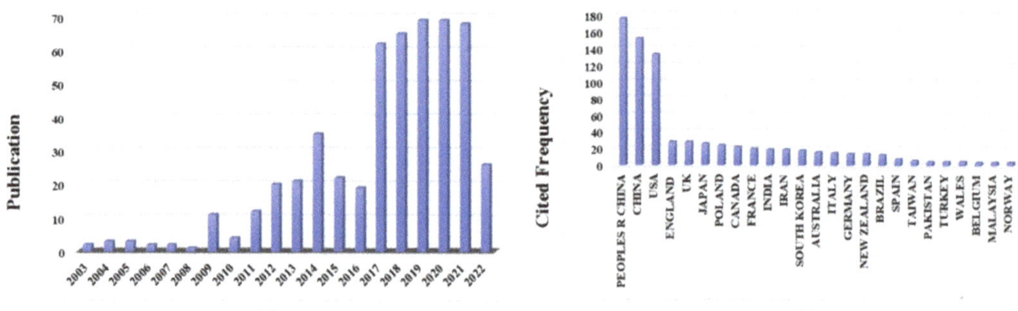

Figure 1. Search results by Web of Science. (**a**) Distribution by time; (**b**) Distribution by country.

The main influencing factor of the vibration isolation frequency band of the vibration isolation system is the system stiffness [37]. Reducing the system stiffness can reduce the natural frequency of the system. The vibration isolation frequency band of the system can be extended to low frequency and ultra-low frequency, so as to obtain a wider vibration isolation frequency band and isolate the harm of vibration [38]. In order to reduce the stiffness of the system, there are usually two technical means: (1) structural optimization design of vibration isolation elements to reduce its stiffness; (2) a negative stiffness element is introduced to reduce the overall stiffness of the system through a quasi-zero stiffness (QZS) mechanism. Due to the limitation of physical size, it is very difficult to effectively reduce the stiffness of the system through structural optimization design [39]. At the same time, the bearing capacity and stability of the system will be limited under low stiffness. The dynamic stiffness of the QZS isolator at the static equilibrium position is zero, which can make the isolator have low stiffness and show excellent low-frequency vibration isolation performance on the premise of ensuring the bearing capacity [40]. Therefore, research on QZS has become a hot topic in the international academic community. The concept of QZS, which is a kind of nonlinear vibration isolator with positive and negative stiffness, was first proposed by Alabuzhev. By optimizing the geometry and stiffness parameters of the negative stiffness mechanism reasonably, the QZS can realize the large bearing capacity of the vibration isolation system and the superior low frequency vibration isolation performance [41,42]. Compared with the traditional passive vibration isolation system, the QZS has the following three advantages: (1) Because of the nonlinear stiffness term, the static bearing capacity is good, and the structural deformation is small. (2) The natural frequency of the system is reduced, and the vibration isolation frequency band is widened. (3) The stiffness characteristics of the system can be flexibly adjusted according to different working conditions. The bearing capacity is determined by the positive stiffness spring, and the negative stiffness element is used to reduce the dynamic stiffness of the system. The positive and negative stiffness in parallel has the characteristics of high static and low dynamic stiffness, which can significantly reduce the system stiffness, broaden the vibration isolation band of the system, and improve the vibration isolation performance of the system on the premise of ensuring the bearing capacity of the system. At the same time, QZS provides a feasible solution for low-frequency vibration energy harvesting due to the very low dynamic stiffness. Therefore, the QZS characteristic can not only widen the vibration isolation region, but also facilitate low-frequency energy harvesting, which has received extensive attention from researchers [43]. In recent years, QZS has been widely studied for simultaneous vibration isolation and energy harvesting. However, there are still many challenges in engineering applications to be overcome. Thus, it is very significant to review and summarize QZS-based simultaneous vibration isolation and energy harvesting. The main purpose of this paper is to review present progress and discuss future directions. The rest of this article is organized as follows. Section 2 describes the development history and basic principles of QZS vibration isolation. Section 3 mainly summarizes the classification of QZS structures. Section 4 summarizes the research progress of QZS synchronous vibration isolation and energy harvesting. Section 5 reviews the challenges and advantages in QZS simultaneous vibration isolation and energy harvesting. Section 6 discusses the shortcomings of current research and provides an outlook for future research. Section 7 makes a brief conclusion.

2. Description of QZS Development

2.1. Basic Principle of QZS

The development of QZS, which is realized by the parallel combination of positive and negative stiffness elements, can be traced back to the 1980s. When a force (ΔF) is applied to a system, the corresponding deformation is denoted as ΔX. Then, if $\Delta F/\Delta X$ is positive, we always call it 'positive stiffness'. Otherwise 'negative stiffness'. The drawback of negative stiffness elastic elements is their instability, so they are rarely used alone in engineering applications. However, it is not the case when positive and negative stiffness elements are connected in parallel. The basic principle of QZS is shown in Figure 2 and the total stiffness of the system can be expressed as $k_0 = k^+ + k^-$. In practice, the positive stiffness k^+ can be adjusted so that $|k^+| \approx |k^-|$ and $|k^+| > |k^-|$. In this case, the total stiffness of the system is infinitely close to zero, so it is called as 'QZS'.

According to the vibration isolation principle, linear vibration isolation devices can only play a role when the excitation frequency is larger than $\sqrt{2}$ times natural frequency of the system. Therefore, the performance of linear vibration isolators is not good under low-frequency or ultra-low frequency vibrations. In order to deal with this problem, it is often necessary to reduce the stiffness of the system. However, reducing the stiffness will reduce the system stability and the loading capacity. To resolve this conflict, nonlinearity is introduced into linear vibration isolators. QZS is just one kind of new nonlinearity, which attracts more and more attention nowadays in the field of vibration isolation due to near-zero dynamic stiffness and strong loading capacity.

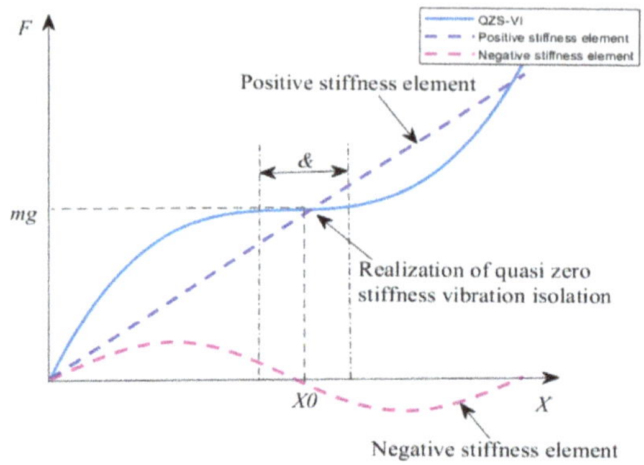

Figure 2. Schematic diagram of QZS.

A simplified schematic diagram of QZS-based vibration isolation is shown in Figure 3. When the mass m is excited by the external force F, the system will generate forced responses, leading to vibration. During the vibration, the positive stiffness element in the QZS plays the key role of loading and damping. The negative stiffness element can adjust the system stiffness according to different working conditions. When the mass is at static equilibrium position, the dynamic stiffness of the system is zero. At the same time, the vibration isolation device has the advantages of low natural frequency and large bandwidth. According to the literature, the current research on QZS-based vibration isolation mainly focuses on two aspects: one is static analysis and the other is stability analysis.

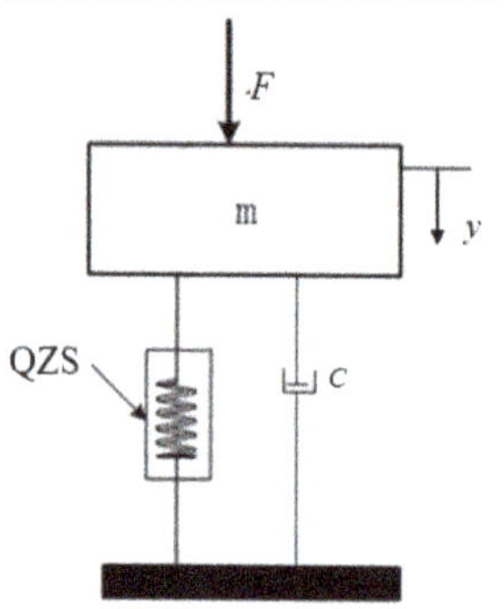

Figure 3. Schematic diagram of QZS-based vibration isolator.

2.2. Development of QZS-Based Vibration Isolation

In 1989, Alabuzhev et al. [44] first presented a comprehensive introduction to the zero stiffness vibration isolation theory. Since then, many implementation schemes of QZS have been proposed. The whole development of QZS-based vibration isolation can be summarized in Figure 4, which is divided into three stages: vibration isolation, vibration isolation and noise reduction, and vibration isolation and energy harvesting. In 1994, Trimboli et al. [45] proposed a new concept of vibration isolation using a magnetic spring and a feedback adjustment system. The magnetic spring exhibits negative stiffness in the limited movement region. The use of magnetic springs in parallel with linear springs leads to ideal vibration isolation characteristics. In 1999, Platus et al. [46] first established a compression spring-rod structure with negative stiffness to counteract the positive stiffness in the suspension. Compared with the state-of-the-art vibration isolation systems, the performance of this new system was greatly improved. In 2003, Takasaki et al. [47] first proposed to use linear actuators for negative stiffness active control in the field of vibration isolation. In 2004, Lee et al. [48] performed dynamic analysis on an elastic device with zero stiffness and then carried out an optimal design of the instability caused by negative stiffness in the system. In 2006, Lee et al. [49] first combined variable structural negative stiffness air damping with positive stiffness spring and applied it for actual seat pneumatic control suspension.

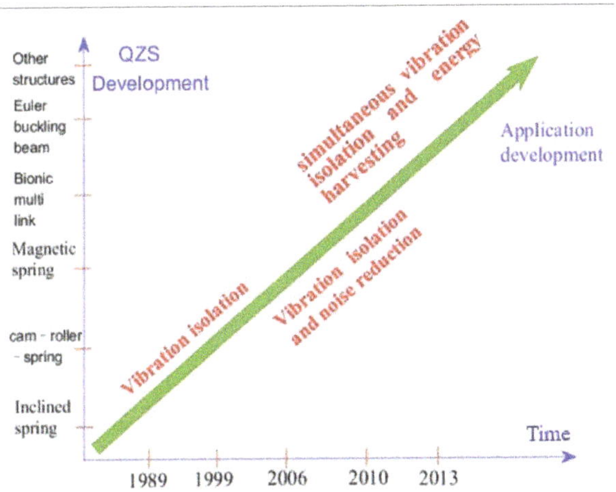

Figure 4. Development of QZS-based vibration isolation.

After 2006, the study of QZS isolation gradually and significantly increased. Carrella et al. [50] analyzed the dynamics of QZS passive vibration isolators by establishing an inclined spring model, and here the concept of high static and low dynamic stiffness was first introduced. In previous studies, it was found that although QZS could be generated at the equilibrium position, the system stiffened once a disturbance occurred. To address this problem, Kovacic et al. [51] used a nonlinear, pre-stressed inclined spring to explore the possible effects of system nonlinearity in 2007. In 2008, Brennan et al. [52] compared the force transmissibility of the QZS vibration isolator with the linear vibration isolator. It was concluded that as long as the system parameters were selected correctly, the QZS vibration isolation performance would be better than that of the linear one. In 2011, Peng et al. [53] first analyzed the stability of a spring-linked QZS isolation device using the L-P regression method. In 2012, Carrella et al. [54] carried out static analysis on nonlinearity in the QZS. Based on this, many researchers subsequently proposed and optimized many kinds of QZS vibration isolators. To date, QZS vibration isolation has been gradually developed to become a research hot-spot in engineering fields. In summary, the utilization of QZS is very promising for vibration isolation and energy harvesting in future [55].

So far, QZS vibration isolators have been widely used in different fields, as shown in Figure 5, such as transportation engineering, marine engineering, aerospace, civil engineering, and so on. As for new energy vehicles, ride comfort, road handling, and battery endurance are main challenges in future. To address these problems, Thanh et al. [56] designed a QZS vibration isolator for a car seat, as shown in Figure 5. Zhou et al. [57] presented a QZS vibration isolator for medical infant cabins. Hu et al. [58] proposed a QZS vibration isolator consisting of three linear springs for pipeline vibration suppression in the field of civil engineering. As for transportation engineering, bridges are threatened by low-frequency vibrations. Attary et al. [59–61] designed a QZS vibration isolation device to reduce the low-frequency vibration of bridges and the simulation results validated its excellent vibration isolation performance. Ishida et al. [62] proposed an origami-type QZS vibration isolator for earthquakes and conducted experimental validations in a simulated seismic zone. In order to protect industrial equipment (e.g., generators, compressors and pipes) from vibration hazards in engineering applications, Burian et al. [63] applied QZS properties to design proper supporting bases for engineering equipment, which could effectively reduce vibration transmission.

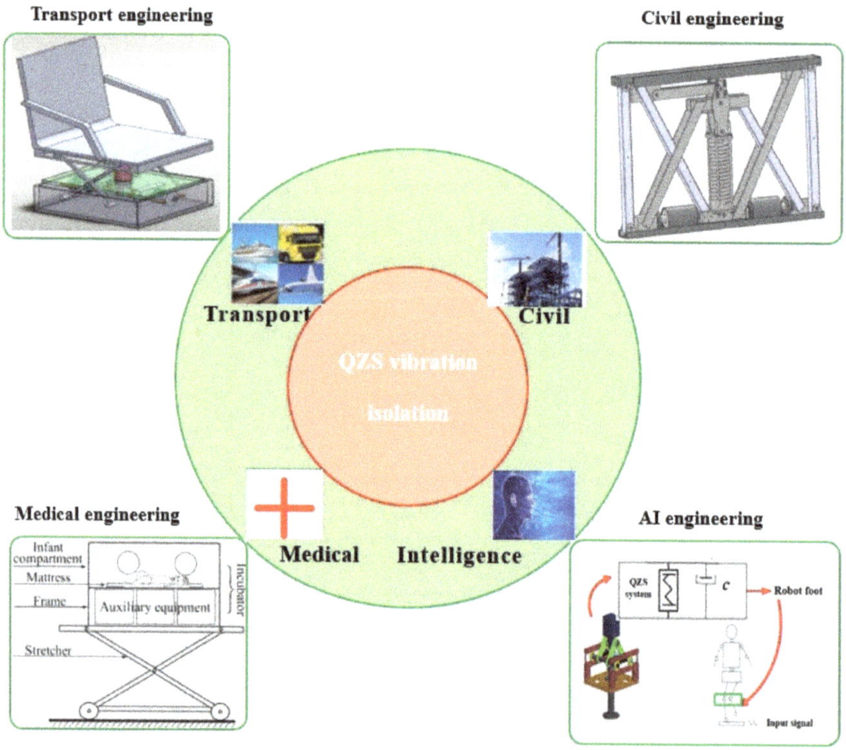

Figure 5. Engineering applications of QZS-based vibration isolation. Reprinted/adapted with permission from Ref. [56]. 2013 Elsevier Ltd. All rights reserved.Thanh Danh Le and Kyoung Kwan Ahn.Reprinted/adapted with permission from Ref. [57].Copyright The Author(s) 2017 Reprints and permissions. Jiaxi Zhou and Kai Wang. Reprinted/adapted with permission from Ref. [59]. Copyright A. S. Elnashai, N.ATTARY and M.SYMANS.

3. Classification of QZS-Based Vibration Isolation Structures

The key point of QZS-based vibration isolation is to design proper negative stiffness structures. Up to now, many scholars have designed various negative stiffness structures for vibration isolation, which can be summarized as the following four classes.

3.1. Classical Structures with Positive and Negative Stiffness

Carrella et al. [50] built a nonlinear vibration isolation system consisting of a vertical spring and two oblique springs connected in parallel. The two inclined springs were used to generate negative stiffness [64–66]. The behaviors of vibration transmission and isolation were studied and the QZS vibration isolation characteristic with high static and low dynamic stiffness was found. At the same time, it was proven that there was a special relationship between the geometry and the QZS stiffness, which provided a theoretical foundation for follow-up research. Later, Xu et al. [67] further increased the number of inclined springs and built a structure consisting of four inclined springs and one vertical spring. Similar to the inclined spring structure, many researchers also studied other related structures. Thanh et al. [56] proposed a novel QZS vibration isolation device shown in Figure 6a, which is composed of two parts. The positive stiffness was provided by the vertical spring and the negative stiffness was generated by two symmetrical structures. Liu et al. [68] proposed a QZS vibration isolator based on the 'spring-rod' structure, as shown in Figure 6b. The isolator was composed of a vertical spring and two identical pre-compressed horizontal springs, where the combination of horizontal spring, slider, and rod generated negative stiffness and the vertical spring generated positive stiffness. The advantages of this kind of isolator include high stability and loading capacity. Finally, a simplified mathematical model of the QZS vibration isolation was established and analyzed.

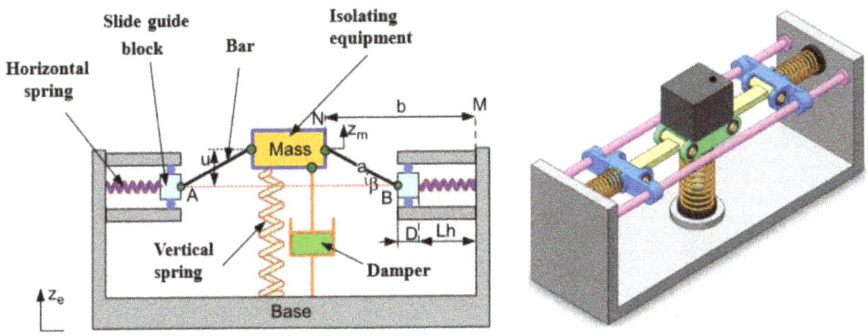

(a) (b)

Figure 6. Schematic diagram of spring-rod structures in (**a**) Spring connecting rod structure. Reprinted/adapted with permission from Ref. [56]. Copyright 2013 Elsevier Ltd., Kyoung KwanAhn and Thanh DanhLe. and (**b**) Three dimensional quasi zero stiffness isolator. Reprinted/adapted with permission from Ref. [57]. Copyright Springer Nature B.V. 2020, Chaoran Liu and Kaiping Yu.

Different from the inclined spring, Liu et al. [69] presented a large negative stiffness under small displacement by utilizing the Euler buckling beam. As shown in Figure 7a, the structure consisted of a linear spring with positive stiffness and two Euler buckling beams with negative stiffness. Huang et al. [70] studied a negative stiffness corrector based on Euler buckling beams. As shown in Figure 7b, two pairs of buckling beams were symmetrically placed at both ends of the mass to provide negative stiffness. Vertical springs provided positive stiffness. This nonlinear isolator performed well under low excitation amplitudes in applications. Carrella et al. [71] proposed a QZS vibration isolation device composed of a magnetic element and a coil spring in parallel. As shown in Figure 7c, it consisted of two linear springs separated by a permanent magnet. Two permanent magnets were symmetrically placed at both ends of the spring and the magnets acted as a negative stiffness. Zheng et al. [72] proposed a QZS shaft coupling, and the repulsive force generated by two coaxial magnetic rings generated negative torsional stiffness.

3.2. Negative-Stiffness Structures with Active Control

A negative stiffness structure with active control is generally established by adding a controller, a sensor, and an external circuit to traditional QZS element. The principle is to introduce a control strategy to adjust dynamic response of the system. Good vibration isolation and stability can be achieved by continuously supplying external electrical energy to the controller.

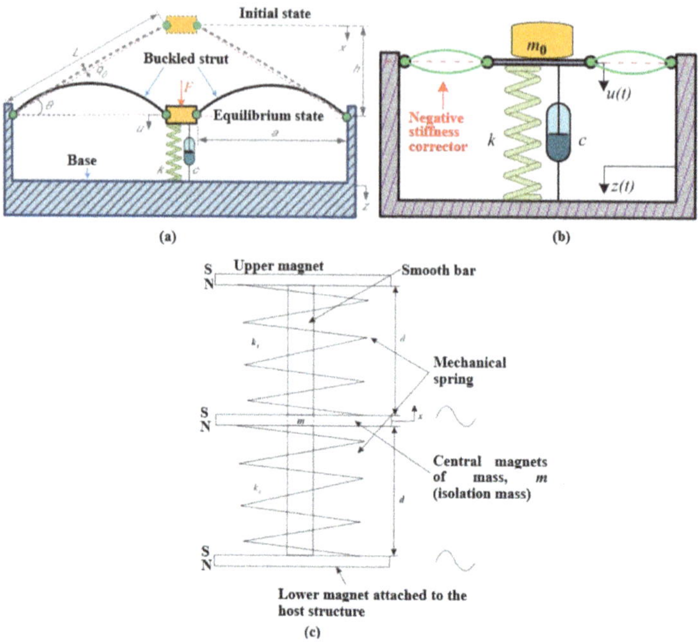

Figure 7. Classical structures with positive and negative stiffness in (**a**) QZS isolator structure of euler buckling beam, (**b**) Double Euler Buckling Beam Symmetrical Distribution QZS Vibration Isolator and (**c**) QZS magnetic spring structure.Reprinted/adapted with permission from Ref. [69]. Copyright 2013 Elsevier Ltd., Xiuchang Huang and Xingtian Liu. Reprinted/adapted with permission from Ref. [70]. Copyright 2013 Elsevier Ltd., Xingtian Liu and Xiuchang Huang. Reprinted/adapted with permission from Ref. [71]. Copyright 2008 Elsevier Ltd. All rights reserved, A. Carrella and M.J. Brennan.

Sun et al. [73] proposed a QZS vibration isolator with simple linear and time-delay active control strategy. As shown in Figure 8a, the structure is connected by two pairs of springs and a rod with length a in the horizontal direction. Each pair of springs is composed of linear stiffness k_1 and nonlinear stiffness k_2. In the vertical direction, the mass is connected to the base by a spring with a stiffness of k_0 and an active control unit. The control unit is composed of a vertical spring and a servo motor, where the motor generates active feedback force. Zhou et al. [74] presented a QZS device with active control, where the positive stiffness was provided by mechanical springs and the negative stiffness was provided by electromagnetic springs. As shown in Figure 8b, the beam acted as a spring and the negative stiffness was determined by the current through the electromagnet. Xu et al. [75] proposed a novel QZS actuator with active control. The actuator was composed of a negative stiffness electromagnetic spring and a positive stiffness flexible link. Thanh et al. [76] proposed a negative stiffness vibration isolator with active control for low-frequency seat vibrations. The isolator introduced an adaptive intelligent controller to obtain high performance of vibration isolation. As shown in Figure 8c, the whole model consisted of a negative stiffness structure, vertical springs, dampers, and a low-friction cylinder controlled by a proportional valve.

3.3. Geometrically Nonlinear Structures

The realization scheme of geometrically nonlinear QZS is supposed to make the force–displacement relation nonlinear due to structural deformation, variable disturbances, etc. The nonlinearity due to geometric instability can be used to generate negative stiffness.

So far, existing QZS vibration isolators with geometric nonlinearity can be classified into cam-like structures and shear-type structures.

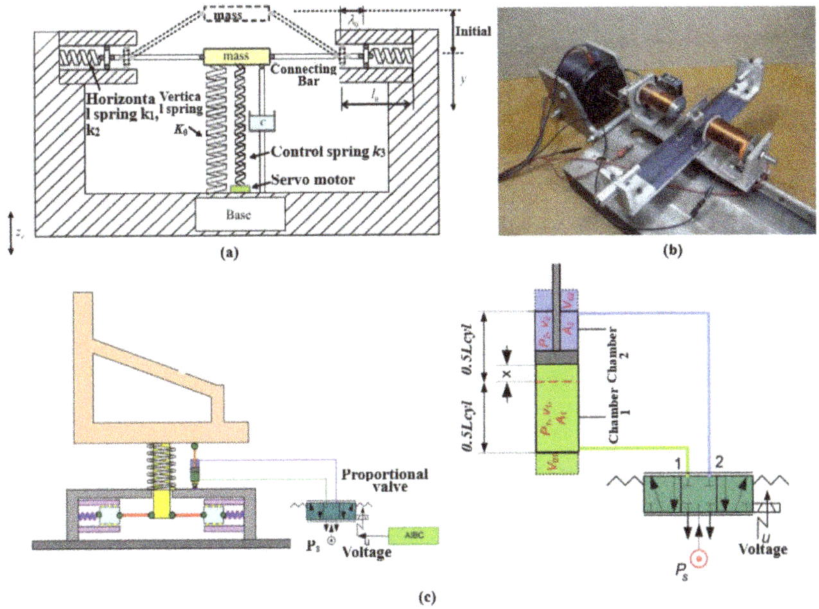

Figure 8. Negative-stiffness structures with active control in (**a**) Simplified schematic diagram of active control, (**b**) Experimental device for active control of negative stiffness structure and (**c**) Negative stiffness structure of active seat control. Reprinted/adapted with permission from Ref. [73]. Copyright 2014 Elsevier Ltd., Xiuting Sun and Jian Xu. Reprinted/adapted with permission from Ref. [74]. Copyright 2009 Elsevier Ltd., N. Zhou and K. Liu. Reprinted/adapted with permission from Ref. [76]. Copyright 2013 Elsevier Ltd., Le Thanh Danh and Kyoung Kwan Ahn.

Li et al. [77] proposed a passive QZS vibration isolator by using a cam structure and the structural profile was designed to achieve the force-displacement relation of QZS. As shown in Figure 9a, the sliding rod and the coil spring connected by the bearing were symmetrically placed in the horizontal direction. The sliding rod and the cam were connected by the bearing in the vertical direction. The experiments demonstrated that the QZS isolator had a low resonance frequency. Zhou et al. [78] proposed a QZS vibration isolator by using a cam-roller-spring structure, as shown in Figure 9b. During vibrations, the cam moved vertically and compressed the rollers connected by the coil spring in the horizontal direction, leading to produce restoring force in different directions. In this structure, the cam-roller-spring generated negative stiffness and the vertical coil springs generated positive stiffness.

Zou et al. [79] proposed a QZS vibration isolator by using a shear structure, as shown in Figure 9c. Then, negative stiffness was provided with by two scissor-like structures and springs. Each scissor-like structure included a coil spring and a linear damper. The geometrically nonlinear stiffness was calculated by parametric analysis and the isolator could be designed to obtain a large frequency band.

3.4. Multi-Direction/Multi-Degree-of-Freedom QZS Structures

In many cases, the vibrations are multi-direction, instead of single-direction. Therefore, it is necessary to design multi-direction QZS vibration isolation devices. Liu et al. [80] proposed an in-plane QZS vibration isolation device, as shown in Figure 10. The device consisted of two concentric magnetic rings and eight cables connected inside the magnetic ring. Interactions between two magnetic rings generated negative stiffness and the cables generated positive stiffness. Vibration in the horizontal direction could be arbitrarily decomposed into orthogonal vibrations, so that vibrations in any direction could be reduced effectively.

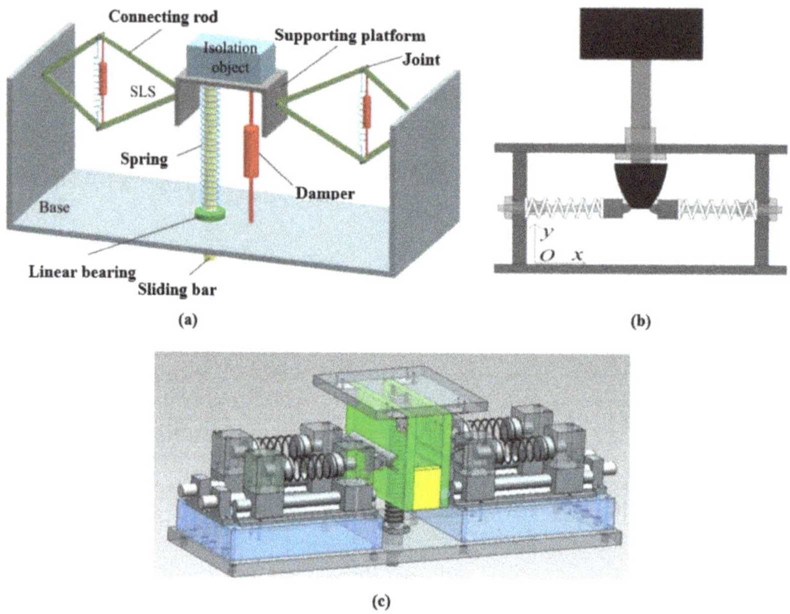

Figure 9. Geometrically nonlinear vibration isolation structure in (**a**) QZS isolator structure of shear type structure, (**b**) QZS isolator for cam mechanism and (**c**) Cam-roller-spring QZS isolator structure. Reprinted/adapted with permission from Ref. [77]. Copyright Springer-V erlag GmbH Germany, part of Springer Nature 2020, Wei Zou and Chun Cheng. Reprinted/adapted with permission from Ref. [78]. Copyright 2020 Published by Elsevier Ltd., Ming Li and Wei Cheng. Reprinted/adapted with permission from Ref. [79]. Copyright 2015 Elsevier Ltd., Jiaxi Zhou and Xinlong Wang.

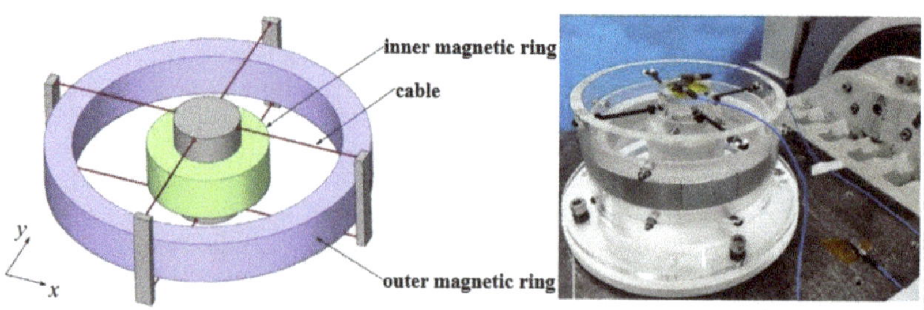

Figure 10. Schematic diagram of multi-direction QZS vibration isolation structure. Reprinted/adapted with permission from Ref. [80]. Copyright 2021 Elsevier Inc., Chaoran Liu and Rui Zhao.

Wu et al. [81] proposed a six-degree-of-freedom passive vibration isolator, which could lead to QZS by adjusting the structural parameters. The whole structure is similar to the Stewart platform, which can achieve passive vibration isolation in six directions. Zhou et al. [82] constructed a six-degree-of-freedom QZS vibration isolation platform by using QZS support poles. Compared with linear vibration isolation devices, this platform had both a wider frequency band of vibration isolation and better isolation performance in the low frequency range.

In summary, the above four types of QZS vibration isolation structures are compared in Tables 1 and 2. It can be seen that: (1) Classical structures with positive and negative stiffness advantages represent the simplest approach, but their stability is poor. (2) The stability of negative stiffness structures with active control is high and their resonance frequencies are low. However, they require a large amount of external electrical energy to maintain, leading to high cost and technical obstacles. (3) Geometrically nonlinear QZS vibration isolators can be more compact than other structures. but they have the disadvantage of limited displacement range and loading capacity. (4) Superior to other structures, multi-direction or multi-degree-of-freedom QZS vibration isolators can reduce multi-direction vibrations and adapt to different working conditions.

Table 1. Comparisons of four QZS implementation structures.

QZS Structure	Representative Literature	Excitation Frequency	Peak Transmissibility
Classical structures with positive and negative stiffness	Thanh et al. [56]	1.5 Hz	9 dB
Negative-stiffness structures with active control	Zhou and Liu et al. [74]	7 Hz	4 dB
Geometrically nonlinear structures	Li and Cheng [77]	10 Hz	17 dB
Multi-direction/Multi-degree-of-freedom structures	Liu et al. [80]	7.23 Hz	11.22 dB

Table 2. Comparisons of four QZS implementation structures.

QZS Structure	Advantages	Disadvantages
Classical structures with positive and negative stiffness	Simple structure	Poor stability
Negative-stiffness structures with active control	low resonance frequency rapid response strong stability	Require external electrical energy High cost
Geometrically nonlinear structures	Compact structure	Limited displacement range Limited loading capacity
Multi-direction/Multi-degree-of-freedom structures	Multi-direction vibration isolation	Complex structure Large space

4. Current Progress of QZS-Based SVIEH

From the viewpoint of energy flow, vibration energy harvesting can also be considered as the counterpart of vibration isolation. As shown in Figure 11, according to different mechanisms, the research of QZS-based SVIEH is mainly composed of electromagnetic, piezoelectric, and hybrid methods.

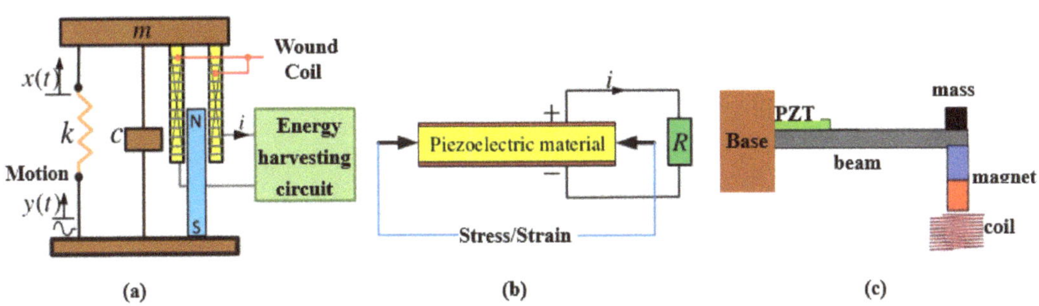

Figure 11. Classification of QZS-based SVIEH [83]: (**a**) Electromagnetic, (**b**) Piezoelectric and (**c**) Hybrid.

4.1. QZS-Based Electromagnetic SVIEH

Electromagnetic vibration energy harvesting was based on the Faraday's law of electromagnetic induction. In recent years, QZS-based electromagnetic SVIEH devices have attracted more and more attention. Yang et al. [83] first proposed a high-order quasi-zero stiffness (HQZS)-based SVIEH structure. As shown in Figure 12a, by adjusting structural parameters, the HQZS can achieve arbitrarily small stiffness at the equilibrium position. Compared with the QZS four-stable system, the HQZS structure had more advantages in vibration isolation and energy harvesting. Li et al. [84] proposed a novel QZS-based bistable electromagnetic energy harvesting device for ocean wave energy. As shown in Figure 12b, the structure consisted of two biomimetic X structures and a power take-off (PTO) system. The new biomimetic X-structure had QZS and bistable properties. There was a mechanical rectifier and electromagnetic DC generator in the PTO system. Compared with linear energy harvesting devices, its energy harvesting efficiency was significantly enhanced [85]. Lu et al. [86] proposed a QZS electromagnetic Stewart platform to achieve SVIEH. As shown in Figure 12c, for various mechanical and electrical parameters, the analytical and numerical results both demonstrate that the frequency band of vibration isolation extended to lower frequencies and produces considerable power output. Moreover, the increase in energy harvesting led to reduced vibration transmissibility under varying some parameters. Yang et al. [87] proposed a multistable and multidirectional (MMD) QZS energy harvesting system based on the former. As shown in Figure 12d, the structure was guided by the slider to realize the relative movement between the permanent magnet and the coil. At the same time, the multi-stable behavior was analyzed. It was verified by experiments that the MMD QZS energy harvesting device can easily realize low-frequency energy harvesting.

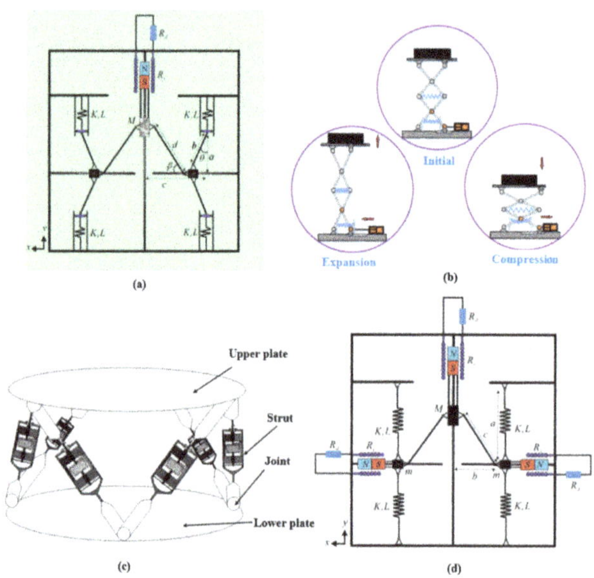

Figure 12. QZS-based SVIEH by using (**a**) the HQZS structure, (**b**) X structure electromagnetic wave energy converter, (**c**) the Stewart Platform, and (**d**) the multistable and multidirectional structure. Reprinted/adapted with permission from Ref. [40]. Copyright 2021 Elsevier Ltd. All rights reserved, Tao Yang and Qingjie Cao. Reprinted/adapted with permission from Ref. [84]. Copyright 2021 Elsevier Ltd. All rights reserved, Meng Li and Xingjian Jing. Reprinted/adapted with permission from Refs. [85,86]. Copyright 2020 Published by Elsevier Inc., Zeqi Lu and Dao Wu. Reprinted/adapted with permission from Ref. [87]. Copyright 2020 Elsevier Ltd. All rights reserved, Tao Yang and Qingjie Cao.

4.2. QZS-Based Piezoelectric SVIEH

Piezoelectric vibration energy harvesting is achieved through the positive piezoelectric effect. When piezoelectric material is deformed by an external force, polarization occurs inside. Then, positive and negative ions in the electrolyte move along the direction of the deformation. By this way, charges will be generated on the two opposite surfaces of the material. Piezoelectric vibration energy harvesting has the advantages of high energy density, simple structure, and easy miniaturization.

Many scholars have combined the advantages of QZS with piezoelectric materials for vibration energy harvesting. Liu et al. [88] proposed piezoelectric beams with QZS for simultaneous low-frequency vibration isolation and energy harvesting. As shown in Figure 13a, the electrical energy output by the deformed piezoelectric material was very high. At the same time, the piezoelectric buckling beam also exhibited negative stiffness. Therefore, the structure had a good vibration isolation effect in the low frequency environment. Compared with traditional linear vibration isolators and cantilever beam piezoelectric devices, it achieved better vibration isolation and energy harvesting in lower frequency environments. Liu et al. [89] further proposed a novel QZS vibration isolation and piezoelectric energy harvesting device for vibration localization in previous research. As shown in Figure 13c, the device consisted of a QZS device and six piezoelectric cantilevers. The device can not only achieve a good low-frequency vibration isolation effect, but also achieve piezoelectric energy harvesting in multiple directions. Sun et al. [90] proposed and analyzed a nonlinear, adjustable, and continuous structure. As shown in Figure 13e, the structure can have large amplitude vibration with adjustable output frequency. Iwaniec et al. [91] designed a QZS dual potential energy well and piezoelectric structure energy harvesting device. As shown in Figure 13b, the QZS effect resonator was able to generate larger amplitude and higher voltage output. The advantage of this device was the wider distribution of potential wells and lower potential barriers. Lu et al. [92] proposed a QZS device of a bistable piezoelectric composite plate structure for SVIEH in a low-frequency environment. As shown in Figure 13d, a feasible analytical model was established through theoretical analysis. The approximate solution of displacement transmissibility and voltage amplitude was obtained. Finally, the accuracy of the approximate solution was judged by numerical simulation. The result show that the energy harvesting increases with the decrease of displacement transfer rate in a certain frequency band range.

4.3. QZS-Based Hybrid SVIEH

Due to the low conversion efficiency of the single-form vibration energy conversion device, researchers designed a hybrid vibration energy harvesting device to overcome the above technical shortcomings [93]. According to current research hybrid QZS vibration energy harvesting devices are not limited to conventional energy conversion methods. In recent years, triboelectric nanogenerators have exhibited excellent energy harvesting performance in low-frequency and ultra-low frequency environments. Therefore, many scholars have proposed coupling triboelectric nanogenerators with electromagnetic generators for energy harvesting in low-frequency environment [94–96]. Wang et al. [97] proposed a nonlinear dual QZS hybrid energy harvesting device. As shown in Figure 14a, the energy conversion part was coupled with a traditional electromagnetic conversion form and a triboelectric nanogenerator to form a hybrid vibration energy conversion device. The effects of QZS and bistable system on energy harvesting efficiency are verified by numerical analysis. Yang et al. [98] designed a time-delay QZS piezoelectric-electromagnetic hybrid energy harvester. As shown in Figure 14b, the parameters that can affect the energy output by external excitation were optimized by designing different types of time-delay control techniques. In this way, the optimal parameters of the energy output efficiency can be obtained. Most importantly, time-delay control techniques may enhance common amplitude values under either random or harmonic excitation. Energy output was significantly enhanced.

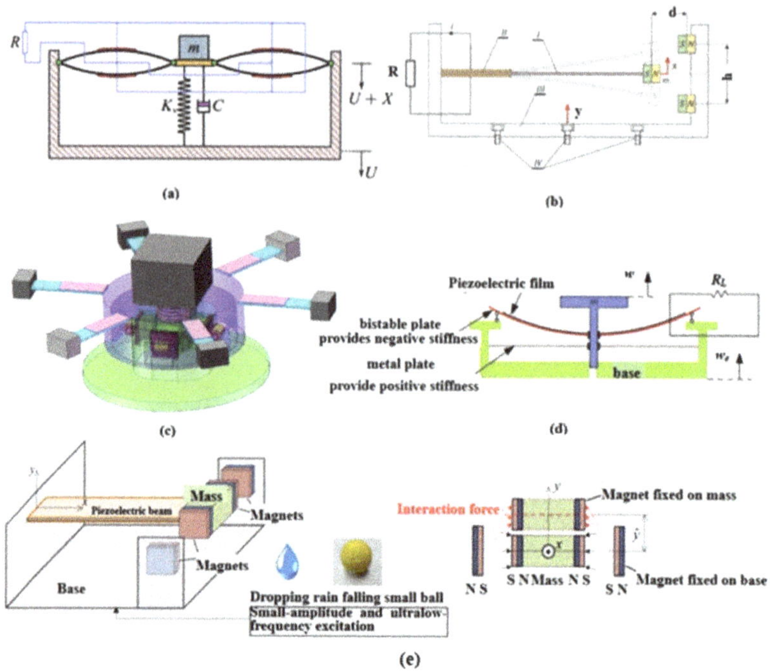

Figure 13. Typical structures based on piezoelectric QZS-SVIEH in (**a**) Simplified diagram of QZS synchronous vibration isolation and energy harvesting, (**b**) Schematic Diagram of Piezoelectric Cantilever Beam Energy Harvesting Structure, (**c**) QZS energy harvesting physical structure, (**d**) Schematic Diagram of Vibration Isolation and Energy Harvesting Integration of Bistable Piezoelectric Composite Plate and (**e**) Piezoelectric energy harvesting structure. Reprinted/adapted with permission from Ref. [88]. Copyright 2021 Elsevier Ltd. All rights reserved, Chaoran Liu and Rui Zhao. Reprinted/adapted with permission from Ref. [91]. Copyright 2021 by the authors. Licensee MDPI, Basel, Switzerland, Joanna Iwaniec and Grzegorz Litak. Reprinted/adapted with permission from Ref. [89]. Copyright 2021 Elsevier Inc. All rights reserved, Chaoran Liu and Rui Zhao. Reprinted/adapted with permission from Ref. [92]. Copyright Journal of Vibration and Control 2019, Vol. 0(0) 1–11, Zeqi Lu and Dong Shao. Reprinted/adapted with permission from Ref. [90]. Copyright Korean Society for Precision Engineering 2019, Xiuting Sun and Feng Wang.

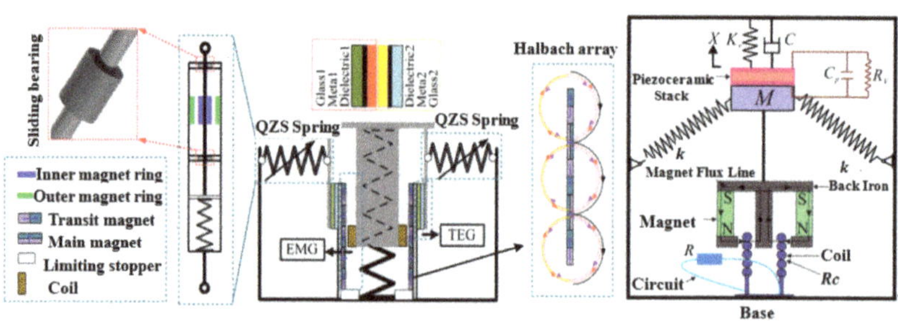

Figure 14. Typical structures based on hybrid QZS-SVIEH in (**a**) Multi directional composite QZS energy harvesting structure and (**b**) Hybrid energy harvesting structure. Reprinted/adapted with

To sum up, the energy density generated by energy harvesting based on QZS is different due to different energy conversion methods. The energy data generated by different energy conversion methods are shown in Table 3.

Table 3. Comparisons of four QZS implementation structures.

Energy Conversion Mode	Representative Literature	Incentive Type	Energy Harvesting Value
Electromagnetic	Yang and Cao et al. [3]	harmonic excitation	0.5 W
	Li and jing et al. [84]	Random excitation	0.5~1 W
	Lu and Wu et al. [86]	harmonic excitation	1.74 W
Piezoelectric	Liu and Zhao et al. [89]	harmonic excitation	2 W
	Sun and Wang et al. [90]	Random excitation	100 mW
Hybrid.	Wang and Zhou et al. [97]	Random excitation	10 mW

5. Key Challenges of QZS-Based SVIEH

Up to now, QZS has been widely studied for vibration isolation and much progress has been achieved [99]. In particular, QZS began to be explored for SVIEH in recent years. It should be pointed out that QZS-based SVIEH is still in its infancy and faces some problems to be solved in engineering applications [100]. According to the literature, several key challenges of QZS-based SVIEH are summarized as follows.

5.1. Improve Energy Harvesting/Weight Ratio

Current negative stiffness is mainly implemented by means of oblique springs, Eulerian flexural beams [101–103], spring-linked bars [104,105], and geometric structures [106]. These structural dimensions make it difficult to carry out the miniaturization of a QZS element. Compared with negative stiffness elements, positive stiffness elements are always realized by using linear helical springs, which also play a major role of load bearing. As a result, the entire QZS device hardly becomes small by changing the positive stiffness element [107,108] and then the power/volume ratio is indeed low. In recent years, magnetic springs have been proposed to realize negative stiffness, which are superior to other structures in miniaturization. However, magnetic springs are prone to magnetic degeneration, so it is difficult keep the bearing capacity of magnetic springs. Therefore, it is urgent to develop QZS-based SVIEH device with simple structure and strong carrying capacity.

5.2. Potential Instability

Most current research on the QZS focuses on the way how to carry out negative stiffness [109,110]. However, negative stiffness itself also leads to the instability of QZS [111]. A QZS system can only have static equilibrium position within a small displacement range. Once the QZS system shifts from its static equilibrium position, the negative stiffness element is most likely to show hardening characteristics, which will cause unstable phenomena, such as jumping [112–114], chaos [115–117], and voltage bifurcation [118,119]. In this case, it is possible that vibration isolation and energy harvesting cannot be achieved simultaneously.

As for QZS-based SVIEH, most studies mainly adopt dynamic modeling, response analysis, sensitivity analysis, performance comparison, and so on [120,121]. The instability in QZS-based SVIEH has rarely been studied widely [122,123]. Nowadays, researchers mainly adjust system parameters to achieve the best performance of the QZS-based SVIEH device and the effects due to the instability of negative stiffness instabilities are ignored. Compared with traditional energy harvesting systems, negative stiffness elements are key components to generate instability. Although multi-stable energy harvesting device can improve the harvesting efficiency, it is prone to jumping back and forth between different energy wells.

5.3. Multi-Stable Phenomenon Due to Nonlinear Negative Stiffness

The characteristic of nonlinear negative stiffness makes a QZS system easy to generate multi-stable phenomenon, so QZS-based energy harvesting system can achieve high harvesting efficiency among a broad frequency band [124,125]. It breaks the limitation that linear energy harvesting devices can only achieve the maximum performance in the resonance region. However, nonlinear QZS systems exhibit multiple resonances among certain frequency ranges. In this case, the advantage of broad band is missing. Especially in the multi-solution region, it is difficult to realize the motion between high-energy wells. At the same time, the distance between potential wells will become narrow and the barrier will become high under ultra-low frequency vibrations, such that it is not easy to transit the barrier.

5.4. Small Range of QZS

In recent years, QZS-based systems have demonstrated good low-frequency vibration isolation characteristics due to high static and low dynamic stiffness. However, there are still low-frequency resonant regions, which receive much attention in the field of low-frequency vibration energy harvesting [126,127]. The researchers obtained the minimum stiffness by adjusting the quasi-zero stiffness structural parameters. The energy output between traps in a low frequency environment is achieved near the static equilibrium position. When the QZS energy harvesting device is excited by low intensity or ultra-low frequency, the realization range of QZS becomes smaller. It is difficult to obtain high-energy inter-well oscillations for QZS devices. Therefore, the small range of QZS realization is still a challenge to be solved in the future.

5.5. Limited Implementation Method

As a key technology to improve QZS-based SVIEH devices, nonlinearity has been recognized by more and more researchers. At present, most studies on the performance of QZS-based SVIEH mainly focus on how to optimally design of nonlinear negative stiffness. Then, the nonlinear negative stiffness element is used in parallel with linear helical springs to achieve high static and low dynamic stiffness. By this way, linear helical springs with positive stiffness are limited to compensate for negative stiffness. Thus, it is difficult to innovatively design QZS-based SVIEH structures and new ideas are much desirable.

6. Future Trend

As for the above challenges of QZS-based SVIEH, new theories, new structures, and new designs should be explored to deal with them. In this paper, the following trends will be discussed.

6.1. Nonlinear Positive Stiffness Compensation

Existing QZS structures for SVIEH are mainly based on symmetrically distributed nonlinear negative stiffness and the positive stiffness is mainly realized by linear helical spring [128]. However, using linear helical springs to compensate for negative stiffness is not perfect for all kinds of QZS [129,130]. From the perspective of dynamics, the QZS is often realized through combining positive and negative stiffness elements. However, the nonlinearity is not limited to only negative stiffness elements. It is also a new idea to realize QZS-based SVIEH by using nonlinear positive stiffness elements to compensate negative stiffness. As shown in Figure 15, Yan et al. [131] proposed to compensate negative stiffness with nonlinear positive stiffness for QZS-based vibration isolation and then analyzed its feasibility. In the future, more nonlinear positive stiffness elements should be studied for QZS-based SVIEH.

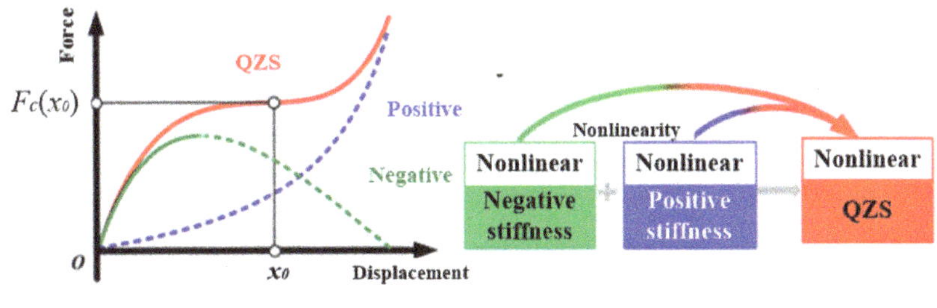

Figure 15. Nonlinear positive stiffness compensates for negative stiffness. (**a**) Force-displacement curve of QZS. (**b**) Analysis Diagram of QZS Realization. Reprinted/adapted with permission from Ref. [131]. Copyright 2022 Elsevier Ltd. All rights reserved, Ge Yan and Zhiyuan Wu.

6.2. QZS Molecular Spring

In recent years, molecular spring isolators have received much attention due to their high static and low dynamic stiffness characteristics. Molecular spring isolators are composed of water and nano hydrophobic microporous materials. This new solid–liquid mixture is like a mechanical spring, which can easily achieve variable stiffness characteristics [132]. Li et al. [133] first proposed a molecular spring isolator for heavy loads and analyzed the vibration isolation performance by means of energy transfer. Nie et al. [134] applied a molecular spring isolator for a car suspension and simulated the smoothness by using MATLAB/Simulink simulation. The results showed that the perturbation of molecular spring isolators was better than that of linear suspensions under external fixed-frequency excitation. Up to now, the QZS molecular spring has been rarely studied for SVIEH, which deserves further study in the future.

6.3. QZS-Based SVIEH Using New Material

In recent years, some metamaterials or synthetic materials have been introduced into vibration energy harvesting. The biggest difference between these new materials and traditional materials is that the new materials easily demonstrate variable stiffness, light weight, and compact structure [135]. For QZS-based SVIEH, structural miniaturization, low-frequency and broad band, nonlinearity, and the efficiency are still the key points in future research. However, natural materials hardly have the above characteristics simultaneously [136]. Thus, the integration of new materials and energy harvesting provides a new idea for QZS-based SVIEH [137]. Although the related studies have not been reported to date, it is worth using them to develop QZS-based SVIEH in the future.

6.4. The Effects of External Circuit Parameters

In recent years, researchers have focused on the effects of structural parameters on QZS-based SVIEH devices, such as the system damping ratio [138], the electromechanical coupling coefficient [139], and the excitation amplitude [140]. When a QZS-based SVIEH system is connected to an external circuit, the external resistance will also affect the performance [141]. Zhang et al. [142] used cantilever piezoelectric structural parameters as quantification. The resistance value with the best energy capture effect was obtained by using the structural parameters. Yan et al. [143] proposed a low-frequency broadband piezoelectric tri-stable external resistive-inductive resonant (RL) circuit energy harvesting device. The results show that the energy harvesting device with the resonant circuit significantly improves the energy harvesting efficiency. At the same time, the vibration response is reduced. Resonant circuits have ushered in new technical means in the field of improving vibration energy capture. Zhou et al. [144] summarized the research on multistable energy harvesting devices while presenting the key challenges for energy harvesting by external circuits. The importance of the conversion circuit is expounded when the AC power obtained by the energy harvesting device under the external excitation is used to supply the DC equipment. For example, circuits, such as synchronous charge extraction techniques [145–147], parallel synchronous switched capture inductor (P-SSHI) circuits [148], and SSHI electronic interface circuits [149], have shown significant improvements in vibration energy harvesting efficiency. At present, much research in the field of QZS vibration isolation and energy harvesting (QZS-VI-EH) mainly focuses on structural design and low-frequency broadband. However, the effect of external circuits on QZS-VI-EH is rarely studied. It is not difficult to find that external circuitry may also have an impact on the QZS-VI-EH device. For future research, the coupling between the external circuit and QZS-VI-EH should be closely combined. An optimal coupling condition is sought to improve the energy harvesting performance.

6.5. QZS-Based SVIEH Double Optimization

Currently, QZS-based SVIEH regards vibration isolation and energy harvesting as two conflicting indicators, and its main work focuses on theoretical modeling, numerical simulation, experimental testing, etc. A tradeoff optimization can be achieved by sacrificing the performance of vibration suppression and energy harvesting, respectively. Therefore, it will be an important direction to realize the dual optimization of vibration isolation and energy harvesting in the future, and the following two aspects can be considered: (1) Mechanism research, according to the application scenario of QZS-based SVIEH, to reveal the relationship between the internal vibration isolation and energy harvesting of the structure, how the performance of QZS-based SVIEH is constrained in structure, etc. (2) A composite QZS-based SVIEH structure is established to realize the decoupling of vibration isolation and energy harvesting from the structure.

7. Conclusions

The QZS devices can not only achieve a good vibration isolation effect in the low frequency or ultra-low frequency environment, but also improve the energy harvesting efficiency. This presents wide applications in the fields of vehicle transportation, aerospace engineering, civil engineering, marine energy equipment engineering, environmental monitoring, and micro-electromechanical engineering. The QZS perfectly fits the two goals of low-frequency vibration isolation and vibration energy harvesting. This paper aims to summarize the research progress of nonlinear QZS in vibration isolation and energy harvesting in recent decades. The theoretical status, basic principle, structural status, and engineering application of zero stiffness vibration isolation are analyzed. The research status of QZS-VI-EH is discussed from the perspective of energy conversion. Finally, according to the problems existing in QZS-VI-EH, the future development trend of QZS-VI-EH is prospected.

Author Contributions: Z.C. (Zhiwen Chen): investigation, methodology, writing—original draft. Z.C. (Zhongsheng Chen): conceptualization, supervision, writing—original draft, funding acquisition. Y.W.: investigation. All authors have read and agreed to the published version of the manuscript.

Funding: This research was funded by the National Natural Science Foundation of China, grant number 51911530202 and 51577189.

Data Availability Statement: Not applicable.

Conflicts of Interest: The authors declare no conflict of interest.

References

1. Yang, T.; Zhou, S.; Fang, S. Nonlinear vibration energy harvesting and vibration suppression technologies: Designs, analysis, and applications. *Appl. Phys. Rev.* **2021**, *8*, 031317. [CrossRef]
2. Cao, D.; Guo, X.; Hu, W. A novel low-frequency broadband piezoelectric energy harvester combined with a negative stiffness vibration isolator. *J. Intell. Mater. Syst. Struct.* **2019**, *30*, 1105–1114. [CrossRef]

3. Yang, T.; Cao, Q.; Hao, Z. A novel nonlinear mechanical oscillator and its application in vibration isolation and energy harvesting. *Mech. Syst. Signal Process.* **2021**, *155*, 107636. [CrossRef]
4. Wang, Q.; Zhou, J.; Xu, D. Design and experimental investigation of ultra-low frequency vibration isolation during neonatal transport. *Mech. Syst. Signal Process.* **2020**, *139*, 106633. [CrossRef]
5. Ye, K.; Ji, J.C.; Brown, T. Design of a quasi-zero stiffness isolation system for supporting different loads. *Sound Vib.* **2020**, *471*, 115198. [CrossRef]
6. Wang, K.; Zhou, J.; Ouyang, H. A semi-active metamaterial beam with electromagnetic quasi-zero-stiffness resonators for ultralow-frequency bandgap tuning. *Int. J. Mech. Sci.* **2020**, *176*, 105548. [CrossRef]
7. Priya, S.; Inman, D.J. *Energy Harvesting Technologies*; Springer: New York, NY, USA, 2009.
8. Zhou, S.; Cao, J.; Erturk, A.; Lin, J. Enhanced broadband piezoelectric energy harvesting using rotatable magnets. *Appl. Phys. Lett.* **2013**, *102*, 173901. [CrossRef]
9. Yan, Z.; Sun, W.; Tan, T. Nonlinear analysis of galloping piezoelectric energy harvesters with inductive-resistive circuits for boundaries of analytical solutions, Commun. *Nonlinear Sci. Numer. Simul.* **2018**, *62*, 90–116. [CrossRef]
10. Margielewicz, J.; Gąska, D.; Litak, G. Nonlinear dynamics of a new energy harvesting system with quasi-zero stiffness. *Appl. Energy* **2022**, *307*, 118159. [CrossRef]
11. Barton, D.A.W.; Burrow, S.G.; Clare, L.R. Energy harvesting from vibrations with a nonlinear oscillator. *J. Vib. Acoust.* **2010**, *132*, 021009. [CrossRef]
12. Abolfathi, A.; Brennan, M.J.; Waters, T.P. On the effects of mistuning a force-excited system containing a quasi-zero-stiffness vibration isolator. *J. Vib. Acoust.* **2015**, *137*, 044502. [CrossRef]
13. Li, S.; Feng, G.; Zhao, Q. Design and research of semiactive quasi-zero stiffness vibration isolation system for vehicles. *Shock Vib.* **2021**, *2021*, 5529509. [CrossRef]
14. Davis, R.B.; McDowell, M.D. Combined Euler column vibration isolation and energy harvesting. *Smart Mater. Struct.* **2017**, *26*, 055001. [CrossRef]
15. Cobb, R.G.; Sullivan, J.M.; Das, A.; Davis, L.P.; Hyde, T.T.; Davis, T.; Rahman, Z.H.; Spanos, J.T. Vibration isolation and suppression system for precision payloads in space. *Smart Mater. Struct.* **1999**, *8*, 798. [CrossRef]
16. Virgin, L.N.; Santillan, S.T.; Plaut, R.H. Vibration isolation using extreme geometric nonlinearity. *J. Sound Vib.* **2008**, *315*, 721–731. [CrossRef]
17. Thakadu, K.; Li, K. A Passive Vibration Isolator Integrated a Dynamic Vibration Absorber with Negative Stiffness Spring. *J. Vib. Eng. Technol.* **2022**, *10*, 71–82. [CrossRef]
18. Valeev, A.; Kharisov, S. Application of vibration isolators with a low stiffness for the strongly vibrating equipment. *Procedia Eng.* **2016**, *150*, 641–646. [CrossRef]
19. Anvar, V.; Alexey, Z.; Artem, T. Study of application of vibration isolators with quasi-zero stiffness for reducing dynamics loads on the foundation. *Procedia Eng.* **2017**, *176*, 137–143. [CrossRef]
20. Ali, S.; Ramaswamy, A. Testing and modeling of MR damper and its application to SDOF systems using integral backstepping technique. *J. Dyn. Syst. Meas. Control* **2009**, *131*, 021009. [CrossRef]
21. Lee, C.T.; Moon, B.Y. Simulation and experimental validation of vehicle dynamic characteristics for displacement-sensitive shock absorber using fluid-flow modelling. *Mech. Syst. Signal Process.* **2006**, *20*, 373–388. [CrossRef]
22. Nie, S.; Zhuang, Y.; Wang, Y. Velocity & displacement-dependent damper: A novel passive shock absorber inspired by the semi-active control. *Mech. Syst. Signal Process.* **2018**, *99*, 730–746.
23. Konieczny, Ł. Analysis of simplifications applied in vibration damping modelling for a passive car shock absorber. *Shock Vib.* **2016**, *2016*, 6182847. [CrossRef]
24. Xu, T.; Liang, M.; Li, C. Design and analysis of a shock absorber with variable moment of inertia for passive vehicle suspensions. *J. Sound Vib.* **2015**, *355*, 66–85. [CrossRef]
25. Bazinenkov, A.M.; Mikhailov, V.P. Active and semi active vibration isolation systems based on magnetorheological materials. *Procedia Eng.* **2015**, *106*, 170–174. [CrossRef]
26. Choi, S.B.; Lee, H.K.; Chang, E.G. Field test results of a semi-active ER suspension system associated with skyhook controller. *Mechatronics* **2001**, *11*, 345–353. [CrossRef]
27. Valášek, M.; Novak, M.; Šika, Z. Extended ground-hook-new concept of semi-active control of truck's suspension. *Veh. Syst. Dyn.* **1997**, *27*, 289–303. [CrossRef]
28. Kamaruzaman, N.A.; Robertson, W.S.P.; Ghayesh, M.H. Vibration Isolation Performance of an LQR-Stabilised Planar Quasi-zero Stiffness Magnetic Levitation System. In *Vibration Engineering for a Sustainable Future*; Springer: Cham, Switzerland, 2021; pp. 237–243.
29. Zhang, Q.; Yang, Z.; Wang, C. Intelligent control of active shock absorber for high-speed elevator car. *Proc. Inst. Mech. Eng. Part C J. Mech. Eng. Sci.* **2019**, *233*, 3804–3815. [CrossRef]
30. Ebrahimi, B.; Bolandhemmat, H.; Khamesee, M.B. A hybrid electromagnetic shock absorber for active vehicle suspension systems. *Veh. Syst. Dyn.* **2011**, *49*, 311–332. [CrossRef]
31. Zheng, P.; Wang, R.; Gao, J. A comprehensive review on regenerative shock absorber systems. *J. Vib. Eng. Technol.* **2020**, *8*, 225–246. [CrossRef]
32. Brennan, M.J.; Tang, B.; Melo, G.P. An investigation into the simultaneous use of a resonator as an energy harvester and a vibration absorber. *J. Sound Vib.* **2014**, *333*, 1331–1343. [CrossRef]

33. Li, Y.; Baker, E.; Reissman, T. Design of mechanical metamaterials for simultaneous vibration isolation and energy harvesting. *Appl. Phys. Lett.* **2017**, *111*, 251903. [CrossRef]
34. Mofidian, S.M.M.; Bardaweel, H. A dual-purpose vibration isolator energy harvester: Experiment and model. *Mech. Syst. Signal Process.* **2019**, *118*, 360–376. [CrossRef]
35. Satpute, N.; Iwaniec, M.; Narina, R. Energy harvesting shock absorber with linear generator and mechanical motion amplification. *Vib. Phys. Syst.* **2020**, *31*, 2020104-1–2020104-8.
36. Shen, W.; Zhu, S.; Zhu, H. Experimental study on using electromagnetic devices on bridge stay cables for simultaneous energy harvesting and vibration damping. *Smart Mater. Struct.* **2016**, *25*, 065011. [CrossRef]
37. Dalela, S.; Balaji, P.S.; Jena, D.P. A review on application of mechanical metamaterials for vibration control. *Mech. Adv. Mater. Struct.* **2022**, *29*, 3237–3262. [CrossRef]
38. Zou, D.; Liu, G.; Rao, Z. A device capable of customizing nonlinear forces for vibration energy harvesting, vibration isolation, and nonlinear energy sink. *Mech. Syst. Signal Process.* **2021**, *147*, 107101. [CrossRef]
39. Wang, K.; Zhou, J.; Ouyang, H. A dual quasi-zero-stiffness sliding-mode triboelectric nanogenerator for harvesting ultralow-low frequency vibration energy. *Mech. Syst. Signal Process.* **2021**, *151*, 107368. [CrossRef]
40. Woodard, S.E.; Housner, J.M. Nonlinear behavior of a passive zero-spring-rate suspension system. *J. Guid. Control. Dyn.* **1991**, *14*, 84–89. [CrossRef]
41. Yan, G.; Zou, H.X.; Wang, S. Large stroke quasi-zero stiffness vibration isolator using three-link mechanism. *J. Sound Vib.* **2020**, *478*, 115344. [CrossRef]
42. Gatti, G.; Shaw, A.D.; Gonçalves, P.J.P. On the detailed design of a quasi-zero stiffness device to assist in the realisation of a translational Lanchester damper. *Mech. Syst. Signal Process.* **2022**, *164*, 108258. [CrossRef]
43. Wang, X.; Shi, Z.; Wang, J. A stack-based flex-compressive piezoelectric energy harvesting cell for large quasi-static loads. *Smart Mater. Struct.* **2016**, *25*, 055005. [CrossRef]
44. Alabuzhev, P.M. *Vibration Protection and Measuring Systems with Quasi-Zero Stiffness*; CRC Press: Boca Raton, FL, USA, 1989.
45. Trimboli, M.S.; Wimmel, R.; Breitbach, E.J. Quasi-active approach to vibration isolation using magnetic springs. In *Smart Structures and Materials 1994, Passive Damping*; SPIE: Bellingham, WA, USA, 1994; Volume 2193, pp. 73–83.
46. Platus, D.L. Negative-stiffness-mechanism vibration isolation systems. In *Vibration Control in Microelectronics, Optics, and Metrology*; International Society for Optics and Photonics: Bellingham, WA, USA, 1992; Volume 1619, pp. 44–54.
47. Mizuno, T.; Toumiya, T.; Takasaki, M. Vibration isolation system using negative stiffness. *JSME Int. J. Ser. C Mech. Syst. Mach. Elem. Manuf.* **2003**, *46*, 807–812. [CrossRef]
48. Lee, C.M.; Goverdovskiy, V.N.; Samoilenko, S.B. Prediction of non-chaotic motion of the elastic system with small stiffness. *J. Sound Vib.* **2004**, *272*, 643–655. [CrossRef]
49. Lee, C.M.; Bogatchenkov, A.H.; Goverdovskiy, V.N. Position control of seat suspension with minimum stiffness. *J. Sound Vib.* **2006**, *292*, 435–442. [CrossRef]
50. Carrella, A.; Brennan, M.J.; Waters, T.P. Static analysis of a passive vibration isolator with quasi-zero-stiffness characteristic. *J. Sound Vib.* **2007**, *301*, 678–689. [CrossRef]
51. Kovacic, I.; Brennan, M.J.; Waters, T.P. A study of a nonlinear vibration isolator with a quasi-zero stiffness characteristic. *J. Sound Vib.* **2008**, *315*, 700–711. [CrossRef]
52. Carrella, A.; Brennan, M.J.; Kovacic, I. On the force transmissibility of a vibration isolator with quasi-zero-stiffness. *J. Sound Vib.* **2009**, *322*, 707–717. [CrossRef]
53. Peng, X.; Zhang, S.X. Nonlinear resonance response analysis of a kind of passive isolation system with quasi-zero stiffness. *J. Hunan Univ.* **2011**, *38*, 34–39.
54. Niu, F.; Meng, L.S.; Wu, W.J. Recent advances in quasi-zero-stiffness vibration isolation systems. *Appl. Mech. Mater. Trans.* **2013**, *397*, 295–303. [CrossRef]
55. Li, B.; Shuai, C.; Xu, W. Review on Low Frequency Control of Quasi-Zero Stiffness. In Proceedings of the 2021 13th International Conference on Computer and Automation Engineering (ICCAE), Melbourne, Australia, 20–22 March 2021; pp. 59–63.
56. Le, T.D.; Ahn, K.K. Experimental investigation of a vibration isolation system using negative stiffness structure. *Int. J. Mech. Sci.* **2013**, *70*, 99–112. [CrossRef]
57. Zhou, J.; Wang, K.; Xu, D. Vibration isolation in neonatal transport by using a quasi-zero-stiffness isolator. *J. Vib. Control* **2018**, *24*, 3278–3291. [CrossRef]
58. Ding, H.; Ji, J.; Chen, L.Q. Nonlinear vibration isolation for fluid-conveying pipes using quasi-zero stiffness characteristics. *Mech. Syst. Signal Process.* **2019**, *121*, 675–688. [CrossRef]
59. Attary, N.; Symans, M.; Nagarajaiah, S. Performance evaluation of negative stiffness devices for seismic response control of bridge structures via experimental shake table tests. *J. Earthq. Eng.* **2015**, *19*, 249–276. [CrossRef]
60. Attary, N.; Symans, M.; Nagarajaiah, S. Numerical simulations of a highway bridge structure employing passive negative stiffness device for seismic protection. *Earthq. Eng. Struct. Dyn.* **2015**, *44*, 973–995. [CrossRef]
61. Attary, N.; Symans, M.; Nagarajaiah, S. Performance assessment of a highway bridge structure employing adaptive negative stiffness for seismic protection. In Proceedings of the Structures Congress 2013, Bridging Your Passion with Your Profession, Pittsburgh, PA, USA, 2–4 May 2013; pp. 1736–1746.

62. Ishida, S.; Suzuki, K.; Shimosaka, H. Design and experimental analysis of origami-inspired vibration isolator with quasi-zero-stiffness characteristic. *J. Vib. Acoust.* **2017**, *139*, 051004. [CrossRef]
63. Burian, Y.A.; Silkov, M.V.; Trifonova, E.N. Support with quasi-zero stiffness effect for processing equipment. *AIP Conf. Proc.* **2019**, *2141*, 03006.
64. Bouna, H.S.; Nbendjo, B.R.; Woafo, P. Isolation performance of a quasi-zero stiffness isolator in vibration isolation of a multi-span continuous beam bridge under pier base vibrating excitation. *Nonlinear Dyn.* **2020**, *100*, 1125–1141. [CrossRef]
65. Liang, H.; Hao, G.; Olszewski, O.Z. A review on vibration-based piezoelectric energy harvesting from the aspect of compliant mechanisms. *Sens. Actuators A Phys.* **2021**, *331*, 112743. [CrossRef]
66. Robertson, W.S.; Kidner, M.R.F.; Cazzolato, B.S. Theoretical design parameters for a quasi-zero stiffness magnetic spring for vibration isolation. *J. Sound Vib.* **2009**, *326*, 88–103. [CrossRef]
67. Xu, D.; Zhang, Y.; Zhou, J. On the analytical and experimental assessment of the performance of a quasi-zero-stiffness isolator. *J. Vib. Control* **2014**, *20*, 2. [CrossRef]
68. Liu, C.; Yu, K. Superharmonic resonance of the quasi-zero-stiffness vibration isolator and its effect on the isolation performance. *Nonlinear Dyn.* **2020**, *100*, 95–117. [CrossRef]
69. Liu, X.; Huang, X.; Hua, H. On the characteristics of a quasi-zero stiffness isolator using Euler buckled beam as negative stiffness corrector. *J. Sound Vib.* **2013**, *332*, 3359–3376. [CrossRef]
70. Huang, X.; Liu, X.; Sun, J. Vibration isolation characteristics of a nonlinear isolator using Euler buckled beam as negative stiffness corrector: A theoretical and experimental study. *J. Sound Vib.* **2014**, *333*, 1132–1148. [CrossRef]
71. Carrella, A.; Brennan, M.J.; Waters, T.P. On the design of a high-static–low-dynamic stiffness isolator using linear mechanical springs and magnets. *J. Sound Vib.* **2008**, *315*, 712–720. [CrossRef]
72. Zheng, Y.; Zhang, X.; Luo, Y. Analytical study of a quasi-zero stiffness coupling using a torsion magnetic spring with negative stiffness. *Mech. Syst. Signal Process.* **2018**, *100*, 135–151. [CrossRef]
73. Sun, X.; Xu, J.; Jing, X. Beneficial performance of a quasi-zero-stiffness vibration isolator with time-delayed active control. *Int. J. Mech. Sci.* **2014**, *82*, 32–40. [CrossRef]
74. Zhou, N.; Liu, K. A tunable high-static–low-dynamic stiffness vibration isolator. *J. Sound Vib.* **2010**, *329*, 1254–1273. [CrossRef]
75. Xu, J.; Yang, X.; Li, W. Design of quasi-zero stiffness joint actuator and research on vibration isolation performance. *J. Sound Vib.* **2020**, *479*, 115367. [CrossRef]
76. Ahn, K.K. Active pneumatic vibration isolation system using negative stiffness structures for a vehicle seat. *J. Sound Vib.* **2014**, *333*, 1245–1268.
77. Li, M.; Cheng, W.; Xie, R. A quasi-zero-stiffness vibration isolator using a cam mechanism with user-defined profile. *Int. J. Mech. Sci.* **2021**, *189*, 1059. [CrossRef]
78. Zhou, J.; Wang, X.; Xu, D. Nonlinear dynamic characteristics of a quasi-zero stiffness vibration isolator with cam–roller–spring mechanisms. *J. Sound Vib.* **2015**, *346*, 53–69. [CrossRef]
79. Zou, W.; Cheng, C.; Ma, R. Performance analysis of a quasi-zero stiffness vibration isolation system with scissor-like structures. *Arch. Appl. Mech.* **2021**, *91*, 117–133. [CrossRef]
80. Liu, C.; Zhao, R.; Yu, K. In-plane quasi-zero-stiffness vibration isolator using magnetic interaction and cables: Theoretical and experimental study. *Appl. Math. Model.* **2021**, *96*, 497. [CrossRef]
81. Wu, Z.; Jing, X.; Sun, B. A 6DOF passive vibration isolator using X-shape supporting structures. *J. Sound Vib.* **2016**, *380*, 90–111. [CrossRef]
82. Zhou, J.; Xiao, Q.; Xu, D. A novel quasi-zero-stiffness strut and its applications in six-degree-of-freedom vibration isolation platform. *J. Sound Vib.* **2017**, *394*, 59–74. [CrossRef]
83. Diala, U. Application of Nonlinear System Frequency Analysis and Design to Vibration Isolation and Energy Harvesting. Ph.D. Thesis, University of Sheffield, Sheffield, UK, 2019.
84. Li, M.; Jing, X. A bistable X-structured electromagnetic wave energy converter with a novel mechanical-motion-rectifier: Design, analysis, and experimental tests. *Energy Convers. Manag.* **2021**, *244*, 114466. [CrossRef]
85. Liang, H.; Hao, G.; Olszewski, O.Z. Ultra-low wide bandwidth vibrational energy harvesting using a statically balanced compliant mechanism. *Int. J. Mech. Sci.* **2022**, *219*, 107130. [CrossRef]
86. Lu, Z.Q.; Wu, D.; Ding, H. Vibration isolation and energy harvesting integrated in a Stewart platform with high static and low dynamic stiffness. *Appl. Math. Model.* **2021**, *89*, 249–267. [CrossRef]
87. Yang, T.; Cao, Q.; Li, Q. A multi-directional multi-stable device: Modeling, experiment verification and applications. *Mech. Syst. Signal Process.* **2021**, *146*, 106986. [CrossRef]
88. Liu, C.; Zhao, R.; Yu, K. A quasi-zero-stiffness device capable of vibration isolation and energy harvesting using piezoelectric buckled beams. *Energy* **2021**, *233*, 121146. [CrossRef]
89. Liu, C.; Zhao, R.; Yu, K. Simultaneous energy harvesting and vibration isolation via quasi-zero-stiffness support and radially distributed piezoelectric cantilever beams. *Appl. Math. Model.* **2021**, *100*, 152–169. [CrossRef]
90. Sun, X.; Wang, F.; Xu, J. Nonlinear piezoelectric structure for ultralow-frequency band vibration energy harvesting with magnetic interaction. *Int. J. Precis. Eng. Manuf.-Green Technol.* **2019**, *6*, 671–679. [CrossRef]
91. Iwaniec, J.; Litak, G.; Iwaniec, M. Response Identification in a Vibration Energy-Harvesting System with Quasi-Zero Stiffness and Two Potential Wells. *Energies* **2021**, *14*, 3926. [CrossRef]

92. Lu, Z.Q.; Shao, D.; Fang, Z.W. Integrated vibration isolation and energy harvesting via a bistable piezo-composite plate. *J. Vib. Control* **2020**, *26*, 779–789. [CrossRef]
93. Zhao, L.C.; Zou, H.X.; Yan, G. A water-proof magnetically coupled piezoelectric-electromagnetic hybrid wind energy harvester. *Appl. Energy* **2019**, *239*, 735–746. [CrossRef]
94. Rajarathinam, M.; Ali, S.F. Energy generation in a hybrid harvester under harmonic excitation. *Energy Convers. Manag.* **2018**, *155*, 10–19. [CrossRef]
95. Karami, M.A.; Inman, D.J. Equivalent damping and frequency change for linear and nonlinear hybrid vibrational energy harvesting systems. *J. Sound Vib.* **2011**, *330*, 5583–5597. [CrossRef]
96. Lallart, M.; Inman, D.J. Mechanical effect of combined piezoelectric and electromagnetic energy harvesting. In *Structural Dynamics and Renewable Energy*; Springer: New York, NY, USA, 2011; Volume 1, pp. 261–272.
97. Wang, K.; Ouyang, H.; Zhou, J. A nonlinear hybrid energy harvester with high ultralow-frequency energy harvesting performance. *Meccanica* **2021**, *56*, 461–480. [CrossRef]
98. Yang, T.; Cao, Q. Time delay improves beneficial performance of a novel hybrid energy harvester. *Nonlinear Dyn.* **2019**, *96*, 1511–1530. [CrossRef]
99. Tao, Y.; Zhou, S.X.; Cao, Q.J. Some advances in nonlinear vibration energy harvesting technology. *Li Xue Xue Bao* **2021**, *53*, 2894–2909.
100. Fulcher, B.A.; Shahan, D.W.; Haberman, M.R. Analytical and experimental investigation of buckled beams as negative stiffness elements for passive vibration and shock isolation systems. *J. Vib. Acoust.* **2014**, *136*, 031009. [CrossRef]
101. Oyelade, A.O. Vibration isolation using a bar and an Euler beam as negative stiffness for vehicle seat comfort. *Adv. Mech. Eng.* **2019**, *11*, 1687814019860983. [CrossRef]
102. Mehreganian, N.; Fallah, A.S.; Sareh, P. Structural mechanics of negative stiffness honeycomb metamaterials. *J. Appl. Mech.* **2021**, *88*, 051006. [CrossRef]
103. Meng, H.; Huang, X.; Chen, Y. Structural vibration absorption in multilayered sandwich structures using negative stiffness nonlinear oscillators. *Appl. Acoust.* **2021**, *182*, 108240. [CrossRef]
104. Kashdan, L.; Seepersad, C.C.; Haberman, M. Design, fabrication, and evaluation of negative stiffness elements using SLS. *Rapid Prototyp. J.* **2012**, *18*, 194–200. [CrossRef]
105. Sarlis, A.A.; Pasala, D.T.R.; Constantinou, M.C. Negative stiffness device for seismic protection of structures. *J. Struct. Eng.* **2013**, *139*, 1124–1133. [CrossRef]
106. Montenbruck, O.; Steigenberger, P.; Darugna, F. Semi-analytical solar radiation pressure modeling for QZS-1 orbit-normal and yaw-steering attitude. *Adv. Space Res.* **2017**, *59*, 2088–2100. [CrossRef]
107. Liu, Y.; Abu-Rub, H.; Ge, B. Front-end isolated quasi-z-source dc–dc converter modules in series for high-power photovoltaic systems—Part II: Control, dynamic model, and downscaled verification. *IEEE Trans. Ind. Electron.* **2016**, *64*, 359–368. [CrossRef]
108. Yan, B.; Wang, X.; Wang, Z. Enhanced lever-type vibration isolator via electromagnetic shunt damping. *Int. J. Mech. Sci.* **2022**, *218*, 107070. [CrossRef]
109. Dong, G.; Zhang, X.; Xie, S. Simulated and experimental studies on a high-static-low-dynamic stiffness isolator using magnetic negative stiffness spring. *Mech. Syst. Signal Process.* **2017**, *86*, 188–203. [CrossRef]
110. Xu, D.; Yu, Q.; Zhou, J. Theoretical and experimental analyses of a nonlinear magnetic vibration isolator with quasi-zero-stiffness characteristic. *J. Sound Vib.* **2013**, *332*, 3377–3389. [CrossRef]
111. Li, H.; Li, Y.; Li, J. Negative stiffness devices for vibration isolation applications: A review. *Adv. Struct. Eng.* **2020**, *23*, 1739–1755. [CrossRef]
112. Zeng, Y.; Ding, H.; Du, R.H. A suspension system with quasi-zero stiffness characteristics and inerter nonlinear energy sink. *J. Vib. Control* **2022**, *28*, 143–158. [CrossRef]
113. Tang, L.; Yang, Y.; Soh, C.K. Improving functionality of vibration energy harvesters using magnets. *J. Intell. Mater. Syst. Struct.* **2012**, *23*, 1433–1449. [CrossRef]
114. Qian, F.; Hajj, M.R.; Zuo, L. Bio-inspired bi-stable piezoelectric harvester for broadband vibration energy harvesting. *Energy Convers. Manag.* **2020**, *222*, 113174. [CrossRef]
115. Zhou, S.; Cao, J.; Lin, J. Exploitation of a tristable nonlinear oscillator for improving broadband vibration energy harvesting. *Eur. Phys. J.-Appl. Phys.* **2014**, *67*, 30902. [CrossRef]
116. Huynh, B.H.; Tjahjowidodo, T.; Zhong, Z.W. Design and experiment of controlled bistable vortex induced vibration energy harvesting systems operating in chaotic regions. *Mech. Syst. Signal Process.* **2018**, *98*, 1097–1115. [CrossRef]
117. Tan, T.; Wang, Z.; Zhang, L. Piezoelectric autoparametric vibration energy harvesting with chaos control feature. *Mech. Syst. Signal Process.* **2021**, *161*, 107989. [CrossRef]
118. Liu, D.; Xu, Y.; Li, J. Randomly-disordered-periodic-induced chaos in a piezoelectric vibration energy harvester system with fractional-order physical properties. *J. Sound Vib.* **2017**, *399*, 182–196. [CrossRef]
119. Huang, D.; Zhou, S.; Han, Q. Response analysis of the nonlinear vibration energy harvester with an uncertain parameter. *Proc. Inst. Mech. Eng. Part K J. Multi-Body Dyn.* **2020**, *234*, 393–407. [CrossRef]
120. Suman, S.; Balaji, P.S.; Selvakumar, K. Nonlinear vibration control device for a vehicle suspension using negative stiffness mechanism. *J. Vib. Eng. Technol.* **2021**, *9*, 957–966. [CrossRef]

121. Nagarajaiah, S.; Pasala, D.T.R.; Reinhorn, A. Adaptive negative stiffness: A new structural modification approach for seismic protection. *Adv. Mater. Res. Trans.* **2013**, *639*, 54–66. [CrossRef]
122. Davoodi, E.; Safarpour, P.; Pourgholi, M. Design and evaluation of vibration reducing seat suspension based on negative stiffness structure. *Proc. Inst. Mech. Eng. Part C J. Mech. Eng. Sci.* **2020**, *234*, 4171–4189. [CrossRef]
123. Cheng, C.; Li, S.; Wang, Y. Force and displacement transmissibility of a quasi-zero stiffness vibration isolator with geometric nonlinear damping. *Nonlinear Dyn.* **2017**, *87*, 2267–2279. [CrossRef]
124. Araki, Y.; Kimura, K.; Asai, T. Integrated mechanical and material design of quasi-zero-stiffness vibration isolator with superelastic Cu–Al–Mn shape memory alloy bars. *J. Sound Vib.* **2015**, *358*, 74–83. [CrossRef]
125. Santhosh, B. Dynamics and performance evaluation of an asymmetric nonlinear vibration isolation mechanism. *J. Braz. Soc. Mech. Sci. Eng.* **2018**, *40*, 1–14. [CrossRef]
126. Han, H.; Sorokin, V.; Tang, L.; Cao, D. A nonlinear vibration isolator with quasi-zero-stiffness inspired by Miura-origami tube. *Nonlinear Dyn.* **2021**, *105*, 1313–1325. [CrossRef]
127. Jin, G.; Wang, Z.; Yang, T. Cascaded quasi-zero stiffness nonlinear low-frequency vibration isolator inspired by human spine. *Appl. Math. Mech.* **2022**, *43*, 813–824. [CrossRef]
128. Yan, G.; Qi, W.H.; Shi, J.W. Bionic paw-inspired structure for vibration isolation with novel nonlinear compensation mechanism. *J. Sound Vib.* **2022**, *525*, 116799. [CrossRef]
129. Yan, G.; Zou, H.X.; Wang, S. Bio-inspired vibration isolation: Methodology and design. *Appl. Mech. Rev.* **2021**, *73*, 020801. [CrossRef]
130. Yan, G.; Zou, H.X.; Wang, S. Bio-inspired toe-like structure for low-frequency vibration isolation. *Mech. Syst. Signal Process.* **2022**, *162*, 108010. [CrossRef]
131. Yan, G.; Wu, Z.Y.; Wei, X.S. Nonlinear compensation method for quasi-zero stiffness vibration isolation. *J. Sound Vib.* **2022**, *523*, 116743. [CrossRef]
132. Yu, M.; Gao, X.; Chen, Q. Study of the mechanical properties and vibration isolation performance of a molecular spring isolator. *Shock Vib.* **2016**, *2016*, 6451829. [CrossRef]
133. Li, Z.; Chen, Q.; Gu, F. Modeling a Mechanical Molecular Spring Isolator with High-Static-Low-Dynamic-Stiffness Properties. *Shock Vib.* **2020**, *2020*, 8853936. [CrossRef]
134. Nie, G.; Li, Z.; Chen, Q. Application of a hydrophobic microporous material in vehicle suspension. *AIP Adv.* **2022**, *12*, 015216. [CrossRef]
135. Chen, Z.; Guo, B.; Yang, Y. Metamaterials-based enhanced energy harvesting: A review. *Phys. B Condens. Matter* **2014**, *438*, 1–8. [CrossRef]
136. Hu, G.; Tang, L.; Liang, J. Acoustic-elastic metamaterials and phononic crystals for energy harvesting: A review. *Smart Mater. Struct.* **2021**, *30*, 085025. [CrossRef]
137. Fan, H.; Yang, L.; Tian, Y. Design of metastructures with quasi-zero dynamic stiffness for vibration isolation. *Compos. Struct.* **2020**, *243*, 112244. [CrossRef]
138. Zhou, Y.; Chen, P.; Mosqueda, G. Analytical and numerical investigation of quasi-zero stiffness vertical isolation system. *J. Eng. Mech.* **2019**, *145*, 04019035. [CrossRef]
139. Hom, C.L.; Pilgrim, S.M.; Shankar, N. Calculation of quasi-static electromechanical coupling coefficients for electrostrictive ceramic materials. *IEEE Trans. Ultrason. Ferroelectr. Freq. Control* **1994**, *41*, 542–551. [CrossRef]
140. Shaw, A.D.; Gatti, G.; Gonçalves, P.J.P. Design and test of an adjustable quasi-zero stiffness device and its use to suspend masses on a multi-modal structure. *Mech. Syst. Signal Process.* **2021**, *152*, 107354. [CrossRef]
141. Hu, X.; Zhou, C. The effect of various damping on the isolation performance of quasi-zero-stiffness system. *Mech. Syst. Signal Process.* **2022**, *171*, 108944. [CrossRef]
142. Zhang, Y.; Cai, C.S. A retrofitted energy harvester for low frequency vibrations. *Smart Mater. Struct.* **2012**, *21*, 075007. [CrossRef]
143. Yan, B.; Zhou, S.; Litak, G. Nonlinear analysis of the tristable energy harvester with a resonant circuit for performance enhancement. *Int. J. Bifurc. Chaos* **2018**, *28*, 1850092. [CrossRef]
144. Zhou, S.; Lallart, M.; Erturk, A. Multistable vibration energy harvesters: Principle, progress, and perspectives. *J. Sound Vib.* **2022**, *2022*, 116886. [CrossRef]
145. Guyomar, D.; Badel, A.; Lefeuvre, E. Toward energy harvesting using active materials and conversion improvement by nonlinear processing. *IEEE Trans. Ultrason. Ferroelectr. Freq. Control* **2005**, *52*, 584–595. [CrossRef]
146. Lefeuvre, E.; Badel, A.; Richard, C. A comparison between several vibration-powered piezoelectric generators for standalone systems. *Sens. Actuators A Phys.* **2006**, *126*, 405–416. [CrossRef]
147. Lallart, M.; Wu, W.J.; Hsieh, Y. Synchronous inversion and charge extraction (SICE): A hybrid switching interface for efficient vibrational energy harvesting. *Smart Mater. Struct.* **2017**, *26*, 115012. [CrossRef]
148. Cheng, C.; Chen, Z.; Xiong, Y. A high-efficiency, self-powered nonlinear interface circuit for bi-stable rotating piezoelectric vibration energy harvesting with nonlinear magnetic force. *Int. J. Appl. Electromagn. Mech.* **2016**, *51*, 235–248. [CrossRef]
149. Shu, Y.C.; Lien, I.C.; Wu, W.J. An improved analysis of the SSHI interface in piezoelectric energy harvesting. *Smart Mater. Struct.* **2007**, *16*, 2253. [CrossRef]

micromachines

MDPI

Article

Scalable, Dual-Band Metasurface Array for Electromagnetic Energy Harvesting and Wireless Power Transfer

Yiqing Wei [1,2], Junping Duan [1,2], Huihui Jing [1,2], Huimin Yang [3], Hao Deng [1,2], Chengwei Song [1,2], Jiayun Wang [1,2], Zeng Qu [1,2] and Binzhen Zhang [1,2,*]

1 School of Instrument and Electronics, North University of China, Taiyuan 030051, China
2 State Key Laboratory of Dynamic Testing Technology, North University of China, Taiyuan 030051, China
3 School of computer and information engineering, Shanxi Technology and Business College, Taiyuan 030072, China
* Correspondence: zhangbinzhen@nuc.edu.cn

Abstract: A dual-band metasurface array is presented in this paper for electromagnetic (EM) energy harvesting in the Wi-Fi band and *Ku* band. The array consists of metasurface unit cells, rectifiers, and load resistors. The metasurface units within each column are interconnected to establish two channels of energy delivery, enabling the transmission and aggregation of incident power. At the terminals of two channels, a single series diode rectifier and a voltage doubler rectifier are integrated into them to rectify the energy in the Wi-Fi band and the *Ku* band, respectively. A 7 × 7 prototype of the metasurface array is fabricated and tested. The measured results in the anechoic chamber show that the RF-to-dc efficiencies of the prototype at 2.4 GHz and 12.6 GHz reach 64% and 55% accordingly, when the available incident power at the surface is 3 dBm and 14 dBm, respectively.

Keywords: ambient energy harvesting (AEH); dual-band rectenna; metasurface array absorber; rectifier; wireless power transfer (WPT)

1. Introduction

Metamaterials are artificial media with negative permittivity and negative permeability and thus have promising applications in electromagnetic (EM) cloaking, anechoic chamber design, and radar cross-section reduction [1–4]. With its unique EM properties, the metamaterial absorber achieves the performance of minimizing reflection coefficient and near-zero transmission coefficient, thereby achieving complete absorption of EM energy [5–8]. However, the energy absorbed in most absorbers is not utilized but rather dissipated in the medium.

Recently, an approach of integrating a rectifier into the metamaterial absorber to fulfill energy conversion has been proposed, which has a promising application in ambient energy harvesting (AEH) and wireless power transfer (WPT). Yet, the very low available power density (μW-mW) in the environment increases the difficulty of energy harvesting [9]. Moreover, the diode rectifier losses have also an influence on efficient energy harvesting. Thus, many approaches have been proposed for reducing energy loss and improving energy collection efficiency. One approach of embedding rectifier diodes directly between the absorber unit cells is proposed to achieve good energy harvesting performance [10–12]. Though the energy transfer loss is eliminated by direct impedance matching between the structure unit cell and the diode, it can only be applied to specific metamaterial resonant structures. In addition, another approach for energy transfer by punching holes in the metamaterial unit cells has been proposed [13–15]. While this approach achieves efficient energy transfer, additional structural layers need to be added to achieve terminal rectification. Similarly, the method of energy transfer through EM coupling between multiple layers is proposed, which obviously requires multiple layers to implement, and therefore inevitably increases the complexity of fabrication [16]. Furthermore, the energy collector usually also requires additional layers to build the rectification network, regardless of the perforation or EM coupling, which also increases the difficulty of efficient energy collection. Additionally, another approach has been proposed to obtain high energy harvesting efficiency by interconnecting metamaterial unit cells and rectifying them at the terminals [17,18]. The energy captured by the metasurface absorber is transferred to the terminals by interconnecting the unit cells, thus the structure of the energy collector is simplified and the energy loss from the multilayer structure is eliminated. However, the size of the unit cell for energy harvesting is still large and the incident power density is still high in order to obtain high harvesting efficiency.

The aim of this paper is to design a novel dual-band scalable periodic metasurface energy harvester based on small unit size, simple structure, and adaptability to different input power levels. Our design is composed of three parts: a periodic meta surface array, a rectifier, and a load. By interconnecting the structural unit cells of a simple sandwich metamaterial absorber, the radio frequency (RF) energy transmission channels are constructed and the incident power at the terminals is aggregated. Therefore, only a low incident power density is required to achieve efficient energy capture. In addition, the cell is constructed with a common three-layer structure, which does not require additional structural layers to deliver energy. Meanwhile, the designed metasurface unit is also relatively small in size and simple in structure with good scalability. Furthermore, the energy of low and high input power is efficiently converted at different frequencies by integrating a single series rectifier network and a voltage doubler rectifier network at the terminal of the array, respectively. Moreover, the metasurface array is able to maintain high energy harvesting efficiency over a wide range of input powers, whether operating at low or high input power levels. In addition, it can be easily adapted to various power density environments and different sizes of spaces for electromagnetic energy harvesting by reconfiguring the number of the array.

Above all, the structural design and performance analysis of the metasurface array are presented in Section 2. Then, a single series rectifier topology and a voltage multiplier rectifier topology are provided in Section 3 and their performance is simulated separately. Next, Section 4 gives the experimental validation of the metasurface array prototype, and conclusions are presented in Section 5.

2. Metasurface Array Analysis and Design

Metamaterial absorbers are used to achieve complete absorption of incident EM waves at resonant frequencies via resonant structures, loss in the dielectric layer, or by adding resistive loads [19,20]. The proposed metasurface array unit cell is a typical sandwich structure, as shown in Figure 1. The resonant structure of the top layer generates EM resonance at the gap on both sides, with resonant frequencies of 2.4 GHz and 12.6 GHz, respectively. Each side of the dielectric layer is etched with a copper of 0.035 mm thickness, forming the top and bottom layers of the metamaterial absorber. The F4B material with thickness t = 3 mm is used as an intermediate dielectric layer with a dielectric constant of $\varepsilon_r = 2.2$ and loss tangent of $\tan\delta = 0.0009$, respectively. At such a low loss tangent, the energy dissipated on the dielectric layer is almost negligible, which makes it possible for efficient collection of the metasurface energy harvester.

The boundary of both sides of the metasurface unit cell is d = 21 mm. Microstrip lines on both sides of the unit cell, with a width w_1 of 0.3 mm, are coincident with the boundary, thus the energy transfer channel of the metasurface array is established. The internal patterns of the resonant structure have a width w_2 of 1 mm and both overlap with the boundary, forming another arm of the energy transfer channel. The distances of both microstrip lines along the y-axis from the boundary are m = 0.5 mm, and the distance between the two sides of the gap causing resonance is also s = 0.5 mm. When the incident EM wave resonates at the gap on both sides, the energy is delivered to the load at the terminal along the microstrip lines on both sides of the unit cell. Then, the incident EM energy is completely captured by the metasurface array and concentrated on the load in the event that suitable load resistance is chosen to allow the array to match the spatial impedance.

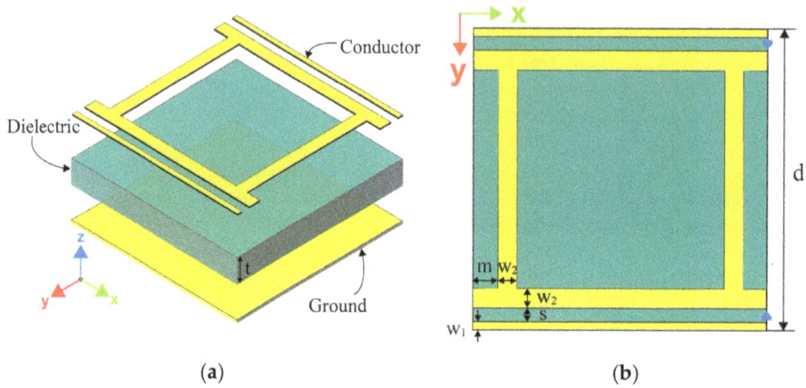

Figure 1. Schematic of the metasurface array unit cell. (**a**) 3-D view. (**b**) Top view.

The proposed metasurface array is simulated by the EM simulation software CST Microwave Studio under periodic boundary conditions with a Floquet port. The unit cell is excited in the xy plane by an incident plane wave along the $-z$ direction. Figure 2 shows the variation curves of the absorptivity and reflectivity as a function of frequency for the metasurface array when the electric field direction of the incident plane wave is parallel to the y-axis. The resistance value of the load located on both sides of the unit cell gap is 900 Ω. It can be observed that the absorptivity of the metasurface array is close to 1 for both resonant frequencies of 2.4 GHz and 12.6 GHz, and the incident EM energy is completely absorbed. It is also demonstrated by the relative input impedance curves of the metasurface array at resonant frequencies shown in Figure 3. It can be noticed that the real part of the relative input impedance of the array is close to 1 and the imaginary part is close to 0 at both 2.4 GHz and 12.6 GHz. It shows that the metasurface array unit cell has good impedance matching at both resonant frequencies when terminated by a 900-Ω load and therefore has good absorption performance.

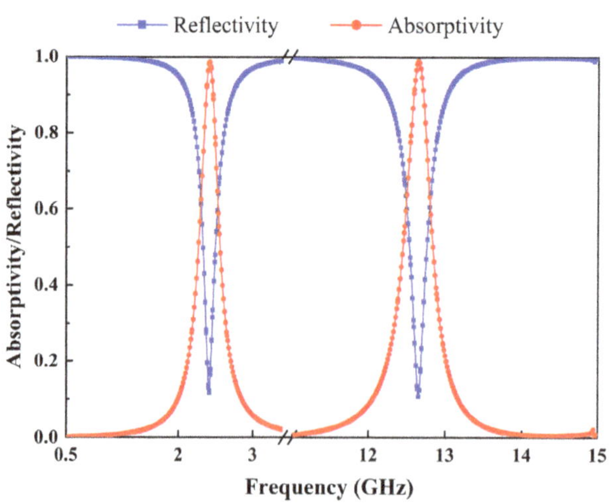

Figure 2. Simulated absorptivity and reflectivity of the metasurface array as a function of frequency.

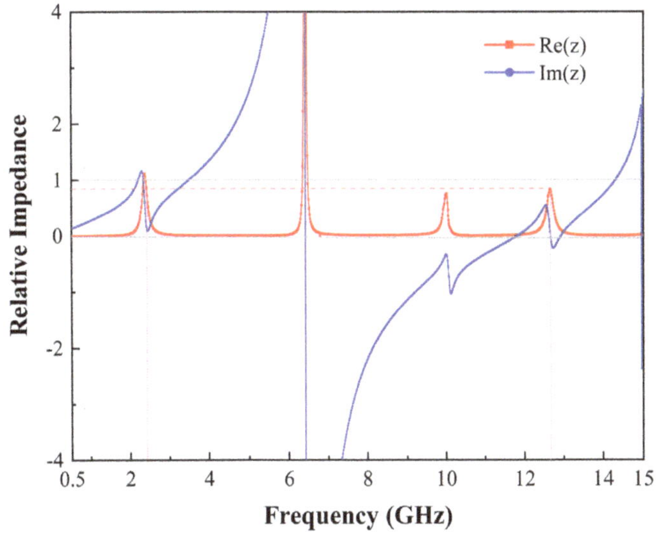

Figure 3. Simulated the real and imaginary parts of the relative input impedance of the metasurface array unit cell.

Next, the dissipated power distribution of each element of the metasurface array is analyzed. Figure 4 shows the overall absorption efficiency of the unit cell and the harvesting efficiency on the metal, dielectric layer, and resistive load when the incident electric field travels along the *y*-axis. The radiation-to-AC conversion efficiency of the metasurface array is determined by

$$\eta_{\text{Rad-ac}} = \frac{P_{\text{ac}}}{P_{\text{rad}}} \tag{1}$$

where P_{ac} is the absorbed power across the resistive load and P_{rad} is the available incident power over the physical area of the unit cell. As can be seen in Figure 4, the energy absorbed by the metasurface array is mainly concentrated in the load resistance achieving 96% and 92.9% at the resonant frequencies of 2.4 GHz and 12.6 GHz, respectively. While the energy dissipated in the metal and the dielectric is almost negligible, which makes it possible to collect the incident energy efficiently.

Finally, we analyzed the electric field, magnetic field, surface current, and power flow distribution of the metasurface array at two resonant frequencies, as shown in Figures 5 and 6. Figure 5 presents the distribution of the electrical properties of the metasurface unit cell at 2.4 GHz. It can be seen that the EM resonance occurs mainly at the gap on both sides of the unit cell at 2.4 GHz, as shown by the electric field distribution in Figure 5a. The strong electric field distribution on both sides of the gap indicates that there is a larger charge accumulation at the edges of the microstrip lines. The microstrip lines on both sides of the gap can be described by the inductance, while the gap can be expressed as the capacitance. Therefore, LC resonant circuit theory can be used to illustrate the resonant absorption of the metasurface array [21]. The magnetic field of the unit cell is mainly distributed on the two branches along the *y*-axis, as shown in Figure 5b. Accordingly, the surface current of the unit cell flows along the two branches of the *y*-axis, as shown in Figure 5c. Finally, the power flow distribution at 2.4 GHz is shown in Figure 5d. It can be observed that the energy captured by the metasurface array at 2.4 GHz is mainly concentrated in the load.

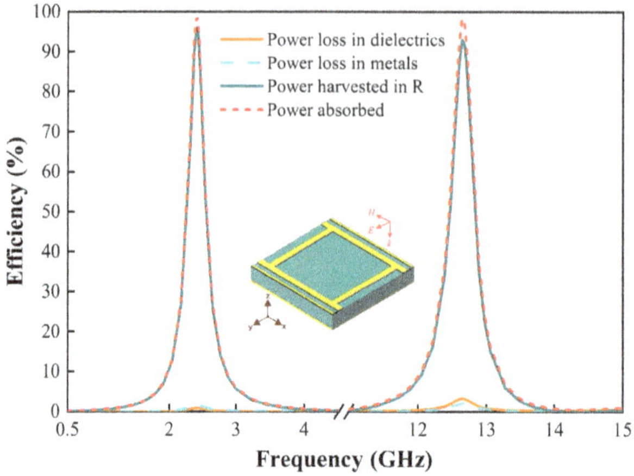

Figure 4. The power absorption, loss, and harvesting efficiency of the metasurface array under incident plane wave irradiation in the $-z$-direction.

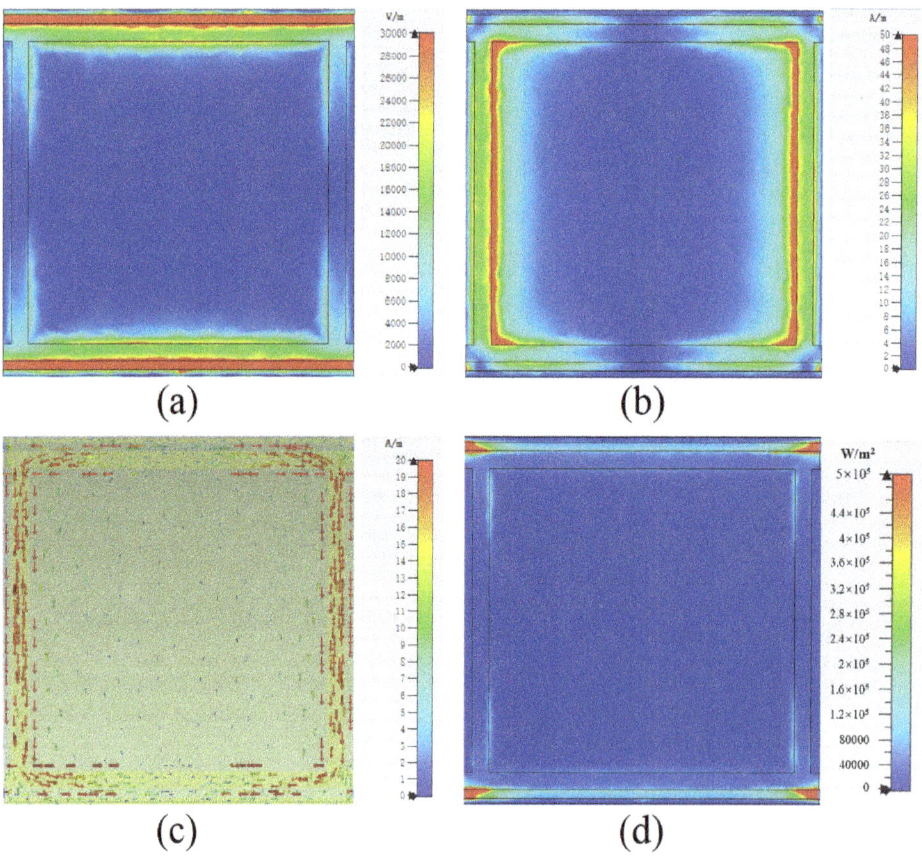

Figure 5. Simulations of (**a**) electric field magnitude, (**b**) magnetic field magnitude, (**c**) surface current distribution, and (**d**) power flow diagram distribution of the metasurface array unit cell at 2.4 GHz with incident plane wave irradiation along the $-z$-direction.

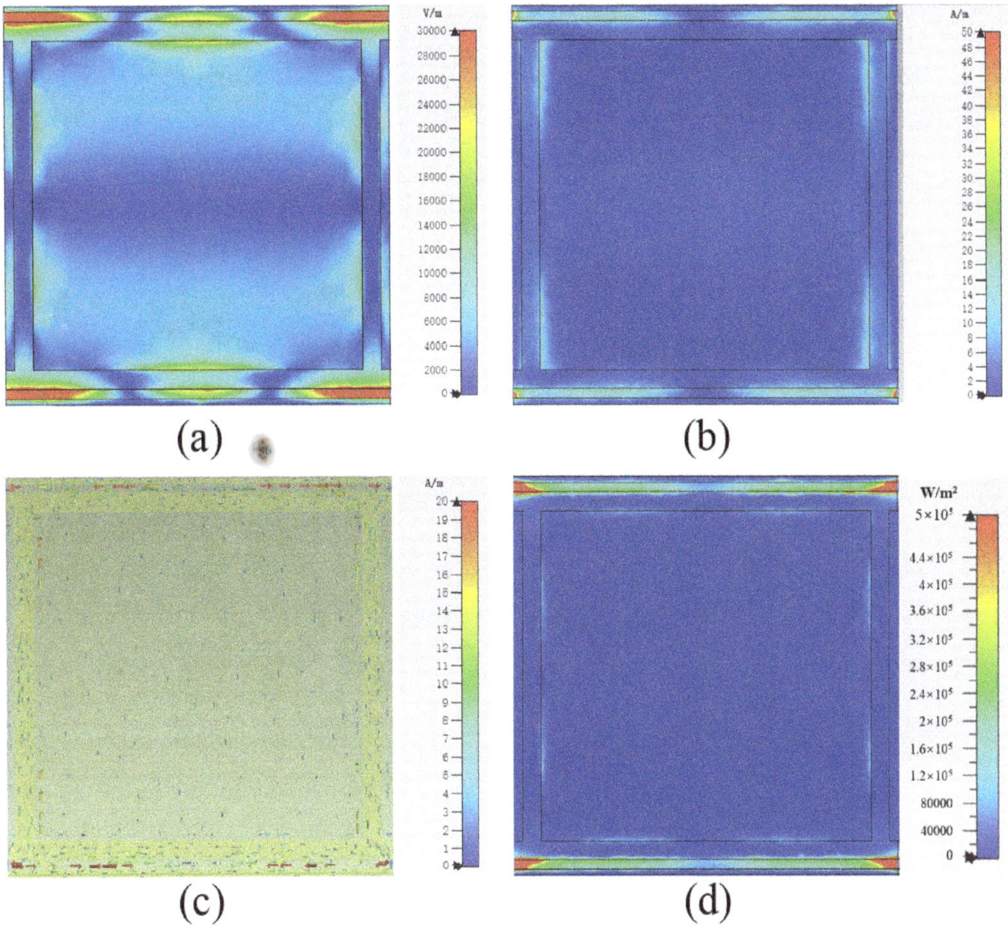

Figure 6. Simulations of (a) electric field magnitude, (b) magnetic field magnitude, (c) surface current distribution, and (d) power flow diagram distribution of the metasurface array unit cell at 12.6 GHz with incident plane wave irradiation along the $-z$-direction.

Figure 6 shows the distribution of electrical properties of the metasurface array at 12.6 GHz. It can be seen from the electric and magnetic field distributions in Figure 6a,b that the EM coupling of the metasurface array is mainly contributed by the gap on both sides of the unit cell in the x-axis direction and the two branches in the y-axis direction. The resonance at higher resonant frequencies can be explained by the fundamental dipole mode [22–24]. Since the dipole resonance has a large local charge distribution on both sides of the junction, the electric field can excite an enhanced local electromagnetic field inside the junction. The surface currents flow along the branches in the x-axis and y-axis directions, respectively, and cross the load resistance, as shown in Figure 6c. Additionally, the power flow distribution is concentrated at both ends of the load resistor in Figure 6d under 12.6 GHz, the same as in Figure 5d, which are all consistent with the findings in Figure 4 where the energy is mainly captured by the resistive load.

3. Dual-Band Rectifier Design

The rectifier, as a component of the energy conversion of the metasurface energy harvester, is usually composed of a rectifier diode and an impedance matching circuit to achieve efficient energy transfer and conversion. According to the connection of the rectifier diodes, rectifiers can be sorted into the single series diode rectifier [25,26], the single parallel diode rectifier [27], the diode bridge rectifier [28], and the voltage doubler rectifier [29], etc. In conventional rectifier designs, the impedance matching networks used are usually different for the different types of diodes, which is caused by the nonlinearity of the diodes and the various packaging technologies. Yet, the proposed metasurface array unit cell is characterized by high input impedance at the terminal, 900 Ω to achieve the perfect matching, and thus can achieve good matching with different types of diode rectifiers. Two energy transfer channels are then established along the x-axis of the metasurface array shown in Figure 1 so that energy harvesting with different frequencies and different input powers can be achieved by connecting different types of diode rectifiers to them.

Figure 7a,b show the topologies of the single series diode rectifier and the voltage doubler rectifier in the Advanced Design System (ADS), respectively. The rectifier shown in Figure 7a is designed to rectify at 2.4 GHz because of its high rectification efficiency at low input power levels. The Schottky diode Avago HSMS-2850 was chosen due to its lower turn-on voltage characteristics, enabling a high rectification efficiency at low input power levels [25,26]. C1 and L1 were used to form an L-shaped impedance matching network for efficient energy transfer. The voltage drop across the load resistor RL can be a monitor of the rectifier's performance. The capacitor C2, connected in parallel, was used to smooth the output waveform and store DC power. Figure 7b shows a voltage doubler rectifier for rectification of incident signals at high input power levels under 12.6 GHz. The Schottky diode Avago HSMS-8202, suitable for operation at high input power, was selected as the main component of the rectifier. As shown in Figure 7b, the topology of L2 and C3 in parallel and L3 and C4 in series constitutes an LC bandpass impedance matching circuit. In addition, C4 also acts as a DC blocker apart from being a part of the impedance matching network. The capacitor C5 is used to smooth the output ripple and ensure that the DC passes to the load.

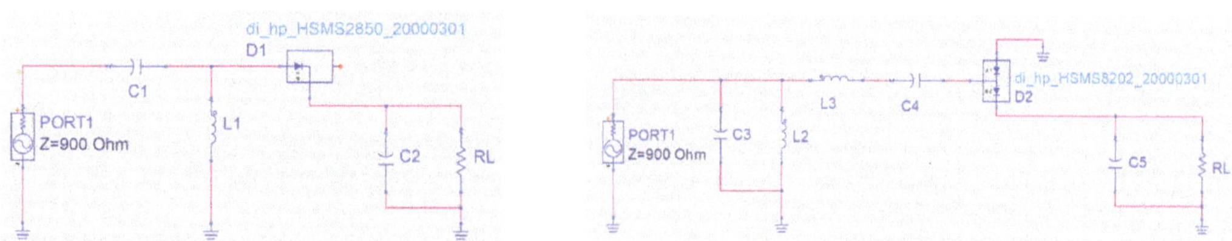

(a) (b)

Figure 7. Schematic of the proposed voltage doubler rectifier in ADS. Schematic diagram of (**a**) the single series diode rectifier and (**b**) the voltage multiplier rectifier proposed in the ADS. Here, C1 = 4 pF, C2 = C5 = 1000 pF, C3 = C4 = 0.2 pF, L1 = 22 nH, L2 = 2.2 nH, L3 = 1.5 nH, (**a**) RL = 1000 Ω, and (**b**) RL = 600 Ω.

The proposed metasurface array is interconnected by energy transfer channels along the x-axis and connected to the rectifier at the terminals, whereby the energy from the end of a row of unit cells is aggregated and captured, resulting in an increase in the incident power. As for the nonlinear variation of the input impedance of the diode with the input power and frequency, we have simulated and optimized the rectifier circuits in Figure 7a,b, respectively, with the harmonic balance (HB) solver in ADS. The overall performance of the rectifier circuits for the metasurface array at two frequencies is shown in Figures 8–10.

The variation curves of rectification efficiency with frequency for different input power levels are shown in Figure 8a,b, respectively. Rectification efficiency can be given as

$$\eta_{\text{ac-dc}} = \frac{P_{dc}}{P_{ac}} \tag{2}$$

where P_{dc} is the output power at the load resistor of the rectifier output terminal and P_{ac} is the incoming power of the rectifier. It can be observed that the rectification efficiency of the single series diode rectifier at 2.4 GHz reaches a maximum of 67.1% when the incident power is 3 dBm, as shown in Figure 8a. While the rectification efficiency of the voltage doubler rectifier at 12.6 GHz reaches the maximum of 60.4% at an incident power of 14 dBm.

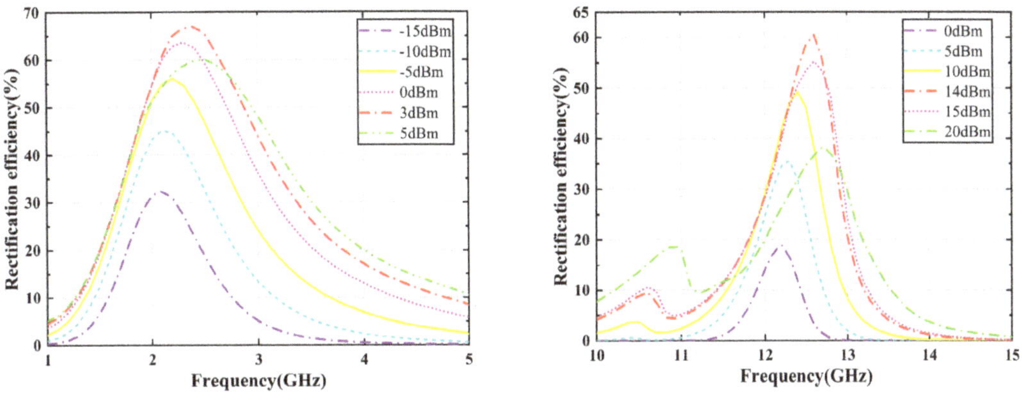

Figure 8. Simulated rectification efficiency curves versus frequency for (**a**) the single series diode rectifier and (**b**) the voltage doubler rectifier at different input power levels.

Furthermore, the variation of rectification efficiency versus input power and load resistance is analyzed as shown in Figure 9a,b. It can be noticed that the rectification efficiency grows with the increase of input power at both 2.4 GHz and 12.6 GHz and reaches the maximum at 3 dBm and 14 dBm, respectively. In this case, the efficiency reaches the maximum at 2.4 GHz and 12.6 GHz for load resistors RL with values of 1000 Ω and 600 Ω, respectively. The variation curve shown in Figure 9 well illustrates the nonlinearity of the rectifier diode against the input power and load resistance, which is directly related to whether the rectifier can be well matched. Finally, the rectification efficiency versus load impedance is analyzed for the given input power, as shown in Figure 10. It can be seen that the rectification efficiency of the two rectifiers gradually reaches its maximum and then gradually drops as the load resistance value increases. Such results are consistent with the maximum efficiency of Figures 8 and 9, which also verify the validity of the impedance matching network and rectifier circuit design.

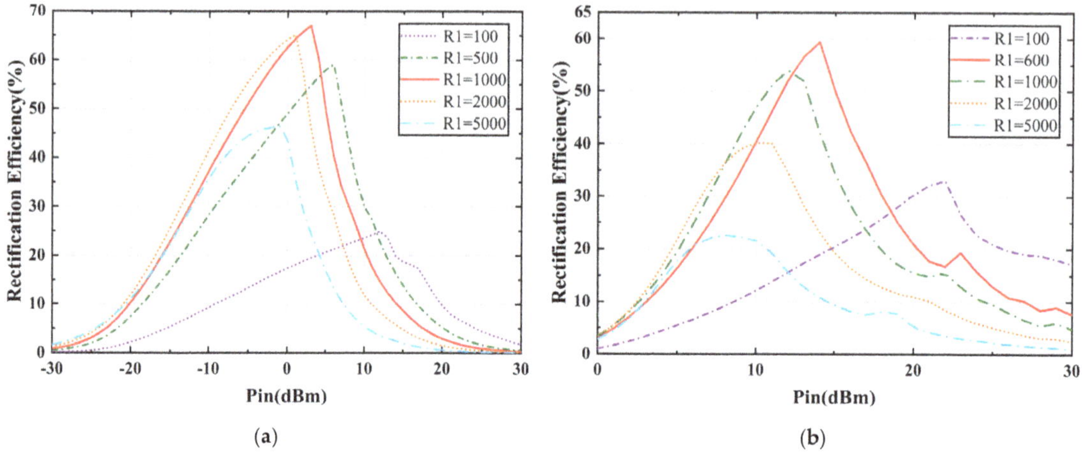

Figure 9. Simulated rectification efficiency of rectifiers versus input power at various loads. (**a**) 2.4 GHz. (**b**) 12.6 GHz.

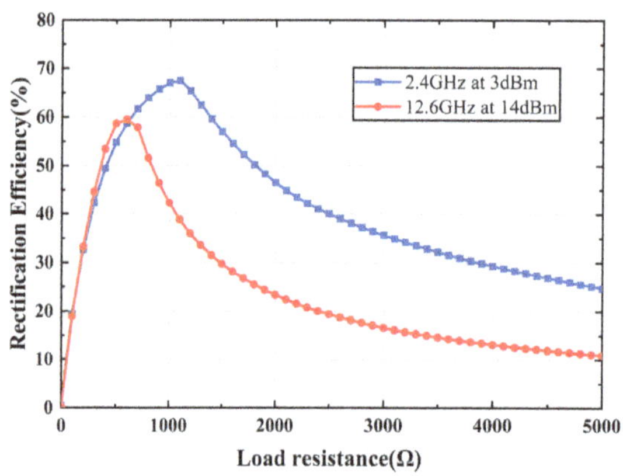

Figure 10. Simulated rectification efficiency curves versus load resistance for the single series diode rectifier and the voltage doubler rectifier.

4. Measurement and Discussion

A 7 × 7 metasurface resonant structure is plated on a 3-mm F4B dielectric substrate with 35-μm copper, as shown in Figure 11. The footprint area of the prototype dielectric layer is 182 mm × 157 mm, while both the top and bottom layers have a copper cladding area of 147 mm × 147 mm. Seven unit cells in each column of the fabricated prototype along the *x*-axis are interconnected to form an energy flow channel. The middle column of the array was tested with a single series diode rectifier and a voltage doubler rectifier, which were integrated with the dual-channel terminals, respectively. The reason for choosing the middle column for testing is to minimize the effect of non-uniform coupling. The remaining six columns of the array are terminated with 900-Ω resistors.

As shown in Figure 12, the measurement of the metasurface array prototype was performed in an anechoic chamber. First, the RF signal produced by the signal generator was transmitted to the standard horn antenna covering the measurement band. Second, the desired incident plane wave was illuminated by the horn antenna onto the prototype situated at the far-field position. Finally, the dc output voltage across the load resistor on the prototype terminal was recorded by a digital multimeter. The overall efficiency of the metasurface array energy harvesting can be exhibited as follows:

$$\eta_{mea.} = \frac{P_{\text{dc}}}{P_{\text{in}}} \tag{3}$$

where P_{dc} is the output power across the rectifier load resistor, and P_{in} is the total available incident power absorbed in the surface region of the array, respectively, shown as follows:

$$P_{\text{dc}} = \frac{V_{\text{out}}^2}{R_{\text{load}}} \tag{4}$$

$$P_{\text{in}} = \frac{G_t \cdot P}{4\pi R^2} \cdot A_s \tag{5}$$

where V_{out} represents the dc voltage landing on the rectifier load R_{load}. G_t denotes the gain of the horn antenna and P represents the power excited by the signal source. R is denoted as the distance from the horn antenna to the prototype. A_s represents the effective receiving aperture of the metasurface array, which can be given by the physical footprint of the middle column under the test.

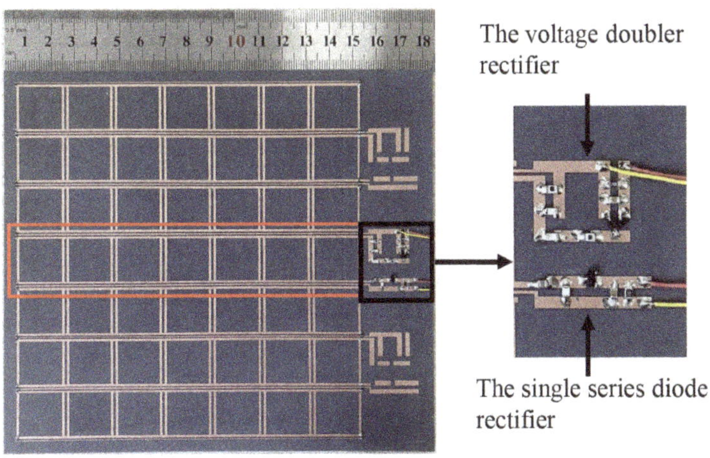

Figure 11. Fabricated metasurface array with integrated rectifier.

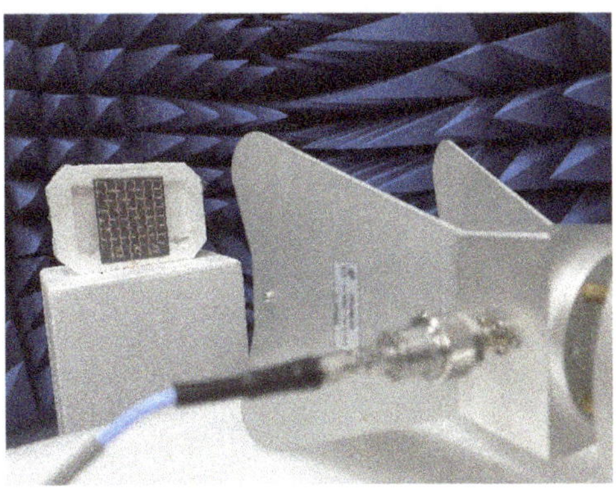

Figure 12. Photograph of metasurface array being measured under broadband horn antenna illumination in the anechoic chamber.

As the fabricated prototype is irradiated by a normal incident plane wave along the $-z$ direction, the RF-to-dc efficiency versus frequency at different available power levels is shown in Figure 13. As can be seen from Figure 13a, the metasurface array with an integrated single series diode rectifier achieves a maximum conversion efficiency of 64% at 2.2 GHz when the incident available power is 3 dBm. When the incident available power is 14 dBm, the conversion efficiency of the metasurface array with an integrated voltage doubler rectifier reaches a maximum of 55% at 12.3 GHz.

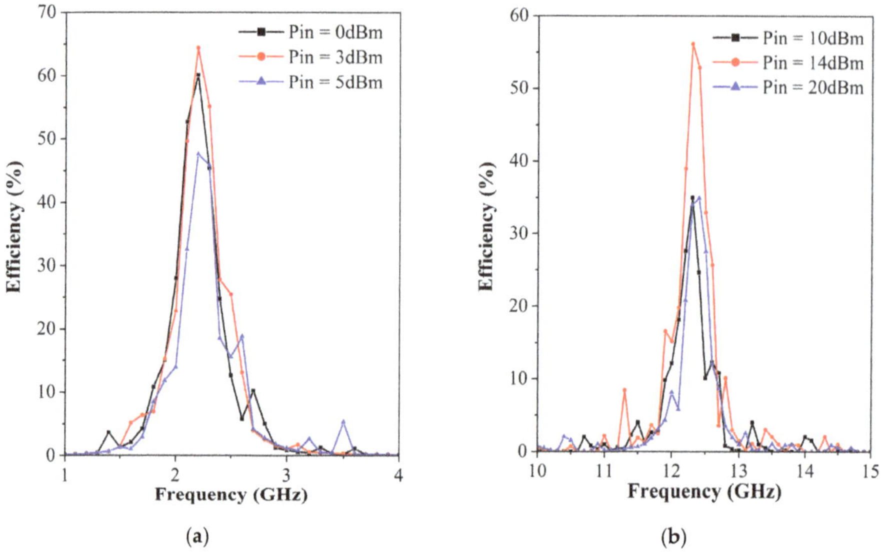

Figure 13. Measured RF-to-dc efficiency of the proposed metasurface array as a function of frequency with (**a**) the single series diode rectifier and (**b**) the voltage doubler rectifier.

It can also be observed in Figure 13 that the manufactured prototype undergoes a frequency shift and a reduction in conversion efficiency. It is firstly due to the conversion efficiency of the prototype being highly dependent on the rectification performance of the rectifier circuit compared to the absorption efficiency of the metasurface array. Secondly, the non-uniform coupling among the finite number of metasurface array units is also a very important factor. Although the middle column of the prototype was tested, it inevitably caused a bias in the test results compared to the simulation of the infinite period metasurface. Finally, manufacturing deviations, whether caused by the 7×7 metasurface array panel or the layout of the rectification network and the soldering of the circuits, will cause frequency shifts and a reduction in conversion efficiency.

Next, the conversion efficiency of the metasurface array at 2.2 GHz and 12.3 GHz was measured separately as a function of the available input power at the surface, as shown in Figure 14. It can be observed that the conversion efficiency of the prototype at 2.2 GHz reaches 64% at 3 dBm when the load of the single series diode rectifier is 1000 Ω. Similarly, the conversion efficiency of the prototype reaches 55% at 14 dBm under 12.6 GHz when the load of the voltage doubler rectifier is 600 Ω. In addition, the metasurface array can still maintain high harvesting efficiency at a wide available power of −10–7 dBm at 2.2 GHz and a wide available power of 8–18 dBm at 12.3 GHz, respectively.

Based on simulations and measurements, the proposed metasurface array is compared with previous energy harvesting works reported in the literature, as shown in Table 1. It can be seen that a new design based on a small unit size, a simple structure, and adaptability to different input powers for a dual-band metasurface is proposed in the article. In comparison with other related work, the metasurface array avoids the integration of rectifier diodes in each cell by a built-in channel approach, which simplifies the number of structural layers and reduces the difficulty of fabrication. Moreover, the feature of interconnecting adjacent units to aggregate energy enables the metasurface array to achieve high harvesting efficiency at low input power. Furthermore, two channels of each column are integrated by a single series diode rectifier and a voltage doubler rectifier, respectively, to achieve high conversion efficiency at different frequencies with different input powers. In addition, the relatively small size of the designed unit makes it easy to expand, which can be easily adapted to different space sizes for energy harvesting. It is also to be noted that the metasurface array can maintain a high energy harvesting efficiency over a wide range of input power, whether it operates at low or high input power levels. Therefore, the proposed metasurface array has better adaptability in AEH and WPT compared to other works.

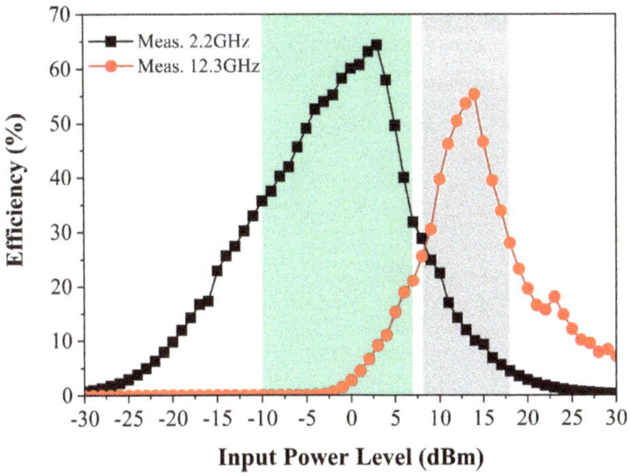

Figure 14. Measured RF-to-dc efficiency of the proposed metasurface array versus input power levels at 2.2 GHz and 12.3 GHz.

Table 1. Comparison of the proposed metasurface array with related works.

Reference	Frequency (GHz)	Diodes to Each Cell	Dimension of Unit Cell (mm)	Prototype Size (mm)	RF-to-dc Efficiency (%)
[16]	Single-band 5.8	No	31.7 × 31.7 (5 layers)	190.2 × 190.2 (array: 6 × 6)	55%@35 μW/cm^2
[17]	Single-band 2.45	Yes	35 × 57.5 × 6.5 (3 layers)	228.6 × 304.8 (array: 5 × 6)	61%@15 dBm
[26]	Dual-band 0.94, 1.84	NA	NA	240 × 240	42%@0.14 μW/cm^2
[28]	Single-band 2.45	No	74 × 67 × 4.769 (5 layers)	229 × 305 (array: 3 × 4)	61%@313 μW/cm^2
[30]	Dual-band 2.4, 5.8	Yes	16 × 15.5 × 1.27 (3 layers)	67 × 64 (array: 4 × 4)	58%, 51%@0 dBm
[31]	Single-band 2.84	Yes	50 × 50 × 3.175 (3 layers)	200 × 200 (array: 4 × 4)	60%@187.5 μW/cm^2
[32]	Single-band 2.45	Yes	20 × 20 × 4.2 (6 layers)	120 × 120 (array: 6 × 6)	66.9%@5000 μW/cm^2
This work	Dual-band 2.4, 12.6	No	21 × 21 × 3.07 (3 layers)	182 × 157 (array: 7 × 7)	64%@3 dBm 2.2 GHz, 55%@14 dBm 12.3 GHz

5. Conclusions

A dual-band metasurface array operating in the Wi-Fi band and Ku band is designed and optimized. Two energy channels are constructed by interconnecting the units, enabling the transmission and aggregation of energy, which in turn increases the power density of the channel terminals. A single series diode rectifier is integrated at the terminal of one channel, thus enabling the efficient collection of low input power at 2.4 GHz. At the terminal of another channel, a voltage doubler rectifier is integrated to achieve high input power collection at 12.6 GHz. Then, a 7 × 7 metasurface array is fabricated. The test results show that the RF-to-dc efficiency of the prototype reaches 64% and 55% at 2.2 GHz and 12.3 GHz, respectively, with the available incident power at the surface of the array correspondingly at 3 dBm and 14 dBm. Moreover, the proposed metasurface array is capable of maintaining a high collection efficiency over a wide range of input power levels at both dual-band resonant frequencies. Additionally, the proposed metasurface array possesses the characteristics of simple design and a scalable number of units, thus allowing efficient energy harvesting in different environments.

Author Contributions: Conceptualization, Y.W.; methodology, J.D.; software, H.J.; validation, H.Y. and Y.W.; formal analysis, H.D.; investigation, C.S.; resources, J.W.; data curation, Z.Q.; writing—original draft preparation, Y.W.; writing—review and editing, Y.W. and B.Z.; visualization, H.Y.; project administration, B.Z.; funding acquisition, B.Z. All authors have read and agreed to the published version of the manuscript.

Funding: This work was supported in part by the National Natural Science Foundation of China under Grant 52175555; the Shanxi Key Research and Development Project (International Cooperation) under Grant 201803D421043; the Innovation Community under Grant 51821003; and the Fundamental Research Program of Shanxi Province under Grant 20210302123074.

Conflicts of Interest: The authors declare no conflict of interest.

References

1. Landy, N.I.; Sajuyigbe, S.; Mock, J.J.; Smith, D.R.; Padilla, W.J. Perfect Metamaterial Absorber. *Phys. Rev. Lett.* **2008**, *100*, 207402. [CrossRef] [PubMed]
2. Schurig, D.; Mock, J.J.; Justice, B.J.; Cummer, S.A.; Pendry, J.B.; Starr, A.F.; Smith, D.R. Metamaterial electromagnetic cloak at microwave frequencies. *Science* **2006**, *314*, 977–980. [CrossRef] [PubMed]
3. Kim, J.; Han, K.; Hahn, J.W. Selective dual-band metamaterial perfect absorber for infrared stealth technology. *Sci. Rep.* **2017**, *7*, 6740. [CrossRef] [PubMed]
4. Iwaszczuk, K.; Strikwerda, A.C.; Fan, K.B.; Zhang, X.; Averitt, R.D.; Jepsen, P.U. Flexible metamaterial absorbers for stealth applications at terahertz frequencies. *OExpr* **2012**, *20*, 635–643. [CrossRef]
5. Li, S.J.; Gao, J.; Cao, X.Y.; Li, W.Q.; Zhang, Z.; Zhang, D. Wideband, thin, and polarization-insensitive perfect absorber based the double octagonal rings metamaterials and lumped resistances. *J. Appl. Phys.* **2014**, *116*, 043710. [CrossRef]
6. Munk, B.A.; Munk, P.; Pryor, J. On designing Jaumann and circuit analog absorbers (CA absorbers) for oblique angle of incidence. *IEEE Trans. Antennas Propag.* **2007**, *55*, 186–193. [CrossRef]
7. Han, Y.; Che, W.Q.; Christopoulos, C.; Xiong, Y.; Chang, Y.M. A Fast and Efficient Design Method for Circuit Analog Absorbers Consisting of Resistive Square-Loop Arrays. *ITEIC* **2016**, *58*, 747–757. [CrossRef]
8. Liang, Y.; Koshelev, K.; Zhang, F.C.; Lin, H.; Lin, S.R.; Wu, J.Y.; Jia, B.H.; Kivshar, Y. Bound States in the Continuum in Anisotropic Plasmonic Metasurfaces. *Nano Lett.* **2020**, *20*, 6351–6356. [CrossRef]
9. Li, L.; Zhang, X.M.; Song, C.Y.; Huang, Y. Progress, challenges, and perspective on metasurfaces for ambient radio frequency energy harvesting. *Appl. Phys. Lett.* **2020**, *116*, 7. [CrossRef]
10. Keyrouz, S.; Perotto, G.; Visser, H.J. Frequency selective surface for radio frequency energy harvesting applications. *IET Microw. Antennas Propag.* **2014**, *8*, 523–531. [CrossRef]
11. Ferreira, D.; Sismeiro, L.; Ferreira, A.; Caldeirinha, R.F.S.; Fernandes, T.R.; Cuinas, I. Hybrid FSS and Rectenna Design for Wireless Power Harvesting. *IEEE Trans. Antennas Propag.* **2016**, *64*, 2038–2042. [CrossRef]
12. Ashoor, A.Z.; Almoneef, T.S.; Ramahi, O.M. A Planar Dipole Array Surface for Electromagnetic Energy Harvesting and Wireless Power Transfer. *IEEE Trans. Microw. Theory Tech.* **2018**, *66*, 1553–1560. [CrossRef]
13. Zhang, X.M.; Liu, H.X.; Li, L. Tri-band miniaturized wide-angle and polarization-insensitive metasurface for ambient energy harvesting. *Appl. Phys. Lett.* **2017**, *111*, 5. [CrossRef]
14. Almoneef, T.S.; Erkmen, F.; Ramahi, O.M. Harvesting the Energy of Multi-Polarized Electromagnetic Waves. *Sci. Rep.* **2017**, *7*, 14. [CrossRef]

15. Zhong, H.T.; Yang, X.X.; Tan, C.; Yu, K. Triple-band polarization-insensitive and wide-angle metamaterial array for electromagnetic energy harvesting. *Appl. Phys. Lett.* **2016**, *109*, 4. [CrossRef]
16. Dinh, M.Q.; Thinh Le, H.; Hong Tien, V.; Nguyen Thanh, T.; Minh Thuy, L. Design, fabrication, and characterization of an electromagnetic harvester using polarization-insensitive metamaterial absorbers. *J. Phys. D Appl. Phys.* **2021**, *54*, 345502. [CrossRef]
17. Erkmen, F.; Almoneef, T.S.; Ramahi, O.M. Scalable Electromagnetic Energy Harvesting Using Frequency-Selective Surfaces. *IEEE Trans. Microw. Theory Tech.* **2018**, *66*, 2433–2441. [CrossRef]
18. Erkmen, F.; Ramahi, O.M. A Scalable, Dual-Band Absorber Surface for Electromagnetic Energy Harvesting and Wireless Power Transfer. *IEEE Trans. Antennas Propag.* **2021**, *69*, 6982–6987. [CrossRef]
19. Ramahi, O.M.; Almoneef, T.S.; AlShareef, M.; Boybay, M.S. Metamaterial particles for electromagnetic energy harvesting. *Appl. Phys. Lett.* **2012**, *101*, 173903. [CrossRef]
20. Alavikia, B.; Almoneef, T.S.; Ramahi, O.M. Electromagnetic energy harvesting using complementary split-ring resonators. *Appl. Phys. Lett.* **2014**, *104*, 163903. [CrossRef]
21. Okamoto, T.; Otsuka, T.; Sato, S.; Fukuta, T.; Haraguchi, M. Dependence of LC resonance wavelength on size of silver split-ring resonator fabricated by nanosphere lithography. *OExpr* **2012**, *20*, 24059–24067. [CrossRef]
22. Perez-Gonzalez, O.; Zabala, N.; Borisov, A.G.; Halas, N.J.; Nordlander, P.; Aizpurua, J. Optical Spectroscopy of Conductive Junctions in Plasmonic Cavities. *Nano Lett.* **2010**, *10*, 3090–3095. [CrossRef]
23. Cheng, Y.Z.; Gong, R.Z.; Cheng, Z.Z. A photoexcited broadband switchable metamaterial absorber with polarization-insensitive and wide-angle absorption for terahertz waves. *Opt. Commun.* **2016**, *361*, 41–46. [CrossRef]
24. Ali, H.O.; Al-Hindawi, A.M.; Abdulkarim, Y.I.; Karaaslan, M. New compact six-band metamaterial absorber based on Closed Circular Ring Resonator (CCRR) for Radar applications. *Opt. Commun.* **2022**, *503*, 10. [CrossRef]
25. Shen, S.; Zhang, Y.; Chiu, C.-Y.; Murch, R. A Triple-Band High-Gain Multibeam Ambient RF Energy Harvesting System Utilizing Hybrid Combining. *ITIE* **2020**, *67*, 9215–9226. [CrossRef]
26. Shen, S.P.; Zhang, Y.J.; Chiu, C.Y.; Murch, R. An Ambient RF Energy Harvesting System Where the Number of Antenna Ports is Dependent on Frequency. *IEEE Trans. Microw. Theory Tech.* **2019**, *67*, 3821–3832. [CrossRef]
27. Hagerty, J.A.; Helmbrecht, F.B.; McCalpin, W.H.; Zane, R.; Popovic, Z.B. Recycling Ambient Microwave Energy With Broad-Band Rectenna Arrays. *IEEE Trans. Microw. Theory Tech.* **2004**, *52*, 1014–1024. [CrossRef]
28. Erkmen, F.; Ramahi, O.M. A Scalable, Dual-Polarized Absorber Surface for Electromagnetic Energy Harvesting and Wireless Power Transfer. *IEEE Trans. Microw. Theory Tech.* **2021**, *69*, 4021–4028. [CrossRef]
29. Song, C.; Huang, Y.; Carter, P.; Zhou, J.; Joseph, S.D.; Li, G. Novel Compact and Broadband Frequency-Selectable Rectennas for a Wide Input-Power and Load Impedance Range. *IEEE Trans. Antennas Propag.* **2018**, *66*, 3306–3316. [CrossRef]
30. Li, L.; Zhang, X.M.; Song, C.Y.; Zhang, W.Z.; Jia, T.Y.; Huang, Y. Compact Dual-Band, Wide-Angle, Polarization- Angle - Independent Rectifying Metasurface for Ambient Energy Harvesting and Wireless Power Transfer. *IEEE Trans. Microw. Theory Tech.* **2021**, *69*, 1518–1528. [CrossRef]
31. Almoneef, T.S.; Erkmen, F.; Alotaibi, M.A.; Ramahi, O.M. A New Approach to Microwave Rectennas Using Tightly Coupled Antennas. *IEEE Trans. Antennas Propag.* **2018**, *66*, 1714–1724. [CrossRef]
32. Duan, X.; Chen, X.; Zhou, L. A metamaterial electromagnetic energy rectifying surface with high harvesting efficiency. *Aip Adv.* **2016**, *6*, 7. [CrossRef]

energies

MDPI

Review

Recent Advances in Energy Harvesting from the Human Body for Biomedical Applications

Ihor Sobianin [1], Sotiria D. Psoma [1] and Antonios Tourlidakis [2,*]

1 School of Engineering & Innovation, The Open University, Walton Hall, Milton Keynes MK7 6AA, UK
2 Department of Mechanical Engineering, University of Western Macedonia, Kozani 50100, Greece
* Correspondence: atourlidakis@uowm.gr

Abstract: Energy harvesters serve as continuous and long-lasting sources of energy that can be integrated into wearable and implantable sensors and biomedical devices. This review paper presents the current progress, the challenges, the advantages, the disadvantages and the future trends of energy harvesters which can harvest energy from various sources from the human body. The most used types of energy are chemical; thermal and biomechanical and each group is represented by several nano-generators. Chemical energy can be harvested with a help of microbial and enzymatic biofuel cells, thermal energy is collected via thermal and pyroelectric nano-generators, biomechanical energy can be scavenged with piezoelectric and triboelectric materials, electromagnetic and electrostatic generators and photovoltaic effect allows scavenging of light energy. Their operating principles, power ratings, features, materials, and designs are presented. There are different ways of extracting the maximum energy and current trends and approaches in nanogenerator designs are discussed. The ever-growing interest in this field is linked to a larger role of wearable electronics in the future. Possible directions of future development are outlined; and practical biomedical applications of energy harvesters for glucose sensors, oximeters and pacemakers are presented. Based on the increasingly accumulated literature, there are continuous promising improvements which are anticipated to lead to portable and implantable devices without the requirement for batteries.

Keywords: bioelectronics; energy harvesting; wearables; implantable electronics

1. Introduction

Continuous monitoring plays a vital role in a timely medical diagnosis and prevents a large spectrum of diseases [1]. In modern medical practice this approach is implemented via wearable biosensors that register any anomalous biomarker deviations [2]. A core problem, however, is that continuous monitoring via wearable biosensors cannot be developed without a sufficient energy source. Typically, energy is provided by either a battery, by an energy harvester, or a combination of both [3,4].

Energy harvesting is a process of collecting, converting and providing ready-to-use energy for an arbitrary device. The circuitry for such a device consists of nanogenerators (NGs) and a power management unit that conditions the scavenged energy into a usable form [5]. Research into energy harvesting requires a multi-disciplinary approach which considers both electrical and biological factors to achieve optimum conversion results. Thermal, biomechanical, biochemical and solar energy (Figure 1) can be scavenged from the ambient environment or the human body [3,6–9]. In addition, the same external stimuli that are converted to energy can effectively be monitored by energy harvesters [10].

The history of energy harvesters can be traced back to 2006 when Zhong Lin Wang published a report on piezoelectric nanogenerator that was comprised of ZnO nanowire arrays [11]. This fundamental article became an instant classic among the scientific community for its breakthrough novelty at that time. The microminiaturization trend goes back to 1959 when Richard Feynman professed a famous lecture with a title "There's Plenty of Room at the Bottom" [12] where he discussed the possibility of manipulating atoms to severely reduce sizes of mechanical and electronic devices. The alluring capabilities of micro-sized world echoed within the research and medical communities [1,13,14] and since then a surge of interest in this area has been developed.

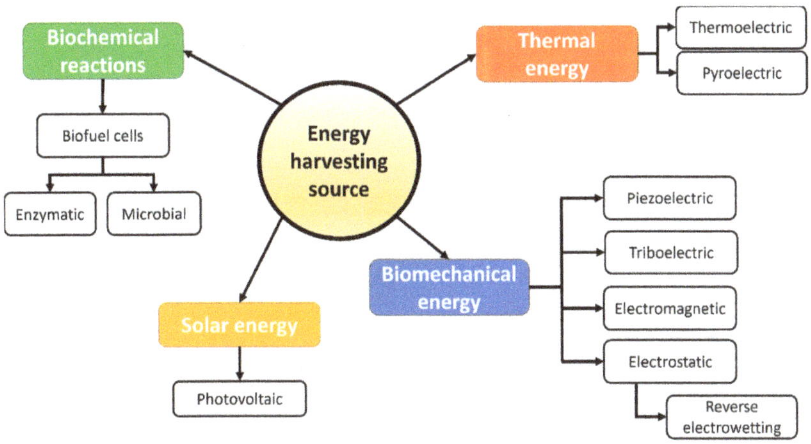

Figure 1. Energy conversion methods based on applied physical principles.

Batteries are quite often used in glucose biosensors, for example, because they represent a reliable and long-lasting source of energy. Typical power values of biosensing electronics are less than 10 μW [15] but due to the sole nature of continuous monitoring, batteries become the bottleneck of the power design. For instance, glucose biosensors have a needle that penetrates the skin in order to collect data (Figure 2) and any time the battery needs to be replaced, the patient must go through a painful procedure of taking the biosensor in and out which involves puncturing the skin. Low battery capacity necessitates regular replacements, every week or two; each time the entire glucose biosensor is replaced, not just the battery. The 1.55 V coin-cell battery, which is mounted inside the Abbot's biosensor, tends to drive higher the end price of the continuous glucose monitoring. Therefore, it is desirable to identify means and methods that allow biosensors to realize continuous monitoring at full potential without the requirement to depend on batteries.

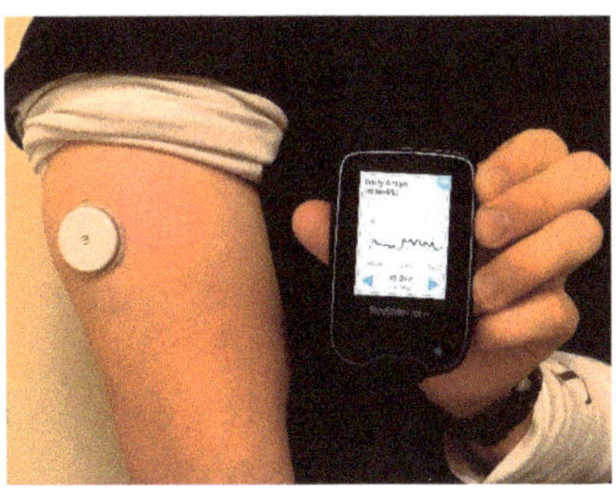

Figure 2. Abbott's continuous glucose monitoring biosensor; https://diatribe.org (accessed on 9 October 2022).

This review is aimed towards offering a basic understanding of what energy harvesters are, of the underlying physical principles, the technological challenges and the future perspectives. Apart from well-known and researched energy harvesting methods such as piezo- or triboelectricity, the review also focuses on some less known or exploited phenomena that are worthy of further investigation thus exposing a wide spectrum of energy scavenging possibilities. The review intends to offer an overview exploring modern energy harvesting techniques and materials and identifying the main trends of energy scavenging. The presentation is kept to a "high-level" style explaining core ideas providing the reader with the necessary references in case he is interested in more detailed descriptions and explanations.

2. Biofuel Cells

The first major entry in nanogenerators are biofuel cells (BFCs) which form a subgroup of fuel cells. Their modus operandi is based on biocatalysts [16–19] that cause a flow of electrons between anode and cathode thus generating energy. BFCs are represented by two main groups: microbial BFCs (Figure 3a) and enzymatic BFCs (Figure 3b). BFCs have two electrodes, anode and cathode, that are submerged in an electrolyte. The oxidation reaction of biofuel occurs at the anode, the oxidants are reduced at the cathode, and this allows electrons to flow through the circuitry. Microbial BFCs exploit microorganisms as biocatalysts forming a bioreactor while enzymatic BFCs utilize enzymes for oxidation and reduction reactions. A major advantage of microbial BFCs is that exoelectrogenic microorganisms operate in the environment that is akin to its natural biome and they have a long life span of up to five years [17,19]. The main advantage of enzymatic BFCs is their ability to effectively utilize glucose and lactate that is present in blood, sweat and tears as a fuel for chemical reactions thus proving to be extremely useful in the field of wearable and implantable electronics [16,20]. Yet, the lifespan of enzymatic BFCs is significantly shorter than of the microbial ones (7–10 days) [19].

Elasticity of BFCs plays an important role in wearable electronics related to the number of stresses and deformations they are subjected to. Electrodes of BFCs need to possess characteristics such as elasticity, electrical conductivity and biocompatibility while simultaneously not impairing the biocatalyst-bioelectrode interface [21]. Carbon-based materials, such as graphene, meet all the criteria. Carbon-based materials can be used in different forms: nanotubes, nanosheets, nanodots, etc. [19]. Combined with flexible substrates, such as elastomers, polymer films, hydrogels, they prove to demonstrate good results [21,22].

There are several materials that can be used as electrolytes. Biofluids such as sweat, urine, tears, saliva, etc. [23] are a common choice for BFCs as they have the necessary fuel to power the system. On the other hand, there are systems where the electrolyte is artificial, such as hydrogel for example [24], and the fuel is supplied from elsewhere.

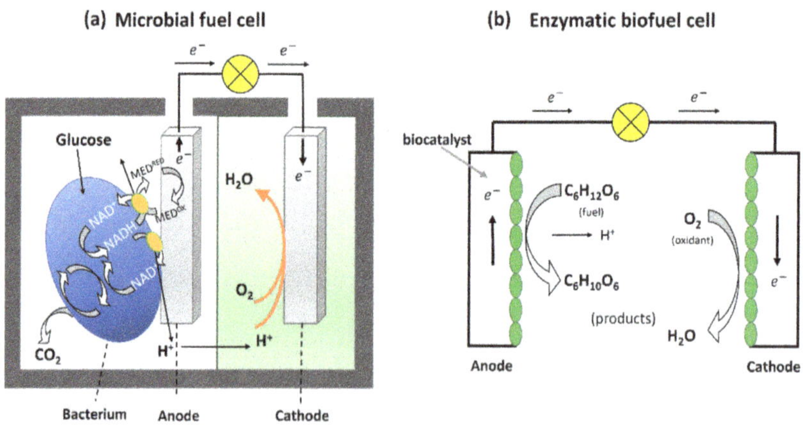

Figure 3. Basic operating principle of (**a**) microbial BFC [25] and (**b**) enzymatic BFC [26].

Zhang et al. [27] were able to exploit an enzymatic BFC, which was integrated into a diaper of diabetic patients, as an effective glucose concentration sensor while simultaneously powering the system. Quantitatively, they managed to approach a power density of 220 μWcm^{-2}. Zhang et al. [28] took a rather unconventional approach and designed an artificial enzyme which mimics its natural competitor with a power density of 149.2 ± 4.0 μW·cm^{-2}. A particularly unique case of microbial BFC was studied by Jayapiriya and Goel [29] where a paper-based 3D printed microbial BFC was designed that was capable of generating as much as 11.8 μW·cm^{-2}. It is also important to add that the manufacturing method employed is important for creating a cost-effective solution for energy harvesters used in low-powered biosensors. An extraordinary power density of 703.55 mW·m^{-2} was achieved by Yan et al. [30] with a nano enhancement of electrode/electrolyte interface. Overall, BFCs are prominent sources of energy for wearable and implantable bioelectronics though they have a number of drawbacks such as low power density, low power output and they are prone to have high cost and short lifetime [31].

3. Thermal Generators

Thermal energy exists in abundance in the ambient environment. Both the human body and the space around it can be seen as energy sources for thermoelectric and pyroelectric devices [32–34]. The physical principle of thermoelectric energy generation is based on the Seebeck effect that describes the generation of an electric current when n and p type conductors are electrically connected under the presence of a temperature gradient (Figure 4) i.e., charge carriers will move from the warmer to the cooler side of the formed thermocouple [34]. Advantages of thermoelectric NGs are that they supply a DC voltage, which means that they do not need to be rectified, thus making the process of energy collection easier at a price of generally having a low voltage output.

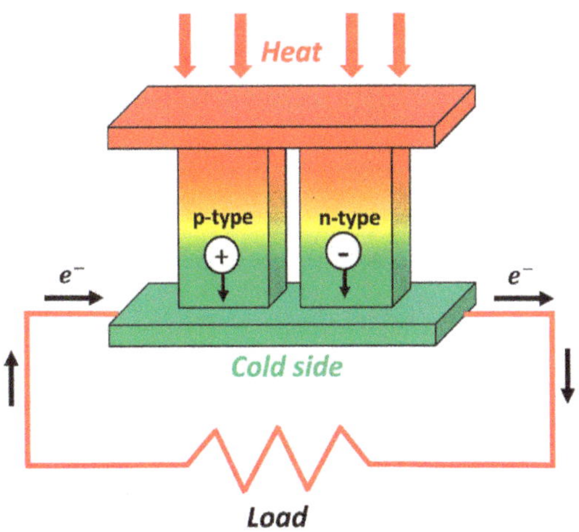

Figure 4. Physical principle of thermoelectric generator [35].

Pyroelectric materials (Figure 5a), are a subclass of piezoelectric materials that are not gradient dependent, but instead they are prone to change polarization due to temperature fluctuations [36]. Pyroelectric materials have an inherent dipole moment that adds up to the spontaneous polarization. The dipole moment remains constant unless there is a temperature fluctuation (Figure 5b). The polarization decreases when the temperature increases (Figure 5c) while a decrease in temperature normalizes the orientation of dipoles and causes the polarization to increase (Figure 5d). The main advantage of pyroelectrics is that they possess piezoelectric properties as well and their main disadvantage is that they depend on constant change of temperature.

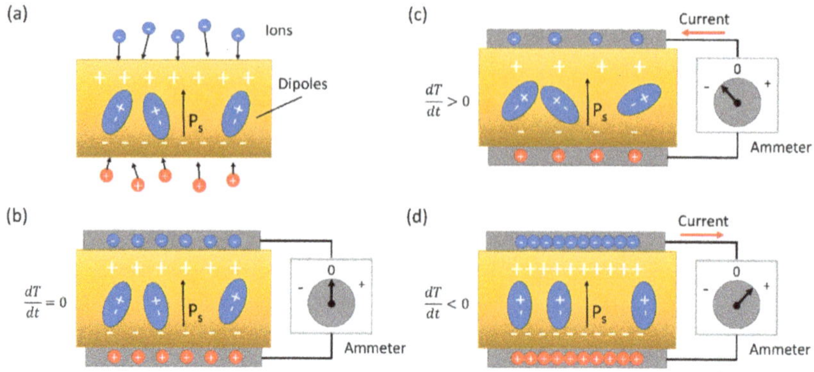

Figure 5. Physical principle of pyroelectric generator: (**a**) ions and dipoles in pyroelectric material, (**b**) without temperature variation, there is no current, (**c**) increase in temperature tends to decrease polarization (**d**) decrease in temperature leads to increase of polarization [37].

Thermal materials can be divided into three main groups: inorganic, organic and hybrid (a combination of both). Inorganic materials consist mainly of bismuth telluride (Bi2Te3) and antimony telluride (Sb2Te3) and their respective alloys [38]. Inorganic materials possess qualities such as high electrical conductivity and they suffer from a mechanically rigid structure and toxicity [39]. Organic materials are represented by carbon-based materials and conductive polymers [38]. They are much more flexible and biocompatible in contrast with their poor electrical parameters. The combination of both types attempts to synthesize materials with the best qualities of both worlds. In practice, thermal materials sit in-between two material types. It is worth noting that usually an organic material is doped by an inorganic one to boost conductive properties of the hybrid.

The human body maintains its temperature constant at approximately 37 °C which allows thermal energy to act as a reliable source for energy harvesting. The ambient temperature and solar energy in particular [40] can be used for conversion of thermal energy. Recent advances in the field of thermoelectric harvesters, indicate that flexibility of the harvester plays one of the most important roles. Klochko et al. [41] used biodegradable polymer nanocellulose and biocompatible semiconductor copper iodide and this method resulted in a flexible design that provided a power density of up to 44 $\mu W \cdot cm^{-2}$. A different approach was followed by Bai et al. [42] who proposed a conductive quasi-solid gel for an electrical interface. Changing flexible solids for a gel had a positive effect on flexibility and proved to offer about 100 μA in parallel and a voltage of 24 mV for 25 generator units. Rösch et al. [43] reported the design of a 3D printable thermoelectric generator with a meander structure. A highlight of this project lies in its 2D layout that was later bent into a 3D origami-like shape which could be easily scaled up. This design provided with an output of 47.8 $\mu W \cdot cm^{-2}$. Ren et al. [44] described a generator that possessed self-healing capabilities, was made of recyclable material and had a modular design which was implemented via making laser cuts in polyimine substrate and inserting in an array of thermoelectric chips. Due to the bond exchange reaction of polyimine, the substate restored its original form even when it was torn apart to a certain degree. Average outputs of 45 and 83 $nW \cdot cm^{-2}$ were achieved while a wearer was sitting and walking respectively. This approach was taken even to a greater extent when Liu et al. [45] described a thermoelectric generator filled with gel that it also had self-healing capabilities meaning it could repair self-inflicted damage to a certain degree and reached output voltage as high as 172.9 nW. The idea of self-repairing generators echoes in the article by Wu et al. [46] which reported the degenerative effect of repeated bending of flexible thermoelectric generators. Pyroelectric harvesters are usually represented by a combination of hybrid piezo- and pyro- effects that take place simultaneously [36]. Mahanty et al. [47] reported their findings from the design of such harvesters with an energy density of 34 $\mu W \cdot cm^{-2}$. The hybridization trend is also reported in Li et al. [40] where an energy harvester that produced 21.3 $mW \cdot m^{-2}$ was designed via a combination of solar, thermal and pyroelectric physical interfaces. The device was integrated onto an outdoor bracelet and was reported to charge after 1 h of exposure to the environment. Moreover, hand movement of the bracelet increased pyroelectric output by more than 12.7 times. Thermoelectric and pyroelectric harvesters represent a huge potential for energy scavenging. While being capable of exploiting the highly stable and accessible energy source of the human body and the ambient temperature, this group possesses good hybridization capabilities, especially when paired with photovoltaic materials.

4. Biomechanical Energy

Biomechanical energy is one of the largest categories and one of the most utilized among all other energy sources [3,7–9,48,49]. The human body is full of flexible joints and points of concentrated pressure (Figure 6) that could be used for energy harvesting from limb movement, respiration, chest and abdomen displacement, eyelid movement, body pressure onto soil while walking, etc [50–53].

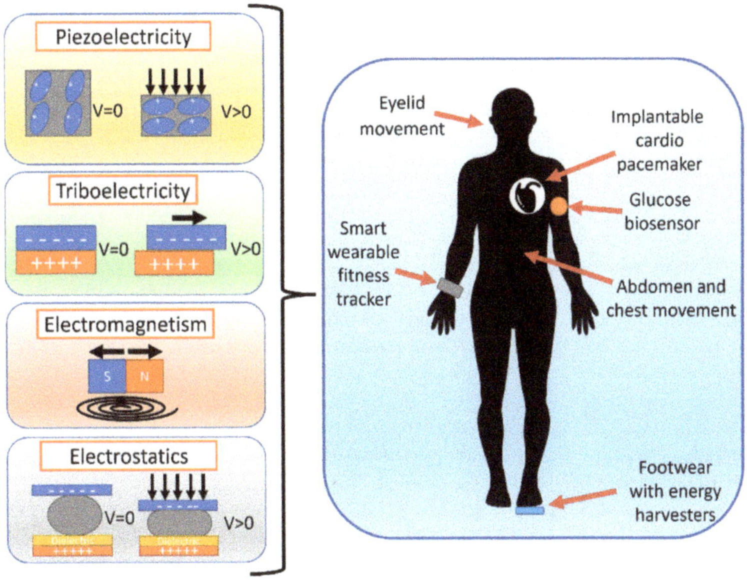

Figure 6. Biomechanical energy in the human body [54].

4.1. *Piezoelectric Nanogenerators*

Piezoelectric nanogenerators (PENGs) are used in many applications such as sensors and actuators [55–58], energy harvesters, micro-electromechanical systems (MEMS) [59] and microfluidics [60]. Energy generation is possible because piezoelectric materials lack central symmetry in their structure. An applied mechanical load induces spontaneous polarization in PENGs and thus electricity is generated (Figure 7).Their popularity is based on their scalability potential, high output power, simplicity of integration [57,61] and availability of "of-the-shelf" energy harvesting solutions (Würth Elektronik, Physik Instrumente, TE connectivity, etc). The downside of PENGs is that their output depends on frequency and optimum actuation values lie in a high-frequency range.

Piezoelectric materials are classified into four main groups: inorganic, organic, composites and biocomposites. Inorganic materials, such as lead zirconate titanate (PZT), barium titanate (BT) and quartz have high values for their electric parameters (piezoelectric strain constant, electromechanical coupling factor and dielectric constant) but they are rigid, brittle and require polling process to function [62]. Organic materials, represented mainly by polyvinylidene difluoride (PVDF) and its copolymer trifluoroethylene (PVDF-TrFE), are on the opposite side of the spectrum and have high elasticity, while having less electrical efficiency. Polymer composites are intended to be the best of both worlds, combining the high piezoelectric parameters of the inorganic materials while staying flexible and biocompatible like the organic materials. Mostly, this is performed by depositing ceramic element, such as BT, on a polymer matrix, such as PVDF [34,63]. Biocomposites are biological tissues that possess inherent piezoelectricity, such as hair, bone, fish bladder, etc. [63].

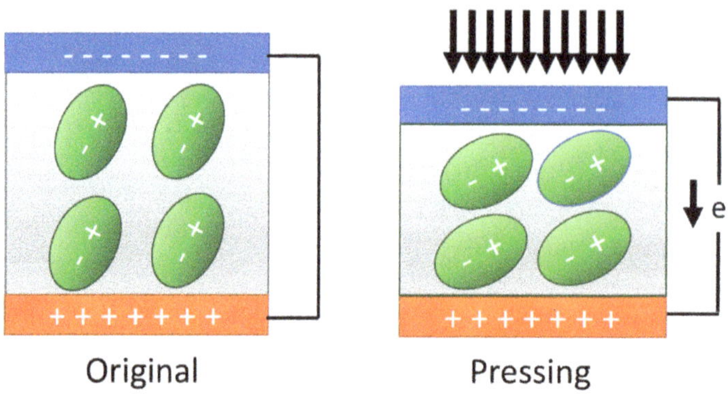

Figure 7. Physical principle of piezoelectric generator [9].

The most commonly used materials are ZnO, polyvinylidene difluoride (PVDF) and lead zirconate titanate (PZT) [57], all of which have inherent pyroelectric properties [36,64].

The recent advances in the field of piezoelectric energy harvesting build on developments in additive manufacturing capabilities [65]. Montero et al. [66] used inkjet printing to deposit an interdigitated electrode which is subsequently coated with PVDF film. The resulted power density was 0.5 $\mu W \cdot cm^{-3}$ with a maximum value of 2.4 nW for the device. Potential applications are varied, for example, a backpack is a quite common form of an attire and thus it is relevant to investigate how much energy can be harvested from it. Zhang et al. [67] used force amplification via a double-bridge frame, studied effects of this method via numerical simulations and experiments and presented a backpack with the maximum output power of 4.13 μW. This power output was reached when the harvester was integrated into a backpack and the person wearing it was jogging. Ruiz et al. [53] demonstrated a respiratory piezoelectric based sensor which, not only provided data for health monitoring, but also could act as an energy harvester. Beyaz [55] described a blood pressure sensing technique which exploits PENGs. Three piezoelectric transducers were used to exhibit and receive acoustic signals to detect geometrical changes in artery tissue. In terms of biodegradable materials, Kumar et al. [68] used reinforced polycarbonate which resulted in a flexible platform that provided high power density up to 2556 $\mu W \cdot cm^{-3}$ at maximum. An unorthodox approach was taken by Sun et al. [69] who reported a PENG made out of a wood sponge. Balsa wood was processed to remove lignin and hemicellulose which resulted in a crystal structure that possessed a high piezoelectric coefficient. Not only does this NG used a biodegradable material, but wood is also present in an abundance thus making this PENG a prominent candidate for wearable harvesters. The scalability potential of the PENG suggests that it can be used for application for wearable electronics and for powering LED lights and LCD screens. The peak power density reached was 0.6 $nW \cdot cm^{-2}$. Due to the way the internal electrostatic field propagates inside the piezoelectric material, it is usually synonymous with relatively low current density and thus lower power density output, especially compared to triboelectric materials. Gu et al. [70] proposed a variant of PENG with high output current density. In order to achieve this, a high piezoelectric coefficient material Sm-PMN-PT was selected, and an intercalation electrode structure was utilized. Essentially, the way to increase piezoelectric energy lies in increasing electrode surface area. The authors of this work proposed a way to enhance PENG by integrating multiple electrodes inside the piezoelectric film and sandwiching them together resulting in a PENG which had as high as 12 units of this structure stacked together. As a result, a maximum peak short-circuit current of 320 μA was obtained by the harvester. Xu et al. [71] designed a PENG that can be attached to one of the leads of a pacemaker. The main material used was microporous P(VDF-TrFE) thin film that was deposited onto Kapton film and sandwiched between Au electrodes. From the engineering point of view, the authors decided to implement a multibeam design. The harvester had a cylindric form and its inner space was filled with piezoelectric films which play role of beams. Then, the harvester was placed onto a lead. A pacemaker lead was fixed by these films and when the lead was subject to movement, its energy was transferred to the PENG. The output energy was 6.5 μJ that was stored in a 120 uF capacitor. It can be deduced that, despite being relatively simple and long-studied subject, PENGs are in a high demand and present a very prominent foundation for energy harvesting. Yet, their cost and dependence on actuation frequency need to be taken into account during the design phase [72].

4.2. Triboelectric Nanogenerators

Triboelectric nanogenerators (TENGs) are used to power many sensors and energy harvesters of different designs [52,73,74]. TENGs harvest kinetic energy of the human body akin to PENGs and exploit the same points of mechanical pressure. The underlying physical principle is the fact that materials have different tribopolarities and when paired properly, spatial manipulation causes triboelectrification and electrostatic induction. Surface porosity plays a vital role in the process of designing TENGs [75]. There are four fundamental operational modes of TENGs as displayed in Figure 8, each utilized under different circumstances [74]. In the vertical contact separation mode (Figure 8a) which features two electrodes and a dielectric film, when the mechanical force is applied, triboelectrification charges the electrodes, and when the force is released and there is a small gap between the electrodes, electrostatic induction causes current to flow. Contact-sliding or lateral-sliding mode (Figure 8b) works similarly to the first mode, yet electrodes move to the sides. In the single-electrode mode (Figure 8c), only one grounded electrode is present which is spatially separated from a dielectric film. While approaching to and departing from the electrode, electrical field changes due to the triboelectrification. The potential difference between the materials cause electrons to flow. In the freestanding triboelectric-layer mode (Figure 8d) the dielectric layer moves, while the electrodes remain still. The major advantage of using TENGs is their low cost per Watt at low frequencies [76], their high output power and that the output voltage is independent of frequency. However, the issue of the overvoltage influences the energy conditioning circuitry and TENGs tend to have small currents.

Triboelectrification occurs between two materials, one of which has a higher electron affinity (acceptor), while the other material has lower electron affinity (donor). According to the results of material selection summary [77], Polytetrafluoroethylene (PTFE), polydimethyl-siloxane (PDMS), fluorinated ethylene propylene (FEP) and Kapton are widely used as electron acceptors. As electron donors, materials such as aluminum, cuprum, skin and Nylon are mostly used. In addition, triboelectric generators can be represented as four (4) layers which are the charge-generating, charge trapping, charge-collecting and charge-storage layers whereas the materials for such complex structure are described elsewhere [78].

In terms of practical application, Zhang et al. [79] presented a paper-based TENG capable of producing up to 171 mW$\cdot$m^{-2}. A highlight of this work is that a single-electrode TENG can be produced relatively economically and fast from waste paper, which serves as a triboelectric layer. Jiang et al. [80] described a TENG with self-healing capabilities. The synthesized elastomer proved to be both extremely elastic (10000%) and could completely repair itself while reaching an output of 450 mW$\cdot$m^{-2}. Another single-electrode mode

TENG for tactile sensors was designed by Wu et al. [81] where the surface porosity was optimized in order to provide the best results possible and a power density of 33.75 $W \cdot m^{-2}$ was achieved. Guan et al. [82] reported a TENG based on a two-electrodes contact-separate mode that had a negative Poisson-ratio fiber inserted into positive Poisson-ratio hallow cylinder. This synergistic structure could provide up to 52.36 $mW \cdot m^{-2}$. Liao et al. [83] proposed an ion-gel based TENG that could heal itself, could effectively work in a wide range of temperatures (−70 to 250 °C) and was transparent. The approach of using ion-gel instead of regular electrodes allows to reach outstanding stretchability parameters (1012%). The self-healing capability was tested by cutting the gel in half and sticking it back together. The demonstrated TENG reached power density of 2.17 $W \cdot m^{-2}$. Another unusual application of gel in TENG was suggested by Sheng et al. [84] who investigated the application of highly stretchable (>10,000%) TENG as an arm training sensor or for helping patients with self-rehabilitation. This harvester reached a power density of 30 $mW \cdot m^{-2}$. Yang et al. [85] indicated that there are still ways to improve the design layout of TENG instead of trying to improve harvesters on molecular level with a report on a honey combed TENG. The outer layer which forms a carcass of the harvester is a 3D printed TPU material, the top three honeycomb walls are coated with conductive fabric and PBAT film while the bottom three walls are coated with Al film. In this case, PBAT and Al work as triboelectric layers when being pressed against each other. Additionally, this design was very easily manufactured making it extremely useful for prototyping and massive production. This design solution reached a power output of 10.79 $W \cdot m^{-2}$ and a high voltage of 1500 V. Ryu et al. [86] described a TENG producing 4.9 $\mu W \cdot cm^{-3}$ which is suitable for use with IMDs and pacemakers. In this work, the TENG consisted of multiple stacked units, it works in free-standing mode and harvests vertical displacement of the chest. The harvested energy was subsequently conditioned and put into a Li-ion battery. This work also featured an in vivo experiment involving an adult mongrel. Over a span of 24 h, 144 mW of power was harvested considering both periods of low and high activity. Biocompatibility was examined by the inflammation response of the animal which proved to be within limits and no signs of infections or fever were observed. Overall, TENGs are a prominent method of energy harvesting that has a considerable amount of output power that is achieved through various intricate design choices. However, the conditioning circuit plays a huge role in this type of harvesters due to the overvoltage [72].

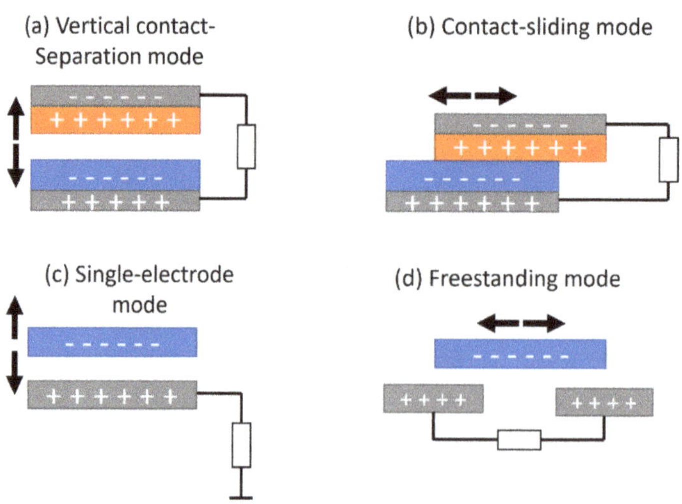

Figure 8. Fundamental operational modes of TENG: (**a**) vertical contact-separation mode, (**b**) contact-sliding mode, (**c**) single-electrode mode, (**d**) freestanding triboelectric-layer mode [9].

4.3. Electromagnetic Generators

Electromagnetic generators (EMGs) follow a variety of constructive implementations that revolve around moving a coil relative to magnets thus exploiting Faraday's law of induction [9,10] which implies that a change in magnetic flux results in an induction of electromotive force. It can be achieved through four main mechanisms: moving the magnet perpendicularly relative to the coil (Figure 9a), moving the magnet in parallel to the coil (Figure 9b), moving the magnet inside the coil (Figure 9c); and moving the coil in-between magnets (Figure 9d). The current trend in EMGs lies in the hybridization of EMGs and TENGs, which proves to have fruitful results. The main advantage of EMGs is that they have low matching resistance values [87] and can be effectively used at low frequencies [88]. On the other hand, they have inherently rigid nature and might have reliability issues [89].

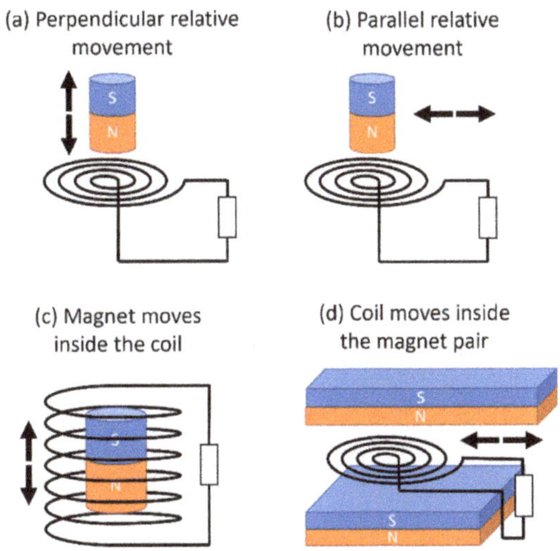

Figure 9. Physical principle of electromagnetic generator: (**a**) perpendicular relative movement, (**b**) lateral or parallel relative movement, (**c**) magnet movement inside the coil, (**d**) coil relative movement in-between magnets [9].

Jiang et al. [90] presented a hybrid harvester where human body pressure from movement is transformed via transmission into radial movement of a generator. The generator consisted of rotating triboelectric blades which worked in the freestanding triboelectric-layer mode, magnets and coils that were placed in the rotor and the stator respectively. In terms of output power, the harvester provided 14.68 mJ when integrated into a shoe. Li et al. [91] developed a magnetized microscale needle array which also possessed triboelectric properties. The device used microneedles that induced triboelectrification when rubbed against the substrate, which also acted as a magnet moving up and down relatively to the coil. This combination of energy harvesting generators achieved up to 16.19 $\mu W \cdot m^{-2}$ for a load of 10^6 Ω. Practical application involved putting the harvester into a shoe and the values of collected voltage for walking and jogging were very different for the TENG (2.8 V and 6 V respectively), whereas the collected current for the EMG indicated less difference (4.5 μA vs. 5.3 μA). Iqbal et al. [87] described PENG mixed with EMG that could be inserted into a shoe insole. As a result, the overall output was better than using PENG or EMG separately and a power of 4.05 $\mu W \cdot cm^{-3}$ was achieved. On the other hand, Lai et al. [92] reported on a hybrid harvester that scavenges electromagnetic energy from the ambient environment such as a mobile phone, laptop or any other electronic device. The mechanical energy was harvested via TENG that worked in the single-electrode mode. The peak power was 360 $\mu W \cdot m^{-1}$. Li et al. [88] designed a hybrid PENG and EMG, where PENG was connected to the EMG via a spring-loaded mechanism. In this harvester, two generators harvested energy separately, but they were combined through the mechanical transmission. A power output of 36.21 mW·cm^{-3} was reached. Electromagnetic energy is an underutilized and unconventional source for biomechanical energy harvesting, but it may prove to be useful in combination with other NGs.

4.4. Electrostatic Nanogenerators

Electrostatic generators (ESGs) are a type of a varying capacitor and their working principle is described by electrostatic induction [8]. The capacitance of such a device changes as a result of changes in the overlap area or in the gap between two electrodes, caused by some external stimuli (Figure 10). The induced electrical charge will result in electric current. There are four main shapes of ESGs: an in-plane gap closing ESG where the electrode moves sideways (Figure 10a), an in-plane overlap ESG where the counter-electrode moves in plane thus changing the overlap (Figure 10b), out-of-plane gap closing converter where the gap between electrodes changes (Figure 10c), in-plane ESG where the overlap varies by moving across one axis (Figure 10d). The application of such varying capacitors ranges from biosensors, textile wearable electronics, biomechanical energy harvesting to solar energy storage. The main advantage of using ESGs is their simple structure and high output at high frequencies however they frequently need bias voltage to function properly and most methods of increasing frequency of a harvester (or self-oscillation feature) includes MEMS [93,94].

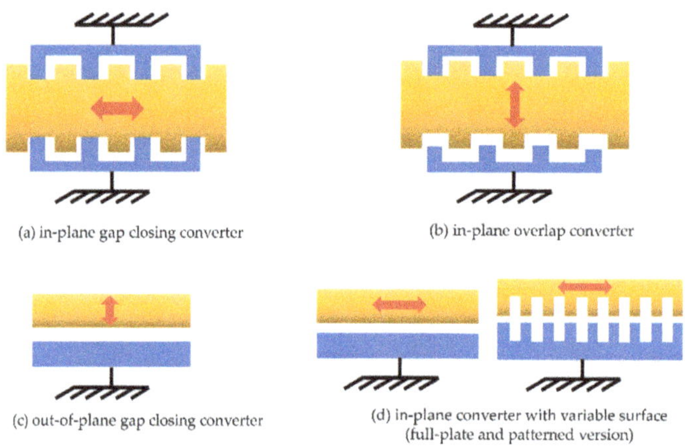

Figure 10. Main shapes of electrostatic generators: (**a**) in-plane gap closing converter, (**b**) in-plane overlap converter, (**c**) out-of-plane gap closing converter, (**d**) in-plane converter with variable surface [95].

Wu et al. [94] continued the trend of hybridization that is so prominent in other NGs and described a rotatory disc-based EMG and ESG. The modus operandi of the EMG followed the principles discussed in the previous section, while ESG consists of an electret film and the respective electrode. The rotatory movement of the electret film relative to the electrode causes a change in the overlap area. The maximum output density is 2.5 W·m^{-3} for EMG and 107.8 W·m^{-3} for ESG. Pourshaban et al. [96] designed an electrostatic generator which is integrated into a contact lens where a tear droplet plays the role of the electret. The average obtained power is 0.265 μW. Erturun et al. [97] proposed a hybrid of PENG and ESG. In this structure, a piezoelectric film is sandwiched between Au and Al electrodes. The Al electrode is coated with Teflon on the opposite side to the PENG and is isolated from the bottom Cu electrode via an air gap confined by two Styrofoam pads. When the harvester is subjected to stress, the electron flow is directed to the Al electrode and outwards, when pressure is released. The maximum attained power density was ~8.8 μW·cm^{-3}. Cao et al. [98] proposed a hybrid of PENG and EMG where vibrations of a piezoelectric cantilever were also used to harvest electrostatic energy. This is due to fluctuations of a beam that has ESG electrode on it which changes the gap distance and induces change in capacitance. Capacitive harvesters represent a stable source of energy and have been thoroughly investigated for a long period of time thus making them a very suitable choice for a hybridization.

4.5. Reverse Electrowetting on Dielectric Phenomenon (REWOD)

Recently, a lot of attention has been drawn to the reverse electrowetting on dielectric phenomenon (REWOD) [99–101]. This is a prominent subclass of ESGs which is based on the direct electrowetting on dielectric (EWOD) phenomenon. EWOD exploits the fact that the liquid-solid (droplet-electrode) interfacial energy can be manipulated by the application of an electric field. This leads to a change in the apparent contact angle between the droplet and the electrode which increases the overlap area (Figure 11). A droplet could potentially be a physiological fluid [102] thus a biomolecular absorption should be taken into account [103]. REWOD works in the opposite manner and when a bubble of conductive liquid is pressed in-between electrodes, the overlap area changes thus significantly increasing the capacitance of the system. Adhikari et al. [101] studied REWOD phenomenon via changing different parameters of the system such as dielectric thickness, electrolyte and dielectric concentrations. When two dissimilar dielectrics are introduced, the system can work without the external bias voltage. A total power density of 53.3 nW$\cdot$cm^{-2} was obtained. Tasneem et al. [104] proposed a wearable motion tracking sensor powered by a REWOD principle. A significant focus of the work was oriented towards the conditioning circuit. In this case, the authors used a rectifier based on a Schottky diode followed up by a booster circuitry and the resulted power density of the harvester was 58 nW$\cdot$cm^{-2}. Hsu et al. [93] described a REWOD harvester that could produce as much as 100 W$\cdot$m^{-2} with a 4.5 V of bias voltage and it instantly became one of the most prominent works in the field. The described harvester heavily relies on multiplication of the number of REWOD units and the frequency of their oscillations. Thus, the REWOD method of harvesting biomechanical energy holds a significant potential in what it can practically be achieved.

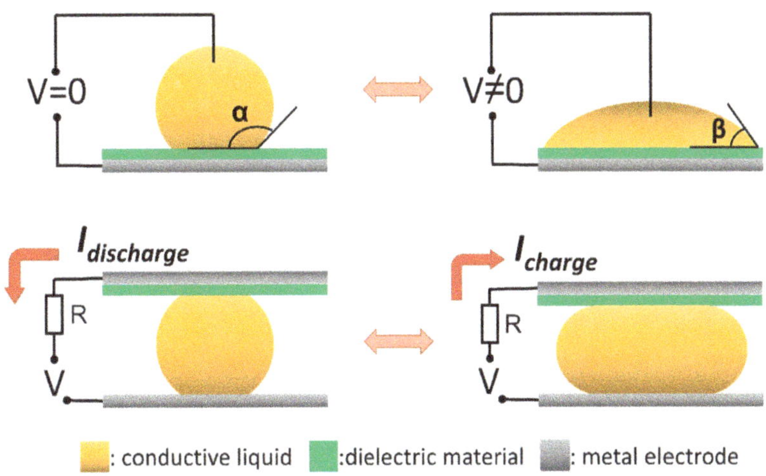

Figure 11. EWOD and REWOD processes [93].

5. Photovoltaic Materials

Solar energy is a renewable energy source with very high potential, it is abundant [105] and can be harvested easily, under appropriate weather conditions [3,9,106,107]. Photovoltaic materials optically collect photons emitted by a light source. The harvested solar energy is then transferred to the electrons, which makes them highly mobile, thus forming an electric current (Figure 12). Merits of using photovoltaic materials are their stable DC output but for wearable electronics they usually need to be hybridized because of limited day-time use.

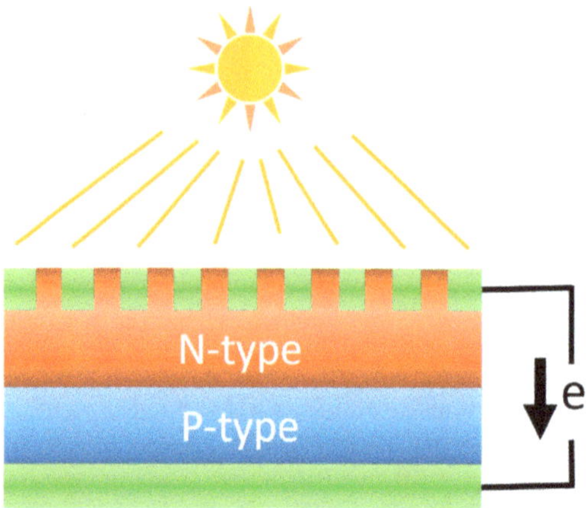

Figure 12. Solar energy harvesting [9].

Zhang et al. [105] reported a photo-thermoelectric hybrid harvester. The seamless integration with the textile fabric creates a possibility for an integration with the continuous monitoring system, while the output power under sunlight reaches 0.24 nW. Zhang et al. [108] described another unconventional application of photo-thermoelectric harvester which is conjugated with hydroelectricity. The hydroelectric effect helps to drastically increase the total output voltage from −0.092 to 1.712 mV; and, in the wet state the harvester reaches 2.45 μW. Moreover, the evaporated body fluid also produces steam, and such a textile-based harvester can also serve as a human body humidity monitor. Yu et al. [109] proposed a gauze kerchief made from photovoltaic textile. In this case, Cu or Cu-coated polymer wires are sequentially coated with Mn substrate, ZnO nanowires, ethanol solution of N719 and CuI forming photoanodes. The synthesized generator was optimized via change of embroidery process parameters, different materials of the counter electrode and how electrodes were connected to one another. 0.79 V were obtained with a light intensity of 100 mW cm^{-2}. Photovoltaic devices represent efficient energy harvesters though it should be viewed as an addition to another NG that can constantly and reliably provide power.

Selected types of NGs cover all the basics needed for design of the effective energy harvester however this is not the end of it. As briefly mentioned in photovoltaic materials, a new hydrovoltaic type of energy attracts a great deal of popularity [110]. Energy that comes from water evaporation can be harvested with the help of a porous conductive structures. The underlying physical principles can be explained through an electrical double layer and electrokinetic effects but the exact mechanism is yet to be described [111]. Recent survey of this topic can be found elsewhere [112].

6. Comparison of Nanogenerators

The energy harvesters that were reviewed in the previous sections and their respective features are compiled and compared in Table 1. Their main features, their electrical properties, their size and their power output are presented and compared.

Table 1. Comparison of Nanogenerators.

NG Type	Authors	Features	Voltage, Load, Frequency	Size	Power Output
enzymatic BFC	Zhang et al. [27]	Integrated into a diaper	300 mV (OC)	6 × 5 cm^2	220 $\mu W \cdot cm^{-2}$
enzymatic BFC	Zhang et al. [28]	artificial enzyme	400 mV (OC)	n/a	149.2 $\mu W \cdot cm^{-2}$
microbial BFC	Jayapiriya and Goel [29]	3D printed, paper-based	480 mV (OC)	Coin size (~2 × 2 cm^2)	11.8 $\mu W \cdot cm^{-2}$
enzymatic BFC	Yan et al. [30]	high-power, nano enhancements of materials	~0.8 V 1000 Ω	~0.3 nm^2	703.55 $mW \cdot m^{-2}$
thermoelectric NG	Klochko et al. [41]	Biodegradable materials	5 V (OC)	3 × 2 cm^2	44 $\mu W \cdot cm^{-2}$
thermoelectric NG	Rösch et al. [43]	3D printed origami-like shape	534 mV (OC)	190 cm^2	47.8 $\mu W \cdot cm^{-2}$
thermoelectric NG	Ren et al. [44]	self-healing recyclable material	5 V	6 × 5 cm^2	83 $nW \cdot cm^{-2}$
thermoelectric NG	Liu et al. [45]	self-healing	4.15 mV 600 Ω	~4 × 2 cm^2	172.9 nW
pyro- and PENG	Mahanty et al. [47]	hybrid	35 V at 3.4 M Ω	8 × 7 cm^2	34 $\mu W \cdot cm^{-2}$
Solar, thermal and pyroelectric NG	Li et al. [40]	hybrid	93.1 (OC)	3 × 9 cm^2	21.3 $mW \cdot m^{-2}$
PENG	Montero et al. [66]	3D printed	1.1 V 60 Hz 100 MΩ	18 × 18 cm^2	0.5 $\mu W \cdot cm^{-3}$
PENG	Zhang et al. [67]	integrated into a backpack	~1.7 V 8 Hz 0.7 MΩ	3.5 × 3.5 × 9 mm^3	4.13 μW
PENG	Kumar et al. [68]	nano enhancements of materials	16.67 V (OC) 5 Hz	12 × 8 mm^2	2556 $\mu W \cdot cm^{-3}$
PENG	Sun et al. [69]	made from wood sponge	0.63 V ~1 Hz 80 MΩ	15 × 15 × 14 mm^3	0.6 $nW \cdot cm^{-2}$
PENG	Xu et al. [71]	multibeam structure implantable	0.3 V 1 Hz	n/a	6.5 μJ
TENG	Zhang et al. [79]	paper-based	~95 V 2 Hz 130 MΩ	4 × 4 cm^2	171 $mW \cdot m^{-2}$
TENG	Jiang et al. [80]	self-healing	75 V (calc) 2 Hz 5 MΩ	2 × 2 cm^2	450 $mW \cdot m^{-2}$
TENG	Wu et al. [81]	optimized surface porosity	78.7 V (OC)	8 × 8 × 0.5 mm^3	33.75 $W \cdot m^{-2}$
TENG	Guan et al. [82]	different Poisson-ratio materials	~25 V 10 MΩ	20 cm × ø 2 mm	52.36 $mW \cdot m^{-2}$
TENG	Liao et al. [83]	ion-gel self-healing	189 V (OC) 1.5 Hz	3 × 3 cm^2	2.17 $W \cdot m^{-2}$
TENG	Yang et al. [85]	3D printed honey-comb structure	1500 V (OC) 3 Hz	68 × 39 mm^2 × 5	10.79 $W \cdot m^{-2}$
TENG	Ryu et al. [86]	implantable stacked TENG structure	10 MΩ	n/a	4.9 $\mu W \cdot cm^{-3}$
EMG and TENG	Li et al. [91]	hybrid Microscale needles	10 V (OC) 1 Hz (TENG)	24 × 24 × 3.2 mm^3	16.19 $\mu W \cdot m^{-2}$
EMG and PENG	Iqbal et al. [87]	hybrid	7.01 V (OC)	3.9 × 3.9 × 2.9 cm^3	4.05 $\mu W \cdot cm^{-3}$
EMG and TENG	Lai et al. [92]	hybrid uses ambient magnetic field	120 V/m 100 MΩ	5 cm (contact length)	360 $\mu W \cdot m^{-1}$ and 8 $\mu W \cdot m^{-1}$

Table 1. *Cont.*

NG Type	Authors	Features	Voltage, Load, Frequency	Size	Power Output
EMG and PENG	Li et al. [88]	hybrid MEMS	18 V (OC) PENG 22 V (OC) EMG	4 × 2.5 × 5.8	36.21 mW·cm^{-3}
ESG and EMG	Wu et al. [94]	hybrid	~40 V (OC) EMG ~0.1 V (OC) ESG 50 rpm	ø 50 mm	2.5 W·m^{-3} and 107.8 W·m^{-3}
ESG	Pourshaban et al. [96]	contact lens	450 mV 100 kΩ	2 × 10 mm^2	0.265 μW
ESG and PENG	Erturun et al. [97]	hybrid	4.2 V 1 MΩ—PENG 17 V 10 MΩ—ESG 20 Hz	ø 3 × 0.65 cm	8.8 μW·cm^{-3}
REWOD NG	Adhikari et al. [101]	no bias voltage	0.103 V 0.15 MΩ 3 Hz	ø 50.5 mm	53.3 nW·cm^{-2}
REWOD NG	Tasneem et al. [104]	conditioning circuit	943 mV pp 1.6 MΩ 10 Hz	4 mm gap between electrodes	58 nW·cm^{-2}
REWOD NG	Hsu et al. [93]	self-oscillation	bias voltage used 0.877 MΩ	40 × 40 mm^2 (628 mm^2)	100 W·m^{-2}
photo-thermoelectric NG	Zhang et al. [105]	hybrid	0.56 mV 1.32 kΩ	4 × 0.3 cm^2	0.24 nW
photo-, thermo-, hydroelectric NG	Zhang et al. [108]	hybrid hydroelectricity	12.5 mV 15.93 Ω	1.5 × 1.5 cm^2	2.45 μW
photovoltaic NG	Yu et al. [109]	photovoltaic textile	0.79 V	4.5 mm pitch	~134 μW

A summary of the advantages and disadvantages of the main classes of NGs is presented in Table 2.

Table 2. Advantages and disadvantages of various types of Nanogenerators.

Energy Harvester Type		Advantages	Disadvantages	Voltage Output	Current Density	Power Density
BFCs	Microbial	No enzymes Long life span (5 years) DC output voltage	Need special mediators Low power density	100–500 mV	10–100 μA×cm^{-2}	1–10 μW×cm^{-2}
	Enzymatic	Effective use of body liquids (sweat, tears, etc) as fuel DC output voltage	Short life span	0.1–1 V	100–500 μA×cm^{-2}	10–100 μW×cm^{-2}
Thermal	Thermoelectric	DC output voltage	Low voltage	0.1–1 V	0.1–10 μA×cm^{-2}	0.1–10 μW×cm^{-2}
	Pyroelectric	Are a sub-class of piezoelectric materials	Require temperature fluctuations	0.1–10 V	0.01–10 μA×cm^{-2}	0.1–1 μW×cm^{-2}
Piezoelectric		Availability "of-the-shelf" Can possess pyroelectric properties	Frequency dependent output	0.1–10 V	0.01–10 μA×cm^{-2}	1–100 μW×cm^{-2}
Triboelectric		High output High elasticity Low cost	Overvoltage Small current	10–1000 V	1–10 μA×cm^{-2}	1–100 mW×m^{-2}
Electromagnetic		Hybridization with TENGs and ESGs Low impedance	MEMS (low reliability) Rigid, non-flexible nature	0.1–10 V	1–10 μA×cm^{-2}	10 μW×cm^{-2} (mostly used as a hybrid with TENGs)
Electrostatic		Simple structure	Practical applications involve MEMS	0.1–1 V	1–10 μA×cm^{-2}	0.1–10 μW×cm^{-2}

Table 2. *Cont.*

Energy Harvester Type	Advantages	Disadvantages	Voltage Output	Current Density	Power Density
REWOD	High output voltage at high frequencies	Needs bias voltage for high output Frequency dependent	10–100 mV	10–100 $nA \times cm^{-2}$	0.01–0.1 $\mu W \times cm^{-2}$
Photovoltaic	DC current	Needs hybridization for constant energy	1–100 mV	1–100 $\mu A \times cm^{-2}$	0.1–10 $\mu W \times cm^{-2}$

7. Biomedical Application of Energy Harvesters

In addition to the aforementioned glucose biosensor, another example of a commonly used biosensor is a pulse oximeter (Figure 13) biosensor that had seen an extensive use throughout the COVID-19 pandemic [113–115]. A pulse oximeter is used to measure the blood oxygen saturation level and the heart rate. Its working principle is based on hemoglobin light absorption [116]. Near infrared and red LEDs are mounted on one side of the device and the opposing side is equipped with photodiodes. Detection of oxygenated and deoxygenated hemoglobin appears at the red-light wavelength while near infrared is used as a reference. Because the heart pumps blood at certain frequency, this pulsative nature or pulse can be monitored with the same method thus making a combined biosensor. Batteries for pulse oxygen biosensors are replaceable and can last for 30 h and more. Biosensors consume approximately 40 mW of total energy during continuous work [117]. Introduction of an energy harvesting module would allow pulse oximeters to continuously monitor hemoglobin without the need to constantly replace or recharge batteries making it fully autonomous.

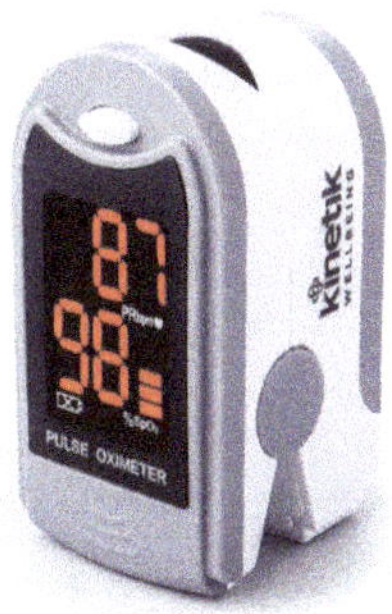

Figure 13. An example of a pulse oximeter from Kinetik wellbeing; https://www.kinetikwellbeing.com/ (accessed on 19 August 2022).

A different application for energy harvesting can be found within cardiovascular implantable medical devices (IMDs) such as implantable cardioverter-defibrillators (ICDs) and pacemakers (Figure 14). ICD represents a combination of a continuous monitoring system and actuator for administration of an electric stimuli if heart rate is higher than a designated threshold (tachycardia) [118]. Pacemakers, depending on their type, can sense intracardiac signal and stimulate the according chambers or ventricles thus pacing heart [119]. Both devices follow a similar design and quite frequently they are combined into one device. A pacemaker, for example, consists of a pulse generator module with leads which are placed into hears chamber or ventricles. The electric stimuli are delivered from the pulse generator to the heart. Typical power consumption values for pacemakers are less than 100 µW while ICDs may reach more than 200 µW [120]. Depending on the frequency of its use, a pacemaker may last up to 12 years [121] but an implantable cardioverter-defibrillator lasts for 9 years [122]. However, when the battery runs out or it depletes prematurely, which happens to 8% of IMDs [123], a new surgery needs to be performed to replace an IMD thus involving health risks and financial implications [86].

Clinical application of biosensors and sensors alike is vast. Apart from glucose and pulse oximeter biosensors, human respiration efficacy, temperature, sweat content, interstitial fluid and heart electrical signal. Such types of sensors and biosensors can be integrated into clothing, wristbands, contact lenses, skin or be implanted into the human body [8,124].

Overall, almost all sensors for monitoring human biomarkers rely on batteries to a certain degree which creates a gap in continuous monitoring and necessitates additional actions to be performed by a patient. Energy harvesting offers a solution which can reduce the size of batteries or get rid of them completely offering more stable and comfortable monitoring platform.

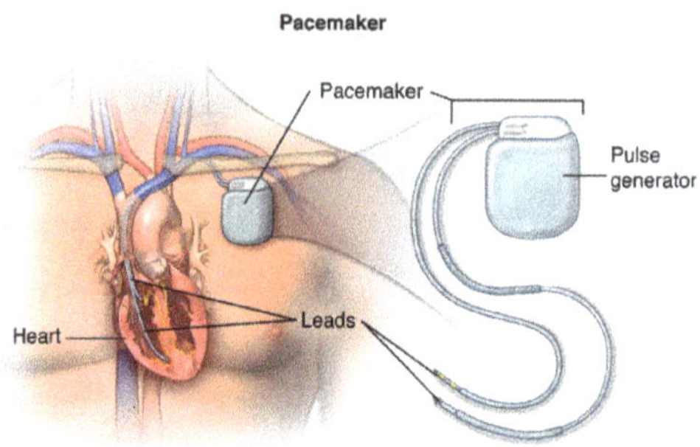

Figure 14. A pacemaker implanted into the human body; https://www.stanfordchildrens.org/ (accessed on 19 August 2022).

8. Discussion on Current Trends

This brief review of energy harvesting identifies two main current trends: hybridization of several NGs and manipulation of material structure. In addition, the overall application of the harvester is of paramount importance because it largely influences the design of harvester. Extensive presence of TENGs and PENGs indicates that biomechanical energy is one of the most used and effective sources for scavenging. Since TENGs and PENGs already represent an established field, prototyping and manufacturing of these devices is lower compared to other NGs. However, the choice between the two is not the simplest decision to make because, as previously mentioned, PENG has lower output than TENG and its impedance is dependent on actuation frequency while TENG have a necessity to come with complex energy conditioning circuitry to combat overvoltage. ESGs and EMGs are essentially MEMS meaning that their mechanical reliability is going to be a prime issue during the design stage [89].

Judging from the comparison presented in Table 1, it can be observed that PENGs and TENGs dominate the area by analyzing how many research publications deal with these types of harvesters and their hybrid derivatives. Moreover, both harvesters seem to be equally frequently chosen for hybridization purposes.

From the same table, it can be discerned that EMGs and ESGs that are based of MEMS are slowly being replaced by solid state NGs such as TENGs and PENGs because of the reliability issues and rigidity. EMGs and ESGs are quite often found in hybrid harvesters as an auxiliary unit while their use as a single harvester is less frequent. This can be expanded as a forecast that MEMS are going to be driven out of this field and solid-state single and hybrid harvesters are going to take their place. This does not mean that EMGs and ESGs are going to cease to exist, they will become more flexible, less dependent on actual MEMS structures and the energy generation is going to take place at nano-scale level, like using magnetized microneedles [91]. Thermoelectric and pyroelectric NGs tend to be hybridized with other NGs more frequently than any other NGs due to nature of the energy source and low power density they exhibit. BFCs are a good choice for specific biosensors, where chemical reactions may be chained in order to harvest energy, but this comes with the cost of manufacturing combined with short life span and low power density.

Looking at the various types of NGs, it is apparent that flexibility plays a major role during the design stage regardless of the chosen type. Due to inherent flexibility of the human body (skin, muscles, joint movements) and unique body structure of every individual, it is imperative to look at flexible materials such as elastomers, polymers, gels. A good description of the challenges of flexible and conductive materials was analyzed in recent works [125–127], where bioinspired by cuit cocoon structure significantly increased foldability of electrodes. In particular, foldability of electrodes was significantly increased by utilizing two principles: firstly, a rigid web structure of a raw cocoon is tough but reeling process forms fibers that can slide one against the other and the layers become separable in a cuit cocoon. Secondly while folding, the cuit cocoon bends into an ε-shape at the crease. These features of the cuit cocoon can be combined with other biological structures, such as leaves. For example, a cockscomb petal has a spring-like structure which allows it to be flexible in terms of compression and tensile and it can be also viewed as a membrane for attachment of substrates, [127]. Combined biological structures of the cuit cocoon and the cockscomb petal were used to design a super-foldable C/NiS nanofiber electrode. In another case [126], a Mimosa leaf structure was used to ensure scatheless foldability of electrodes by addition of cone-like arrays (bottom diameter was ~100 nm and top diameter was ~20 nm) on top of the C-fiber/FeOOH electrode.

A more general trend is emerging when the overall use of electronic sensors and electronics in general is expected to increase in the following years [128]. The same trend also correlates with the rising interest in the field of biosensors as a whole [129]. This in turn will increase the total energy consumption which can be balanced out with the use of energy harvesters. Therefore, the multi-disciplinary field of energy harvesting will highly likely experience even bigger influx of research articles and industry interest in the following years.

9. Conclusions

Continuous monitoring of biomarkers and reliable supply of energy to implantable bioelectronics are topics that are closely interrelated with energy harvesting. As this review indicated, there are numerous biosensors that are used in medical practice today that can be improved by integrating them with efficient energy harvesters.

The human body produces several types of energy such as biochemical, thermal and mechanical. Biomechanical energy is the largest and most well-refined energy type that is currently used for harvesting. The most prominent type of harvesters employed here are piezoelectric and triboelectric harvesters. Photovoltaic materials had also seen an increase in popularity in a form of wearable energy scavengers.

The main problem of energy harvesters is their low power density and there are two main design philosophies that try to solve this issue. It is possible to alternate material structure in nanoscale employing different manufacturing processes. This usually increases complexity of the material while improving its output parameters. Another way of maximizing the power output is through the hybridization of different harvesters in one compact structure. A resulting harvester offers a better output power at the cost of more complex external conditioning circuitry. Flexibility and foldability are important material properties and increasing research accumulates in order to accomplish this requirement.

Author Contributions: Conceptualization, I.S. and S.D.P.; methodology, I.S., S.D.P. and A.T.; formal analysis, I.S.; investigation, I.S., S.D.P. and A.T.; resources, S.D.P.; data curation, I.S. and S.D.P.; writing—original draft preparation, I.S. and A.T.; writing—review and editing, I.S., S.D.P. and A.T.; visualization, I.S., S.D.P. and A.T.; supervision S.D.P. and A.T.; project administration, S.D.P.; funding acquisition, S.D.P. All authors have read and agreed to the published version of the manuscript.

Funding: The present research was supported by the Faculty of Science, Technology, Engineering and Mathematics of the Open University and received no external funding.

Institutional Review Board Statement: Not applicable.

Informed Consent Statement: Not applicable.

Data Availability Statement: The reported data in this review article are obtained from the references which are included in the paper.

Conflicts of Interest: The authors declare that they do not have any conflict of interest.

Abbreviations

BFC	Biofuel cell
EMG	Electromagnetic generators
ESG	Electrostatic generators
EWOD	Electrowetting on dielectric
ICD	Implantable cardioverter-defibrillators
IMD	Implantable medical devices
MEMS	Micro-electromechanical system
NG	Nanogenerator
OC	Open-circuit
PENG	Piezoelectric nanogenerator
PVDF	Polyvinylidene fluoride
PZT	Lead zirconate titanate
REWOD	Reverse electrowetting on dielectric
TENG	Triboelectric nanogenerator

References

1. Gambhir Sanjiv, S.; Ge, T.J.; Vermesh, O.; Spitler, R.; Gold Garry, E. Continuous health monitoring: An opportunity for precision health. *Sci. Transl. Med.* **2021**, *13*, eabe5383. [CrossRef] [PubMed]
2. Leenen, J.P.L.; Leerentveld, C.; van Dijk, J.D.; van Westreenen, H.L.; Schoonhoven, L.; Patijn, G.A. Current Evidence for Continuous Vital Signs Monitoring by Wearable Wireless Devices in Hospitalized Adults: Systematic Review. *J. Med. Internet Res.* **2020**, *22*, e18636. [CrossRef] [PubMed]
3. Song, Y.; Mukasa, D.; Zhang, H.; Gao, W. Self-Powered Wearable Biosensors. *Acc. Mater. Res.* **2021**, *2*, 184–197. [CrossRef]
4. Tan, P.; Zou, Y.; Fan, Y.; Li, Z. Self-powered wearable electronics. *Wearable Technol.* **2020**, *1*, e5. [CrossRef]
5. Molina Arias, L.; Iwaniec, J.; Iwaniec, M. Modeling and Analysis of the Power Conditioning Circuit for an Electromagnetic Human Walking-Induced Energy Harvester. *Energies* **2021**, *14*, 3367. [CrossRef]
6. Guo, X.; Liu, L.; Zhang, Z.; Gao, S.; He, T.; Shi, Q.; Lee, C. Technology evolution from micro-scale energy harvesters to nanogenerators. *J. Micromech. Microeng.* **2021**, *31*, 093002. [CrossRef]
7. He, T.; Guo, X.; Lee, C. Flourishing energy harvesters for future body sensor network: From single to multiple energy sources. *iScience* **2021**, *24*, 101934. [CrossRef] [PubMed]
8. Rong, G.; Zheng, Y.; Sawan, M. Energy Solutions for Wearable Sensors: A Review. *Sensors* **2021**, *21*, 3806. [CrossRef] [PubMed]
9. Zhang, T.; Yang, T.; Zhang, M.; Bowen, C.R.; Yang, Y. Recent Progress in Hybridized Nanogenerators for Energy Scavenging. *iScience* **2020**, *23*, 101689. [CrossRef] [PubMed]
10. Xu, C.; Song, Y.; Han, M.; Zhang, H. Portable and wearable self-powered systems based on emerging energy harvesting technology. *Microsyst. Nanoeng.* **2021**, *7*, 25. [CrossRef] [PubMed]
11. Wang, Z.L.; Song, J. Piezoelectric nanogenerators based on zinc oxide nanowire arrays. *Science* **2006**, *312*, 242–246. [CrossRef] [PubMed]
12. Feynman, R.P. There's Plenty of Room at the Bottom. *Eng. Sci.* **1960**, *23*, 63–76.
13. Schüssler-Fiorenza Rose, S.M.; Contrepois, K.; Moneghetti, K.J.; Zhou, W.; Mishra, T.; Mataraso, S.; Dagan-Rosenfeld, O.; Ganz, A.B.; Dunn, J.; Hornburg, D.; et al. A longitudinal big data approach for precision health. *Nat. Med.* **2019**, *25*, 792–804. [CrossRef] [PubMed]
14. Sharma, A.; Badea, M.; Tiwari, S.; Marty, J.L. Wearable Biosensors: An Alternative and Practical Approach in Healthcare and Disease Monitoring. *Molecules* **2021**, *26*, 748. [CrossRef] [PubMed]
15. Lee, P.M.; Xiong, Z.; Ho, J. Methods for powering bioelectronic microdevices. *Bioelectron. Med.* **2018**, *1*, 201–217. [CrossRef]
16. Hao, S.; Sun, X.; Zhang, H.; Zhai, J.; Dong, S. Recent development of biofuel cell based self-powered biosensors. *J. Mater. Chem. B* **2020**, *8*, 3393–3407. [CrossRef] [PubMed]
17. Katz, E.; Bollella, P. Fuel Cells and Biofuel Cells: From Past to Perspectives. *Isr. J. Chem.* **2021**, *61*, 68–84. [CrossRef]
18. Nasar, A.; Perveen, R. Applications of enzymatic biofuel cells in bioelectronic devices—A review. *Int. J. Hydrogn Energy* **2019**, *44*, 15287–15312. [CrossRef]
19. Sharma, A.; Singh, G.; Arya, S.K. Biofuel cell nanodevices. *Int. J. Hydrogn Energy* **2021**, *46*, 3270–3288. [CrossRef]
20. Yimamumaimaiti, T.; Lu, X.; Zhang, J.-R.; Wang, L.; Zhu, J.-J. Efficient Blood-toleration Enzymatic Biofuel Cell via In Situ Protection of an Enzyme Catalyst. *ACS Appl. Mater. Interfaces* **2020**, *12*, 41429–41436. [CrossRef]

21. Wu, H.; Zhang, Y.; Kjøniksen, A.-L.; Zhou, X.; Zhou, X. Wearable Biofuel Cells: Advances from Fabrication to Application. *Adv. Funct. Mater.* **2021**, *31*, 2103976. [CrossRef]
22. Buaki-Sogó, M.; García-Carmona, L.; Gil-Agustí, M.; García-Pellicer, M.; Quijano-López, A. Flexible and Conductive Bioelectrodes Based on Chitosan-Carbon Black Membranes: Towards the Development of Wearable Bioelectrodes. *Nanomaterials* **2021**, *11*, 2052. [CrossRef] [PubMed]
23. Zeng, X.; Peng, R.; Fan, Z.; Lin, Y. Self-powered and wearable biosensors for healthcare. *Mater. Today Energy* **2022**, *23*, 100900. [CrossRef]
24. Chen, Z.; Yao, Y.; Lv, T.; Yang, Y.; Liu, Y.; Chen, T. Flexible and Stretchable Enzymatic Biofuel Cell with High Performance Enabled by Textile Electrodes and Polymer Hydrogel Electrolyte. *Nano Lett.* **2021**, *22*, 196–202. [CrossRef] [PubMed]
25. Seselj, N.; Engelbrekt, C.; Zhang, J. Graphene-supported platinum catalysts for fuel cells. *Sci. Bull.* **2015**, *60*, 864–876. [CrossRef]
26. Ivan, I.; Vidaković-Koch, T.; Sundmacher, K. Recent Advances in Enzymatic Fuel Cells: Experiments and Modeling. *Energies* **2010**, *3*, 803. [CrossRef]
27. Zhang, J.; Liu, J.; Su, H.; Sun, F.; Lu, Z.; Su, A. A wearable self-powered biosensor system integrated with diaper for detecting the urine glucose of diabetic patients. *Sens. Actuators B Chem.* **2021**, *341*, 130046. [CrossRef]
28. Zhang, H.; Huang, L.; Chen, J.; Liu, L.; Zhu, X.; Wu, W.; Dong, S. Bionic design of cytochrome c oxidase-like single-atom nanozymes for oxygen reduction reaction in enzymatic biofuel cells. *Nano Energy* **2021**, *83*, 105798. [CrossRef]
29. Jayapiriya, U.S.; Goel, S. Influence of cellulose separators in coin-sized 3D printed paper-based microbial fuel cells. *Sustain. Energy Technol. Assess.* **2021**, *47*, 101535. [CrossRef]
30. Yan, Y.; Hou, Y.; Yu, Z.; Tu, L.; Qin, S.; Lan, D.; Chen, S.; Sun, J.; Wang, S. B-doped graphene quantum dots implanted into bimetallic organic framework as a highly active and robust cathodic catalyst in the microbial fuel cell. *Chemosphere* **2022**, *286*, 131908. [CrossRef]
31. Kano, K.; Shirai, O.; Kitazumi, Y.; Sakai, K.; Xia, H.-Q. Applications to Biofuel Cells and Bioreactors. In *Enzymatic Bioelectrocatalysis;* Kano, K., Shirai, O., Kitazumi, Y., Sakai, K., Xia, H.-Q., Eds.; Springer: Singapore, 2021; pp. 115–131. [CrossRef]
32. Khan, S.; Kim, J.; Acharya, S.; Kim, W. Review on the operation of wearable sensors through body heat harvesting based on thermoelectric devices. *Appl. Phys. Lett.* **2021**, *118*, 200501. [CrossRef]
33. Li, X.; Cai, K.; Gao, M.; Du, Y.; Shen, S. Recent advances in flexible thermoelectric films and devices. *Nano Energy* **2021**, *89*, 106309. [CrossRef]
34. Soleimani, Z.; Zoras, S.; Ceranic, B.; Cui, Y.; Shahzad, S. A comprehensive review on the output voltage/power of wearable thermoelectric generators concerning their geometry and thermoelectric materials. *Nano Energy* **2021**, *89*, 106325. [CrossRef]
35. Abedlhafd, M.; Izman, S.; Noor, A.; Basheer, U.; Rajoo, S. A review of thermoelectric p-type $Ca_3Co_4O_9$ nanostructured ceramics for exhaust energy recovery. In Proceedings of the 2nd International Symposium on Engine Boosting and Energy Recovery, Kuala Lumpur, Malaysia, 11–13 September 2017.
36. Surmenev, R.A.; Chernozem, R.V.; Pariy, I.O.; Surmeneva, M.A. A review on piezo- and pyroelectric responses of flexible nano- and micropatterned polymer surfaces for biomedical sensing and energy harvesting applications. *Nano Energy* **2021**, *79*, 105442. [CrossRef]
37. Kishore, R.; Priya, S. A Review on Low-Grade Thermal Energy Harvesting: Materials, Methods and Devices. *Materials* **2018**, *11*, 1433. [CrossRef] [PubMed]
38. Hasan, M.N.; Wahid, H.; Nayan, N.; Mohamed Ali, M.S. Inorganic thermoelectric materials: A review. *Int. J. Energy Res.* **2020**, *44*, 6170–6222. [CrossRef]
39. Caballero-Calero, O.; Ares, J.; Martín-González, M. Environmentally Friendly Thermoelectric Materials: High Performance from Inorganic Components with Low Toxicity and Abundance in the Earth. *Adv. Sustain. Syst.* **2021**, *5*, 2100095. [CrossRef]
40. Li, H.; Koh, C.S.L.; Lee, Y.H.; Zhang, Y.; Phan-Quang, G.C.; Zhu, C.; Liu, Z.; Chen, Z.; Sim, H.Y.F.; Lay, C.L.; et al. A wearable solar-thermal-pyroelectric harvester: Achieving high power output using modified rGO-PEI and polarized PVDF. *Nano Energy* **2020**, *73*, 104723. [CrossRef]
41. Klochko, N.P.; Barbash, V.A.; Petrushenko, S.I.; Kopach, V.R.; Klepikova, K.S.; Zhadan, D.O.; Yashchenko, O.V.; Dukarov, S.V.; Sukhov, V.M.; Khrypunova, A.L. Thermoelectric textile devices with thin films of nanocellulose and copper iodide. *J. Mater. Sci. Mater. Electron.* **2021**, *32*, 23246–23265. [CrossRef]
42. Bai, C.; Wang, Z.; Yang, S.; Cui, X.; Li, X.; Yin, Y.; Zhang, M.; Wang, T.; Sang, S.; Zhang, W.; et al. Wearable Electronics Based on the Gel Thermogalvanic Electrolyte for Self-Powered Human Health Monitoring. *ACS Appl. Mater. Interfaces* **2021**, *13*, 37316–37322. [CrossRef]
43. Rösch, A.G.; Gall, A.; Aslan, S.; Hecht, M.; Franke, L.; Mallick, M.M.; Penth, L.; Bahro, D.; Friderich, D.; Lemmer, U. Fully printed origami thermoelectric generators for energy-harvesting. *Npj Flex. Electron.* **2021**, *5*, 1. [CrossRef]
44. Ren, W.; Sun, Y.; Zhao, D.; Aili, A.; Zhang, S.; Shi, C.; Zhang, J.; Geng, H.; Zhang, J.; Zhang, L.; et al. High-performance wearable thermoelectric generator with self-healing, recycling, and Lego-like reconfiguring capabilities. *Sci. Adv.* **2021**, *7*, eabe0586. [CrossRef] [PubMed]
45. Liu, Z.; Wang, X.; Wei, S.; Lv, H.; Zhou, J.; Peng, P.; Wang, H.; Chen, G. A Wavy-Structured Highly Stretchable Thermoelectric Generator with Stable Energy Output and Self-Rescuing Capability. *CCS Chem.* **2021**, *3*, 2404–2414. [CrossRef]
46. Wu, Z.X.; Wang, B.L.; Hou, S.H.; Zheng, L. Degeneration of power output of a flexible and wearable thermoelectric module under bending fatigue. *Mech. Mater.* **2021**, *161*, 104027. [CrossRef]

47. Mahanty, B.; Ghosh, S.K.; Maity, K.; Roy, K.; Sarkar, S.; Mandal, D. All-fiber pyro- and piezo-electric nanogenerator for IoT based self-powered health-care monitoring. *Mater. Adv.* **2021**, *2*, 4370–4379. [CrossRef]
48. Afroz, A.S.; Romano, D.; Inglese, F.; Stefanini, C. Towards Bio-Hybrid Energy Harvesting in the Real-World: Pushing the Boundaries of Technologies and Strategies Using Bio-Electrochemical and Bio-Mechanical Processes. *Appl. Sci.* **2021**, *11*, 2220. [CrossRef]
49. Zou, Y.; Bo, L.; Li, Z. Recent progress in human body energy harvesting for smart bioelectronic system. *Fundam. Res.* **2021**, *1*, 364–382. [CrossRef]
50. Li, J.; Long, Y.; Yang, F.; Wang, X. Respiration driven triboelectric nanogenerators for biomedical applications. *EcoMat* **2020**, *2*, e12045. [CrossRef]
51. Pourshaban, E.; Karkhanis, M.U.; Deshpande, A.; Banerjee, A.; Ghosh, C.; Kim, H.; Mastrangelo, C.H. A Magnetically-Coupled Micromachined Electrostatic Energy Harvester Driven by Eye Blinking Motion. In Proceedings of the 2021 21st International Conference on Solid-State Sensors, Actuators and Microsystems (Transducers), Orlando, FL, USA, 20–24 June 2021; pp. 960–963.
52. Pu, X.; An, S.; Tang, Q.; Guo, H.; Hu, C. Wearable triboelectric sensors for biomedical monitoring and human-machine interface. *iScience* **2021**, *24*, 102027. [CrossRef] [PubMed]
53. Ruiz, L.L.; Lin, V.; Fitzgerald, L.; Zhu, J.; Borish, L.; Quinn, D.; Lach, J. Piezoelectric-Based Respiratory Monitoring: Towards Self-Powered Implantables for the Airways. In Proceedings of the 2021 IEEE 17th International Conference on Wearable and Implantable Body Sensor Networks (BSN), Athens, Greece, 27–30 July 2021; pp. 1–5.
54. Khalid, S.; Raouf, I.; Khan, A.; Kim, N.; Kim, H.S. A Review of Human-Powered Energy Harvesting for Smart Electronics: Recent Progress and Challenges. *Int. J. Precis. Eng. Manuf.-Green Technol.* **2019**, *6*, 821–851. [CrossRef]
55. Beyaz, M.İ. An acoustic blood pressure sensing scheme using time of flight and shear wave elastography techniques. *Sens. Actuators A Phys.* **2021**, *330*, 112865. [CrossRef]
56. Chen, Y.; Yang, Y.; Li, M.; Chen, E.; Mu, W.; Fisher, R.; Yin, R. Wearable Actuators: An Overview. *Textiles* **2021**, *1*, 283–321. [CrossRef]
57. Mahapatra, S.D.; Mohapatra, P.C.; Aria, A.I.; Christie, G.; Mishra, Y.K.; Hofmann, S.; Thakur, V.K. Piezoelectric Materials for Energy Harvesting and Sensing Applications: Roadmap for Future Smart Materials. *Adv. Sci.* **2021**, *8*, 2100864. [CrossRef] [PubMed]
58. Veronica, A.; Hsing, I.m. An Insight into Tunable Innate Piezoelectricity of Silk for Green Bioelectronics. *ChemPhysChem* **2021**, *22*, 2266–2280. [CrossRef] [PubMed]
59. Nagaraj, M.J.; Shantha, V.; Nishanth, N.; Parthsarathy, V. Study and Optimization of Piezoelectric Materials for MEMS Biochemical Sensor Applications. In *Advances in Renewable Energy and Electric Vehicles*; Sanjeevikumar, P., Nagesh, p., Suryanarayana, K., Eds.; Springer: Singapore, 2022; Volume 767, pp. 419–425.
60. Wang, D.-H.; Peng, Y.-H.; Tang, L.-K.; Yu, H.-Q. A multi-chamber piezoelectric pump based on pumping unit with double circular piezoelectric unimorph actuators. *Smart Mater. Struct.* **2021**, *30*, 095023. [CrossRef]
61. Tong, Z.; Hu, H.; Wu, Z.; Xie, S.; Chen, G.; Zhang, S.; Lou, L.; Liu, H. An Ultrasonic Proximity Sensing Skin for Robot Safety Control by Using Piezoelectric Micromachined Ultrasonic Transducers (PMUTs). *IEEE Sens. J.* **2021**, *22*, 17351–17361. [CrossRef]
62. Vijayakanth, T.; Liptrot, D.J.; Gazit, E.; Boomishankar, R.; Bowen, C.R. Recent Advances in Organic and Organic–Inorganic Hybrid Materials for Piezoelectric Mechanical Energy Harvesting. *Adv. Funct. Mater.* **2022**, *32*, 2109492. [CrossRef]
63. Lay, R.; Deijs, G.S.; Malmström, J. The intrinsic piezoelectric properties of materials—A review with a focus on biological materials. *RSC Adv.* **2021**, *11*, 3657–3673. [CrossRef] [PubMed]
64. Sultana, A.; Alam, M.M.; Middya, T.R.; Mandal, D. A pyroelectric generator as a self-powered temperature sensor for sustainable thermal energy harvesting from waste heat and human body heat. *Appl. Energy* **2018**, *221*, 299–307. [CrossRef]
65. Koroglu, L.; Ayas, E.; Ay, N. 3D Printing of Polyvinylidene Fluoride Based Piezoelectric Nanocomposites: An Overview. *Macromol. Mater. Eng.* **2021**, *306*, 2100277. [CrossRef]
66. Montero, K.L.; Laurila, M.M.; Mäntysalo, M. Fully Printed Unobtrusive and Skin-conformable Piezoelectric Energy Harvester. In Proceedings of the 2021 IEEE International Conference on Flexible and Printable Sensors and Systems (FLEPS), Manchester, UK, 20–23 June 2021; pp. 1–4.
67. Zhang, J.; Yu, X.; Zhao, W.; Qu, D. A piezoelectric vibration energy harvester based on the reverse-rhombus double-bridge force amplification frame. *J. Phys. D Appl. Phys.* **2021**, *54*, 365501. [CrossRef]
68. Kumar, N.; Mahale, B.; Muzata, T.S.; Ranjan, R. Energy harvesting with flexible piezocomposite fabricated from a biodegradable polymer. *Int. J. Energy Res.* **2021**, *45*, 19395–19404. [CrossRef]
69. Sun, J.; Guo, H.; Ribera, J.; Wu, C.; Tu, K.; Binelli, M.; Panzarasa, G.; Schwarze, F.W.M.R.; Wang, Z.L.; Burgert, I. Sustainable and Biodegradable Wood Sponge Piezoelectric Nanogenerator for Sensing and Energy Harvesting Applications. *ACS Nano* **2020**, *14*, 14665–14674. [CrossRef] [PubMed]
70. Gu, L.; Liu, J.; Cui, N.; Xu, Q.; Du, T.; Zhang, L.; Wang, Z.; Long, C.; Qin, Y. Enhancing the current density of a piezoelectric nanogenerator using a three-dimensional intercalation electrode. *Nat. Commun.* **2020**, *11*, 1030. [CrossRef] [PubMed]
71. Xu, Z.; Jin, C.; Cabe, A.; Escobedo, D.; Hao, N.; Trase, I.; Closson, A.B.; Dong, L.; Nie, Y.; Elliott, J.; et al. Flexible Energy Harvester on a Pacemaker Lead Using Multibeam Piezoelectric Composite Thin Films. *ACS Appl. Mater. Interfaces* **2020**, *12*, 34170–34179. [CrossRef]

72. Thainiramit, P.; Yingyong, P.; Isarakorn, D. Impact-Driven Energy Harvesting: Piezoelectric Versus Triboelectric Energy Harvesters. *Sensors* **2020**, *20*, 5828. [CrossRef] [PubMed]
73. Song, Y.; Wang, N.; Hu, C.; Wang, Z.L.; Yang, Y. Soft triboelectric nanogenerators for mechanical energy scavenging and self-powered sensors. *Nano Energy* **2021**, *84*, 105919. [CrossRef]
74. Zou, Y.; Raveendran, V.; Chen, J. Wearable triboelectric nanogenerators for biomechanical energy harvesting. *Nano Energy* **2020**, *77*, 105303. [CrossRef]
75. Pharino, U.; Sinsanong, Y.; Pongampai, S.; Charoonsuk, T.; Pakawanit, P.; Sriphan, S.; Vittayakorn, N.; Vittayakorn, W. Influence of pore morphologies on the mechanical and tribo-electrical performance of polydimethylsiloxane sponge fabricated via commercial seasoning templates. *Radiat. Phys. Chem.* **2021**, *189*, 109720. [CrossRef]
76. Ahmed, A.; Hassan, I.; Helal, A.S.; Sencadas, V.; Radhi, A.; Jeong, C.K.; El-Kady, M.F. Triboelectric Nanogenerator versus Piezoelectric Generator at Low Frequency (<4 Hz): A Quantitative Comparison. *iScience* **2020**, *23*, 101286. [CrossRef] [PubMed]
77. Zhang, R.; Olin, H. Material choices for triboelectric nanogenerators: A critical review. *EcoMat* **2020**, *2*, e12062. [CrossRef]
78. Kim, D.W.; Lee, J.H.; Kim, J.K.; Jeong, U. Material aspects of triboelectric energy generation and sensors. *NPG Asia Mater.* **2020**, *12*, 6. [CrossRef]
79. Zhang, Z.; Jie, Y.; Zhu, J.; Zhu, Z.; Chen, H.; Lu, Q.; Zeng, Y.; Cao, X.; Wang, N.; Wang, Z. Paper triboelectric nanogenerator designed for continuous reuse and quick construction. *Nano Res.* **2021**, *15*, 1109–1114. [CrossRef]
80. Jiang, J.; Guan, Q.; Liu, Y.; Sun, X.; Wen, Z. Abrasion and Fracture Self-Healable Triboelectric Nanogenerator with Ultrahigh Stretchability and Long-Term Durability. *Adv. Funct. Mater.* **2021**, *31*, 2105380. [CrossRef]
81. Wu, M.; Gao, Z.; Yao, K.; Hou, S.; Liu, Y.; Li, D.; He, J.; Huang, X.; Song, E.; Yu, J.; et al. Thin, soft, skin-integrated foam-based triboelectric nanogenerators for tactile sensing and energy harvesting. *Mater. Today Energy* **2021**, *20*, 100657. [CrossRef]
82. Guan, X.; Xu, B.; Huang, J.; Jing, T.; Gao, Y. Fiber-shaped stretchable triboelectric nanogenerator with a novel synergistic structure of opposite Poisson's ratios. *Chem. Eng. J.* **2022**, *427*, 131698. [CrossRef]
83. Liao, W.; Liu, X.; Li, Y.; Xu, X.; Jiang, J.; Lu, S.; Bao, D.; Wen, Z.; Sun, X. Transparent, stretchable, temperature-stable and self-healing ionogel-based triboelectric nanogenerator for biomechanical energy collection. *Nano Res.* **2021**, *15*, 2060–2068. [CrossRef]
84. Sheng, F.; Yi, J.; Shen, S.; Cheng, R.; Ning, C.; Ma, L.; Peng, X.; Deng, W.; Dong, K.; Wang, Z.L. Self-Powered Smart Arm Training Band Sensor Based on Extremely Stretchable Hydrogel Conductors. *ACS Appl. Mater. Interfaces* **2021**, *13*, 44868–44877. [CrossRef] [PubMed]
85. Yang, Y.; Chen, L.; He, J.; Hou, X.; Qiao, X.; Xiong, J.; Chou, X. Flexible and Extendable Honeycomb-Shaped Triboelectric Nanogenerator for Effective Human Motion Energy Harvesting and Biomechanical Sensing. *Adv. Mater. Technol.* **2021**, *7*, 2100702. [CrossRef]
86. Ryu, H.; Park, H.-m.; Kim, M.-K.; Kim, B.; Myoung, H.S.; Kim, T.Y.; Yoon, H.-J.; Kwak, S.S.; Kim, J.; Hwang, T.H.; et al. Self-rechargeable cardiac pacemaker system with triboelectric nanogenerators. *Nat. Commun.* **2021**, *12*, 4374. [CrossRef] [PubMed]
87. Iqbal, M.; Nauman, M.M.; Khan, F.U.; Abas, P.E.; Cheok, Q.; Iqbal, A.; Aissa, B. Multimodal Hybrid Piezoelectric-Electromagnetic Insole Energy Harvester Using PVDF Generators. *Electronics* **2020**, *9*, 635. [CrossRef]
88. Li, Z.; Luo, J.; Xie, S.; Xin, L.; Guo, H.; Pu, H.; Yin, P.; Xu, Z.; Zhang, D.; Peng, Y.; et al. Instantaneous peak 2.1 W-level hybrid energy harvesting from human motions for self-charging battery-powered electronics. *Nano Energy* **2021**, *81*, 105629. [CrossRef]
89. Zhou, W.; He, J.; Peng, P.; Chen, L.; Cao, K. Reliability of Microelectromechanical Systems Devices. In *Reliability and Maintenance: An Overview of Cases*; IntechOpen: London, UK, 2020.
90. Jiang, D.; Ouyang, H.; Shi, B.; Zou, Y.; Tan, P.; Qu, X.; Chao, S.; Xi, Y.; Zhao, C.; Fan, Y.; et al. A wearable noncontact free-rotating hybrid nanogenerator for self-powered electronics. *InfoMat* **2020**, *2*, 1191–1200. [CrossRef]
91. Li, Y.; Chen, Z.; Zheng, G.; Zhong, W.; Jiang, L.; Yang, Y.; Jiang, L.; Chen, Y.; Wong, C.-P. A magnetized microneedle-array based flexible triboelectric-electromagnetic hybrid generator for human motion monitoring. *Nano Energy* **2020**, *69*, 104415. [CrossRef]
92. Lai, Y.-C.; Lu, H.-W.; Wu, H.-M.; Zhang, D.; Yang, J.; Ma, J.; Shamsi, M.; Vallem, V.; Dickey, M.D. Elastic Multifunctional Liquid–Metal Fibers for Harvesting Mechanical and Electromagnetic Energy and as Self-Powered Sensors. *Adv. Energy Mater.* **2021**, *11*, 2100411. [CrossRef]
93. Hsu, T.-H.; Manakasettharn, S.; Taylor, J.A.; Krupenkin, T. Bubbler: A Novel Ultra-High Power Density Energy Harvesting Method Based on Reverse Electrowetting. *Sci. Rep.* **2015**, *5*, 16537. [CrossRef] [PubMed]
94. Wu, Z.; Cao, Z.; Ding, R.; Wang, S.; Chu, Y.; Ye, X. An electrostatic-electromagnetic hybrid generator with largely enhanced energy conversion efficiency. *Nano Energy* **2021**, *89*, 106425. [CrossRef]
95. Boisseau, S.; Despesse, G.; Seddik, B. Electrostatic Conversion for Vibration Energy Harvesting. In *Small-Scale Energy Harvesting*; BoD–Books: Norderstedt, Germany, 2012. [CrossRef]
96. Pourshaban, E.; Karkhanis, M.U.; Deshpande, A.; Banerjee, A.; Ghosh, C.; Kim, H.; Mastrangelo, C.H. Flexible Electrostatic Energy Harvester Driven by Cyclic Eye Tear Wetting and Dewetting. In Proceedings of the 2021 IEEE International Conference on Flexible and Printable Sensors and Systems (FLEPS), Manchester, UK, 20–23 June 2021; pp. 1–4.
97. Erturun, U.; Eisape, A.A.; Kang, S.H.; West, J.E. Energy harvester using piezoelectric nanogenerator and electrostatic generator. *Appl. Phys. Lett.* **2021**, *118*, 063902. [CrossRef]

98. Cao, Y.; Nie, D.; Zhang, J.; Wang, Y.; He, R.; Zhao, Z.; Wu, J.; Ji, B.; Xie, J.; Tao, K. Investigation of electrostatic-piezoelectric hybrid vibrational power generators with different frequency broadening schemes. In Proceedings of the 2021 IEEE 16th International Conference on Nano/Micro Engineered and Molecular Systems (NEMS), Xiamen, China, 25–29 April 2021; pp. 528–532.
99. Boroujeni, F.G.; Raissi, B.; Jafarabadi-Ashtiani, S.; Riahifar, R.; Sahba-Yaghmaee, M. Droplet-based energy harvester considering electrowetting phenomena. *Eng. Res. Express* **2020**, *2*, 045028. [CrossRef]
100. Wu, H.; Mendel, N.; van der Ham, S.; Shui, L.; Zhou, G.; Mugele, F. Charge Trapping-Based Electricity Generator (CTEG): An Ultrarobust and High Efficiency Nanogenerator for Energy Harvesting from Water Droplets. *Adv. Mater.* **2020**, *32*, 2001699. [CrossRef]
101. Adhikari, P.R.; Tasneem, N.T.; Reid, R.C.; Mahbub, I. Electrode and electrolyte configurations for low frequency motion energy harvesting based on reverse electrowetting. *Sci. Rep.* **2021**, *11*, 5030. [CrossRef] [PubMed]
102. Srinivasan, V.; Pamula, V.K.; Fair, R.B. An integrated digital microfluidic lab-on-a-chip for clinical diagnostics on human physiological fluids. *Lab. A Chip* **2004**, *4*, 310–315. [CrossRef] [PubMed]
103. Yoon, J.-Y.; Garrell, R.L. Preventing Biomolecular Adsorption in Electrowetting-Based Biofluidic Chips. *Anal. Chem.* **2003**, *75*, 5097–5102. [CrossRef] [PubMed]
104. Tasneem, N.T.; Biswas, D.K.; Mahbub, I.; Adhikari, P.R.; Reid, R. Self-Powered Motion Tracking Sensor Integrated with Low-Power CMOS Circuitry. In Proceedings of the 2021 IEEE International Symposium on Circuits and Systems (ISCAS), Daegu, Korea, 22–28 May 2021; pp. 1–5.
105. Zhang, X.; Shiu, B.C.; Li, T.-T.; Liu, X.; Ren, H.-T.; Wang, Y.; Lou, C.-W.; Lin, J.-H. Photo-thermoelectric nanofiber film based on the synergy of conjugated polymer and light traps for the solar-energy harvesting of textile solar panel. *Sol. Energy Mater. Sol. Cells* **2021**, *232*, 111353. [CrossRef]
106. Singh, J.; Kaur, R.; Singh, D. Energy harvesting in wireless sensor networks: A taxonomic survey. *Int. J. Energy Res.* **2020**, *45*, 118–140. [CrossRef]
107. Xie, L.; Song, W.; Ge, J.; Tang, B.; Zhang, X.; Wu, T.; Ge, Z. Recent progress of organic photovoltaics for indoor energy harvesting. *Nano Energy* **2021**, *82*, 105770. [CrossRef]
108. Zhang, X.; Shiu, B.C.; Li, T.-T.; Liu, X.; Ren, H.-T.; Wang, Y.; Lou, C.-W.; Lin, J.-H. Synergistic work of photo-thermoelectric and hydroelectric effects of hierarchical structure photo-thermoelectric textile for solar energy harvesting and solar steam generation simultaneously. *Chem. Eng. J.* **2021**, *426*, 131923. [CrossRef]
109. Yu, H.; Xiang, S.; Tao, H.; Xue, J.; Tao, C.; Li, C.; Zhang, N.; Fan, X. Embroidering a Light and Foldable Photovoltaic Gauze Kerchiefs. *Energy Technol.* **2021**, *9*, 2100285. [CrossRef]
110. Zhang, Z.; Li, X.; Yin, J.; Xu, Y.; Fei, W.; Xue, M.; Wang, Q.; Zhou, J.; Guo, W. Emerging hydrovoltaic technology. *Nat. Nanotechnol.* **2018**, *13*, 1109–1119. [CrossRef]
111. Yin, J.; Zhou, J.; Fang, S.; Guo, W. Hydrovoltaic Energy on the Way. *Joule* **2020**, *4*, 1852–1855. [CrossRef]
112. Tan, J.; Fang, S.; Zhang, Z.; Yin, J.; Li, L.; Wang, X.; Guo, W. Self-sustained electricity generator driven by the compatible integration of ambient moisture adsorption and evaporation. *Nat. Commun.* **2022**, *13*, 3643. [CrossRef] [PubMed]
113. Ganesh, K.V.S.S.; Ruhan Bevi, A. Low-Cost Pulse Oximeter & Heart Rate Measurement for COVID Diagnosis. *J. Phys. Conf. Ser.* **2021**, *1964*, 62035. [CrossRef]
114. Michard, F.; Shelley, K.; L'Her, E. COVID-19: Pulse oximeters in the spotlight. *J. Clin. Monit. Comput.* **2020**, *35*, 11–14. [CrossRef] [PubMed]
115. Akhtaruzzaman, A.K.M.; Kamal, M.M.; Parveen, M.; Rabbi, M.; Dhali, R.; Islam, M.S.; Bhowmick, D.K. Remote monitoring of COVID–19 patients using home pulse oximetry and virtual platform: An observational study. *Anaesth. Pain Intensive Care* **2022**, *26*, 89–95. [CrossRef]
116. Yoon, J.-Y. Spectrophotometry and Optical Biosensor. In *Introduction to Biosensors*; Springer: New York, NY, USA, 2013; pp. 121–139. [CrossRef]
117. Elsamnah, F.; Bilgaiyan, A.; Affiq, M.; Shim, C.H.; Ishidai, H.; Hattori, R. Comparative Design Study for Power Reduction in Organic Optoelectronic Pulse Meter Sensor. *Biosensors* **2019**, *9*, 48. [CrossRef] [PubMed]
118. Kossaify, A. Sensing and Detection Functions in Implantable Cardioverter Defibrillators: The Good, the Bad and the Ugly. *Acta Cardiol. Sin.* **2020**, *36*, 308–317. [CrossRef]
119. Mulpuru, S.K.; Madhavan, M.; McLeod, C.J.; Cha, Y.-M.; Friedman, P.A. Cardiac Pacemakers: Function, Troubleshooting, and Management: Part 1 of a 2-Part Series. *J. Am. Coll. Cardiol.* **2017**, *69*, 189–210. [CrossRef]
120. Moerke, C.; Wolff, A.; Ince, H.; Ortak, J.; Öner, A. New strategies for energy supply of cardiac implantable devices. *Herzschrittmachertherapie Elektrophysiologie* **2022**, *33*, 224–231. [CrossRef]
121. Dong, L.; Han, X.; Xu, Z.; Closson, A.; Liu, Y.; Wen, C.; Liu, X.; Escobar, G.; Oglesby, M.; Feldman, M.; et al. Flexible Porous Piezoelectric Cantilever on a Pacemaker Lead for Compact Energy Harvesting. *Adv. Mater. Technol.* **2018**, *4*, 1800148. [CrossRef]
122. Ammannaya, G.K.K. Implantable cardioverter defibrillators-the past, present and future. *Arch. Med. Sci. Atheroscler. Dis.* **2020**, *5*, e163–e170. [CrossRef]
123. Dong, L.; Closson, A.; Jin, C.; Nie, Y.; Cabe, A.; Escobedo, D.; Huang, S.; Trase, I.; Xu, Z.; Chen, Z.; et al. Multifunctional Pacemaker Lead for Cardiac Energy Harvesting and Pressure Sensing. *Adv. Healthc. Mater.* **2020**, *9*, 53. [CrossRef] [PubMed]
124. Rodrigues, D.; Barbosa, A.I.; Rebelo, R.; Kwon, I.K.; Reis, R.L.; Correlo, V.M. Skin-Integrated Wearable Systems and Implantable Biosensors: A Comprehensive Review. *Biosensors* **2020**, *10*, 79. [CrossRef] [PubMed]

125. Zan, G.; Wu, T.; Zhu, F.; He, P.; Cheng, Y.; Chai, S.; Wang, Y.; Huang, X.; Zhang, W.; Wan, Y.; et al. A biomimetic conductive super-foldable material. *Matter* **2021**, *4*, 3232–3247. [CrossRef]
126. Zan, G.; Wu, T.; Zhang, Z.; Li, J.; Zhou, J.; Zhu, F.; Chen, H.; Wen, M.; Yang, X.; Peng, X.; et al. Bioinspired Nanocomposites with Self-Adaptive Stress Dispersion for Super-Foldable Electrodes. *Adv. Sci.* **2022**, *9*, 2103714. [CrossRef] [PubMed]
127. Zan, G.; Wu, T.; Dong, W.; Zhou, J.; Tu, T.; Xu, R.; Chen, Y.; Wang, Y.; Wu, Q. Two-Level Biomimetic Designs Enable Intelligent Stress Dispersion for Super-Foldable C/NiS Nanofiber Free-Standing Electrode. *Adv. Fiber Mater.* **2022**, *4*, 1177–1190. [CrossRef]
128. Bonnaud, O. The technological challenges of microelectronics for the next generations of connected sensors. *Int. J. Plasma Environ. Sci. Technol.* **2020**, *14*, e01002.
129. Singh, A.; Sharma, A.; Ahmed, A.; Sundramoorthy, A.K.; Furukawa, H.; Arya, S.; Khosla, A. Recent Advances in Electrochemical Biosensors: Applications, Challenges, and Future Scope. *Biosensors* **2021**, *11*, 336. [CrossRef]

energies

MDPI

Review

Modelling of Flow-Induced Vibration of Bluff Bodies: A Comprehensive Survey and Future Prospects

Ying Wu, Zhi Cheng, Ryley McConkey, Fue-Sang Lien * and Eugene Yee

Mechanical and Mechatronics Engineering, University of Waterloo, 200 University Avenue West, Waterloo, ON N2L 3G1, Canada
* Correspondence: fue-sang.lien@uwaterloo.ca

Abstract: A comprehensive review of modelling techniques for the flow-induced vibration (FIV) of bluff bodies is presented. This phenomenology involves bidirectional fluid–structure interaction (FSI) coupled with non-linear dynamics. In addition to experimental investigations of this phenomenon in wind tunnels and water channels, a number of modelling methodologies have become important in the study of various aspects of the FIV response of bluff bodies. This paper reviews three different approaches for the modelling of FIV phenomenology. Firstly, we consider the mathematical (semi-analytical) modelling of various types of FIV responses: namely, vortex-induced vibration (VIV), galloping, and combined VIV-galloping. Secondly, the conventional numerical modelling of FIV phenomenology involving various computational fluid dynamics (CFD) methodologies is described, namely: direct numerical simulation (DNS), large-eddy simulation (LES), detached-eddy simulation (DES), and Reynolds-averaged Navier–Stokes (RANS) modelling. Emergent machine learning (ML) approaches based on the data-driven methods to model FIV phenomenology are also reviewed (e.g., reduced-order modelling and application of deep neural networks). Following on from this survey of different modelling approaches to address the FIV problem, the application of these approaches to a fluid energy harvesting problem is described in order to highlight these various modelling techniques for the prediction of FIV phenomenon for this problem. Finally, the critical challenges and future directions for conventional and data-driven approaches are discussed. So, in summary, we review the key prevailing trends in the modelling and prediction of the full spectrum of FIV phenomena (e.g., VIV, galloping, VIV-galloping), provide a discussion of the current state of the field, present the current capabilities and limitations and recommend future work to address these limitations (knowledge gaps).

Keywords: flow-induced vibration; mathematical modelling; numerical modelling; machine learning techniques; fluid energy harvesting

1. Introduction

Flow-induced vibration (FIV) of bluff body structures is a classical bidirectional flow–structure interaction problem, which is linked to various fluid dynamics phenomena (e.g., boundary-layer separation, vortex formation and shedding, hydrodynamic loading on the structures) as well as structure vibrations. These phenomena occur widely in various structures (e.g., bridges, transmission lines, marine cables, riser pipes) exposed to wind, tidal waves, or river flow. As a consequence, FIV phenomena have attracted increasing attention over the past few decades owing to complex physics and important practical value [1].

Flow-induced vibration involves the strong interaction between flow and structure. This interaction consists of highly non-linear dynamical phenomena which pose a great challenge for its modelling. FIV itself is also a multi-parameter problem, involving a number of flow properties (e.g., Reynolds number Re, Strouhal number S_t, inflow speed, turbulent intensity), the geometry of the moving body (e.g., cross-sectional shape), and the elasticity of the supporting structure (e.g., mass ratio, damping ratio, spring stiffness). Furthermore, the complete analysis of FIV phenomena must include not only the oscillatory (vibrational) motion of the body (e.g,. displacement, frequency, phase angle), but also the complex dynamics of the flow past the body responsible for this motion (e.g., vortex mode, wake evolution, boundary-layer separation and reattachment). Needless to say, FIV is a very complex physical phenomenon that poses a grand challenge for modelling.

Flow-induced vibration is a general term encompassing various structure oscillations stimulated by the fluid flow, such as vortex-induced vibration, galloping, flutter, buffeting, and wake galloping. Generally speaking, the inducing mechanisms and the response characteristics of these FIV phenomena are distinctive. A key component of research in FIV concerns the underlying mechanisms leading to the structural oscillations—the mechanisms facilitate the identification of the type of FIV and, as a result, lead to improvements in our current understanding of this important physical phenomenon [2].

To study FIV of a bluff body, the most direct and effective approach is to undertake wind-tunnel or water-channel experimental investigations. However, these investigations need to be supplemented with the mathematical or numerical modelling of the phenomenon—this modelling effort provides the required insights and intuition into the physical mechanisms underpinning the observed FIV phenomenon. General modelling methods are based on physical theory or numerical investigations. The main difference between these two approaches concerns the dynamical behavior of fluid—the hydrodynamic forces acting on the structure is either represented (approximated) using a simplified dynamical system model (mathematical modelling) or numerically computed using various approximations of the Navier–Stokes (NS) equations (numerical modelling). In general, mathematical modelling is more efficient and plays a critical role in our understanding of the underlying physical nature of the FIV phenomenon. By contrast, numerical modelling can obtain more accurate predictions of the FIV response and provide much more detailed information of flow field, albeit at much greater computational cost especially for a fully three-dimensional (3D) numerical simulation. In addition to these more traditional approaches for modelling FIV phenomena, there are newly emerging (nascent) machine learning procedures that can potentially be utilized to address the FIV problem. Indeed, ML has been applied recently in the context of fluid dynamics to address a number of different problems of engineering and practical interest [3–5].

In terms of practical applications, the suppression of FIV in structures in order to mitigate structural damage has been the dominant area of research. However, in light of an emerging trend toward sustainability goals for clean energy generation in recent years, there is increasing interest in the utilization of FIV phenomena for clean energy production, especially as it relates to the development of small-scale (portable) wind generators—indeed, the energy that can be harnessed from an oscillating body undergoing FIV can be effectively utilized and converted into usable forms of energy, such as electricity, by certain piezoelectric or electromagnetic materials [6–8]. In consequence, there is a growing body of research whose objective is to amplify and harness the oscillations of a bluff body immersed in a flow, rather than trying to suppress these oscillations (which is the focus of more traditional research to maintain structural integrity).

This review will focus on providing an extensive and systematic overview of traditional as well as novel (emerging) approaches for modelling FIV phenomena. To this purpose, we elaborate on how these different modelling methodologies can be applied to specific engineering applications of FIV phenomena which we believe will be of interest to both the general and specialist reader. The end goal is to provide a comprehensive survey of traditional (mathematical and numerical) and data-driven (machine-learning) approaches for addressing the modelling of FIV phenomena which we hope will serve as a reference point for both researchers and practitioners (both in and out of the field) not only by providing a categorization of the existing methods, but also providing recommendations for their applicability to various problems. Towards this objective, the application of FIV phenomena to energy harvesting is used as a representative application case to discuss how these various modelling methods can be implemented to address a specific engineering problem. Finally, an outlook focused on the challenges and future directions for research in FIV modelling is provided.

To reiterate, we believe that a comprehensive review of the principal prevailing trends in the modelling and prediction of all aspects of FIV phenomena (e.g., VIV, galloping, VIV-galloping) presented from an integrated and unified perspective will provide an important impetus to advance our current state of knowledge in this field. Indeed, we expect that this new clarity can improve our current understanding of FIV phenomena and enable robust progress—indeed, in this work, we present current limitations in the modelling of FIV phenomena and, based on these limitations, we propose various settings within the current modelling progress of FIV phenomena that are under-explored but are potentially critical for real-world applications of FIV (e.g., energy harvesting).

The paper is organized as follows. Section 2 provides the fundamental concepts and necessary background for the general reader to understand FIV phenomena which includes key definitions and the classification and the characteristics of critical response characteristics of FIV. Two conventional (traditional) modelling approaches based on mathematical (semi-analytical) modelling and numerical modelling (Navier–Stokes equations) are summarized in Sections 3 and 4. This is followed by a survey of the use of emerging artificial intelligence (or data-driven) approaches for modelling FIV phenomena in Section 5. Section 6 describes the application of these modelling approaches to an FIV energy harvester. Finally, the challenges and prospects of FIV modelling (and their applications to energy harvesting) and new research directions are discussed in Section 7.

2. Fundamental Concepts and Background of FIV

Flow-induced vibration refers to the alternating motion of a structure due to the aerodynamic force exerted by a moving fluid over the body surface. The flow regime that results is significantly affected by this structural motion—in consequence, FIV is a classical bidirectional FSI problem.

In 1977, Blevins [9] first coined the term "FIV" and classified FIV mechanisms according to steady and unsteady flow conditions. In a steady flow, the interaction of the fluid and structure always has a fixed pattern which is predictable. On the other hand, an unsteady FSI can involve either random patterns dominated by transient forces (e.g., turbulence and sudden changes of the flow), or very regular patterns generated by a known excitation (e.g., forced vibration). Nakamura et al. [1] further identified two-phase flow where the changes of flow momentum and pressure over time are the major causes of the structural vibration. Among these flow regimes, the two-way interactive impact of fluid and structure in a steady single-phase flow is the most common case investigated, where the mechanisms responsible for FIV can be further classified into two general categories based on the source of the unsteady forces (viz., resonance-type and instability-type) [1].

The resonance-type structural oscillation is essentially a kind of forced vibration, which is typically driven by external oscillatory forces arising from multiple sources, such as vortex shedding behind the structure (VIV) and vibrations in the incoming (incident) flow (buffeting). When the fluctuation frequency of external excitation is very close to the natural frequency of the structure, a resonance response may occur resulting in a much larger amplitude in the oscillations. By comparison, if some initial unsteady flow-induced forces are applied to the body to make it move first, the moving body will induce a periodically hydrodynamic force on itself. The motion-induced force and oscillating system mutually reinforce each other until a dynamic equilibrium is reached—a phenomenon that is referred to as an instability-type oscillation. In other words, the structure undergoing an instability response requires an initial small disturbance to begin the motion and then this motion is gradually converted to significant periodic oscillations autonomously—this is similar to a self-excited motion with no upper limit in the amplitude of vibration. These representative instabilities include galloping and flutter [10]. These four common forms of FIV will be explained in detail below.

2.1. Vortex-Induced Vibration

As a resonance-type response, the well-known vortex-induced vibration is generated by unsteady aerodynamic forces exerted by the fluid on an immersed structure. When a viscous flow passes over an elastically-supported rigid body, a boundary layer forms and then separates from near the rear of the body. This action results in vortex shedding from either side of the body producing a von Kármán vortex street in the wake. This alternate vortex shedding from the back of the body causes an asymmetric pressure distribution on body surface giving rise to the fluid forces responsible for VIV of the bluff body [9].

"Lock-in" or synchronization occurs when the vortex-shedding frequency f_{vs} is approximately equal to the structural natural frequency f_n. This results in a resonant oscillation of the body that is characterized by a larger but, nevertheless, limited vibration amplitude. The phenomenon of lock-in occurs for a certain range of inflow (incident) wind speeds (cf. left panel of Figure 1). From these considerations, VIV is an inherently self-governed, self-regulated, and self-limiting phenomenon which can occur in one- or multi-degree-of-freedom dynamical systems. In addition to the "lock-in" phenomenon, VIV also exhibits various forms of stronger non-linear oscillatory behavior. These include resonance delay (where the amplitude of vibration reaches its maximum at a velocity greater than the resonance velocity $U_R \equiv f_n D/S_t$ where S_t is the Strouhal number and D is a characteristic length of the structure), hysteresis (producing different amplitude responses as the flow speed is increased or decreased), and a multi-valued response (where a given fixed velocity can result in multiple values for the vibration amplitude) [11].

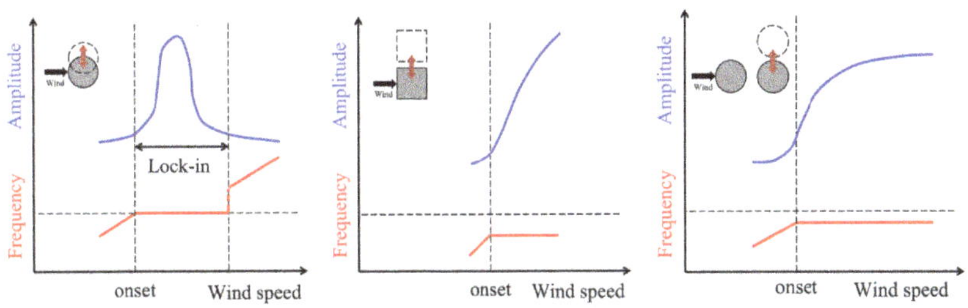

Figure 1. Comparison of amplitude-velocity and frequency-velocity of FIV phenomenon: VIV (**left**); instability like galloping and flutter (**middle**); and, wake galloping (**right**).

The dynamical characteristics of lock-in makes VIV an ideal choice for fluid energy harvesting. More specifically, a VIV-based energy converter exhibits a better power performance and a higher energy efficiency when the incoming (incident) wind speed lies in the synchronization range [7].

2.2. Buffeting

Buffeting is also a resonance-type response, but the resonance results from a fluctuating incoming (incident) flow rather than from a vortex-related instability as in the "lock-in" phenomenon. As an example, the fluctuations in the incident flow can arise from the natural atmospheric turbulence as is commonly observed in long-span bridges [12] or from the oscillating wake generated by structures upstream of a bluff body as in multi-body system—this phenomenon is also referred to as wake galloping [10] (cf. right panel of Figure 1). Investigations undertaken with regard to bridge structures and aeronautics have shown that a buffeting response may occur even at low wind speeds, accompanied by a smaller oscillation amplitude and a wider frequency range than that obtained from VIV [10].

These characteristics suggest that buffeting can be utilized to harness fluid energy in a multi-body design and the power performance from such a design is expected be strongly dependent on the precise layout (arrangement) of the various oscillating bodies.

2.3. Galloping

As a representative instability response, a body subjected to galloping will first undergo a very small oscillatory motion induced by an initial perturbation. These motions result subsequently in significant oscillations once a critical incident flow velocity is exceeded (cf. middle panel of Figure 1). Galloping can occur in the transverse direction for an elastically-mounted body or in torsion for an hinged body. Torsional galloping is a much more complex phenomenon in terms of the angle of attack, the angular displacement and velocity, the phase difference and other complex dynamics involving rotational motion. It is noted that torsional galloping is not as common as transverse galloping in the context of actual engineering applications [13]. In consequence, only transverse galloping will be discussed in this paper—for simplicity, transverse galloping will be referred to as galloping hereafter.

After the onset of galloping, the vibration amplitude will increase monotonically with the increasing flow velocity and will not come to rest again even at very large flow velocities. This is the most distinct difference between instability and resonance. Galloping is associated with a much lower vibration frequency than that of vortex shedding. The characteristics of a monotonically increasing amplitude with increasing flow velocity amplitude and a frequency that is not determined by any form of lock-in is what fundamentally distinguishes a galloping (or instability) phenomenon from a resonance (VIV) phenomenon.

In general, galloping is considered to have a greater energy potential in terms of energy harvesting than VIV owing to its much larger vibration amplitudes and extended range of wind speeds for which galloping occurs, once the critical velocity for onset is exceeded. Moreover, galloping occurs only for cylinders with specific cross-sectional shapes (e.g., square, rectangular, *D*-section) or for cylinder–appendage systems (e.g., cylinder with an attached splitter plate)—indeed, galloping does not occur for flow past a circular cylinder. As a consequence, an energy converter utilizing galloping will require a special geometrical design in order to allow the occurrence of galloping or of the interaction of VIV and galloping.

2.4. Flutter

Flutter is also a typical non-self-limited and self-sustained fluid instability usually applicable to dynamical systems involving two or more degree of freedoms. This phenomenon is closely related to the coupling of resonant bending and torsion deformation of a body, and has some common features with galloping.

Figure 1 compares the oscillation characteristics of VIV, galloping, flutter, and wake galloping in terms of the amplitude and frequency response. As shown, all of these oscillatory phenomena have a threshold flow speed for onset—the wind speed for VIV onset is lower than that for galloping and flutter. Only VIV exhibits a lock-in of frequency and a limited displacement—the other forms of flow-induced vibration are non-self-limiting with respect to the oscillation amplitude and are associated with a frequency lower than the natural structural frequency. As discussed previously, FIV exhibits a complex taxonomy—each category in the classification corresponds to a different generating mechanism and is associated with different response modes. Furthermore, the vibration of a given bluff body can exhibit a number of different resonances and/or instabilities [1].

3. Mathematical Modelling of FIV

Mathematical modelling of FIV involves the formulation of set of ordinary differential equations (ODEs) that describe the complex physical processes associated with this phenomenon. This set of ODEs is solved numerically when given appropriate initial and boundary conditions. The success of this modelling approach depends critically on an accurate and deep intuitive understanding of the physics underlying FIV phenomena. This is the key to formulating a good theoretical model that meets the following conditions [11]: (i) simplicity of formulation; (ii) inclusion of all essential (important) characteristics of the underlying physical phenomena; (iii) utilization of concepts and quantities of interest that are physically meaningful (interpretable and explainable); (iv) capability to provide accurate and universal predictions; (v) wide applicability especially to complex cases; and (vi) possibility for model to be further developed and generalized. It should be noted that frequently the terminology of mathematical modelling as it relates to FIV concerns the theoretical analysis of the phenomenon and, in this context, it is also referred as a theoretical or analytical model. Generally speaking, a mathematical model of FIV cannot be directly derived from first principles but needs to be supplemented with various semi-empirical relationships that are obtained from experimental investigations. From this perspective, a mathematical model for FIV can also be referred to as a semi-empirical or a phenomenological model.

3.1. Classification

Presently, there are two common methods used to classify mathematical models of FIV. One was proposed by Gabbai and Benaroya [14] and involves the following three classes of models: namely, a wake oscillator model, a single-degree-of-freedom (SDOF) model and a force decomposition model. The other classification scheme was proposed by Paidoussis et al. [15]—the scheme involves also three classes of models: namely, a forced system model, a fluid–elastic system model and a coupled system model. Based on this background, we present a new formalism for the classification of FIV models that we believe makes the ordering of FIV models more legible to researchers and practitioners and makes discussing new research directions easier. From the perspective of a bidirectional FSI problem, a structural oscillation can be modelled in terms of the motion displacement using a mass–spring–damper system. However, there are a number of different strategies that can be used the model the dynamics of the fluid (in the fluid–structure interaction). Depending on the modelling methodology used for the fluid and for the coupling between the fluid and the structure, mathematical models of FIV can be classified unambiguously into two main classes: namely, a coupled wake-oscillator model and a single degree-of-freedom model as exhibited in Figure 2.

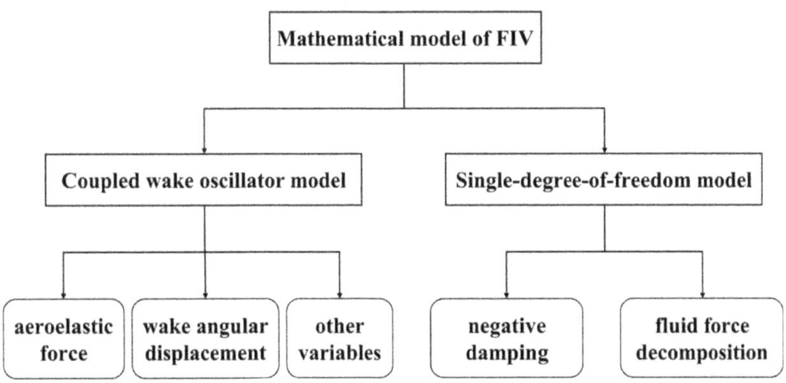

Figure 2. Proposed classification of mathematical modelling of FIV used to predict the flow-induced vibration of a bluff body.

The coupled wake-oscillator model involves the development of a fully non-linear wake model (also referred to as a wake- or fluid-oscillator model) using either a Rayleigh-type [16] or a van der Pol-type [17] oscillator. The model includes a negative damping term in order to represent the excitation of self-sustained oscillations that are characteristic of FIV. This model is then coupled to a structure equation through a forcing term that is dependent on the state of motion of the oscillatory system. As a consequence, the coupled wake-oscillator model is composed of two ordinary differential equations whose solution can be obtained either analytically or numerically in order to predict the FIV response. In contrast to the more intuitive representation of the structural motion using the readily interpretable concepts of displacement, velocity or acceleration of a moving body, the wake-oscillator model is more abstract in the sense that it is characterized by state variables that are more difficult to interpret. Finally, we note that the interactions between the various components of the coupled wake-oscillator model involve the following processes—the aerodynamic force acts on the structure whose motion determines the angular displacement of near-wake field generated by the structure, as well as other dynamical state variables describing the dynamics of the wake.

As the name implies, the single-degree-of-freedom model consists of only one ordinary differential equation with a forcing term that is used to describe the dynamic behavior of the structure. This model is applied primarily to predict the amplitude of the FIV response [18]. The basic strategy that underpins the SDOF model is the decomposition of the hydrodynamic force (viz., the forcing term) as well as the phenomenological representation of all the force components—the self-excited response is induced on the structure by these external forces. In accordance to the nature of the forcing term, the SDOF model can be divided into two sub-classes: namely, one that is based on a mathematical formulation of negative damping and another that is based on force data obtained from experiments.

The mathematical models of FIV mentioned above will be described in detail in Sections 3.2 and 3.3 in relation to their modelling of various FIV phenomena—VIV, galloping, and combined VIV-galloping.

3.2. *Vortex-Induced Vibration*

3.2.1. Coupled Wake-Oscillator Model

(1) Aerodynamic force coefficient.

All the available coupled wake-oscillator models that use a physical quantity related to the aerodynamic force as the fluid variable is a derivative of the seminal ideas of Bishop and Hassan [19]. These investigators were the first to advocate using a fluctuating lift force acting on a circular cylinder to model the dynamic behavior of the near-wake region. Figure 3 displays a schematization of this kind of model—here, the instantaneous displacement of the cylinder is $y(t)$ and the wake dynamics are characterized by a specific physical quantity $q(t)$ (e.g., oscillating fluid force $C_L(t)$). Following on from this effort, Hartlen and Currie [20] formulated a widely used Rayleigh-type wake oscillator in terms of the fluctuating lift coefficient C_L and coupled this model with a damped linear system through the cylinder velocity $\dot{y}$ in order to predict the transverse VIV of an elastically-mounted rigid circular cylinder. The dimensionless governing equations for this model are given by

$$\ddot{Y} + 2\zeta_c\dot{Y} + Y = m^*\nu^2 C_L\,, \tag{1}$$

$$\ddot{C_L} - \alpha\nu\left(1 - \frac{4}{3{C_{L0}}^2\nu^2}\dot{C_L}^2\right)\dot{C_L} + \nu^2 C_L = b\dot{Y}\,, \tag{2}$$

where the dot over a quantity refers to the derivative with respect to the dimensionless time $\tau \equiv \omega_n t$; $Y \equiv y/D$, $\dot{Y}$, and $\ddot{Y}$ are the normalized transverse displacement, velocity, and acceleration, respectively; $\nu \equiv \omega_{vs}/\omega_n$ is the ratio of the vortex-shedding frequency ω_{vs} to the natural structural frequency ω_n; $m^* \equiv \rho D^2/(8\pi^2 {S_t}^2 M_c)$ is the structural mass ratio where M_c and D are the cylinder mass per unit length and cylinder diameter, respectively; S_t is the Strouhal number; ρ is the fluid density; the structural damping ratio $\zeta_c \equiv C_c/(2\omega_n M_c)$ is defined as the ratio of the structural damping coefficient C_c to critical damping; C_{L0} is lift coefficient for the stationary cylinder; and α and b are constant coefficients that define the model.

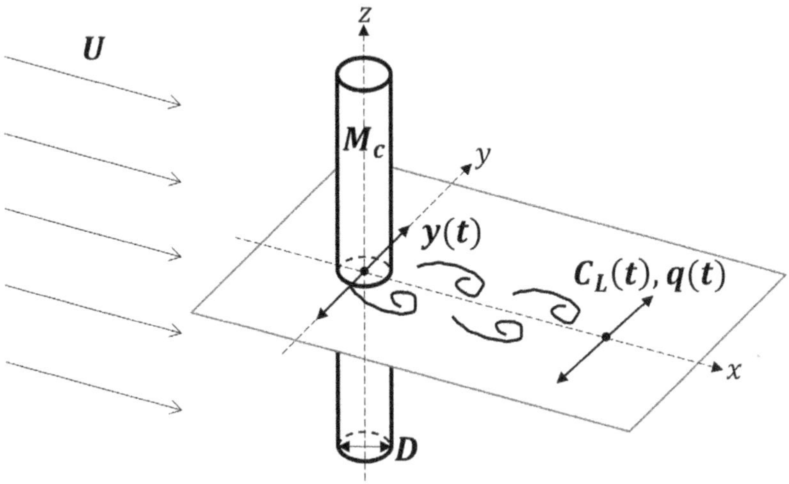

Figure 3. A schematic model of an elastically mounted circular cylinder undergoing cross-flow VIV due to aerodynamic fluid forces acting on the oscillating body.

A number of models (both variants and generalizations) have been formulated based on the Hartlen–Currie model [20]. Skop and Griffin [21] and Griffin et al. [22] applied a modified van der Pol-type wake oscillator with concomitant model parameters describing structural properties, such mass and damping ratio and showed that the model agreed very well with experimental data. Landl [23] reformulated the Hartlen–Currie wake oscillator using a van der Pol-type oscillator and introduced a fifth-order non-linear damping term (viz., $\gamma {C_L}^4 \dot{C}_L$ where γ is a model parameter) into the model in order to account the hysteresis effect of VIV. It is noted that these model variants were developed to predict the VIV of a rigid bluff body. Skop and Griffin [24] extended their earlier model [21] to predict the VIV arising from a *flexible* circular cylinder (e.g., a slender cable). This was accomplished by expressing the sectional vibration displacement and lift coefficient using a modal expansion (normal mode) and formulating the governing equations for the various terms in this expansion. Following on from this effort, Skop and Balasubramanian [25] introduced a so-called stall component (viz., $-2\alpha\dot{Y}/\omega_{vs}$ where α is a model parameter) as part of the forcing term in the structure equation. To this purpose, the transverse fluid force consists of two components: namely, one component (modelled using a van der Pol wake oscillator) that is used to cause the body to move and another component (which proportional to the negative structural velocity) which is used to reduce the amplitude of the lift coefficient for large (significant) structural motion (stall term). This innovative modification of the lift force is significant because it confers on the model the capability to predict the asymptotic VIV response in the vicinity of zero structural damping—this has, up until then, never been realized. Skop and Luo [26] further refined the stall term (viz., $-2\alpha\dot{Y}(\omega_{vs}/\omega_n)^k/\omega_{vs}$) in order to ensure the accuracy of induced asymptotic behavior.

The models described above (whether for a rigid or flexible body) have the following common characteristic: namely, the wake variable is explicitly expressed in terms of the fluctuating (instantaneous) lift coefficient C_L. Facchinetti et al. [27] first introduced a generalized dimensionless wake variable q in order to characterize the oscillating wake (cf. Equations (3) and (4)). Although in their case the action of the fluid on the structure was still considered as a lift force with q representing the reduced vortex lift coefficient ($q \equiv q_0 C_L/C_{L0}$ where $q_0 = 2$ is the finite amplitude of a stable quasi-harmonic oscillation), Facchinetti et al. [27] demonstrated that q can be chosen to be other physical quantities that are capable of describing the fluctuation characteristics of the near wake. The dimensionless governing equations for this model can be expressed as follows:

$$\ddot{Y} + \left(2\zeta_c \nu + \frac{C_D}{4\pi S_t m^*}\right)\dot{Y} + \nu^2 Y = \frac{C_{L0}}{8\pi^2 {S_t}^2 m^* q_0} q\,, \tag{3}$$

$$\ddot{q} + \varepsilon\left(q^2 - 1\right)\dot{q} + q = A\ddot{Y}/\dot{Y}/Y\,. \tag{4}$$

The dimensionless quantities in this model have different definitions than those of the Hartlen–Currie model [20]. These include the dimensionless time $\tau \equiv \omega_{vs} t$, the structural mass ratio $m^* \equiv M/(\rho D^2)$ with M being the sum of the cylinder mass M_c and added mass M_a, and the frequency ratio which is now defined as $\nu \equiv \omega_n/\omega_{vs}$. Finally, ε and A are constant coefficients that define the model.

Another contribution arising from the model proposed by Facchinetti et al. [27] and inspired by the effort of Skop and Balasubramanian [25] for a flexible body was the inclusion of the stall term (i.e., a fluid-added damping term) in the structure equation of motion using a drag coefficient C_D—as a reminder to the reader, this term is given by $C_D\dot{Y}/(4\pi S_t m^*)$. These investigators also examined different forcing terms for the wake oscillator (displacement, velocity, and acceleration) and found that the use of the acceleration provided the best conformance with the experimental data. The model formulated by Facchinetti et al. [27] is so popular that many variants of this model have been used in subsequent theoretical studies. For example, Ogink and Metrikine [28] introduced an acceleration–velocity coupling term ($\widetilde{A}(\omega)\ddot{Y} + \widetilde{B}(\omega)\dot{Y}$) that is dependent on the oscillation frequency ω of the cylinder in order to provide good predictions for both forced and free vibrations.

It should be noted that all the models described above are designed to predict the one-degree-of-freedom (1DOF) transverse VIV motion of a bluff body. To be more realistic, some theoretical studies focus on modelling both the in-line and cross-flow vibrations. A number of two-degree-of-freedom (2DOF) phenomenological models have been developed based on using two governing equations for the x- and y- directions of motion in order to simulate the possible coupling between these directions of oscillatory motion. In comparison to the 1DOF models, the 2DOF models for VIV are still rather limited and will not described in detail in this paper. Interested readers are referred [29–33] for more information concerning 2DOF models.

(2) Wake angular displacement.

Rather than representing the dynamic wake using an instantaneous lift force coefficient, Birkhoff [34] developed a linear wake oscillator based on the concept of a wake rotation angle (α) in order to simulate the oscillations in the near wake region behind a stationary cylinder (see Figure 4a). This model assumes the following mathematical form:

$$2\rho\bar{l}^3 h\ddot{\alpha} + \pi\rho U^2 \bar{l}^2 \alpha = 0\,, \tag{5}$$

where $\bar{l}$ and h are half-length and width of the near-wake lamina which can be determined using experimental data. It is noteworthy that Birkhoff's wake oscillator is derived using physically-based concepts, such as the Kutta–Joukowsky lift theory [35,36] and Newton's second law of motion. In this sense, Birhoff's wake oscillator model is more of a physics-based model than models that directly incorporate a van der Pol-type governing equation.

Funakawa [37] modified Birkhoff's wake oscillator slightly based on experimental data in the following two aspects: (i) the near wake develops from the rear (back) of cylinder rather than from the cylinder center; and (ii) the location of the restoring force coincides with the center of gravity as shown in Figure 4b. With these modifications, the governing equation is given by

$$2\rho\bar{l}\left(\frac{D}{2} + \bar{l}\right)^2 h\ddot{\alpha} + 2\pi\rho U^2 \bar{l}\left(\frac{D}{2} + \bar{l}\right)\alpha = 0\,. \tag{6}$$

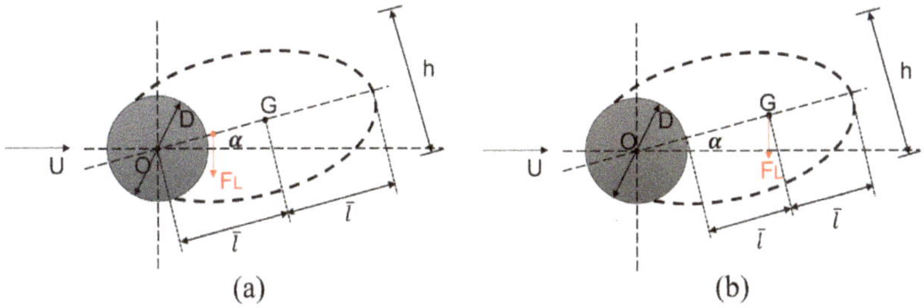

Figure 4. A schematic of the wake oscillator model developed by (**a**) Birkhoff [34] and (**b**) Funakawa [37].

Tamura [38] further assumed that the wake angular displacement $\alpha(t)$ and wake length $l(t)$ are dependent on the time t and replaced the constant wake length in Funakawa's wake oscillator with time-dependent wake length. The time dependence of $\alpha(t)$ and $l(t)$ were obtained from measurements of the wake dynamics behind a fixed cylinder. Furthermore, Tamura included a viscous force using a negative damping term into the ordinary differential equation and derived the associated damping coefficient using the assumption that the work done by the lift and viscous forces is equal. With these modifications, the governing equation for Tamura's wake oscillator is expressed as

$$2\rho\bar{l}\left(\frac{D}{2}+\bar{l}\right)^2\ddot{\alpha}-\bar{C}\left(1-\frac{4{f_m}^2}{{C_{L0}}^2}\alpha^2\right)\dot{\alpha}+2\pi\rho U^2\bar{l}\left(\frac{D}{2}+\bar{l}\right)\alpha=0\,, \tag{7}$$

where $\bar{C}$ denotes the damping coefficient of the near wake and f_m is the Magnus factor which has a value of 1.16 for a circular cylinder [38]. We note that Equation (7) reduces essentially to a non-linear and self-excited van der Pol-type equation and, as a consequence, it can potentially be coupled to a structural equation to predict self-sustained FIV phenomena just like the lift oscillator described above. The consistency in the structure of the lift oscillator and the wake-angle oscillator suggests that the van der Pol equation provides a reasonable and effective paradigm for modelling the wake dynamics behind a bluff body experiencing FIV.

Tamura and Matsui [39] extended Tamura's non-linear wake oscillator for a stationary cylinder to a vibrating system by coupling it to a linear mass–spring–damper system. The external driving force in the structure equation is obtained from the transverse component of the lift and drag forces, so $F_y = F_{Ly} + F_{Dy}$. The forcing term of the wake oscillator is a linear function of the velocity and acceleration of the oscillation. The dimensionless governing equations for this model are given by

$$\ddot{Y}+\left(2\zeta_c+\frac{m^*\nu C_D}{2\pi S_t}\right)\dot{Y}+Y=-\frac{f_m m^*\nu^2}{(2\pi S_t)^2}\left(\alpha+\frac{2\pi S_t}{\nu}\dot{Y}\right), \tag{8}$$

$$\ddot{\alpha}-2\zeta_w\nu\left(1-\frac{4{f_m}^2}{{C_{L0}}^2}\alpha^2\right)\dot{\alpha}+\nu^2\alpha=-\frac{\ddot{Y}}{(0.5+\bar{l}^*)}-(2\pi S_t)\nu\dot{Y}\,. \tag{9}$$

The quantities in these governing equations basically follow the nomenclature for the Hartlen–Currie model [20], except that mass ratio per unit length m^* is redefined as $m^* \equiv \rho D^2/(2M_c)$, a damping ratio ζ_w of oscillating wake is included, and the normalized wake length is defined as $\bar{l}^* \equiv \bar{l}/D$. The characteristic wake variable α is related to the lift coefficient through the Magnus effect, so $C_L = -f_m(\alpha + 2\pi S_t\dot{Y}/\nu)$. An obvious difference between the Tamura–Matsui model [39] (which uses the angular displacement α in the wake oscillator) and previous models (which use the lift (either C_L or q) in wake oscillator) is that all the parameters in the former model have a clear physical meaning (interpretation) and can be determined from direct measurements in wind-tunnel experiments, while the latter set of models includes a number non-physical (tuning) constants (e.g., α and b in Equation (2) or ε and A in Equation (4)) that can only be determined from curve fitting of experimental data. The reason for this is that the Tamura–Matsui model [39] is derived from series of physically-based concepts (as noted earlier) instead of assuming a priori the applicability of a van der Pol-type oscillator. The Tamura–Matsui model [39] is designed to predict 1DOF transverse VIV for a two-dimensional (2D) rigid circular cylinder and, as a result, this model can reproduce the key non-linear characteristic of VIV for this case. Finally, Tamura and Amano [40] extended the Tamera–Matsui model to a three-dimensional circular cylinder. This was accomplished by utilizing the modal expansion method for the sectional vibration displacement of the cylinder and the angular displacement of the wake.

In addition to predicting the VIV for 2D and 3D circular cylinders, Tamura and his collaborators also combined the non-linear wake oscillator in terms of the angular displacement with an equation of motion for a square cylinder in order to predict galloping and VIV-galloping. This generalization of the model will be described in detail in Sections 3.3 and 3.4. The series of models developed by Tamura and his collaborators (which spanned the period from 1979 to 1987) are representative of mathematical models that apply the rotation angle of the near wake in order to characterize the flow dynamics behind a bluff body undergoing FIV. The interested reader can refer to a comprehensive review paper by Tamura [11] for a broad overview of the motivation, development, and derivation of these types of model.

(3) Other wake variables.

Among the theoretical models that use wake variables other than the lift coefficient and the wake angular displacement, the most representative one is that proposed by Iwan and Blevins [41]. This particular model is derived from the basic fluid mechanics underpinning the von Kármán vortex street behind a cylinder (momentum equation in y-direction) and involves the introduction of a hidden fluid variable z denoting the weighted average of the transverse flow motion (e.g., velocity, acceleration) in order to formulate the vertical momentum in a control volume. Interestingly, the Iwan–Blevins wake oscillator (which is based on the physics of the flow field) also assumes the form of a van der Pol equation which, in turn, provides further support for the use of the latter model from a purely physics-based perspective.

Another interesting coupled model was proposed by Krenk and Nielsen [42], who utilized a generalized fluid oscillation displacement w of the equivalent fluid mass as the state variable in a wake oscillator. The non-linear negative damping incorporated in this model was defined as the quadratic form

$$-\left[1-\left(w^2+\dot{w}^2/\omega_{vs}^2\right)w_0^2\right]\dot{w}\,, \tag{10}$$

where $\omega_{vs} = 2\pi f_{vs}$ denotes the circular frequency of vortex shedding and ω_0 is the vibration amplitude of fluid oscillator over a stationary cylinder. In consequence, the wake oscillator defined by this model is actually a combination of a van der Pol and Rayleigh oscillator.

3.2.2. Single-Degree-of-Freedom Model

The key to the formulation of a SDOF model is to correctly represent the forcing term so that it can induce finite-amplitude vibrations on the bluff body. A forcing term based on a negative damping is generally expressed as a forcing function F, for which the aeroelastic damping term may be offset by positive mechanical damping and causes the total system damping to be zero for a certain range of inflow (incident) wind speeds. This is the reason why the energy of the fluid is transferred from the wake flow to the bluff body, initiating the oscillation of the body. Concerning the mathematical formulation, F is a function of the status of the body motion (e.g., displacement, velocity, acceleration, oscillation frequency) and can be regarded as a "stack" of multiple vibration sources each multiplied by a corresponding prefactor that is dependent on the reduced frequency defined as $K \equiv \omega_n D/U$ (U is the inflow wind speed).

Motivated by an earlier linear model [43] as the basis for development, a representative non-linear SDOF model was proposed by Simiu and Scanlan [44]. This model is given by

$$\begin{aligned} F &= M_c\left(\ddot{y} + 2\zeta_c\omega_n\dot{y} + {\omega_n}^2 y\right) \\ &= \frac{1}{2}\rho U^2(2D)\left(\underbrace{Y_1(K)\left(1-\varepsilon(K)\frac{y^2}{D^2}\right)\frac{\dot{y}}{D} + Y_2(K)\frac{y}{D}}_{\text{motion-induced force}} \right. \\ &\qquad \left. + \underbrace{\frac{1}{2}C_L(K)\sin(\omega_{vs}t+\varphi)}_{\text{vortex-shedding force}}\right). \end{aligned} \tag{11}$$

In this model, the hydrodynamic force that drives VIV on the structure depends on two components. The first component is the motion-induced force that is expressed in terms of y and $\dot{y}$, which is characterized by the following non-linear damping term:

$$Y_1(K)\left(1-\varepsilon(K)\frac{y^2}{D^2}\right)\frac{\dot{y}}{D}. \tag{12}$$

The second component is a linear stiffness term $Y_2(k)y/D$ where $Y_1(k)$, $Y_2(k)$, and $\epsilon(K)$ are model parameters. The combination of these two terms, especially the inclusion of van der Pol-type negative damping with a non-constant coefficient, induces the self-excited and self-limited VIV response. The other force component in the model is the vortex-shedding force that is directly related to the lift coefficient $C_L(k)$—a term which forces the structure to vibrate at the vortex-shedding circular frequency ω_{vs}.

Ehsan et al. [45] further extended the Simiu–Scanlan model [44] for a rigid body to a long-span flexible bridge. Inspired by the coupled wake-oscillator model proposed by Billah [46], Goswami et al. [47] modified this model by replacing the direct forcing term which is dependent on $C_L(K)$ in Simiu–Scanlan model with a parametric stiffness term given by $J_2(K)\cos(2\omega_{vs}t)y/D$. Here, $J_2(K)$ is a model parameter that excites the body to oscillate at the frequency $2\omega_{vs}$. This model can reproduce the non-linear characteristics of VIV, such as lock-in and hysteresis.

Another type of SDOF model decomposes the fluid force according to its phase difference relative to the motion of the structure. More specifically, the fluid inertial force is in-phase with the acceleration, while the fluid damping force inducing the transfer of energy from the fluid to the body (giving rise to structural oscillation) is in-phase with the velocity and displacement. Rather than deriving a specific mathematical relationship, the composed forces are obtained directly from measurements obtained from forced-vibration experiments in which the lift force is measured at a given frequency. As a result, this type of SDOF model is more suitable for the calculation of the vibration at a single distinct frequency and can be used to obtain accurate predictions of maximum response. Interested readers are referred to references [48–51] for more details regarding the development of SDOF models based on fluid force measurements.

As an example of the application of the mathematical models described in this section, Figure 5 shows the predictions obtained from three mathematical models of the normalized maximum oscillation amplitude A_{max}/D as a function of the reduced velocity U_r for a VIV system consisting of the flow past a circular cylinder of diameter D. The three mathematical models are those proposed by Iwan and Blevins [41] (I&B), Tamura and Matsui [39] (T&M) and Facchinetti et al. [27]. These model predictions are compared to some experimental data reported by Feng [52] for VIV of a circular cylinder with a mass ratio of 0.00257 and damping ratio of 0.00181. As shown, model proposed by Iwan and Blevins (which is based on a hidden flow variable) predicts a quite narrow lock-in and a slightly larger amplitude.

The model proposed by Tamura and Matsui (which is based on the wake inclination angle) generally provides a good agreement with experimental data, although the model provides a premature termination of the synchronization. Finally, Facchinetti et al.'s model with regard to the lift coefficient gives a reasonably accurate prediction of the lock-in range, but the onset of VIV occurs at a smaller value of U_r. It should be noted that the mathematical model predictions depend critically on the appropriate selection of the parameters for each model, and the values of these parameters can vary from case to case.

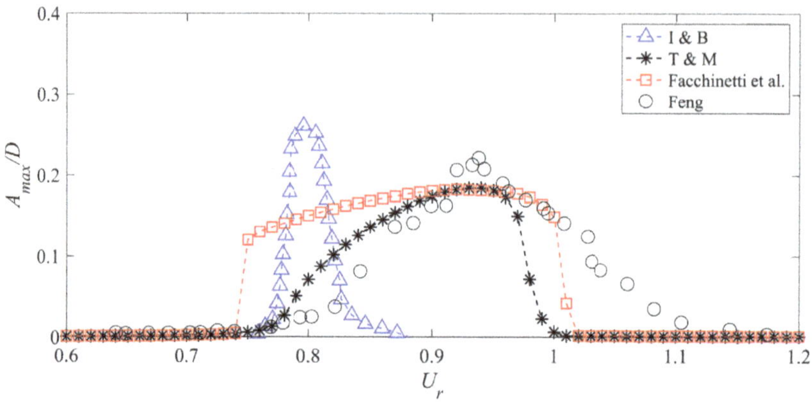

Figure 5. Comparison of predictions of the maximum oscillation amplitude A_{max} as a function of the reduced velocity U_r of the VIV of a circular cylinder with diameter D obtained using three mathematical models with a benchmark experiment conducted by Feng [52], Iwan and Blevins [41] (I&B), Tamura and Matsui [39] (T&M) and Facchinetti et al. [27].

3.3. *Galloping*

Galloping is an important aeroelastic response of a bluff body—its analytical modelling is generally based on the so-called quasi-steady (QS) hypothesis which states that the instantaneous aeroelastic force exerted by the fluid on a vibrating structure is the same as that on a stationary body at the same angle of attack.

Figure 6 displays the simplest physical model for the analysis of the galloping behavior of a cylindrical body restrained to oscillate only in the transverse direction and subjected to an uniform streamwise inflow (incident) wind speed of U [53]. As shown, the damped spring-mounted system is characterized by the mass per unit length M_c, the mechanical damping coefficient C_c, the spring stiffness k, and the characteristic length D of the body normal to incoming flow. The structural motion is governed by the following dimensional second-order ordinary differential equation:

$$M_c\ddot{y} + C_c\dot{y} + ky = F_y\,, \tag{13}$$

where F_y is the lateral aerodynamic force inducing the galloping which can be calculated using QS theory. Once the body begins to move in the transverse y-direction (viz., $\dot{y} \neq 0$), the relative velocity between the oscillating body and the fluid is $U_{rel} = (U^2 + \dot{y}^2)^{1/2}$ and the angle of attack (AOA) is determined using AOA $= \arctan(\dot{y}/U)$ which varies periodically over time owing to the transverse oscillation of the structure. The time variation of the AOA induces a time variation in the lift and drag forces given by $F_L(\text{AOA}) = \frac{1}{2}\rho U_{rel}^2 DC_L(\text{AOA})$ and $F_D(\text{AOA}) = \frac{1}{2}\rho U_{rel}^2 DC_D(\text{AOA})$. The lift and drag coefficients can be determined from conducting static experiments.

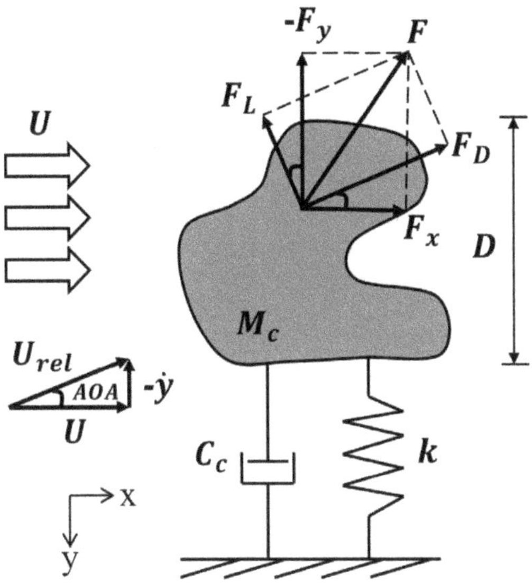

Figure 6. Schematic of a bluff body experiencing transverse galloping in a fluid flow.

The transverse fluid force F_y is a function of AOA and will vary with time. This force component can be derived by projecting the lift and drag forces F_L and F_D along transverse y-axis [2]:

$$F_y(\mathrm{AOA}) = -\frac{1}{2}\rho U_{rel}^2 D\Big[C_L(\mathrm{AOA})\cos(\mathrm{AOA}) + C_D(\mathrm{AOA})\sin(\mathrm{AOA})\Big]. \tag{14}$$

The force component F_y can also be formulated in terms of the transverse force coefficient C_{Fy} as follows:

$$F_y(\mathrm{AOA}) = \frac{1}{2}\rho U^2 D C_{Fy}(\mathrm{AOA}). \tag{15}$$

It should be noted that the velocity in Equation (15) is the inflow (incident) velocity U (rather than U_{rel}) which is referenced to a direction perpendicular to the direction used to define C_{Fy}. Equations (13)–(15) can be combined to obtain the transverse force coefficient given by

$$C_{Fy}(\mathrm{AOA}) = -\frac{C_L(\mathrm{AOA}) + C_D(\mathrm{AOA})\tan(\mathrm{AOA})}{\cos(\mathrm{AOA})}. \tag{16}$$

To model the incipient instability at the onset to galloping (viz., $\dot{y} \approx 0$ implying that $\alpha \approx \dot{y}/U \approx 0$), one can apply a one-dimensional Taylor series expansion for $F_y(\mathrm{AOA})$ about the point AOA $= 0$. The Taylor series expansion of $F_y(\mathrm{AOA})$ truncated at the first-order term is given by

$$F_y(\mathrm{AOA}) = F_y\big|_{\mathrm{AOA}=0} + \frac{dF_y(\mathrm{AOA})}{d\mathrm{AOA}}\bigg|_{\mathrm{AOA}=0}\mathrm{AOA}. \tag{17}$$

Combining Equations (13) to (17), the equation of motion can be written as follows:

$$M_c\ddot{y} + C_c\dot{y} + ky = F_y(0) - \frac{1}{2}\rho U^2 D\left(\frac{dC_L(\mathrm{AOA})}{d\mathrm{AOA}} + C_D(\mathrm{AOA})\right)\bigg|_{\mathrm{AOA}=0}\frac{\dot{y}}{U}. \tag{18}$$

The second term on the right-hand side can be treated as the contribution to the effective system damping (also called fluid dynamic damping) [9] arising from the aerodynamic forces. This term can be merged with the mechanical damping term (second term on the left-hand-side), so that the total damping ratio ζ_{total} assumes the following form:

$$\zeta_{\text{total}} = \zeta_c + \frac{\rho U D}{4 M_c \omega_n} \left(\frac{d C_L(\text{AOA})}{d\text{AOA}} + C_D(\text{AOA}) \right) \bigg|_{\text{AOA}=0} . \tag{19}$$

The prerequisite for galloping is the presence of a negative system damping (viz., $\zeta_{\text{total}} < 0$). Because the mechanical damping is positive ($\zeta_c = C_c/(2\omega_n M_c) > 0$), the second term on the right-hand-side of Equation (19) is less than zero, so

$$\left(\frac{d C_L(\text{AOA})}{d\text{AOA}} + C_D(\text{AOA}) \right) \bigg|_{\text{AOA}=0} < 0 . \tag{20}$$

Equation (20) is the well-known Glauert–den Hartog galloping criterion [54] setting the necessary condition for aerodynamically unstable behavior of a SDOF oscillator. This criterion can also be used to evaluate the galloping onset U_g by setting the mechanical damping to be equal to the fluid dynamic damping giving

$$U_g = -\frac{4 M_c \zeta_c \omega_n}{\rho D \left(\frac{d C_L(\text{AOA})}{d\text{AOA}} + C_D(\text{AOA}) \right) \Big|_{\text{AOA}=0}} . \tag{21}$$

The rationale underlying the Glauert–den Hartog galloping criterion is related to the boundary-layer separation and reattachment phenomenon at a certain angle of attack [55–57].

Parkinson and Brooks [58] combined the aerodynamic force derived from QS theory with a linear dynamical system to develop a non-linear self-excited oscillation model for a bluff body experiencing 1DOF "plunging" oscillation (viz., galloping). The variation of the aerodynamic force coefficient in the transverse direction (C_{Fy}) as a function of AOA (obtained from static test data of the lift and drag coefficients (C_L and C_D) for different angles of attack) was fitted with a fifth-order polynomial. This polynomial approximation for C_{Fy} as a function of AOA was then incorporated as a forcing term in the governing equation for a linear mass–spring–damper system. Subsequently, a more accurate seventh-order polynomial fitting of $C_{Fy}(\text{AOA})$ (see Equation (22)) was used in order to reproduce hysteresis effects observed in wind-tunnel experiments [59]. The seventh-order polynomial approximation for $C_{Fy}(\text{AOA})$ has the following form:

$$C_{Fy} = A_1 \frac{\dot{y}}{U} + A_3 \left(\frac{\dot{y}}{U} \right)^3 + A_5 \left(\frac{\dot{y}}{U} \right)^5 + A_7 \left(\frac{\dot{y}}{U} \right)^7 , \tag{22}$$

where A_1, A_3, A_5, and A_7 are fitting coefficients for the model. The corresponding dimensionless equation of motion for a square cylinder can be written as follows:

$$\begin{aligned} \ddot{Y} - m^* A_1 \Bigg[& \left(U_r - \frac{2\zeta}{m^* A_1} \right) + \left(\frac{A_3}{A_1 U_r} \right) \dot{Y}^2 \\ & + \left(\frac{A_5}{A_1 {U_r}^3} \right) \dot{Y}^4 + \left(\frac{A_7}{A_1 {U_r}^5} \right) \dot{Y}^6 \Bigg] \dot{Y} + Y = 0 , \end{aligned} \tag{23}$$

where $U_r \equiv U/f_n D$ is the reduced velocity. The Parkinson–Brooks model [58] can be solved analytically using various approximation methods when the mass ratio is small (e.g., in air m^* is typically of order 10^{-3}). As an example of this strategy, Parkinson and Brooks [58] used the Krylov–Bogoliubov asymptotic method to solve Equation (23) and found that the (approximate) solution provided a good agreement with some wind-tunnel measurements. Furthermore, Equation (23) can be solved numerically (using, for example, the fourth-order

Runge–Kutta method). The form of Equation (23) resembles the SDOF model for VIV, described previously in Section 3.2.2, in the sense of having one governing equation where the forcing term is related to the aerodynamic force acting on structure—the only difference between these two models is that the force coefficient used to model galloping involves a polynomial that has been fitted to experimental data of the static force as a function of the angle of attack.

The result provided by the QS model is highly dependent on the polynomial fitting $C_{Fy}(\mathrm{AOA})$. Luo et al. [60] and Ng et al. [61] compared different high-order polynomials and found the minimal order of the fitting polynomial needs to be seven in order to capture the key characteristics of C_{Fy} as a function of AOA and to predict certain features of the non-linear dynamics of galloping, such as hysteresis. Using a ninth- or eleventh-order polynomial may provide a better fit to the variation of C_{Fy} with AOA, but this does not change the characteristics of the governing equation, such as the number of positive real roots representing the stationary oscillation amplitude of a bluff body, implying that the predicted galloping hysteresis would be the same as that obtained from a seventh-order polynomial approximation for $C_{Fy}(\mathrm{AOA})$. Barrero-Gil et al. [62] and Joly et al. [63] also incorporated the Reynolds number (Re) into the classical QS galloping model by expressing the polynomial coefficients as a function of Re in order to investigate the occurrence of galloping and the hysteresis effect of a square cylinder in the laminar flow regime.

From a physical and mathematical viewpoint, the underlying QS assumption is that the transverse galloping force depends on the instantaneous position of the body, as well as on the instantaneous relative velocity between the body and the fluid [64]—this is true in certain circumstances, such as when vortex shedding frequency is much larger than the structural oscillation frequency (viz., $f_c \ll f_{vs} \equiv S_t U/D$), implying that the wake dynamics are essentially uncoupled from the body motion so that the influence of the vortex force on the galloping driving force is negligible. In addition, the applicable velocity range of the QS model is determined (approximately or better) by the condition $U_r = U/(f_c D) \gg (U/D)(D/(S_t U)) = 1/S_t$ where the Strouhal number S_t often has an order of magnitude of 0.1, suggesting that the QS assumption can be satisfied at large values of U_r [62,64–66]. Even with these limitations, the easy-to-use QS model still plays a fundamental role in the mathematical modelling of galloping, especially when few experimental or numerical data are available.

The den Hartog criterion and QS galloping models are limited to 1DOF translational galloping. There also exists some QS-based theoretical models for 2DOF galloping (vertical and torsional) [13,67,68] or for 3DOF galloping (vertical, lateral, and torsional) [64,69,70].

3.4. Combined VIV and Galloping

Instead of undergoing VIV or galloping only, some engineering structures with aerodynamically unstable cross-sectional shapes can experience the mutual effect of both types of FIV phenomena. Parkinson and Sullivan [53] and Parkinson and Wawzonek [71] conducted some wind-tunnel experiments involving square and rectangular towers, which demonstrated that the synergy between VIV and galloping can provoke large-amplitude transverse vibrations that are not predicted using models for either galloping or VIV. As a consequence, it is of practical importance to construct a simple and efficient mathematical model for a (combined) VIV-galloping response. Based on the discussion in Sections 3.2 and 3.3, the essential problem still reduces to correct simulation of the unsteady aerodynamic force responsible for the dynamic response—the most straightforward way to accomplish this is to simply superimpose the VIV and galloping force components.

An early theoretical work following this methodology was proposed by Parkinson and Bouclin [72] who used the Hartlen–Currie lift oscillator [20] to calculate the fluctuating lift force due to vortex shedding that induces the VIV (C_L) and the Parkinson–Smith quasi-steady model [59] to represent the perturbational transverse force that triggers galloping ($C_{Fy}(\dot{Y})$). In this approach, the sum of both these forces is incorporated as the forcing term in the structural equation of motion and used to predict either VIV or galloping response of a bluff body.

Corless and Parkinson [73] modified the Parkinson–Bouclin model [72] by adding an acceleration coupling term in the wake oscillator. With the inclusion of this term, the Corless–Parkinson model assumes the form

$$\ddot{Y}+2\zeta_c\dot{Y}+Y=m^*\nu^2\left[C_L+A_1\left(\frac{2\pi S_t}{\nu}\right)\dot{Y}+A_3\left(\frac{2\pi S_t}{\nu}\right)^3\dot{Y}^3+A_5\left(\frac{2\pi S_t}{\nu}\right)^5\dot{Y}^5+A_7\left(\frac{2\pi S_t}{\nu}\right)^7\dot{Y}^7\right], \tag{24}$$

$$\ddot{C_L}-a(\nu)\left(1-\frac{4}{3{C_{L0}}^2\nu^2}\dot{C_L}^2\right)\dot{C_L}+\nu^2C_L=b(\nu)\dot{Y}+d(\nu)\ddot{Y}, \tag{25}$$

where the coefficients $a(\nu)$, $b(\nu)$ and $d(\nu)$ are functions of ν. These coefficients can be determined by curve fitting of experimental data. The other quantities in Equations (24) and (25) are consistent with those described earlier for the Hartlen–Currie lift oscillator [20] (see Equations (1) and (2)). The system of ODEs here were solved using the Multiple Time Scales (MTS) method.

Following the same strategy, Han et al. [74] incorporated the quasi-steady galloping force into another representative lift-type wake-oscillator model—Facchinetti et al.'s model [27]—to give a coupled mathematical model for VIV-galloping of a square prism. Using the same dimensionless terms as in the model proposed by Facchinetti et al. [27], the formulation that incorporates acceleration coupling and a seventh-order polynomial fitting of the galloping force reduces to the following form:

$$\ddot{Y}+2\zeta_c\nu\dot{Y}+\nu^2Y=\frac{1}{8\pi^2{S_t}^2m^*}\left(\frac{C_{L0}}{q_0}q+A_1(2\pi S_t\dot{Y})+A_3(2\pi S_t\dot{Y})^3+A_5(2\pi S_t\dot{Y})^5+A_7(2\pi S_t\dot{Y})^7\right), \tag{26}$$

$$\ddot{q}+\varepsilon\left(q^2-1\right)\dot{q}+q=A\ddot{Y}. \tag{27}$$

It is possible to combine not only the lift oscillator model with the QS model for galloping, but also the wake oscillator model with the wake rotation angle as the state variable (see Section 3.2.1). Tamura and Shimada [75] proposed a mathematical model for the interaction of the VIV resonance with galloping for 2D and 3D rigid square cylinders. In this model, the unsteady vortex shedding and galloping forces were simulated using the Tamura–Matsui VIV model [39] and the QS model developed by Parkinson and colleagues [58,59], respectively. Following the nomenclature used earlier in Equations (8) and (9), the model recast in dimensionless form is given by

$$\ddot{Y}+2\zeta_c\dot{Y}+Y=-\frac{m^*\nu^2}{(2\pi S_t)^2}\left(\underbrace{f_m\left(\alpha+\frac{2\pi S_t}{\nu}\dot{Y}\right)}_{\text{VIV transverse lift coefficient}}+\underbrace{C_D\frac{2\pi S_t}{\nu}\dot{Y}}_{\text{VIV transverse drag coefficient}}\right)-\frac{m^*\nu^2}{(2\pi S_t)^2}\left(\underbrace{A_1\left(\frac{2\pi S_t}{\nu}\right)\dot{Y}+A_3\left(\frac{2\pi S_t}{\nu}\right)^3\dot{Y}^3+A_5\left(\frac{2\pi S_t}{\nu}\right)^5\dot{Y}^5+A_7\left(\frac{2\pi S_t}{\nu}\right)^7\dot{Y}^7}_{\text{galloping transverse force coefficient}}\right), \tag{28}$$

$$\ddot{\alpha} - 2\zeta_w \nu \left(1 - \frac{4 f_m{}^2}{C_{L0}{}^2} \alpha^2\right) \dot{\alpha} + \nu^2 \alpha = -\frac{\ddot{Y}}{0.5 + \bar{I}^*} - (2\pi S_t) \nu \dot{Y} \,. \tag{29}$$

The models described above have been important mathematical tools for the prediction of combined VIV and galloping for a square cylinder. These basic models have been validated and extended in a number of ways. An important contribution in this respect is the body of work conducted by Mannini and colleagues [76–79]. In particular, Mannini et al. [76] compared the Corless–Parkinson model [73] and the Tamura–Shimada model [75] for the prediction of the FIV response of a rectangular cylinder and validated their model performance using their wind-tunnel test data. A key conclusion of this investigation was that the Tamura–Shimada model [75] is more accurate than the Corless–Parkinson model [73] in terms of their predictions of the correlation between the oscillation amplitude and the wind speed over the range where galloping occurs. Furthermore, Mannini et al. [77] modified the Tamura–Shimada model [75] by replacing Funakawa's wake oscillator [80] (see Figure 4b) with Birkchoff's wake oscillator [34] (see Figure 4a) and used an improved tuning process of the model parameters. These investigators also incorporated the effects of turbulence into the modified the Tamura–Shimada model [78]. Finally, Chen et al. [79] proposed a more reasonable definition of the near-wake laminar for a rectangular cylinder—the modified model was shown to provide better predictions of the (combined) VIV-galloping instability.

4. Numerical Modelling

Owing to the continuous advancement of computational fluid dynamics (CFD) and computer-aided engineering (CAE) in recent years, numerical modelling of the complicated FSI problems such as the FIV of a bluff body is now practical for real-world engineering and industrial problems. The various forms of CFD that can be used to model numerically the FIV of a bluff body include direct numerical simulation (DNS), large-eddy simulation (LES), detached-eddy simulation (DES), delayed detached-eddy simulation (DDES), and Reynolds-averaged Navier–Stokes (RANS) modelling, among others. The numerical discretization approaches used to solve the fluid dynamics equations for these various approaches encompass the finite-element method (FEM), the finite-volume method (FVM), the finite-differencing method (FDM), spectral elements method (SEM) which is essentially a Galerkin-based FEM with Lagrange shape functions, the discrete vortex method (DVM), the entropy–viscosity method (EVM), the lattice-Boltzmann method (LBM), among others. The fluid–structure interaction schemes include either a moving-mesh or a fixed-mesh strategy. The commonly-used numerical codes for addressing FIV of a bluff body include the various commercial programs, such as Ansys Fluent [81] (which uses FEM) from Ansys Inc. and STAR-CCM+ [82] from Siemens PLM, as well as open-source software packages, such as OpenFOAM [83] (which uses FVM and unstructured grids).

For the benefit of those readers who are unfamiliar with CFD—the key-enabling technology underpinning the numerical simulation of FIV phenomena—we will provide a brief overview of the key concepts. The basic idea of CFD is to use appropriate algorithms to solve some form of the Navier–Stokes (NS) equations that describe the fluid flow. To this purpose, the NS equations that govern a viscous, incompressible flow, whether laminar or turbulent, express the conservation of mass (incompressibility of the fluid volume) and the conservation of momentum, respectively, as follows:

$$\frac{\partial u_i}{\partial x_i} = 0 \,, \tag{30}$$

and

$$\frac{\partial u_i}{\partial t} + u_j \frac{\partial u_i}{\partial x_j} = -\frac{1}{\rho} \frac{\partial p}{\partial x_i} + \nu \frac{\partial^2 u_i}{\partial x_j \partial x_j} \,, \tag{31}$$

where x_i is the Cartesian coordinate associated the i-th direction (so, $(x_1, x_2, x_3) = (x, y, z)$); u_i is the Cartesian fluid velocity component in the x_i direction; p is the pressure; ρ is the fluid density; ν is the fluid kinematic viscosity; and, t is time. In the above equations, the Einstein summation convention is assumed to apply to repeated indices.

In direct numerical simulation (DNS), the NS equations is solved numerically with the entire band of scales of the flow from the large energy-producing to the small dissipative motions fully resolved in space and time. Here, no ad hoc models are required to represent the unresolved motions of the flow. The application of DNS is currently restricted to low and moderate Reynolds numbers (laminar or weakly turbulent flows). For larger Reynolds number flows (strongly turbulent flows which are prevalent in engineering and geophysical applications), a simpler level of description is required. To this purpose, one can apply a filtering (averaging) operation to the NS equations. The resulting filtered (averaged) NS equations give rise to more unknowns than the number of equations—unknown subgrid stresses or Reynolds stresses in the momentum transport equation—resulting in the so-called turbulence closure problem. In large-eddy simulations (LESs), only the large-scale portion of the turbulence is explicitly computed while the small-scale portion (subgrid scale or SGS) of the fluid motions are represented using a so-called SGS model. Alternatively, in the statistical simulation of the fluid flow based on the Reynolds-averaged NS (RANS) equation, the averaged effect of the turbulence on the flow (embodied in the Reynolds stress tensor) is modelled. Finally, there are hybrid methods of fluid flow simulation (e.g., detached-eddy simulation or DES) in which a RANS model is used to represent some parts of the flow (e.g., near solid walls) while LES (with a SGS model) is used elsewhere, hence providing a compromise between the computational expense of LES and the computational efficiency of RANS. For more information on CFD, the reader is referred to [84].

The numerical modelling of FIV is capable of providing accurate predictions, as well as deeper physical insights into the fundamental processes underpinning the FIV response. In consequence, numerical simulations are a invaluable tool that complements physical experiments for the practical design and analysis of a bluff body subjected to FIV. Even so, we stress that the FIV problem involves very complex flow dynamics and non-linear bidirectional interactions between the fluid and the structure—these complexities limit the predictive capabilities of FIV simulations, especially in the high-Reynolds number (and strongly turbulent) flows in the range of from 10^4 to 10^6. It should be stressed that real-world FIV problems almost always involve flows in this flow regime where either the three-dimensional flow transitions from laminar to turbulent or the flow is fully turbulent [85]. These limitations prevent CFD being used exclusively to address real-world engineering and industrial problems involving FIV phenomena. Nevertheless, it can be stated that improvements of accuracy and reliability of FIV numerical simulations for flows in the subcritical and high-Reynolds number range is essential in order to provide confidence in the use of numerical modelling of FIV for addressing real-world applications.

4.1. Numerical Techniques for FSI Simulation

The numerical simulation of a two-way FSI problem, such as FIV involves both the effect of the fluid flow on the motion of the structure and, conversely, the effect of the structure on the local flow. As shown in Section 3, the precise mathematical description of FIV consists of the governing equations for the fluid flow dynamics (NS equations) which are then coupled to the equation of motion for the structure (which generally is modelled as a mass–spring–damper system). This coupled set of equations needs to be solved numerically. The use of an appropriate fluid–structure interaction strategy is critical for the simulation of the coupled dynamics of the fluid–structure interaction. Depending on the treatment of the computational mesh, the numerical methodologies for the simulation of FSI can be stratified into two categories: namely, a moving mesh (or conforming) method and a fixed mesh (or non-conforming) method [86].

(1) Moving mesh.

The moving mesh method involves a dynamic re-meshing (or mesh updating) of the computational grid at each time step (viz., the mesh moves to accommodate the dynamically changing topology of the spatial fluid domain owing to the motion of the structure which, in turn, necessitates the tracking of the fluid–structure interface at each time step). As a consequence, an efficient moving mesh module and a fast and automatic mesh deformation procedure are of great importance for the successful simulation of a FSI using this approach.

An often-used mesh diffusion scheme is exhibited in Figure 7. In this scheme, the mesh around the moving body (green) moves with the body without any deformation as if the mesh was "rigid" (viz., attached to the body). The outer boundary cells (white) in this scheme remain fixed—only the portion of the computational grid lying between the inner and outer boundaries (orange) are deformed and re-generated at each time step. This moving mesh scheme has been demonstrated to give almost no projection error in the solution and, furthermore, improves the accuracy of the simulations by preserving the gradients in the vicinity of the moving body [87,88]. Another key issue is the mesh morphing algorithm that deals with the motion of the body. As an example, the displacement and velocity of the mesh in the orange region of Figure 7 for each time step is obtained from the solution of a Laplacian equation. The reader is referred to Ref. [89] for more details of mesh deformation methods (remeshing or adaptive meshing strategies).

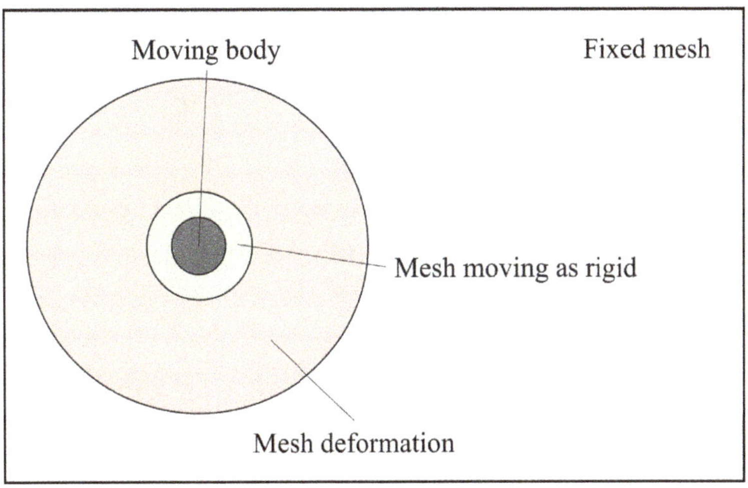

Figure 7. Schematic describing the components of a mesh moving scheme.

A classical and popular moving mesh technique is the arbitrary Lagrangian–Eulerian (ALE) algorithm, which can both follow the moving boundary and preserve the mesh grid shape even for large structural displacements at the same time. This algorithm accomplishes this by defining three domains: namely, a material domain to provide a Lagrangian formulation of the moving body, a spatial domain described in an Eulerian (fixed) reference frame, and an arbitrary domain involving an arbitrary motion that determines the deformation of the moving mesh. The mapping of ALE to the NS equations involves explicitly incorporating an arbitrary mesh velocity into the convective term of the momentum transport equations, so that the body motion and the mesh deformations required to maintain an adequate spatial discretization can be taken into account. It should be noted that in the ALE scheme, the computational grids are allowed to move independently of the fluid velocity in order to avoid an excessive distortion of the mesh. Another common moving mesh technique is the so-called Deforming-Spatial-Domain/Stabilized Space-Time (DSD/SST) algorithm, which has been used by Mittal and Kumar [90] and by Singh and Mittal [87] in their numerical simulations of FIV.

In general, the numerical computation involved in the application of a moving mesh approach consists of three basic steps [91]: (1) the fluid dynamics computation; (2) the structure dynamics computation; and (3) the mesh movement computation. A detailed flowchart showing how these steps are implemented is displayed in Figure 8 for the ALE algorithm (left panel). For examples of the application of the ALE algorithm to address the FIV problem, the interested reader is referred to references [85,92–97]. The moving-mesh method using a body-fitted coordinate system can generally achieve a high-mesh resolution near the moving boundary and, as a result, can provide excellent simulations in that critical portion of the flow field (viz., where the fluid and the structure interact). However, this methodology requires frequent updating of the mesh (usually at every time step) which may be computationally prohibitive—hence, limiting the wider acceptance of the methodology for addressing FSI problems (particularly those involving very large amplitudes in the motion of the body giving rise to large distortions of the mesh).

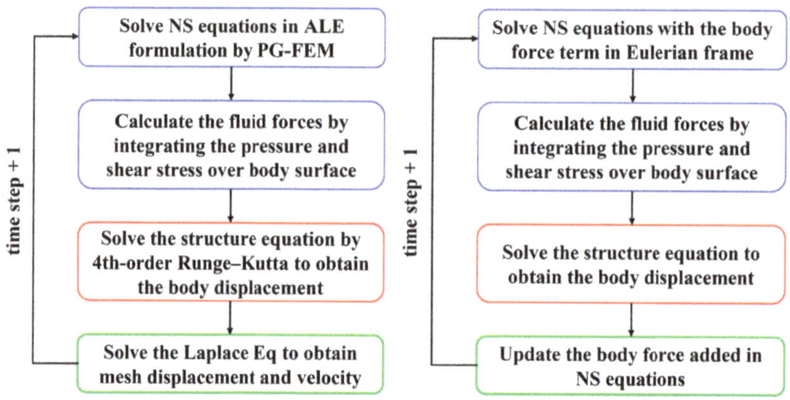

Figure 8. Numerical computational procedure for the simulation of FIV of a bluff body using the ALE algorithm for a moving mesh scheme (left panel) and the immersed boundary method for a fixed mesh scheme (right panel).

(2) Fixed mesh.

To circumvent some of the problems associated with using a moving and regenerating mesh in the body-conforming formulation, a fixed mesh method can alternatively be used for FIV simulation. This methodology uses the notion of an immersed method that employs an overlapping description of the fluid (in an Eulerian frame) and the structure (in a Lagrangian frame) and a non-conforming fluid–structure interface in order to avoid the movement of the mesh (and concomitant re-meshing) at each time step [86,98]. In this case, the fluid–structure interaction can be taken into account by including an additional inertial forcing term into the NS equations, resulting in the transformation from the dynamic Lagrangian coordinate system (for a moving body computation) to a fixed Eulerian coordinate system (for the fluid computation). Needless to say, the method used to determine the body force and how to distribute this force from the "immersed" points into the surrounding Cartesian grid will significantly affect the accuracy of the immersed method [99].

The immersed boundary method (IBM) is a popular non-conforming mesh technique that is based on a fixed Cartesian grid that is not fitted to the geometry of the moving body [100]. This methodology considers the moving structure as a moving boundary that is represented using a set of discrete Lagrangian points and resolves the fluid dynamics on a fixed Eulerian mesh. In order to couple the Eulerian and Lagrangian variables, IBM formulates the body force term (included in the momentum transport equations) as an integral representation involving Dirac delta function kernels—these are generally replaced with regularized delta function kernels in the numerical treatment. In comparison to the moving mesh scheme, IBM is advantageous in that there is no need to regenerate the computational mesh at each time step as a result of the motion of the body. This dramatically improves the computational efficiency for a complex FSI problem, especially in cases involving a changing topology. However, it should be stressed that this computational efficiency comes with a price: namely, the solution obtained using this methodology is probably less accurate in the vicinity of the fluid–structure interface. The steps involved in the application of IBM for FIV simulation are summarized in Figure 8 (right panel).

Apart from IBM, there are other fixed mesh schemes such as the fictitious domain (FD) method [101], the embedded boundary (EB) method [102,103], the immersed interface method, and the immersed body method. These methodologies are useful for the case where the geometric complexity of the fluid–structure interface is so complicated that it would be difficult to address using a moving mesh technique.

In summary, this section provides a brief review of two commonly used numerical methodologies for the simulation of FIV of a bluff body. For a more detail description of these methodologies, the interested reader is directed to references [86,98,104].

4.2. Two-Dimensional or Three-Dimensional Numerical Simulations of FIV

It is known that the FIV response of a bluff body is inherently three-dimensional at a high-Reynolds number in the turbulent flow regime. As a consequence, 2D numerical simulations of FIV for this case are not adequate to capture all the dynamical characteristics of this three-dimensional turbulent flow. Moreover, it should be stressed that even fully 3D turbulence simulation or modelling methodologies (e.g., 3D numerical simulations using LES, DES, RANS) can potentially not fully reproduce the complex flow dynamics in this case, especially for a laminar–turbulence transition that can occur in wakes, shear layers, and boundary layers [105]. In principle, the only numerical methodology that can potentially simulate every aspect of turbulence is DNS which involves solving the three-dimensional NS equations exactly. Unfortunately, a full numerical simulation must provide a complete solution for both the fluid dynamics and fluid–structure interaction which for high-Re turbulent flows is computationally, prohibitive (even insuperable using current computational technology). Nevertheless, if we are interested in the modelling (approximation) of the principal characteristics of the flow field, both the 2D and 3D numerical solutions obtained using either LES or RANS do provide a reasonably accurate approximation of the FIV response of a bluff body.

To this purpose, 3D numerical simulations should be used if the predictive accuracy of the fluid forces and structural response is required, as in a research problem. However, for addressing engineering and industrial FIV problems where parametric studies involving a large number of test cases are required, 2D numerical simulations with modest computational resources may be sufficient (indeed attractive). On the one hand, it has been demonstrated that 2D simulations of the FIV of an elastically-mounted cylinder may be sufficient owing to the fact that oscillations of the cylinder result in a significant increase in the wake correlation length—an increase which drastically diminishes the 3D effects of the vortex shedding [106]. This implies that 2D simulations can be used to provide good predictions for various integrated parameters, such as the aerodynamic forces coefficients. On the other hand, the largest discrepancy between 2D and 3D simulations is expected for high-Reynolds number flows involving strong turbulence effects which are not completely accounted for in a 2D simulation. As an example, the lift and drag coefficients obtained from 2D simulations in this case are overpredicted. Indeed, a number of features in the upper branch of the VIV amplitude response at high values of *Re* as obtained from experimental data are not present in 2D low-*Re* simulations—and, moreover, the corresponding vortex modes are different. In view of this, it is of great importance to ascertain whether a 2D simulation is sufficient to capture the relevant physical mechanisms of FIV in the problem of interest or whether it is necessary to resort to more computationally prohibitive 3D simulations [106] to address the problem.

To determine whether a 2D numerical simulation can be used to address an inherently 3D FIV problem to give useful results, it is necessary to consider firstly whether the wake structures for the problem are essentially two-dimensional (viz., it is necessary to ascertain the threshold Reynolds number for the onset of three-dimensional instabilities for the bluff body under consideration). It is well known that with increasing Reynolds number, the wake of a cylindrical body undergoes a transition from a two-dimensional flow to a three-dimensional perturbation. The critical Reynolds number at which this occurs is determined by the geometric shape and other constraints on the cylinder. A large body of research has demonstrated that the threshold Reynolds number for the flow transition to three-dimensionality over a stationary body occurs at a Reynolds number of about $Re_c = 190$ for a circular cylinder [107,108] and about $Re_c = 160$ for a square cylinder [109]. For other cross-sectional shapes, such as a rectangular cylinder, an elliptic cylinder, and a flat plate, the critical Reynolds number Re_c is a function of the aspect ratio [110,111]. If the cylinder is free to move, the value of the critical Reynolds number Re_c is generally larger than that for the stationary cylinder. For example, for a transversely oscillating circular cylinder, $Re_c \approx$ 250–300 [96,112]. Because the flow is essentially two-dimensional (approximately or better) for $Re \lesssim Re_c$, it is advantageous (in terms of both accuracy and computational efficiency) to perform 2D simulations of FIV for flow Reynolds numbers less than Re_c. For these cases, it is expected that the results are reliable and effectively identical to the 3D simulations of FIV (at least for the primary dynamic responses, although some discrepancies may exist in the predictions for the vortex-shedding patterns). Another consideration concerning whether a 2D numerical simulation can be used to address a FIV problem stems from the fact that the inducing mechanism of relevant dynamics of the flow and structure, such as the branching behavior, the frequency lock-in, the phase jump and the wake mode has no direct relevance to the presence of three-dimensionality in the flow, so that a knowledge of these 2D characteristics can provide valuable information for the more complex 3D flow due to the qualitative similarities between them. To this point, Leontini et al. [112] analyzed in detail the genesis of 3D higher-Re flow behaviors present in a 2D flow in terms of the vibration amplitude, the frequency, and the phase difference.

If the above two considerations have been addressed, one efficient strategy to take full advantage of both 2D and 3D numerical simulations would be to first carry out a large number of 2D numerical simulation cases covering the whole spectrum of relevant parameters at low-Reynolds number, and then conduct a few representative 3D high-*Re* numerical simulations to investigate in detail the cases of interest from the perspective of the three-dimensional flow structures.

4.3. Direct Numerical Simulation (DNS) of FIV

4.3.1. Two-Dimensional DNS in the Laminar Regime

Owing to the prohibitive computational cost and uncertainties in turbulence modelling, most numerical simulations of FIV of a bluff body are conducted at a low-Reynolds number for laminar flow. This involves solving the two-dimensional NS equations coupled with the structure equation of motion. Indeed, a 2D DNS is an indispensable tool for exploring the physical mechanisms and obtaining a fundamental understanding of FIV phenomena.

(1) Single cylinder.

Mittal and various collaborators conducted a series of seminal 2D simulations of the VIV response of a single circular cylinder at a low-Reynolds number. Mittal and Kumar [90] reported the "soft-lock-in" phenomenon (where $f_{vs} < f_n$) in a 2DOF-VIV of a circular cylinder at a low mass ratio for $Re = 325$—a phenomenon that appears to be related to a self-limiting mechanism of the cylinder. Using 2D DNS, Singh and Mittal [87] found that the Reynolds number ($Re =$ 50–500) has a significant influence on the VIV amplitude and vortex shedding mode. More specifically, the latter transitions from a 2S mode for $Re < 300$ to a P + S mode for $Re > 300$. Furthermore, blockage has an important impact on the non-linearity of 2DOF-VIV-like hysteretic behavior at a low-Reynolds number [113]. Prasanth et al. [114] investigated the combined effect of mass ratio and blockage on the transverse oscillation of a circular cylinder and found that the hysteretic behavior appears when the critical mass ratio exceeds a value of 10.11. Prasanth and Mittal [115] conducted a comprehensive numerical study of 2DOF-VIV of a circular cylinder at a low-Reynolds number of $Re = 60$–200 and discovered that a phase jump of 180° occurring in the middle of the lock-in range is related to the self-limiting mechanism of VIV. Mittal [116] conducted 2D numerical simulations for flow past a circular cylinder under a number of different conditions (e.g., increasing/decreasing U_r, unsteady/steady flow) and reported a multi-branch VIV response of the cylinder at $Re = 100$—these results were used to define a new branch which was referred to as Initial Branch II.

Leontini et al. [117] reported two branches in the VIV amplitude response at $Re = 200$ and suggested that a 2D low-Re numerical simulation can be used to garner deeper physical insights into the results obtained from a 3D high-Re simulation (at least as it concerns the nature of the underlying physical mechanisms responsible for important VIV response features). Ji et al. [101] demonstrated the feasibility of using the fictitious domain method for the simulation of a 2DOF-VIV of a circular cylinder at a low-Reynolds number of $Re = 200$. These researchers focused on how the Skop–Griffin parameter ($S_G \equiv 2\pi^3 S_t^2 m^* \zeta_c$), which encapsulates the combined effect of mass and damping on the structure, determines the nature of the VIV response. Their numerical simulations showed that the response amplitude decreased rapidly with increasing values of S_G. Li et al. [118] conducted numerical simulations at low-Reynolds number and at small mass-damping parameter in order to analyze various non-linear phenomena associated with 1DOF- and 2DOF-VIV (e.g., lock-in, beating, phase-switching, limit cycles, bifurcation) at low-Reynolds number and low mass-damping parameter. These investigators concluded that the streamwise vibrations influence the transverse vibrations. Étienne and Pelletier [119] utilized results obtained from numerical simulations of a zero mass and zero damping vibrating circular cylinder in order to define the regions in the (Re, U_r) plane where vibrations are either absent or present—these results showed that transverse vibrations occur for a reduced velocity $U_r > 40$. Bourguet and Jacono [120] performed a parametric study using 2D numerical simulations to investigate the transverse VIV of a rotating cylinder with a given rotation rate at $Re = 100$. These researchers reported a bell-shaped amplitude response, the maximum amplitude of which was three times that of the VIV of a non-rotating cylinder. In addition, the autorotation of the cylinder was found to affect the wake pattern, the lift frequency and the phase. Zhao et al. [121] conducted numerical simulations of the 2DOF-VIV of a rotating circular cylinder at a low rotation rate at $Re = 150$ and found that the forced rotation of the cylinder significantly enhances the in-line oscillation amplitude and lock-in range.

Some recent numerical studies of the VIV characteristics of a single circular cylinder in laminar flow include the following: VIV of a cooled circular cylinder in the presence of thermal buoyancy [122]; some unusual behavior in the time history of the displacement, frequency, and lift coefficient of VIV [123]; and the presence of a three-branch VIV response for a low mass-damping cylinder at low $Re = 300$ [124].

A number of researchers have conducted 2D simulations of laminar flow past elastically-mounted non-circular cylinders (e.g., elliptical, square, rectangular, triangular). A circular cylinder can be considered as a special case of an elliptic cylinder with an aspect ratio of unity. As a consequence, the effect of the aspect ratio (viz., ratio of the cylinder dimension in the transverse direction to that in the streamwise direction) on the FIV response of an elliptic cylinder can be studied numerically. For example, Yogeswaran et al. [125] systematically investigated the dynamics of a freely vibrating elliptic cylinder at eight different aspect ratios AR ($0.7 < \mathrm{AR} < 1.43$) in the laminar flow regime ($60 < Re < 140$). These researchers examined the VIV response (amplitude, phase lag, fluid force), the flow pattern in the wake, and the hysteresis phenomenon and found that the peak amplitude increases with an increasing AR. Chen et al. [126] found that the aspect ratio has a significant effect on the VIV response and on the heat transfer to the elliptical cylinder. Leontini et al. [127] investigated the FIV response of an elliptic cylinder with a fixed aspect ratio of 1.5 for various angles of attack AOA (viz., the angle between flow direction and major axis of ellipse) in the range $0^\circ < \text{AOA} < 90^\circ$. These researchers observed that the VIV for the elliptic cylinder is very similar to that of a circular cylinder for $\text{AOA} < 30^\circ$. However, it was found that a galloping-type oscillation may be triggered when $\text{AOA} > 30^\circ$, which may be due to an enhancement of the lift force by the stronger wake vortices.

It is well known that the flow past a square cylinder can exhibit both VIV and galloping owing to its asymmetry. Research conducted for the flow past a square cylinder has focused primarily on the specific conditions that trigger galloping, such as the critical values of the Reynolds number [128–131], the mass ratio [63,129–132], and the angle of attack [95,133]. In general, the consensus is that a light square cylinder can undergo galloping for $\text{AOA} = 0^\circ$, but exhibits VIV for $\text{AOA} = 45^\circ$ and for intermediate angles of attack provided the Reynolds number exceeds some critical value. Two-dimensional simulations at low-Reynolds number can also be utilized to generate a large dataset of lift and drag coefficients which, in turn, can be used to determine the parameters of that determine a quasi-steady galloping model [63]. The FIV characteristics of a square cylinder under other conditions can also be investigated using 2D numerical simulation (e.g., shear flow past a square cylinder [134]). A rectangular cylinder is the generalization of a square cylinder having different aspect ratios (viz., an aspect ratio other than unity). Presently, low-*Re* 2D numerical simulations of the flow past a rectangular cylinder have focused primarily on the effect of the aspect ratio on the dynamic response of the vibrating system. For example, Zhao [97] investigated the flow past a rectangular cylinder for five different aspect ratios (namely, 0.3, 0.5, 0.7, 1, and 1.25) and observed that both VIV and galloping occur provided the aspect ratio is larger than 0.3. Similar to square and rectangular cylinders, a triangular cylinder is also prone to galloping due to the presence of sharp corners in the structure, to fixed separation points and to the afterbody in the separated wake. Wang et al. [135] investigated the streamwise and cross-flow FIV responses of a triangular cylinder under three typically low incidence angles (namely, 0°, 30°, and 60°) at $Re = 100$. They found the dynamic response at the small incidence angles of 0° and 30° exhibited VIV similar to that for the flow past a circular cylinder, while for the larger incidence angle of 60° the dynamic response exhibited galloping.

(2) Passive turbulence control cylinder.

Two-dimensional numerical simulations can be used to study the flow past a cylinder with passive turbulence control (PTC) which is, perhaps, the simplest flow control strategy employed to control the intensity of motion of a body. Passive turbulence control involves either attaching to or placing around the body one or more small device(s) (e.g., rod, splitter plate, roughness plate, fairing) to control the intensity of motion of the body.

Small controlling rods are one of the most commonly used PTC devices. Zhao et al. [95] used numerical simulation to investigate the dynamical characteristics of a PTC cylinder consisting of a circular cylinder of diameter D and a small rod (with diameter $0.2\ D$) attached at different angular locations on the cylindrical surface. Liu et al. [136] fixed the control rod upstream of a cylinder of diameter D and studied the effects of rod diameter ($0.4\ D$, $0.6\ D$, and $0.8\ D$) and center-to-center spacing between the cylinder and rod ($1.2\ D$ and $2.8\ D$) on the dynamical response of the system. Liu et al. [137] used the numerical simulation of a cylinder–rod system, consisting of a cylinder of diameter D and six detached controlling rods of diameter $0.25\ D$ in order to investigate the effect of the separation (ranging from $0.1\ D$–$2\ D$) between the cylinder and the multiple rods on the dynamical response.

The splitter plate is another easy-to-use flow control device. Various configurations, such as the relative location of the splitter plate with respect to the main cylinder (e.g., upstream, downstream, side-by-side), the type of connection (e.g., attached or detached), the nature of the plate material (e.g., flexible or rigid), the constraints applied to the cylinder and/or plate (e.g., fixed or free) and the plate length can be investigated using 2D numerical simulations in laminar flow. For these various configurations, the effect of the mass ratio, of the damping ratio, of the Reynolds number, of the plate length, and of the separation between the cylinder and the detached plate can be studied [138–140].

(3) Array of cylinders.

An array of cylinders is a more realistic and practical system to study than a single cylinder owing to the fact that this configuration is found frequently in the form of offshore platforms (floating or fixed), multiple marine risers, and a fluid energy harvester system. We consider the simpler case where each cylinder in the cylinder array has the same diameter. Needless to say, the dynamic response of a cylinder array is expected to be much more complex than that of a single cylinder—an observation which highlights the challenges that need to be confronted if such a system is to be investigated numerically using 3D flow simulation at high-Reynolds number. In view of this, it is expected that 2D low-Reynolds number numerical simulations are useful for providing the much needed physical insights into the dynamic response and the underlying physics of a multi-cylinder system.

Carmo et al. [106] applied 2D numerical simulations at $Re = 150$ to study the effect of the separation between two cylinders on the FIV response of the system. In this study, the separation between the two cylinders is carefully selected such that the downstream freely-vibrating cylinder experiences the transverse FIV (wake galloping) that results due to the wake shed from the upstream fixed cylinder. The numerical results show that the tandem two-cylinder system provokes FIV with a wider synchronization boundary, a larger maximum amplitude, and an overall larger vibration amplitude at high speeds in comparison with that obtained for a single cylinder. This may be due to the presence of an oscillatory flow between the two cylinders that is absent for a single cylinder. Other numerical studies have focused on a number of different configurations of the cylinder array—key parameters that determine the configuration include the number of cylinders, the relative location of the cylinders (e.g., tandem, staggered, side-by-side), and the constraints imposed on each cylinder (e.g., fixed or free) [94,97,141–143].

Table 1 summarizes the 2D numerical studies of FIV for various bluff bodies conducted for low-Reynolds number flows. A 2D DNS is expected to provide detailed flow structures and ameliorate the uncertainties associated with the turbulence modelling at low-Reynolds numbers less than 300 where the 3D turbulent characteristics are not relevant. Moreover, a 2D DNS permits large simulations to be conducted at affordable computational cost—hence, it is perhaps the most widely used numerical technique used to study the interactions between the flow dynamics and the FIV response of a bluff body, especially for those cases involving complex geometries and/or dynamics.

Table 1. Summary of representative 2D DNS studies of flow-induced vibration of a bluff body for low-Reynolds number flows in the laminar flow regime.

Author/Year	Motion	Shape	m^*	ζ	Re	CFD	FSI Scheme
Mittal and Kumar [90] [1]/1999	2DOF-VIV	circular	4.72	0.0003	325	SST-FEM	DSD/SST
Singh and Mittal [87]/2005	2DOF-VIV	circular	10	0	50–500	SST-FEM	DSD/SST
Prasanth et al. [113]/2006	2DOF-VIV	circular	10	0	60–125	SST-FEM	DSD/SST
Prasanth and Mittal [115]/2008	2DOF-VIV	circular	10	0	60–200	SST-FEM	DSD/SST
Prasanth et al. [114]/2011	2DOF-VIV	circular	1–100	0–0.1	60–150	SST-FEM	DSD/SST
Mittal [116]/2017	1DOF-VIV	circular	70	0	100	SST-FEM	DSD/SST
Leontini et al. [117]/2006	1DOF-VIV	circular	10	0.01	200	SEM	fixed mesh
Ji et al. [101]/2011	2DOF-VIV	circular	1.27	0.003–3.3	60–300	FDM	FD
Li et al. [118]2011	1DOF-VIV	circular	1	0.003	200	ST-FEM	moving mesh
Étienne et al. [119]/2012	2DOF-VIV	circular	0	0	15–45	FEM	ALE
Bourguet and Jacono [120]/2014	1DOF-VIV	circular	10	0	100	SEM	fixed mesh
Zhao et al. [121]/2014	2DOF-VIV	circular	2	0	150	PG-FEM	ALE
Garg et al. [122]/2019	1DOF-VIV	circular	2	0	150	FDM	IBM
Dorogi and Baranyi [124]/2020	1DOF-VIV	circular	10	0–0.05	300	FDM	fixed mesh
Shen and Chen [123]/2022	1DOF-VIV	circular	2–50	0.041	150	FVM	IBM
Yogeswaran et al. [125]/2014	2DOF-VIV	elliptic	10	0	60–140	SST-FEM	DSD/SST
Leontini et al. [127]/2018	1DOF-FIV	elliptic	1	0	200	SEM	fixed mesh
Chen et al. [126]/2021	1DOF-VIV	elliptic	10	0	150	FVM	moving mesh
Su et al. [133]/2007	1DOF-FIV	square	3	0	100	FEM	ALE

Table 1. *Cont.*

Author/Year	Motion	Shape	m^*	ζ	Re	CFD	FSI Scheme
Sen and Mittal [128]/2011	2DOF-FIV	square	10	0	60–250	SST-FEM	DSD/SST
Joly et al. [63]/2012	DOF-FIV	square	1–20	0–0.01	150	FEM	moving mesh
He et al. [129]/2012	2DOF-FIV	square	10, 20, 40	0	60–250	FEM	ALE
Sen and Mittal [130]/2015	2DOF-FIV	square	1, 5, 10, 20	0	60–250	SST-FEM	DSD/SST
Zhao et al. [95]/2013	2DOF-VIV	square	3	0	100	PG-FEM	ALE
Sun et al. [134]/2017	2DOF-FIV	square	2	0	80,160	FEM	ALE
Sourav and Sen [131]/2019	2DOF-FIV	square	3–4	0	60–250	SST-FEM	DSD/SST
Zhao [97]/2015	2DOF-VIV	rectangular	10	0	200	PG-FEM	ALE
Wang et al. [135]/2014	2DOF-FIV	triangular	2	0	100	FEM	ALE
Zhao et al. [95]/2013	2DOF-FIV	circular-rod	2	0	250	PG-FEM	ALE
Liu et al. [136]/2020	2DOF-FIV	circular-rod	2	0.007	150	FEM	ALE
Liu et al. [137]/2022	1DOF-FIV	circular-rod	2	0	200	FEM	ALE
Sahu et al. [138]/2019 [2]	1DOF-FIV	circular-plate	2–1000	0	92–150	SST-FEM	moving mesh
Zhu et al. [139]/2020 [3]	2DOF-FIV	circular-plate	6.9	0.01	120	FVM	moving mesh
Wang et al. [140]/2018 [4]	2DOF-FIV	circular-plate	10	0	100, 200	FEM	ALE
Borazjani and Sotiropoulos [141]/2009	2DOF-VIV	2-circular	2	0	200	IBM	IBM
Carmo et al. [106]/2010	1DOF-FIV	2-circular	2	0.007	150	SEM	ALE
Griffith et al. [142]/2017	1DOF-FIV	2-circular	2.546	0	200	FDM	IBM
Zhao [94]/2013	1DOF-FIV	2-circular	2	0	150	PG-FEM	ALE
Han et al. [143]/2015	2DOF-VIV	4-circular	6	0	80–160	CBS-FEM [5]	ALE
Zhao [97]/2015	1DOF-FIV	36-circular	2.5	0	100	PG-FEM	ALE

[1] The new moving mesh scheme was utilized to remesh only as needed instead of frequent remeshing. SST: stabilized space–time. [2] A circular cylinder with a splitter plate attachment. [3] circular cylinder with upstream and/or downstream splitter plate attachment(s). [4] The flexible splitter plate was subjected to the wake from the upstream fixed circular cylinder. [5] CBS-FEM: characteristic-based split finite-element method.

4.3.2. Three-Dimensional DNS in the Turbulent Regime

Three-dimensional DNS is needed in order to conduct the most realistic simulations of the FIV of a bluff body and, indeed, is required for cases where 3D flow effects are significant, such as spanwise wake correlations and the interaction of streamwise and spanwise vortices in strongly turbulent flows. Newman and Karniadakis [144] used 3D DNS to investigate the oscillation of an infinite flexible cable at $Re = 200$ (early transitional regime) and compared these results with 2D numerical simulations conducted at $Re = 100$ (laminar flow). These researchers found that the transverse amplitude predicted by the 3D numerical simulation agreed well with experimental data. However, the transverse amplitude was found to be larger than that obtained from the 2D numerical simulation. Newman and Karniadakis [144] also used the spectral element-Fourier method for the spatial discretization of the coupled system consisting of the fluid NS equations and the cable equation. This numerical methodology has also been used in many subsequent DNS studies [145–148].

Evangelinos et al. [145] used 3D DNS to simulate the transverse VIV of both rigid and flexible circular cylinders at a subcritical Reynolds number of $Re = 1000$ and presented results for the detailed force distribution along the cylinder span. These investigators also modified the semi-empirical model proposed by Skop and Griffin [21] by using their DNS data to provide improved model coefficients (using a least-squares fitting of the coefficients of the model to their data). Lucor et al. [146,148] conducted 3D DNS of the transverse VIV of a rigid circular cylinder in a uniform flow at three different Reynolds numbers ($Re = 1000$, 2000, and 3000) with a low mass-damping parameter and a large aspect ratio of 26. These investigators successfully reproduced the three-branch VIV and found that the large-amplitude response in upper branch (lock-in range) was strongly dependent on the Reynolds number. Furthermore, they confirmed the occurrence of a sharp drop in the spanwise correlation of the force and wake near the transition from the upper to the lower branch (still at a large amplitude)—a result which appears to be related to the phase difference between the displacement and lift force.

Carmo et al. [106] and Zhao et al. [96] compared 2D and 3D DNS results for an oscillating cylinder. Carmo et al. [106] also conducted a number of 3D numerical simulations in the transitional regime ($Re = 300$) in order to elucidate the relationship between the 3D flow structures and the VIV of a rigid circular cylinder. By comparing these results with 2D numerical simulations performed *Re* = 150, these researchers found that the 2D and 3D simulation results were very similar to each other in terms of the amplitude response—the difference in the maximum amplitude between the two results was small, supporting the notion that a 2D numerical simulation in the transitional regime can yield accurate results. Zhao et al. [96] investigated the transverse VIV of a rigid circular cylinder at different Reynolds numbers using a 3D DNS implemented with the Petrov–Galerkin Finite Element Method (PG-FEM). A number of important results were reported in this investigation. Firstly, the wake flow behind a circular cylinder undergoing VIV was shown to be two-dimensional at *Re* = 150 (laminar regime) as the 2D and 3D numerical simulations yielded almost identical results in terms of the vibration amplitude, the frequency, and the vortex flow patterns. Secondly, 3D numerical simulations of VIV at *Re* = 1000 (turbulent regime) were undertaken to study the variation of the flow along the cylinder span. Thirdly, the numerical simulation of flows with Reynolds numbers varying from 200 to 800 was performed at a constant reduced velocity in the middle of lock-in for both *Re* = 150 and 1000 in order to determine the critical Reynolds number for the wake flow to transition from a 2D to a 3D flow—for a vibrating cylinder, the critical Reynolds number was shown lie in the range from 250 to 300.

Mittal [88] carried out a 3D DNS study of the streamwise and transverse VIV of a rigid circular cylinder at *Re* = 1000. To this purpose, he used the stabilized space–time FEM (SST-FEM) technique to solve the coupled set governing equations (fluid and structure), as well as a moving mesh scheme for handling the fluid–structure interaction. A number of interesting phenomena were reported in the work including the three-branch VIV response, the beating of vibration amplitude and fluid force within the initial branch, the soft-lock-in (viz., the vortex shedding frequency matches the vibration frequency, but is different than the natural frequency) in the lower branch, the hysteresis effect in initial-upper transition, and the intermittency in upper-lower transition. In particular, Mittal [88] observed a 2S vortex mode in the upper branch which differs from the 2P vortex mode obtained from experiments at higher *Re*.

Gsell et al. [149] used 3D DNS (based on FVM) to investigate the 2DOF-VIV of a rigid circular cylinder at the moderate Reynolds number of *Re* = 3900. These researchers reported a three-branch VIV response and, in particular, they examined the fluid force in terms of the phase difference and the spectral content. In a recent effort, Chen et al. [150] used 3D DNS to investigate 2DOF-VIV corresponding to the flow past an elastically-mounted circular cylinder at *Re* = 500 with the cylinder positioned near a stationary wall—a scenario that mimics an engineering application, such as a marine pipeline above the seabed. These investigators reported a two-branched (initial and lower) VIV response accompanied by two desynchronization regions. They observed a significant modification of the three-dimensional wake structures, the vortex dynamics and the flow physics at each velocity range due to the presence of the stationary wall and the concomitant boundary layer.

Table 2 presents a concise summary of the details of the 3D DNS studies of VIV of a circular cylinder described above. It should be stressed that 3D DNS is only applicable for low- and moderate-Reynolds number flow in the transitional regime owing to its prohibitive computational cost. For high-*Re* flows (e.g., laminar–turbulent transition, fully turbulent), various advanced turbulent modelling techniques have been developed and can be used to predict the turbulent flow at an affordable computational cost (albeit at a lower fidelity).

Table 2. Summary of 3D DNS of flow-induced vibration of a bluff body in the turbulent regime.

Author/Year	Motion	Shape	m^*	ζ	*Re*	CFD	FSI Scheme
Newman and Karniadakis [144]/1997	1DOF-VIV	circular	2	0	200	SEM	fixed mesh
Evangelinos et al. [145]/2000	1DOF-VIV	circular	2	0	1000	SEM	fixed mesh
Lucor et al. [148]/2005	1DOF-VIV	circular	2	0	1000–3000	SEM	fixed mesh
Carmo et al. [106]/2010	2DOF-VIV	2-circular	2	0.007	300	SST-FEM	ALE
Mittal [88]/2013	2DOF-VIV	circular	10	0	1000	SST-FEM	ALE
Zhao et al. [96]/2014	1DOF-VIV	circular	2	0	1000	PG-FEM	ALE
Gsell et al. [149]/2016	2DOF-VIV	circular	2	0	3900	FVM	fixed mesh
Chen et al. [150]/2022	2DOF-VIV	circular	2	0	500	FDM	IBM

4.4. Large-Eddy Simulation (LES) of FIV

The main idea of large-eddy simulation (LES) is to resolve the larger turbulent eddies whose scales are larger or comparable to the size of a computational grid cell and to model the smaller turbulent eddies whose scales are smaller than the grid cell size and so cannot be resolved adequately. These smaller eddies are modelled using a so-called subgrid scale (SGS) model. Typically, a spatial filter is defined either explicitly or implicitly and used is to separate the resolved and unresolved turbulent flow scales. Naturally, the particular spatial filter used and the choice of the SGS model will affect the accuracy of the LES results. When using LES to simulate the FIV response of a bluff body, the boundary-layer flow near the cylinder wall needs to be modelled carefully because the flow separation and the vortex formation and shedding typically occur in this region—a region characterized by large velocity gradients and an increased anisotropy in the turbulent fluctuations. When compared to DNS, LES is computationally less costly and is capable of capturing the key unsteady flow features. In consequence, LES provides the most accurate predictions of FIV apart from DNS (where no modelling of the small-scale turbulence is required).

Large-eddy simulation using the Smagorinsky (SM) SGS model has been used to investigate the free vibration of a 2D circular cylinder [105,151–153] and the forced vibration of a 3D circular cylinder [154] for high-Reynolds number flows. These studies demonstrated that LES using the Smagorinsky SGS model provided reasonably accurate predictions of the FIV phenomenon. More recently, Pastrana et al. [155] investigated the transverse and streamwise VIV of a rigid circular cylinder with low mass-damping for flows with Reynolds numbers in the range *Re* =3900–11,000 using a high-fidelity LES with a wall-adapting local-eddy (WALE) viscosity model that accounts for the near-wall scaling of the eddy viscosity. These researchers concluded that LES provided predictions that are in good conformance with the experimental data—the VIV maximum amplitude, the characteristic flow features, and the non-linear VIV behavior in the super-upper branch (e.g., hysteresis, transition delay) are well predicted using LES. In order to assess the reliability of strip theory for the semi-empirical prediction of VIV, Wang et al. [156] conducted an LES study involving the free vibration of a 3D flexible cylinder and the forced vibration of a 2D rigid cylinder. A key result of this study was that the hydrodynamic force coefficients obtained in the two cases were comparable, but the wake patterns predicted for the two cases were different. Janocha et al. [157] performed a number of 3D LES calculations to investigate the VIV of a circular cylinder at a Reynolds number of *Re* = 3900 for representative reduced velocities within the initial ($U_r = 3$), upper ($U_r = 5$) and lower ($U_r = 7$) branches. These investigators applied the multi-scale proper orthogonal decomposition (mPOD) technique to quantify the three-dimensional structures in the wake flow. A very good agreement between the predictions provided by the 3D LES and the experimental data were reported—providing a validation for the accuracy of 3D LES for the prediction of VIV of a bluff body in the subcritical flow regime.

Daniels et al. [158] used 3D LES to investigate the free motion (namely, heaving and pitching) of a rectangular bridge deck subjected to the uniform and turbulent flow. These simulations provided a detailed visualization of the complex vortex formation and development in this flow and yielded accurate predictions of the peak loading. However, it was also shown that the turbulence intensity in the incident flow significantly decreased the amplitude of oscillation, while the turbulent integral length scales in the free stream enhanced the spanwise wake correlation and, as a result, increased the amplitude of oscillation of the body.

Table 3 summarizes the key properties (e.g., physical parameters, Reynolds number, numerical scheme) of a number of LES studies of the FIV of a bluff body. It should be stressed that the body of work here appears to suggest that LES can be used to obtain high-fidelity simulations of the FIV response of a bluff body—albeit at a computational cost that is still prohibitive for real-world engineering and industrial applications of practical relevance.

Table 3. Summary of LES studies of the flow-induced vibration of a bluff body.

Author/Year	Motion	Shape	m^*	ζ	Re	Scale	SGS	CFD	FSI Scheme
Zhang and Dalton [151] [1]/1996	1DOF-VIV	circular	10	0.02	13,000	2D	SM	FDM	fixed mesh
Al-Jamal and Dalton [105] [2]/2004	1DOF-VIV	circular	7.85	0.02	8000	2D	SM	FDM	fixed mesh
Tutar and Holdo [154] [3]/2000	1DOF-forced	circular	/	/	24,000	3D	SM	FEM	fixed mesh
Pastrana et al. [155] [4]/2018	2DOF-VIV	circular	2.6	0	3900–11,000	2D	WALE	FVM	ALE
Wang et al. [156] [5]/2021	1DOF-VIV	circular	4	0.087	550–900	3D	/	EVM	fixed mesh
Janocha et al. [157] [6]/2022	1DOF-VIV	circular	/	/	3900	3D	WALE	FVM	ALE
Daniels et al. [158]/2016	1DOF-VIV	rectangular	57	0.0021	40,000	3D	MTS	FVM	ALE

[1] The 2D LES predictions were in good agreement with the experimental data. [2] The 2D LES accurately predicted the VIV response, but did not capture the vortex modes well. [3] The 3D LES provided more realistic predictions of the flow field and the aerodynamic forces than those provided by the 2D LES. [4] LES results agreed well with the experimental data for the VIV response and the flow pattern. [5] Comparison of LES results for the free vibration of a flexible body and the forced vibration of a rigid body. [6] LES reproduces correctly the three branches of the VIV response and the corresponding vortex modes: namely, the initial (2S), upper (2P), and lower (2P) branches.

4.5. Reynolds-Averaged Navier–Stokes (RANS) Modelling of FIV

RANS modelling decomposes an instantaneous flow quantity (e.g., velocity) into the mean and fluctuating components using the Reynolds decomposition—the former component is determined using from the time-averaged (or ensemble-averaged) NS equations and the latter component is modelled using some form of semi-empirical approximations (e.g., a turbulence closure model). With an acceptable computational cost while providing a reasonable accuracy, RANS has become the primary numerical tool used in engineering applications—and, more particularly, for the numerical modelling of the VIV of a circular cylinder at Reynolds numbers in the subcritical flow regime (viz., $300 < Re < 3 \times 10^5$) that consists of a completely turbulent wake flow and a laminar boundary-layer flow on the cylinder surface.

In 1996, Khalak and Williamson [159] undertook a series of experimental investigations in order to elucidate the dynamic characteristics of transverse VIV of a rigid circular cylinder for Reynolds-number flows in the range from 2000 to 12,000. These experiments provided benchmark data that have been used since for the validation of various RANS modelling studies of FIV [92,160–162]. As an example of this effort, Figure 9 displays the maximum vibration amplitude A_{max} (normalized by the cylinder diameter D) as a function of the reduced velocity U_r—the RANS modelling of an elastically-mounted circular cylinder that can undergo transverse VIV conducted by Guilmineau and Queutey [160] (G&Q), Pan et al. [161], and Wanderley et al. [162] are compared with some experimental data obtained by Khalak and Williamson [159] (K&W).

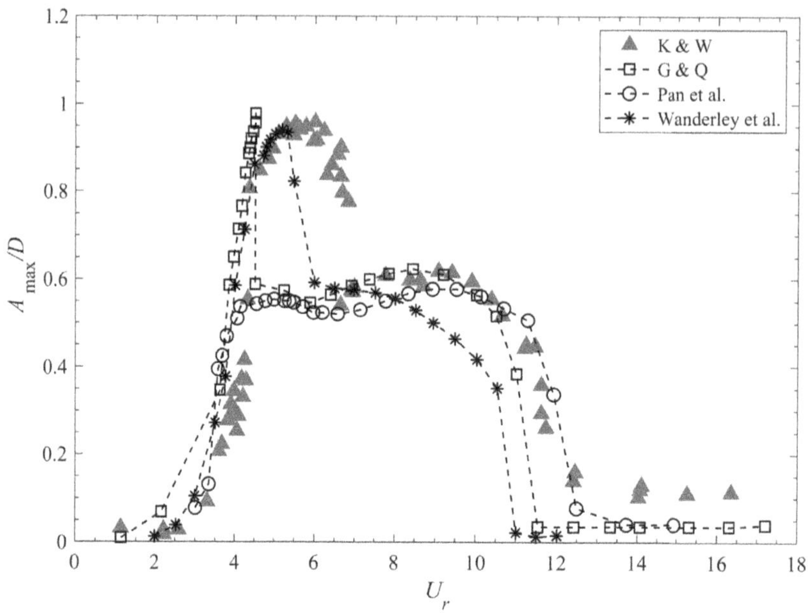

Figure 9. The variation with the reduced velocity U_r of the normalized maximum structural amplitude (A_{max}/D) of FIV consisting of an elastically-mounted circular cylinder of diameter D with a mass ratio of 2.4 and mass-damping ratio of 0.013. The results of three RANS modelling studies are compared with some experimental data reported by Khalak and Williamson [159] (K&W). The RANS modelling studies shown here were conducted by Guilmineau and Queutey [160] (G&Q), Pan et al. [161], and Wanderley et al. [162].

Guilmineau and Queutey [160] applied RANS with the shear stress transport (SST) k-ω model to investigate the influence of initial conditions on the transverse VIV of a circular cylinder for flows with Reynolds number ranging from 900 to 15,000. A key result reported in this modelling study is that initial conditions of rest and of decreasing velocity reproduced the two-branch VIV response (initial and lower), while that of an increasing velocity yielded a very narrow upper branch with an amplitude that is comparable to the experimental data. Applying the same numerical methodology as used in Guilmineau and Queutey [160], Pan et al. [161] focused their attention on the reason for failure of RANS modelling to correctly predict the upper branch of the VIV response. These researchers suggested that the reason was related to the random (stochastic) behavior of the vortex-shedding process which cannot be achieved using a time-averaged (or ensemble-average) RANS modelling strategy. Wanderlay et al. [162] applied RANS with the k-ϵ turbulence closure model and demonstrated that this approach was able to provide very accurate predictions of the three branches of VIV response, of the force coefficients, and of the lock-in range.

Zhao and Cheng [92] conducted RANS modelling of the streamwise and transverse VIV of a rigid circular cylinder of diameter D and correctly predicted the characteristic behavior of the super-upper branch of the VIV response—a maximum amplitude of 1.5D when the flow velocity is increasing. These researchers also reported three vortex shedding modes of VIV in the near-wall region based on their RANS modelling study. Some numerical studies apply RANS modelling to investigate the influence of the Reynolds number on the VIV response of a circular cylinder. Wanderlay and Soares [163] used RANS to model the 1DOF-VIV response of a circular cylinder at a small mass-damping parameter for three different ranges of the Reynolds number (namely, Re = 2000–6000, 2000–12,000, and 2000–24,000) and for six fixed values of the Reynolds number (namely, Re = 100, 200, 300, 400, 1000, and 1200). It was found that the Reynolds number has a significant effect on the amplitude response, the lift coefficient, and the vibration frequency for a low mass-damping system and that the upper branch of the VIV response only appears as the value of Re is varied, but not for a fixed value of Re. This study suggests that the combined influence of Re and U_r on the VIV response requires more investigation.

Recently, Gu et al. [164] applied RANS with the SST k-ω model to study the effect of Reynolds number on the VIV response at small and moderate values of Re. A non-dimensional parameter $\alpha \equiv f_n D^2/\nu$ was defined in order to deal with the combined influence of Re and U_r on the VIV response—a parameter that can be varied by changing the flow velocity, the cylinder diameter and the structural stiffness while ensuring that the values of the mass and the damping ratio are fixed. These researchers concluded that the VIV response is sensitive to the Reynolds number for moderate values of Re and, indeed, it was shown that the maximum amplitude of the VIV response is strongly dependent on Re. Kang et al. [165] used RANS to study Re effects for a 2DOF-VIV consisting of a circular cylinder at small and large mass ratios for three different Re ranges: namely, Re = 1450–10,200, 5800–40,800, and 13,050–91,800. With respect to the influence of mass and damping on the VIV response, Khan et al. [166] and Anwar et al. [167] used RANS with the SST k-ω turbulence closure to model the transverse VIV response of a circular cylinder with a mass ratio of 11 at Re = 10,000. These researchers demonstrated that RANS can be used to correctly predict the three branches and vortex modes in the VIV response for higher mass ratio at a high-Reynolds number—albeit the peak amplitude within lock-in range was underestimated. Martins and Avila [168] investigated different combinations of the structural damping ratio and the Reynolds number on the VIV response using the RANS SST k-ω model.

The studies summarized above involve RANS modelling of the VIV of an elastically-mounted circular cylinder. However, RANS modelling has also been utilized to predict the FIV of bluff bodies with more complex (non-circular) cross-sectional shapes. Cui et al. [169] used 2D RANS modelling to investigate the transverse FIV of square and rectangular cylinders for different inflow angles at moderate values of the Reynolds number. These investigators reported that galloping is induced on the square cylinder at an inflow angle of 0° and on the rectangular cylinder at inflow angles of 0° and 90°. Han et al. [170] employed 2D RANS to model the power performance of a low-mass square energy harvester based on an 1DOF-FIV (VIV and galloping) at Re = 24,000–160,000 and investigated the effect of the inflow angle and the structural damping ratio on the FIV response. One interesting phenomenon observed in this study was that a large-amplitude response occurs at large value of U_r for the square cylinder, regardless of the value of the inflow angle (0°, 22.5°, and 45°) or mode of response (viz., whether VIV or galloping). Other notable studies using 2D RANS modelling to investigate the FIV response of non-cylindrical bluff bodies include Ding et al. [171] (for square, triangular, quasi-trapezoidal shaped bodies), Zhang et al. [172] (for triangular, square, pentagonal, circular-shaped bodies), Zhang et al. [173] (for triangular, square, hexagonal, octagonal, 24-sided, circular shaped bodies), and Wang et al. [174] (for different rear edges such as semicircular, triangular, trapezoidal, rectangular). In these studies, the RANS modelling provided predictions that agreed well with experimental data in terms of the FIV response. Most of these studies also determined the power production arising from the FIV of different shapes of the bluff bodies in order to provide practical guidance for the design of a FIV energy harvester.

RANS modelling can be used to predict the dynamic response of a cylinder with PTC involving, in particular, attached or detached appendages having more complex and/or irregular geometric structure. To this purpose, Zhu and Yao [175] used 2D RANS to show that the correct placement of multiple small rods around an elastically mounted circular cylinder can result in a significant suppression of the streamwise and transverse VIV response. With respect to the incitement of oscillatory motion, Ding et al. [176] used 2D RANS with the Spalart–Allmaras (SA) turbulence closure to model the FIV response of a rigid circular cylinder of diameter D with two roughness strips for flows at high-Reynolds numbers in the range from 30,000 to 110,000. These researchers showed the presence of a five-branch FIV response for this configuration (viz., static, VIV initial, VIV upper, transition from VIV to galloping, and galloping). Furthermore, it was shown that the maximum amplitude was increased due to the presence of the roughness strips—indeed, the maximum amplitude was as large as 3.5 D. Another parametric study was conducted by Wang et al. [177] who showed that attaching two symmetrical small-diameter cylindrical rods at angular locations of 55° to 65° around the circumference of the cylinder significantly extended the range of aeroelastic instability of the vibrating system.

Zhao and Cheng [93] used RANS with the SST k-ω model to study a 2DOF-VIV system consisting of a rigidly connected four-cylinder square configuration arranged with a fixed center-to-center separation. The modelling was undertaken for flows with Reynolds numbers in the range Re = 1000–20,000. A key result of this study was that the inflow angle of the incident flow has a significant effect on the lock-in range, the mode of vibration of the amplitude response, and the hydrodynamic force coefficients. RANS modelling has been used to investigate the transverse VIV of a two-cylinder system, such as two rigidly coupled side-by-side cylinders with different diameters [178], side-by-side and tandem arrangements of cylinders [179], and two elastically coupled side-by-side cylinders in symmetric and asymmetric configurations [180].

Table 4 provides a summary of a number of important studies involving the use of RANS modelling to predict the FIV response for single or multiple bluff bodies of various shapes (single cylinder, cylinder with PTC, array of cylinders). Some studies included here involve very complicated FIV responses, especially those for flows in the subcritical Reynolds number regime.

Table 4. Summary of a number of representative RANS modelling studies of flow-induced vibration of bluff bodies.

Author/Year	Motion	Shape	m^*	ζ	Re [1]	Turbulence	CFD
Guilmineau and Queutey [160]/2004	1DOF-VIV	circular	2.4	0.0054	900–15,000	SST k-ω	FVM
Pan et al. [161]/2007	1DOF-VIV	circular	2.4	0.0054	2500–13,000	SST k-ω	FVM
Wanderlay et al. [162]/2008	1DOF-VIV	circular	1.88	0.00542	2000–12,000	k-ϵ	FDM
Zhao and Cheng [92]/2011	2DOF-VIV	circular	2.6	0.005	1000–15,000	SST k-ω	PG-FEM
Wanderlay and Soures [163]/2015	1DOF-VIV	circular	1.88	0.00542	100–24,000	k-ϵ	FDM
Gu et al. [164]/2022	1DOF-VIV	circular	1.28, 2.4	0.01, 0.006	200–27,000	SST k-ω	FEM
Kang et al. [165]/2019	2DOF-VIV	circular	2.6, 13	0.005	1450–91,800	SST k-ω	FVM
Anwar et al. [167]/2022	1DOF-VIV	circular	2.4, 11	/	10,000	SST k-ω	FEM
Martins and Avila [168] [2]/2019	2DOF-VIV	circular	2.4	0.005–0.01	750–130,000	SST k-ω	FEM
Cui et al. [169]/2016	1DOF-FIV	rectangular	2.4	0.0054	385–19,250	SST k-ω	PG-FVM
Han et al. [170]/2021	1DOF-FIV	square	1.45	0.01–0.7	24,000–160,000	SST k-ω	FEM
Ding et al. [171] [3]/2015	1DOF-FIV	various	varied	0.017	10,000–130,000	SA	FVM
Zhang et al. [172]/2017	1DOF-FIV	various	0.93	0.1076	2800–12,000D	SST k-ω	FEM
Zhang et al. [173]/2019	1DOF-FIV	various	2.4	0.0054	16,100–242,000	SST k-ω	FEM
Wang et al. [174]/2020	1DOF-FIV	various	2.6	0.002	8000–56,000	SST k-ω	FEM
Zhu and Yao [175] [4]/2015	2DOF-FIV	cir-rods	2.4	0.0054	1631–6,387.1	SST k-ω	FEM
Ding et al. [176]/2016	1DOF-FIV	cir-strips	1.896	0.04	30,000–110,000D	SA	FVM
Wang et al. [177] [5]/2020	1DOF-FIV	cir-rods	1.68	0.016	30,000–10,0000	SST k-ω	FVM
Zhao and Cheng [93]/2012	2DOF-VIV	4-circular	2	0.001	1000–20,000	SST k-ω	PG-FVM
Rahmanian et al. [178]/2014	1DOF-VIV	2-circular	5	0.0008	5000D	SST k-ω	PG-FVM
Cui et al. [180]/2014	1DOF-VIV	2-circular	2	0	5000	SST k-ω	PG-FVM
Zhao et al. [179]/2016	1DOF-VIV	2-circular	2.5	0	5000	SST k-ω	PG-FVM

[1] The Reynolds number Re was varied by changing the reduced velocity U_r. [2] Three-dimensional RANS modelling. [3] The mass ratio varies with the cross-sectional shape with the mass per unit length fixed at a value of 11.76 kg m^{-1}. [4] A circular cylinder with multiple rods placed around it. [5] An overset mesh is used in the numerical solution of the FSI problem. All other studies summarized in this table use a moving mesh in the numerical solution of the FSI problem.

4.6. Detached-Eddy Simulation (DES) of FIV

To strike a balance between simulating (resolving) and modelling turbulence, detached-eddy simulation is a hybrid LES/RANS approach that can be used to predict the flow over a stationary or vibrating cylinder. DES involves modelling the small-scale (unresolved) turbulence in the near-wall regions of the flow with a sufficiently fine grid using RANS and simulating (resolving) the turbulent flow in the regions away from the wall using a grid that is appropriate for LES. The governing equations of DES can be written in a general form applicable for both RANS and LES depending on how the mean quantities are defined (viz., ensemble-averaging for RANS and spatial filtering for LES). DES is more applicable for the simulation of highly separated flows in the high-Reynolds number regime at a computational cost that is intermediate between those of LES and RANS.

Nguyen and Nguyen [181] used a DES based on the SST k-ω model to simulate an 1DOF-VIV system consisting of a single circular cylinder subjected to free (undisturbed) incident flow and a 2DOF-VIV system consisting of a downstream cylinder subjected to the wake flow generated in the upstream direction by a fixed cylinder. Both cases involve a high-Reynolds number flow. These investigators reported a very good agreement of their predictions with experimental data and suggested that DES yielded more accurate predictions of the fluid forces, the VIV response and the phase difference than a comparable 2D RANS model. Joshi and Jaiman [182] applied the delayed detached-eddy simulation methodology to simulate the VIV of a slender flexible riser immersed in a turbulent flow at a subcritical Reynolds number. In this study, the response amplitudes and the motion trajectories were found to be in good agreement with the experimental measurements. Recently, Ma et al. [183] used DDES to investigate FIV associated with a cylinder with passive turbulence control (involving, as such, attachments of two strips to the cylinder of different thickness at different locations to give different coverage). Their numerical results agreed very well with the experiment data and successfully reproduced almost all the key features of the FIV response, such as the maximum amplitude, the multiple branches in the response, and the vortex shedding modes. Another similar DES study was conducted by Ma et al. [184] but for a flow at the lower Reynolds number of $Re = 3900$.

Table 5 summarizes the details of these DES studies of FIV phenomena. Generally speaking, this body of work suggests that DES is a very promising methodology for the prediction of the FIV response for rigid/flexible circular cylinders and and for cylinder–appendage system at high-Reynolds numbers.

Table 5. Summary of some key DES studies of the flow-induced vibration of bluff bodies.

Author/Year	Motion	Shape	m^*	ζ	Re	Dimension	SGS
Nguyen and Nguyen [181]/2016	2DOF-VIV	circular	2.6, 11	0.001, 0.005	3000–30,000	2D	SST k-ω
Joshi and Jaiman [182]/2017	2DOF-VIV	circular	2.23	/	4,000	3D	SST-DDES
Ma et al. [184]/2021	1DOF-FIV	cir-strips	2.4	0.0054	3900	3D	SST-DDES
Ma et al. [183]/2022	1DOF-FIV	cir-strips	2.4	0.0054	3000-19,000	3D	SST-DDES

All studies in this table use the FVM numerical method and the ALE FSI technique.

5. Machine Learning and FIV

The numerical modelling described in Section 4 is a full-order model (FOM) based on the numerical solution of various forms of the Navier–Stokes equations coupled with an equation of motion for the structure (body). This is a useful strategy to study the FIV of a bluff body, but is limited to low-Reynolds number flow (if DNS is used) or requires significant computational resources for flows in the strongly turbulent regime at very high Reynolds number. Furthermore, as noted by Hollenbach et al. [185] and by Sanders [186], the traditional CFD technology has no ability to provide physical insight into the mechanism underpinning the FIV behavior.

Fortunately, the recent advancements in data-driven modelling based on machine learning make it possible to address the issue of the high computational cost of conducting CFD simulation/modelling of FIV phenomena. Data-driven modelling using machine learning can be applied in two different contexts: (1) construction of a reduced-order model (ROM) of the FIV phenomena (viz., a simple mathematical representation of the principal characteristics of the full-order model based on CFD) as discussed in Section 5.1; and (2) use of a deep neural network (DNN) either to accelerate the conventional numerical solution of CFD or to map the non-linear relationship between inputs (e.g., flow and structure properties) and outputs (e.g., dynamic response, FIV forces, wake patterns) as described in Section 5.2.

For readers who may not be familiar with machine learning, we will provide a brief introduction to certain machine learning techniques that have been applied to modelling FIV phenomena (described later in this section). Machine learning is a rapidly developing discipline that lies at the intersection of computer science, statistics, and applied mathematics. Machine learning algorithms are data-driven in the sense that these algorithms are used to generate models from data—as more data are used, the models can improve. From this perspective, machine learning can be used in any data-rich field (more specifically, for the modelling of FIV phenomena considered herein) that generates a great amount of data (big data).

In the application of machine learning to FIV modelling, the non-linear dynamical system that generates the relationship between the inputs (e.g., flow) and output (e.g., transverse displacement) is characterized through the training data. Machine learning algorithms can be used to build input–output models using time series data obtained from either experimental measurements or numerical simulations of the dynamical system. Examples of such machine learning algorithms are the autoregressive with exogenous input model (and various extensions), the eigensystem realization algorithm (ERA), and the dynamic mode decomposition (DMD)—all of which will be described below as it applies to unravelling the dynamics associated with FIV phenomena. These methodologies can be interpreted as model reduction methods that generate a low-dimensional representation of the complex input–output relationship of the dynamical system from black-box data streams (e.g., a series of "snapshots" or solutions obtained from a numerical simulation of the dynamical system).

Other machine learning techniques include artificial neural networks (ANNs) (and, more specifically, deep neural networks (DNNs)) as depicted in Figure 10a which consist of a number of individual computational units (neurons) organized in a number of successive layers. The computational units of the network are connected in a graph structure. A DNN (unlike a shallow neural network) contains multiple computational layers—the intermediate layers between the input and output layers are referred to as hidden layers with the nodes in one layer connected to those in the next layer (in a architecture known as a feedforward network). The output of the j-th computational unit (neuron) in the l-th layer of a DNN (each layer of which consists of n neurons) is given by

$$a_j^l = \sigma\left(\sum_{i=i}^{n} a_i^{l-1} w_{i,j}^l + b_j^l\right), \tag{32}$$

where a_r^s is output of the r-th neuron in layer s, $w_{i,j}^l$ is the weight that connects the i-th neuron in layer $(l-1)$ to the j-th neuron in layer l, b_j^l is the bias imposed on the j-th neuron of layer l, and $\sigma(v)$ is a non-linear activation function. Typical activation functions include the sigmoid function $\sigma(v) = 1/(1+\exp(-v))$ and the rectified linear unit (ReLU) function $\sigma(v) = \max(0, v)$. The neural network has a number of unknown parameters, namely the weights and biases, whose values are obtained by the neural network model so that it provides a good fit to the training data (without overfitting these data so that the model generalizes well—viz., gives good predictions of data not used in the training process). Usually, these parameters (weights and biases) are determined by modifying their values using a gradient descent algorithm to minimize a loss (or error) function that provides a quantitative measure of the performance (or accuracy) of the neural network predictions on the data.

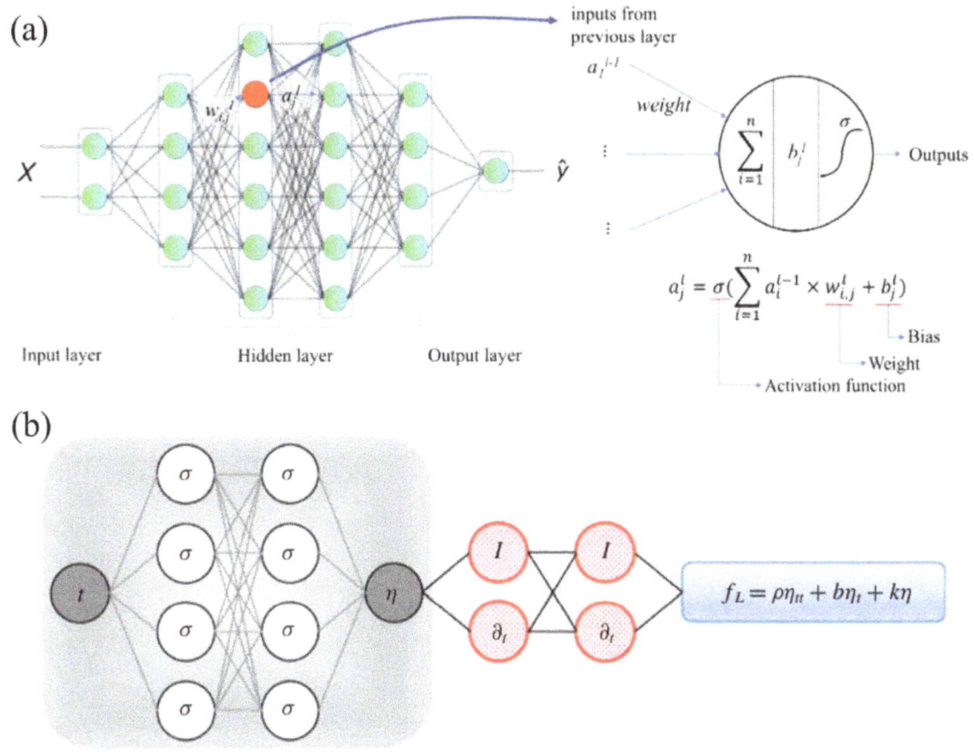

Figure 10. The architecture of a deep neural network (DNN). (**a**) A general deep neural network consisting of a number of computational units (neurons) organized into a number of layers and connected in a graph structure [187]. In this example, the neural network of an input layer having two neurons (viz., input X is a two-dimensional vector), an output layer with one neuron (giving a prediction of the output $\hat{y}$) and four hidden layers that pass information from a previous layer into the next layer. (**b**) A physics-informed neural network (PINN) with an input t, an output η, and two hidden layers with four neurons each [188].

A variant of DNNs is the so-called physics-informed neural networks (PINNs), an example of which is exhibited in Figure 10b. A PINN is an example of a scientific machine learning technique which integrates both data and mathematical physics models (physical constraints and domain knowledge) in the learning algorithm. The mathematical physics model (usually expressed as a partial differential equation) is embedded into the loss function of the DNN using automatic differentiation. More specifically, the loss function of a PINN consists typically of the weighted losses of the data and the partial differential equations encoding the underlying physics of the problem. The PINN is then trained by minimizing the loss using a gradient-based optimizer just as for the physics-uninformed neural network (see Figure 10a).

5.1. Data-Driven Reduced-Order Models (ROM) of FIV

Reduced-order modelling in the context of fluid dynamics involves the development of a low-dimensional mathematical model from a large flow dataset obtained either from experimental measurements or from full-scale CFD simulations. The unsteady aerodynamic models can be divided into three classes [189]. Firstly, there are the purely linear models where the static and dynamic aerodynamic forces acting on the body are linear—these models are most useful for steady or quasi-steady flows. Secondly, there are dynamically linear models where the flow is non-linear in space, but adheres to a linear relationship in time (also referred to as a time linearization model). A projection-based ROM and a system identification-based ROM are examples of this kind of model. The latter type of ROM is typically used to characterize an unsteady flow past a structure subject to a small disturbance (perturbation). Thirdly, there are the dynamic non-linear models. With an increase in the amplitude of motion, the aerodynamic characteristics of the dynamic non-linearity reveals itself and the implicit assumption of a small disturbance (perturbation) is no longer valid. Dynamically linear and non-linear models are frequently used to represent the FIV response of bluff bodies.

5.1.1. Projection-Based ROM

Proper orthogonal decomposition (POD) [190] and dynamic mode decomposition (DMD) are two methodologies that are used to construct a projection-based ROM. The datasets that are used to construct this type of ROM are sequences of snapshots of the flow field (e.g., velocity, pressure). This type of ROM is frequently used to study the characteristics of coherent wake structures in a flow.

The proper orthogonal decomposition (as referred to as principal components analysis in the statistical literature) has been used in conjunction with the Galerkin projection method to construct a reduced-order model. This consists of two steps: namely, (1) determine a low-dimensional subspace where the dominant dynamics occur by decomposing a high-dimensional flow field (assembled as a data matrix $\mathbf{X}$) into a set of orthogonal modes; and (2) project the high-order Navier–Stokes equations (partial differential equation, PDEs) that determine this high-dimensional flow into this low-rank subspace to obtain a system of ODEs using a Galerkin projection methodology [191,192]. It is noted that POD is an efficient method for model order reduction in the flow dynamics of a system—it involves computing the singular value decomposition (SVD) of the data matrix $\mathbf{X}$ and using the information contained in the decomposition to construct the optimal basis modes (optimal POD modes) that can be used to extract a low-rank, low-dimensional subspace of dynamic flow evolution. Following from this analysis, it is possible to approximate the flow field evolution given the NS equations using a Galerkin projection of the NS equations onto the optimal POD modes to give the dimensionally reduced evolution equations.

One common shortcoming of the conventional POD/Galerkin methodology is the lack of a clear separation of the individual dynamics embodied in the optimal basis modes—this makes it difficult to build a high-quality ROM [193]. Moreover, certain low-energy components that do not contribute to POD modes can have a large influence on the underlying mechanisms responsible for the flow features, implying that the POD-based ROM may provide poor predictions in this case. As a result, many variants of the conventional POD/Galerkin methodology have been developed to mitigate these shortcomings, such as the balanced POD (BPOD) [194] and the spectral POD (SPOD) [193].

Recently, Schubert et al. [195] developed a data-driven ROM based on SPOD and applied it to reconstruct the turbulent wake dynamics of a circular cylinder undergoing VIV at $Re = 4000$. These researchers applied a sparsification procedure to determine the temporal coefficients corresponding to the POD modes and enhanced the reliability and robustness of the ROM.

Rather than basing a decomposition on the energy content of orthogonal modes while ignoring the temporal information as in POD, DMD provides a modal decomposition of the high-dimensional flow structures, where each mode in the decomposition embodies spatially correlated structures that have the same linear behavior in time—in so doing, the DMD provides not only a dimensionality reduction as in POD but also gives a model for the time evolution of these modes. Each DMD mode corresponds to a single complex frequency (oscillation that grows or decay) which is useful for providing a deeper understanding of the flow mechanism, as well as for stability analysis. Moreover, the eigenvalue/eigenvector decomposition provided by the DMD modes provides the flow evolution, obviating the need to apply the Galerkin projection to obtain the dimensionally reduced evolution equations for the dynamical system. As an example, Kou et al. [196] employed DMD to investigate the temporal evolution of the purely fluid modes at Reynolds numbers between 12 and 55—this information was used to explain the occurrence and disappearance of VIV phenomenon in the subcritical regime.

5.1.2. System Identification-Based ROM

Another time-linearized ROM uses system identification to unravel the dynamical relationship between the input and output data—two common data-driven methodologies based on this approach are the autoregressive with exogenous input (ARX) model and the eigensystem realization algorithm (ERA).

The key steps for the development of a system identification-based ROM are: (1) perform full-order CFD simulations in order to obtain the required high-fidelity input-output datasets; (2) train the unsteady fluid dynamics ROM, using either the ARX or ERA methodology based on the data collected in the previous step; (3) couple the equation of motion for the structure with the data-driven model to give an unsteady coupled ROM for the (total) FSI system; and (4) conduct a linear stability analysis (LSA) of the dynamical system using the coupled ROM. Note that either ARX or ERA is used to derive the ROM for the fluid dynamics in step (2)—the other three steps are the same for both methodologies.

The ARX model is a linear time series model based on a discrete difference equation for the input–output data, in which the current system output is expressed as the linear superposition of the current input, as well as previous (in time) inputs and outputs, so

$$y_F(k) = \sum_{i=1}^{n_a} A_i y_F(k-i) + \sum_{j=0}^{n_b-1} B_j u(k-j)\,, \tag{33}$$

where the input u and output y_F are obtained from full-order CFD simulations. The time delay order parameters n_a and n_b are specified by the user and the model coefficients A_i $(i = 1, 2, \ldots, n_a)$ and B_j $(j = 0, 1, \ldots, n_b)$ need to be estimated (either through a least-squares fitting process or application of the maximum likelihood procedure). The development of an unsteady fluid ROM based on ARX consists of four steps: (1) compute the steady-state base flow for the FIV system (subject to a specific input signal) using a numerical solver; (2) estimate the ARX model coefficients A_i and B_i defined in Equation (33); (3) construct the reduced matrices using the estimated values for A_i and B_j; and (4) recast the fluid dynamics ROM into a standard state-space format.

In general, a unit step signal can be used as the input (training) signal for computation of the steady-state base flow. The requirement is that the prespecified training signal should be persistently exciting (viz., its spectrum should contain a sufficiently large number of harmonics) so that the input can trigger the unsteady linear flow dynamics. In the context of the FIV response of a bluff body, the single input and output can be taken to be the transverse displacement of the cylinder and the instantaneous lift coefficient acting on the cylinder, respectively. The application of ARX to better understand FIV phenomena has described in [197–200]. More specifically, Zhang et al. [197] constructed a fluid ROM using ARX and this data-driven model is then coupled to the structure equation. This coupled ROM for FSI was used to investigate the physical mechanisms underlying the frequency lock-in in VIV for a flow at Re = 60. Utilizing linear stability analysis of VIV based on the coupled ROM, these researchers reported two lock-in patterns: namely, a "resonance-induced" lock-in (only the wake mode is unstable) and a "flutter-induced" lock-in (both the wake mode and the structure mode are unstable). Moreover, Zhang et al. [197] also investigated the effect of the natural frequency, the mass, and the damping of the cylinder on the VIV dynamics using the FSI ROM. The main results concerning lock-in of VIV reported by Zhang et al. [197] were further confirmed by other global LSA studies (e.g., Navrose and Mittal [201] applied LSA using a linearized Navier–Stokes equations in which the equations for the steady flow and disturbance field were obtained separately).

The intrinsic mechanisms of various FIV response have been investigated using ARX in conjunction with LSA. These include investigations of the transverse galloping of a square cylinder in laminar flow (at Re = 140–180) [199], the wake-induced vibration of two identical tandem cylinders at Re = 60 [198] and the VIV suppression of a circular cylinder by a tuned mass damper [200].

The eigensystem realization algorithm (ERA) is a data-driven methodology originally presented by Juan and Pappa [202] who used it to identify modal parameters relevant to modelling structures on the space shuttle. ERA is a linear system identification methodology that uses the impulse response of the system to construct a ROM for the dynamics. The development of a fluid ROM using ERA consists of the following key steps: (1) calculate a steady-state base flow subject to an input impulse response using a numerical solver; (2) use this impulse response data to construct a generalized Hankel matrix H obtained by stacking shifted time sequences of the data; (3) compute the singular value decomposition (SVD) of H; (4) calculate the reduced system state matrices; and (5) recast the fluid ROM into a state-space format. In ERA, the dimension of the Hankel matrix H is a critical parameter that controls the convergence of the constructed ROM.

In greater detail, ERA is applied with reference to a discrete-time state-space system with the form

$$x(i+1) = Ax(i) + Bu(i)\,, \tag{34}$$

$$y(i) = Cx(i) + Du(i)\,, \tag{35}$$

where $x(i)$, $u(i)$, and $y(i)$ are the state vector, input vector, and output vector at discrete time step i; and, A, B, C, and D are the system, input, output, and feed-through matrices, respectively. ERA uses measurements of the impulse response of a dynamical system to infer the matrices (A, B, C, and D) that define the state-space model. To this purpose, a discrete Kronecker delta function input of the form

$$u^{\delta}(i) \equiv u^{\delta}(i\Delta t) = \begin{cases} I, & i = 0; \\ 0, & i = 1,2,3,\dots\,, \end{cases} \tag{36}$$

(I is the identify matrix and Δt is the time-step size) gives rise to the impulse response (output of the dynamical system) in the form

$$y^{\delta}(i) \equiv y^{\delta}(i\Delta t) = \begin{cases} D\,, & i = 0 \\ CA^{i-1}B\,, & i = 1,2,3,\dots \end{cases}. \tag{37}$$

This impulse response data are used to form a Hankel matrix H obtained by stacking shifted time series of the data to give

$$H = \begin{bmatrix} y^{\delta}(1) & y^{\delta}(2) & \cdots & y^{\delta}(s) \\ y^{\delta}(2) & y^{\delta}(3) & \cdots & y^{\delta}(s+1) \\ \vdots & \vdots & \ddots & \vdots \\ y^{\delta}(r) & y^{\delta}(r+1) & \cdots & y^{\delta}(s+r-1) \end{bmatrix}. \tag{38}$$

Next, we take the SVD of H, truncate the small singular values in this decomposition (viz., retain only the dominant singular values), and create a second shifted Hankel matrix H' (obtained by shifting the columns of H one time step to the left). The information in the truncated SVD and in H' can be used to construct estimates for (A, B, C, and D) that define the discrete-time state-space model (reduced-order model) of the dynamical system.

Jaiman and colleagues conducted number of investigations of the FIV response of a bluff body using the ROM/ERA methodology [203–207]. Yao and Jaiman [203] conducted LSA on the transverse VIV of a circular cylinder. These researchers then investigated the stability characteristics of FIV of a cylinder having different cross-sectional shapes (e.g., ellipse, forward triangle, diamond, and squares with various rounding radii). Generally speaking, these results demonstrated the effectiveness of using ROM/ERA to predict various features of the FIV phenomenon. Indeed, ROM/ERA has been used to conduct stability analysis of various vibrating systems, such as the FIV of a cylinder with different passive suppression devices (e.g., fairings, splitter plate, connected-C) [206], of two cylinders in either a tandem or side-by-side layout [205], and of active control of the VIV of a circular cylinder [204] and a sphere [207].

The most useful aspect of the ARX and ROM/ERA methodologies is that once the ROM is obtained using the data-driven process, it is straightforward to perform a linear stability analysis of the dynamical system—an analysis that facilitates deeper insights into the underlying mechanisms responsible for FIV through the examination of eigenvalue trajectories (which allows one to better understand the competition between structure and wake modes). For example, Cheng et al. [208] investigated the modal interaction and transformation of a VIV system consisting of the flow past an elastically-mounted circular cylinder for Reynolds numbers in the range from 20 to 180. These researchers used ERA/ROM the explain the beating phenomenon observed in VIV which occurs in the initial branch of the response and the significant lag time that arises between the initial branch and the occurrence of a fully developed response in the lower branch. Other relevant investigations utilizing LSA in the context of ROM/ERA in order to better understand the FIV phenomenon include Bukka et al. [206], Yao and Jaiman [204], Chizfahm and Jaiman [207], and Li et al. [199].

It is important to stress that the linear ROMs constructed using ARX or ERA cannot be employed to study the non-linear characteristics of the flows associated with the FIV phenomenon.

5.1.3. Non-Linear ROM

The projection-based ROMs based on either POD or DMD and the system identification-based ROMs based on either ARX or ERA are linear models of the system dynamics (viz., they are time-linearized, dynamically linear or statically non-linear [209]). One common weakness of these approaches is that they are difficult to generalize to high-dimensional non-linear systems involving large-amplitude motions of the structure such as the FIV response of a bluff body subjected to a strongly turbulent flow.

In order to predict the non-linear dynamics of a complex flow system using a ROM, a general strategy is to incorporate within the linear model some mechanism to account for non-linearities in the dynamics, such as the Koopman operator [210] and the discrete empirical interpolation method (DEIM) [211]. In particular, Miyanawala and Jaiman [211] combined POD and DEIM to present a non-linear ROM for the 2DOF-VIV of a sharp-cornered square cylinder immersed in either a laminar or turbulent flow. These researchers successfully used this non-linear ROM to interpret the most significant wake features and their contributions to the forces on the vibrating body.

An alternative for the generation of a reduced-order non-linear model of a high-dimensional flow field uses machine learning, especially deep neural networks. Some DNN architectures can be interpreted as non-linear generalizations of either POD or DMD in the sense that they decompose the high-dimensional flow fields into low-dimensional non-linear modes. An example is provided by the convolutional neural network (CNN) autoencoder proposed by Murata et al. [212]. Another strategy is to combine a DNN with a traditional linear model decomposition in order to construct a hybrid ROM that integrates the linear and non-linear characteristics of an unsteady flow. The conventional POD performs well in dimensionality reduction based on a set of time-invariant orthogonal modes. As a generalization, the temporal evolution of the POD modes (represented as modal coefficients) can be learned using a deep neural network—examples include the CNN [213] and the encoder–decoder recurrent neural network (RNN) using a long short-term memory (LSTM) cell [214]. The effectiveness of these hybrid POD-DNN-based ROMs has been demonstrated by their successful prediction of various aspects of the FIV of a bluff body for flows at low- and high-Reynolds numbers.

In order to identify non-linear low-dimensional features of a flow, Bukka et al. [215] incorporated a CNN with an RNN to give a convolutional recurrent autoencoder network (CRAN) model. In this framework, a low-dimensional embedding of the state variables was learned by the CNN and the transient evolution of the dynamics was learned using the RNN-LSTM. A conventional linear system identification technique can also be combined with a DNN and used to investigate the underlying mechanism and stability characteristics of a strongly non-linear dynamical system. To this purpose, Kou and Zhang [209,216] proposed a hybrid ROM to accurately predict the integrated flutter and buffeting of an airfoil. Here, an ARX model was trained and used to approximate the linear dynamics while the non-linear deviations between the real outputs for large-amplitude inputs and the ARX model was represented using a radial basis function neural network (RBFNN). In the context of predicting the VIV of a freely vibrating sphere in the transverse direction, Chizfahm and Jaiman [217] used a predictive RNN-LSTM as a DNN-ROM to provide the non-linear mapping of the input-output dynamics of the FSI system. This was integrated with a ROM/ERA to provide a linear approximation of the non-linear model for the stability prediction.

5.2. Data-Driven Deep Neural Network Applied to FIV

Recent advancements in using machine learning within the broader field of fluid dynamics [3] can be applied for the specific problem of simulating FIV. The possible application modes can be divided into three types: (1) use machine learning to augment the numerical solving procedure in standard CFD solvers; (2) use deep neural network to infer the flow field and FIV response; and (3) use deep neural network to model the fluid force of FIV. The above three applications are all data-driven methods, which will be reviewed in detail below.

Deep learning is a subset of ML that refers to the use of highly multi-layered neural networks to understand a complex dataset, with the intention of predicting certain features in the dataset.

5.2.1. DNN-Accelerated Fluid Modelling

The first scenario integrates the machine learning technique into the standard CFD solver to accelerate the key calculation steps. This is a hybrid application of traditional CFD simulation and learned neural network in fluid dynamics. Duraisamy et al. [218] reviewed various machine learning methods for augmenting numerical simulation techniques at various levels of fidelity.

High-fidelity simulations of FIV require the use of DNS or of LES. While DNS simulations of high-Reynolds number turbulent flows is prohibitively expensive, recent advancements in applying machine learning to perform accurate DNS on a coarser grid greatly reduce the computational expense [219]. In DNS, a major computational burden is resolving the smallest physical scales in the flow through the use of extremely fine grids. In order to accelerate the computation of DNS and LES without loss of accuracy, Kochkov et al. [219] embedded a neural network interpolator into the standard CFD solver on a coarser grid—specifically, in order to approximate the convective flux term for DNS and model the residual correction for LES. The training dataset is high-resolution simulations for Kolmogorov flow at $Re = 1000$. In this application, the isotropy of Kolmogorov flow particularly lends itself to this methodology and the training procedure and model structure can benefit from numerous invariances in this flow. In principle, the ML-based CFD solver proposed by Kochkov et al. [219] can be applied to any flow since the output of the neural network only depends on local features, which was validated by a test case with large-domain size, decaying turbulent flow and a higher Reynolds number of 4000.

The traditional turbulence modellings in CFD apply some approximations for turbulence effects at high-Reynolds numbers, such as a SGS model in LES or the SST k-ω model in RANS, which is the main source of numerical errors. Therefore, an alternative way to improve the accuracy of simulations and to speed up the calculation of LES and

RANS is to modify the Reynolds stress or SGS stress through the application of machine learning techniques. A notable machine learning technique within the field of LES is the neural network SGS model of Maulik et al. [220], which allows for a more accurate closure within LES. Compared to LES, machine-learning augmented RANS is a broad and rapidly developing field. Several techniques have been developed to accelerate RANS modelling and increase their accuracy. A popular approach is to integrate a richer eddy viscosity closure relationship. For example, Ling et al. [221] used a tensor basis neural network (TBNN) to predict the anisotropy tensor, a highly versatile closure relationship. Another notable work in augmented eddy viscosity modelling includes the framework proposed by Wu et al. [222].

The application of machine learning mentioned above is interpreted from an overall perspective of fluid dynamics, focusing more on the general approaches instead of application to a specific problem. There is no doubt that these ML-based techniques, regardless whether it is the ML-accelerated DNS/LES, or the ML-modified turbulence modelling of LES/RANS, could in principle be used to address numerically the FIV response of a bluff body, increase the accuracy of LES/RANS simulation, or develop a highly-tuned LES/RANS model for turbulent FIV simulations.

5.2.2. Physics-Informed Neural Network of FIV

Contrary to the ML-based DNS or LES or RANS models, the second application of machine learning in fluid dynamics is completely out of the traditional CFD simulations—that is to say, no numerical calculations are needed during the inference procedure by a trained deep neural network. It should be noted that the training dataset can be experimental data or numerical results—if it is the latter, the necessary CFD simulations would be carried out during the data preparation period. This data-driven deep learning technique has been employed explicitly for modelling the FIV response of a circular cylinder.

Raissi et al. [188] conducted some seminal work to apply a physics-informed neural network (PINN) model to accurately infer the lift and drag forces acting on a circular cylinder undergoing a 1DOF-VIV from scattered simulation data of the velocity field for a laminar flow at $Re = 100$. It should be noted that the PINN presented by Raissi et al. [188] is not a black-box model like traditional data-driven neural networks based on large training data, because it encodes the NS equations coupled with the structural motion equation into the loss function as regularization (constraints on the output of the neural network), which makes it have potential to infer target variables based on the governing physical laws and on a limited training dataset. Inspired by the introduction of the eddy viscosity ν_t in conventional turbulence models used in CFD (e.g., LES, DES, RANS) for the closure of either the time-averaged or spatially-filtered NS equations, Xu et al. [223] introduced an artificial viscosity ν_n in the RANS framework to formulate the parameterized Navier–Stokes equations, which was treated as the physical constraint of the established PINN. For the case of flow over a stationary cylinder at a high-Reynolds number of 10^5, the physics-informed deep learning successfully reconstructed the missing flow dynamics and inferred the lift and drag coefficient profiles. As an extension of the study reported by Raissi et al. [188] and by Xu et al. [223], Bai and Zhang [224] further used both the parameterized Navier–Stokes equations-based PINN (PNS-PINN) and the Navier–Stokes equations-based PINN (NS-PINN) to infer the unsteady velocity field and force coefficients for the transverse VIV of a circular cylinder in a turbulent flow. The training data for neural network models consisted of 2D RANS data at Re = 10,000 and employed the SST k-ω model and the SST-SAS turbulence closure models. These researchers also tested two kinds of datasets based on the flow velocity and the dye trace, respectively. The results demonstrated the predictive capability of the PNS-PINN trained by a velocity dataset for turbulent VIV at subcritical Reynolds numbers, in terms of correctly predicting the flow field and the lift and drag coefficients. Cheng et al. [225] applied a similar PNS-PINN approach to study the 2DOF-VIV of a single circular cylinder at Re = 2889 and wake galloping of two rigidly-connected circular cylinders at $Re = 9000$ and reported good success.

The above four representative studies have demonstrated that the data-driven deep learning technique underlying the PINN framework is capable of inferring the complicated FIV response of a bluff body from a limited training dataset, in terms of dynamic parameters (damping and stiffness), the entire velocity and pressure fields, and the lift and drag forces acting on the vibrating system. Moreover, PINN can be applied in various scenarios, from laminar to turbulent and for a single cylinder or multiple (array of) cylinders.

5.2.3. DNN-Based FIV Force Model

The last machine learning model is specialized in predicting the FIV modelling of a bluff body, which involves some small-scale numerical calculations, not for the NS equations (multiple PDEs) governing the fluid dynamics, but for the mass–spring–damper (one ODE) system controlling the structural oscillation.

As an example of this approach, we focus on a recent work by Zhang et al. [187] which will be explained in detail in order to help the reader understand what role machine learning plays in the VIV prediction for a single circular cylinder. As shown in Equation (13), the forcing term on the right-hand side represents the fluid force responsible for the VIV response of a circular cylinder, which is closely related to the flow properties and the vibration parameters. Zhang et al. [187] established a deep neural network to map the relationship between these two, which has as the aerodynamic force coefficient as the single output and multiple input features, such as the Reynolds number, the reduced velocity, the oscillation amplitude and the velocity of the circular cylinder, as well as a few parameters characterizing the oscillating incident flow. The neural network was trained using a large dataset from forced-vibration experiments. To predict the transverse VIV of flexible pipes, the trained neural network was applied to calculate the fluid force at each time instant, which was then substituted into Equation (13), so that this second-order ODE can be numerically solved. This was accomplished by using the Newmark-β method combined with a finite-element discretization of the long flexible pipe. Another similar case will be given in Section 6 for a VIV energy harvester.

It can be inferred that the excitation force coefficient model based on a deep neural network is computationally efficient owing to the fact that the calculation of the mass–spring–damper system is much simpler than the solution of the NS equations in the first scenario.

6. Energy Harvesting Based on FIV

Sections 2–5 have provided a comprehensive overview of the various methodologies that can be used to model the FIV response, such as mathematical modelling, numerical modelling, and modelling based on machine learning approaches. Flow-induced vibration is essentially a FSI phenomenon and a deeper understanding of this phenomenon is important for real-world engineering and industrial applications—the most common is perhaps the control (suppression) of FIV in engineering structures such as offshore production risers [226], heat exchanger tubes [227], suspension bridges [228], and power transmission lines [229]. The efficient harvesting of the energy potential of FIV is a key application arising from the growing demand for clean and renewable energy, as well as the rapid development of related technologies, such as wireless sensor networks and micro-electromechanical systems [7,230]. This section focuses on recent advancements in flow-induced vibration energy harvesting (FIVEH)—this application is used to illustrate how the various modelling methodologies described earlier can be used to address this important problem.

The concept of FIVEH is to harness wind and/or water energy from the surrounding environment by utilizing structures that oscillate in the flow (providing various FIV responses) and then converting this energy of oscillation into usable energy (e.g., electricity). Therefore, FIVEH primarily involves two forms of coupling: namely, (1) the bidirectional fluid–structure interaction converting fluid energy to mechanical energy and (2) the electromechanical coupling transforming the mechanical energy to electrical energy. The former specifically refers to various FIV phenomena (see Section 2) consisting primarily of VIV and galloping. The oscillatory energy arising from the FIV response can be converted by various means—electromagnetic, piezoelectric, electrostatic, dielectric, triboelectric, interface circuit, and power management [6]—to a usable form of energy.

Like other energy generators, the core research objective of FIVEH is to obtain the optimal power performance—specifically, to enhance the oscillation by widening the effective flow speed range and reducing the cut-in velocity of FIV so as to maximize the energy harvesting efficiency and power generation. The performance of FIVEH can be directly evaluated by the output voltage, or indirectly analyzed by examining critical oscillation properties, such as the displacement amplitude, the vibration frequency, and the lift force acting on the moving body.

Numerous efforts have been made to enhance the power performance of a FIV energy harvester in a number of ways. The most direct and easiest way is to change the geometrical shape of the bluff body which can include utilizing a cylinder with a non-circular cross-section (e.g., square cylinder [231], rectangular cylinder [232], triangular cylinder [233], *D*-section cylinder [234]) and incorporating a PTC device to change the surface roughness of the cylinder (e.g., adding attachment(s) to a cylinder, such as roughness strips [176], trip wires [235], a splitter plate [236], small rods [237], and a free-to-rotate impeller [238]). The underlying strategy here is to obtain an asymmetrical geometry of the rigid body which induces an instability in the shear layer to produce a negative damping in the transverse (crosswind) direction and, hence, trigger galloping on the vibrating body. As mentioned in Section 2.3, an energy harvester driven by galloping can potentially yield a larger power production than that driven by VIV. In addition to incorporating appendages on an oscillating body, the non-linear spring stiffness and damping can be changed on the supporting structure to improve the harvested power and efficiency [239,240]. The mechanical part of the energy harvester can also be designed as a dual-mass system (i.e., one mass undergoing FIV is used to drive another mass with an attached magnet that moves relative to a coil in order to produce electricity [241]), as a multi-degree-of-freedom system [242], and as a multi-directional system [243]. All these strategies have been shown to improve the extraction of fluid energy from the ambient flow. Moreover, an hybrid energy harvester based either on two or more FIV phenomena (e.g., VIV-galloping) or on utilizing multiple forms of renewable energy simultaneously (e.g., wind, water, photovoltaic) based on multiple energy conversion mechanisms [244] can generally yield higher energy production. Lastly, an array of cylinders with an appropriate (optimal) layout can significantly increase the total power generation for the system.

Figure 11 summarizes some common FIV-based energy harvesters. The interested reader can refer to the excellent review paper by Lv et al. [8] for more details concerning the operating principles of various energy harvesters and for examples of various methodologies that have been used to improve the power production of a FIV energy harvester.

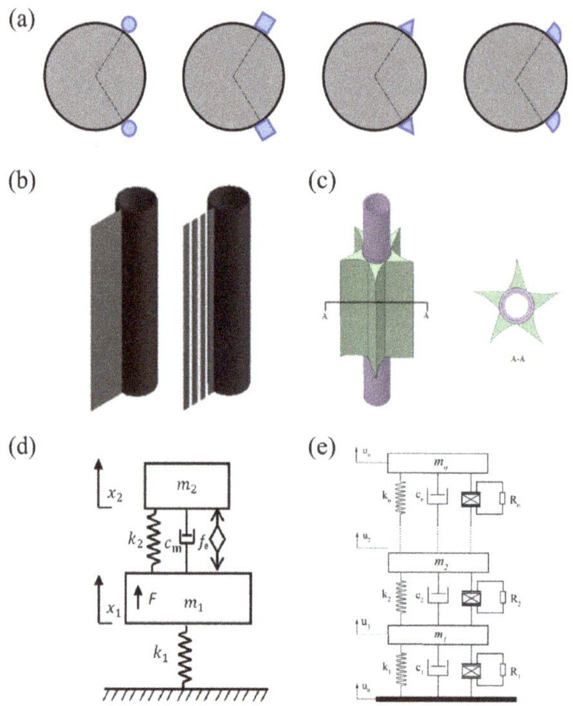

Figure 11. Strategies to improve the power production from a vibrating cylinder undergoing FIV for energy harvesting: (**a**) attaching small rods (circular, square, triangular, fin-shaped); (**b**) attaching a splitter plate (solid and slotted) [245]; (**c**) attaching a free-to-rotate pentagram impeller [238]; (**d**) incorporating a dual-mass system [241]; and (**e**) utilizing a multi-degree-of-freedom system with multiple piezoelectric elements [242].

6.1. Mathematical and Numerical Modelling of FIVEH

Mathematical and numerical modellings are important tools to study FIVEH, since accurate theoretical models and CFD simulations can predict the dynamic response of the system and the power production of the energy harvester. To provide a comprehensive example of how mathematical and numerical modelling can be applied to FIVEH, we consider, specifically, a piezoelectric energy converter for this illustration.

Figure 12 depicts the main components of a simple piezoelectric energy harvester. As shown, the cylindrical rigid body connected to the free end of a cantilever beam undergoes FIV as a result of the excitation from the incident flow. This motion deforms the piezoelectric sheet bonded to the root of the cantilever beam—the deformation converts the mechanical vibration of the body to electrical energy in the interface circuit with a load resistance of R. Note that the fluid, the bluff body, and the piezoelectric transducer are considered as a single dynamical system that can be modelled using a set of governing equations—these equations can be solved to determine the vibration response and the power output of the energy harvester.

The semi-empirical model described in Section 3 provides the coupling of a fluid motion equation (wake oscillator) and an equation of motion for the body (mass–spring–damper system). In addition, a governing equation for the piezoelectric circuit is required to model the system. In accordance to Erturk and Inman [246], a SDOF lumped-parameter model that describes the coupled electro-mechanical system can be expressed as

$$M_c\ddot{y} + C_c\dot{y} + K_c y - \theta V = F_y\,, \tag{39}$$

$$\theta \dot{y} + C_P \dot{V} + \frac{V}{R} = 0\,. \tag{40}$$

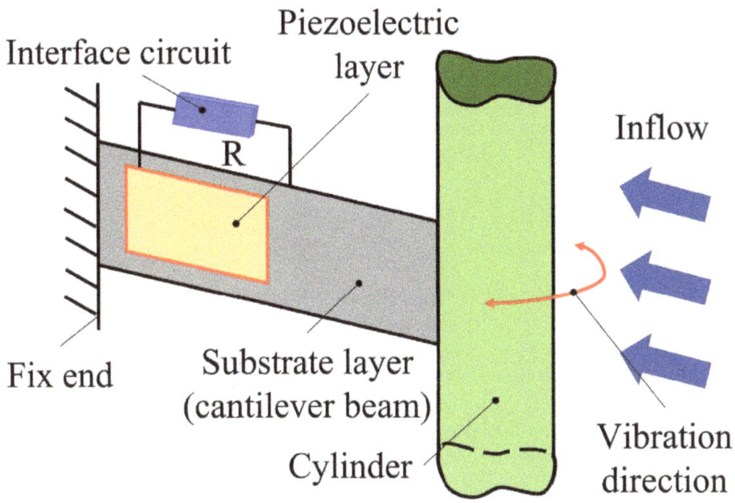

Figure 12. Sketch of a simple piezoelectric energy harvester based on flow-induced vibration.

If we compare these equations to the mathematical model (Equation (13)) in Section 3, the equation of motion of the structure given by Equation (39) has been modified to include the mechanical force which is modelled as being proportional to the voltage $V(t)$ across the external electrical load resistance R with the coefficient of proportionality for the coupling to the electro-mechanical system given by θ. Other physical quantities, such as M_c, C_c, K_c, y, and F_y, have the same meaning as those in Equation (13). Equation (40) is the circuit equation where $V(t)$ arises from the capacitance of the piezoelectric layer C_p and the electro-mechanical coupling θ.

If the rigid body in Figure 12 is replaced with a smooth circular cylinder, the system becomes a VIV-based piezoelectric energy harvester. Dai et al. [247] developed a reduced-order model for this device by utilizing the Euler–Lagrange principle and a Galerkin discretization based on the wake-oscillator model proposed by Facchinetti et al. [27]. The governing equations are given by

$$\begin{aligned} \ddot{r}_i + 2\zeta_i\omega_i\dot{r}_i + \frac{1}{2}C_D\rho DUL_c\left[\varphi_i(L_s) + \frac{D}{2}\varphi_i{}'(L_s)\right]\left[\sum_{j=1}^{n}\left(\varphi_j(L_s) + \frac{D}{2}\varphi_j{}'(L_s)\right)\dot{r}_j\right] \\ + \omega_i{}^2 r_i - \theta_i V = \frac{1}{4}C_{L0}\rho DU^2 L_0\left[\varphi_i(L_s) + \frac{D}{2}\varphi_i{}'(L_s)\right]q\,, \end{aligned} \tag{41}$$

$$\ddot{q} + \lambda\omega_v s\left(q^2 - 1\right)\dot{q} + \omega_v s^2 q = \frac{A}{D}\sum_{i=1}^{n}\left[\left(\varphi_i(L_s) + \frac{D}{2}\varphi_i{}'(L_s)\right)\ddot{r}_i\right], \tag{42}$$

$$C_P\dot{V} + \frac{V}{R} + \sum_{i=1}^{n}\theta_i\dot{r}_i = 0\,. \tag{43}$$

Equation (41) models the motion of the free end of the piezoelectric beam where r_i $(i = 1,2,\ldots,n)$ denotes the modal coordinates of the displacement; L_c and L_s are the lengths of the cylinder and the substrate layer, respectively; $\varphi_i(L_s)$ is the displacement of the end of the cantilever beam. Equation (42) encapsulates the dynamics of the near wake behind the cylinder based on Facchinetti et al.'s wake oscillator model with an acceleration coupling term—here, $\lambda = 0.3$ and $A = 1.2$. Equation (43) is the governing equation for the piezoelectric circuit, similar to Equation (40). Dai et al. [247] validated their non-linear mathematical model using some experimental results and reported a very good agreement. These investigators applied this model to study the effect of the tip mass and the piezoelectric sheet length on the maximum generated voltage and harvested power of the piezoelelectri energy harvester.

The linear benchmark model [247] is a common theoretical form for the FIV piezoelectric energy harvester. A number of researchers has modified this basic form and tailored it to solve various engineering problems. For example, Hou et al. [248] introduced a non-linear magnetic force term F_m into Equation (41) with the following form:

$$F_m = \frac{3\mu_0 M^2}{2\pi \varphi_1{}^3(L_s)} r \left[r^2 + \left(\frac{d_0}{\varphi_1(L_s)} \right)^2 \right]^{-5/2}. \tag{44}$$

Here, μ_0, M and d_0 are the vacuum permeability, the magnetization intensity, and the distance between the magnets, respectively.

Rather than formulating the VIV aerodynamic force using a wake-oscillator model, Wang et al. [249] presented a new semi-empirical model for the VIV piezoelectric energy harvester using the Scanlan–Ehsan non-linear representation for the forcing term F_y in Equation (39). This formulation was compared to some wind-tunnel data and found to yield accurate predictions. A significant benefit of this model is that it can be used not only for a smooth circular cylinder, but also for cylinders having different cross-sectional shapes and for a cylinder–attachment system.

Galloping is provoked when a square or rectangular cylinder is incorporated in the beam. For a piezoelectric energy harvester exhibiting galloping, the external excitation can be modelled using quasi-steady theory—the modelling here is similar to that for the polynomial expansion of the transverse force coefficient (see Section 3.3 for more details). While the electro-mechanical coupling term is different, the model can be formulated in a number of ways, such as: (1) using a simple SDOF mass–spring–damper system [250,251]; (2) using a distributed parameter model with Rayleigh–Ritz discretization [252]; and (3) using a Euler–Bernoulli distributed parameter model [253]. A mathematical model can also be derived for the piezoelectric energy harvester based on other types of FIV response, such as a flutter converter [253] and a wake-galloping converter [254]. As a final note, the use of a SDOF mass–spring–damper system [250,251] mentioned here involves simply using an odd-order polynomial approximation for F_y in Equation (39) which results in

$$M_c\ddot{y} + C_c\dot{y} + K_c y - \theta V = \frac{1}{2}\rho D L_c U^2 \left[A_1 \left(\frac{\dot{y}}{U} \right)^1 + A_3 \left(\frac{\dot{y}}{U} \right)^3 + \ldots \right]. \tag{45}$$

In addition to mathematical modelling (described above), numerical modelling based on CFD can be employed to predict the dynamics of the FIV piezoelectric energy harvester. To this purpose, numerical modelling can be used in several different ways to address problems related to FIVEH. Firstly, high-fidelity CFD predictions can be used to validate various theoretical models of FIVEH (e.g., the analytical and semi-empirical models mentioned above). Secondly, the distributed parameters in a semi-empirical model of FIVEH can be determined using reliable data obtained from high-fidelity numerical simulations (e.g., DNS or LES), especially when wind-tunnel experimental data are not available. Thirdly, CFD can be used to predict the detail characteristics of the flow and the power efficiency of a FIV energy harvester [255–261]. This is of particular importance when it is difficult and/or impossible to develop a detailed mathematical model for an energy harvester (e.g., a harvester that utilizes a bluff body having a very complex geometry). Generally, CFD simulations generate more accurate predictions than those provided by any analytical and/or semi-empirical mathematical model, frequently at only a fraction of the cost required for conducting a detailed experimental investigation. As a result, numerical modelling occupies a central role in research concerning FIVEH.

Recently, Ramirez [262] proposed a new modelling approach which couples CFD with an electro-mechanical model to obtain the flow dynamics and the electric power of multiple interacting energy harvesters. The approach advocated here is to integrate the NS equations for incompressible fluid flow with the Gauss law for the piezoelectric equations and the mass–spring–damper system for the oscillating rigid body. The proposed model uses the NS equations to make detailed predictions of the flow, rather than employing a simple (semi-empirical) wake-oscillator model, so it can generate more accurate predictions of the FIV displacement in the energy harvester.

In summary, employing reliable mathematical and numerical modelling can effectively predict the aerodynamic response of a bluff body, such as the vibration displacement, the frequency, and the lift force, as well as the power and output voltage of an energy harvester of a specified design. Moreover, these powerful mathematical and numerical tools can be used to undertake a parametric study of the influence of various aero-electro-mechanical parameters relevant to the system on the performance of the FIV energy harvester.

6.2. Machine Learning for Modelling FIVEH

Until now, traditional research methods, such as theoretical analysis (mathematical modelling), CFD simulations (numerical modelling), and wind-tunnel experiments, are still the principal approaches used to investigate FIVEH, but these approaches have some limitations. For example, a mathematical model is not always available for a complicated FIV energy harvester that has an irregular cross-sectional shape, or attached with small devices, or consists of a multi-cylinder system. On the other hand, CFD simulations and experimental investigations are time-consuming and expensive. As a consequence, machine learning approaches capable of addressing the FSI problem (see Section 5) is another efficient and economical alternative methodology that can be used to predict the dynamic response intensity and power generation capacity of a FIV energy harvester.

Some researchers employ machine learning and neural networks to wind assessment (e.g., wind speed, wind pressure) and power forecasting of a large-scale horizontal-axis wind turbine or a wind farm [263–266], which is instructional to the implementation of these artificial intelligence (AI) approaches on a FIV energy harvester. However, to the authors' knowledge, there are very few studies that apply ML or NN in the prediction of the dynamics associated with a FIV energy harvester. Two representative works will be reviewed here in detail in order to help the reader understand and appreciate the possible and feasible application of ML to support the design of current and future energy harvesters for exploiting FIV.

Zhang et al. [267] investigated wake galloping in a piezoelectric energy harvester consisting of an upstream stationary obstacle and a downstream piezoelectric cantilever beam connected with a cuboid bluff body on the free end, as shown in Figure 13. These researchers first conducted a series of wind-tunnel experiments to study the influence of shapes and sizes of upstream obstacles, as well as the distances between the fixed and vibrating body on the wake galloping response. Subsequently, they employed three different algorithms (viz., a decision-tree regression (DTR), a random forest (RF), and gradient boosting regression trees (GBRT)) to train a machine learning model based on the large datasets collected from their experimental measurements, including multiple input signals (e.g., wind speed and geometric variables), as well as two output signals (root-mean-square voltage V_{rms} and the maximum displacement y_{max}). These researchers found the GBRT model with best hyperparameters optimized using a ten-fold cross-validation method can obtain the most accurate predictions of V_{rms} and y_{max}. This example demonstrates the possibility of training a ML model with experimental data to predict the power output and vibration intensity of a wake-galloping energy harvester. The modelling logic of this novel approach is summarized in Figure 14 (left panel).

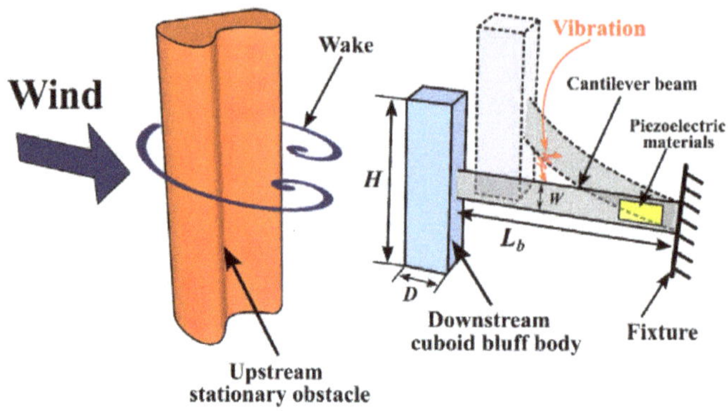

Figure 13. Schematic of wake galloping in a piezoelectric energy harvester [267].

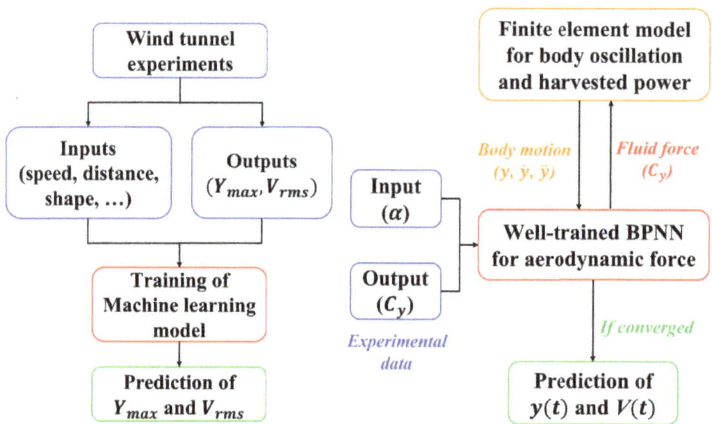

Figure 14. Machine-learning modelling procedure for a piezoelectric energy harvester based on wake galloping [267] (**left panel**) and galloping [268] (**right panel**).

By contrast, Hu et al. [268] applied AI to model a galloping piezoelectric energy harvester in a more mathematical way. As described in Section 3.3, the most-commonly used theoretical model of galloping uses the quasi-steady assumption of the aerodynamic force. Instead of representing the external excitation by an odd-order polynomial function fitting the discretized experimental data points like others, Hu et al. [268] applied a well-trained three-layer back propagation neural network (BPNN) to establish the non-linear relationship between the force coefficient C_y (output) and the angle of attack α (input). To demonstrate the superiority of using a neural network, these researchers first replaced the polynomial expression of aerodynamic force on the right-hand side of the structural equation (Equation (45)) with the output of the trained BPNN, and combined this with the electrical circuit equation (Equation 40)) to form a BPNN SDOF model. Furthermore, the authors proposed an architecture integrating the finite-element modelling of the electro-mechanical structure with a trained BPNN of the aerodynamic force. Specifically, during each time step of simulation, the body motion calculated from the finite-element model is first passed as an input to the trained BPNN, which then predicts the aerodynamic force acting on the body at the current moment and updates the fluid force in the finite-element model in real-time. The simulation converges after an iterative process. In this case, the use of BPNN only involves the action of the fluid on the body (viz., prediction of the aerodynamic force), while the body motion and the electro-mechanical conversion are addressed by the finite-element model. A flowchart of the application logic is shown in Figure 14 (right panel).

Obviously, the above two studies use completely different AI approaches in addressing the problem of FIVEH. The first case [267] developed a non-linear black-box model for the dynamical system without knowing the particular mathematical structure or underlying physics, in which the relationship between the input and the output are obtained by training a machine learning model using a large dataset. On the other hand, the second case [268] actually lies at the intersection of machine learning and numerical modelling, in which the former is used for prediction of the aerodynamic force and the latter is responsible for calculating the motion of the bluff body and the output voltage of the energy harvester. In addition, the second application mode of AI has a more solid theoretical foundation because the flow information and the energy conversion are obtained by numerically solving the NS equations and the electro-mechanical circuit equation—in this sense, it is more useful and practical when there is a lack of training data.

In summary, AI approaches applied to the FIV energy harvesting is nascent, but as these new techniques mature and become more widespread, they could certainly aid in the design of FIV energy harvesting systems. Some of the previously mentioned PINN techniques for flow field reconstruction in Section 5 could also be applied to in situ tuning and control of a FIV energy harvester. By allowing for accelerated and more accurate simulations of FIV devices, the design process will be better informed.

7. Outlook

Flow-induced vibration of a bluff body is a classical FSI problem with deep theoretical significance and practical value. The objective of this comprehensive survey is to provide a systematic overview on the current modelling approaches for FIV—a full spectrum of approaches is reviewed including mathematical modelling, numerical simulation/modelling based on CFD and application of machine learning. These approaches are illustrated through an application to a real-world problem—design of a FIV energy harvester. Each modelling approach has its own strengths and weaknesses. In this section, we consider some of the outstanding challenges (future issues) that remain to be addressed and summarize some directions for future work and perspectives for the modelling of FIV phenomena.

7.1. Future Issues

(1) Mathematical modelling.

- Mathematical modelling plays an indispensable role in our understanding of the underlying physics of the FIV phenomenon. However, formulating an accurate mathematical model for a highly non-linear dynamical system, such as FIV phenomena is highly challenging. The mathematical models proposed to date based on experimental data and/or engineering experience are not purely analytical models derived from first principles using a comprehensive theory and, as a result, these models have a limited range of applicability. Is it possible to develop a more complete theory of various aspects of FIV phenomena that will allow the formulation of more general analytical models?
- Most theoretical models for FIV, whether a coupled wake-oscillator model or single-degree-of-freedom model, are semi-empirical in nature—implying that the model has unknown parameters whose values need to be determined (fitted) to available experimental data. As a result, these models tend to have a very limited range of applicability (viz., a model that has been developed for VIV for a particular bluff body shape cannot be used for VIV for another body shape or for galloping for the same body shape without changing either the structure of the model and/or re-fitting the free parameters that define the model). More specifically, the need for reformulating the structure of terms and/or inclusion of new terms in the model along with a re-

fitting of the free coefficients using available experimental data need to be conducted for new scenarios.

- Most of the current mathematical models for FIV phenomena have been developed for VIV, which can also include galloping behavior in the formulation through use of the quasi-steady assumption—this usually involves inclusion of an odd-order polynomial approximation of the aerodynamic force responsible for galloping in the mathematical model. Although some other methods (e.g., neural network mappings) have been proposed to improve the curve fitting accuracy of the forcing term responsible for galloping, it is stressed that the quasi-steady assumption has inherent limitations and is applicable only under certain conditions.

(2) Numerical modelling.

- Owing to its moderate computational cost, RANS modelling will remain the workhorse of CFD applications to engineering and industrial FIV problems of practical relevance into the foreseeable future. The essential element for the successful use of RANS modelling is formulation of turbulence closure models. Is it possible to simulate FIV for these engineering and industrial applications using lower-fidelity numerical models such as unsteady RANS?
- Even so, for FIV problems that require greater accuracy or detailed information on the spatial and temporal fluctuations of the flow field, hybrid methods may be the numerical tool of choice (at least in the near future) given that these methods are not as computationally prohibitive as either LES or DNS. Hybrid methods represent the amalgamation of closure modelling with eddy-resolving simulation. The key question here is what blending and/or interpolation functions should be used in hybrid methods to give the best prediction of FIV phenomena?

(3) Machine learning technique.

- In machine learning, the availability of large public datasets has spurred rapid progress in computer vision, speech recognition, and other applications. Many sophisticated machine learning techniques (including those for numerical simulation of turbulent flows) require large datasets consisting of high-quality information. Perhaps the biggest challenge for machine learning techniques within FIV simulation is the lack of availability of data. Every investigator that applies a state-of-the-art machine learning method to FIV simulation must generate their own dataset. Currently, it is impossible to benchmark various machine learning techniques, such as PINN or augmented RANS modelling as applied to FIV, because there exists no suitable public dataset for this purpose.
- Aside from the lack of available data, there are major physics-based challenges for applying machine learning technique to FIV simulation. Changes in the Reynolds number, in the freestream turbulence, in vibration parameters, and in the body shape all greatly affect the FIV response. Even without incorporating machine learning techniques, traditional physics-based numerical methods, such as RANS and LES, often struggle to reproduce the complicated interactions within FIV. When applying a machine learning technique for FIV, the design of the training dataset will be of utmost importance. Generalizability issues within machine learning greatly affect the model's ability to perform outside of the parameters given in the training dataset. For example, a model trained at $Re = 2000$ may produce erroneous results at $Re = 30{,}000$, because transitional turbulence regions in the former case become fully turbulent in the latter case [269]. These small changes in turbulence can have large effects on the vibration response. Therefore, the training dataset should be designed to incorporate a wide variety of FIV parameters, so that the model can perform well for a variety of test cases. The issue of generalizability within machine learning augmented turbulence modelling is currently under investigation. Machine learning techniques may be better suited to provide accelerated and higher-accuracy simulation results within a known

range of parameters, rather than attempting to uncover new physics (for example, at a higher-Reynolds number than any of the cases included in the training dataset).

- While several PINN techniques have been developed and applied to the unsteady FIV problem, RANS-based machine learning techniques for unsteady cases are still in development. The overwhelming majority of machine learning augmented RANS investigations use a steady state flow for training and testing. Some techniques within augmented RANS (e.g., iterative methods) are more suited for unsteady flows, but have not been rigorously tested. Therefore, another key issue is the ability of augmented turbulence models to be used for transient cases. Machine learning augmented LES closures are more immediately applicable for unsteady flow simulation due to the nature of LES. However, the suitability of these closure for the FIV problem is yet to be demonstrated.

7.2. Future Work and Perspectives

In spite of important advances, much remains to be done with respect to research concerning the modelling and prediction of the myriad of physical processes underlying various FIV phenomena—there is much future work to be done. In light of the challenges in current FIV modelling efforts identified in Section 7.1, we summarize directions for future work.

- With respect to the mathematical modelling of FIV phenomena, the development of a general theory for one or more aspects of FIV phenomena from first principles and the comprehensive analysis following from the application of such a general theory is a worthwhile (albeit extremely difficult) avenue of investigation.
- To allow mathematical models for FIV phenomena to have greater flexibility/generality (and lacking a general theory for FIV at present), research into the application of improved optimization methods for determination of model parameters along with an uncertainty quantification of the prediction provided by such models would be extremely beneficial.
- Work on the development of an improved mathematical representation for the inclusion of galloping in mathematical models for FIV could potentially improve significantly the prediction accuracy of the current generation of models for a bluff body experiencing either pure galloping or a combined FIV that includes galloping.
- More work needs to be done to improve the predictive accuracy of low-fidelity numerical methodologies, such as unsteady RANS modelling for routine application to FIV problems of industrial relevance. Schemes to provide improved closure models for addressing the FIV problem need to be developed. From this perspective, the application of data-driven approaches using ML (e.g., neural nets, gradient boosting machines [270]) or inverse modelling (e.g., Bayesian optimization [271], likelihood-free inference [272]) can be used to design application specific closure models for FIV phenomena.
- More effort needs to be focused on the proper application of hybrid methods to the prediction of FIV phenomena. To this purpose, more research should be conducted on the appropriate selection of the mode of computation (whether RANS modelling or eddy-resolving simulation) that is specifically tailored to address the peculiarities of FIV phenomena (e.g., development of a specific hybrid blending or interpolation function that more intelligently selects whether RANS or eddy-resolving simulation should be used in various regions of the FIV flows).
- Methods of machine learning can be applied potentially to provided improved closure models for numerical simulation of FIV phenomena. However, for this to occur, a considerable effort should be made to obtain high-quality FIV datasets from high-fidelity numerical simulation (DNS and/or LES). The authors strongly encourage the future publication of public-domain FIV datasets that can be used for bench marking machine learning techniques.

- Work should be undertaken to better understand the synergy in mathematical, numerical, and machine learning techniques for the modelling of FIV phenomena (and, more specifically, on how to exploit this synergy to advance future modelling efforts in this field of endeavor). For example, high-fidelity numerical simulation methodologies (DNS and/or LES) provide the data for machine learning algorithms that can potentially lead to improved closure models for low-fidelity RANS models. The predictions obtained from these models provide the information required to obtain deeper insights and a better understanding of FIV phenomenology—this better understanding can potentially lead to the development of improved and more principled and general mathematical models of FIV phenomena.

The key takeaway of this review is that modelling is arguably the most important tool for investigation of various aspects of FIV phenomena. Advances in the modelling of FIV phenomena will undoubtedly impact a number of engineering and industrial applications of practical real-world relevance. A deeper understanding of FIV phenomena will allow engineers to design potentially significantly improved fluid kinetic energy harvesting devices (viz., enhance the energy harvesting efficiency of such devices). However, this application will require development of a system-of-systems modelling approach involving coupling models of FIV, electromechanics, and complex interface circuits. Alternatively, modelling advances in FIV will lead to a better understanding of fatigue failure arising from vibration in structures exposed to high energy fluid flow (e.g., transmission lines critical to modern electricity generation infrastructure, fuel rod assembly in conventional or small modular nuclear reactors and implications for nuclear safety, risers for deep water applications, such as oil and gas production). An under-explored area of research in FIV is multi-physics coupling in the simulation. For example, transmission lines in cold climates during winter storms are subject to icing as well as to wind-induced (aeolian) vibration that can cause outages—how the vibration response of the transmission line is altered with ice accretion has not been studied to the authors' knowledge.

Author Contributions: Conceptualization, F.-S.L. and E.Y.; methodology, Y.W., Z.C. and R.M.; software, Y.W., Z.C.; validation, Y.W., Z.C., R.M., F.-S.L. and E.Y.; formal analysis, Y.W.; investigation, Y.W., Z.C., R.M., F.-S.L. and E.Y.; resources, F.-S.L.; data curation, Y.W. and Z.C.; writing—original draft preparation, Y.W., Z.C. and R.M.; writing—review and editing, E.Y. and F.-S.L.; visualization, Y.W. and Z.C.; supervision, F.-S.L. and E.Y.; project administration, F.-S.L. and E.Y.; funding acquisition, F.-S.L. All authors have read and agreed to the published version of the manuscript.

Funding: This research was funded by the Natural Sciences and Engineering Research Council of Canada (NSERC) Discovery Grants program Grant Nos. 50503-10234.

Data Availability Statement: Not applicable.

Conflicts of Interest: The authors declare no conflict of interest.

Abbreviations

The following abbreviations are used in this manuscript:

AI	Artificial Intelligence
ALE	Arbitrary Lagrangian–Eulerian
AOA	Angle of Attack
ARX	Autoregressive with Exogenous Input
BPNN	Back Propagation Neural Network
BPOD	Balanced POD

CAE	Computer-Aided Engineering
CFD	Computational Fluids Dynamics
CNN	Convolutional Neural Network
CRAN	Convolutional Recurrent Autoencoder Network
DDES	Delayed Detached-Eddy Simulation
DES	Detached-Eddy Simulation
DEIM	Discrete Empirical Interpolation Method
DMD	Dynamic Mode Decomposition
DNN	Deep Neural Network
DNS	Direct Numerical Simulation
DSD/SST	Deforming-Spatial-Domain/Stabilized Space-Time
DTR	Decision-Tree Regression
DVM	Discrete Vortex Method
EB	embedded boundary
ERA	Eigensystem Realization Algorithm
EVM	Entropy–Viscosity Method
FD	Fictitious Domain
FDM	Finite Differencing Method
FEM	Finite Element Method
FIV	Flow-Induced Vibration
FIVEH	Flow-Induced Vibration Energy Harvesting
FOM	Full-Order Model
FSI	Flow-Structure Interaction
FVM	Finite Volume Method
GBRT	Gradient Boosting Regression Trees
IBM	Immersed Boundary Method
LBM	Lattice-Boltzmann Method
LES	Large-Eddy Simulation
LSA	Linear Stability Analysis
LSTM	Long Short-Term Memory
ML	Machine Learning
MTS	Multiple Time Scales
NS	Navier–Stokes
NS-PINN	Navier–Stokes equations-based PINN
ODE	Ordinary Differential Equation
PDE	Partial Differential Equation
PG-FEM	Petrov–Galerkin Finite Element Method
PINN	Physics-Informed Neural Networks
PNS-PINN	Parameterized Navier–Stokes equations-based PINN
POD	Proper Orthogonal Decomposition
PTC	Passive Turbulence Control
QS	Quasi-Steady
RANS	Reynolds-averaged Navier–Stokes
RBFNN	Radial Basis Function Neural Network
RF	Random Forest
RNN	Recurrent Neural Network
ROM	Reduced-Order Model
SA	Spalart–Allmaras
SDOF	Single Degree-Of-Freedom
SEM	Spectral Elements Method
SGS	Sub-grid Scale
SM	Smagorinsky
SPOD	Spectral POD
SST	Shear Stress Transport
SST-FEM	Stabilized Space–Time FEM
SVD	Singular Value Decomposition
TBNN	Tensor Basis Neural Network

VIV	Vortex-Induced Vibration
WALE	Wall-Adapting Local-Eddy
mPOD	Multiscale POD
1DOF	One-Degree-Of-Freedom
2DOF	Two-Degree-Of-Freedom
2D	Two-Dimensional
3D	Three-Dimensional

References

1. Nakamura, T.; Kaneko, S.; Inada, F.; Kato, M.; Ishihara, K.; Nishihara, T.; Mureithi, N.W.; Langthjem, M.A. *Flow-Induced Vibrations: Classifications and Lessons from Practical Experiences*; Butterworth-Heinemann: Portsmouth, NH, USA, 2013.
2. Parkinson, G. Phenomena and modelling of flow-induced vibrations of bluff bodies. *Prog. Aerosp. Sci.* **1989**, *26*, 169–224. [CrossRef]
3. Brunton, S.L.; Noack, B.R.; Koumoutsakos, P. Machine learning for fluid mechanics. *Annu. Rev. Fluid Mech.* **2020**, *52*, 477–508. [CrossRef]
4. Kutz, J.N. Deep learning in fluid dynamics. *J. Fluid Mech.* **2017**, *814*, 1–4. [CrossRef]
5. Usman, A.; Rafiq, M.; Saeed, M.; Nauman, A.; Almqvist, A.; Liwicki, M. Machine learning computational fluid dynamics. In Proceedings of the IEEE 2021 Swedish Artificial Intelligence Society Workshop (SAIS), Stockholm, Sweden, 14–15 June 2021; pp. 1–4.
6. Ma, X.; Zhou, S. A review of flow-induced vibration energy harvesters. *Energy Convers. Manag.* **2022**, *254*, 115223. [CrossRef]
7. Wang, J.; Geng, L.; Ding, L.; Zhu, H.; Yurchenko, D. The state-of-the-art review on energy harvesting from flow-induced vibrations. *Appl. Energy* **2020**, *267*, 114902. [CrossRef]
8. Lv, Y.; Sun, L.; Bernitsas, M.M.; Sun, H. A comprehensive review of nonlinear oscillators in hydrokinetic energy harnessing using flow-induced vibrations. *Renew. Sustain. Energy Rev.* **2021**, *150*, 111388. [CrossRef]
9. Blevins, R.D. *Flow-Induced Vibration*; Van Nostrand Reinhold Company: New York, NY, USA, 1977.
10. Rostami, A.B.; Armandei, M. Renewable energy harvesting by vortex-induced motions: Review and benchmarking of technologies. *Renew. Sustain. Energy Rev.* **2017**, *70*, 193–214. [CrossRef]
11. Tamura, Y. Mathematical models for understanding phenomena: Vortex-induced vibrations. *Jpn. Archit. Rev.* **2020**, *3*, 398–422. [CrossRef]
12. Su, Y.; Di, J.; Li, S.; Jian, B.; Liu, J. Buffeting response prediction of long-span bridges based on different wind tunnel test techniques. *Appl. Sci.* **2022**, *12*, 3171. [CrossRef]
13. Luongo, A.; Piccardo, G. Linear instability mechanisms for coupled translational galloping. *J. Sound Vib.* **2005**, *288*, 1027–1047. [CrossRef]
14. Gabbai, R.D.; Benaroya, H. An overview of modeling and experiments of vortex-induced vibration of circular cylinders. *J. Sound Vib.* **2005**, *282*, 575–616. [CrossRef]
15. Païdoussis, M.P.; Price, S.J.; De Langre, E. *Fluid-Structure Interactions: Cross-Flow-Induced Instabilities*; Cambridge University Press: Cambridge, UK, 2010.
16. Rayleigh, J.W.S.B. *The Theory of Sound*; Macmillan: New York, NY, USA, 1896; Volume 2.
17. van der Pol, B.; van der Mark, J. LXXII. The heartbeat considered as a relaxation oscillation, and an electrical model of the heart. *Lond. Edinb. Dublin Philos. Mag. J. Sci.* **1928**, *6*, 763–775. [CrossRef]
18. Goswami, U. Toward an interactive analogy model of reading development: Decoding vowel graphemes in beginning reading. *J. Exp. Child Psychol.* **1993**, *56*, 443–475. [CrossRef]
19. Bishop, R.E.D.; Hassan, A. The lift and drag forces on a circular cylinder oscillating in a flowing fluid. *Proc. R. Soc. Lond. Ser. Math. Phys. Sci.* **1964**, *277*, 51–75.
20. Hartlen, R.T.; Currie, I.G. Lift-oscillator model of vortex-induced vibration. *J. Eng. Mech. Div.* **1970**, *96*, 577–591. [CrossRef]
21. Skop, R.; Griffin, O. A model for the vortex-excited resonant response of bluff cylinders. *J. Sound Vib.* **1973**, *27*, 225–233. [CrossRef]
22. Griffin, O.; Skop, R.; Koopmann, G. The vortex-excited resonant vibrations of circular cylinders. *J. Sound Vib.* **1973**, *31*, 235–IN3. [CrossRef]
23. Landl, R. A mathematical model for vortex-excited vibrations of bluff bodies. *J. Sound Vib.* **1975**, *42*, 219–234. [CrossRef]
24. Skop, R.; Griffin, O. On a theory for the vortex-excited oscillations of flexible cylindrical structures. *J. Sound Vib.* **1975**, *41*, 263–274. [CrossRef]
25. Skop, R.; Balasubramanian, S. A new twist on an old model for vortex-excited vibrations. *J. Fluids Struct.* **1997**, *11*, 395–412. [CrossRef]
26. Skop, R.; Luo, G. An inverse-direct method for predicting the vortex-induced vibrations of cylinders in uniform and nonuniform flows. *J. Fluids Struct.* **2001**, *15*, 867–884. [CrossRef]
27. Facchinetti, M.L.; De Langre, E.; Biolley, F. Coupling of structure and wake oscillators in vortex-induced vibrations. *J. Fluids Struct.* **2004**, *19*, 123–140. [CrossRef]
28. Ogink, R.; Metrikine, A. A wake oscillator with frequency dependent coupling for the modeling of vortex-induced vibration. *J. Sound Vib.* **2010**, *329*, 5452–5473. [CrossRef]

29. Furnes, G.K.; Sørensen, K. Flow induced vibrations modeled by coupled non-linear oscillators. In Proceedings of the Seventeenth International Offshore and Polar Engineering Conference, Lisbon, Portugal, 1–6 July 2007; OnePetro: Richardson, TX, USA, 2007.
30. Ge, F.; Long, X.; Wang, L.; Hong, Y. Flow-induced vibrations of long circular cylinders modeled by coupled nonlinear oscillators. *Sci. China Ser. Phys. Mech. Astron.* **2009**, *52*, 1086–1093. [CrossRef]
31. Srinil, N.; Zanganeh, H. Modelling of coupled cross-flow/in-line vortex-induced vibrations using double Duffing and van der Pol oscillators. *Ocean Eng.* **2012**, *53*, 83–97. [CrossRef]
32. Bai, X.; Qin, W. Using vortex strength wake oscillator in modelling of vortex induced vibrations in two degrees of freedom. *Eur. J. Mech. B Fluids* **2014**, *48*, 165–173. [CrossRef]
33. Postnikov, A.; Pavlovskaia, E.; Wiercigroch, M. 2DOF CFD calibrated wake oscillator model to investigate vortex-induced vibrations. *Int. J. Mech. Sci.* **2017**, *127*, 176–190. [CrossRef]
34. Birkhoff, G. Formation of vortex streets. *J. Appl. Phys.* **1953**, *24*, 98–103. [CrossRef]
35. Kutta, W. Auftriebskräfte in strömenden Flüssigkeiten. *Illus. Aeronaut. Mitteilung.* **1902**, *6*, 133–135.
36. Joukowsky, N. Über die konturen der Tragflächen der Drachenflieger. *Z. Flugtech. Mot.* **1910**, *1*, 281–285.
37. Funakawa, M. Excitation mechanism of elastically supported circular cylinder in the flow. *Bull. Jpn. Soc. Mech. Eng.* **1970**, *36*, 303–312.
38. Tamura, Y. Vortex induced vibration of circular cylinder (Part I Wake oscillator model). *Trans. Archit. Inst. Jpn.* **1978**, *266*, 87–95. [CrossRef]
39. Tamura, Y.; Matsui, G. *Wind Engineering, Proceedings of the 5th International Conference on Wind Engineering, Fort Collins, CO, USA, 8–13 July 1979*; Cermak, J.E., Ed.; Pergamon Press Ltd.: Amsterdam, The Netherlands, 1979; Volume 266, pp. 1085–1094.
40. Tamura, Y.; Amano, A. Mathematical model for vortex-induced oscillations of continuous systems with circular cross section. *J. Wind. Eng. Ind. Aerodyn.* **1983**, *14*, 431–442. [CrossRef]
41. Iwan, W.; Blevins, R. A model for vortex induced oscillation of structures. *J. Appl. Mech.* **1974**, *41*, 581–586. [CrossRef]
42. Krenk, S.; Nielsen, S.R. Energy balanced double oscillator model for vortex-induced vibrations. *J. Eng. Mech.* **1999**, *125*, 263–271. [CrossRef]
43. Scanlan, R.H. *State-of-the-Art Methods For Calculating Flutter, Vortex-Induced, and Buffeting Response of Bridge Structures*; Technical Report No. FHWA/RD-80-050; Federal Highway Administration: Washington, DC, USA, 1981.
44. Simiu, E.; Scanlan, R. *Wind Effects on Structures*, 2nd ed.; John Wiley and Sons: Hoboken, NJ, USA, 1986.
45. Ehsan, F.; Scanlan, R.H.; Bosch, H.R. Modeling spanwise correlation effects in the vortex-induced response of flexible bridges. *J. Wind. Eng. Ind. Aerodyn.* **1990**, *36*, 1105–1114. [CrossRef]
46. Billah, K.Y.R. A Study of Vortex-Induced Vibration. Ph.D. Thesis, Princeton University, Princeton, NJ, USA, 1989.
47. Goswami, I.; Scanlan, R.H.; Jones, N.P. Vortex-induced vibration of circular cylinders. II: New model. *J. Eng. Mech.* **1993**, *119*, 2288–2302. [CrossRef]
48. Sarpkaya, T. Fluid forces on oscillating cylinders. *J. Waterw. Port, Coast. Ocean. Div.* **1978**, *104*, 275–290. [CrossRef]
49. Staubli, T. Calculation of the vibration of an elastically mounted cylinder using experimental data from forced oscillation. *J. Fluids Eng.* **1983**, *105*, 225–229. [CrossRef]
50. Stäubli, M.; Bircher, J.; Galeazzi, R.; Remund, H.; Studer, H. Serum concentrations of amiodarone during long term therapy. Relation to dose, efficacy and toxicity. *Eur. J. Clin. Pharmacol.* **1983**, *24*, 485–494. [CrossRef]
51. Griffin, O.; Koopmann, G. The vortex-excited lift and reaction forces on resonantly vibrating cylinders. *J. Sound Vib.* **1977**, *54*, 435–448. [CrossRef]
52. Feng, C. The Measurement of Vortex Induced Effect in Flow Past Stationary and Oscillating Circular and D-Section Cylinders. Master's Thesis, University of British Columbia, Vancouver, BC, Canada, 1968.
53. Parkinson, G.; Sullivan, P. Galloping response of towers. *J. Wind. Eng. Ind. Aerodyn.* **1979**, *4*, 253–260. [CrossRef]
54. den Hartog, J. Transmission line vibration due to sleet. *Trans. Am. Inst. Electr. Eng.* **1932**, *51*, 1074–1076. [CrossRef]
55. Meseguer Ruiz, J.; Sanz Andres, A.P.; Alonso Rodrigo, G. Determination of maximum mechanical energy efficiency in energy galloping systems. *J. Eng. Mech.* **2015**, *141*, 1–8. [CrossRef]
56. Alonso, G.; Meseguer, J.; Pérez-Grande, I. Galloping stability of triangular cross-sectional bodies: A systematic approach. *J. Wind. Eng. Ind. Aerodyn.* **2007**, *95*, 928–940. [CrossRef]
57. Simiu, E.; Yeo, D. *Wind Effects on Structures: Modern Structural Design for Wind*; John Wiley & Sons: Hoboken, NJ, USA, 2019.
58. Parkinson, G.; Brooks, N. On the aeroelastic instability of bluff cylinders. *J. Appl. Mech.* **1961**, *28*, 252–258. [CrossRef]
59. Parkinson, G.; Smith, J. The square prism as an aeroelastic non-linear oscillator. *Q. J. Mech. Appl. Math.* **1964**, *17*, 225–239. [CrossRef]
60. Luo, S.; Chew, Y.; Ng, Y. Hysteresis phenomenon in the galloping oscillation of a square cylinder. *J. Fluids Struct.* **2003**, *18*, 103–118. [CrossRef]
61. Ng, Y.; Luo, S.; Chew, Y. On using high-order polynomial curve fits in the quasi-steady theory for square-cylinder galloping. *J. Fluids Struct.* **2005**, *20*, 141–146. [CrossRef]
62. Barrero-Gil, A.; Sanz-Andres, A.; Roura, M. Transverse galloping at low Reynolds numbers. *J. Fluids Struct.* **2009**, *25*, 1236–1242. [CrossRef]
63. Joly, A.; Etienne, S.; Pelletier, D. Galloping of square cylinders in cross-flow at low Reynolds numbers. *J. Fluids Struct.* **2012**, *28*, 232–243. [CrossRef]

64. Piccardo, G.; Pagnini, L.C.; Tubino, F. Some research perspectives in galloping phenomena: Critical conditions and post-critical behavior. *Contin. Mech. Thermodyn.* **2015**, *27*, 261–285. [CrossRef]
65. Bearman, P.; Gartshore, I.; Maull, D.; Parkinson, G. Experiments on flow-induced vibration of a square-section cylinder. *J. Fluids Struct.* **1987**, *1*, 19–34. [CrossRef]
66. Norberg, C. Flow around rectangular cylinders: Pressure forces and wake frequencies. *J. Wind. Eng. Ind. Aerodyn.* **1993**, *49*, 187–196. [CrossRef]
67. Desai, Y.; Shah, A.; Popplewell, N. Galloping analysis for two-degree-of-freedom oscillator. *J. Eng. Mech.* **1990**, *116*, 2583–2602. [CrossRef]
68. Jones, K.F. Coupled vertical and horizontal galloping. *J. Eng. Mech.* **1992**, *118*, 92–107. [CrossRef]
69. He, M.; Macdonald, J.H. An analytical solution for the galloping stability of a 3 degree-of-freedom system based on quasi-steady theory. *J. Fluids Struct.* **2016**, *60*, 23–36. [CrossRef]
70. He, M.; Macdonald, J. 3DOF galloping analysis based on quasi-steady theory with reference to the aerodynamic stiffness coupling. In Proceedings of the 7th European and African Conference on Wind Engineering, Liege, Belgium, 4–7 July 2017.
71. Parkinson, G.; Wawzonek, M. Some considerations of combined effects of galloping and vortex resonance. *J. Wind. Eng. Ind. Aerodyn.* **1981**, *8*, 135–143. [CrossRef]
72. Parkinson, G.; Bouclin, D. Hydroelastic oscillations of square cylinders. In Proceedings of the International Research Seminar on Safety of Structures under Dynamic Loading, Trondheim, Norway, 25–30 June 1977.
73. Corless, R.M.; Parkinson, G. A model of the combined effects of vortex-induced oscillation and galloping. *J. Fluids Struct.* **1988**, *2*, 203–220. [CrossRef]
74. Han, P.; Hémon, P.; Pan, G.; de Langre, E. Nonlinear modeling of combined galloping and vortex-induced vibration of square sections under flow. *Nonlinear Dyn.* **2021**, *103*, 3113–3125. [CrossRef]
75. Tamura, Y.; Shimada, K. A mathematical model for the transverse oscillations of square cylinders. In Proceedings of the International Conference on Flow Induced Vibrations, Bowness-Windermere, UK, 12–14 May 1987; pp. 12–14.
76. Mannini, C.; Massai, T.; Marra, A.M.; Bartoli, G. Modelling the interaction of VIV and galloping for rectangular cylinders. In Proceedings of the 14th International Conference on Wind Engineering, Porto Alegre, Brazil, 21–26 June 2015; pp. 1–20.
77. Mannini, C.; Massai, T.; Marra, A.M. Modeling the interference of vortex-induced vibration and galloping for a slender rectangular prism. *J. Sound Vib.* **2018**, *419*, 493–509. [CrossRef]
78. Mannini, C. Incorporation of turbulence in a nonlinear wake-oscillator model for the prediction of unsteady galloping response. *J. Wind. Eng. Ind. Aerodyn.* **2020**, *200*, 104141. [CrossRef]
79. Chen, C.; Mannini, C.; Bartoli, G.; Thiele, K. Wake oscillator modeling the combined instability of vortex induced vibration and galloping for a 2: 1 rectangular cylinder. *J. Fluids Struct.* **2022**, *110*, 103530. [CrossRef]
80. Funakawa, M. The vibration of a cylinder caused by wake force in a flow. *Bull. JSME* **1969**, *12*, 1003–1010. [CrossRef]
81. ANSYS Homepage. Available online: http:www.ansys.com (accessed on 30 August 2022).
82. Siemens PLM STAR-CCM+ Homepage. Available online: https://mdx.plm.automation.siemens.com/star-ccm-plus (accessed on 30 August 2022).
83. OpenFOAM Project Homepage. Available online: http://www.openfoam.org (accessed on 30 August 2022).
84. Selvam, R.P. *Computational Fluid Dynamics for Wind Engineering*; John Wiley & Sons: Hoboken, NJ, USA, 2022.
85. Zhao, M.; Cheng, L. Vortex-induced vibration of a circular cylinder of finite length. *Phys. Fluids* **2014**, *26*, 2548. [CrossRef]
86. Hou, G.; Wang, J.; Layton, A. Numerical methods for fluid-structure interaction—A review. *Commun. Comput. Phys.* **2012**, *12*, 337–377. [CrossRef]
87. Singh, S.; Mittal, S. Vortex-induced oscillations at low Reynolds numbers: Hysteresis and vortex-shedding modes. *J. Fluids Struct.* **2005**, *20*, 1085–1104. [CrossRef]
88. Mittal, S. Free vibrations of a cylinder: 3-D computations at Re = 1000. *J. Fluids Struct.* **2013**, *41*, 109–118.
89. Alauzet, F. A changing-topology moving mesh technique for large displacements. *Eng. Comput.* **2014**, *30*, 175–200. [CrossRef]
90. Mittal, S.; Kumar, V. Finite element study of vortex-induced cross-flow and in-line oscillations of a circular cylinder at low Reynolds numbers. *Int. J. Numer. Methods Fluids* **1999**, *31*, 1087–1120. [CrossRef]
91. Longatte, E.; Bendjeddou, Z.; Souli, M. Application of arbitrary Lagrange Euler formulations to flow-induced vibration problems. *J. Press. Vessel Technol.* **2003**, *125*, 411–417. [CrossRef]
92. Zhao, M.; Cheng, L. Numerical simulation of two-degree-of-freedom vortex-induced vibration of a circular cylinder close to a plane boundary. *J. Fluids Struct.* **2011**, *27*, 1097–1110. [CrossRef]
93. Zhao, M.; Cheng, L. Numerical simulation of vortex-induced vibration of four circular cylinders in a square configuration. *J. Fluids Struct.* **2012**, *31*, 125–140. [CrossRef]
94. Zhao, M. Flow induced vibration of two rigidly coupled circular cylinders in tandem and side-by-side arrangements at a low Reynolds number of 150. *Phys. Fluids* **2013**, *25*, 123601. [CrossRef]
95. Zhao, M.; Cheng, L.; Zhou, T. Numerical simulation of vortex-induced vibration of a square cylinder at a low Reynolds number. *Phys. Fluids* **2013**, *25*, 023603. [CrossRef]
96. Zhao, M.; Cheng, L.; An, H.; Lu, L. Three-dimensional numerical simulation of vortex-induced vibration of an elastically mounted rigid circular cylinder in steady current. *J. Fluids Struct.* **2014**, *50*, 292–311. [CrossRef]

97. Zhao, M.; Cheng, L.; An, H.; Tong, F. Flow and flow-induced vibration of a square array of cylinders in steady currents. *Fluid Dyn. Res.* **2015**, *47*, 045505. [CrossRef]
98. Griffith, B.E.; Patankar, N.A. Immersed methods for fluid-structure interaction. *Annu. Rev. Fluid Mech.* **2020**, *52*, 421. [CrossRef]
99. Chen, W.; Ji, C.; Xu, W.; Liu, S.; Campbell, J. Response and wake patterns of two side-by-side elastically supported circular cylinders in uniform laminar cross-flow. *J. Fluids Struct.* **2015**, *55*, 218–236. [CrossRef]
100. Peskin, C.S. *Flow Patterns Around Heart Valves: A Digital Computer Method for Solving the Equations of Motion*; Yeshiva University: Manhattan, NY, USA, 1972.
101. Ji, C.; Xiao, Z.; Wang, Y.; Wang, H. Numerical investigation on vortex-induced vibration of an elastically mounted circular cylinder at low Reynolds number using the fictitious domain method. *Int. J. Comput. Fluid Dyn.* **2011**, *25*, 207–221. [CrossRef]
102. Löhner, R.; Baum, J.D.; Mestreau, E.; Sharov, D.; Charman, C.; Pelessone, D. Adaptive embedded unstructured grid methods. *Int. J. Numer. Methods Eng.* **2004**, *60*, 641–660. [CrossRef]
103. Gonzalez Cornejo, F.A.; Cruchaga, M.A.; Celentano, D.J. Modelling low Reynolds number vortex-induced vibration problems with a fixed mesh fluid-solid interaction formulation. *Phys. Fluids* **2017**, *29*, 113601. [CrossRef]
104. Wall, W.A.; Gerstenberger, A.; Gamnitzer, P.; Förster, C.; Ramm, E. Large deformation fluid-structure interaction–Advances in ALE methods and new fixed grid approaches. In *Fluid-Structure Interaction*; Springer: Berlin/Heidelberg, Germany, 2006; pp. 195–232.
105. Al-Jamal, H.; Dalton, C. Vortex induced vibrations using large-eddy simulation at a moderate Reynolds number. *J. Fluids Struct.* **2004**, *19*, 73–92. [CrossRef]
106. Carmo, B.S.; Sherwin, S.J.; Bearman, P.W.; Willden, R. Flow-induced vibration of a circular cylinder subjected to wake interference at low Reynolds number. *J. Fluids Struct.* **2011**, *27*, 503–522. [CrossRef]
107. Williamson, C.H.; Roshko, A. Vortex formation in the wake of an oscillating cylinder. *J. Fluids Struct.* **1988**, *2*, 355–381. [CrossRef]
108. Barkley, D.; Henderson, R.D. Three-dimensional Floquet stability analysis of the wake of a circular cylinder. *J. Fluid Mech.* **1996**, *322*, 215–241. [CrossRef]
109. Jiang, H.; Cheng, L.; An, H. Three-dimensional wake transition of a square cylinder. *J. Fluid Mech.* **2018**, *842*, 102–127. [CrossRef]
110. Ryan, K.; Thompson, M.C.; Hourigan, K. Three-dimensional transition in the wake of bluff elongated cylinders. *J. Fluid Mech.* **2005**, *538*, 1–29. [CrossRef]
111. Leontini, J.S.; Jacono, D.L.; Thompson, M.C. Stability analysis of the elliptic cylinder wake. *J. Fluid Mech.* **2015**, *763*, 302–321. [CrossRef]
112. Leontini, J.S.; Thompson, M.C.; Hourigan, K. Three-dimensional transition in the wake of a transversely oscillating cylinder. *J. Fluid Mech.* **2007**, *577*, 79–104. [CrossRef]
113. Prasanth, T.; Behara, S.; Singh, S.; Kumar, R.; Mittal, S. Effect of blockage on vortex-induced vibrations at low Reynolds numbers. *J. Fluids Struct.* **2006**, *22*, 865–876. [CrossRef]
114. Prasanth, T.; Premchandran, V.; Mittal, S. Hysteresis in vortex-induced vibrations: Critical blockage and effect of m. *J. Fluid Mech.* **2011**, *671*, 207–225. [CrossRef]
115. Prasanth, T.; Mittal, S. Vortex-induced vibrations of a circular cylinder at low Reynolds numbers. *J. Fluid Mech.* **2008**, *594*, 463–491. [CrossRef]
116. Mittal, S. A new regime of multiple states in free vibration of a cylinder at low Re. *J. Fluids Struct.* **2017**, *68*, 310–321.
117. Leontini, J.S.; Thompson, M.C.; Hourigan, K. The beginning of branching behaviour of vortex-induced vibration during two-dimensional flow. *J. Fluids Struct.* **2006**, *22*, 857–864. [CrossRef]
118. Li, T.; Zhang, J.; Zhang, W. Nonlinear characteristics of vortex-induced vibration at low Reynolds number. *Commun. Nonlinear Sci. Numer. Simul.* **2011**, *16*, 2753–2771. [CrossRef]
119. Étienne, S.; Pelletier, D. The low Reynolds number limit of vortex-induced vibrations. *J. Fluids Struct.* **2012**, *31*, 18–29. [CrossRef]
120. Bourguet, R.; Jacono, D.L. Flow-induced vibrations of a rotating cylinder. *J. Fluid Mech.* **2014**, *740*, 342–380. [CrossRef]
121. Zhao, M.; Cheng, L.; Lu, L. Vortex induced vibrations of a rotating circular cylinder at low Reynolds number. *Phys. Fluids* **2014**, *26*, 073602. [CrossRef]
122. Garg, H.; Soti, A.K.; Bhardwaj, R. Vortex-induced vibration of a cooled circular cylinder. *Phys. Fluids* **2019**, *31*, 083608. [CrossRef]
123. Shen, L.; Chen, J. Irregular vortex-induced vibrations of a two-dimensional circular cylinder at a low Reynolds number. *Eur. J. Mech. B Fluids* **2022**, *94*, 128–147. [CrossRef]
124. Dorogi, D.; Baranyi, L. Identification of upper branch for vortex-induced vibration of a circular cylinder at *Re* = 300. *J. Fluids Struct.* **2020**, *98*, 103135. [CrossRef]
125. Yogeswaran, V.; Sen, S.; Mittal, S. Free vibrations of an elliptic cylinder at low Reynolds numbers. *J. Fluids Struct.* **2014**, *51*, 55–67.
126. Chen, G.; Liang, X.F.; Li, X.B.; Zhou, D.; Lien, F.S. Numerical investigation of vortex induced vibration effects on the heat transfer for various aspect ratios ellipse cylinder. *Int. J. Therm. Sci.* **2021**, *170*, 107138. [CrossRef]
127. Leontini, J.S.; Griffith, M.; Jacono, D.L.; Sheridan, J. The flow-induced vibration of an elliptical cross-section at varying angles of attack. *J. Fluids Struct.* **2018**, *78*, 356–373. [CrossRef]
128. Sen, S.; Mittal, S. Free vibration of a square cylinder at low Reynolds numbers. *J. Fluids Struct.* **2011**, *27*, 875–884. [CrossRef]
129. He, T.; Zhou, D.; Bao, Y. Combined interface boundary condition method for fluid–rigid body interaction. *Comput. Methods Appl. Mech. Eng.* **2012**, *223*, 81–102. [CrossRef]

130. Sen, S.; Mittal, S. Effect of mass ratio on free vibrations of a square cylinder at low Reynolds numbers. *J. Fluids Struct.* **2015**, *54*, 661–678. [CrossRef]
131. Sourav, K.; Sen, S. Transition of VIV-only motion of a square cylinder to combined VIV and galloping at low Reynolds numbers. *Ocean Eng.* **2019**, *187*, 106208. [CrossRef]
132. Abdelkefi, A.; Hajj, M.R.; Nayfeh, A.H. Power harvesting from transverse galloping of square cylinder. *Nonlinear Dyn.* **2012**, *70*, 1355–1363. [CrossRef]
133. Su, Z.; Liu, Y.; Zhang, H.; Zhang, D. Numerical simulation of vortex-induced vibration of a square cylinder. *J. Mech. Sci. Technol.* **2007**, *21*, 1415–1424. [CrossRef]
134. Sun, W.; Zhou, D.; Tu, J.; Han, Z. Vortex-induced vibrations of a square cylinder under linear shear flow. *Fluid Dyn. Res.* **2017**, *49*, 025502. [CrossRef]
135. Wang, H.; Zhao, D.; Yang, W.; Yu, G. Numerical investigation on flow-induced vibration of a triangular cylinder at a low Reynolds number. *Fluid Dyn. Res.* **2014**, *47*, 015501. [CrossRef]
136. Liu, M.; Jin, R.; Tang, G.; Li, C. Numerical investigation of vortex-induced vibration for two tandem circular cylinders with different diameters. *Proc. Inst. Mech. Eng. Part M J. Eng. Mar.Environ.* **2020**, *234*, 676–685. [CrossRef]
137. Ming-Ming, L.; Hao-Cheng, W.; Fei-Fei, S.; Xin, J.; Tang, G.Q.; Fan, Y. Numerical investigation on vortex-induced vibration of an elastically mounted circular cylinder with multiple control rods at low Reynolds number. *Appl. Ocean. Res.* **2022**, *118*, 102987.
138. Sahu, T.R.; Furquan, M.; Jaiswal, Y.; Mittal, S. Flow-induced vibration of a circular cylinder with rigid splitter plate. *J. Fluids Struct.* **2019**, *89*, 244–256. [CrossRef]
139. Zhu, H.; Li, G.; Wang, J. Flow-induced vibration of a circular cylinder with splitter plates placed upstream and downstream individually and simultaneously. *Appl. Ocean. Res.* **2020**, *97*, 102084. [CrossRef]
140. Wang, H.; Zhai, Q.; Zhang, J. Numerical study of flow-induced vibration of a flexible plate behind a circular cylinder. *Ocean Eng.* **2018**, *163*, 419–430. [CrossRef]
141. Borazjani, I.; Sotiropoulos, F. Vortex-induced vibrations of two cylinders in tandem arrangement in the proximity–wake interference region. *J. Fluid Mech.* **2009**, *621*, 321–364. [CrossRef]
142. Griffith, M.D.; Jacono, D.L.; Sheridan, J.; Leontini, J.S. Flow-induced vibration of two cylinders in tandem and staggered arrangements. *J. Fluid Mech.* **2017**, *833*, 98–130. [CrossRef]
143. Han, Z.; Zhou, D.; He, T.; Tu, J.; Li, C.; Kwok, K.C.; Fang, C. Flow-induced vibrations of four circular cylinders with square arrangement at low Reynolds numbers. *Ocean Eng.* **2015**, *96*, 21–33. [CrossRef]
144. Newman, D.J.; Karniadakis, G.E. A direct numerical simulation study of flow past a freely vibrating cable. *J. Fluid Mech.* **1997**, *344*, 95–136. [CrossRef]
145. Evangelinos, C.; Lucor, D.; Karniadakis, G.E. DNS-derived force distribution on flexible cylinders subject to vortex-induced vibration. *J. Fluids Struct.* **2000**, *14*, 429–440. [CrossRef]
146. Lucor, D.; Foo, J.; Karniadakis, G. Correlation length and force phasing of a rigid cylinder subject to VIV. In Proceedings of the IUTAM Symposium on Integrated Modeling of Fully Coupled Fluid Structure Interactions Using Analysis, Computations and Experiments, New Brunswick, NJ, USA, 2–6 June 2003; Springer: Berlin/Heidelberg, Germany, 2003; pp. 187–199.
147. Lucor, D.; Karniadakis, G.E. Effects of oblique inflow in vortex-induced vibrations. *Flow Turbul. Combust.* **2003**, *71*, 375–389. [CrossRef]
148. Lucor, D.; Foo, J.; Karniadakis, G. Vortex mode selection of a rigid cylinder subject to VIV at low mass-damping. *J. Fluids Struct.* **2005**, *20*, 483–503. [CrossRef]
149. Gsell, S.; Bourguet, R.; Braza, M. Two-degree-of-freedom vortex-induced vibrations of a circular cylinder at *Re* = 3900. *J. Fluids Struct.* **2016**, *67*, 156–172. [CrossRef]
150. Chen, W.; Ji, C.; Xu, D.; Zhang, Z. Three-dimensional direct numerical simulations of vortex-induced vibrations of a circular cylinder in proximity to a stationary wall. *Phys. Rev. Fluids* **2022**, *7*, 044607. [CrossRef]
151. Zhang, J.; Dalton, C. Interactions of vortex-induced vibrations of a circular cylinder and a steady approach flow at a Reynolds number of 13,000. *Comput. Fluids* **1996**, *25*, 283–294. [CrossRef]
152. Dalheim, J. Numerical prediction of VIV on deepwater risers subjected to shear currents and waves. In Proceedings of the Offshore Technology Conference, Houston, TX, USA, 5–7 November 1999.
153. Herfjord, K.; Drange, S.O.; Kvamsdal, T. Assessment of vortex-induced vibrations on deepwater risers by considering fluid-structure interaction. *J. Offshore Mech. Arct. Eng.* **1999**, *121*, 207–212. [CrossRef]
154. Tutar, M.; Holdo, A.E. Large eddy simulation of a smooth circular cylinder oscillating normal to a uniform flow. *J. Fluids Eng.* **2000**, *122*, 694–702. [CrossRef]
155. Pastrana, D.; Cajas, J.; Lehmkuhl, O.; Rodríguez, I.; Houzeaux, G. Large-eddy simulations of the vortex-induced vibration of a low mass ratio two-degree-of-freedom circular cylinder at subcritical Reynolds numbers. *Comput. Fluids* **2018**, *173*, 118–132. [CrossRef]
156. Wang, Z.; Fan, D.; Triantafyllou, M.S.; Karniadakis, G.E. A large-eddy simulation study on the similarity between free vibrations of a flexible cylinder and forced vibrations of a rigid cylinder. *J. Fluids Struct.* **2021**, *101*, 103223. [CrossRef]
157. Janocha, M.J.; Ong, M.C.; Yin, G. Large eddy simulations and modal decomposition analysis of flow past a cylinder subject to flow-induced vibration. *Phys. Fluids* **2022**, *34*, 045119. [CrossRef]

158. Daniels, S.J.; Castro, I.P.; Xie, Z.T. Numerical analysis of freestream turbulence effects on the vortex-induced vibrations of a rectangular cylinder. *J. Wind. Eng. Ind. Aerodyn.* **2016**, *153*, 13–25. [CrossRef]
159. Khalak, A.; Williamson, C. Dynamics of a hydroelastic cylinder with very low mass and damping. *J. Fluids Struct.* **1996**, *10*, 455–472. [CrossRef]
160. Guilmineau, E.; Queutey, P. Numerical simulation of vortex-induced vibration of a circular cylinder with low mass-damping in a turbulent flow. *J. Fluids Struct.* **2004**, *19*, 449–466. [CrossRef]
161. Pan, Z.; Cui, W.; Miao, Q. Numerical simulation of vortex-induced vibration of a circular cylinder at low mass-damping using RANS code. *J. Fluids Struct.* **2007**, *23*, 23–37. [CrossRef]
162. Wanderley, J.B.; Souza, G.H.; Sphaier, S.H.; Levi, C. Vortex-induced vibration of an elastically mounted circular cylinder using an upwind TVD two-dimensional numerical scheme. *Ocean Eng.* **2008**, *35*, 1533–1544. [CrossRef]
163. Wanderley, J.B.; Soares, L.F.N. Vortex-induced vibration on a two-dimensional circular cylinder with low Reynolds number and low mass-damping parameter. *Ocean Eng.* **2015**, *97*, 156–164. [CrossRef]
164. Gu, J.; Fernandes, A.C.; Han, X.; Kuang, X.; Chen, W. Numerical investigation of Reynolds number effects on vortex-induced vibrations at low and moderate *Re* regimes. *Ocean Eng.* **2022**, *245*, 110535. [CrossRef]
165. Kang, Z.; Zhang, C.; Chang, R.; Ma, G. A numerical investigation of the effects of Reynolds number on vortex-induced vibration of the cylinders with different mass ratios and frequency ratios. *Int. J. Nav. Archit. Ocean. Eng.* **2019**, *11*, 835–850. [CrossRef]
166. Khan, N.B.; Ibrahim, Z.; Nguyen, L.T.T.; Javed, M.F.; Jameel, M. Numerical investigation of the vortex-induced vibration of an elastically mounted circular cylinder at high Reynolds number (*Re* = 104) and low mass ratio using the RANS code. *PLoS ONE* **2017**, *12*, e0185832. [CrossRef] [PubMed]
167. Anwar, M.U.; Lashin, M.; Khan, N.B.; Munir, A.; Jameel, M.; Muhammad, R.; Guedri, K.; Galal, A.M. Effect of variation in the mass ratio on vortex-induced vibration of a circular cylinder in crossflow directio n at Reynold Number = 104: A numerical study using RANS model. *J. Mar. Sci. Eng.* **2022**, *10*, 1126. [CrossRef]
168. Martins, F.; Avila, J. Effects of the Reynolds number and structural damping on vortex-induced vibrations of elastically-mounted rigid cylinder. *Int. J. Mech. Sci.* **2019**, *156*, 235–249. [CrossRef]
169. Cui, Z.; Zhao, M.; Teng, B.; Cheng, L. Two-dimensional numerical study of vortex-induced vibration and galloping of square and rectangular cylinders in steady flow. *Ocean Eng.* **2015**, *106*, 189–206. [CrossRef]
170. Han, P.; Huang, Q.; Pan, G.; Wang, W.; Zhang, T.; Qin, D. Energy harvesting from flow-induced vibration of a low-mass square cylinder with different incidence angles. *AIP Adv.* **2021**, *11*, 025126. [CrossRef]
171. Ding, L.; Zhang, L.; Wu, C.; Mao, X.; Jiang, D. Flow induced motion and energy harvesting of bluff bodies with different cross sections. *Energy Convers. Manag.* **2015**, *91*, 416–426. [CrossRef]
172. Zhang, B.; Song, B.; Mao, Z.; Tian, W.; Li, B. Numerical investigation on VIV energy harvesting of bluff bodies with different cross sections in tandem arrangement. *Energy* **2017**, *133*, 723–736. [CrossRef]
173. Zhang, B.; Song, B.; Mao, Z.; Li, B.; Gu, M. Hydrokinetic energy harnessing by spring-mounted oscillators in FIM with different cross sections: From triangle to circle. *Energy* **2019**, *189*, 116249. [CrossRef]
174. Wang, W.; Song, B.; Mao, Z.; Tian, W.; Zhang, T.; Han, P. Numerical investigation on vortex-induced vibration of bluff bodies with different rear edges. *Ocean Eng.* **2020**, *197*, 106871. [CrossRef]
175. Zhu, H.; Yao, J. Numerical evaluation of passive control of VIV by small control rods. *Appl. Ocean. Res.* **2015**, *51*, 93–116. [CrossRef]
176. Ding, L.; Zhang, L.; Bernitsas, M.M.; Chang, C.C. Numerical simulation and experimental validation for energy harvesting of single-cylinder VIVACE converter with passive turbulence control. *Renew. Energy* **2016**, *85*, 1246–1259. [CrossRef]
177. Wang, J.; Zhao, W.; Su, Z.; Zhang, G.; Li, P.; Yurchenko, D. Enhancing vortex-induced vibrations of a cylinder with rod attachments for hydrokinetic power generation. *Mech. Syst. Signal Process.* **2020**, *145*, 106912. [CrossRef]
178. Rahmanian, M.; Cheng, L.; Zhao, M.; Zhou, T. Vortex induced vibration and vortex shedding characteristics of two side-by-side circular cylinders of different diameters in close proximity in steady flow. *J. Fluids Struct.* **2014**, *48*, 260–279. [CrossRef]
179. Zhao, M.; Murphy, J.M.; Kwok, K. Numerical simulation of vortex-induced vibration of two rigidly connected cylinders in side-by-side and tandem arrangements using RANS model. *J. Fluids Eng.* **2016**, *138*, 1257. [CrossRef]
180. Cui, Z.; Zhao, M.; Teng, B. Vortex-induced vibration of two elastically coupled cylinders in side-by-side arrangement. *J. Fluids Struct.* **2014**, *44*, 270–291. [CrossRef]
181. Nguyen, V.T.; Nguyen, H.H. Detached eddy simulations of flow induced vibrations of circular cylinders at high Reynolds numbers. *J. Fluids Struct.* **2016**, *63*, 103–119. [CrossRef]
182. Joshi, V.; Jaiman, R.K. A variationally bounded scheme for delayed detached eddy simulation: Application to vortex-induced vibration of offshore riser. *Comput. Fluids* **2017**, *157*, 84–111. [CrossRef]
183. Ma, C.; Zhao, W.; Wan, D. Numerical investigations of the flow-induced vibration of a three-dimensional circular cylinder with various symmetric strips attached. *Phys. Fluids* **2022**, *34*, 065102. [CrossRef]
184. Ma, C.; Zhao, W.; Wan, D. Numerical investigations of three-dimensional flows around a cylinder attaching with symmetric strips. *Phys. Fluids* **2021**, *33*, 075101. [CrossRef]
185. Hollenbach, R.; Kielb, R.; Hall, K. Extending a van der Pol-based reduced-order model for fluid-structure interaction applied to non-synchronous vibrations in turbomachinery. *J. Turbomach.* **2021**, *144*, 031006. [CrossRef]

186. Sanders, A.J. Non-synchronous vibration (NSV) due to a flow-induced aerodynamic instability in a composite fan stator. *Turbo Expo Power Land Sea Air* **2004**, *6*, 507–516. [CrossRef]
187. Zhang, M.; Fu, S.; Ren, H.; Ma, L.; Xu, Y. A hybrid FEM-DNN-based vortex-induced vibration prediction method for flexible pipes under oscillatory flow in the time domain. *Ocean Eng.* **2022**, *246*, 110488. [CrossRef]
188. Raissi, M.; Wang, Z.; Triantafyllou, M.S.; Karniadakis, G.E. Deep learning of vortex-induced vibrations. *J. Fluid Mech.* **2019**, *861*, 119–137. [CrossRef]
189. Dowell, E.; Curtiss, H.; Scanlan, R.; Sisto, F. *A Modern Course in Aeroelasticity*; Springer: Berlin/Heidelberg, Germany, 2021; Volume 264.
190. Lumley, J. Coherent Structures in Turbulence. In *Transition and Turbulence*; Meyer, R.E., Ed.; Academic Press: London, UK, 1981; pp. 215–242. [CrossRef]
191. Star, S.; Stabile, G.; Rozza, G.; Degroote, J. A POD-Galerkin reduced order model of a turbulent convective buoyant flow of sodium over a backward-facing step. *Appl. Math. Model.* **2021**, *89*, 486–503. [CrossRef]
192. Lorenzi, S.; Cammi, A.; Luzzi, L.; Rozza, G. POD-Galerkin method for finite volume approximation of Navier–Stokes and RANS equations. *Comput. Methods Appl. Mech. Eng.* **2016**, *311*, 151–179. [CrossRef]
193. Sieber, M.; Paschereit, C.O.; Oberleithner, K. Spectral proper orthogonal decomposition. *J. Fluid Mech.* **2016**, *792*, 798–828. [CrossRef]
194. Rowley, C.W. Model reduction for fluids, using balanced proper orthogonal decomposition. *Int. J. Bifurc. Chaos* **2005**, *15*, 997–1013. [CrossRef]
195. Schubert, Y.; Sieber, M.; Oberleithner, K.; Martinuzzi, R. Towards robust data-driven reduced-order modelling for turbulent flows: Application to vortex-induced vibrations. *Theor. Comput. Fluid Dyn.* **2022**, *36*, 1–27. [CrossRef]
196. Kou, J.; Zhang, W.; Liu, Y.; Li, X. The lowest Reynolds number of vortex-induced vibrations. *Phys. Fluids* **2017**, *29*, 041701. [CrossRef]
197. Zhang, W.; Li, X.; Ye, Z.; Jiang, Y. Mechanism of frequency lock-in in vortex-induced vibrations at low Reynolds numbers. *J. Fluid Mech.* **2015**, *783*, 72–102. [CrossRef]
198. Li, X.; Zhang, W.; Gao, C. Proximity-interference wake-induced vibration at subcritical Re: Mechanism analysis using a linear dynamic model. *Phys. Fluids* **2018**, *30*, 033606. [CrossRef]
199. Li, X.; Lyu, Z.; Kou, J.; Zhang, W. Mode competition in galloping of a square cylinder at low Reynolds number. *J. Fluid Mech.* **2019**, *867*, 516–555. [CrossRef]
200. Luo, F.; Gao, C.; Zhang, W. The key to suppress vortex-induced vibration: Stability of the structural mode. *J. Fluids Struct.* **2022**, *113*, 103692. [CrossRef]
201. Navrose; Mittal, S. Lock-in in vortex-induced vibration. *J. Fluid Mech.* **2016**, *794*, 565–594. [CrossRef]
202. Juang, J.N.; Pappa, R.S. An Eigensystem Realization Algorithm (ERA) for Modal Parameter Identification, NASA. In Proceedings of the JPL Workshop on Identification and Control of Flexible Space Structures, Pasadena, CA, USA, 4–6 July 1984.
203. Yao, W.; Jaiman, R.K. Model reduction and mechanism for the vortex-induced vibrations of bluff bodies. *J. Fluid Mech.* **2017**, *827*, 357–393. [CrossRef]
204. Yao, W.; Jaiman, R. Feedback control of unstable flow and vortex-induced vibration using the eigensystem realization algorithm. *J. Fluid Mech.* **2017**, *827*, 394–414. [CrossRef]
205. Reddy, S.B.; Magee, A.R.; Jaiman, R.K. A data-driven approach for the stability analysis of vortex-induced vibration. In Proceedings of the International Conference on Offshore Mechanics and Arctic Engineering, Madrid, Spain, 17–22 June 2018; American Society of Mechanical Engineers: New York, NY, USA, 2018; Volume 51210, p. V002T08A004.
206. Bukka, S.; Magee, A.; Jaiman, R. Stability analysis of passive suppression for vortex-induced vibration. *J. Fluid Mech.* **2020**, *886*, A12. [CrossRef]
207. Chizfahm, A.; Jaiman, R. Data-driven stability analysis and near-wake jet control for the vortex-induced vibration of a sphere. *Phys. Fluids* **2021**, *33*, 044104. [CrossRef]
208. Cheng, Z.; Lien, F.S.; Yee, E.; Zhang, J.H. Mode transformation and interaction in vortex-induced vibration of laminar flow past a circular cylinder. *Phys. Fluids* **2022**, *34*, 033607. [CrossRef]
209. Kou, J.; Zhang, W. A hybrid reduced-order framework for complex aeroelastic simulations. *Aerosp. Sci. Technol.* **2019**, *84*, 880–894. [CrossRef]
210. Rowley, C.W.; Dawson, S.T. Model reduction for flow analysis and control. *Annu. Rev. Fluid Mech.* **2017**, *49*, 387–417. [CrossRef]
211. Miyanawala, T.P.; Jaiman, R.K. Decomposition of wake dynamics in fluid—Structure interaction via low-dimensional models. *J. Fluid Mech.* **2019**, *867*, 723–764. [CrossRef]
212. Murata, T.; Fukami, K.; Fukagata, K. Nonlinear mode decomposition with convolutional neural networks for fluid dynamics. *J. Fluid Mech.* **2020**, *882*, A13. [CrossRef]
213. Miyanawala, T.; Jaiman, R.K. A hybrid data-driven deep learning technique for fluid-structure interaction. In Proceedings of the International Conference on Offshore Mechanics and Arctic Engineering, Glasgow, Scotland, 9–14 June 2019; American Society of Mechanical Engineers: New York, NY, USA, 2019; Volume 58776, p. V002T08A004.
214. Reddy, S.B.; Magee, A.R.; Jaiman, R.K.; Liu, J.; Xu, W.; Choudhary, A.; Hussain, A. Reduced order model for unsteady fluid flows via recurrent neural networks. In Proceedings of the International Conference on Offshore Mechanics and Arctic Engineering

Glasgow, Scotland, 9–14 June 2019; American Society of Mechanical Engineers: New York, NY, USA, 2019; Volume 58776, p. V002T08A007.
215. Bukka, S.R.; Gupta, R.; Magee, A.R.; Jaiman, R.K. Assessment of unsteady flow predictions using hybrid deep learning based reduced-order models. *Phys. Fluids* **2021**, *33*, 013601. [CrossRef]
216. Kou, J.; Zhang, W. Layered reduced-order models for nonlinear aerodynamics and aeroelasticity. *J. Fluids Struct.* **2017**, *68*, 174–193. [CrossRef]
217. Chizfahm, A.; Jaiman, R. Deep Learning for Stability Analysis of a Freely Vibrating Sphere at Moderate Reynolds Number. *arXiv* **2021**, arXiv:abs/2112.09858.
218. Duraisamy, K.; Iaccarino, G.; Xiao, H. Turbulence modeling in the age of data. *Annu. Rev. Fluid Mech.* **2019**, *51*, 357–377. [CrossRef]
219. Kochkov, D.; Smith, J.A.; Alieva, A.; Wang, Q.; Brenner, M.P.; Hoyer, S. Machine learning accelerated computational fluid dynamics. *Proc. Natl. Acad. Sci. USA* **2021**, *118*, e2101784118. [CrossRef]
220. Maulik, R.; San, O.; Rasheed, A.; Vedula, P. Subgrid modelling for two-dimensional turbulence using neural networks. *J. Fluid Mech.* **2019**, *858*, 122–144. [CrossRef]
221. Ling, J.; Kurzawski, A.; Templeton, J. Reynolds averaged turbulence modelling using deep neural networks with embedded invariance. *J. Fluid Mech.* **2016**, *807*, 155–166. [CrossRef]
222. Wu, J.L.; Xiao, H.; Paterson, E. Physics-informed machine learning approach for augmenting turbulence models: A comprehensive framework. *Phys. Rev. Fluids* **2018**, *7*, 1–28. [CrossRef]
223. Xu, H.; Zhang, W.; Wang, Y. Explore missing flow dynamics by physics-informed deep learning: The parameterized governing systems. *Phys. Fluids* **2021**, *33*, 095116. [CrossRef]
224. Bai, X.D.; Zhang, W. Machine learning for vortex induced vibration in turbulent flow. *Comput. Fluids* **2022**, *235*, 105266. [CrossRef]
225. Cheng, C.; Meng, H.; Li, Y.Z.; Zhang, G.T. Deep learning based on PINN for solving 2 DOF vortex induced vibration of cylinder. *Ocean Eng.* **2021**, *240*, 109932. [CrossRef]
226. Nikoo, H.M.; Bi, K.; Hao, H. Three-dimensional vortex-induced vibration of a circular cylinder at subcritical Reynolds numbers with low-Re correction. *Mar. Struct.* **2019**, *66*, 288–306. [CrossRef]
227. Mosaferi, A.A.; Esmaeili, M.; Rabiee, A.H. Effect of aligned magnetic field on the 2DOF VIV suppression and convective heat transfer characteristics of a circular cylinder. *Int. Commun. Heat Mass Transf.* **2022**, *130*, 105807. [CrossRef]
228. Ma, C.; Wang, J.; Li, Q.; Qin, H.; Liao, H. Vortex-induced vibration performance and suppression mechanism for a long suspension bridge with wide twin-box girder. *J. Struct. Eng.* **2018**, *144*, 04018202. [CrossRef]
229. Sui, J.; Wang, J.; Liang, S.; Tian, Q. VIV suppression for a large mass-damping cylinder attached with helical strakes. *J. Fluids Struct.* **2016**, *62*, 125–146. [CrossRef]
230. Wang, J.; Yurchenko, D.; Hu, G.; Zhao, L.; Tang, L.; Yang, Y. Perspectives in flow-induced vibration energy harvesting. *Appl. Phys. Lett.* **2021**, *119*, 100502. [CrossRef]
231. Andrianne, T.; Aryoputro, R.P.; Laurent, P.; Colson, G.; Amandolese, X.; Hémon, P. Energy harvesting from different aeroelastic instabilities of a square cylinder. *J. Wind. Eng. Ind. Aerodyn.* **2018**, *172*, 164–169. [CrossRef]
232. Yu, H.; Zhang, M. Effects of side ratio on energy harvesting from transverse galloping of a rectangular cylinder. *Energy* **2021**, *226*, 120420. [CrossRef]
233. Javed, U.; Abdelkefi, A.; Akhtar, I. An improved stability characterization for aeroelastic energy harvesting applications. *Commun. Nonlinear Sci. Numer. Simul.* **2016**, *36*, 252–265. [CrossRef]
234. Sirohi, J.; Mahadik, R. Harvesting wind energy using a galloping piezoelectric beam. *J. Vib. Acoust.* **2012**, *134*, 011009. [CrossRef]
235. Quadrante, L.A.R.; Nishi, Y. Amplification/suppression of flow-induced motions of an elastically mounted circular cylinder by attaching tripping wires. *J. Fluids Struct.* **2014**, *48*, 93–102. [CrossRef]
236. Song, J.; Hu, G.; Tse, K.; Li, S.; Kwok, K. Performance of a circular cylinder piezoelectric wind energy harvester fitted with a splitter plate. *Appl. Phys. Lett.* **2017**, *111*, 223903. [CrossRef]
237. Hu, G.; Tse, K.T.; Wei, M.; Naseer, R.; Abdelkefi, A.; Kwok, K.C. Experimental investigation on the efficiency of circular cylinder-based wind energy harvester with different rod-shaped attachments. *Appl. Energy* **2018**, *226*, 682–689. [CrossRef]
238. Zhu, H.; Zhao, Y.; Zhou, T. CFD analysis of energy harvesting from flow induced vibration of a circular cylinder with an attached free-to-rotate pentagram impeller. *Appl. Energy* **2018**, *212*, 304–321. [CrossRef]
239. Ramlan, R.; Brennan, M.; Mace, B.; Kovacic, I. Potential benefits of a non-linear stiffness in an energy harvesting device. *Nonlinear Dyn.* **2010**, *59*, 545–558. [CrossRef]
240. Zhang, B.; Li, B.; Fu, S.; Mao, Z.; Ding, W. Vortex-induced vibration (VIV) hydrokinetic energy harvesting based on nonlinear damping. *Renew. Energy* **2022**, *195*, 1050–1063. [CrossRef]
241. Tang, X.; Zuo, L. Enhanced vibration energy harvesting using dual-mass systems. *J. Sound Vib.* **2011**, *330*, 5199–5209. [CrossRef]
242. Xiao, H.; Wang, X.; John, S. A multi-degree of freedom piezoelectric vibration energy harvester with piezoelectric elements inserted between two nearby oscillators. *Mech. Syst. Signal Process.* **2016**, *68*, 138–154. [CrossRef]
243. Wu, Y.; Qiu, J.; Zhou, S.; Ji, H.; Chen, Y.; Li, S. A piezoelectric spring pendulum oscillator used for multi-directional and ultra-low frequency vibration energy harvesting. *Appl. Energy* **2018**, *231*, 600–614. [CrossRef]
244. Yang, B.; Lee, C.; Kee, W.L.; Lim, S.P. Hybrid energy harvester based on piezoelectric and electromagnetic mechanisms. *J. Micro Nanolithogr. Mems Moems* **2010**, *9*, 023002. [CrossRef]

245. Assi, G.R.; Bearman, P.W. Transverse galloping of circular cylinders fitted with solid and slotted splitter plates. *J. Fluids Struct.* **2015**, *54*, 263–280. [CrossRef]
246. Erturk, A.; Inman, D.J. Issues in mathematical modeling of piezoelectric energy harvesters. *Smart Mater. Struct.* **2008**, *17*, 065016. [CrossRef]
247. Dai, H.; Abdelkefi, A.; Wang, L. Theoretical modeling and nonlinear analysis of piezoelectric energy harvesting from vortex-induced vibrations. *J. Intell. Mater. Syst. Struct.* **2014**, *25*, 1861–1874. [CrossRef]
248. Hou, C.; Shan, X.; Zhang, L.; Song, R.; Yang, Z. Design and modeling of a magnetic-coupling monostable piezoelectric energy harvester under vortex-induced vibration. *IEEE Access* **2020**, *8*, 108913–108927. [CrossRef]
249. Wang, J.; Tang, L.; Zhao, L.; Hu, G.; Song, R.; Xu, K. Equivalent circuit representation of a vortex-induced vibration-based energy harvester using a semi-empirical lumped parameter approach. *Int. J. Energy Res.* **2020**, *44*, 4516–4528. [CrossRef]
250. Yang, Y.; Zhao, L.; Tang, L. Comparative study of tip cross-sections for efficient galloping energy harvesting. *Appl. Phys. Lett.* **2013**, *102*, 064105. [CrossRef]
251. Zhang, M.; Zhao, G.; Wang, J. Study on fluid-induced vibration power harvesting of square columns under different attack angles. *Geofluids* **2017**, *2017*, 1–18. [CrossRef]
252. Sirohi, J.; Mahadik, R. Piezoelectric wind energy harvester for low-power sensors. *J. Intell. Mater. Syst. Struct.* **2011**, *22*, 2215–2228. [CrossRef]
253. Abdelkefi, A.; Yan, Z.; Hajj, M.R. Modeling and nonlinear analysis of piezoelectric energy harvesting from transverse galloping. *Smart Mater. Struct.* **2013**, *22*, 025016. [CrossRef]
254. Hu, Y.; Yang, B.; Chen, X.; Wang, X.; Liu, J. Modeling and experimental study of a piezoelectric energy harvester from vortex shedding-induced vibration. *Energy Convers. Manag.* **2018**, *162*, 145–158. [CrossRef]
255. Song, R.; Shan, X.; Lv, F.; Li, J.; Xie, T. A novel piezoelectric energy harvester using the macro fiber composite cantilever with a bicylinder in water. *Appl. Sci.* **2015**, *5*, 1942–1954. [CrossRef]
256. Zhang, M.; Liu, Y.; Cao, Z. Modeling of piezoelectric energy harvesting from freely oscillating cylinders in water flow. *Math. Probl. Eng.* **2014**, *2014*, 985360. [CrossRef]
257. Dai, H.; Abdelkefi, A.; Wang, L. Piezoelectric energy harvesting from concurrent vortex-induced vibrations and base excitations. *Nonlinear Dyn.* **2014**, *77*, 967–981. [CrossRef]
258. Franzini, G.R.; Bunzel, L.O. A numerical investigation on piezoelectric energy harvesting from vortex-induced vibrations with one and two degrees of freedom. *J. Fluids Struct.* **2018**, *77*, 196–212. [CrossRef]
259. Wang, J.; Zhang, C.; Gu, S.; Yang, K.; Li, H.; Lai, Y.; Yurchenko, D. Enhancement of low-speed piezoelectric wind energy harvesting by bluff body shapes: Spindle-like and butterfly-like cross-sections. *Aerosp. Sci. Technol.* **2020**, *103*, 105898. [CrossRef]
260. Wang, J.; Gu, S.; Zhang, C.; Hu, G.; Chen, G.; Yang, K.; Li, H.; Lai, Y.; Litak, G.; Yurchenko, D. Hybrid wind energy scavenging by coupling vortex-induced vibrations and galloping. *Energy Convers. Manag.* **2020**, *213*, 112835. [CrossRef]
261. Sun, W.; Zhao, D.; Tan, T.; Yan, Z.; Guo, P.; Luo, X. Low velocity water flow energy harvesting using vortex induced vibration and galloping. *Appl. Energy* **2019**, *251*, 113392. [CrossRef]
262. Ramirez, J.M. A coupled formulation of fluid-structure interaction and piezoelectricity for modeling a multi-body energy harvester from vortex-induced vibrations. *Energy Convers. Manag.* **2021**, *249*, 114852. [CrossRef]
263. Barbounis, T.G.; Theocharis, J.B.; Alexiadis, M.C.; Dokopoulos, P.S. Long-term wind speed and power forecasting using local recurrent neural network models. *IEEE Trans. Energy Convers.* **2006**, *21*, 273–284. [CrossRef]
264. Kusiak, A.; Zheng, H.; Song, Z. Short-term prediction of wind farm power: A data mining approach. *IEEE Trans. Energy Convers.* **2009**, *24*, 125–136. [CrossRef]
265. Hu, G.; Kwok, K.C. Predicting wind pressures around circular cylinders using machine learning techniques. *J. Wind. Eng. Ind. Aerodyn.* **2020**, *198*, 104099. [CrossRef]
266. Clifton, A.; Kilcher, L.; Lundquist, J.; Fleming, P. Using machine learning to predict wind turbine power output. *Environ. Res. Lett.* **2013**, *8*, 024009. [CrossRef]
267. Zhang, C.; Hu, G.; Yurchenko, D.; Lin, P.; Gu, S.; Song, D.; Peng, H.; Wang, J. Machine learning based prediction of piezoelectric energy harvesting from wake galloping. *Mech. Syst. Signal Process.* **2021**, *160*, 107876. [CrossRef]
268. Hu, G.; Wang, J.; Lan, C.; Tang, L.; Liang, J. Deep-learning assisted finite element model of a galloping piezoelectric energy harvester. In Proceedings of the Active and Passive Smart Structures and Integrated Systems XV. SPIE, Online, 22–26 March 2021; Volume 11588, pp. 58–66.
269. Zdravkovich, M.M. Conceptual overview of laminar and turbulent flows past smooth and rough circular cylinders. *J. Wind. Eng. Ind. Aerodyn.* **1990**, *33*, 53–62. [CrossRef]
270. Mohri, M.; Rostamizadeh, A.; Talwalker, A. *Foundations of Machine Learning*; The MIT Press: Cambridge, MA, USA, 2018.
271. Shan, S.; Wang, G. Survy of modeling and optimization strategies to solve high-dimensional design problems with computationally-expensive black-box functions. *Struct. Multidiscip. Optim.* **2010**, *52*, 219–241. [CrossRef]
272. Gutmann, M.U.; Corander, J. Bayesian optimization for likelihood-free inference of simulator-based statistical models. *J. Mach. Learn. Res.* **2016**, *17*, 1–47.

energies

MDPI

Article

Aerodynamic Performance Analysis of Trailing Edge Serrations on a Wells Turbine

Abdullah Saad Alkhalifa, Mohammad Nasim Uddin and Michael Atkinson *

Mechanical Engineering Department, North Carolina A&T State University, Greensboro, NC 27411, USA
* Correspondence: mdatkinson@ncat.edu; Tel.: +1-(336)-285-2202

Abstract: The primary objective of this investigation was to explore the aerodynamic impact of adding trailing edge serrations to a Wells turbine. The baseline turbine consists of eight NACA 0015 blades. The blade chord length was 0.125 m and the span was 0.100 m. Two modified serrated blade configurations were studied: (1) full-span, and (2) partial-span covering 0.288c of the trailing edge. The numerical simulations were carried out by solving the three-dimensional, incompressible steady-state Reynolds Averaged Navier-Stokes (RANS) equations using the k-ω SST turbulence model in ANSYS™ (CFX). The aerodynamic performance of the modified Wells turbine was compared to the baseline by calculating non-dimensional parameters (i.e., torque coefficient, pressure drop coefficient, and turbine efficiency). A comparison of the streamlines was performed to analyze the flow topology around the turbine blades for a flow coefficient range of $0.075 \leq \phi \leq 0.275$, representing an angle of attack range of $4.29° \leq \alpha \leq 15.3°$. The trailing edge serrations generated a substantial change in surface pressure and effectively reduced the separated flow region, thus improving efficiency in most cases. As a result, there was a modest peak efficiency increase of 1.51% and 1.22%, for the partial- and full-span trailing edge serrations, respectively.

Keywords: renewable energy; wave energy conversion; CFD

1. Introduction

The negative impacts of carbon-emitting energy sources have hastened the need to develop alternative forms of energy production. The Office of Energy Efficiency & Renewable Energy reports that renewable energy accounted for approximately 20% of all U.S. electricity in 2021, and that percentage is expected to grow [1]. The International Energy Agency (IEA) is expecting the renewable capacity to increase over 8% in 2022 [2]. Popular forms of eco-friendly renewable energy include land- and sea-based solar and wind applications. Novel devices that utilize marine energy offer the possibility of harnessing the kinetic energy of ocean waves. One such design uses the oscillating water column principle and has received renewed attention. The Oscillating Water Column (OWC) is a well-studied marine renewable energy harvesting technique that has successfully produced electricity. A significant component of the conventional OWC is the Wells turbine. However, an OWC equipped with a Wells turbine has a limited operating range, low aerodynamic efficiency, high noise levels, and poor starting characteristics. Among the wide range of renewable energy sources, wave energy could become economically viable in the near future. The conversion of wave energy is an attractive proposition for countries surrounded by the sea, such as the British Isles, Japan, India, and others. Globally, ocean waves offer huge potential for harnessing energy. Locally, in the U.S., more than 50% of the population lives within 50 miles of the coast, making wave energy a logical and ideal form of power generation [3].

In the late 1970s, Prof. Alan Arthur Wells of Queen's University Belfast developed the Wells turbine, an effective air turbine used for bi-directional air flow. The Wells turbine rotor is illustrated in Figure 1. It consists of several symmetrical airfoil blades arranged around a central hub with their chord lines normal to the rotation axis. This turbine is more efficient than other turbines because it doesn't require rectifying air valves and can extract power at low mass flow rates. An axial flow air turbine such as the Wells is considered one of the simplest and most economical ways to harness wave energy

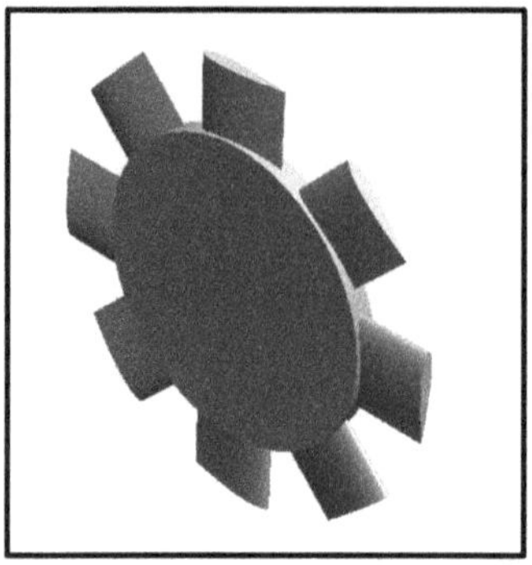

Figure 1. Wells Turbine.

It is proven that the Wells turbine is an adequate wave energy converter, even with its well-known aerodynamic performance limitations [4]. This research will investigate the aerodynamic effect of adding trailing edge serrations to Wells turbine blades.

Ocean waves create a reciprocating airflow that transfers energy to the turbine blades through the OWC. Air, which is the circulating fluid, reverses its direction with the oscillating wave (i.e., inhalation and exhalation), but the Wells turbine rotation direction remains the same. Researchers have investigated the effect of varying Wells turbine design parameters through experimental and numerical analyses. In the following section, we briefly discuss prior studies whose aim was to assess and improve its design for performance.

Using NACA0015 profiles, Ghisu et al. [5] compared experimental and numerical performance characteristics of a high-solidity Wells turbine at different non-dimensional piston frequencies. The unsteady numerical simulations were carried out using ANSYS™ Fluent, a commercial computational fluid dynamics (CFD) flow solver. The computations focused on the flow upstream and downstream of the rotor and investigated hysteresis during piston acceleration and deceleration. Examination of the flow field differences, between the intake and exhaust strokes, were result of the machine's asymmetry. Moreover, this study shows that turbine performance can be solely attributed to delays introduced by the internal flow field correlating the turbine's performance with flow parameters (i.e., velocity and pressure) near the blade. In a similar study, Ghisu et al. [6] showed that compressibility effects determine the hysteresis between acceleration and deceleration, while differences in flow parameters and secondary structures near the rotor are negligible under similar flow conditions.

Taha et al. [7] compared the performance of a NACA0020 blade profile with various uniform tip and non-uniform [8] tip clearances using CFD. The results showed that the turbine performance is significantly influenced by tip clearance. It was shown that a turbine with non-uniform tip clearance performs similarly to one that is uniform. However, a turbine with non-uniform tip clearance, however, seems to have a better overall performance. One of the most noticeable features is that a turbine with a larger tip clearance is able to operate at a greater flow rate without stalling. Additionally, this research showed that turbine performance was greatly influenced by tip leakage flow. The study concluded that increased tip clearance to chord length ratios led to reduced peak efficiency and wider stall margins of the turbine.

As part of the effort to improve Wells turbine performance, different researchers considered different methodologies. Takao et al. [9] improved the original Wells turbine design by adding end plates. Das et al. [10] obtained improvement of a Wells turbine through an automated optimization technique using OPAL++. In this study, a turbine with a ring-type endplate was optimized using four geometric properties. The modified geometry increased the peak torque by 120%, the efficiency by 9%, and delayed the stall point. Shaaban et al. [11] used a duct geometry on Wells turbine to improve the performance by optimizing the duct geometry. By optimizing the duct geometry, turbine power and efficiency are increased by 14% and 9%, respectively. Halder et al. [12] improved the design of a Wells turbine by increasing the operating range. In order to find the optimal design, Reynolds-averaged Navier-Stokes (RANS) equations were solved and a surrogate approximation model was constructed. For this design, the peak efficiency dropped by 3.1%. However, compared to the reference case, both relative power output and relative stall point increased by 29% and 18%, respectively.

Mohamed et al. [13] studied multi-objective optimization of the turbine blade's airfoil to increase power and improve efficiency. The methodology explored using a monoplane Wells turbine with symmetric airfoil blades to optimize efficiency by increasing the tangential force. The performance of the proposed solution is demonstrated by detailed comparisons with the classical Wells turbine with symmetric airfoils. A new design of the airfoil shape increased power output by 11.3%, and efficiency by 1% throughout the operating range.

Setoguchi et al. [14] used 2-D and 3-D models with guide vanes to improve turbine performance. This study showed that design parameters as well as the nature of the flow downstream and upstream to determine the efficiency of a turbine. Furthermore, the study showed the upstream guide vanes are primarily responsible for improving the turbine running characteristic, while the downstream guide vanes are improving the turbine starting characteristic.

Halder et al. [15] completed a detailed review of Wells turbine design modifications and some key selected design changes include variation of airfoil profiles, blade sweep, pitch angle, and geometry. However, the research focused on multi-objective optimization of blade sweep. This study performed numerical analysis by solving the RANS equations and k-ω SST turbulence model for the flow field. Based on the computational results, the peak of torque coefficient was increased by 28.28% and the efficiency was decreased by 13.5%

Watterson et al. [16] studied the effect of solidity calculations on Wells turbine performance. A three-dimensional RANS flow solver was used to determine the performance of a Wells turbine. Using NACA 0015 blades, a monoplane device performance has been calculated with Reynolds number 8×10^5, tip Mach number 0.4, hub-to-tip ratio 0.6, and a tip clearance of 2%. For different blade counts, calculations were performed to study the effects of solidity. The results showed that by increasing turbine solidity and thus blade-to-blade fluid interaction, the turbine is less susceptible to stall.

In the present work, a Wells turbine blade design was changed by adding the Trailing Edge Serration (TES) patterns. Several theoretical [17,18], numerical [19], and experimental [20,21] studies have shown that TES can mitigate negative aeroacoustics impacts caused by the interaction of turbulent boundary layers at the trailing edge of the blade. The results of these studies verified the positive effect of trailing edge serrations on noise reduction. Recent studies have demonstrated that the practical implementation of the serration design alters the flow streamlines at the teeth's roots [22,23]. While trailing edge serrations reduce noise, the aerodynamic impacts on turbine performance have not been studied thoroughly.

2. Materials and Methods

In this work, we assess the aerodynamic performance of different trailing-edge serration designs, by solving the RANS equations using ANSYS™ CFX. Fourteen three-dimensional incompressible simulations were completed. The turbine's performance was evaluated by its torque coefficient, efficiency, and pressure drop coefficient under steady-state conditions. The system of equations was closed by coupling the RANS equations with the k-ω SST turbulence model. In order to ensure numerical accuracy, we used the grid metrics carried out in previous work [24]. Furthermore, the computational model was validated by comparing the results to available numerical [25] and experimental [26] data. A summary of the grid independence study is provided below.

2.1. Geometry Modeling

A Wells turbine with symmetrical NACA 0015 airfoil blades fixed at a 90-degree staggered angle was utilized. The demonstrated turbine model consists of eight blades around the hub and perpendicular to the axial fluid flow direction. Each blade has a 0.125 m chord length, with hub and tip radius of 0.200 m and 0.300 m, respectively. To precisely assess the turbine performance, the tip clearance has been taken into consideration. For fluid flow in turbomachinery, the tip leakage flow is a significant secondary flow which was chosen to be 1% of the chord length.

The turbine rotor blade solidity is used to account for the wake effect on the blades. During the rotation of a turbine, solidity indicates the amount of obstructions in the airflow. Furthermore, it is a measurement of a turbine's blades' mutual interaction and a crucial design element for self-starting [27–31]. Solidity is constant at 0.67 which leads to having eight blades around the rotor. As a result of numerous studies [32], it has been demonstrated that this solidity value of 0.67 is highly suitable for Wells turbines. As a result of the Wells turbine's symmetry, we have thoroughly analyzed only one of its eight blades.

CFD analysis requires discretization of the governing equations. Also, consequently, discretization errors can appear in the CFD results. In numerical analysis, errors can be primarily caused by the quality of the computational grids. ANSYS meshing was used to discretize the flow domain by forming an unstructured mesh for the presented designs (see Figure 2). The boundary layer flow of the rotor blades must be sufficiently resolved to predict the dynamic stall point of a turbine. In order to accurately predict the viscous sublayer, because of their sensitivity to meshes, the k-ω SST turbulence model requires high resolution with $y^+ < 1$. Therefore, to capture the boundary layer flow, inflation layers of closely spaced hexahedrons were generated on the walls of the turbine blades.

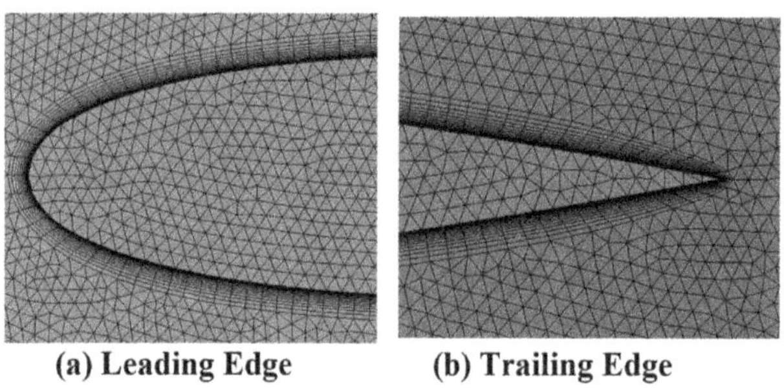

Figure 2. ANSYS Meshing Generated Around the Baseline Blade.

An inviscid flow occurs away from the airfoil surface, and a viscous flow occurs adjacent to it, where friction dissipates energy (i.e., the boundary layer). CFD has traditionally measured boundary layer thickness by δ, which is a dimensionless distance normal to the airfoil surface. By adjusting this parameter, the boundary layer develops correctly. Fluid density, absolute distance from wall, friction velocity, and viscosity are all used to calculate y^+.

$$y^+ = \frac{\rho y u_\tau}{\mu} \tag{1}$$

Based on equation or $y^+ < 1$, the first layer height was obtained as 1.1×10^{-5} with a growth factor of 1.2. The total number of inflation layers was 20. As shown below, friction velocity can be calculated using the square root of the wall shear stress to fluid density ratio.

$$u_\tau = \sqrt{\frac{\tau_w}{\rho}} \tag{2}$$

At the velocity inlet, there is a 5% turbulent intensity distributed uniformly across the velocity profile. Output boundary conditions are imposed by static pressure of 101.3 KPa at the outlet. There are four and six times the blade chord lengths upstream and downstream of the computational domain length. At the hub, tip, and blade surface, no-slip boundary conditions are specified. Considering that Wells turbines have symmetrical geometry, calculations are restricted to one blade-to-blade passage with periodic boundary conditions, as illustrated in Figure 3. Also, the turbine was set to rotate at 2000 rpm and the assigned convergence criterion for the solver was 1×10^{-5}.

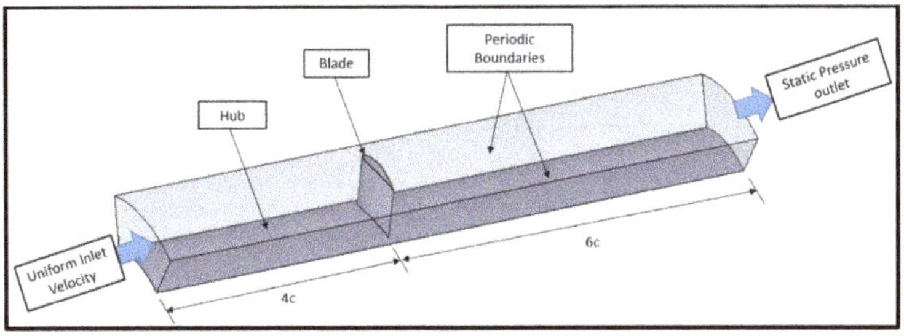

Figure 3. Computational Domain and the Boundary Conditions.

In the fluid zone, the Moving Reference Frame (MRF) rotates at a speed equivalent to the Wells turbine rotors. The governing equations incorporate centripetal and Coriolis components to take into account the MRF. ANSYS™ CFX solved the following governing equations:

The conservation of mass

$$\frac{\partial \rho}{\partial t} + \nabla.\rho\vec{v_r} = 0 \tag{3}$$

The conservation of momentum

$$\frac{\partial\left(\rho\vec{v}\right)}{\partial t} + \nabla.\left(\rho\vec{v_r}\vec{v}\right) + \rho\left[\left(\omega\hat{a}\right) \times \left(\vec{v} - \vec{v_t}\right)\right] = -\nabla p + \nabla.\underline{\tau} \tag{4}$$

In this equation, $\vec{v_t}$, $\vec{v}$, ω, $\underline{\tau}$, and $\hat{a}$ respectively define transitional velocity, absolute velocity, angular velocity, viscous stress, and axis of rotation.

2.2. Grid Independance Study

In a prior study, four different cell-size grids were analyzed [24]. The simulations were carried out at $\phi = 0.225$ and the torque coefficient was compared. A change in grid size resulted in discrepancies of approximately 24%, 4%, and 0.5% between medium, fine, and extra-fine grids, respectively on Table 1. For the fine grid, variation in the results was negligible, resulting in reasonable accuracy. However, after examining the flow topology there was a qualitative difference between the fine and extra-fine grid. This is typical of RANS turbulence modeling of separated flows. As a result, subsequent simulations were performed using the extra-fine grids; grid dependency statistics are shown in Figure 4.

Table 1. Grid Sizes.

Mesh	No. of Grids (Millions)	Grid Size (mm)
Coarse	3.6	10.0
Medium	7.6	5.0
Fine	18.1	2.5
Extra Fine	32	1.25

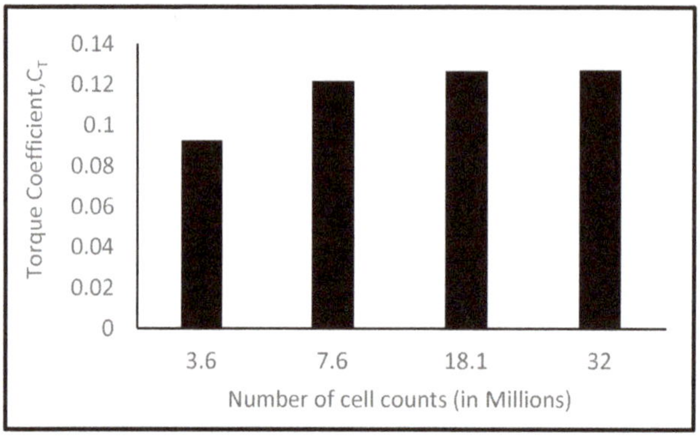

Figure 4. Grid Independence Study of Torque Coefficient at ϕ = 0.225.

2.3. Validation

Upon completion of the grid independence study, we compared the torque coefficient (C_T), pressure drop coefficient (ΔP_0^*), and efficiency (η) at various flow coefficients conducted in prior work [24]. As shown in Figures 5–7, the CFD results are in good agreement until the flow coefficient reached $\phi = 0.200$. Here the flow coefficient is representative of a high angle of attack. Therefore, flow separation starts to form at the leading edge and causes a rapid decrease in the torque coefficient and efficiency. The numerical results deviate from the experiment because of RANS modeling inability to adequately resolve highly separated flows. Therefore, it tends to over predict torque after stall [25]. However, the present work corresponds well to other CFD results [25], given the same geometrical features.

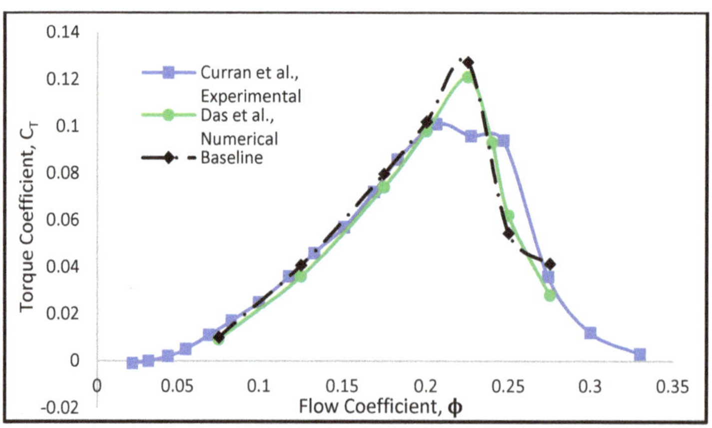

Figure 5. Comparison of numerical and experimental torque coefficient, C_T.

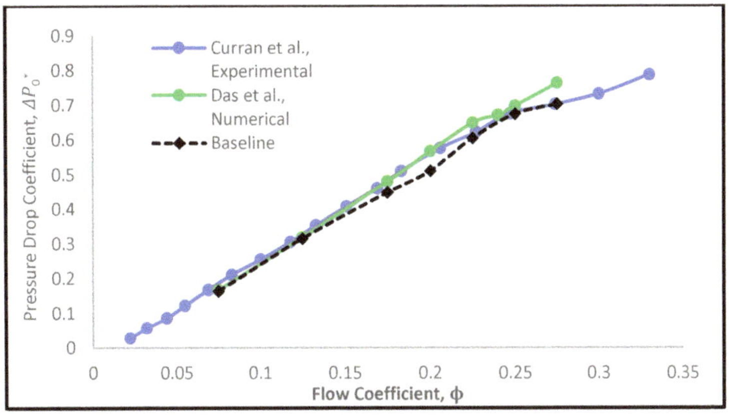

Figure 6. Comparison of numerical and experimental pressure drop coefficient, ΔP_0^*.

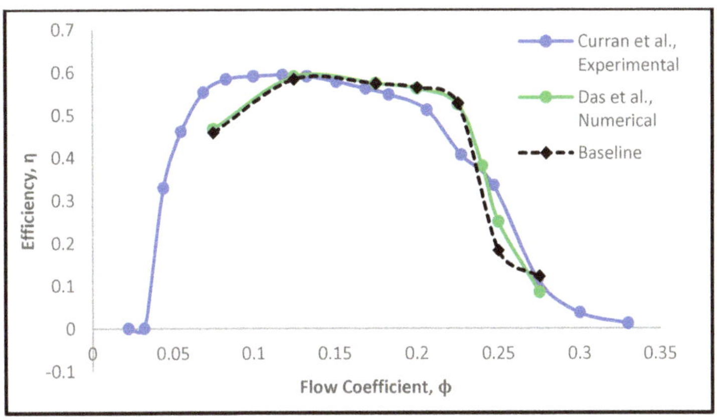

Figure 7. Comparison of numerical and experimental efficiency, η.

Traditionally, turbine performance is characterized by non-dimensional coefficients such as:

The torque coefficient, $\mathbf{C_T}$

$$\mathbf{C_T} = \frac{\mathbf{T}}{\rho\omega^2\mathbf{R}^5} \quad (5)$$

The Pressure drop coefficient, $\Delta\mathbf{P}_0^*$

$$\Delta\mathbf{P}_0^* = \frac{\Delta\mathbf{p}_0}{\rho\omega^2\mathbf{R}^2} \quad (6)$$

The efficiency, η

$$\eta = \frac{\mathbf{T}\omega}{\Delta\mathbf{p}_0\mathbf{Q}} \quad (7)$$

The terms $\mathbf{T}$, $\mathbf{Q}$, $\Delta\mathbf{p}_0, \mathbf{U_R}$, $\mathbf{b}, \mathbf{c}, \mathbf{z}, \mathbf{v}$, ω, **and** ρ represent the blade torque, volumetric flow rate, static pressure drop across the turbine, the circumferential velocity at blade mean radius, blade height, blade chord length, number of blades, axial inlet velocity, the angular velocity of the turbine, and air density, respectively.

2.4. Serrations Setup

The modified blade model corresponds to a NACA 0015 airfoil, with chord length of c = 0.125 m and a span of L = 0.100 m. The NACA 0015 blade's profile was designed in the x-axis and y-axis, then extruded in the z-axis direction. The serrations were made by cutting triangular patterns on the Trailing Edge (TE) with assigned parameters. For our serration parameters, we have considered the existing modeling criterion for noise attenuation, even though we are primarily investigating the aerodynamic characteristics of the serrated trailing edges.

At a Strouhal number of $\sim \mathbf{1}$ noise reduction occurs in the low-frequency region when $\frac{f\delta}{U_0} < \boldsymbol{St_\delta}$. On the other hand, noise can increase when $\frac{f\delta}{U_0} > \boldsymbol{St_\delta}$, reguardless of angle of attack and mean flow velocity [33]. Furthermore, according to Das et al. [25] a Wells turbine normally operates at a reduced frequency in the order of 10^{-3}, which makes it fall under the condition of $\boldsymbol{St_\delta} < 1$. However, Gruber et al. [33] have considered critical values of either $\frac{h}{\delta} \sim \mathbf{0.5}$ or $\frac{h}{\lambda} \sim \mathbf{1}$, beyond which significant noise reduction can occur. Our assigned serration depth, $\boldsymbol{h} = \mathbf{0.02}\boldsymbol{c}$ and width is $\boldsymbol{\lambda} = \mathbf{0.096}\boldsymbol{c}$ (See Figure 8). By the assigned dimensions; therefore, our serration pattern can meet the noise attenuation condition of:

$$\frac{h}{\delta} = \mathbf{0.83} \tag{8}$$

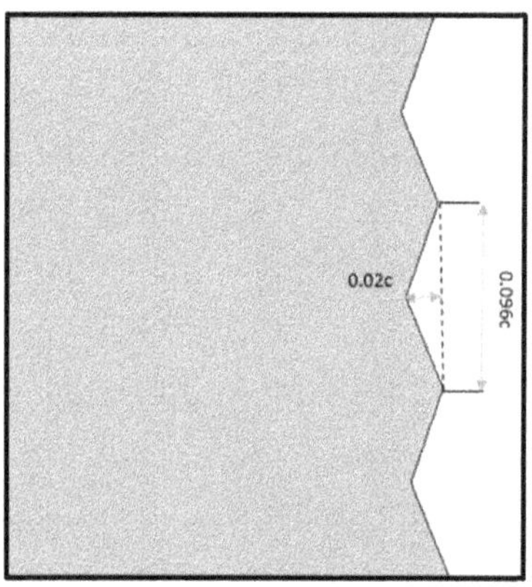

Figure 8. Trailing edge serration pattern of the NACA 0015 airfoil.

To obtain an optimal design of the TE serration. The first model consisted of a fully serrated trailing edge (FSTE), shown in Figure 9.

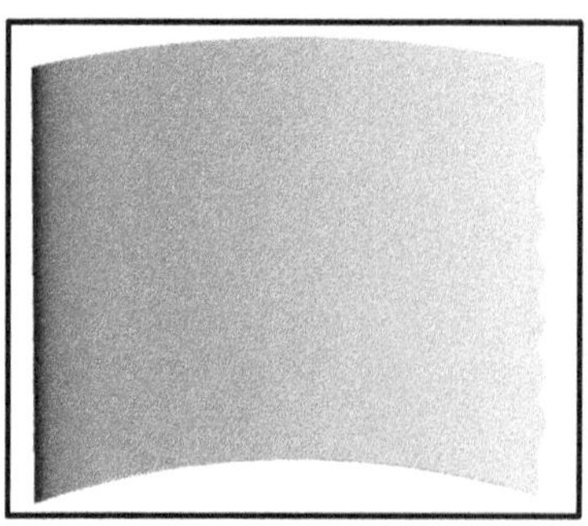

Figure 9. Fully Serrated Trailing Edge Model on NACA0015 airfoil.

While observing the baseline turbine flow field, through range of $0.075 \leq \phi \leq 0.275$, it was observed that a small area of recirculation region begins in near hub location of the span. Therefore, we designed a second model which is a partially serrated trailing edge (PSTE), where the serration pattern covers 0.288c (36 **mm**) of the trailing edge as shown in Figure 10. Since, the serrations attenuate the noise level by mitigating the vortex shedding at the trailing edge we introduced the partial trailing edge serrations close to hub [34].

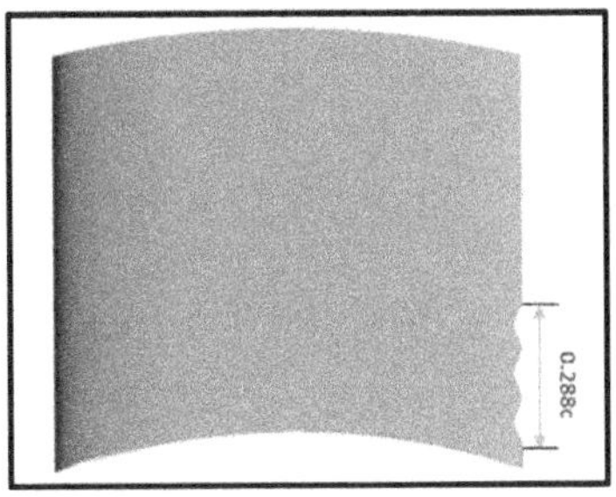

Figure 10. Partially Serrated Trailing Edge Model on NACA0015 airfoil.

After creating the serration models, the meshes were generated using ANSYS meshing by specifying same criterion as the baseline model face sizing and inflation layers. In addition, the boundary conditions remain constant for all three models. Figure 11 show the final meshing results for the fully serrated blade model. Figure 12 shows the generated mesh around the partially serrated trailing edge.

1. Fully Serrated Trailing Edge:

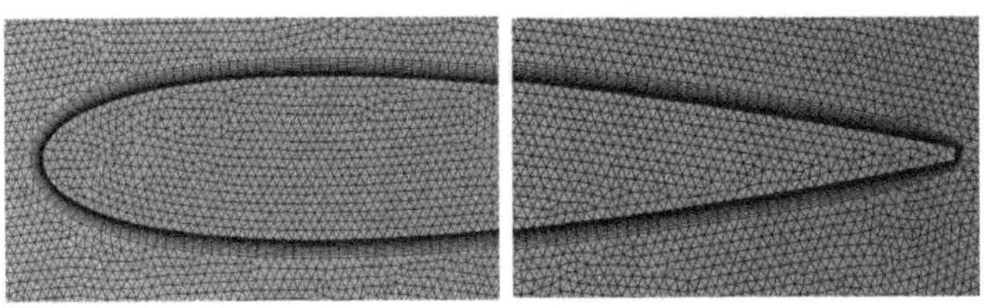

(a) Leading Edge (b) Trailing Edge

Figure 11. ANSYS Meshing Generated Around the Fully Serrated Blade.

2. Partially Serrated Trailing Edge:

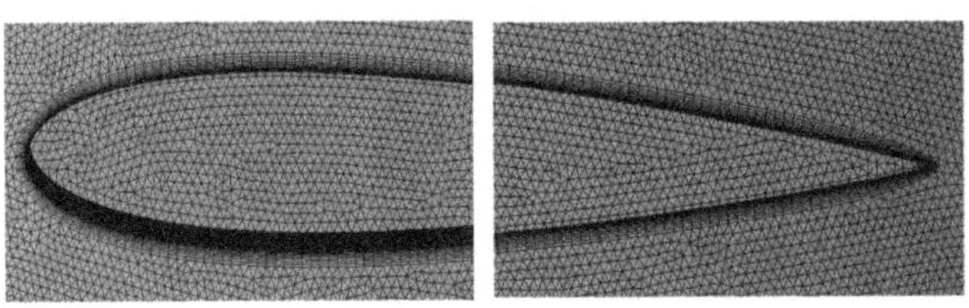

(a) Leading Edge (b) Trailing Edge

Figure 12. ANSYS Meshing Generated Around the Partially Serrated Blade.

3. Results

The CFD simulations of the modified Wells turbine were carried out at different flow coefficients that represent pre-stall, stall, and post-stall conditions. In this section, we discuss the results quantitatively and qualitatively by comparing the performance coefficients and flow topology to the baseline configuration.

3.1. Comparison of Performance Coefficients and Turbine Efficiency

The baseline has the highest torque coefficient prior to the stall point followed by the PSTE design. From Figure 13, it is shown, for all three designs, that the torque coefficient increases up to $\phi = 0.225$, from there it decreases. Hence, the turbine blades stall at a flow coefficient of 0.250, i.e., angle of attack of 14.03°. Mostly, the FSTE design produced lower torque compared to the other designs, except for, $\phi = 0.200$. This may be a result of the presence of flow separation closer to the hub and trailing edge, unlike the baseline and PSTE models. The PSTE mostly produced higher torque compared to the FSTE and lower torque compared to the baseline blade.

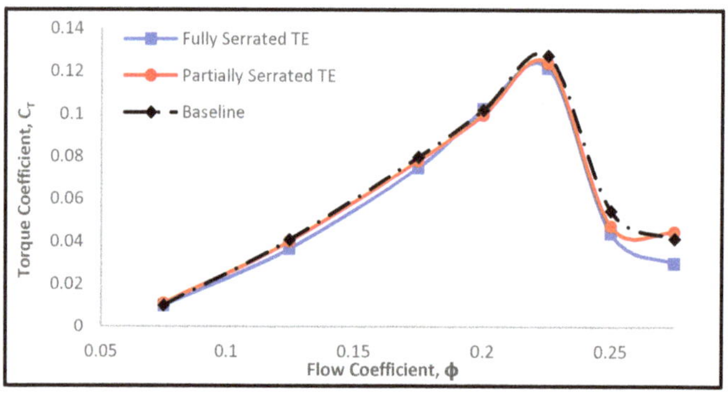

Figure 13. Torque Coefficient, C_T.

Figure 14 shows a comparison of the pressure drop across the turbine. Here the results show a linear increase with an increasing flow coefficient. The reference (i.e., baseline) case has a higher pressure drop coefficient for our range of ϕ, the FSTE has the lowest, while the PSTE design fluctuates between the baseline and FSTE configuration.

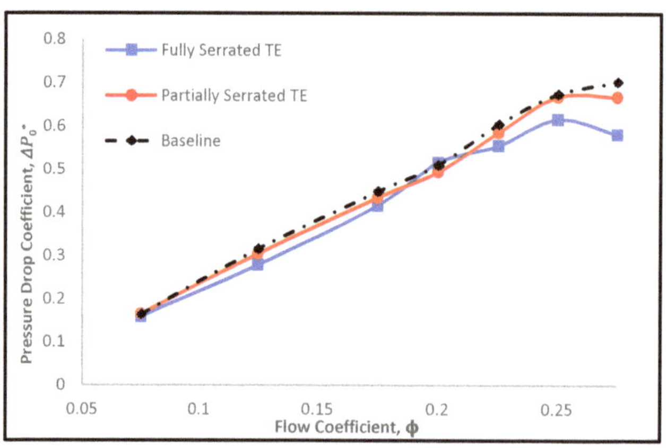

Figure 14. Pressure Drop Coefficient, ΔP_0^*.

The efficiency plot, shown in Figure 15, shows that the PSTE has the highest efficiency prior to the stall condition. PSTE increases the efficiency at the peak point by 1.51%. Secondly, the FSTE model shows a 1.22% improvement in peak efficiency compared to the baseline model. At the stall point, flow recirculation appears and more negatively impacts the serrated models. Therefore, at $\phi = 0.250$, the baseline model is more efficient because the serration pattern allows the air on the pressure side to bleed to the low-pressure suction side.

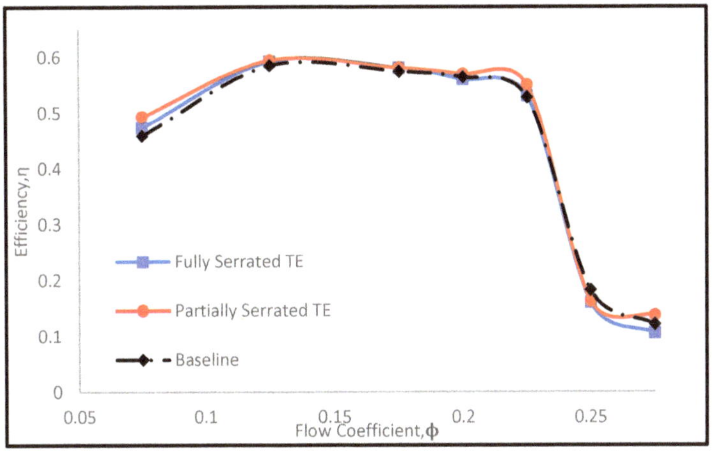

Figure 15. Efficiency, η.

3.2. Comparison of Flow Topology and Pressure Coefficient

Discussed in this section are the differences in the pressure distribution between the baseline and serrated blades at pre-stall, $\phi = 0.225$, stall, $\phi = 0.250$, and post-stall, $\phi = 0.275$ points. This section studies the flow field behavior by describing sectional views (see Figure 16) of the streamlines, C_p contours, and C_p plots at different span locations.

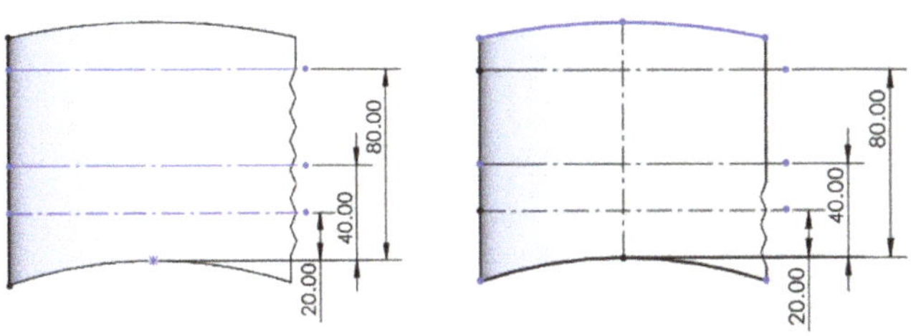

Figure 16. Sectional plane locations for FSTE on the right side, and PSTE on the left side. Pre-Stall, $\phi = 0.225$.

Figure 17 shows the streamlines at $\phi = 0.225$. We observe that the streamlines start to detach across the span near the trailing edge. For the serrated models, the saw-tooth trailing edge reduces the strength of the trailing edge wake and encourages upstream mixing and induces separation near the hub. For the FSTE, at the near-hub location, the flow bleeds over on to the suction side from the higher-pressure side [Figure 17]. Further from the hub, the flow detaches near the trailing edge at the mid- and near-tip locations for all configurations. For the PSTE configuration, while not as substantial as the FSTE, the flow moves to the suction side near-hub and reattaches on the suction side and reduces the size of the wake, but results in induced separation on the suction side, thus reducing torque. However, due to a lower pressure drop for the serrated configurations, there is a modest improvement in efficiency. The PSTE has higher efficiency compared to the FSTE configuration because the flow attaches closer to the TE.

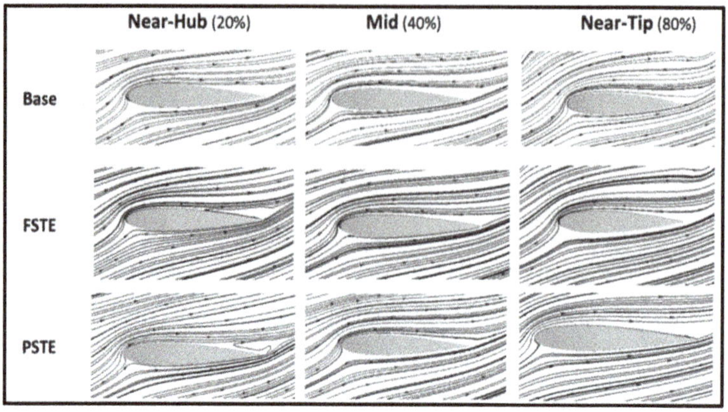

Figure 17. Streamlines at ϕ = 0.225.

Figure 18 presents the C_p contours of three models along the span. The baseline model has the largest low-pressure zone close to the leading edge on the suction side. We observe higher pressure starting to bleed onto the suction side at the mid-span location. The FSTE configuration has the largest suction side trailing edge low-pressure zone among the presented designs. Creating lower pressure on the suction side is undesired in this case because it reduces the aerodynamic performance of the turbine. Additionally, the FSTE design creates a large adverse pressure gradient which leads to flow separation. It can be shown that the FSTE model shows greater bleed compared to the other designs. The PSTE decreased the bleed of air from the pressure to the suction side at the trailing edge.

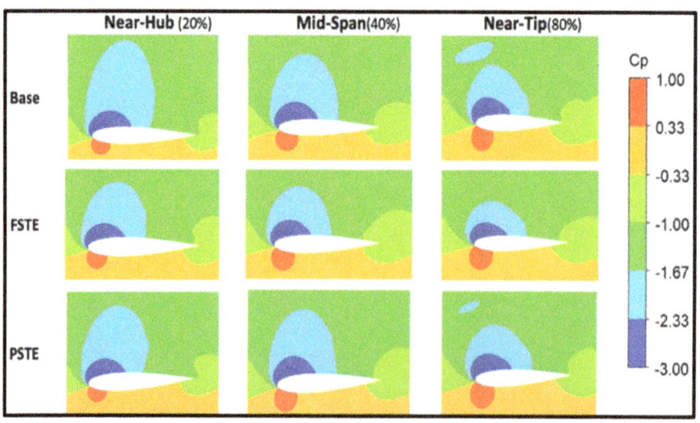

Figure 18. Contours of C_p. for ϕ = 0.225.

The fully serrated trailing edge has a higher-pressure gradient at the trailing edge because of the flow bleeding from the pressure side to the suction side, as shown in Figure 19. Further, the high-pressure gradient at the trailing edge acts as resistance to turbine rotation which is responsible for lower torque generation. Therefore, the FSTE model has the lowest efficiency.

At the stall point, ϕ = 0.250, a recirculation zone appears, for all models, as shown in Figure 20. For the baseline configuration, recirculation appears near the mid-span at the trailing edge. Near the tip location, the recirculation has increased in size and moved towards the leading edge. For the FSTE model, minor recirculation is observed at the near hub and becomes larger at the mid-span of the trailing edge. At the near tip location, the recirculation field moved towards the leading edge. The PSTE model, recirculation appears near the hub due to the higher pressure at the trailing edge. At the near-tip location, the circulation is close to the leading edge similar to the baseline model.

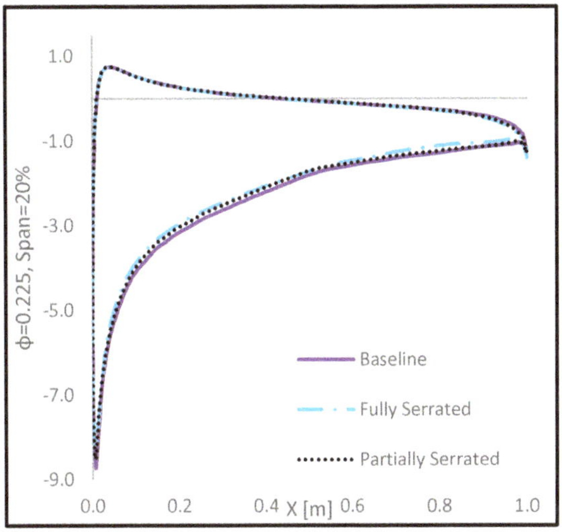

(a)

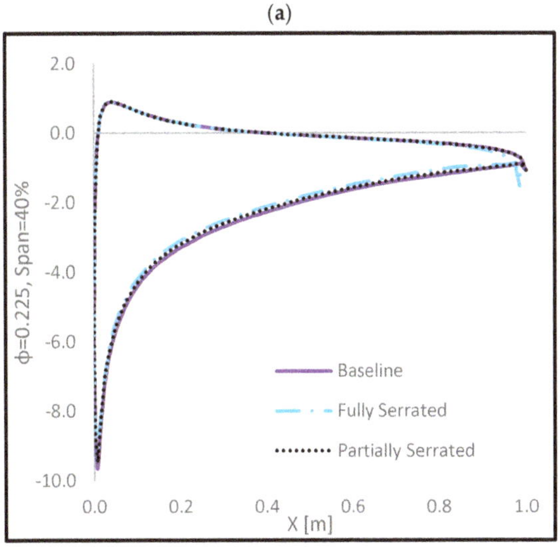

(b)

Figure 19. *Cont.*

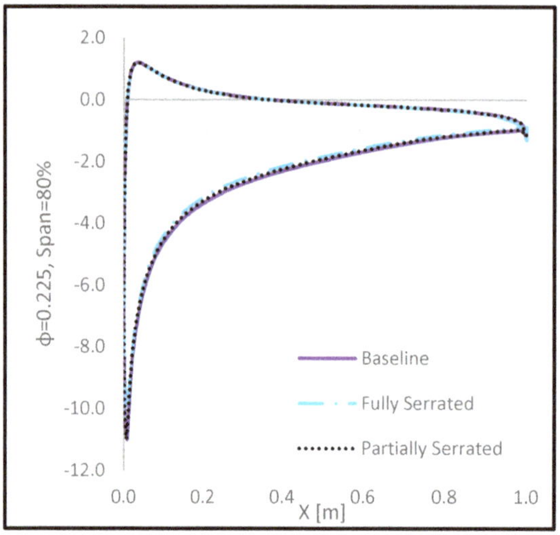

(c)

Figure 19. Plot of C_p, ϕ = 0.225. (**a**) 20% span, (**b**) 40% span, (**c**) 80% span. Stall, ϕ = 0.250.

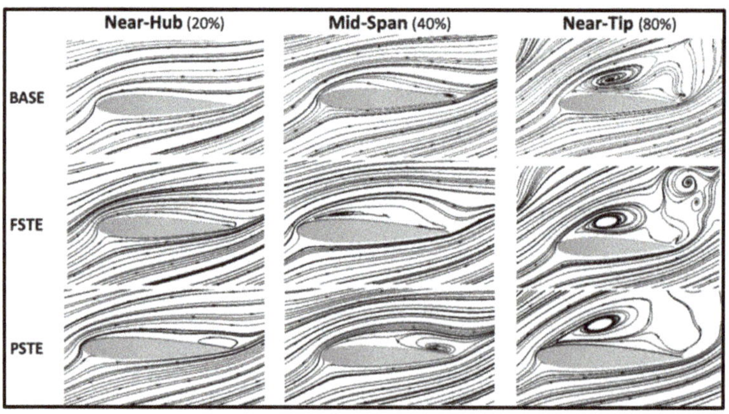

Figure 20. Streamlines at ϕ = 0.250.

Figure 21 presents the C_p contours at the stalling point for all designs. It can be shown that the baseline model has a greater lower pressure region across the chord compared to the pre-stall condition. This low-pressure creates suction and causes a recirculation zone by attracting the higher-pressure trailing edge fluid thus leading to stall. Due to the partial span serrations of the PSTE, has a localized low-pressure zone near-hub compared to the other designs as shown in Figure 21. As a result, it has minor recirculation at the trailing edge due to bleed from the pressure side. Near the mid-span location, unlike the previous locations, the FSTE model has two distinct low-pressure zones distributed close to the leading and trailing edges on the suction side of the blade. Therefore, at the near mid-span location, this design tends to have multiple recirculation zones. The PSTE model has a greater lower pressure zone compared to the pre-stall condition. Recirculation appeared starting at the near-hub location. The flow detached near the mid-span due to higher pressure around the trailing edge. Near-tip, the lower pressure zone increased which allows additional high-pressure air to bleed onto the suction side.

Figure 21. Contours of C_p. for $\phi = 0.250$. 20% span, 40% span, 80% span.

The FSTE model reduced the low-pressure region along the suction side of the blade. This is shown in Figure 22. The C_p plots clearly show higher suction side pressure that results in a reduction of pressure drop and torque coefficient. However, this results in a slight improvement of efficiency. At the near-hub, the PSTE demonstrates better performance even though there is evidence of trailing edge separation. Due to the serrations, the pressure drop is less severe compared to the baseline configuration, thus improving efficiency. However, at the near mid-span location, the low-pressure zone extends approximately 80% of the airfoil which possibly represents the turbine experiencing stall sooner than the other configurations. Again, the low-pressure zone appears at 80% chord and continues until the trailing edge causing a large recirculation zone for the FSTE. As a result, the FSTE produced the lowest aerodynamic performance. When it comes to the near-tip location, the baseline and PSTE models show similar trends. Although, the low-pressure region is slightly higher for the baseline, near the leading edge, up to the mid-chord position. From 30% of the chord to the near trailing edge, the FSTE has the smallest pressure gradient. However, at the trailing edge, the FSTE model has the highest flow bleed resulting in less torque and lower efficiency.

In the end, the post-stall condition, $\phi = 0.275$, is shown in Figure 23. The baseline model, near-hub, does not have recirculation but at the near mid-span recirculation appears. Also, near-tip, the flow is highly separated, and a recirculation zone appears. For the FSTE model, recirculation appeared at all observed span locations. The recirculation at the mid-span has increased in size and shows substantial flow separation near the trailing edge. The PSTE model, shows no detached flow near the hub; however, recirculation appears near the mid-span on the leading and trailing edges. Lastly, even though the near tip location shows less recirculation for the PSTE model, the turbine is fully stalled.

Figure 24 presents the C_p contours for the post-stall condition. It is obvious that the lower pressure region has expanded for all three models compared to previous angles of attack (i.e., flow coefficient). By increasing the lower pressure area, the suction of higher-pressure flow will increase, as a result, multiple recirculation zones are observed.

The FSTE model produced the lowest suction side pressure on the airfoil, except at the trailing edge, for all the span locations [Figure 25]. At the near-hub location, the PSTE demonstrates better performance regarding flow separation compared to the FSTE and baseline models. However, at the near mid-span location, the low-pressure zone for the baseline configuration extends to around 80% of the airfoil chord which typically represents turbine stall. Again, the low-pressure zone appears at 80% of the chord and extends to the trailing edge which causes huge recirculation in the FSTE model. As a result, the FSTE showed the lowest aerodynamic performance. When it comes to the near-tip location, the baseline and PSTE models show similar trends. Although, the low-pressure region is slightly higher for the baseline model near the leading edge up to the mid-chord. From the airfoil chord position of 30% to near the trailing edge, the FSTE model has the smallest pressure gradient. However, at the trailing edge FSTE model, the flow bleed causes less torque production and lower efficiency.

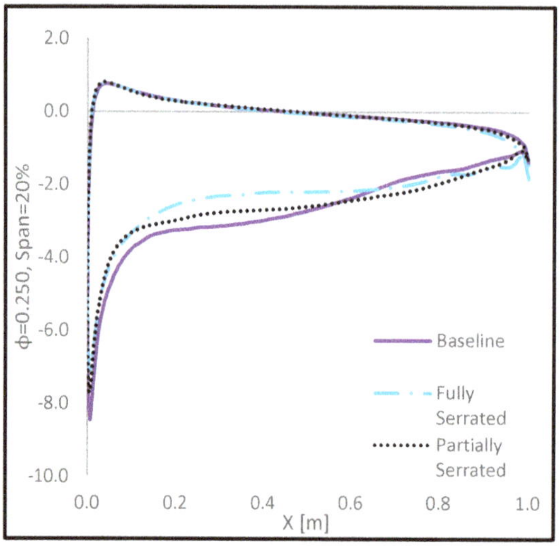

(**a**)

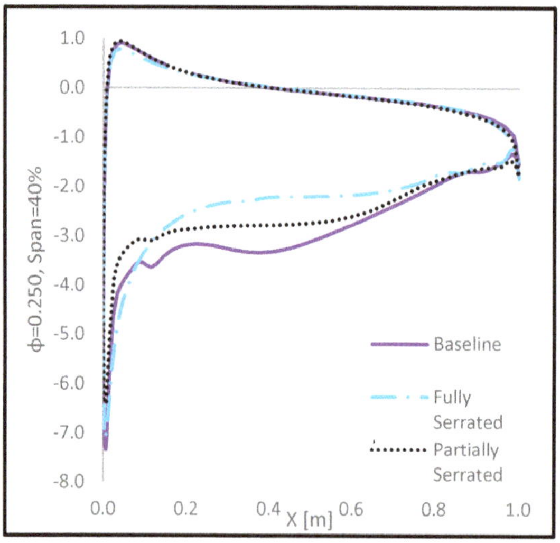

(**b**)

Figure 22. *Cont.*

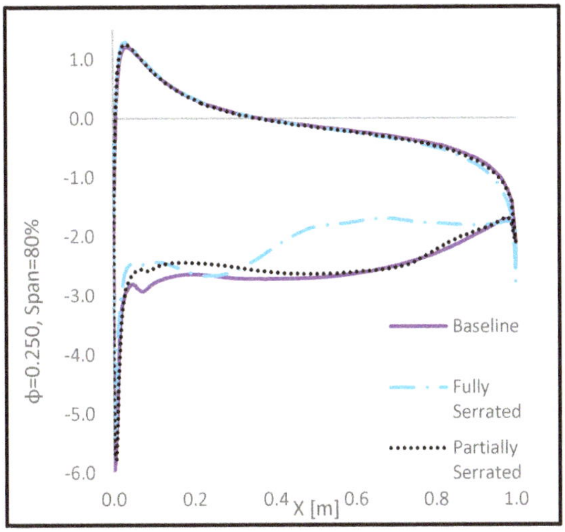

(c)

Figure 22. Plot of C_p, ϕ = 0.250. (a) 20% span, (b) 40% span, (c) 80% span. Post-Stall, ϕ = 0.275.

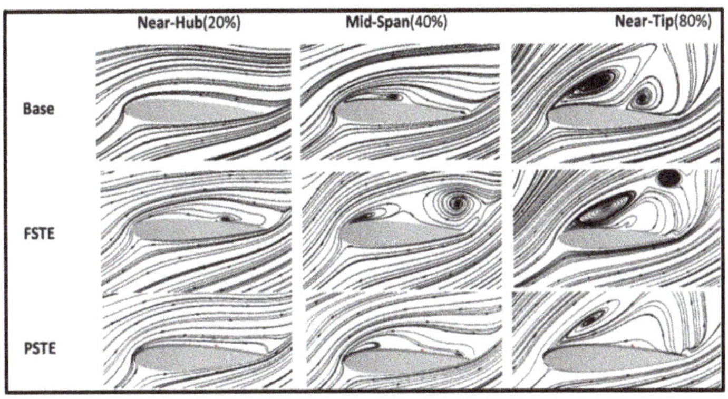

Figure 23. Streamlines at the post-stall condition, ϕ = 0.275.

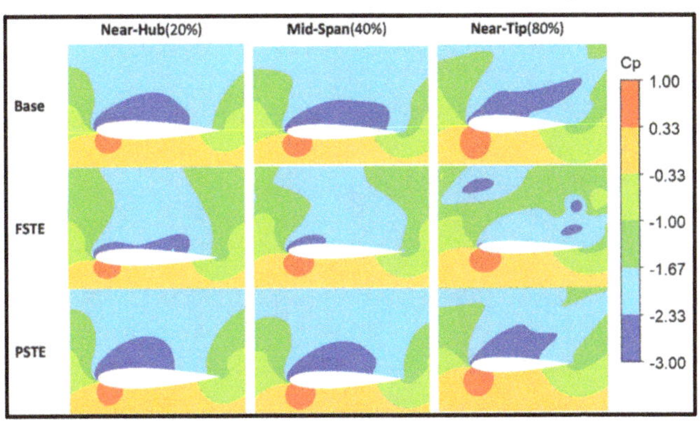

Figure 24. Contours of C_p. for ϕ = 0.275.

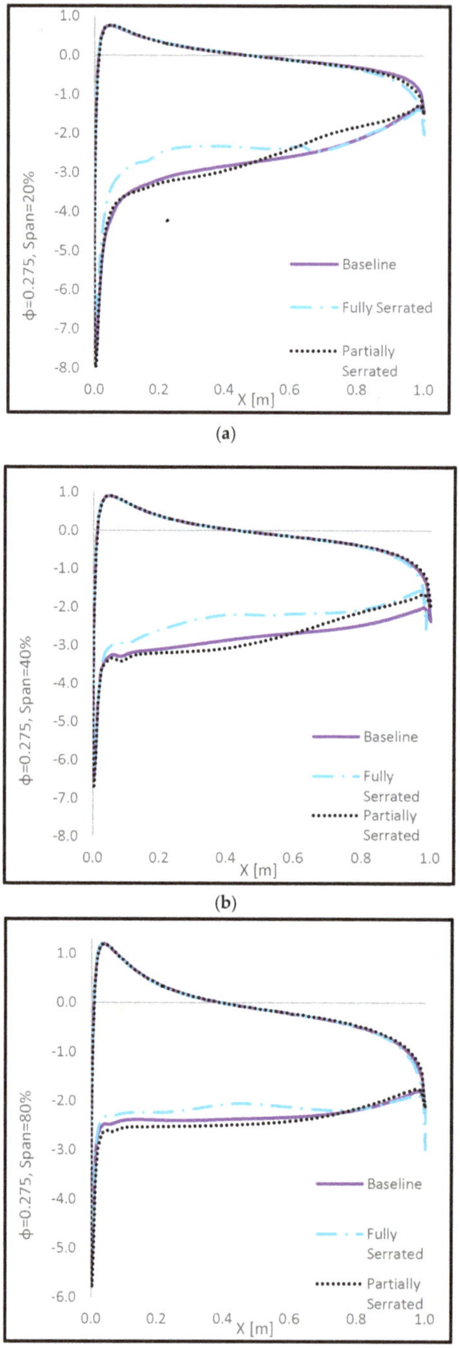

(**a**)

(**b**)

(**c**)

Figure 25. Plot of C_p, ϕ = 0.275. (**a**) 20% span, (**b**) 40% span, (**c**) 80% span.

4. Discussion

This study investigated the influence of trailing edge serrations on a NACA- 0015 Wells turbine. Two serration patterns were explored: (1) a fully-serrated trailing edge, and (2) a partially-serrated trailing edge. The flow topology and aerodynamic performance were assessed and compared to a baseline (non-serrated) NACA 0015 airfoil at varying flow coefficients. Three-dimensional incompressible RANS simulations showed that the PSTE model has the highest efficiency under unstalled conditions. Also, at low flow coefficients, the PSTE increases the peak efficiency by 1.51%; The FSTE had a 1.22% increase. Due to massive flow separation close to the stall point, the RANS solver overpredicts torque. This is typical of such turbulence models and it is notoriously difficult to resolve such flow fields. Furthermore, our results are consistent with other numerical simulations. However, at unstalled conditions the solver computes accurate results and demonstrates good agreement with the experimental data for $\phi \leq 0.200$. Because the efficiency improves in the range where the solver shows excellent agreement, we can say that the computed results are accurate and show a modest improvement. The FSTE and PSTE designs tend to have a high degree of trailing edge induced separation due to higher air bleed, thus resulting in reduced torque. This results in a reduced pressure drop and in a modest increase in efficiency. Future studies should include higher-fidelity simulations to resolve the highly separated flow. Furthermore, as the PSTE model a have higher peak efficiency, parametric studies and shape optimization of the turbine blades with partial serrations should be explored to determine the impact on turbine performance. Variation of serration dimensions and locations can be studied to determine the optimum partial serration pattern.

Author Contributions: Conceptualization, A.S.A., M.A. and M.N.U.; methodology, A.S.A.; software, A.S.A.; validation, M.N.U. and A.A.; formal analysis, A.S.A.; investigation, A.S.A.; resources, M.A.; data curation, A.S.A. and M.N.U.; writing—original draft preparation, A.S.A.; writing—review and editing, A.S.A., M.N.U. and M.A.; visualization, A.S.A.; supervision, M.A.; project administration, M.A.; funding acquisition, M.A. All authors have read and agreed to the published version of the manuscript.

Funding: This research was funded by North Carolina Renewable Ocean Energy Program grant number 111110-740106-72000-0000.

Institutional Review Board Statement: Not applicable.

Informed Consent Statement: Not applicable.

Data Availability Statement: Not applicable.

Conflicts of Interest: The authors declare no conflict of interest.

Nomenclature

a	acceleration (m/s^2)
τ	shear stress
ρ	density (kg/m^3)
c	chord length (m)
C_T	torque coefficient
λ	nondimensional width
h	nondimensional depth
St_δ	Strouhal number
L	length scale (m)
Δp_0	static pressure drop (Pa)
$\Delta \mathbf{P}_0^*$	pressure drop coefficient
Q	volumetric flow rate (m^3/s)
R	rotor radius (m)
Re	Reynolds number
T	blade torque (Nm)

u_t	friction velocity (m/s)
v	axial velocity (m/s)
y	first layer distance from the wall (m)
y^+	dimensionless wall distance
f	frequency
ν_t	turbulent viscosity (kg/m-s)
ϕ	flow coefficient
η	efficiency
μ	dynamic viscosity (Ns/m^2)
ω	angular velocity (rad/s)
δ	boundary layer thickness (m)

References

1. EIA Expects Renewables to Account for 22% of U.S. Electricity Generation in 2022. Available online: https://www.eia.gov/todayinenergy/detail.php?id=53459 (accessed on 2 November 2022).
2. Renewable Energy Market Update—May 2022—Analysis, IEA. Available online: https://www.iea.org/reports/renewable-energy-market-update-may-2022 (accessed on 2 November 2022).
3. Marine Energy Basics. Energy.gov. Available online: https://www.energy.gov/eere/water/marine-energy-basics (accessed on 2 November 2022).
4. Lekube, J.; Garrido, A.J.; Garrido, I.; Otaola, E.; Maseda, J. Flow Control in Wells Turbines for Harnessing Maximum Wave Power. *Sensors* **2018**, *18*, 535. [CrossRef]
5. Ghisu, T.; Puddu, P.; Cambuli, F. Numerical analysis of a wells turbine at different non-dimensional piston frequencies. *J. Therm. Sci.* **2015**, *24*, 535–543. [CrossRef]
6. Ghisu, T.; Puddu, P.; Cambuli, F. A detailed analysis of the unsteady flow within a Wells turbine. *Proc. Inst. Mech. Eng. Part A J. Power Energy* **2017**, *231*, 197–214. [CrossRef]
7. Taha, Z.; Sawada, T. A comparison of computational and experimental results of Wells turbine performance for wave energy conversion. *Appl. Ocean Res.* **2010**, *32*, 83–90. [CrossRef]
8. Taha, Z.; Ya, T.T.; Sawada, T. Numerical investigation on the performance of Wells turbine with non-uniform tip clearance for wave energy conversion. *Appl. Ocean Res.* **2011**, *33*, 321–331. [CrossRef]
9. Takao, M.; Setoguchi, T.; Kinoue, Y.; Kaneko, K. Wells Turbine with End Plates for Wave Energy Conversion. *Ocean Eng.* **2007**, *34*, 1790–1795. [CrossRef]
10. Das, T.K.; Kerikous, E.; Venkatesan, N.; Janiga, G.; Thevenin, D.; Samad, A. Performance Improvement of a Wells Turbine through an Automated Optimization Technique. *Energy Convers. Manag. X* **2022**, *16*, 100285. [CrossRef]
11. Shaaban, S.; Hafiz, A.A. Effect of duct geometry on Wells turbine performance. *Energy Convers. Manag.* **2012**, *61*, 51–58. [CrossRef]
12. Halder, P.; Samad, A.; Thévenin, D. Improved design of a Wells turbine for higher operating range. *Renew. Energy* **2017**, *106*, 122–134. [CrossRef]
13. Mohamed, M.; Janiga, G.; Pap, E.; Thévenin, D. Multi-objective optimization of the airfoil shape of Wells turbine used for wave energy conversion. *Energy* **2011**, *36*, 438–446. [CrossRef]
14. Setoguchi, T.; Santhakumar, S.; Takao, M.; Kim, T.; Kaneko, K. Effect of guide vane shape on the performance of a Wells turbine. *Renew. Energy* **2001**, *23*, 1–15. [CrossRef]
15. Halder, P.; Rhee, S.H.; Samad, A. Numerical optimization of Wells turbine for wave energy extraction. *Int. J. Nav. Arch. Ocean Eng.* **2017**, *9*, 11–24. [CrossRef]
16. Watterson, J.K.; Raghunathan, S. Computed Effects of Solidity on Wells Turbine Performance. *JSME Int. J. Ser. B* **1998**, *41*, 177–183. [CrossRef]
17. Howe, M. A review of the theory of trailing edge noise. *J. Sound Vib.* **1978**, *61*, 437–465. [CrossRef]
18. Howe, M. Aerodynamic noise of a serrated trailing edge. *J. Fluids Struct.* **1991**, *5*, 33–45. [CrossRef]
19. Jones, L.; Sandberg, R. *Direct Numerical Simulations of Noise Generated by the Flow over an Airfoil with Trailing Edge Serrations*; American Institute of Aeronautics and Astronautics: Reston, VA, USA, 2009. [CrossRef]
20. Oerlemans, S.; Fisher, M.; Maeder, T.; Kögler, K. Reduction of Wind Turbine Noise Using Optimized Airfoils and Trailing-Edge Serrations. *AIAA J.* **2009**, *47*, 1470–1481. [CrossRef]
21. Gruber, M.; Azarpeyvand, M.; Joseph, P.F. Airfoil Trailing Edge Noise Reduction by the Introduction of Sawtooth and Slitted Trailing Edge Geometries. In Proceedings of the 20th International Congress on Acoustics, Sydney, Australia, 23–27 August 2010; p. 9.
22. Ragni, D.; Avallone, F.; van der Velden, W.C.P.; Casalino, D. Measurements of near-wall pressure fluctuations for trailing-edge serrations and slits. *Exp. Fluids* **2018**, *60*, 6. [CrossRef]
23. Avallone, F.; Arce León, C.; Pröbsting, S.; Lynch, K.; Ragni, D. Tomographic-PIV Investigation of the Flow over Serrated Trailing-Edges. In Proceedings of the 54th AIAA Aerospace Sciences Meeting, San Diego, CA, USA, 4–8 January 2016.
24. Uddin, M.N.; Atkinson, M.; Das, T.K.; Esau, S. *Computational Fluid Dynamics Investigation of a Novel Bi-Directional Axial Flow Turbine for Wave Energy Conversion*; American Institute of Aeronautics and Astronautics: Reston, VA, USA, 2022. [CrossRef]
25. Das, T.K.; Samad, A. Influence of stall fences on the performance of Wells turbine. *Energy* **2019**, *194*, 116864. [CrossRef]

26. Curran, R.; Gato, L. The energy conversion performance of several types of Wells turbine designs. *Proc. Inst. Mech. Eng. Part A J. Power Energy* **1997**, *211*, 133–145. [CrossRef]
27. Liu, Z. Self-Starting Analysis of an OWC Axial Impulse Turbine in Constant Flows: Experimental and Numerical Studies. *Appl. Ocean Res.* **2018**, *82*, 458–469.
28. Raghunathan, S. The wells air turbine for wave energy conversion. *Prog. Aerosp. Sci.* **1995**, *31*, 335–386. [CrossRef]
29. Raghunathan, S.; Tan, C.P.; Wells, N. Wind Tunnel Tests on Airfoils in Tandem Cascade. *AIAA J.* **1981**, *19*, 1490–1492. [CrossRef]
30. Raghunathan, S.; Tan, C. Aerodynamic performance of a Wells air turbine. *J. Energy* **1983**, *7*, 226–230. [CrossRef]
31. Suzuki, M.; Arakawa, C.; Tagori, T. Fundamental Studies on Wells Turbine for Wave Power Generator; 1st Report, The Effect of Solidity, and Self-Starting. *Bull. JSME* **1984**, *27*, 1925–1931. [CrossRef]
32. Setoguchi, T.; Santhakumar, S.; Takao, M.; Kim, T.; Kaneko, K. A modified Wells turbine for wave energy conversion. *Renew. Energy* **2003**, *28*, 79–91. [CrossRef]
33. Gruber, M.; Joseph, P.; Chong, T. On the Mechanisms of Serrated Airfoil Trailing Edge Noise Reduction. In Proceedings of the 17th AIAA/CEAS Aeroacoustics Conference (32nd AIAA Aeroacoustics Conference), Portland, OR, USA, 5–8 June 2011; American Institute of Aeronautics and Astronautics: Reston, VA, USA, 2011.
34. Intravartolo, N.; Sorrells, T.; Ashkharian, N.; Kim, R. Attenuation of Vortex Noise Generated by UAV Propellers at Low Reynolds Numbers. In Proceedings of the 55th AIAA Aerospace Sciences Meeting, Grapevine, TX, USA, 9–13 January 2017; American Institute of Aeronautics and Astronautics: Reston, VA, USA, 2017.

micromachines

MDPI

Review

Towards a Highly Efficient ZnO Based Nanogenerator

Mohammad Aiman Mustaffa [1], Faiz Arith [1,*], Nur Syamimi Noorasid [1], Mohd Shahril Izuan Mohd Zin [1], Kok Swee Leong [1], Fara Ashikin Ali [2], Ahmad Nizamuddin Muhammad Mustafa [2,3] and Mohd Muzafar Ismail [2]

1 Faculty of Electronic and Computer Engineering, Universiti Teknikal Malaysia Melaka (UTeM), Hang Tuah Jaya, Melaka 76100, Malaysia
2 Faculty of Electrical and Electronic Engineering Technology, Universiti Teknikal Malaysia Melaka (UTeM), Hang Tuah Jaya, Melaka 76100, Malaysia
3 Department of Materials, Faculty of Engineering, Imperial College London, London SW7 2AZ, UK
* Correspondence: faiz.arith@utem.edu.my

Abstract: A nanogenerator (NG) is an energy harvester device that converts mechanical energy into electrical energy on a small scale by relying on physical changes. Piezoelectric semiconductor materials play a key role in producing high output power in piezoelectric nanogenerator. Low cost, reliability, deformation, and electrical and thermal properties are the main criteria for an excellent device. Typically, there are several main types of piezoelectric materials, zinc oxide (ZnO) nanorods, barium titanate ($BaTiO_3$) and lead zirconate titanate (PZT). Among those candidate, ZnO nanorods have shown high performance features due to their unique characteristics, such as having a wide-bandgap semiconductor energy of 3.3 eV and the ability to produce more ordered and uniform structures. In addition, ZnO nanorods have generated considerable output power, mainly due to their elastic nanostructure, mechanical stability and appropriate bandgap. Apart from that, doping the ZnO nanorods and adding doping impurities into the bulk ZnO nanorods are shown to have an influence on device performance. Based on findings, Ni-doped ZnO nanorods are found to have higher output power and surface area compared to other doped. This paper discusses several techniques for the synthesis growth of ZnO nanorods. Findings show that the hydrothermal method is the most commonly used technique due to its low cost and straightforward process. This paper reveals that the growth of ZnO nanorods using the hydrothermal method has achieved a high power density of 9 μWcm^{-2}.

Keywords: nanogenerator; zinc oxide (ZnO) nanorods; hydrothermal method; higher output power; piezoelectric effect

1. Introduction

Urbanization and rapid industrial growth drive the demand for energy resources for human civilization. Since the beginning of the industrial revolution, fossil fuels such as natural gas, petroleum and coal have been the main sources of energy generation. However, the reliance on the use of fossil fuels, which are non-renewable resources, is appalling as they will be depleted in the future. In addition, a lot of pollution and waste from the effects of the use of fossil fuels [1]. Therefore, generating new energy sources as an alternative to traditional fossil fuels is critical and coveted with the increase in the human population. Nuclear, geothermal, tidal, wind and solar energy are viable alternative sources to replace fossil fuels. In general, renewable energy usage mitigates the adverse effects of fossil fuels on the environment, such as environmental contamination and the greenhouse effect [2]. Energy resource constraints can be reduced with the discovery, strategy and generation of these new energy sources but are still not sufficient to meet the demands of social progress [1,3–7]. Therefore, energy generation from portable microelectronic devices is also an alternative to energy generation. These energy sources have the potential to provide humans with a safer, more reliable and stable, and continuous supply of energy [1].

Energy generation technologies have been extensively established with the introduction of miniaturized devices such as electrodynamic, photovoltaic and thermoelectric effects. Some are based on novel concepts, while others leverage advancements in micro-electro-mechanical Systems (MEMS) and nanotechnology, occasionally paired with sensor technology to provide genuinely self-powered sensors [8]. Energy harvesting (EH) is a technique that collects and generates energy from various sources, including mechanical loads, vibrations, temperature and light gradients, generating relatively low force levels in the nW to mW range [9,10]. Mechanical energy harvesting technology has garnered considerable attention due to the pervasiveness of motion and vibration. Research areas include miniaturization, effective batch-production techniques, and wide frequency ranges [8]. Battery replacement is impractical, especially when sensors are placed in inhospitable or difficult locations. Wearable electronics and sensors for on-demand monitoring may result from substantial advancements in low-power electronics. On the other hand, as the Internet of Things (IoT) evolves, it is anticipated that hundreds of millions of low-power sensors will be used globally, reducing or ultimately removing their reliance on battery power via the EH approach [11,12]. Recently, EH such as piezoelectric, electrostatic, electromagnetic, flexoelectric and triboelectric generators are among the technologies that enable mechanical load and vibration conversion [13–19]. Each nanogenerator (NGs) has its own advantages and limitations. For example, a triboelectric nanogenerator (TENGs) can generate high-output voltage but with a relatively small amount of current, due to its high resistance. In general, the internal resistance of a piezoelectric nanogenerator (PENG) is lower than that in TENGs and thus generates much higher voltage and current output, mostly due to its sensitivity. However, for sensing functions, PENGs are more practical and preferred. However, the main disadvantage of these two NGs is degradation and mechanical damage due to external cyclic mechanical stress. The cyclic mechanical stress in TENGs due to the frictional heat generated between the contact layers causes the thermal structure and mechanical properties to change. To overcome that problem, researchers use self-healing materials in NGs. In addition, triboelectricity also suffers from a durability issue, limited short-circuit output current, structural changes and post-stress state [20,21]. On the other hand, pyroelectric nanogenerators (PyNGs) have a high pyroelectric coefficient and low dielectric loss, but it is difficult to achieve a crystalline structure [22].

The piezoelectric effect, which occurs in noncentrosymmetric crystals, is defined as a linear electromechanical interaction between mechanical and electrical conditions; thus, an electric charge accumulates in response to the applied mechanical stress, as shown in Figure 1. The direct piezoelectric effect can be reversed to generate mechanical strain through the application of an electric charge, thus forming an inverted piezoelectric effect [23–27]. Piezoelectric materials, a subset of ferroelectric materials, display a localized charge separation called an electric dipole because of their noncentrosymmetric structure [14,15,17,18,28,29].

Piezoelectric materials are highly desirable for energy-harvesting applications due to the direct conversion of mechanical vibrations to electricity via piezoelectric effects [14,15,18,23–29]. Generally, piezoelectric materials can be classified into two categories, lead-containing (Pb) and lead-free (Pb-free). Lead zirconate titanate ($PbxZ{-}1{-}xTiO_3$, abbreviated PZT) and different modified forms are the most common lead-containing piezoelectric materials, and they have exceptional piezoelectric characteristics. However, due to the harmful effects of Pb elements on humans and the environment, the material's future application is limited. As a result, it is vital to study high-performance materials that are also environmentally friendly (Pb-free). Numerous piezoelectric materials containing no lead have been studied, including barium titanate ($BaTiO_3$), zinc oxide (ZnO) and polyvinylidene fluoride (PVDF). These materials possess better features in piezoelectricity, structural simplicity, synthesis capabilities, low manufacturing costs and appropriateness for mass production and the application of piezoelectric materials [30–37].

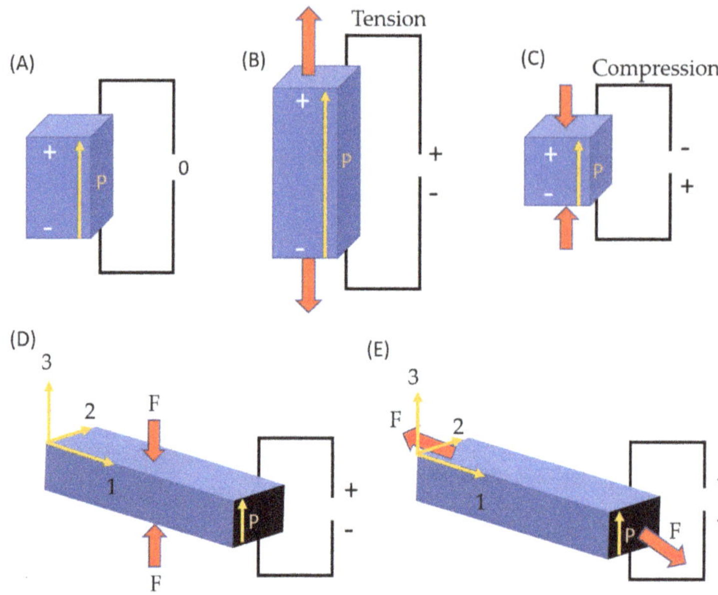

Figure 1. Direct piezoelectric action schematic: (**A**) electrical charge creation in a piezoelectric material in the absence of an external force, (**B**) tension and (**C**) compression. The two most often used operating modes for (**D**) 33 (stock configuration) and (**E**) 31 (bending configuration); the pole (producing the net polarization ion) direction is in the order of "3" for both configurations.

2. Piezoelectric Nanogenerator (PENGs)

In 1880, Pierre and Jacques Curie realized the presence of the piezoelectric effect, which occurs in bulk or nanostructured semiconductor crystals, where the central symmetry is broken under the action of an external force, and thus produces the potential for a piezoelectric generation [38]. The PENG is one of the most promising portable energy-harvester devices that was developed by Z. Wang in 2006 [34]. It is a straightforward device that converts mechanical energy to electrical energy through piezoelectric material, and a schematic diagram is shown in Figure 2a [39]. The principle work of PENGs is based on the piezoelectric effect, which means that the electricity is generated under mechanical stress. As a weak force bends a nanowire, piezoelectric potential is generated at its top and bottom. The method of using nanowires (NWs) to generate electricity under the action of external forces represents the power generation function of materials at the nanometer scale. It provides an experimental and theoretical basis for the design of self-powered nanodevices. In 2007, Z. Wang et al. [40] successfully developed a PENG driven by ultrasonic waves to continuously produces output and work in a standalone fashion, as shown in Figure 2b [41]. It consists of a zigzag electrode, a ZnO NWs, a fixed substrate and an external load. This work lays the foundation for the technical transformation and application of nanogenerators, making it a milestone in the field of nanomaterials [41]. Typically, PENGs comprise a piezoelectric layer, a substrate and two electrodes. PENGs use the piezoelectric effect to capture green energy from ocean waves, wind, biomechanical movements and mechanical vibrations in the environment. The output voltage of this type of nanogenerator is influenced by the mechanical deformation and parameters of the piezoelectric layer. Mechanical vibrations in the environment can cause varying deformations in the piezoelectric nanogenerator that generates the AC-output voltage. Nevertheless, a polymeric encapsulation layer technique is proposed to cover the nanostructure layer, providing physical protection, as well as granting chemical and solution stability to the nanodevice [42]. In general, the PENGs have shown appropriate features due to their structural design, simple performance, simple construction method, high stability and low cost [38].

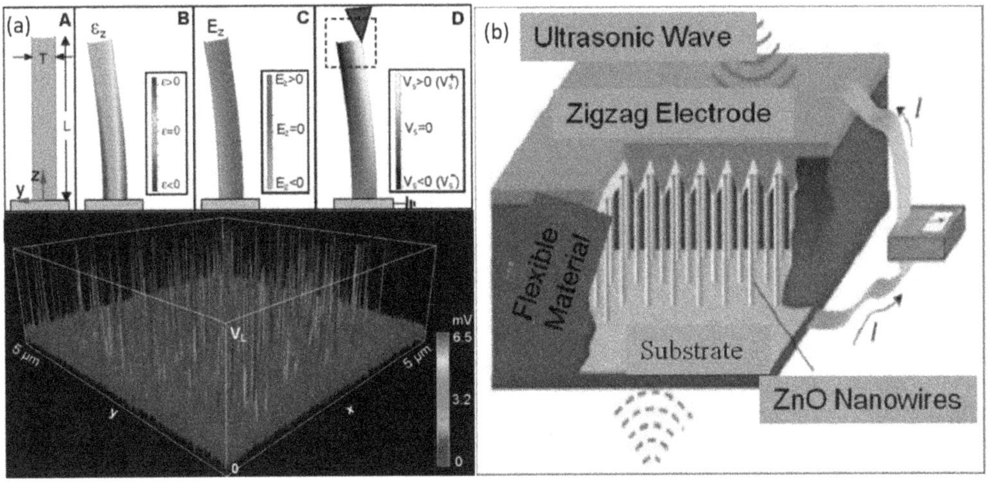

Figure 2. (**a**) Schematic diagram of (**A**) PENG nanostructure under (**B**) longitudinal strain, (**C**) corresponding longitudinal electric field and (**D**) potential distribution and (**b**) the typical structure of PENGs. Reproduced with permission from reference [41].

3. Types of Piezoelectric Material in NGs

3.1. Zinc Oxide (ZnO)

ZnO has been widely used as a piezoelectric nanogenerator due to its excellent structural properties, wherein the lack of a center of symmetry combined with a large electromechanical coupling allows the production of a large piezoelectric response [43]. Kirubaveni Savarimuthu et al. [44] successfully synthesized zinc oxide (ZnO) nanorods (NRs) on a Kapton substrate with a low-temperature hydrothermal method. The ZnO NRs with a silver-electrode-coated insulation layer are purposely to reduce the screening effect and thus increase the output signal strength. Annealing conditions play an important role during the synthesis, wherein the structure of the porous effect can be controlled so as to increase the number of rods that allow a larger contact surface with the counter electrode. This indirectly improves the charge carrier flow mechanism and thus increases the power generated by the device.

Apart from that, Majid S. Al-Ruqeishi et al. [43] fabricated a ZnO NRs-based piezoelectric nanogenerator using an in-tube chemical vapor deposition approach. This technique allows the growth of a high number of ZnO NRs over a large area. It was found that the average length and diameter of ZnO NRs were 3.9 μm and 57 nm, respectively, as shown in Figure 3, and produced as high as 8.97 μWcm^{-2} power density. However, the dimensions of the NRs are non-uniform and rather scattered, resulting in dubious structural quality.

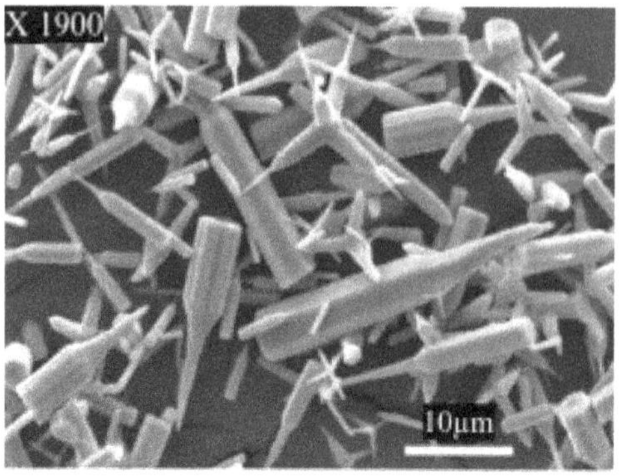

Figure 3. Morphology of ZnO NRs. Reproduced with permission from reference [43].

A polymer coated with a layer of indium tin oxide (ITO) is also widely used as a substrate for ZnO NRs [45]. The polymer substrate offers flexibility in generating energy from the vibration mechanism. In [45], Joe Briscoe et al., synthesis ZnO NRs on /poly(methyl methacrylate) (PMMA) and poly(3,4-ethylenedioxythiophene)poly(styrenesulfonate) (PEDOT:PSS) and found that the internal impedance affects the output power generation, which indicates that the interface state between the substrate and ZnO NRs and the flexibility of the structure cannot be underestimated [45]. However, both ZnO NGs have grown orderly dense ZnO NRs on the polymers, as shown in Figure 4. It is challenging to ensure that ZnO is designed consistently throughout the device manufacturing process, and more ordered structures are required to produce high-performance devices. Compared to other types of ZnO nanostructures, NRs are more promising in the formation of more ordered structures. ZnO NRs can also be produced on various types of substrates with a large surface area. Various deposition techniques have been tried by researchers to create the best arrays of highly oriented ZnO NRs for NG applications [46]. The basic properties of ZnO are shown in Table 1.

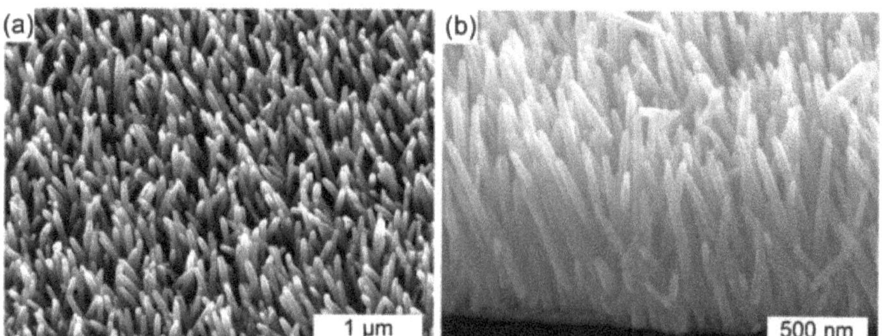

Figure 4. Morphology images of (**a**) surface and (**b**) cross-sectional view of the growth of ZnO NRs. Reproduced with permission from reference [45].

Table 1. Fundamental Properties of ZnO [47].

Properties	Value
Lattice parameters at 300 K	
a_0	0.325 nm
c_0	0.521 nm
a_0/c_0	1.602 (ideal hexagonal structure shows 1.633)
u	0.345
Density	5.606 g cm^{-3}
Stable phase at 300 K	Wurtzite
Melting point	1975 °C
Thermal conductivity	0.6, 1–1.2
Linear expansion coefficient (/C)	a_0: 6.5×10^{-6} c_0: 3.0×10^{-6}
Static dielectric constant	8.656
Refractive index	2.008, 2.029
Energy gap	3.4 eV, direct
Intrinsic carrier concentration	<106 cm^{-3}
Exciton binding energy	60 meV
Electron effective mass	0.24
Electron Hall mobility at 300 K for low n-type conductivity	200 cm^2 $V^{-}1$ s−1
Hole effective mass	0.59
Hole Hall mobility at 300 K for low p-type conductivity	5–50 cm^2 V^{-1} s^{-1}

3.2. *Barium Titanate ($BaTiO_3$)*

On the other hands, Aneesh Koka et al. [48] developed a mechanical nanogenerator system utilizing vertically aligned $BaTiO_3$ nanowire arrays (NWs) as the main semiconducting material. The $BaTiO_3$ NWs arrays were synthesized on fluoride-doped tin oxide (FTO) glass by a two-step hydrothermal method resulting in a well-structured form, as shown in Figure 5. The aligned nanowire arrays have a higher strain than the bulk layer and thereby improve the power conversion output. It is also found that the vertically aligned $BaTiO_3$ NGs produce 16 times greater power density compared to the bulk ZnO-based NGs [48]. This shows that the nanorod structure significantly improves the nanogenerator performance, not limited to ZnO materials but also for other semiconductor materials.

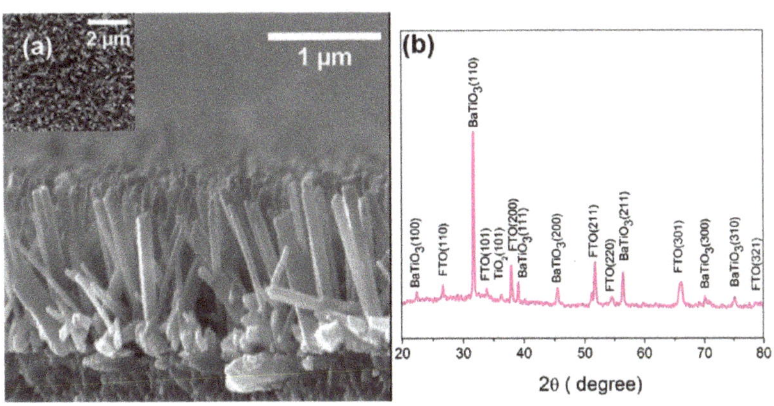

Figure 5. (**a**) $BaTiO_3$ NWs arrays in top view. (**b**) $BaTiO_3$ constitutes the majority of the peaks in the X-ray diffraction spectra of $BaTiO_3$ NW arrays. Reproduced with permission from reference [48].

3.3. *Lead Zirconate Titanate (PZT)*

Ceramics are also used as the main piezoelectric material for NG applications. Weiwei Wu et al. [49] used a robust lead zirconate titanate (PZT) layer on polyethylene terephthalate (PET), which is suitable for flexible and wearable applications, as shown in Figure 6. The developed PZT film is randomly oriented, shrinks and crystallizes after the calcination process, which leads to a fully functional nanogenerator. Compared to ZnO nanowire-based nanogenerators, the PZT-based device has been shown to exceed the output power density by threefold, indicating the suitability of the PZT-based nanogenerator.

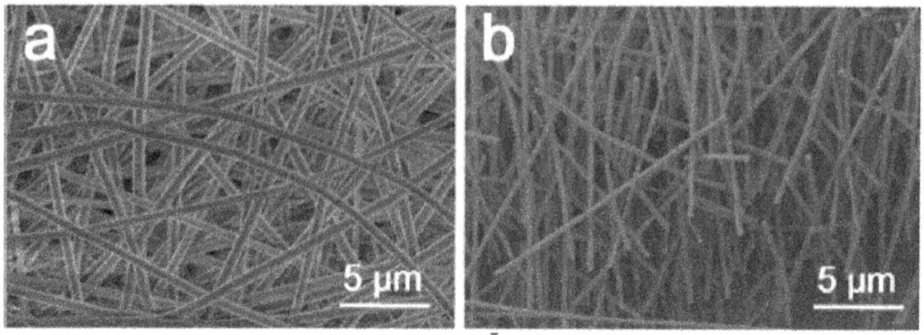

Figure 6. Morphology of PZT NWs (**a**,**b**). Reproduced with permission from reference [49].

3.4. Indium Nitride (InN)

In 2017, Nai-Jen Ku et al. [50] introduced indium nitride (InN) NWs as the nanogenerator material for a mechanical energy harvester. Obliquely aligned InN NWs have successfully produced adequate output power, exceeding that achieved by ZnO NWs counterparts. However, indium nitride (InN) nanowires are grown on silicon substrates and have not yet been demonstrated on flexible substrates. Recently, first-principles calculations have reported that 2-D InN has high out-of-plane vertical polarization between interlayers in InN, resulting in piezoelectric polarization features. This proves the potential of 2-D InN material to be used as a high-performance nanogenerator [51].

3.5. Other Piezoelectric Materials

In recent years, great efforts have been involved in developing nanoscale energy harvesters or NGs and led to the emergence of many new piezoelectric materials. In 2021, Siju Mishra et al. [52] reported the generation of 2-D zinc sulfide (ZnS) nanosheets for the first time using a simple hydrothermal step. ZnS has high piezoelectric material characteristics, as well as being bio-compatible and non-toxic [53]. In addition, nanocomposite piezoelectric nanogenerator fibers consisting of a combination of poly(vinylidene fluoride) (PVDF) with other materials are also promising [54–57]. A number of PVDF composites have successfully generated large output power and have the potential to be used as nanogenerators, thereby replacing the role of batteries in generating energy. Apart from that, lead-free (K, Na)NbO_3 has also been introduced in energy-harvesting nanodevices [42]. The KNN nanorods were vertically grown in a (100)-orientation on a $SrTiO_3$ substrate using a hydrothermal technique at low-temperature conditions, which allows the possibility of a ferroelectric nanorods-based device.

Table 2 shows a summary of NGs with different piezoelectric materials, growth conditions, output power and others. Compared to other materials, ZnO NRs generated a large power density as high as 9 μWcm^{-2}, as tabulated in Table 2. Based on the findings, ZnO NRs produced high power density with the formation of an organized structure and a large surface area by performing an optimization process. In addition, various deposition techniques have been reported to form highly oriented ZnO NR arrays for PENG applications.

Table 2. A summary of various piezoelectric materials based on nanogenerators.

No	Material	Substrate	Method	Geometry D: Diameter L: Length	Aspect Ratio	Output Voltage [V]	Power Density [Wcm^{-2}]	Ref	Year
1	ZnO NRs	Kapton	Hydrothermal	D: 64 nm L: 1.7 μm	27	400 m	9 μ	[44]	2017
2	ZnO NRs	Quartz tube	Chemical vapor deposition	D: 57 nm, L: 3.9 μm	68	0.74	8.97 μ	[43]	2019
3	$BaTiO_3$ NWs	FTO Glass	Hydrothermal	D: 90 nm, L: 1 μm	11	311.5 m	6.27 μ	[48]	2014
4	ZnO NRs/PEDOT:PSS	PET	Thermal evaporation	D: 64 nm, L: 1 μm	16	90 m	5.2 μ	[45]	2013
5	ZnO NRs	Sapphire	Vapor liquid solid	D: 100 nm, L: 1 μm	10	42.5 m	0.4 μ	[34]	2014
6	ZnO NRs/PMMA	PET	Spin coating	D: 64 nm, L: 1.2 μm	19	252 m	0.02 μ	[45]	2013
7	ZnO NRs	Electrospun PVDF fibers	Hydrothermal	D: 30 nm, L: 183 nm	6	79.95 m	0.91 n	[58]	2018
8	ZnO NRs	Terylene-fabric	Hydrothermal	D: 200 nm, L: 5 μm	25	10 m		[59]	2014
9	$BaTiO_3$ nanotube	Titanium	Hydrothermal	D: 144 nm, L: 6 μm	42	6.3 m		[60]	2020
10	ZnO NRs	Cotton Fabric	Aqueous chemical growth	D: 200 nm		9.5 m		[61]	2012
11	PZT nanofibers	Silicon	Electrospinnig process	D: 100 nm, L: 85 μm		0.4 m		[62]	2009
12	Indium nitride nanowire	Silicon		D: 43 nm		825 μ	2.9 n	[50]	2017
13	Indium nitride					0.195	73 m	[51]	2021
14	Zinc sulfide nanosheets	flexible aluminum	Hydrothermal			600 m	219.5 n	[52]	2021
15	Zinc sulfide nanofibers		Hydrothermal	D: 600 nm		3		[53]	2020

4. Effect of Dopant on ZnO Nanorods (NRs) Piezoelectric Nanogenerator (PENGs)

The ZnO PENGs, PZT and $BaTiO_3$ have distinct characteristics. The performance of PENGs can be enhanced by altering the micromorphology of the piezoelectric material. However, the improvement is very marginal. Researchers have introduced chemical techniques to increase the piezoelectric coefficients in piezoelectric materials, thereby improving the piezoelectric properties of the material and the performance of the piezoelectric nanogenerators [41,63]. The piezoelectric coefficients and dielectric constants of the piezoelectric material can be altered by the addition of a small amount of doping into the piezoelectric main material, resulting in a more productive energy generation process. For instance, ZnO, which is a high-performance semiconductor material, can be adapted with two distinct doping strategies: N-type doping and P-type doping. Doping with n-type elements provides an increase in the crystalline lattice tension along the ZnO crystal axis and increases the piezoelectric coefficient, thus improving the output performance of the PENGs. However, these dopant ions have the feature of being overly permeable and produce excessively high doping concentrations. As a result, more lattice defects can arise in this circumstance, obstructing charging on external circuits and reducing energy generation performance. The effects of free electron protection caused by doping in ZnO can be minimized by the use of the p-type dopant [41].

4.1. Silver-Doped (Ag-Doped) ZnO Nanorods (NRs)

Sumera Rafique et al. [64] developed a PENG based on Ag-doped ZnO NRs on cotton fabrics, as shown in Figure 7. In general, the potential is generated due to the piezoelectric action originated from the relative dislocation of Zn^{+2} cations and O^{-2} anions in the wurtzite crystal structure. Both undoped and Ag-doped ZnO nanorods-based nanogenerators exhibit good arrangement synchronization, implying that the maximum number of NRs generates piezoelectric potential simultaneously and in the same direction. However, it was found that the Ag-doped device produced three times the power generation compared to the undoped ZnO NRS-based PENGs [64]. This is possibly due to the reduction of free charge concentration by the Ag-doped passivation. The grown Ag-doped ZnO NRs are dense, oriented and have formed an appropriate atomic diameter of Ag-doped ZnO NRs, as shown in Figure 8, which allow the formation of excellent interfaces with the hydroxyl groups of cotton fabrics.

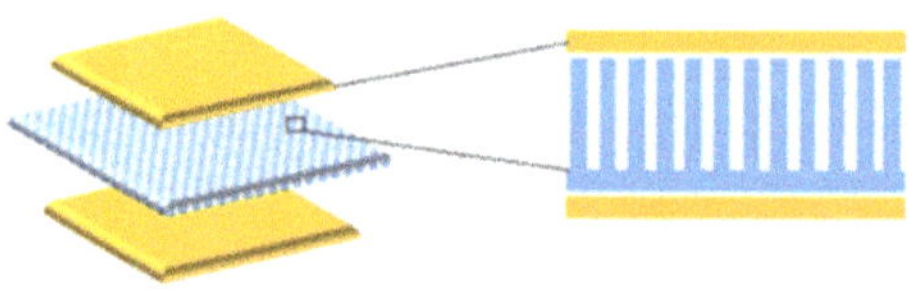

Figure 7. Schematic diagram of a piezoelectric nanogenerator based on Ag-doped ZnO NRs. Reproduced with permission from reference [64].

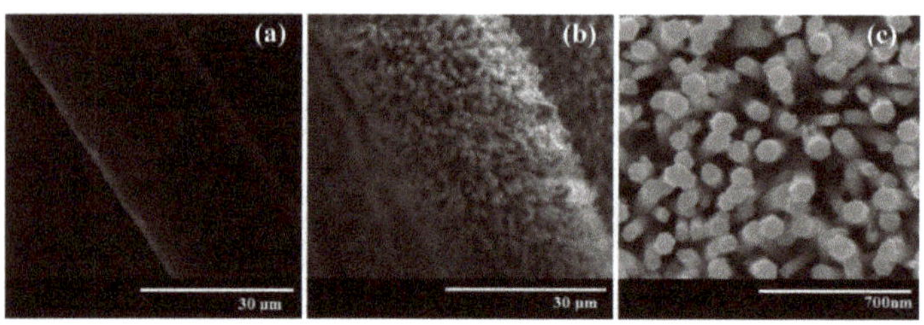

Figure 8. (**a**) Observation of a cotton fiber without ZnO development, (**b**) picture demonstrating the presence of Ag-doped ZnO NRs on a cotton fiber, (**c**) better magnification image of Ag-doped ZnO NRs on a cotton fiber. Reproduced with permission from reference [64].

Furthermore, Ag doping assists in reducing the growth of nucleation activation energy in ZnO NRs and also results in an increase in the NR diameter [64]. The addition of the Ag dopant also keeps the growth process at a lower temperature, allowing the fabrication of devices on flexible substrates.

4.2. Aluminium Doped (Al-Doped) ZnO Nanorods (NRs)

Wen Yang Chang et al. [65] developed a PENG based on Al-doped ZnO (AZO) nanorods with a V-zigzag layer, as depicted in Figure 9. The AZO layer is grown at a low temperature on ITO glass, using an aqueous solution approach. The V-zigzag structure improves the bending and compression deformation characteristics of the NGs. Interestingly, the AZO NGs successfully generate output power at an ultra-low temperature, near to liquid helium state. This proves that the AZO NGs are reliable and can be used as an energy harvester even under extreme conditions.

In addition, the concentration of the Al dopant in ZnO NRs also affects device performance [66]. Aqueous solutions provide an excellent medium for the process of dopant concentration optimization. This process is promising and has attracted much attention due to being able to produce controlled, highly compact and highly oriented ZnO nanostructures [66,67].

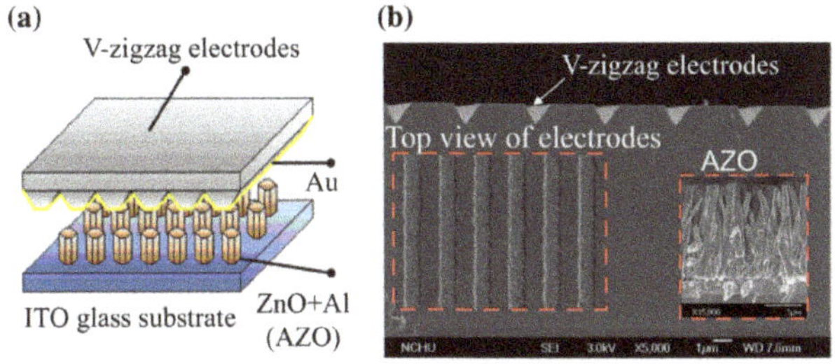

Figure 9. Mechanism energy harvesting based on V-zigzag electrodes: (**a**) AZO and V-zigzag structure, (**b**) SEM pictures of AZO NRs. Reproduced with permission from reference [65].

4.3. Nickel Doped (Ni-Doped) ZnO Nanorods (NRs)

Nickel (Ni) has been widely used in semiconductor materials as a dopant for various applications in improving charge transfer performance and also in passivating defects in the host materials. There are many similarities between the valence of Zn^{2+} and Ni^{2+}, allowing the vacancy defects of Zn^{2+} to be easily passivated by Ni^{2+}. This is also due to the ionic radii for Zn^{2+} and Ni^{2+} being almost similar, 0.074 nm and 0.069 nm, respectively, allowing the transport and separation of charges to be enhanced [68]. In 2020, Yen-Lin Chu et al. [68] introduced the Ni-doped ZnO NRs NGs that were grown by chemical bath deposition on the ITO substrate. Compared to the pure ZnO NRs NGs, the Ni-doped device shows a dense nanorod structure as shown in Figure 10. In addition, the output power is also enhanced by three-fold than that obtained for the pure ZnO NRs NGs. Ni-doped ZnO has also been used as a nanocomposite filler and has improved the piezoelectric characteristics of NGs [69].

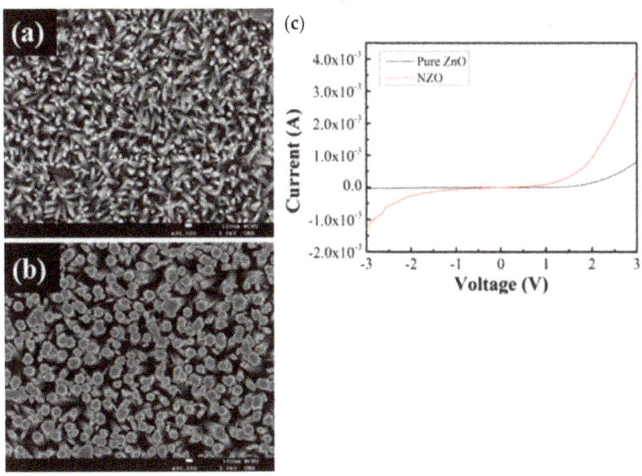

Figure 10. Shows the top view of pure ZnO NRs and Ni-doped ZnO NRs structure (**a**,**b**) using FESEM and the IV characteristic of Pure ZnO and Ni-doped ZnO NRs (**c**). Reproduced with permission from reference [68].

4.4. Rare Earth Material ZnO Nanorods (NRs)

The rare-earth-doped ZnO NRs NGs significantly improved the piezoelectric properties and response. Recently, several researchers adopted the idea and reported promising results. Nd-doped [70], La-doped [71], Tb-doped [72] and Ce-doped [73] ZnO nanorod nanogenerators have shown very encouraging findings. It was documented that the rare-earth-doped ZnO nanorod nanogenerator devices increased the output power by at least three times compared to pure ZnO NRs. This clearly shows the upright effect obtained by adding rare earth ions into the ZnO NRs in increasing the performance of generated energy.

4.5. Other Dopants in ZnO Nanorods (NRs)

Researchers also devoted a great deal of effort in other dopant material in ZnO NR NGs to improve the output power generation. The chemical modification of the ZnO NRs has proven to be a successful strategy to increase energy-harvester performance. Various research groups have reported vanadium (V5+)- [74], barium (Ba)- [75], gallium (Ga)- [76], Cr- [77], and S [78]-doped ZnO NRs NGs. Overall, the results show a positive effect provided by all dopants. Modifications to the bandgap, ZnO matrix and passivation defects are believed to occur with the presence of the dopant. Among the dopants, Ni and La elements have shown remarkable power density output compared to others when incorporated with ZnO material as nanogenerator. This is most probably due to the appropriate ionic radii diameter with Zn^{2+}. It is well known that the Zn^{2+} vacancies are the main site defects that contribute the charge carrier deficiency and thus degrade the device performance. Thus, incorporating dopants with almost similar ionic radii diameters will effectively replace or passivate the Zn^{2+} vacancies, thereby increasing the power density generation.

Table 3 shows the summary of all dopants in ZnO NRs NGs. Based on the findings, Ni-doping is promising and has achieved high output power up to 735 mWcm^{-2}. This is possibly due to the highly magnetic behavior of nickel [79] and also the higher surface area covered by Ni-doping over the ZnO nanorods layer [80].

Table 3. A summary of various dopants in ZnO nanogenerators.

No	Material	Substrate	Method	Geometry D: Diameter L: Length	Aspect Ratio	Output Voltage [V]	Power Density [Wcm^{-2}]	Ref	Year
1	Ni-Doped ZnO NRs	Glass	Chemical bath deposition	L: 1.72 μm		0.07	735 m	[68]	2020
2	Ag-doped ZnO NRs	cotton fabric	Hydrothermal	D: 86 nm		6.85	1.45 m	[64]	2019
3	Al-doped ZnO NRs	Glass	Chemical aqueous solution	L: 1.81 μm		1.35 m	1.026 n	[65]	2015
4	Al-doped ZnO NRs	Glass	Chemical solution	D: 70 nm, L: 1.75 μm	25	60 m	0.84 n	[81]	2010
5	Ag-doped ZnO NRs	PET	Hydrothermal	D: 110 nm, L: 2.02 μm	18	5.2 m		[82]	2017
6	PVDF-HFP/Ni-doped ZnO nanocomposites		Hydrothermal			1.2		[69]	2017
7	Neodymium (Nd)-doped ZnO NRs	PET	Wet chemical coprecipitation	D: 101 nm L: 412 nm	4	31		[70]	2018
8	La-doped ZnO		Wet chemical	D: 60 nm		1.6	50 m	[71]	2019
9	Tb-doped ZnO nanotapers		Wet chemical coprecipitation			2.3		[72]	2020
10	Ce-doped ZnO nanoparticles		Wet chemical solution	D: 18 nm L: 100 nm	6	2.5		[73]	2014
11	Vanadium-doped ZnO nanosheet	PET	Cost-effective seed-assisted solution	D: 100 nm L: 1.5 μm	15			[74]	2013
12	Ba-doped ZnO NRs		Wet chemical coprecipitation	D: 73.25 nm L: 476.51 nm	7	10.5		[75]	2021
13	Ga-doped ZnO NRs	PES	An aqueous solution	D: 75 nm				[76]	2012
14	Cr-doped ZnO NRs		Wet chemical solution	D: 300 nm		8 m		[77]	2014
15	S-doped ZnO NRs		Hydrothermal	L: 2 μm		150 m	24 n	[78]	2022

5. Synthesis Techniques of Growth Process ZnO

Over few past decades, ZnO NRs have attracted a lot of attention among researchers for energy-harvester materials, owing to its excellent physical properties [83]. Properties, such as piezoelectric features, high thermal conductivity, high catalytic capabilities, UV filtering capability, semiconductor and antifungal and antibacterial effects have marked ZnO as a well-established material in the electronic, cosmetic and medical fields alike [84]. The synthesis of ZnO nanostructures can be approached by a variety of different methods, resulting in a wide range of different nano-structures. The piezoelectric properties of ZnO NRs are greatly influenced by the synthesis technique used in growing ZnO NRs. The surface morphology of ZnO NRs [85,86], structure orientation [87], crystallinity [88], and majority charge carriers of ZnO nanorods [89,90] can be highly tuned by choosing an accurate synthesis method. In this section, the synthesis methods that have formed the finest ZnO NRs for PENGs applications are discussed, among which are chemical vapor deposition (CVD) and hydrothermal and electrochemical deposition.

5.1. Chemical Vapor Deposition (CVD) Technique

Chemical vapor deposition is a common technique to grow ZnO NRs for large-scale production. ZnO NRs are grown using zinc powder and zinc oxide powder precursors in a horizontal tube furnace, as illustrated in Figure 11. Typically, the synthesis temperature is in a range between 450 °C to 900 °C for zinc powder [91,92], while 1200 °C and above for zinc oxide powder [93]. Both precursors release zinc vapor phase, which is then adsorbed on the surface of the heated substrate and simultaneously reacts with oxygen for the growth of ZnO NRs. Factors that affect the morphology and piezoelectric characteristics of ZnO NRs developed with this technique include the type of substrate used [94], [95], vacuum pressure [96], type of carrier gas [97], and synthesis temperature [98]. In this way, ZnO nanorods are grown vertically with a uniform geometry, thus allowing more electrical contact with the electrode. Vertically aligned ZnO NRs are well grown on substrates with minimal lattice mismatch with ZnO, such as α-Al_2O_3, TiN epilayer, GaN/Al_2O_3, graphene/Al_2O_3 and 4H-SiC epilayer. Typically, ZnO NRs produce high crystal quality by using the CVD technique due to relatively high synthesis temperatures [98]. However, the CVD method is a vacuum-based technique that incurs high operating costs and also requires a high synthesis temperature reaching over 450 °C, which is not suitable for polymer substrates.

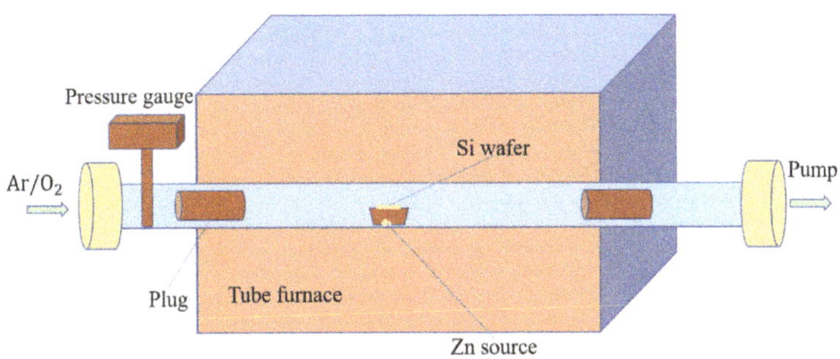

Figure 11. Schematic diagram of CVD process for ZnO NR growth in a horizontal tube furnace.

5.2. Hydrothermal Technique

Recently, researchers have focused on synthesis techniques via a wet chemical path in forming the ZnO NRs. Generally, the ZnO NRs can be grown on a flexible polymer substrate, which requires a relatively low synthesis temperature, i.e., below 100 °C. The hydrothermal technique is the most widely used for ZnO NRs synthesis method and has demonstrated a highly oriented ZnO NRs [99]. A large number of ZnO NRs have been successfully grown on flexible substrates for NG application such as polyethylene terephthalate (PET) [100], polycarbonate (PC) [101–103], carbon fiber [104], poly(methyl methacrylate) (PMMA) [105], textile fabric [106–108], Ti foil [109], graphene-coated polyethylene terephthalate [110] and polyethylene naphthalate (PEN) [111]. There are two steps involved in this synthesisL seeding on the substrate, followed by the growth of ZnO NRs on the seed layer, as shown in Figure 12. A seed layer of ZnO is pre-deposited on the substrate in order to facilitate the growth process of ZnO NRs on the substrate. In addition, the ZnO seed layer is one of the factors that significantly affect the morphology of ZnO NRs. Therefore, it is one of the crucial factors that must be considered in order to produce the best ZnO NRs. There are several methods that can be used to deposit the ZnO seed layer, such as radio frequency (RF) sputtering [63,101,112–115], spin coating [116], electro-spraying [117] and the atomic layer deposition method [118]. After the seed layer is deposited, the substrate is placed in an autoclave filled with a zinc precursor solution. The pH of the zinc precursor solution can be adjusted with sodium hydroxide and/or hydrochloric acid, since the optimal pH for the growth of ZnO NRs on seeded substrates is between 7 and 10 [119,120]. Thus, for the ZnO NRs growth process, the ZnO NRs synthesis process is started first by dissolving zinc nitrate hexahydrate into distilled water. Generally, the growth temperature of ZnO by using hydrothermal method is between 85 to 100 °C, which is suitable for polymer substrates, and the duration of the growth process of ZnO NRs varied from 3 to 12 h.

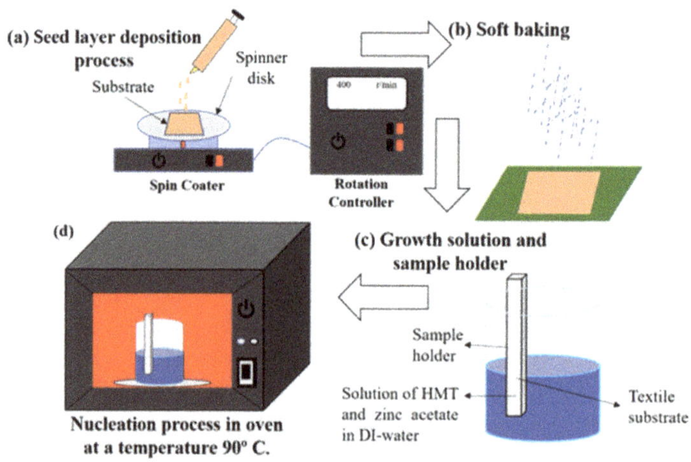

Figure 12. Schematic diagram of hydrothermal process for ZnO NRs.

Tao et al. [121] found that seed layers prepared under different oxygen partial pressure sputtering parameters and annealing treatments had a great influence on the morphology of ZnO NR arrays. It also stated that the average length and diameter of ZnO NRs grew as the pH solution increased, and the optimal pH range for well-aligned ZnO NRs growth was between 10 to 10.4 for a 40 nm-thick zinc seed layer [122]. In conclusion, the alignment of ZnO NR synthesis through the hydrothermal technique is highly reliant on the preferred crystal orientation of the ZnO seed layer. The hydrothermal technique has attracted the attention of many researchers due to its numerous advantages. It is a simple procedure that generates little heat and yields high results at a low cost. Furthermore, it is a fast and controlled technique that can produce structures that are good in terms of shape and clarity [123].

5.3. Electrochemical Deposition Technique (ECD)

Electrochemical deposition (ECD) is a surface modification process in which a thin and strongly adherent coating of metal, oxide, salt or macromolecules is deposited onto a conducting substrate by the simple electrolysis of a solution containing the desired substance. Besides ECD, this technique is also known as electro-deposition, electroplating or electrolytic deposition [124]. ECD has several advantages over other deposition techniques, including its simplicity, low cost, low temperature, high deposition rate and suitability for substrates with a large surface area. This method is therefore one of the most effective methods for growing ZnO thin films. A thin layer of ZnO can be made on any conductive substrate, including transparent conducting oxide or any other metal plate through this ECD technique, wherein merely low cathode voltage or current is required [125]. In addition, ECD is similar to hydrothermal deposition, comprising a two-step process and has been widely used in the fabrication of ZnO-based mechanical energy-harvesting devices [126,127].

The schematic flow process of ECD ZnO NRs is depicted in Figure 13. ZnO NRs are grown on top of a ZnO seed layer placed on the conductive substrate (working electrode) as the seed layer to provide a nucleation site for the growth of ZnO NRs [127]. The ZnO seed layer is pre-deposited on an ITO substrate by using RF magnetron sputtering in an inert gas environment. Next, a double junction of Ag/AgCl (reference electrode) and platinum (Pt) foil (counter electrode) is linked to the cathode through an external circuit. Then, both electrodes are immersed in the zinc precursor solution, while magnetic stirring and heating are continuously maintained. The electrochemical process occurs when an external voltage is applied between two electrodes in order to speed up the driving force of the deposition. The electrochemical process begins when the ions mix with ions in the solution, then diffuses to the working electrode, thus resulting in the deposition of ZnO on the ZnO seed layer substrate. Afterward, ZnO NRs are produced by dehydrating $Zn(OH)_2$ units at a temperature between 75 to 80 °C [127].

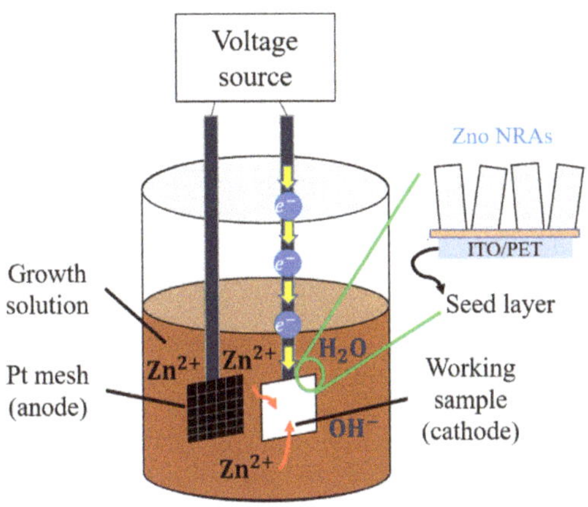

Figure 13. Schematic illustration of ECD.

In this technique, the diameter and length of ZnO NRs can be controlled by the growth time, concentration of the precursor and the external applied voltage. Although the ECD technique has many advantages over other deposition techniques, it is difficult to produce good quality crystals of ZnO NRs due to its low synthesis temperature. Moreover, previous results showed that, with the application of the ECD technique, the ZnO NRs are randomly grown, which causes poor rod termination and poor electrical contact at the counter electrode. Furthermore, the array of ZnO NRs grown on the working electrode has a larger diameter distribution and lower charge carrier density compared to the hydrothermal technique [128]. Therefore, these disadvantages limit its use for PENG fabrication.

6. Recent Technology in Piezoelectric Nanogenerators (PENG)

In recent years, researchers have conceived four methods for improving the performance of ZnO-based PENG output; (i) doping, (ii) increasing the area density of ZnO NRs,

(iii) interface modification and (iv) combination methods. All of these methods have their own characteristics wherein they can compete and complement each other in improving output performance [129].

6.1. Doping

The P-type doping of ZnO NRs is a method to improve the performance of ZnO-based PENGs. It is also a simple approach to increase the piezoelectric output, because p-type doping can reduce the filtering effect in ZnO NRs by replacing the majority of the carriers (electrons) with holes, thus causing the excessive contraction of free electrons. There are several p-type dopants that can be used, such as silver (Ag) [130], lanthanum (La) [131], lithium (Li) [132] and antimony (Sb) [133]. L. Kang et al., investigated the effect of doping on piezoelectric performance using La-doped and undoped ZnO NWs. The study found that, when 5 mol% La-doped ZnO NWs was added, the output voltage increased from 2.1 V to 3.0 V. These results proved that the P-type doping method boosts piezoelectric performance [131]. In addition, the piezoelectric performance of p-type doping has also been proven in [132]. Y. Chang et al. [132] have successfully grown p-type ZnO NWs on Si substrates and claimed that p-type doped ZnO NWs are among the best candidates for nano-generating devices. This is based on an analysis that was carried out on the piezoelectric performance between p-type doped ZnO NWs and pure ZnO NWs through the piezoelectric response. The results show that P-type doped ZnO NWs produce high mobility and effective hole carrier concentration, thus leading to higher piezoelectric output current, power and voltage than pure ZnO NWs.

6.2. Areal Density

As is well known, doping is one of the feasible strategies to overcome the piezoelectric quality of ZnO nanorods. However, the output performance of PENG devices can be further improved by incorporating ZnO NR arrays into complex structures with higher effective NR area density. DM Shin et al., fabricated a two-sided heterostructure of free-standing ZnO NRs/graphene/ZnO NRs. They claim that the double-sided heterostructure produces a coupling of the piezoelectric effect from both the upward and downward growing nanorods, thus resulting in twice the total output voltage and different current density compared to the single heterostructure [119]. M. R. Hassan et al., formed a heterojunction with a Si micropillar (MP) array as the enhancement process in order to enhance the performance of piezoelectric nanogenerators [118]. It is stated that the length of the SiMP array influences the growth and crystalline quality of ZnO NRs and the piezoelectric output voltage. As the length of the SiMP array increases from 0 to 20 mm, the output voltage also increases from 0.7 to 4.0 V. Furthermore, this enhancement process has also been applied to flexible polyester (PS) for both sides [134] and stainless steel (SS) substrates [135]. This improvement process is explained based on the ZnO NRs series connection, which is considered a source of piezoelectric potential.

6.3. Interfacial Modification

Interfacial modifications have been adopted to increase the piezoelectric output power by addressing the main problem of internal and external filtering effects. In this technique, a p–n junction or Schottky contact between ZnO and metal or metal oxide is built, respectively. The interfacial modification technique improves the output performance through two mechanisms. Firstly, the built-in electric field near the p–n junction efficiently consumes the excess electrons in ZnO, thereby reducing the internal screening effect for generated piezoelectric potentials. Secondly, the output performance can be improved through the barrier interface between the Schottky metal and ZnO, as this interfacial modification reduces the carrier concentration and leakage current [136]. Yin et al. [136] formed a (NiO/ZnO) heterojunction at the interface instead of Schottky barrier in order to enhance the piezoelectric performance. The design based on the p–n junction increased the output voltage to 430 mV, which was 21 times higher while increasing the output current density to 40 nA, which was 13 times higher than the ZnO film nanogenerator.

6.4. Combination Method

The combination method is another method that can be implemented to achieve a good performance of ZnO PENG. The combination method is a method that combines the doping method and the interfacial modification method [63,101]. Liu et al. [63] have combined the doping and the interfacial modification method to synthesize a CI-doped ZnO/CuO PENGs device. The result of the study, as illustrated in Figure 14, shows that, by using this combination method, the output voltage and current have increased by 2.2 V and 1000 nAcm^{-2}. This enhancement is caused by the Cl dopant-induced lattice strain along the c-axis of ZnO and the reduction of free electrons at the p–n ZnO/CuO interface. Additionally, tuning the lattice strain along the ZnO c-axis from a compressive to a tensile state can enhance piezoelectric performance. Due to the difference in ionic size between dopants and oxygen elements, this tuning process can be done by substituting halogen dopants from fluorine with other halogen elements like F, I and Br [101].

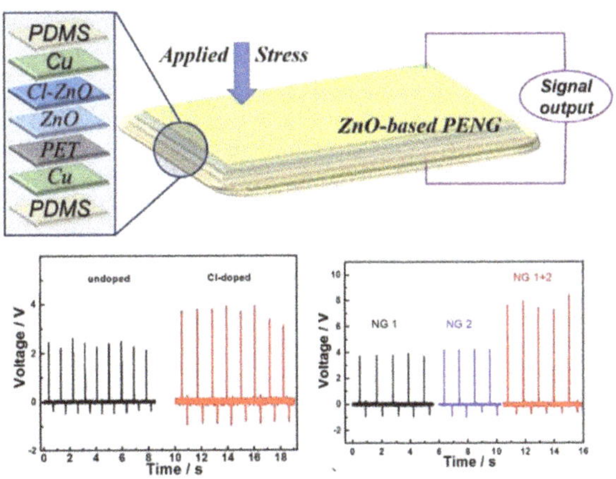

Figure 14. Schematic of Cl-doped ZnO PENGs and its output voltage and the current of the undoped and Cl doped ZnO. Reproduced with permission from reference [63].

7. Discussion

Herein, the progress of PENGs, focusing on piezoelectric semiconductor materials, has been discussed in detail. Based on the description discussed, it has been proven that ZnO NRs have shown good performance compared to other piezoelectric materials. The fact that ZnO NRs can be grown in an orderly and uniform structure. In addition, ZnO NRs have a large surface area allowing the motion to escalate, thus generating more energy. Another piezoelectric material discussed in this paper is lead zirconate titanate (PZT), which has a higher piezoelectric coefficient than other piezoelectric materials, such as barium titanate and ZnO nanorods. However, the PZT nanogenerator is less efficient than the ZnO NGs due to the high piezoelectric coefficient of PZT, causing energy conversion to be less efficient. Furthermore, PZT is not suitable for use at high temperatures, which results in instability and rigidity. The piezoelectric material barium titanate ($BaTiO_3$) is also promising, with further structural design significantly increasing the piezoelectric response through domain engineering. However, this material has a low piezoelectric coefficient, and no satisfactory application has been found for related piezoelectric devices. Various synthesis techniques have been proposed and applied in producing highly oriented ZnO NRs arrays. The hydrothermal method has proven to be the best route in forming ZnO NRs. It is a simple procedure that requires low temperatures but produces satisfactory results at a reasonable cost. In addition, it is a rapid and controlled technique that results in a crystal-clear structure with excellent shape and clarity. In short, the alignment of ZnO NRs produced by the hydrothermal technique strongly relies on the crystal orientation of the ZnO seed layer used. A hydrothermal technique was established and can produce ZnO NRs on a large scale in a cost-effective manner. In addition, ZnO NRs have a relatively wide bandgap of 3.3 eV, which is an advantage to ZnO in allowing more energy to be generated. Based on these findings, ZnO possesses exceptional semiconducting and piezoelectric capabilities, relative abundance, low cost, chemical stability in air, bio separation, complexity and several crystal growth techniques. On the other hand, several synthesis techniques have been discussed in this paper. Figure 15 shows the correlation of output power with the aspect ratio based on previously reported synthesis techniques, including hydrothermal, chemical vapor deposition (CVD), thermal evaporation, vapor liquid solid (VLS) and spin coating techniques. The aspect ratio can be controlled by changing the molar concentration of the growth solution and the growth duration. Among the best device performance was obtained through the hydrothermal method, resulting in a high power density output for a piezoelectric nanogenerator. Furthermore, the aspect ratio related to the length and diameter of ZnO NRs is also an important factor in obtaining high performance. Thinner diameters and longer lengths with good alignment tend to bend more easily under external pressure, resulting in efficient piezoelectric generation.

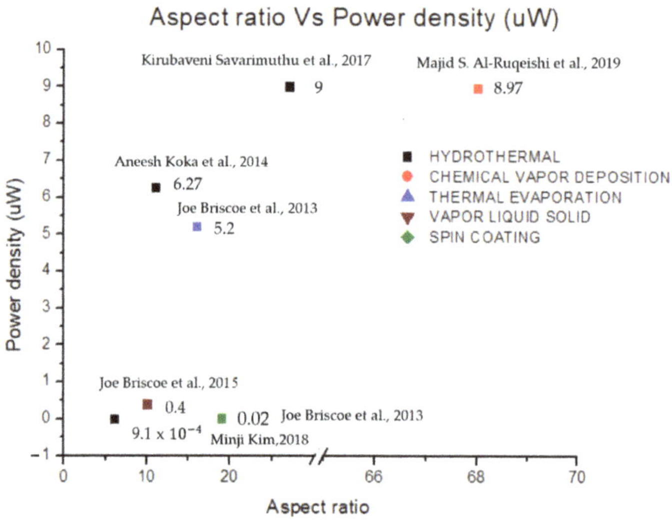

Figure 15. Power density vs. aspect ratio based on several techniques [34,43–45,48,58].

8. Conclusions

The PENGs show high potential in the evolution of the NGs industry. However, continuous studies and modifications are needed to achieve the high performance of PENGs in order to commercialize and compete with other PENGs. Utilizing various micro-morphologies of piezoelectric materials and developing composite thin-film materials are the key strategies for enhancing the performance of piezoelectric nanogenerators. In this paper, several types of piezoelectric materials were discussed, including ZnO NRs, $BaTiO_3$ and PZT. The ZnO NRs advantages showed good performance due to the wide bandgap semiconductor energy of 3.3 eV, high exciton binding energy (60 meV) and low thermal energy at room temperature (25 meV), which allows stable exciton existence at room temperature. These distinctive properties make ZnO suitable for a number of prospective applications, such as transparent electrodes in optoelectronic and solar cell devices, laser diodes and light-emitting diodes, etc. Apart from that, doping the ZnO NRs also enhanced performance. Based on this research, Ni-doped ZnO NRs are shown to achieve higher output power compared to other dopings, as well as leading to a higher surface area. The exceptional chemical stability of nickel (Ni) on zinc (Zn) sites qualifies nickel (Ni) as an effective doping element in ZnO for improving its different properties. Besides, several approaches for the synthesis techniques of ZnO nanostructures have been discussed, such as hydrothermal technique, CVD and ECD. In the finding, the hydrothermal technique has attracted considerable interest due to its numerous advantages compared to other techniques. It is a straightforward procedure under a low-temperature condition and results in a high yield at a reasonable cost. In addition, it is a quick and controlled technique that results in a crystal-clear structure with excellent shape and clarity.

Author Contributions: Conceptualization, M.A.M., F.A. and F.A.A.; methodology, M.A.M., F.A., F.A.A., N.S.N., M.S.I.M.Z. and A.N.M.M.; fabricate, M.A.M., F.A. and F.A.A.; validation, M.A.M., F.A., M.M.I. and F.A.A.; investigation, M.A.M., F.A. and F.A.A.; resources, K.S.L. and F.A.; data curation, M.A.M., F.A. and F.A.A.; writing—original draft preparation, M.A.M. and F.A.; writing—review and editing, M.A.M., F.A., F.A.A., N.S.N., M.S.I.M.Z. and A.N.M.M.; visualization, M.M.I.; supervision, F.A. and F.A.A.; project administration, F.A.; funding acquisition, F.A., F.A.A., M.M.I. and A.N.M.M. All authors have read and agreed to the published version of the manuscript.

Funding: This research was funded by Universiti Teknikal Malaysia Melaka, Malaysia under grant PJP/2020/FTKEE/PP/S01748.

Data Availability Statement: Not applicable.

Conflicts of Interest: The authors declare no conflict of interest.

References

1. Noorasid, N.; Arith, F.; Mustafa, A.; Azam, M.; Mahalingam, S.; Chelvanathan, P.; Amin, N. Current advancement of flexible dye sensitized solar cell: A review. *Optik* **2022**, *254*, 168089. [CrossRef]
2. Owusu, P.A.; Asumadu-Sarkodie, S. A review of renewable energy sources, sustainability issues and climate change mitigation. *Cogent Eng.* **2016**, *3*, 1167990. [CrossRef]
3. Nguyen, V.; Zhu, R.; Yang, R. Environmental effects on nanogenerators. *Nano Energy* **2015**, *14*, 49–61. [CrossRef]
4. Lin, Z.; Chen, J.; Yang, J. Recent Progress in Triboelectric Nanogenerators as a Renewable and Sustainable Power Source. *J. Nanomater.* **2016**, *2016*, 5651613. [CrossRef]
5. Fan, F.-R.; Tian, Z.-Q.; Wang, Z.L. Flexible triboelectric generator. *Nano Energy* **2012**, *1*, 328–334. [CrossRef]
6. Wang, Z.J.; Wang, D.-J. An RF field pattern with improved B1 amplitude homogeneity. *Concepts Magn. Reson. Part B* **2005**, *24*, 1–5. [CrossRef]
7. Hu, Y.; Wang, Z.L. Recent progress in piezoelectric nanogenerators as a sustainable power source in self-powered systems and active sensors. *Nano Energy* **2015**, *14*, 3–14. [CrossRef]
8. Bogue, R. Energy harvesting: A review of recent developments. *Sens. Rev.* **2015**, *35*, 1–5. [CrossRef]
9. Soin, N. *Magnetic Nanoparticles-Piezoelectric Polymer Nanocomposites for Energy Harvesting*; Elsevier Inc.: Amsterdam, The Netherlands, 2018.
10. Bayramol, D.V.; Soin, N.; Shah, T.; Siores, E.; Matsouka, D.; Vassiliadis, S. *Energy Harvesting Smart Textiles*; Springer: Berlin/Heidelberg, Germany, 2017; pp. 199–231.
11. Harstad, S.; D'Souza, N.; Soin, N.; El-Gendy, A.A.; Gupta, S.; Pecharsky, V.K.; Shah, T.; Siores, E.; Hadimani, R.L. Enhancement of β-phase in PVDF films embedded with ferromagnetic Gd_5Si_4 nanoparticles for piezoelectric energy harvesting. *AIP Adv.* **2017**, *7*, 056411. [CrossRef]
12. Soin, N.; Shah, T.H.; Anand, S.C.; Geng, J.; Pornwannachai, W.; Mandal, P.; Reid, D.; Sharma, S.; Hadimani, R.L.; Bayramol, D.V.; et al. Novel "3-D spacer" all fibre piezoelectric textiles for energy harvesting applications. *Energy Environ. Sci.* **2014**, *7*, 1670–1679. [CrossRef]
13. Volk, J.; Nagata, T.; Erdélyi, R.; Bársony, I.; Tóth, A.L.; Lukács, I.E.; Czigány, Z.; Tomimoto, H.; Shingaya, Y.; Chikyow, T. Highly Uniform Epitaxial ZnO Nanorod Arrays for Nanopiezotronics. *Nanoscale Res. Lett.* **2009**, *4*, 699–704. [CrossRef]
14. Dagdeviren, C.; Li, Z.; Wang, Z.L. Energy Harvesting from the Animal/Human Body for Self-Powered Electronics. *Annu. Rev. Biomed. Eng.* **2017**, *19*, 85–108. [CrossRef]
15. Soin, N.; Anand, S.; Shah, T. *Energy Harvesting and Storage Textiles*, 2nd ed.; Elsevier Ltd.: Amsterdam, The Netherlands, 2016.
16. Bowen, C.R.; Kim, H.A.; Weaver, P.M.; Dunn, S. Piezoelectric and ferroelectric materials and structures for energy harvesting applications. *Energy Environ. Sci.* **2013**, *7*, 25–44. [CrossRef]
17. Wan, C.; Bowen, C.R. Multiscale-structuring of polyvinylidene fluoride for energy harvesting: The impact of molecular-, micro- and macro-structure. *J. Mater. Chem. A* **2017**, *5*, 3091–3128. [CrossRef]

18. Ramadan, K.S.; Sameoto, D.; Evoy, S. A review of piezoelectric polymers as functional materials for electromechanical transducers. *Smart Mater. Struct.* **2014**, *23*, 033001. [CrossRef]
19. Xu, Y.; Yu, Q.; Li, J.-F. A facile method to fabricate vertically aligned (K,Na)NbO_3 lead-free piezoelectric nanorods. *J. Mater. Chem.* **2012**, *22*, 23221–23226. [CrossRef]
20. Singh, S.; Tripathi, R.K.; Gupta, M.K.; Dzhardimalieva, G.I.; Uflyand, I.E.; Yadav, B. 2-D self-healable polyaniline-polypyrrole nanoflakes based triboelectric nanogenerator for self-powered solar light photo detector with DFT study. *J. Colloid Interface Sci.* **2021**, *600*, 572–585. [CrossRef]
21. Singh, S.; Yadav, P.; Gupta, M.K.; Dzhardimalieva, G.I.; Yoon, J.; Maiti, C.; Yadav, B.C. Gigantic stimulation in response by solar irradiation in self-healable and self-powered LPG sensor based on triboelectric nanogenerator: Experimental and DFT computational study. *Sens. Actuators B: Chem.* **2022**, *359*, 131573. [CrossRef]
22. Thakre, A.; Kumar, A.; Song, H.-C.; Jeong, D.-Y.; Ryu, J. Pyroelectric Energy Conversion and Its Applications—Flexible Energy Harvesters and Sensors. *Sensors* **2019**, *19*, 2170. [CrossRef]
23. Li, X.; Liang, R.; Tao, J.; Peng, Z.; Xu, Q.; Han, X.; Wang, X.; Wang, C.; Zhu, J.; Pan, C.; et al. Flexible Light Emission Diode Arrays Made of Transferred Si Microwires-ZnO Nanofilm with Piezo-Phototronic Effect Enhanced Lighting. *ACS Nano* **2017**, *11*, 3883–3889. [CrossRef]
24. Li, X.; Chen, M.; Yu, R.; Zhang, T.; Song, D.; Liang, R.; Zhang, Q.; Cheng, S.; Dong, L.; Pan, A.; et al. Enhancing Light Emission of ZnO-Nanofilm/Si-Micropillar Heterostructure Arrays by Piezo-Phototronic Effect. *Adv. Mater.* **2015**, *27*, 4447–4453. [CrossRef] [PubMed]
25. Pandey, R.K.; Dutta, J.; Brahma, S.; Rao, B.; Liu, C.-P. Review on ZnO-based piezotronics and piezoelectric nanogenerators: Aspects of piezopotential and screening effect. *J. Phys.: Mater.* **2021**, *4*, 044011. [CrossRef]
26. Wen, X.; Wu, W.; Pan, C.; Hu, Y.; Yang, Q.; Wang, Z.L. Development and progress in piezotronics. *Nano Energy* **2015**, *14*, 276–295. [CrossRef]
27. Lu, J.; Xu, C.; Li, F.; Yang, Z.; Peng, Y.; Li, X.; Que, M.; Pan, C.; Wang, Z.L. Piezoelectric Effect Tuning on ZnO Microwire Whispering-Gallery Mode Lasing. *ACS Nano* **2018**, *12*, 11899–11906. [CrossRef]
28. Jiji, G. A retrospect on the role of piezoelectric nanogenerators in the development of the green world. *RSC Adv.* **2017**, *7*, 33642–33670. [CrossRef]
29. Song, H. *Fabrication and Characterisation of Electrospun Polyvinylidene Fluoride (PVDF) Nanocomposites for Energy Harvesting Applications*; Brunel University London: London, UK, 2016; p. 147.
30. Ma, J.; Hu, J.; Li, Z.; Nan, C.-W. Recent Progress in Multiferroic Magnetoelectric Composites: From Bulk to Thin Films. *Adv. Mater.* **2011**, *23*, 1062–1087. [CrossRef]
31. Hu, J.-M.; Li, Z.; Wang, J.; Ma, J.; Lin, Y.H.; Nan, C.W. A simple bilayered magnetoelectric random access memory cell based on electric-field controllable domain structure. *J. Appl. Phys.* **2010**, *108*, 043909. [CrossRef]
32. Ma, J.; Shi, Z.; Nan, C.-W. Magnetoelectric Properties of Composites of Single Pb(Zr,Ti)O_3 Rods and Terfenol-D/Epoxy with a Single-Period of 1-3-Type Structure. *Adv. Mater.* **2007**, *19*, 2571–2573. [CrossRef]
33. Fujii, I.; Nakashima, K.; Kumada, N.; Wada, S. Structural, dielectric, and piezoelectric properties of $BaTiO_3$-Bi(Ni1/2Ti1/2)O_3 ceramics. *J. Ceram. Soc. Jpn.* **2012**, *120*, 30–34. [CrossRef]
34. Briscoe, J.; Dunn, S. Piezoelectric nanogenerators—A review of nanostructured piezoelectric energy harvesters. *Nano Energy* **2015**, *14*, 15–29. [CrossRef]
35. Zhu, G.; Wang, A.C.; Liu, Y.; Zhou, Y.; Wang, Z.L. Functional Electrical Stimulation by Nanogenerator with 58 V Output Voltage. *Nano Lett.* **2012**, *12*, 3086–3090. [CrossRef] [PubMed]
36. Qi, Y.; McAlpine, M.C. Nanotechnology-enabled flexible and biocompatible energy harvesting. *Energy Environ. Sci.* **2010**, *3*, 1275–1285. [CrossRef]
37. Wang, Z.L.; Wu, W. Nanotechnology-Enabled Energy Harvesting for Self-Powered Micro-/Nanosystems. *Angew. Chem. Int. Ed.* **2012**, *51*, 11700–11721. [CrossRef]
38. Delgado-Alvarado, E.; Elvira-Hern, E.A.; Huerta-Chua, J.; Vázquez-Leal, H.; Mart, J.; Garc, P.J. Recent Progress of Nanogenerators for Green Energy Harvesting: Performance, Applications and Challenges. *Nanomaterials* **2022**, *12*, 2549. [CrossRef]
39. Wang, Z.L.; Song, J. Piezoelectric Nanogenerators Based on Zinc Oxide Nanowire Arrays. *Science* **2006**, *312*, 242–246. [CrossRef]
40. Wang, X.; Song, J.; Liu, J.; Wang, Z.L. Direct-Current Nanogenerator Driven by Ultrasonic Waves. *Science* **2007**, *316*, 102–105. [CrossRef]
41. Hu, D.; Yao, M.; Fan, Y.; Ma, C.; Fan, M.; Liu, M. Strategies to achieve high performance piezoelectric nanogenerators. *Nano Energy* **2019**, *55*, 288–304. [CrossRef]
42. Bouvet-Marchand, A.; Graillot, A.; Volk, J.; Dauksevicius, R.; Sturm, C.; Grundmann, M.; Saoutieff, E.; Viana, A.; Christian, B.; Lebedev, V.; et al. Design of UV-crosslinked polymeric thin layers for encapsulation of piezoelectric ZnO nanowires for pressure-based fingerprint sensors. *J. Mater. Chem. C* **2017**, *6*, 605–613. [CrossRef]
43. Al-Ruqeishi, M.; Mohiuddin, T.; Al-Habsi, B.; Al-Ruqeishi, F.; Al-Fahdi, A.; Al-Khusaibi, A. Piezoelectric nanogenerator based on ZnO nanorods. *Arab. J. Chem.* **2019**, *12*, 5173–5179. [CrossRef]
44. Savarimuthu, K.; Rajamanickam, G.; Shankararajan, R.; Perumal, R.; Rayarfrancis, A. Experimental Study on Flexible ZnO Based Nanogenerator Using Schottky Contact for Energy Harvesting Applications. *IEEE Trans. Nanotechnol.* **2017**, *16*, 469–476. [CrossRef]

45. Briscoe, J.; Jalali, N.; Woolliams, P.; Stewart, M.; Weaver, P.M.; Cain, M.; Dunn, S. Measurement techniques for piezoelectric nanogenerators. *Energy Environ. Sci.* **2013**, *6*, 3035–3045. [CrossRef]
46. Cao, P.; Yang, Z.; Navale, S.; Han, S.; Liu, X.; Liu, W.; Lu, Y.; Stadler, F.; Zhu, D. Ethanol sensing behavior of Pd-nanoparticles decorated ZnO-nanorod based chemiresistive gas sensors. *Sens. Actuators B Chem.* **2019**, *298*, 126850. [CrossRef]
47. Pearton, S.; Norton, D.; Ip, K.; Heo, Y.; Steiner, T. Recent progress in processing and properties of ZnO. *Superlattices Microstruct.* **2003**, *34*, 3–32. [CrossRef]
48. Koka, A.; Zhou, Z.; Sodano, H.A. Vertically aligned $BaTiO_3$nanowire arrays for energy harvesting. *Energy Environ. Sci.* **2013**, *7*, 288–296. [CrossRef]
49. Wu, W.; Bai, S.; Yuan, M.; Qin, Y.; Wang, Z.L.; Jing, T. Lead Zirconate Titanate Nanowire Textile Nanogenerator for Wearable Energy-Harvesting and Self-Powered Devices. *ACS Nano* **2012**, *6*, 6231–6235. [CrossRef]
50. Ku, N.-J.; Liu, G.; Wang, C.-H.; Gupta, K.; Liao, W.-S.; Ban, D.; Liu, C.-P. Optimal geometrical design of inertial vibration DC piezoelectric nanogenerators based on obliquely aligned InN nanowire arrays. *Nanoscale* **2017**, *9*, 14039–14046. [CrossRef]
51. Islam, S.; Zamil, Y.; Mojumder, R.H.; Stampfl, C.; Park, J. Strong tribo-piezoelectric effect in bilayer indium nitride (InN). *Sci. Rep.* **2021**, *11*, 18669. [CrossRef]
52. Mishra, S.; Supraja, P.; Sankar, P.; Kumar, R.; Prakash, K.; Haranath, D. Controlled synthesis of luminescent ZnS nanosheets with high piezoelectric performance for designing mechanical energy harvesting device. *Mater. Chem. Phys.* **2021**, *277*, 125264. [CrossRef]
53. Parangusan, H.; Bhadra, J.; Al-Thani, N. Flexible piezoelectric nanogenerator based on [P(VDF-HFP)]/PANI-ZnS electrospun nanofibers for electrical energy harvesting. *J. Mater. Sci. Mater. Electron.* **2021**, *32*, 6358–6368. [CrossRef]
54. Banerjee, S.; Bairagi, S.; Ali, S.W. A lead-free flexible piezoelectric-triboelectric hybrid nanogenerator composed of uniquely designed PVDF/KNN-ZS nanofibrous web. *Energy* **2022**, *244*, 123102. [CrossRef]
55. Zhao, B.; Chen, Z.; Cheng, Z.; Wang, S.; Yu, T.; Yang, W.; Li, Y. Piezoelectric Nanogenerators Based on Electrospun PVDF-Coated Mats Composed of Multilayer Polymer-Coated $BaTiO_3$ Nanowires. *ACS Appl. Nano Mater.* **2022**, *5*, 8417–8428. [CrossRef]
56. He, Y.; Wang, H.; Sha, Z.; Boyer, C.; Wang, C.-H.; Zhang, J. Enhancing output performance of PVDF-HFP fiber-based nanogenerator by hybridizing silver nanowires and perovskite oxide nanocrystals. *Nano Energy* **2022**, *98*, 107343. [CrossRef]
57. Mallick, Z.; Saini, D.; Sarkar, R.; Kundu, T.K.; Mandal, D. Piezo-phototronic effect in highly stable lead-free double perovskite Cs2SnI6-PVDF nanocomposite: Possibility for strain modulated optical sensor. *Nano Energy* **2022**, *100*, 107451. [CrossRef]
58. Kim, M.; Wu, Y.S.; Kan, E.C.; Fan, J. Breathable and Flexible Piezoelectric ZnO@PVDF Fibrous Nanogenerator for Wearable Applications. *Polymers* **2018**, *10*, 745. [CrossRef] [PubMed]
59. Qiu, Y.; Lei, J.; Yang, D.; Yin, B.; Zhang, H.; Bian, J.; Ji, J.; Liu, Y.; Zhao, Y.; Luo, Y.; et al. Enhanced performance of wearable piezoelectric nanogenerator fabricated by two-step hydrothermal process. *Appl. Phys. Lett.* **2014**, *104*, 113903. [CrossRef]
60. Aleman, C.K.; Narvaez, J.A.B.; Lopez, G.D.B.; Mercado, C.C. Array pattern effects on the voltage output of vertically aligned $BaTiO_3$ nanotubular flexible piezoelectric nanogenerator. *MRS Commun.* **2020**, *10*, 500–505. [CrossRef]
61. Khan, A.; Abbasi, M.A.; Hussain, M.; Ibupoto, Z.H.; Wissting, J.; Nur, O.; Willander, M. Piezoelectric nanogenerator based on zinc oxide nanorods grown on textile cotton fabric. *Appl. Phys. Lett.* **2012**, *101*, 193506. [CrossRef]
62. Chen, X.; Xu, S.; Yao, N.; Xu, W.; Shi, Y. Potential measurement from a single lead ziroconate titanate nanofiber using a nanomanipulator. *Appl. Phys. Lett.* **2009**, *94*, 253113. [CrossRef]
63. Liu, C.; Yu, A.; Peng, M.; Song, M.; Liu, W.; Zhang, Y.; Zhai, J. Improvement in the Piezoelectric Performance of a ZnO Nanogenerator by a Combination of Chemical Doping and Interfacial Modification. *J. Phys. Chem. C* **2016**, *120*, 6971–6977. [CrossRef]
64. Rafique, S.; Kasi, A.K.; Kasi, J.K.; Aminullah; Bokhari, M.; Shakoor, Z. Fabrication of silver-doped zinc oxide nanorods piezoelectric nanogenerator on cotton fabric to utilize and optimize the charging system. *Nanomater. Nanotechnol.* **2020**, *10*, 1847980419895741. [CrossRef]
65. Chang, W.-Y.; Fang, T.-H.; Tsai, J.-H. Electromechanical and Photoluminescence Properties of Al-doped ZnO Nanorods Applied in Piezoelectric Nanogenerators. *J. Low Temp. Phys.* **2014**, *178*, 174–187. [CrossRef]
66. Vayssieres, L.; Keis, K.; Lindquist, S.-E.; Hagfeldt, A. Purpose-Built Anisotropic Metal Oxide Material: 3D Highly Oriented Microrod Array of ZnO. *J. Phys. Chem. B* **2001**, *105*, 3350–3352. [CrossRef]
67. Fang, T.-H.; Kang, S.-H. Surface and physical characteristics of ZnO:Al nanostructured films. *J. Appl. Phys.* **2009**, *105*, 113512. [CrossRef]
68. Chu, Y.-L.; Young, S.-J.; Ji, L.-W.; Chu, T.-T.; Chen, P.-H. Synthesis of Ni-Doped ZnO Nanorod Arrays by Chemical Bath Deposition and Their Application to Nanogenerators. *Energies* **2020**, *13*, 2731. [CrossRef]
69. Parangusan, H.; Ponnamma, D.; AlMaadeed, M.A.A. Flexible tri-layer piezoelectric nanogenerator based on PVDF-HFP/Ni-doped ZnO nanocomposites. *RSC Adv.* **2017**, *7*, 50156–50165. [CrossRef]
70. Batra, K.; Sinha, N.; Goel, S.; Yadav, H.; Joseph, A.J.; Kumar, B. Enhanced dielectric, ferroelectric and piezoelectric performance of Nd-ZnO nanorods and their application in flexible piezoelectric nanogenerator. *J. Alloy. Compd.* **2018**, *767*, 1003–1011. [CrossRef]
71. Pandey, R.; Khandelwal, G.; Palani, I.A.; Singh, V.; Kim, S.-J. A La-doped ZnO ultra-flexible flutter-piezoelectric nanogenerator for energy harvesting and sensing applications: A novel renewable source of energy. *Nanoscale* **2019**, *11*, 14032–14041. [CrossRef]
72. Batra, K.; Sinha, N.; Kumar, B. Tb-doped ZnO:PDMS based flexible nanogenerator with enhanced piezoelectric output performance by optimizing nanofiller concentration. *Ceram. Int.* **2020**, *46*, 24120–24128. [CrossRef]

73. Sinha, N.; Ray, G.; Bhandari, S.; Godara, S.; Kumar, B. Synthesis and enhanced properties of cerium doped ZnO nanorods. *Ceram. Int.* **2014**, *40*, 12337–12342. [CrossRef]
74. Gupta, M.K.; Lee, J.-H.; Lee, K.Y.; Kim, S.-W. Two-Dimensional Vanadium-Doped ZnO Nanosheet-Based Flexible Direct Current Nanogenerator. *ACS Nano* **2013**, *7*, 8932–8939. [CrossRef]
75. Batra, K.; Sinha, N.; Kumar, B. Ba-doped ZnO nanorods: Efficient piezoelectric filler material for PDMS based flexible nanogenerator. *Vacuum* **2021**, *191*, 110385. [CrossRef]
76. Lee, J.-H.; Lee, K.Y.; Kumar, B.; Kim, S.-W. Synthesis of Ga-Doped ZnO Nanorods Using an Aqueous Solution Method for a Piezoelectric Nanogenerator. *J. Nanosci. Nanotechnol.* **2012**, *12*, 3430–3433. [CrossRef]
77. Sinha, N.; Ray, G.; Godara, S.; Gupta, M.K.; Kumar, B. Enhanced piezoelectric output voltage and Ohmic behavior in Cr-doped ZnO nanorods. *Mater. Res. Bull.* **2014**, *59*, 267–271. [CrossRef]
78. Lam, K.-T.; Chu, Y.-L.; Ji, L.-W.; Hsiao, Y.-J.; Chu, T.-T.; Huang, B.-W. Characterization of nanogenerators based on S-doped zinc oxide nanorod arrays. *Microsyst. Technol.* **2020**, *28*, 395–401. [CrossRef]
79. Ng, Z.-N.; Chan, K.-Y.; Tohsophon, T. Effects of annealing temperature on ZnO and AZO films prepared by sol–gel technique. *Appl. Surf. Sci.* **2012**, *258*, 9604–9609. [CrossRef]
80. Jain, S.; Karmakar, N.; Shah, A.; Shimpi, N.G. Development of Ni doped ZnO/polyaniline nanocomposites as high response room temperature NO2 sensor. *Mater. Sci. Eng. B* **2019**, *247*, 114381. [CrossRef]
81. Fang, T.-H. Physical Properties of ZnO: Al Nanorods for Piezoelectric Nanogenerator Application. *Curr. Nanosci.* **2010**, *6*, 505–511. [CrossRef]
82. Rajabi, M.; Ghorbani, M. Performance evaluation of nanogenerators based on Ag doped ZnO nanorods. *Sens. Actuators A Phys.* **2017**, *266*, 338–344. [CrossRef]
83. Hasanpoor, M.; Aliofkhazraei, M.; Delavari, H. In-situ study of mass and current density for electrophoretic deposition of zinc oxide nanoparticles. *Ceram. Int.* **2016**, *42*, 6906–6913. [CrossRef]
84. Kołodziejczak-Radzimska, A.; Jesionowski, T. Zinc Oxide—From Synthesis to Application: A Review. *Materials* **2014**, *7*, 2833–2881. [CrossRef]
85. Qin, C.; Gu, Y.; Sun, X.; Wang, X.; Zhang, Y. Structural dependence of piezoelectric size effects and macroscopic polarization in ZnO nanowires: A first-principles study. *Nano Res.* **2015**, *8*, 2073–2081. [CrossRef]
86. Cauda, V.; Stassi, S.; Lamberti, A.; Morello, M.; Pirri, C.F.; Canavese, G. Leveraging ZnO morphologies in piezoelectric composites for mechanical energy harvesting. *Nano Energy* **2015**, *18*, 212–221. [CrossRef]
87. Tlemcani, T.S.; Justeau, C.; Nadaud, K.; Poulin-Vittrant, G.; Alquier, D. Deposition Time and Annealing Effects of ZnO Seed Layer on Enhancing Vertical Alignment of Piezoelectric ZnO Nanowires. *Chemosensors* **2019**, *7*, 7. [CrossRef]
88. Joshi, S.; Nayak, M.; Rajanna, K. Effect of post-deposition annealing on transverse piezoelectric coefficient and vibration sensing performance of ZnO thin films. *Appl. Surf. Sci.* **2014**, *296*, 169–176. [CrossRef]
89. Sohn, J.I.; Cha, S.N.; Song, B.G.; Lee, S.; Kim, S.M.; Ku, J.; Kim, H.J.; Park, Y.J.; Choi, B.L.; Wang, Z.L.; et al. Engineering of efficiency limiting free carriers and an interfacial energy barrier for an enhancing piezoelectric generation. *Energy Environ. Sci.* **2012**, *6*, 97–104. [CrossRef]
90. Pham, T.T.; Lee, K.Y.; Lee, J.; Kim, K.H.; Shin, K.S.; Gupta, M.K.; Kumar, B.; Kim, S. Reliable operation of a nanogenerator under ultraviolet light via engineering piezoelectric potential. *Energy Environ. Sci* **2013**, *6*, 841–846. [CrossRef]
91. Cha, S.N.; Seo, J.-S.; Kim, S.M.; Kim, H.J.; Park, Y.J.; Kim, S.-W.; Kim, J.M. Sound-Driven Piezoelectric Nanowire-Based Nanogenerators. *Adv. Mater.* **2010**, *22*, 4726–4730. [CrossRef]
92. Barreca, D.; Bekermann, D.; Comini, E.; Devi, A.; Fischer, R.A.; Gasparotto, A.; Maccato, C.; Sberveglieri, G.; Tondello, E. 1D ZnO nano-assemblies by Plasma-CVD as chemical sensors for flammable and toxic gases. *Sens. Actuators B Chem.* **2010**, *149*, 1–7. [CrossRef]
93. Phan, T.-L.; Yu, S.; Vincent, R.; Dan, N.; Shi, W. Photoluminescence properties of various CVD-grown ZnO nanostructures. *J. Lumin* **2010**, *130*, 1142–1146. [CrossRef]
94. Lee, D.J.; Park, J.Y.; Yun, Y.S.; Hong, Y.S.; Moon, J.H.; Lee, B.-T.; Kim, S.S. Comparative studies on the growth behavior of ZnO nanorods by metalorganic chemical vapor deposition depending on the type of substrates. *J. Cryst. Growth* **2004**, *276*, 458–464. [CrossRef]
95. Kumar, B.; Lee, K.Y.; Park, H.; Chae, S.J.; Lee, Y.H.; Kim, S. Controlled Growth of Semiconducting Nanowire, Nanowall, and Hybrid Nanostructures on Graphene for Piezoelectric Nanogenerators. *Acs Nano* **2011**, *5*, 4197–4204. [CrossRef] [PubMed]
96. Zhang, D.; Wang, C.; Zhang, F. Oxygen pressure and measurement temperature dependence of defects related bands in zinc oxide films. *Vacuum* **2010**, *85*, 160–163. [CrossRef]
97. Menzel, A.; Subannajui, K.; Bakhda, R.; Wang, Y.; Thomann, R.; Zacharias, M. Tuning the Growth Mechanism of ZnO Nanowires by Controlled Carrier and Reaction Gas Modulation in Thermal CVD. *J. Phys. Chem. Lett.* **2012**, *3*, 2815–2821. [CrossRef]
98. Ghosh, M.; Rao, M.G. Growth mechanism of ZnO nanostructures for ultra-high piezoelectric d33 coefficient. *Mater. Express* **2013**, *3*, 319–327. [CrossRef]
99. Vayssieres, L. Growth of Arrayed Nanorods and Nanowires of ZnO from Aqueous Solutions. *Adv. Mater.* **2003**, *15*, 464–466. [CrossRef]

100. Poulin-Vittrant, G.; Oshman, C.; Opoku, C.; Dahiya, A.; Camara, N.; Alquier, D.; Hue, L.-P.T.H.; Lethiecq, M. Fabrication and Characterization of ZnO Nanowire-based Piezoelectric Nanogenerators for Low Frequency Mechanical Energy Harvesting. *Phys. Procedia* **2015**, *70*, 909–913. [CrossRef]
101. Zhang, Y.; Liu, C.; Liu, J.; Xiong, J.; Liu, J.; Zhang, K.; Liu, Y.; Peng, M.; Yu, A.; Zhang, A.; et al. Lattice Strain Induced Remarkable Enhancement in Piezoelectric Performance of ZnO-Based Flexible Nanogenerators. *ACS Appl. Mater. Interfaces* **2016**, *8*, 1381–1387. [CrossRef]
102. Stassi, S.; Cauda, V.; Ottone, C.; Chiodoni, A.; Pirri, C.F.; Canavese, G. Flexible piezoelectric energy nanogenerator based on ZnO nanotubes hosted in a polycarbonate membrane. *Nano Energy* **2015**, *13*, 474–481. [CrossRef]
103. Nour, E.; Sandberg, M.; Willander, M.; Nur, O. Handwriting enabled harvested piezoelectric power using ZnO nanowires/polymer composite on paper substrate. *Nano Energy* **2014**, *9*, 221–228. [CrossRef]
104. Liao, Q.; Zhang, Z.; Zhang, X.; Mohr, M.; Zhang, Y.; Fecht, H.-J. Flexible piezoelectric nanogenerators based on a fiber/ZnO nanowires/paper hybrid structure for energy harvesting. *Nano Res.* **2014**, *7*, 917–928. [CrossRef]
105. Saravanakumar, B.; Kim, S.-J. Growth of 2D ZnO Nanowall for Energy Harvesting Application. *J. Phys. Chem. C* **2014**, *118*, 8831–8836. [CrossRef]
106. Zhang, Z.; Chen, Y.; Guo, J. ZnO nanorods patterned-textile using a novel hydrothermal method for sandwich structured-piezoelectric nanogenerator for human energy harvesting. *Phys. E: Low-Dimens. Syst. Nanostructures* **2018**, *105*, 212–218. [CrossRef]
107. Khan, A.; Hussain, M.; Nur, O.; Willander, M. Mechanical and piezoelectric properties of zinc oxide nanorods grown on conductive textile fabric as an alternative substrate. *J. Phys. D Appl. Phys.* **2014**, *47*, 345102. [CrossRef]
108. Khan, A.; Hussain, M.; Nur, O.; Willander, M. Fabrication of zinc oxide nanoneedles on conductive textile for harvesting piezoelectric potential. *Chem. Phys. Lett.* **2014**, *612*, 62–67. [CrossRef]
109. Zang, W.; Wang, W.; Zhu, D.; Xing, L.; Xue, X. Humidity-dependent piezoelectric output of Al–ZnO nanowire nanogenerator and its applications as a self-powered active humidity sensor. *RSC Adv.* **2014**, *4*, 56211–56215. [CrossRef]
110. Hussain, M.; Abbasi, M.A.; Ibupoto, Z.H.; Nur, O.; Willander, M. The improved piezoelectric properties of ZnO nanorods with oxygen plasma treatment on the single layer graphene coated polymer substrate. *Phys. Status Solidi* **2014**, *211*, 455–459. [CrossRef]
111. Opoku, C.; Dahiya, A.S.; Oshman, C.; Cayrel, F.; Poulin-Vittrant, G.; Alquier, D.; Camara, N. Fabrication of ZnO Nanowire Based Piezoelectric Generators and Related Structures. *Phys. Procedia* **2015**, *70*, 858–862. [CrossRef]
112. Lee, C.-T.; Chiu, Y.-S. Piezoelectric ZnO-nanorod-structured pressure sensors using GaN-based field-effect-transistor. *Appl. Phys. Lett.* **2015**, *106*, 073502. [CrossRef]
113. Lei, J.; Yin, B.; Qiu, Y.; Zhang, H.; Chang, Y.; Luo, Y.; Zhao, Y.; Ji, J.; Hu, L. Flexible piezoelectric nanogenerator based on Cu_2O–ZnO p–n junction for energy harvesting. *RSC Adv.* **2015**, *5*, 59458–59462. [CrossRef]
114. Wang, Q.; Yang, D.; Qiu, Y.; Zhang, X.; Song, W.; Hu, L. Two-dimensional ZnO nanosheets grown on flexible ITO-PET substrate for self-powered energy-harvesting nanodevices. *Appl. Phys. Lett.* **2018**, *112*, 063906. [CrossRef]
115. Zhang, X.; Yang, D.; Qiu, Y.; Nie, Y.; Chu, H.; Zhang, H.; Hu, L. Improvement in the Piezoelectric Performance of a ZnO Nanogenerator by a ZnO/Spiro-MeOTAD ps-n Heterojunction. *Phys. Status Solidi* **2019**, *216*, 1800717. [CrossRef]
116. Lu, S.; Liao, Q.; Qi, J.; Liu, S.; Liu, Y.; Liang, Q.; Zhang, G.; Zhang, Y. The enhanced performance of piezoelectric nanogenerator via suppressing screening effect with Au particles/ZnO nanoarrays Schottky junction. *Nano Res.* **2015**, *9*, 372–379. [CrossRef]
117. Pan, C.; Chen, Y.; Hsieh, C.; Lin, C.-H.; Su, C.; Yen, C.; Liu, Z.; Wang, W. Ultrasonic sensing device with ZnO piezoelectric nanorods by selectively electrospraying method. *Sens. Actuators A: Phys.* **2014**, *216*, 318–327. [CrossRef]
118. Hasan, R.; Baek, S.-H.; Seong, K.S.; Kim, J.H.; Park, I.-K. Hierarchical ZnO Nanorods on Si Micropillar Arrays for Performance Enhancement of Piezoelectric Nanogenerators. *ACS Appl. Mater. Interfaces* **2015**, *7*, 5768–5774. [CrossRef]
119. Shin, D.-M.; Tsege, E.L.; Kang, S.H.; Seung, W.; Kim, S.-W.; Kim, H.K.; Hong, S.W.; Hwang, Y.-H. Freestanding ZnO nanorod/graphene/ZnO nanorod epitaxial double heterostructure for improved piezoelectric nanogenerators. *Nano Energy* **2015**, *12*, 268–277. [CrossRef]
120. Shin, C.; Heo, J.; Park, J.; Lee, T.; Ryu, H.; Shin, B.; Lee, W.; Kim, H.-K. The effect of pH on ZnO hydrothermal growth on PES flexible substrates. *Phys. E: Low-Dimens. Syst. Nanostructures* **2010**, *43*, 54–57. [CrossRef]
121. Tao, Y.; Fu, M.; Zhao, A.; He, D.; Wang, Y. The effect of seed layer on morphology of ZnO nanorod arrays grown by hydrothermal method. *J. Alloy. Compd.* **2010**, *489*, 99–102. [CrossRef]
122. Tak, Y.; Yong, K. Controlled Growth of Well-Aligned ZnO Nanorod Array Using a Novel Solution Method. *J. Phys. Chem. B* **2005**, *109*, 19263–19269. [CrossRef]
123. Kim, H.; Kim, E.; Kim, S. Growth of ZnO Nanorods on ITO Film for Piezoelectric Nanogenerators. *Materials* **2021**, *14*, 1461. [CrossRef]
124. Krukiewicz, K.; Zak, J.K. Electrochemistry role in production of nano-coatings. In *Comprehensive Guide for Nanocoatings Technology*; Nova Science Publishers: Hauppauge, NY, USA, 2015; pp. 1–33.
125. Kumar, M.; Sasikumar, C. Electrodeposition of Nanostructured ZnO Thin Film: A Review. *Am. J. Mater. Sci. Eng.* **2014**, *2*, 18–23. [CrossRef]
126. Tamvakos, D.; Lepadatu, S.; Antohe, V.-A.; Tamvakos, A.; Weaver, P.M.; Piraux, L.; Cain, M.G.; Pullini, D. Piezoelectric properties of template-free electrochemically grown ZnO nanorod arrays. *Appl. Surf. Sci.* **2015**, *356*, 1214–1220. [CrossRef]

127. Nagaraju, G.; Ko, Y.H.; Yu, J.S. Effect of diameter and height of electrochemically-deposited ZnO nanorod arrays on the performance of piezoelectric nanogenerators. *Mater. Chem. Phys.* **2015**, *149–150*, 393–399. [CrossRef]
128. Leprince-Wang, Y. *Piezoelectric ZnO Nanostructure for Energy Harvesting*; John Wiley & Sons: Hoboken, NJ, USA, 2015; Volume 1, pp. 1–124. [CrossRef]
129. Le, A.T.; Ahmadipour, M.; Pung, S.-Y. A review on ZnO-based piezoelectric nanogenerators: Synthesis, characterization techniques, performance enhancement and applications. *J. Alloy. Compd.* **2020**, *844*, 156172. [CrossRef]
130. Lee, S.; Lee, J.; Ko, W.; Cha, S.; Sohn, J.; Kim, J.; Park, J.; Park, Y.; Hong, J. Solution-processed Ag-doped ZnO Nanowires grown on Flexible Polyester for Wearable Nanogenerator Applications. *Nanoscale* **2013**, *5*, 207890. [CrossRef]
131. Kang, L.; An, H.; Park, J.Y.; Hong, M.H.; Nahm, S.; Lee, C.G. La-doped p-type ZnO nanowire with enhanced piezoelectric performance for flexible nanogenerators. *Appl. Surf. Sci.* **2019**, *475*, 969–973. [CrossRef]
132. Chang, Y.-T.; Chen, J.-Y.; Yang, T.-P.; Huang, C.-W.; Chiu, C.-H.; Yeh, P.-H.; Wu, W.-W. Excellent piezoelectric and electrical properties of lithium-doped ZnO nanowires for nanogenerator applications. *Nano Energy* **2014**, *8*, 291–296. [CrossRef]
133. Laurenti, M.; Canavese, G.; Sacco, A.; Fontana, M.; Bejtka, K.; Castellino, M.; Pirri, C.F.; Cauda, V. Nanobranched ZnO Structure: P-Type Doping Induces Piezoelectric Voltage Generation and Ferroelectric-Photovoltaic Effect. *Adv. Mater.* **2015**, *27*, 4218–4223. [CrossRef]
134. Hu, Y.; Lin, L.; Zhang, Y.; Wang, Z.L. Replacing a Battery by a Nanogenerator with 20 V Output. *Adv. Mater.* **2011**, *24*, 110–114. [CrossRef]
135. Jung, D.-Y.; Baek, S.-H.; Hasan, R.; Park, I.-K. Performance-enhanced ZnO nanorod-based piezoelectric nanogenerators on double-sided stainless steel foil. *J. Alloy. Compd.* **2015**, *641*, 163–169. [CrossRef]
136. Yin, B.; Qiu, Y.; Zhang, H.; Lei, J.; Chang, Y.; Ji, J.; Luo, Y.; Zhao, Y.; Hu, L. Piezoelectric performance enhancement of ZnO flexible nanogenerator by a NiO–ZnO p–n junction formation. *Nano Energy* **2015**, *14*, 95–101. [CrossRef]

energies

MDPI

Article

Numerical Simulation on Electromagnetic Energy Harvester Oscillated by Speed Ripple of AC Motors

Masayuki Kato

Department of Electrical and Electronics Systems, Faculty of Engineering, Ibaraki University, Hitachi 3168511, Japan; masayuki.kato.actuator@vc.ibaraki.ac.jp; Tel.: +81-294-38-5182

Abstract: The suppression of torque ripples in an interior permanent magnet synchronous motor (IPMSM) is essential to improve its efficiency and responsiveness. Here, we report on the development of an electromagnetic energy harvester incorporated into an IPMSM to suppress its torque ripples. The proposed harvester is driven to oscillations by the speed ripple of the AC motor. We derived the motion and circuit equations for the motor and the harvester according to Euler–Lagrange equations. We discussed the principle of electrical power generation and used MATLAB/Simulink numerical simulations to investigate the dynamic behavior of the proposed harvester. Our findings revealed that the active Coriolis force unnecessarily reduces the motor's original torque, leading to unsuccessful power generation. Nevertheless, our results demonstrated that the reactive Coriolis force successfully suppresses the motor torque ripple.

Keywords: AC motors; electromagnetic energy harvester; Coriolis force; torque ripple; speed ripple

1. Introduction

Interior permanent magnet synchronous motors (IPMSMs) are widely used in industrial applications due to their high efficiency and high power density [1]. However, their intrinsic operational principle leads to inevitable torque ripples caused by the space harmonics of the magnetic flux and the time harmonics of the output voltage. IPMSMs contain various kinds of space harmonics of the magnetic flux in the air gap due to the saliency of the stator and the rotor. Moreover, the pulse width modulation (PWM) inverter and square-wave drives include odd-order time harmonics of the winding current. These harmonic components generate torque ripples that increase the noise and vibration and reduce the motor's responsiveness without contributing to the motor's output.

The suppression of torque ripples in IPMSMs has been reported in the literature considering the motor's structure and control. The structural approach includes methods based on skewing the rotor [2] and optimizing the shapes of the permanent magnet and the iron core in the rotor [3–5]. One method was reported as a control approach based on the estimation and the control of instantaneous torque ripples using motor constants [6–8].

We proposed a different approach to suppress torque ripples using the Coriolis force generated by the radial vibration of a linear oscillatory actuator (LOA) [9,10]. The Coriolis force is caused by the resonant behavior of the LOA's mover, which oscillates radially in the rotational coordinate of the rotor. The LOA resonates by matching the frequency of the centrifugal force ripple received by the LOA mover to its natural frequency. Furthermore, the control of the LOA's current can be used for actively controlling the Coriolis force. However, this method requires a set of elements, including an external power supply to drive the LOA, a controller, and slip rings for power and signal transmission to the LOA installed on the rotor's surface. The combination of these elements results in a large motor drive system. The present study proposes a new method to eliminate the external power supply by operating the IPMSM as an electromagnetic energy harvester. Energy harvesting techniques have been developed in the past decade, including piezoelectric [11–13] and electromagnetic [14–16]. However, no literature has noted that the torque ripple of a motor can be a source of energy for vibration power generation. This paper investigates and clarifies the feasibility of the energy harvester oscillated by torque ripple of the AC motor. The proposed suppression method in this paper is advantageous because it can be combined with existing suppression methods.

This paper is organized as follows. In Section 2, a mathematical model is derived based on the Euler–Lagrange equations. In Section 3, numerical simulations using MATLAB/Simulink reveal that the electrical power generation and the suppression of torque ripples can be achieved concurrently. In Section 4, energy flow of the proposed harvester is discussed.

2. Electromagnetic Energy Harvester Oscillated by Speed Ripple of Motors

2.1. Mathematical Model

A simplified schematic of the proposed energy harvester system is shown in Figure 1; the stator of the motor and other components are not depicted. Two energy harvesters are mounted symmetrically on the surface of the rotor. The motion of their movers in the radial direction is synchronized with the rotational motion of the rotor. The stator and the coil of the LOA are stationary in the rotating coordinate of the rotor. The permanent magnets on the movers oscillate laterally relative to the coil, resulting in a power-generating back electromotive force (EMF). The axial distance between the energy harvesters and the rotor is assumed to be sufficiently large to prevent flux interferences. In contrast to the systems studied in previous work [9,10], this energy-harvesting system requires neither an external power supply nor a controller. Nonetheless, a slip ring is still necessary for externally extracting the generated electrical power.

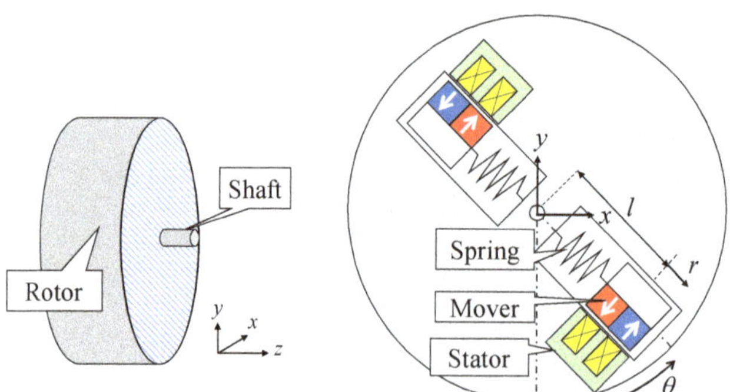

Figure 1. Configuration of the proposed energy harvester.

The equations of motion for the energy harvester and the AC motor are given by the Lagrangian equations as follows:

$$J\ddot{\theta} + m(l+r)^2\ddot{\theta} + c_\theta\dot{\theta} + 2m(l+r)\dot{r}\dot{\theta} = T_e - T_l \quad (1)$$

$$m\ddot{r} + c_r\dot{r} + k_r r = m(l+r)\dot{\theta}^2 + K_t I. \quad (2)$$

Here, J is the rotor's moment of inertia, θ is the rotational angle of the rotor, l is the initial length from the origin, r is the radial displacement of the harvester, c_θ is the viscous damping coefficient for the rotational motion, T_e is the output torque of the AC motor, T_l is the load torque, c_r is the viscous damping coefficient for the radial oscillation, k_r is the spring stiffness of the harvester, K_t is the thrust constant of the harvester, and I is the current caused by electromagnetic induction. The detailed mathematical model can be found in [10,17]. In (1), the multiple harvester oscillators are treated as single equivalent inertia m. The fourth term on the left side of (1) is the Coriolis force acting on the rotor, and the first term on the right side of (2) is the centrifugal force acting on the harvester's oscillator. The electrical circuit equation of the harvester's coil is

$$(R_c + R_l)I + L\frac{dI}{dt} + K_e\dot{r} = 0 \tag{3}$$

where R_c is the coil resistance of the harvester and R_l is the load resistance. For simplicity, the position and current dependence of the coil inductance L and the generation constant K_e were ignored.

2.2. Operational Principle of Power Generation

In this section, we compare the power generation conditions of the proposed energy harvester with those reported in the authors' previous study [10]. The power generation principle in the proposed harvester is illustrated in Figure 2. A rotational vibration system was assumed in [10], in which the rotor was connected to a fixed surface via a torsional spring. Hence, the motion of the rotor was a rotational vibration rather than a steady rotation. The ratio of the torsional motion's frequency (see the second row in Figure 2) to the natural frequency of the oscillator (see the fourth row in Figure 2) was set to 1:2 to induce a parametric excitation and suppress the torsional vibration effectively. A different condition is required for the power generation in the energy harvester proposed in this study. Only the sixth-order component (i.e., the largest component) is assumed for the torque ripple of the AC motor [18], which induces a speed ripple at the same frequency in the rotor. The centrifugal force ripple caused by this speed ripple drives the harvester's movers into oscillation in the radial direction. The frequency of the centrifugal force ripple matches that of the speed ripple because the DC component (constant rotation) is dominant compared with the AC component (ripple). Therefore, matching the natural frequency of the harvester with that of the centrifugal force ripple (i.e., a frequency ratio of 1:1) induces resonance and maximizes the vibration displacement of the harvester. Electric power is generated in the harvester by electromagnetic induction resulting from the relative motion of the permanent magnet and the coil. At the same time, the torque ripple is suppressed by the Coriolis force acting on the rotor due to the harvester's vibration. The harvester's mass and spring constant should be adjusted to match the frequency of the torque ripple and the natural frequency of the harvester.

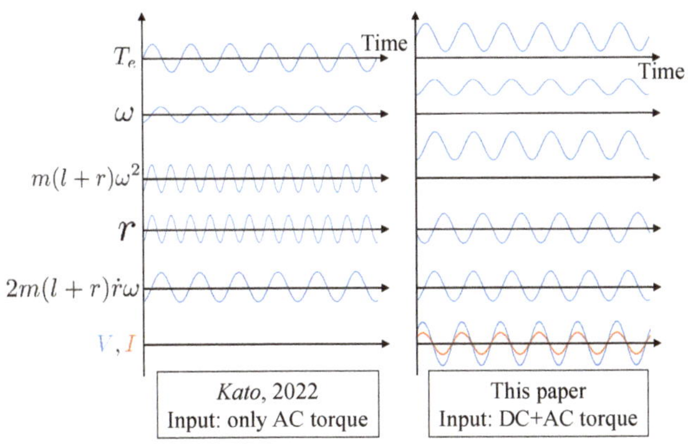

Figure 2. Principle of the electrical power generation in the proposed harvester compared with that reported in [10].

3. Numerical Simulation Results

3.1. MATLAB/Simulink Model and Condition

The mathematical model in Equations (1)–(3) contains several adjustable parameters. As the motion equations are nonlinear differential equations due to the centrifugal force and the Coriolis force, they should be solved numerically rather than analytically. For this reason, we performed numerical simulations using MATLAB/Simulink to clarify the effect of each parameter on power generation. The 1D model of the IPMSM (JMAG-RT model) [19] shown in Table 1 was selected as the target motor. Only the torque constant and moment of inertia in Table 1 were used as reference values. Table 2 lists the additional calculation conditions.

Table 1. Specifications of the selected IPMSM model.

Parameter (Unit)	Value
Power (W)	1000
Number of poles	4
Number of slots	24
Diameter of stator (mm)	54
Diameter of rotor (mm)	27.5
Stack length (mm)	51
Winding type	Distributed winding
Number of turns	46
Inertia J (kgm^2)	1.8×10^{-5}
Torque constant (Nm/A)	0.126

Table 2. Simulation conditions.

Parameter (Unit)	Value
Rotational speed (rpm)	500
DC component of motor torque (Nm)	3
AC component of motor torque (Nm$_{0-p}$)	0.3
Frequency of torque ripple (Hz)	100
Load torque T_l (Nm)	3
Mass of mover m (kg)	0.1
Spring constant k_r (N/mm)	39.5
Viscous damping coefficient c_r (Ns/m)	6.3
Thrust constant K_t (= K_e) (N/A)	1
Arm length l (mm)	50
Coil resistance R_c (Ω)	0.1
Load resistance R_l (Ω)	0.1

3.2. Verification of the Operational Principle

The power generation principle described in Section 2 was verified numerically, as shown in Figure 3, using the conditions listed in Table 2. The obtained waveforms are qualitatively similar to those shown in Figure 2. The simulation results indicated an average power generation of 1.9 mW and an amplitude value for the oscillator displacement equal to 0.41 mm$_{p-p}$, which is acceptable for the design of a harvester with feasible size.

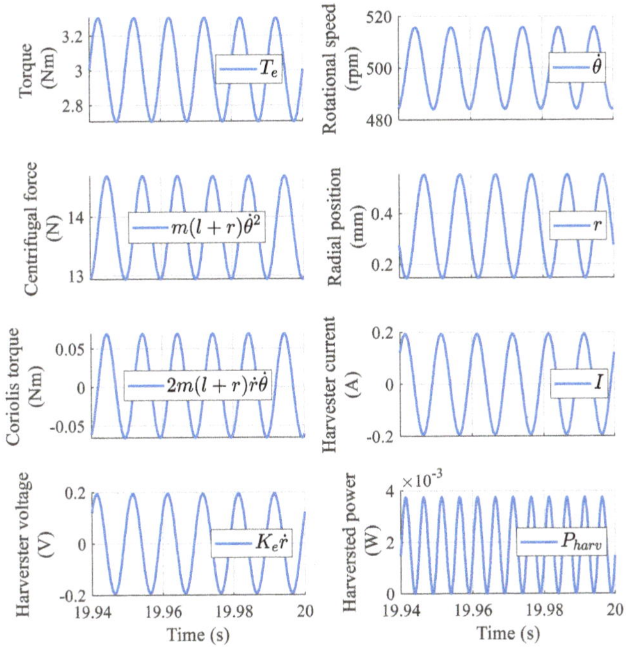

Figure 3. Numerical verification of power generation.

3.3. Parametric Study of the Power Generation

3.3.1. Oscillator Mass and Frequency Ratio

The oscillator mass of the harvester is an independently adjustable parameter. Moreover, the spring constant can be used to adjust the ratio of the torque ripple's frequency to the natural frequency of the harvester. Here, we investigate the effects of these two parameters on power generation through numerical simulations, as shown in Figure 4. The oscillator mass m was increased from 0.01 kg to 1 kg while the arm length l was simultaneously decreased to maintain the equivalent moment of inertia ml^2 constant (2.5×10^{-4} kgm^2). When the mass of the oscillator m = 0.01 kg, the arm length l = 158 mm, which is considerably larger than the stator outer diameter. In this case, Figure 4 shows that more power is generated as the oscillator mass decreases (i.e., the arm length increases). Although the principle explained in Section 2.2 suggested that the amount of power generation is maximized for a frequency ratio of 1:1, the simulation results indicated a maximum power generation for ratios smaller than 1. We attributed this difference to the effect of the harvester's viscosity.

The corresponding results of the torque ripple's reduction rate are shown in Figure 5. Positive and negative values correspond to a decrease and an increase in the torque ripple, respectively. The results indicated that the torque ripple could be suppressed mainly in the region where the frequency ratio was greater than 1, which deviates from the region of maximized power generation, as shown in Figure 4. Consequently, a frequency ratio larger than one was selected as a reasonable condition for concurrent power generation and torque ripple suppression.

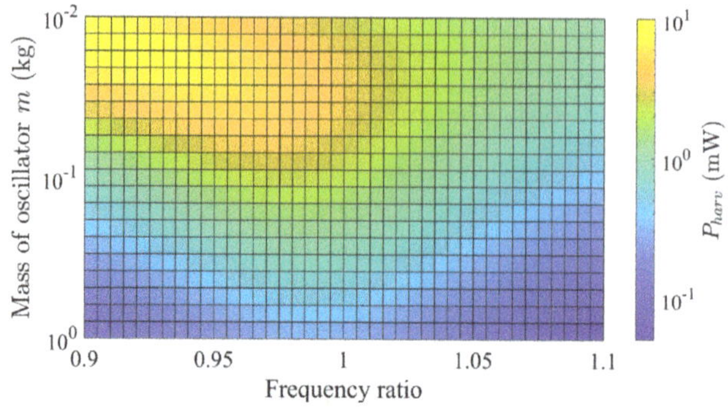

Figure 4. Simulated results of harvested power as a function of mass and frequency ratio.

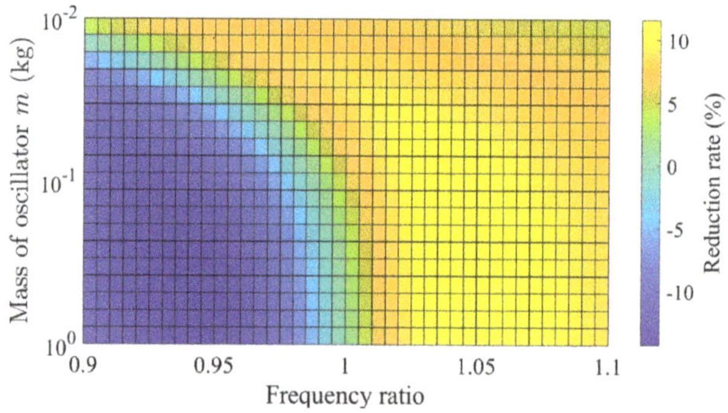

Figure 5. Simulated results of the torque ripple reduction rate as a function of the mass and frequency ratio.

3.3.2. Rotational Speed and Torque Ripple

Motor speed is essential when selecting a motor for the harvesters. Here, we numerically investigated the effects of the motor's speed and the torque ripple's amplitude on power generation. The computed results are shown in Figure 6, where the motor's speed was varied from 50 rpm to 1000 rpm, and the torque ripple's amplitude was varied from 0.3 Nm_{0-p} to 3 Nm_{0-p}. The result showed that the amount of power generated decreased as the rotation speed increased. According to (2), the mechanical response of the motor's rotor is a first-order lag system, and the amplitude of the velocity ripple decreases with increasing speed (i.e., increasing the torque ripple's frequency). It follows that the centrifugal force ripple, which is the source of the harvester's excitation and the resultant power generation, decrease. Therefore, this harvester is suitable for use in the low-speed operation regime of the motor.

Furthermore, we performed a regression analysis on the results in Figure 6, which showed that the amount of power generation increased proportionally to the square of the torque ripple's amplitude. Moreover, the amplitude of the oscillator's displacement increased linearly with that of the torque ripple. Consequently, the amount of generated power increased proportionally to the square of the linearly increasing voltage (i.e., if the generated voltage is V, the amount of generated power is V^2/R_l). Figure 7 shows the computed results of the torque ripple's reduction rate for the conditions shown in Figure 6. The results indicate that this reduction rate is independent of the torque ripple's amplitude. In addition, they show that it decreased in the low-speed range, where an increase in power generation was expected, taking negative values in some regions. To further examine the cause of these observations, we performed calculations using a rotational speed of

50 rpm and a torque ripple amplitude of 3 Nm_{0-p}. In this case, the AC (ripple) component was dominant over the DC (constant rotation) component of the rotation speed. As a result, higher-order components (12th and 18th order), rather than the assumed 6th-order rotational component, were superimposed on the centrifugal force ripple. Therefore, the torque ripple could not be suppressed, which does not satisfy the original power generation principle described in Sections 2 and 3.2. The numerical simulation revealed that the proposed harvester deviates from the assumed power generation principle and loses its suppression effects when the allowable torque ripple amplitude is exceeded.

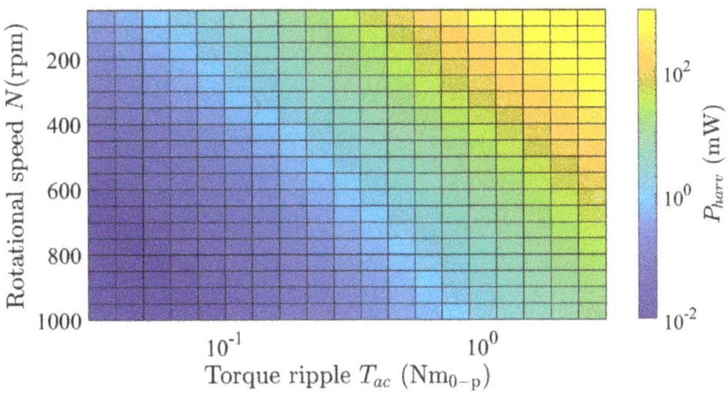

Figure 6. Simulated results of harvested power as a function of rotational speed and torque ripple amplitude.

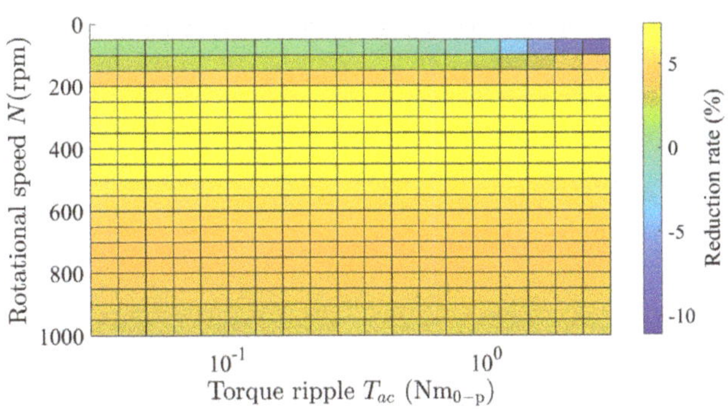

Figure 7. Simulated results of the torque ripple's reduction rate as a function of rotational speed and torque ripple amplitude.

3.3.3. EMF Constant and Impedance Matching

The energy harvester's EMF constant can be adjusted by designing the flux linkage through the coil. Several factors determine the flux linkage, including the permanent magnets' grade, the number of turns, and the magnetic path's design to achieve a high air gap flux density. Moreover, the maximum power transfer theorem [20] is expected to play an important role. It states that power generation is maximized when the coil and load resistance are equal. Nonetheless, the relationship between this theorem and the torque ripple suppression remains unclear. Here, we numerically investigated the effects of the EMF constant and impedance matching on power generation.

Figure 8 shows the simulated result for power generation. Our results showed that as the power generation constant increased, the corresponding maximum point deviated from the theoretical value of impedance matching, R_l/R_c = 1. Moreover, Figure 9 shows the calculation results of the torque ripple's reduction rate for the power generation conditions shown in Figure 8. Although the reduction rate was positive in all regions, the region of maximum value conflicted with that of maximum power generation. Therefore, these characteristic maps may be useful for the designing this harvester's magnetic circuit.

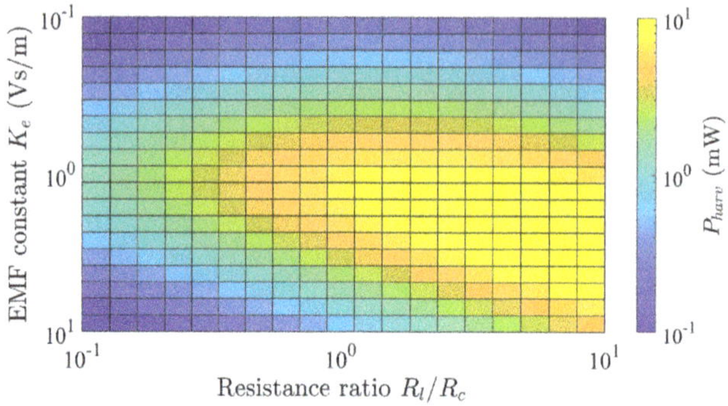

Figure 8. Simulated results of harvested power as a function of load resistance and EMF constant.

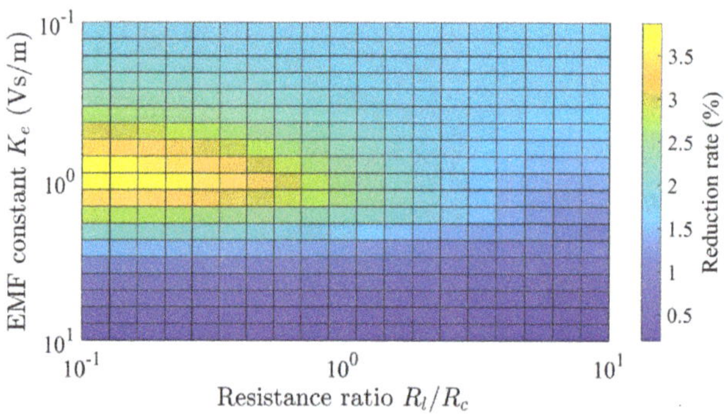

Figure 9. Simulated results of the torque ripple's reduction rate as a function of load resistance and EMF constant.

4. Discussion on Energy Flow

In Section 3, we evaluated the effects of the motor's and harvester's parameters on power generation. However, the mechanism by which the reactive mechanical output, which is the product of the torque and speed ripples, transfers energy to the harvester for power recovery was unclear. Here, we numerically simulated the energy flow to clarify active and reactive power transfer. To convert the differential Equations (1)–(3) into the power dimension, we multiplied both sides of each equation by the rotor's angular velocity d/dt, oscillator velocity dr/dt, and current I flowing in the coil of the harvester, respectively. The following three equations are obtained:

$$\frac{d}{dt}\left(\frac{1}{2}\{J+m(l+r)^2\}\dot{\theta}^2\right)+c_\theta\dot{\theta}^2+2m(l+r)\dot{r}\dot{\theta}^2=T_e\dot{\theta}-T_l\dot{\theta} \tag{4}$$

$$\frac{d}{dt}\left(\frac{1}{2}m\dot{r}^2\right)+c_r\dot{r}^2+\frac{d}{dt}\left(\frac{1}{2}k_r r^2\right)=m(l+r)\dot{\theta}^2\dot{r}+K_t I\dot{r} \tag{5}$$

$$(R_c+R_l)I^2+\frac{d}{dt}\left(\frac{1}{2}LI^2\right)+K_e\dot{r}I=0 \tag{6}$$

Figures 10–12 show the time variation of the energy flow calculated from the numerical results in Figure 3, based on Equations (4)–(6), respectively. Our results (Figure 10) indicated that the active mechanical power of the motor (157 W) was converted into active mechanical power due to the load torque and mechanical loss (i.e., viscous friction). The active power due to the Coriolis force was approximately 100 mW on average. In other words, the active mechanical power due to the Coriolis force acted as a load torque, whereas the reactive mechanical power ($35W_{0-p}$) was distributed to the load torque, rotor inertia, equivalent inertia of the oscillator, viscosity, and Coriolis force. Moreover, Equations (4) and (5) suggest that the active and reactive mechanical powers due to the Coriolis force are transferred to the harvester as active and reactive powers via the centrifugal force. In our simulations, the active mechanical power (50 mW) was distributed into mechanical power (3.8 mW) due to the harvester's electromagnetic force and a mechanical loss (46.2 mW) due to viscous friction. In contrast, the reactive power due to centrifugal force was distributed to all other terms. It is clear from Equations (5) and (6) that the mechanical power due to the electromagnetic force is equivalent to the electrical power due to the power generation by electromagnetic induction. This mechanical output (3.8 mW) is converted into Joule losses due to the coil and load resistances, yielding 3.8/2 = 1.9 mW of electric power.

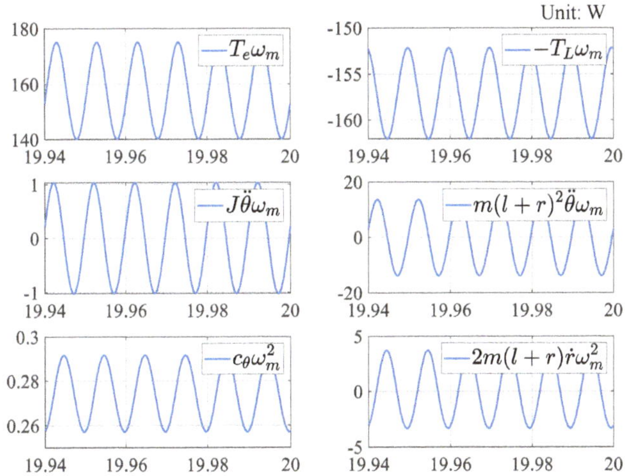

Figure 10. Mechanical energy flow for motor.

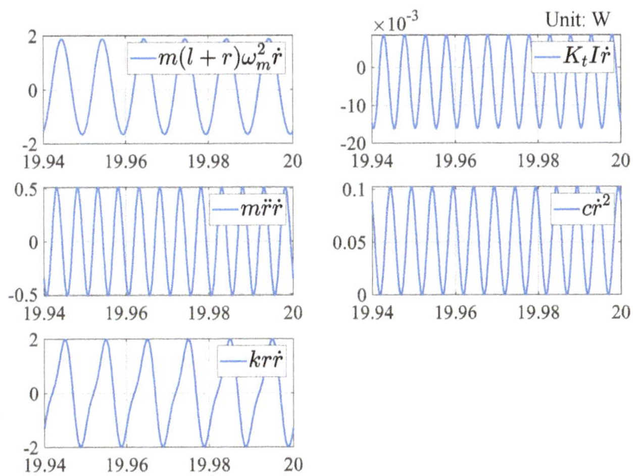

Figure 11. Mechanical energy flow for harvester.

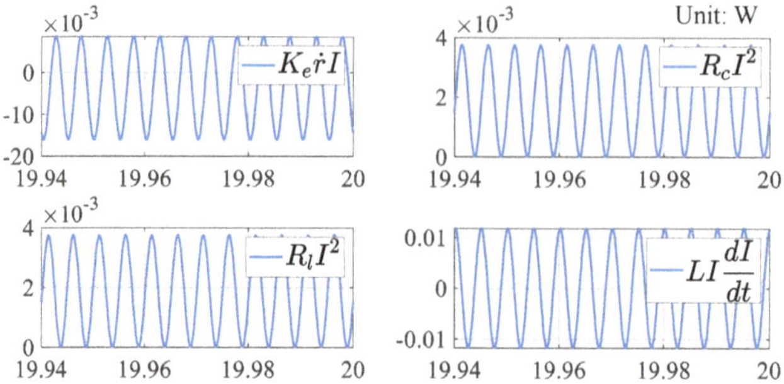

Figure 12. Electrical energy flow for harvester.

The above energy transfer mechanism suggests that the proposed harvester merely receives a part of the motor's active mechanical power, and cannot regenerate the reactive power of the motor caused by the torque ripple. However, because this harvester uses the velocity ripple caused by the torque ripple as an excitation source, energy is transferred from the mechanical system of the motor to the mechanical and electrical systems of the harvester, triggered by the reactive power. The Coriolis force is successfully generated in the process of energy transfer and works to suppress the torque ripple. When this harvester is installed in an AC motor, the Coriolis force's effective output is part of the mechanical load. It reduces the motor's effective output, whereas the reactive output of the Coriolis force suppresses the torque ripple.

5. Conclusions

In this study, we proposed an energy harvester that uses the centrifugal force ripple caused by the speed ripple of an AC motor as the excitation source. We showed, through numerical simulations, that the reactive mechanical power of the motor could not be regenerated as electrical power. Furthermore, we demonstrated that the reactive mechanical power indirectly excites the mover of the harvester, which produces the Coriolis force suppressing the torque ripple.

The proposed suppression method in this paper is advantageous because it can be combined with existing suppression methods. Condition monitoring of AC motors can be a possible application because the proposed energy harvester is able to generate electrical power directly from the torque ripple. In the future, we plan to optimize an energy harvester by using finite-element-based electromagnetic analysis and develop a prototype for experimental verification.

Funding: This research was supported by JKA and its promotion funds from KEIRIN RACE.

Data Availability Statement: Not applicable.

Conflicts of Interest: The authors declare no conflict of interest.

References

1. Sul, S.K.; Kim, S. Sensorless Control of IPMSM: Past, Present, and Future. *IEEJ J. Ind. Appl.* **2012**, *1*, 15–23. [CrossRef]
2. Jahns, T.M.; Soong, W.L. Pulsating Torque Minimization Techniques for Permanent Magnet AC Motor Drives-A Review. *IEEE Trans. Ind. Electron.* **1996**, *43*, 321–330. [CrossRef]
3. Totoki, E.; Yamaguchi, S.; Tanaka, T.; Ito, K.; Daikoku, A.; Morimoto, S. Torque Ripple Reduction for Permanent Magnet Motor using New Configuration for Concentrated Windings. *IEEJ J. Ind. Appl.* **2022**, *11*, 531–537. [CrossRef]
4. Hiramoto, K.; Takeda, Y.; Sanada, M.; Morimoto, S. Torque Ripple Reduction of Reluctance Torque Assisted Motors Using Asymmetric Flux Barriers. *IEEJ J. Ind. Appl.* **2004**, *124*, 208–214. [CrossRef]
5. Yamamoto, Y.; Morimoto, S.; Sanada, M.; Inoue, Y. Reduction of Torque Ripple in Synchronous Reluctance Motor by Combining Different Flux Barrier Structures. *IEEJ J. Ind. Appl.* **2019**, *8*, 430–436. [CrossRef]
6. Nakamura, K.; Fujimoto, H.; Fujitsuna, M. Torque Ripple Suppression Control for PM Motor Considering the Bandwidth of Torque Meter. *IEEJ J. Ind. Appl.* **2010**, *130*, 1241–1247. [CrossRef]

7. Nakao, N.; Akatsu, K. Torque Ripple Control Based on Instantaneous Torque Estimation in PMSMs. *IEEJ J. Ind. Appl.* **2011**, *131*, 1120–1227. [CrossRef]
8. Qu, J.; Jatskevich, J.; Zhang, C.; Zhang, S. Torque Ripple Reduction Method for Permanent Magnet Synchronous Machine Drives With Novel Harmonic Current Control. *IEEE Trans. Energy Convers.* **2021**, *36*, 2502–2513. [CrossRef]
9. Kato, M.; Naito, S.; Kitayama, F. Basic Study on Reduction of Rotational Vibration Using Linear Oscillatory Actuator. *J. Jpn. Soc. Appl. Electromagn. Mech.* **2021**, *29*, 309–314. [CrossRef]
10. Kato, M.; Kitayama, F. Reduction of Rotational Vibration Using Coriolis Force Generated by Electromagnetic Oscillatory Actuator Moving in Radial Direction. *IEEE Trans. Magn.* **2022**, *58*, 8200205. [CrossRef]
11. Sahoo, S.; Ratha, S.; Rout, C.S.; Nayak, S.K. Self-charging supercapacitors for smart electronic devices: A concise review on the recent trends and future sustainability. *J. Mater. Sci.* **2022**, *57*, 4399–4440. [CrossRef]
12. Zhao, K.; Zhang, Q.; Wang, W. Optimization of Galloping Piezoelectric Energy Harvester with V-Shaped Groove in Low Wind Speed. *Energies* **2019**, *12*, 4619. [CrossRef]
13. Ducharne, B.; Gupta, B.; Litak, G. Simulation of Synchronized-Switching Method Energy Harvester Including Accurate Piezoceramic Nonlinear Behavior. *Energies* **2019**, *12*, 4466. [CrossRef]
14. Wang, H.; Shi, G.; Han, C. A Free-Standing Electromagnetic Energy Harvester for Condition Monitoring in Smart Grid. *Wirel. Power Transf.* **2021**, *2021*, 6685308. [CrossRef]
15. Ibrahim, P.; Arafa, M.; Anis, Y. An Electromagnetic Vibration Energy Harvester with a Tunable Mass Moment of Inertia. *Sensors* **2021**, *21*, 5611. [CrossRef]
16. Foisal, A.-R.-M.; Hong, C.; Chung, G.-S. Multi-frequency electromagnetic energy harvester using a magnetic spring cantilever. *Sens. Actuators A Phys.* **2012**, *182*, 106–113. [CrossRef]
17. Viet, L.D.; Anh, N.D.; Matsuhisa, H. The effective damping approach to design a dynamic vibration absorber using Coriolis force. *J. Sound Vib.* **2011**, *330*, 1904–1916. [CrossRef]
18. Su, T.; Hattori, S.; Ishida, M.; Hori, T. Suppression Control Method for Torque Vibration of AC Motor Utilizing Repetitive Controller with Fourier Transform. *IEEE Trans. Ind. Appl.* **2002**, *39*, 1316–1325.
19. [RTML-001] PMSM/IPM Constant Rating 1000(W) 3-Phase. Available online: https://www.jmag-international.com/modellibrary/001/ (accessed on 30 November 2022).
20. Thompson, S.P. *Dynamo-Electric Machinery; A Manual for Students of Electrotechnics*; Bibliolife: Charleston, SC, USA, 2009.

sensors

MDPI

Article

Enhancing the Bandwidth and Energy Production of Piezoelectric Energy Harvester Using Novel Multimode Bent Branched Beam Design for Human Motion Application

Iresha Erangani Piyarathna [1,*], Yee Yan Lim [2], Mahesh Edla [3], Ahmed Mostafa Thabet [4], Mustafa Ucgul [1] and Charles Lemckert [1]

1 Faculty of Science and Engineering, Southern Cross University, Military Road, East Lismore, NSW 2480, Australia
2 Sri Emas International School, Shah Alam, Selangor 40000, Malaysia
3 Department of Research and Development, MYRA Corporate, UAG Bionutrients, Alstonville Industrial Area, Alstonville, NSW 2477, Australia
4 Fortescue Future Industries Pty Ltd., 160 Lakes Rd, Hazelmere, WA 6055, Australia
* Correspondence: i.rathnayaka.mudiyanselage.10@student.scu.edu.au

Abstract: In recent years, harvesting energy from ubiquitous ultralow-frequency vibration sources, such as biomechanical motions using piezoelectric materials to power wearable devices and wireless sensors (e.g., personalized assistive tools for monitoring human locomotion and physiological signals), has drawn considerable interest from the renewable energy research community. Conventional linear piezoelectric energy harvesters (PEHs) generally consist of a cantilever beam with a piezoelectric patch and a proof mass, and they are often inefficient in such practical applications due to their narrow operating bandwidth and low voltage generation. Multimodal harvesters with multiple resonances appear to be a viable solution, but most of the previously proposed designs are unsuitable for ultralow-frequency vibration. This study investigated a novel multimode design, which included a bent branched beam harvester (BBBH) to enhance PEHs' bandwidth output voltage and output power for ultralow-frequency applications. The study was conducted using finite element method (FEM) analysis to optimize the geometrical design of the BBBH on the basis of the targeted frequency spectrum of human motion. The selected design was then experimentally studied using a mechanical shaker and human motion as excitation sources. The performance was also compared to the previously proposed V-shaped bent beam harvester (VBH) and conventional cantilever beam harvester (CBH) designs. The results prove that the proposed BBBH could harness considerably higher output voltages and power with lower idle time. Its operating bandwidth was also remarkably widened as it achieved three close resonances in the ultralow-frequency range. It was concluded that the proposed BBBH outperformed the conventional counterparts when used to harvest energy from ultralow-frequency sources, such as human motion.

Keywords: human motion; piezoelectric energy harvesting; bent branched beam harvester; macro-fiber composite (MFC)

1. Introduction

Over the past few decades, energy-harvesting systems have experienced remarkable growth due to the rapid emergence of the renewable energy industry. In addition to macroscale energy-harvesting technologies, such as solar and wind energy harvesting, innovations and the popularity of handheld electronic devices, medical devices, wireless sensors, and microelectromechanical (MEMS) systems have spurred the demand for micro-energy-harvesting techniques to minimize the reliance on conventional batteries. Using electrochemical batteries in devices such as intelligent electronics, smart wearables for biomedical monitoring, medical implants, wireless sensors, and other ultralow-power transferable electronics suffers from several shortcomings. This includes limited battery lifespan, practical difficulty with battery replacement, and environmental pollution arising from the disposal of batteries. In response to this, researchers have conducted numerous studies to explore alternative micro-energy-harvesting solutions to replace the need for conventional batteries or to extend the battery's lifespan. Mechanical motion, particularly vibration or random displacements, is abundant and ubiquitous, from manmade systems to human bodies. In recent years, mechanical motion has emerged as a core energy source and has been widely studied for micro-energy harvesting.

Some popular mechanical energy-based transduction mechanisms are electrostatic, electromagnetic, triboelectric, and piezoelectric transduction [1]. Among these variant technologies for vibration-dependent energy harvesting, the piezoelectric transduction mechanism has attracted significant interest owing to its simplicity in design and implementation, good scalability, high power density, and minimal damping [2]. Moreover, piezoelectric devices can be manufactured on a microscale in flexible and stretchable miniature devices, marking a significant milestone in powering implantable medical devices [3]. The conventional linear piezoelectric energy harvester (PEH) is generally designed in unimorph or bimorph straight cantilevers, consisting of one or two piezoelectric transducers, respectively, often bonded on a metallic cantilever beam [4]. These piezoelectric patches are usually attached to the fixed end where the highest strain occurs. A proof mass is often attached at the free end to tune the system's resonance frequency and enhance the vibration-induced stress [4].

Various studies on mechanical design improvement of piezoelectric energy-harvesting systems associated with high and low-frequency vibrations have been conducted [1,5–8]. Moreover, there have been recent developments in electrical design improvements, material improvements, and hybrid structural improvements with related analytical and theoretical frameworks of PEHs [9–13]. However, studies focusing on the ultralow excitation frequency [14] (i.e., 1 Hz to 10 Hz), such as human-induced motions, are relatively limited. The main reason is the inherent high natural frequency response of most conventional linear PEH systems, which significantly limits the effective energy harvesting from ultralow-frequency excitations [15].

The energy of ambient vibrations widely exists in nature and is often randomly distributed over a broad frequency spectrum. However, conventional PEH devices are usually designed to operate at their first resonance frequency within a narrow bandwidth. It is well understood that the PEH generates the highest output power when the entire system, or specifically the cantilever beam, is excited at its natural frequency. These have been identified as significant drawbacks of the device. Various inventive solutions have been proposed to widen the operating bandwidth, including tunable linear and nonlinear harvesters, impact-driven harvesters, cantilever arrays, and multimodal systems which possess multiple resonances across a broader frequency spectrum [16].

Multimodal PEH designs are relatively advanced systems with more than one proof mass that can achieve two or more close resonance peaks, resulting in a wider operating bandwidth [1]. Wu et al. [17] proposed a two-degree-of-freedom PEH consisting of inner and outer unimorph beams. To obtain multiple resonances, the weight of the tip mass was altered accordingly. Another widely used concept to achieve multiple close resonance peaks is PEH with multiple branched beams. Branches attached to the main beam can be tuned to achieve several close resonance peaks, leading to a widening operating bandwidth. Zhang and Hu [18] studied a PEH with several branched beams attached to the main beam with a single piezoelectric transducer. Tip masses were attached to each branched beam to increase the vertical deflection of the main harvester beam. Upadrashta et al. [19] continued experimenting with the branched beam concept and introduced a PEH capable of performing in a wider operating bandwidth of 15 to 20 Hz. The harvester comprises three sub-beams with various sizes attached to a main unimorph beam. Following their study, Izadgoshasb et al. [16] proposed a multi-resonance PEH by optimizing the beam shape, which was a unimorph beam with two triangular branches. Most of these designs were superior in terms of power density to the conventional linear PEH due to the ability of a single transducer to attain multiple resonance peaks. However, these energy harvesters are less effective when operating under multidirectional ambient vibration sources with a broader spectrum [20], such as those induced by human motion.

To tackle this problem, multidirectional harvesters with multiple resonances were introduced by a few researchers. Erturk et al. [21] proposed an L-shaped multidirectional design harvester with two close resonance peaks. Su et al. [22] further modified this L-shaped design into a V-shaped bent beam harvester and investigated its performance with varying angles. However, the output power generated was much smaller and, thus, not suitable to act as a standalone harvester for real-life applications. Zhao et al. [20] took another step to improve Su et al.'s design. With the aid of the distributed parameter model, they investigated the multimodal behavior of the V-shaped bent beam harvester. Two close resonance peaks were achieved within the 0 Hz to 60 Hz frequency spectrum, which considerably improved the output power. However, the design produced only a single resonance below 10 Hz when the angle was fixed at 45°. Since the design needed to be manually tuned, its practical applicability was deemed limited and unsuitable for ultralow-frequency vibration sources.

Further investigation was carried out by Jiang et al. [23] by introducing an impact stopper to the V-shaped bent beam harvester. Compared to other designs, the design seems to be improved with a broader operating bandwidth in low-frequency regions. Nevertheless, the lowest resonance peak lay between 8 Hz to 15 Hz, which was much higher than the frequency of human motion (1 Hz to 5 Hz). Furthermore, the power density of the harvester was low since the harvester consisted of two bimorphs, a connector, an impact stopper, and a proof mass in total. In addition, several design criteria must be addressed while downscaling the harvester for practical applications.

Several research groups have attempted different technical approaches to design PEH with multiple resonance peaks in a low-frequency spectrum to widen the operating bandwidth as discussed above [16–23]. However, it was found that it is very challenging to produce a design that achieves improvements in both power efficiency and bandwidth widening, particularly for a complex vibration source such as human motion. Moreover, a simple, standalone PEH design with fewer accessories is preferred considering the end-user's comfort (human). Many preceding designs could hardly be miniaturized into microscale due to their complex design features, which are essential for wearable devices.

This study proposed a simple and effective multimodal PEH design, incorporating the V-shaped bent beam harvester and branched beam harvester for harvesting energy from ultralow-frequency excitation sources, particularly human motion. The key objective of this design was to improve the output voltage and power of the multimodal PEH merely through geometrical improvement, without the need for additional accessories (i.e., no weight penalty). The other objective was to broaden the operating bandwidth by achieving two or more close resonance peaks in the ultralow-frequency range.

The proposed multimodal PEH was designed to produce high power density using the branched beam concept. The designed bent branched beam harvester (BBBH) consisted of a unimorph beam with two bent branches attached at the free end. Finite element method (FEM) analysis was conducted to optimize the geometrical parameters of the BBBH based on the targeted frequency spectrum of human motion, while experimental tests were conducted using a mechanical shaker and human motion as sources of ultralow-frequency excitation. The performance of the proposed BBBH, in terms of the output voltage, the output power, the idle time, and the operating bandwidth, was compared with the existing designs, namely, the conventional cantilever beam harvester (herein denoted as CBH) and the V-shaped bent beam harvester (herein denoted as VBH). It is worth noting that a theoretical solution can be obtained by combining theoretical and FEA (including piezoelectric elements and electromechanical coupling) studies. However, this paper is focused on an experimental study in order to prove the proposed concept; hence, the theoretical study will be conducted in future research instead of in this study.

2. Materials and Methods

In this section, the materials and approaches used in the study are discussed.

2.1. Beam Harvesters Used in the Study

The proposed BBBH consisted of one main cantilever beam and two branches folded in a V shape at 45° (Figure 1). A macro-fiber composite (MFC; M2814-P2) patch of 28 mm × 14 mm × 0.350 mm was attached near the fixed end of the main cantilever beam, and the two branches were attached to the other end through a joint. This study also considered two different beam designs, the CBH (Figure 2a) and the VBH (Figure 2b).

For comparison, the size and properties of the main cantilever beam (horizontal portion) were the same for all three designs. CBH consisted only of the main cantilever beam and the joint, acting as the proof mass. As for the VBH, the main cantilever beam was equivalent to the CBH, with a bent beam added at the free end through the joint, forming a V shape. Aluminum alloy (Grade 6063-T5) was used for various components of the beams, whereas stainless steel (Grade 304) was used for the joints. The overall volume of the proposed BBBH and the VBH were kept identical for a fair comparison.

2.2. Finite Element Method (FEM) Analysis

In order to understand the mechanical behavior of beam harvesters, FEM analysis was conducted using ANSYS Multiphysics [16–19] software. Three-dimensional (3D) hexahedral elements were used to model all the beams and the joints. This element contained 20 nodes, and each node possessed three translational degrees of freedom in the polar coordinate system. The element size selected was 1 mm, sufficiently acceptable to obtain accurate results while not incurring excessively long computational time [16].

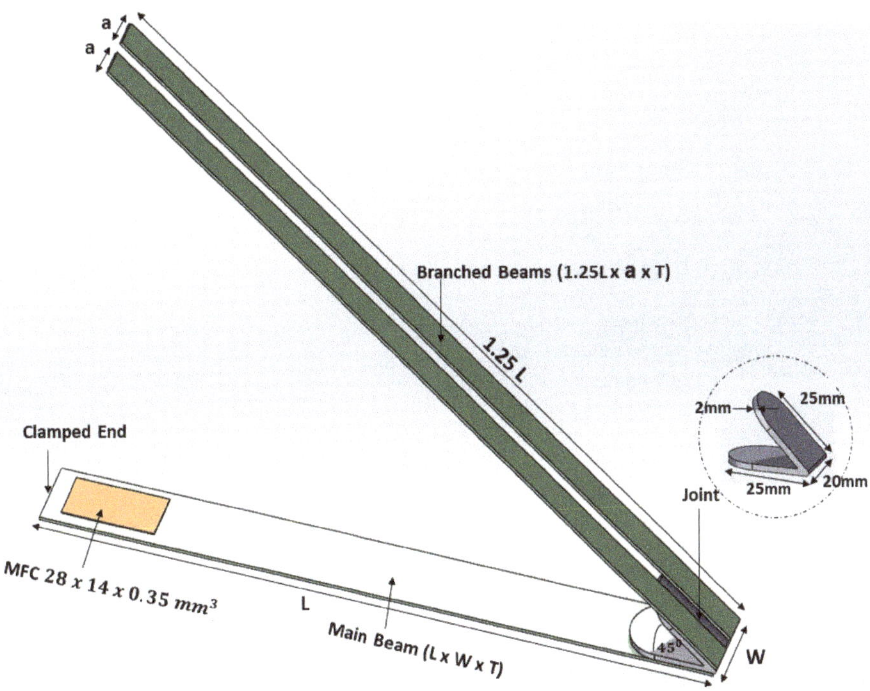

Figure 1. Schematic of proposed BBBH.

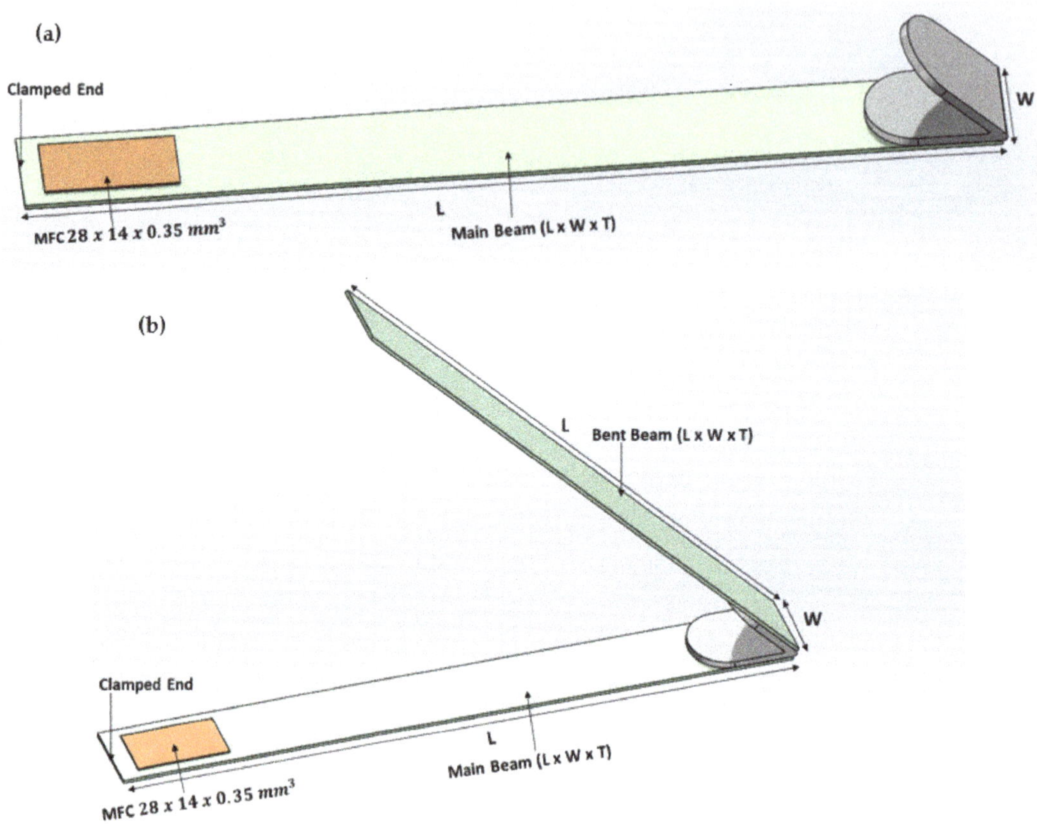

Figure 2. Schematics of two PEH designs; (**a**) CBH and (**b**) VBH.

The optimum dimensions of the main beam were then determined through a parametric study using the constructed FEM models. The main criterion is to achieve two or more close resonances in the ultralow-frequency range. Different geometries of the main cantilever beam were considered, directly affecting the vibratory response, such as the resonance frequency. Moreover, the suitability of the geometric properties for human motion application was also considered to achieve a user-friendly design.

Once the dimensions of the main cantilever beam were identified (denoted as L, W, and T standing for length, width, and thickness, respectively), it was used as the main beam (i.e., horizontal portion) for all three designs. For the VBH, the same dimensions (i.e., L, W, and T) were selected for the bent section. For the proposed BBBH, the length of the two branches was fixed at 1.25 L. This adjustment was made to ensure the total volume of BBBH was the same as VBH for a fair comparison. A gap was preserved between the two branches (width of each branch = a). Furthermore, 45° was adopted as the angle between the horizontal and the bent sections of the BBBH and the VBH, which was the optimal angle recommended in the previous study [20]. All symbolic dimensions are shown in Figure 1.

In the FEA, the main cantilever beams of all three designs were fixed at one end and free at the other to simulate the experimental condition. A V-shaped joint (weight: 15 g; size: 25 mm × 20 mm × 2 mm) was used as a proof mass for CBH and as a connector for VBH and BBBH. The material properties of the aluminum beam and the joint are summarized in Table 1. In addition, the material properties of the MFC actuator used in this study, are summarised in Table 2. These values have been obtained from the manufacturer's profile (Smart Materials Corporation).

Table 1. Material properties of aluminum beam and joint.

Parameters	Aluminum Beam	Joint
Materials	Aluminum alloy 6063-T5	Stainless steel 304
Elastic modulus (GPa)	68	190
Poisson's ratio	0.33	0.26
Density (kg/m^3)	2700	8100

Table 2. MFC actuator material properties.

Property	Young's Modulus E_1 (GPa)	Young's Modulus E_2 (GPa)	Poisson's Ratio V_{12}	Shear Modulus G_{12} (GPa)	Piezoelectric Charge Constants d_{33} (pC/N)	Piezoelectric Charge Constants d_{31} (pC/N)	Active Area Density (g/cm^3)
Value	30.336	15.857	0.31	5.515	4.6×10^2	-2.1×10^2	5.44

2.2.1. Geometrical Optimization of Main Beam

Modal analysis was first conducted using FEM analysis to determine the optimum geometrical properties for the main cantilever beam with the first resonance frequency ranging between 1 Hz and 10 Hz. It is worth noting that during the FEM analysis, discretization and convergence criteria were considered to identify the most suitable number of nodes (12,200) and element size (1 mm) in the model. From this point onwards, there was no change in result accuracy by increasing the number of nodes in the system. When the number of nodes was greater than the current consideration (12,200<), the computational time required was excessively long. Hence, the element size was kept at 1 mm for all components of the model. The resonance frequencies and modal characteristics can be tuned by changing the main cantilever beam, including the length, thickness, width, and the attached proof mass. For this particular study, only the beam's length and width were changed, while the proof mass and the thickness of the beam (1 mm) were kept constant. The minimum length considered in the parametric study was 75 mm, considering the length of the MFC patch (30 mm) and the length of the joint (25 mm).

For the first scenario, the width of the beam was kept at 20 mm, and the length was varied. Five trial lengths labeled Design L1 to L5 were considered (Table 3). The first six resonance frequencies (NF) of each length obtained from the modal analysis are presented in the table.

Table 3. First six NF of main cantilever beam with varying length (width = 20 mm).

Design	Length (mm)	NF (Hz)					
		1	2	3	4	5	6
L1	75	30.43	160.11	317.55	870.29	1686.80	2511.40
L2	100	18.68	124.60	209.00	543.28	833.72	1947.50
L3	150	9.54	91.69	115.67	282.77	346.57	793.72
L4	200	5.39	65.47	69.27	168.85	186.32	388.08
L5	250	4.15	47.98	62.57	128.83	138.92	276.79

It can be observed that the first natural frequencies of Design L1 and Design L2 were much higher than the desired frequency range (i.e., <10 Hz) and, hence, were not suitable for this study. The first resonance of Design L3 was also considered too high for human motion. On the other hand, as stated previously, the volumes of the proposed BBBH and

VBH were kept the same. For a main beam length of 250 mm (Design L5), the required branch length would be 315 mm for BBBH, which was relatively too long and undesirable. Therefore, Design L4 (L = 200 mm) was the most suitable design for this study.

To further tune the natural frequency according to the targeted frequency spectrum, the width of the beam was adjusted with the length fixed at 200 mm, based on Design L4. Considering that the width of the MFC was 15 mm, the minimum width for the trial was 20 mm. Three different widths (20 mm, 25 mm, and 30 mm) were analyzed, and the corresponding natural frequencies are presented in Table 4.

Table 4. First six NFs of a main cantilever beam with varying width (length = 200 mm).

Design	Width (mm)	NF (Hz)					
		1	2	3	4	5	6
W1	20	5.39	65.47	69.27	168.85	186.32	388.08
W2	25	6.57	73.83	83.43	212.28	236.63	437.95
W3	30	7.09	75.07	89.40	218.52	293.88	446.65

It was observed that the resonance frequency increased with increasing width. Note that the increase in width also remarkably affected the stress of the beam. For instance, when the width of the beam was 200 mm, the stress gain near the fixed end was 8.73×10^5 Pa. The stress gain gradually decreases to 7.29×10^5 Pa, and then 6.08×10^5 Pa when the width is increased to 250 mm and then 300 mm, respectively. In addition, a lower natural frequency was preferred for human motion application; hence, the dimension of the main cantilever beam was selected as 200 mm × 20 mm × 1 mm for this study. For VBH, an identical section (200 mm × 20 mm × 1 mm) was attached at the free end of the main beam, forming an angle of 45° with the horizontal plane. As for BBBH, two identical branches (250 mm × 8 mm × 1 mm) were attached to the free end of the main beam at 45°. The gap between the two branches was 4 mm. Consequently, the volumes of VBH and BBBH were the same. However, it is worth noting that these dimensions were chosen as a proof of concept, and with the special characteristics of piezo and the simplicity of the device, it can be scalable to use as microstructures for real-time applications.

2.2.2. Modal Analysis of CBH, VBH, and BBBH

Modal analysis was further conducted for all three PEH designs to identify their resonance frequencies in the ultralow range. The first six resonance frequencies of each PEH are presented in Table 5.

Table 5. First six NFs of each PEH design.

Design	NF (Hz)					
	1	2	3	4	5	6
CBH	5.39	65.47	69.27	168.85	186.32	388.08
VBH	4.50	11.80	14.24	72.09	130.68	132.43
BBBH	4.09	8.72	10.56	14.47	55.71	90.63

It can be noted that the CBH possessed only one resonance in the targeted frequency spectrum (<10 Hz). In addition, the first six resonance frequencies of the CBH spread over a wide frequency range of 5.39 Hz to 388.08 Hz. The VBH showed slightly improved behavior compared to the CBH, with a lower first resonance frequency of 4.50 Hz. Furthermore, the spread of natural frequencies was reduced from 4.50 Hz to 132.43 Hz. However, VBH also only possessed one resonance within the ultralow-frequency range.

The proposed BBBH, on the other hand, achieved a significant improvement compared to the CBH and the VBH. The first three resonance frequencies were 4.09 Hz, 8.72 Hz, and

10.56 Hz, indicating that the proposed BBBH could achieve three close resonance peaks in the targeted range (1 Hz–10 Hz), indicating the widening of operating bandwidth. This suggested that the proposed PEH could potentially harness higher output voltage and power than the other two designs in the frequency range of interest.

The mode shapes and vertical displacements of the BBBH for the first six resonances were simulated as illustrated in Figure 3. It can be noted that both branches were vibrating in the same direction for the first, second, third, and fifth resonances which would generate maximum strain in MFC [19]. For the fourth and sixth modes, the branches were vibrating in opposite directions, which was expected to yield minimum strain in the MFC [19,24]. For this study, only the first three mode shapes were relevant since they fell within the desired frequency range.

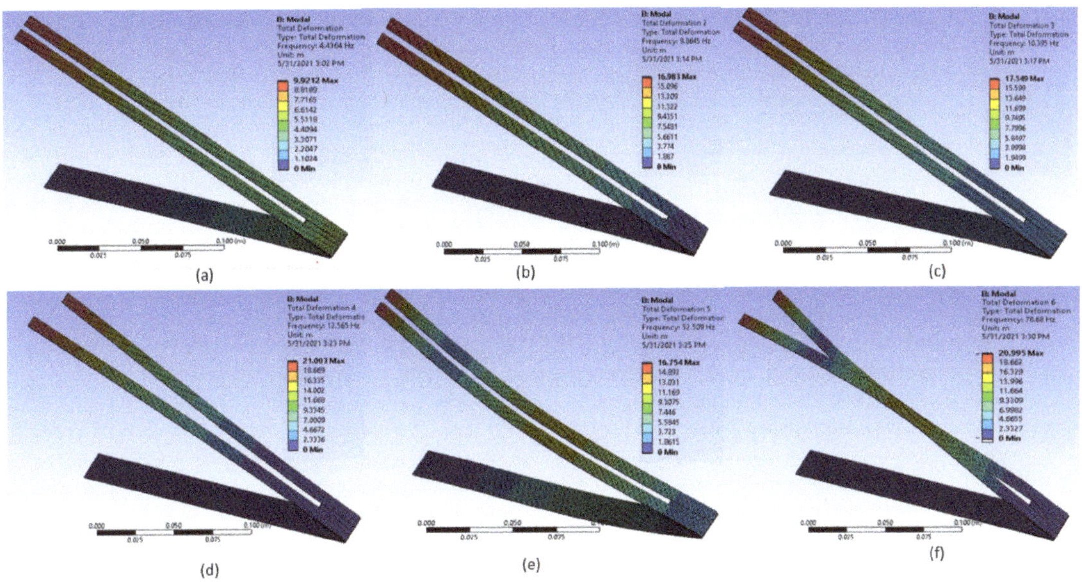

Figure 3. Deformed shapes of the first six resonances of BBBH obtained from modal analysis. (**a**) NF1, (**b**) NF2, (**c**) NF3, (**d**) NF4, (**e**) NF5, (**f**) and NF6.

2.2.3. Static Analysis of CBH, VBH, and BBBH

The energy produced from PEH is directly related to the strain gained in the main cantilever beam. The study's purpose of the static analysis was to elucidate the highest stress and strain achievable by various PEH designs. In the static analysis, the material properties were the same as those simulated for modal analysis. The bolt mass (1.2 g) was simulated as a point load [19]. The material properties of the beams and the proof mass (joint) are listed in Table 1. The simulation outcomes, in terms of highest strain and stress, are tabulated in Table 6.

Table 6. Highest stress and strain achieved by various PEH under static FEM.

PEH Types	Parameters	Highest Recorded Values
CBH	Strain (m/m)	1.28×10^{-5}
	Stress (Pa)	8.73×10^{5}
VBH	Strain (m/m)	1.29×10^{-5}
	Stress (Pa)	8.77×10^{5}
BBBH	Strain (m/m)	1.43×10^{-5}
	Stress (Pa)	9.74×10^{5}

As expected, the highest strain and stress of the cantilever beam occurred near the fixed end, where the piezoelectric patch was attached. It was noted that the CBH achieved the lowest stress of 8.73×10^5 Pa, while the VBH attained a fairly similar stress level of 8.77×10^5 Pa. The proposed BBBH yielded the highest stress of 9.74×10^5 Pa, which was relatively higher than the other designs. A similar trend was observed with the induced strain. These results indicated that the proposed BBBH could generate relatively higher stress and strain, potentially generating higher voltage and power when dynamically excited.

2.3. Experimental Study

A series of experimental tests were conducted in two stages to study the performance of the proposed BBBH. In the first stage, the three designs were excited using a mechanical shaker to generate controlled and stable vibrations at the ultralow-frequency range. In the second stage, they were excited by human walking motions to generate real-life complex vibrations within the ultralow-frequency range. The performance of the three designs was studied and compared in terms of output voltage, output power, and idle time.

2.3.1. Fabrication of the Harvesters

The size and design of the CBH, VBH, and BBBH were the same as the model developed in FEM analysis. All beam sections were fabricated using aluminum grade 6063-T5. For the CBH, a proof mass of 15 g was added at the free end of the main cantilever beam. The material properties of beams and joints are presented in Table 1. An MFC (M2814-P2) patch was attached near the fixed end of the main beam using two-part epoxy. The weight of the MFC patch was 0.5 g, and the size was 28 mm × 14 mm × 0.35 mm. However, it is worth noting that the size of the deployable energy harvester should be small; however, to demonstrate the concept and underlying working principle, a mesoscale-sized prototype was tested. With the special characteristics of piezo, the device can be scalable to use as microstructures for real-time applications.

2.3.2. Stage 1: Shaker Test

In Stage 1, all three PEH designs were tested with a mechanical shaker as the source of excitation. One end of the main cantilever beam was mounted on an electrodynamic shaker (APS-113, APS Dynamics, Inc. San Juan Capistrano, CA, USA) capable of emulating low-frequency vibrations. A power amplifier (APS125, APS Dynamics) was used to drive the mechanical shaker according to the input frequency from the function generator (Agilent 33210A). An accelerometer (Dytran 3305A2, 0.3 to 5000 Hz, ±5%) was fixed at the base of each PEH. The acceleration and the output voltage were measured using NI DAQ modules, NI 9234 and NI 9229, respectively, with the aid of Signal Express software. Next, the output voltage (V) was used to calculate the generated power under open circuit conditions. Considering that the load resistance (Rload) of the open circuit was close to infinity, a considerable load resistance (1 MΩ) [2,3] was assumed in the calculation. The output power was calculated as V^2/R_{load}. The idle time of each harvester was then calculated by analyzing the duration of damping between each impact. A shorter duration of damping indicated a shorter idle time and, thus, high efficiency. Figure 4 illustrates the schematic of the experimental setup. Each PEH design was tested across a frequency range of 1 Hz to 5 Hz, at a step of 1 Hz. Two accelerations (i.e., 1 g and 2 g) were sequentially applied for each frequency.

2.3.3. Stage 2: Human Motion Test

Human motion is a complex movement that can occur in various directions with different amplitudes and accelerations, depending on the source and intention of movements. In general, the amplitude is high, and the frequency is low. This study selected typical human motions with relatively steady and repetitive movements, namely, the walking and the running motions (as sources of excitation). A wearable, customized mechanical fixture [25] was used to attach the three PEH designs to the leg of a female weighing 54 kg, as illustrated in Figure 5. As explained previously in Section 2.3.1, the size of the proposed design can be easily downsized to fit into a wearable device, i.e., without harming the user [26]. Each PEH was attached to the mechanical fixture and tightened with screws to preserve the fixed end conditions. Previous studies found that the impact generated during foot strikes predominantly induced vertical acceleration [27–30]. Thus, for ease of comparison, the outcome was governed by vertical acceleration, which was considered the main motion with the largest amplitude.

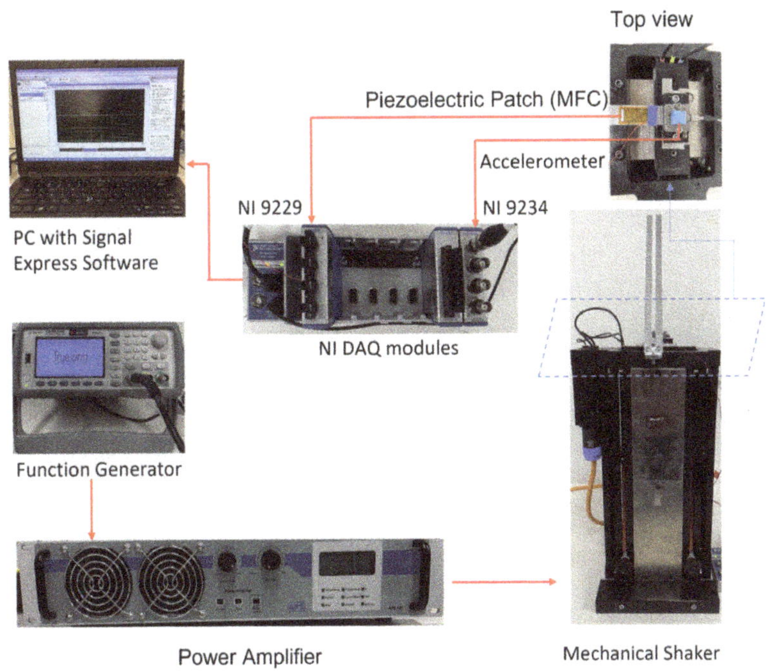

Figure 4. Experimental setup.

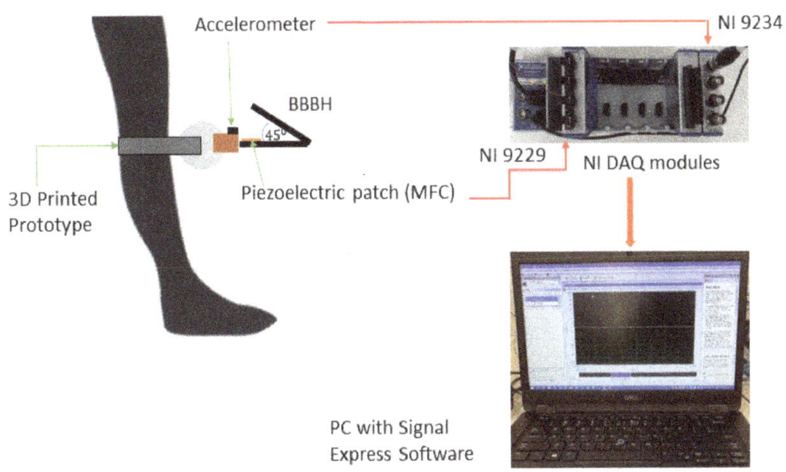

Figure 5. Schematic of experimental setup for the human motion test.

It is worth noting that the mechanical fixture was affixed to the bony landmark location [31] to reduce the relative motion between the bone and the PEH system to minimize the damping effect caused by the skin [25]. The accelerometer was mounted at the base of the main cantilever beam to capture the acceleration induced by walking and running motions. Acceleration generated by different subjects could vary according to their weight, height, and habit of walking or running.

In order to obtain repeatable and stable outcomes, the test subject performed the walking and running motions on a treadmill at constant speeds of 2 km/h and 4 km/h, respectively. The frequencies measured were 1 Hz and 1.5 Hz, respectively. Table 7 gives a summary of all variables involved in the shaker test and the human motion test. In addition, the test subject was required to walk or run continuously for at least 5 min to reach a relatively stable state before data acquisition commenced. For each test scenario, three measurements were taken to ensure repeatability. Similarly, the acceleration and the output voltage were measured using the same NI DAQ modules. The test was repeated for all three designs.

Table 7. Variables engaged in shaker and human motion tests.

Tests	Variables	
	Acceleration (g)	**Frequency (Hz)**
Mechanical Shaker Test	1 and 2	1, 2, 3, 4, and 5
Human Motion Test	0.2 and 0.45	1 and 1.5

3. Experimental Results and Discussions

Results obtained from mechanical shaker test and human motion test are discussed in this section.

3.1. Shaker Test

3.1.1. Output Voltage

In the shaker test, all three designs were subjected to two sets of base accelerations (1 g and 2 g) and five sets of frequencies (1, 2, 3, 4, and 5 Hz). Figure 6 illustrates the acceleration and output voltages against time measured from all designs under 1 Hz and 1 g base excitation. The highest output voltages obtained from CBH (Figure 5b) and VBH (Figure 5c) were about 10 V and 12 V, respectively. A slightly improved performance can be noticed in the VBH compared to the CBH. On the other hand, the voltage generated by BBBH (nearly 20 V in Figure 5d) was remarkably higher than the existing designs (i.e., CBH and VBH), which was approximately two times the CBH and one and a half times the VBH. This was deemed a significant improvement in output voltage for ultralow-frequency vibrations, achieved purely through geometrical design/modification, without additional components such as proof masses, magnets, shock vibrations, or impact stoppers.

Next, the effect of varying excitation frequency was further investigated to understand the operating bandwidth of harvesters. For illustration, the peak output voltage recorded from all three designs under 1 g acceleration for 1 Hz to 10 Hz is presented in Figure 7. The results indicated that the proposed BBBH design was superior to the existing designs, producing the highest output voltage of 44 V at 4 Hz within the frequency range of interest (1 Hz–5 Hz). As per the FEM analysis, the first natural frequency of BBBH was 4.09 Hz. As expected, the higher output voltage was achieved when the excitation frequency was close to the natural frequency. Similarly, the highest voltage gained by VBH was 16 V at about 4 Hz, which was also close to the natural frequency obtained for VBH in the FEA analysis (i.e., 4.5 Hz). The CBH achieved its highest voltage of 23 V at about 5 Hz since the excitation frequency was close to its natural frequency (i.e., 5.39 Hz from FEA).

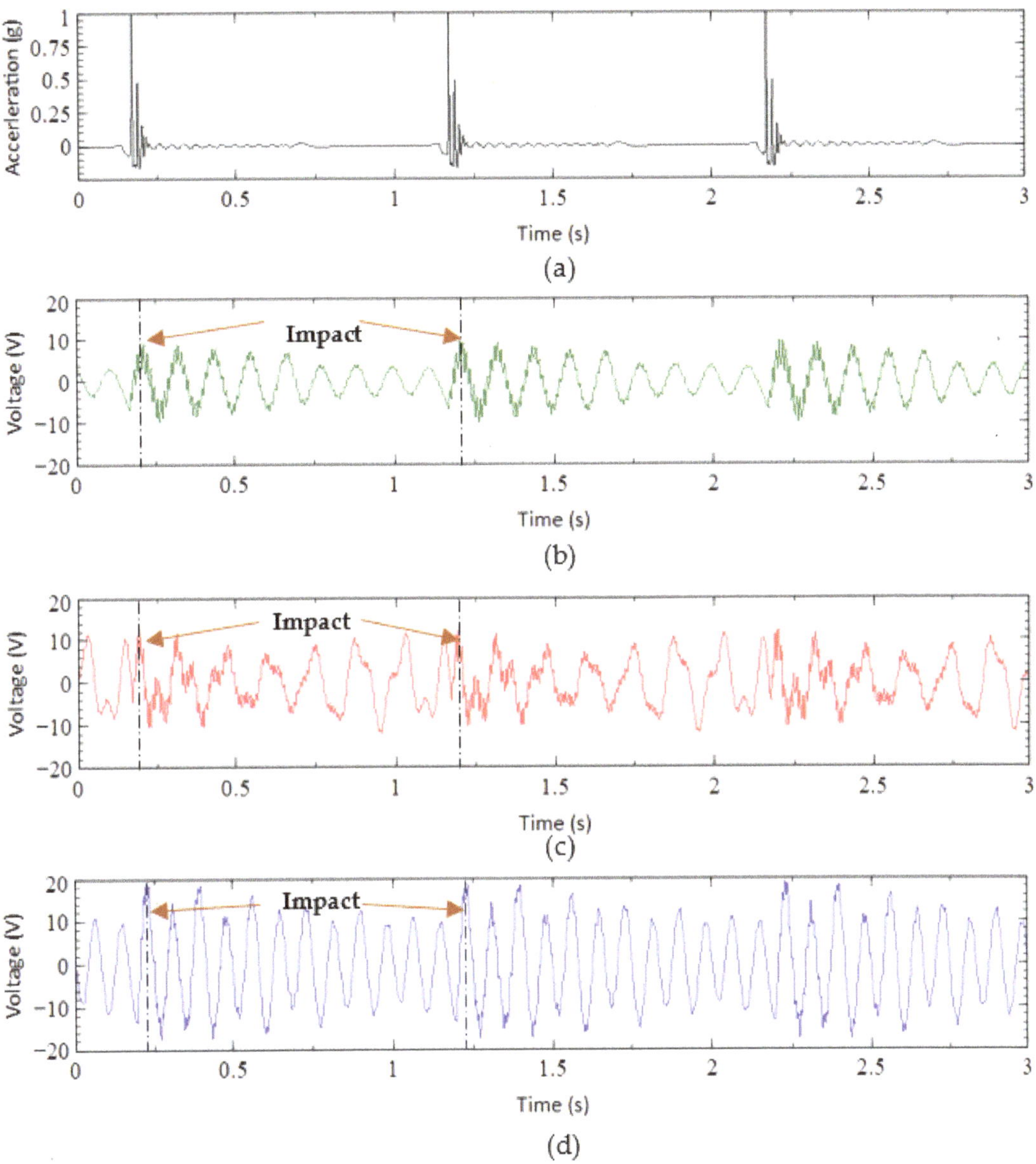

Figure 6. (**a**) Base acceleration and voltage output of (**b**) CBH, (**c**) VBH, and (**d**) BBBH under 1 Hz and 1 g acceleration.

Another observation in Figure 7 was that the BBBH could perform remarkably at its second natural frequency, which was around 8 Hz (FEA 8.72 Hz), with a voltage output of 62 V. As noted in the FEM analysis study and experiments, CBH and VBH could achieve only one natural frequency within 1–10 Hz. Hence, this further proves that the proposed BBBH widened the operating bandwidth, achieving two natural frequencies below 10 Hz. The superiority of BBBH in voltage generation was due to the higher displacement caused by the amplification of the main cantilever beam by both branched beams.

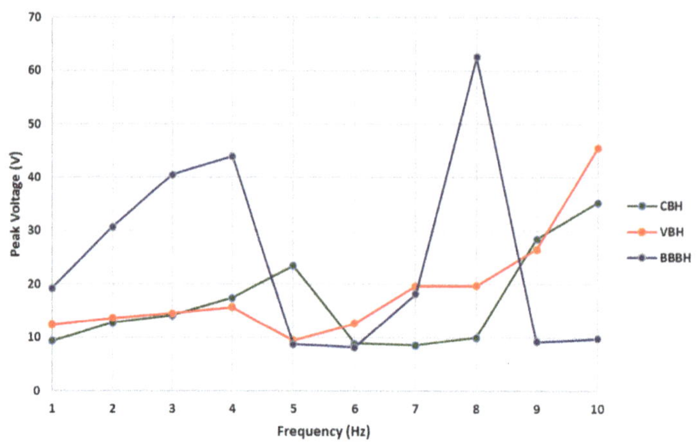

Figure 7. Output voltage of CBH, VBH, and BBBH under excitation frequency of 1 Hz–10 Hz under 1 g acceleration.

Next, the same set of experimental tests was conducted for 1 Hz to 5 Hz under 2 g base acceleration. When the acceleration was increased to 2 g for 1 Hz, the general behavior of the three designs and the observations made were similar to the case of 1 Hz under 1 g acceleration. However, the magnitude of the voltage was higher due to higher acceleration. For instance, the BBBH was able to generate a maximum of 20 V for 1 Hz under 1 g acceleration. Under 2 g acceleration, BBBH produced 50 V, almost 2.5 times higher than the case of 1 g acceleration for 1 Hz. Considering that the acceleration was doubled, this outcome was expected. Similarly, for other frequencies (2 Hz to 5 Hz), a similar trend was also observed, although the magnitude of the voltage peak and the number of peaks per cycle were different. For simplicity, graphical representations of other cases were omitted.

3.1.2. Idle Time

The idle time can be used as an indicator of the efficiency of the PEH design. In this study, idle time referred to the period in which the energy harvested was relatively low. This usually occurs between each impact cycle and is particularly critical in the case of human motion, which is characterized by relatively large amplitude but a very low frequency. Generally, when the idle time is shorter, more spikes (i.e., voltage peaks) arbitrate in a motion cycle, resulting in higher output voltage and output power. If a PEH is not designed properly, the idle time is significant, resulting in low efficiency.

In this study, the idle time of each harvester in a motion cycle was analyzed. In the case of CBH presented in Figure 8a, the peak voltage obtained was about 10 V during a single motion cycle (i.e., one second), representing a single impact given at the base of CBH by the mechanical shaker. Figure 6b illustrates that the peak voltage was achieved at the moment of impact. However, the voltage dramatically dropped immediately after the impact due to damping. Despite the spike in voltage within a short timeframe, approximately 0.5 s, or nearly half of the motion cycle, was idle or mainly wasted. For instance, if the peak voltage (10 V) of CBH at 1 Hz under 1 g acceleration was considered as the baseline. Approximately 0.5 s after the impact, the voltage generated reduced to 4 V, reflecting nearly 60% of the voltage drop. In the next half-motion cycle, the voltage further reduced to 2 V, reflecting nearly 80% of the voltage reduction compared to the baseline. This latter 0.5 s was referred to as the idle time of the CBH. During the idle time, the voltage generated was remarkably smaller than the peak voltage obtained. When the voltage was low, the average power or energy produced was also expected to be low. This is one of the major drawbacks of conventional CBH.

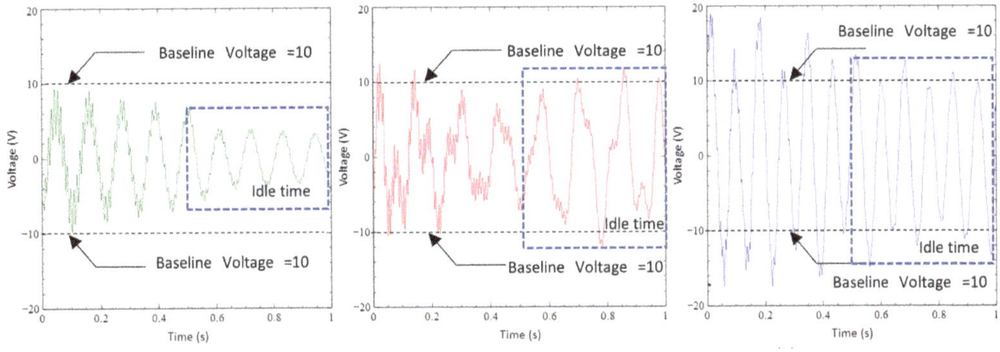

Figure 8. Idle time for (**a**) CBH, (**b**) VBH, and (**c**) BBBH at 1 Hz under 1 g acceleration.

The peak voltage output of VBH was nearly 12 V (Figure 8b), which was a slight improvement compared to CBH. If the same baseline (peak voltage of CBH) was applied to VBH, the number of voltage spikes reaching or exceeding the baseline improved considerably. For instance, after 0.5 s of the impact (first half motion cycle) of VBH, the voltage reduced to 7 V, which was a 30% of voltage reduction compared to the baseline. During the next half motion cycle, VBH still produced voltage spikes of 10 V (a gain of 100% compared to the baseline). This can be due to the increased vertical displacement of the bent section of VBH, which amplifies the main beam's vertical displacement. Compared to the CBH, the idle time of VBH was reduced since the voltage reduction of the whole motion cycle was only about 30%. Hence, VBH was capable of producing a higher voltage output with idle time in each motion cycle compared to CBH.

In the case of BBBH, the peak voltage generated by the harvester was about 20 V, which was two times higher than the peak voltage of CBH (the baseline voltage). If the same baseline was considered for BBBH, all the voltage spikes obtained for BBBH were equal to or larger than the peak voltage of CBH. To be further explicit, the whole motion cycle is presented in Figure 8c. The peak voltage of 20 V was generated at the impact instance, which was 200% of the baseline. Then, after 0.5 s, BBBH could still generate 16 V, which was 160% of the baseline voltage. During the next half-motion cycle, BBBH produced a minimum of 10 V equal to the baseline voltage (100%). Therefore, the voltage drop relative to the peak voltage generated by CBH was negligible for the case of BBBH. This reflected that the idle time of BBBH was remarkably improved compared to CBH and VBH. The reduction in idle time greatly enhanced the efficiency of the harvester.

3.1.3. Number of Voltage Peaks/Spikes

It was noticed that the number of voltage spikes generated during one motion cycle varied across different harvesters. This number of voltage spikes, and their corresponding magnitude, was also a useful indicator of the performance of different harvesters. In general, more spikes indicate less idle time and, thus, higher efficiency. For such a purpose, the output voltage of CBH, VBH, and BBBH at 1 Hz under 1 g acceleration within one motion cycle was considered. Again, the peak output voltage of CBH (10 V) was considered as the baseline for the comparison. With such conditions, the CBH could only produce one 10 V voltage spike within one motion cycle. For the VBH, approximately six voltage spikes reached or exceeded the baseline (10 V) within 1 s, which was a considerable improvement.

On the other hand, the BBBH was able to produce 24 voltage spikes which were equal to or larger than the baseline voltage within one motion cycle, implicating a significant reduction in idle time. The presence of stronger voltage spikes was mainly due to the two branched beams, which amplified the main beam vibration due to the higher vertical displacement experienced by the branches in a single impact. The improvement of BBBH, in terms of the number of voltage spikes, was 24 times and 4 times more than CBH and VBH, respectively. Table 8 tabulates the baseline voltage and the number of peaks attained for all three designs under various excitation conditions. A similar trend was also observed for all the five frequencies under 2 g acceleration; however, for simplicity, representations of these cases were omitted.

Table 8. Number of peaks obtained for different frequencies within one motion cycle under 1 g.

Acceleration	Frequency	Harvester Design	Baseline Voltage	The Number of Peaks Reached the Baseline Voltage
1 g	1 Hz	CBH	10 V (peak voltage of CBH)	1
		VBH		6
		BBBH		24
	2 Hz	CBH	13 V (peak voltage of CBH)	2
		VBH		8
		BBBH		24
	3 Hz	CBH	14 V (peak voltage of CBH)	3
		VBH		9
		BBBH		24
	4 Hz	CBH	16 V (peak voltage of VBH)	16
		VBH		8
		BBBH		24
	5 Hz	CBH	8 V (peak voltage of BBBH)	20
		VBH		10
		BBBH		5

3.1.4. Harvested Power

The output power was calculated to further analyze the proposed design's productivity. Because the open circuit's load resistance (Rload) was close to infinity, a considerable 1 MΩ load resistance was assumed in the calculation [16,19]. The output power was calculated as V^2/R_{load}. Figure 9 illustrates the power output by various designs during three motion cycles under the impact of 1 Hz from the mechanical shaker at 1 g acceleration. The output power from different designs varied remarkably, while their trends were similar to the output voltage. Figure 9a shows that the highest power generated by CBH was 90 µW. The power between each motion cycle decreased rapidly due to damping after each impact.

The power generated by the VBH was considerably higher than the CBH in one complete cycle, as presented in Figure 9b. The highest power of nearly 160 µW was achieved. However, similar to the case of voltage, the power remained relatively low between each impact. The power generated by BBBH is illustrated in Figure 9c. BBBH generated a maximum power of 360 µW, which was more than twice the VBH's highest power. In addition, the BBBH could produce power ranging from 90 µW to 300 µW during each cycle. It is worth mentioning that the average low-power peaks generated by BBBH were higher than or equal to the highest power achieved by the CBH.

Thus, the BBBH was identified as the most efficient PEH in power generation among the three designs. This superiority was explained earlier in the study of output voltage, where the higher displacement caused the amplification of the main cantilever beam by both branched beams. As per the observations made during the experiment, the influence of multidirectional vibration experienced by the branches can be another cause for this high-power generation. Further study is recommended to understand better the interaction between the branches and the main beam, as well as the mechanisms involved in voltage and power amplification.

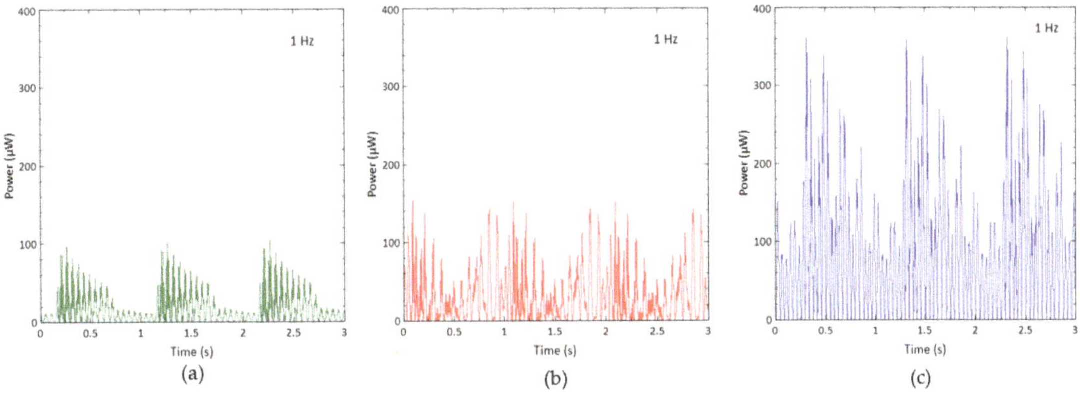

Figure 9. Output power of (**a**) CBH, (**b**) VBH, and (**c**) BBBH when excited under 1 Hz and 1 g acceleration.

Graphical representations of other frequencies (2 Hz to 5 Hz) for BBBH under 1 g acceleration are provided in Figure 10. BBBH was capable of producing at least 500 µW for the cases of 2 Hz and 3 Hz. Overall, the optimum performance of the BBBH was observed at 4 Hz, producing power ranging from 1400 µW to 1900 µW. However, the output power at 5 Hz decreased dramatically as the excitation frequency (5 Hz) was away from the natural frequency.

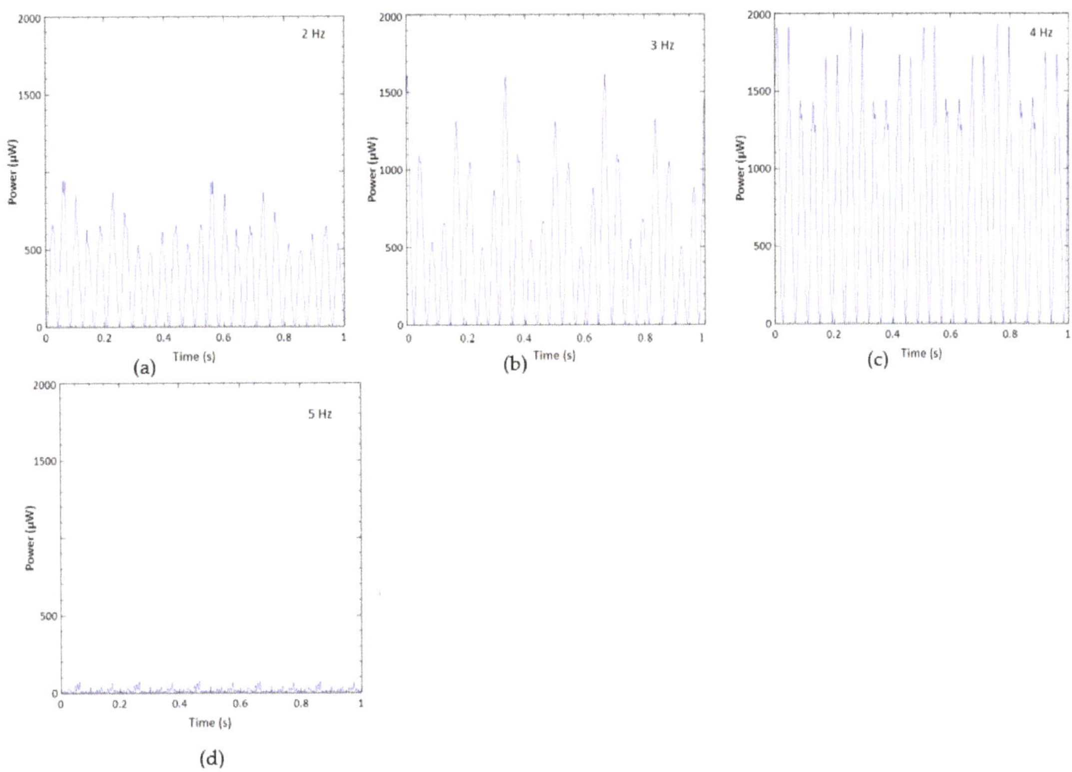

Figure 10. Output power generated by BBBH at (**a**) 2 Hz, (**b**) 3 Hz, (**c**) 4 Hz, and (**d**) 5 Hz under 1 g acceleration.

The superior performance of BBBH over the counterparts was also observed in these frequency ranges, and the details were omitted herein for simplicity. A similar trend was also noticed for various excitation frequencies under 2 g acceleration.

3.1.5. Average Output Power

To further understand the power generation of each PEH design under varying acceleration and frequency, the average output power produced across one cycle was calculated and illustrated in Figure 11. The average output power generated from the CBH gradually increased from 15.67 μW to 146.95 μW when the frequency was increased from 1 Hz to 5 Hz. A slight improvement of 31.04 μW in average output power was noticed for the VBH at 1 Hz under 1 g acceleration compared to the CBH. For higher frequencies (2 Hz to 4 Hz), the average output power gradually increased from 37.89 μW to 41.78 μW; however, for these frequencies, the performance of VBH was similar to CBH. The lowest average power of 11.82 μW was obtained at 5 Hz for VBH when the excitation frequency was away from its natural frequency.

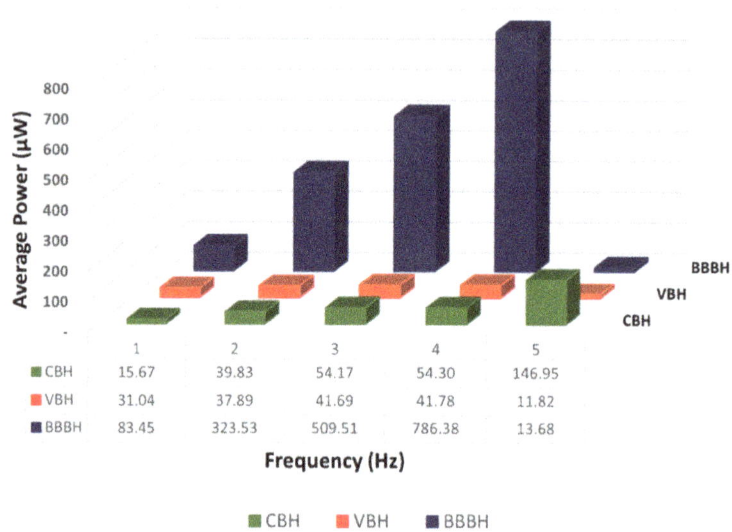

Figure 11. Comparison of average output power against frequency generated by three PEH designs under 1 g acceleration.

On the other hand, the BBBH showed a dramatic improvement in the average output power from 1 Hz to 4 Hz, with values ranging from 83.45 μW to 786.38 μW, respectively. The lowest average output power of BBBH, 13.68 μW, achieved at 5 Hz, coincided with the lowest average output power achieved by CBH at 1 Hz.

3.2. Human Motion Test

The human motion test was further conducted to study the proposed design's suitability in practical applications. Once again, the results obtained from the BBBH were compared against the CBH and the VBH. The output voltage and power produced during walking (1 Hz) and running (1.5 Hz) motions were recorded, analyzed, and discussed. Furthermore, the efficiency of the harvester in terms of idle time and the average output power was discussed.

3.2.1. Output Voltage

Figure 12 presents the output voltage generated by the three designs during walking with an impact frequency of 1 Hz. This impact occurred when the heel of the leg hit the ground. The acceleration of each impact was controlled at approximately 0.2 g for all designs. The highest output voltage generated by the CBH under such circumstances was about 20 V (Figure 12a) at the time of impact. Like the shaker test, the voltage decreased dramatically due to damping. Figure 12b represents the output voltage of VBH under the same excitation condition. The highest output voltage from VBH for walking motion was about 22 V, slightly improving than the CBH. This observation matched the trend observed in the mechanical shaker test. Figure 12c shows the output voltage generated by the BBBH during the walking motion. The BBBH produced nearly 40 V during the impact, with almost double the highest voltage produced by the CBH and the VBH. Several voltage peaks (10–30 V) were present between each impact, while the period of most peaks was also relatively longer.

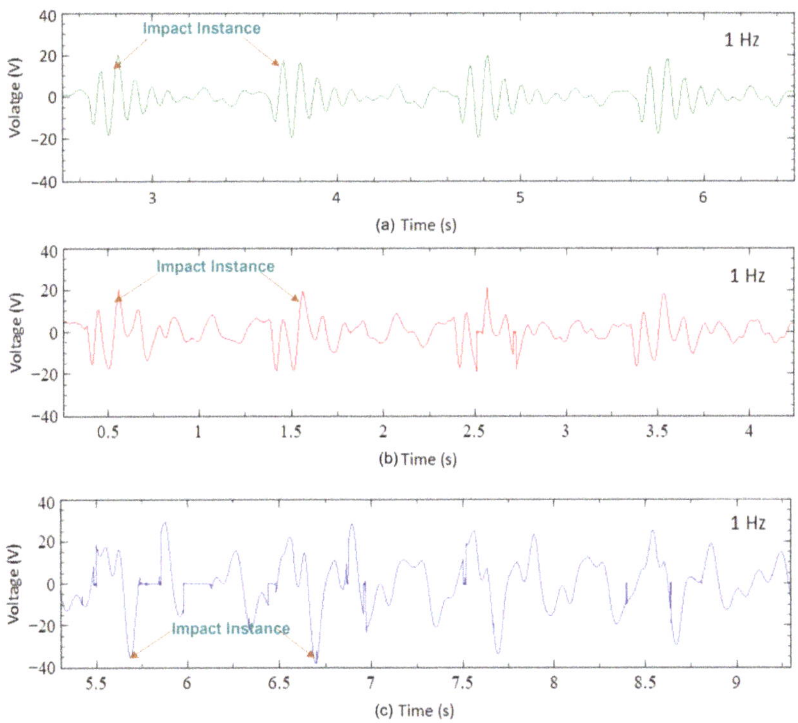

Figure 12. Voltage output of (**a**) CBH, (**b**) VBH, and (**c**) BBBH for human walking (1 Hz).

Similarly, a running test was conducted to further investigate the proposed design's performance. Figure 13a shows the voltage harvested by the CBH during running with an impact frequency of 1.5 Hz. The acceleration was maintained at approximately 0.45 g. For each design, four motion cycles were considered where six impacts were encountered. The highest output voltage generated by the CBH during running motion was 37 V, which was considerably higher than the walking motion. Once the impact was released, the output voltage decreased dramatically, similar to the walking and shaker test cases. Figure 13b represents the voltage output produced by VBH. The highest voltage obtained was nearly 40 V. In the remainder of the period, a minimum of 10 V was often achieved in one cycle. This was a notable improvement over the CBH.

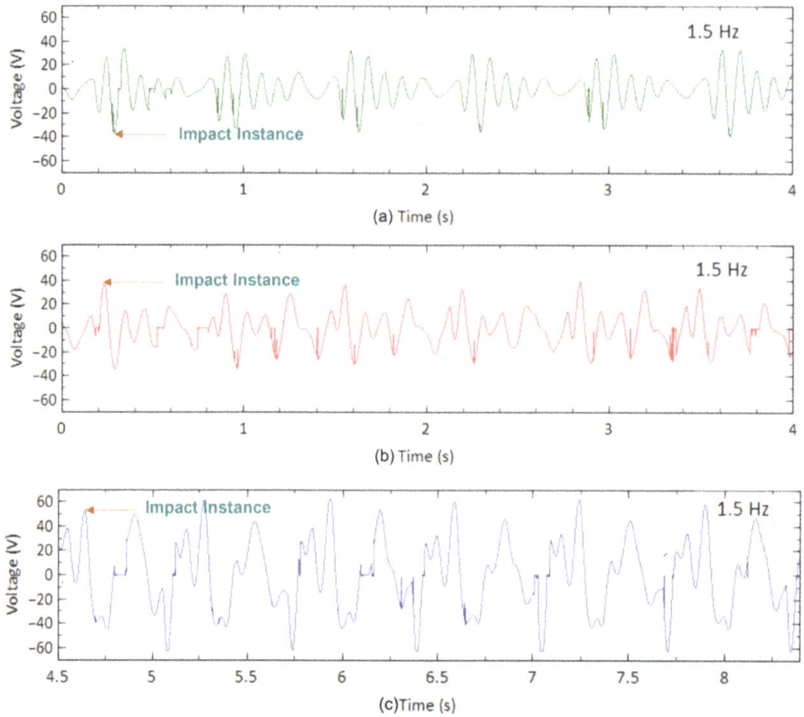

Figure 13. Voltage output of (**a**) CBH, (**b**) VBH, and (**c**) BBBH for human running motion (1.5 Hz).

In contrast, the BBBH (Figure 13c) showed a significant voltage improvement in running motion. In each cycle, the highest voltage achieved was 60 V. Between each impact, multiple voltage peaks ranging from 30 V to 60 V originated (further discussed in the next section). This was a vast improvement compared to the performance of CBH and VBH.

3.2.2. Idle Time

To illustrate the idle time variation among the harvesters for human motion, a walking motion was considered. In the case of CBH, the peak voltage achieved was 20 V at each impact during a single motion cycle (Figure 12a). However, the voltage generated was heavily damped following each impact. Apart from the spike in voltage within a short period, approximately 0.5 s, or nearly half of the motion cycle, was idle or wasted. In contrast, if the peak voltage (20 V) of CBH was considered as the baseline, nearly 0.5 s after the impact instance, the voltage generated was reduced to 5 V, illustrating approximately a 75% voltage drop. In the next half-motion cycle, the voltage further reduced to 2 V, indicating a 90% reduction. The voltage generated during the idle time was remarkably smaller than the peak voltage. This relatively long idle time rendered the CBH rather inefficient. It is worth noting that this behavior was similar to the observation from the mechanical shaker test.

The peak voltage achieved for VBH (22 V, illustrated in Figure 12b) was slightly improved compared to CBH. For a fair comparison of idle time, if the same baseline (peak voltage of CBH) was considered for VBH, after 0.5 s of the impact, the voltage dropped to 10 V, emphasizing a 50% voltage reduction. During the next half cycle, the voltage slightly reduced to 7 V and persisted until the next impact. Thus, the voltage reduction at the end of the latter half-cycle of the motion was about 75%. Compared to CBH, this was a slight improvement.

Figure 12c depicts the voltage generation of BBBH during the walking motion. The peak voltage generated by BBBH was nearly 40 V, approximately twice that of CBH and VBH. Thus, if the same baseline (peak voltage of CBH) was considered for BBBH to maintain a fairground, the peak voltage achieved by BBBH at the impact instance was 200% compared to the baseline voltage. After 0.5 s from the impact, BBBH still produced a maximum of 30 V, about 150% of the baseline voltage. During the latter half of the motion cycle, the voltage varied between 15 V and 20 V, indicating a maximum of 25% reduction from the baseline voltage. This was deemed a significant improvement compared to the existing designs in idle time, and the observations were similar to the shaker test.

For running motion, a similar trend was noticed. However, the idle time in walking motion was longer than in running motion, most likely caused by the higher frequency and acceleration in the running along with the complexity of human motion. In addition to the peak voltage generated during the motion cycle, the number of voltage spikes that appeared was also different for each harvester. Table 9 summarizes the number of voltage peaks generated by each harvester under different motion types. The table summarizes the results of Figures 12 and 13 in various forms. For instance, a higher number of voltage spikes exceeding the baseline voltage indicates that less time in a motion cycle is wasted (i.e., less idle time).

Table 9. The number of voltage peaks obtained in one motion cycle for walking and running motion.

Motion Type	Harvester Design	Number of Peaks Reaching Baseline Voltage (20 V) in One Cycle
Walking (1 Hz)	CBH	1
	VBH	2
	BBBH	6
Motion type	**Harvester design**	**Number of peaks reaching baseline voltage (37 V) in two cycles**
Running (1.5 Hz)	CBH	3
	VBH	5
	BBBH	15

For this case, the number of voltage spikes achieved for walking motion was considered as an example. As above, the peak voltage (20 V) obtained for CBH was considered the baseline voltage. During a single motion cycle, the CBH was able to produce only one voltage peak. On the other hand, VBH managed to generate two voltage peaks during a single motion cycle. In the case of BBBH, six noticeable voltage peaks were in a single motion. This further certified the superior improvement of BBBH against CBH and VBH in terms of idle time. As stated earlier, reducing idle time could significantly influence the efficiency of a harvester. A similar comparison could be made for running motion. Since the frequency of running motion was considered as 1.5 Hz in this study, instead of a single motion cycle, two motion cycles had to be considered when assuming that three impacts were made during two motion cycles. The baseline voltage for running motion was 37 V. For simplicity, the results are presented in Table 9.

Even though the idle time was improved in BBBH among the other two designs, higher idle time was noticed for human motion compared to the shaker test due to the higher damping present in human motion. This could have been a result of the difficulty in the rigidity of fixing the PEH on the human leg, as muscles and skin could dampen the generated vibration.

3.2.3. Harvested Power

Similar to the shaker test, 1 MΩ resistance was used to represent the open-circuit condition. Figure 14 shows the power generated during normal walking, while Figure 15 shows the results obtained from running motion. For both scenarios, the CBH could produce only one power peak in one motion cycle with a magnitude of approximately 375 µW and 1400 µW for walking and running motions, respectively. On the other hand, the VBH produced slightly higher power peaks in a single motion cycle, achieving about 400 µW and 1500 µW for walking and running motions, respectively. This trend was very similar to the shaker test. Compared to CBH, VBH managed to produce a minimum of 50 µW throughout the period for walking and a minimum of 200 µW throughout the period of running. Once again, the BBBH harvested the highest amount of power, reaching nearly 1400 µW and 4000 µW (Figures 14c and 15c) for walking and running motions, respectively. In contrast, BBBH generated a minimum of 150 µW and 400 µW for walking and running motions, respectively, throughout the period of interest.

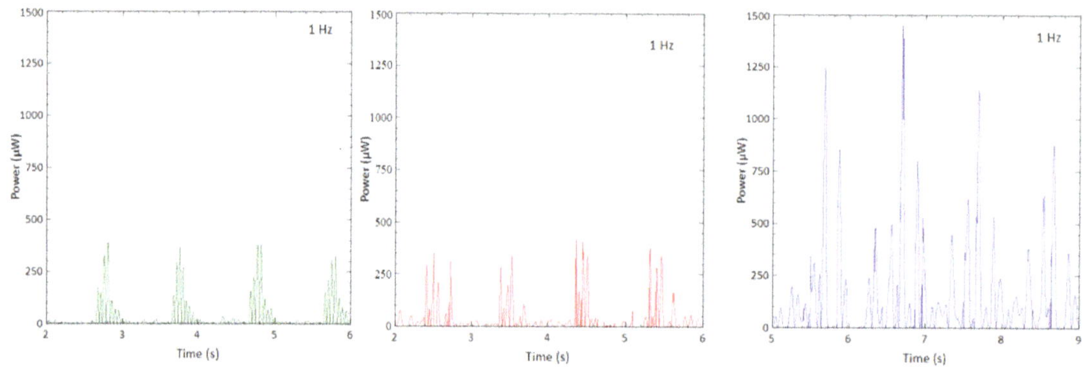

(a) (b) (c)

Figure 14. Power generated by (**a**) CBH, (**b**) VBH, and (**c**) BBBH for human walking (1 Hz).

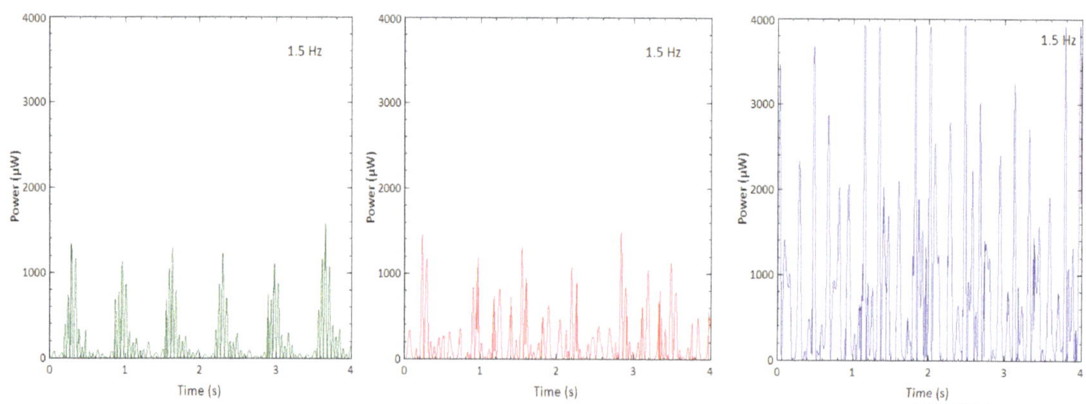

(a) CBH (b) VBH (c) BBBH

Figure 15. Power generated by (**a**) CBH, (**b**) VBH, and (**c**) BBBH for human running motion (1.5 Hz).

3.2.4. Average Output Power

The average output power generated for the human motion was calculated and is presented in Figure 16. For both walking and running motions, the average power output increased slightly from the CBH to the VBH and then considerably to the BBBH, highlighting the superior performance of the BBBH. The CBH and the VBH produced rather low average power for walking motion due to their relatively low output power and considerably long idle time. In contrast, the BBBH produced much higher output power with a higher number of power peaks and shorter idle time. Thus, the BBBH could produce a remarkably higher average power output of 168 μW compared to 33 μW and 43 μW of CBH and VBH, respectively.

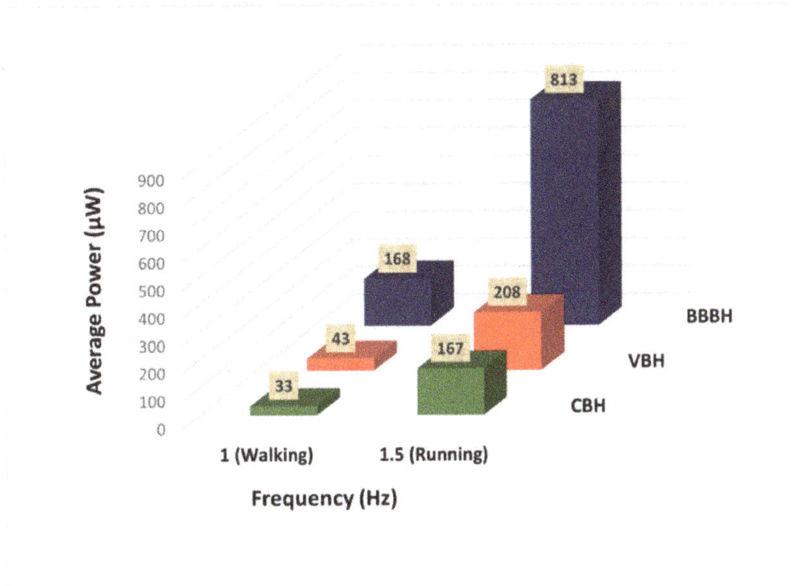

Figure 16. Comparison of average power generated by three different designs of harvesters for human motion.

For running motion, the CBH produced 167 μW of average output power, similar to the average power production of the BBBH during the walking motion. The VBH produced an average power of nearly 208 μW during running motion, slightly better than the CBH. On the other hand, the average output power produced by the BBBH was dramatically higher, achieving 813 μW. This outcome again confirmed the superiority of the proposed BBBH design in the ultralow-frequency range. The BBBH was also identified to perform the best during the running motion, as the average or highest output power generated was several orders higher than the other two designs.

With remarkably higher output voltage, output power, and average output power, as well as shorter idle time, the BBBH was identified to be the best performer among the three designs under all testing scenarios considered in this study.

3.2.5. Performance Comparison with Existing Designs

In terms of average output power, a performance comparison was made between the proposed BBBH and selected designs for human motion applications available in the literature, as summarized in Table 10. All the harvester designs were attached to the human leg and tested during the human walking motion. The proposed design (BBBH) achieved 168 μW of average power compared to the VBH design (which existed in the literature), which generated 100 μW. In addition, the proposed BBBH showed superior power output compared to the other existing designs.

There are a few studies in the literature focusing on the development of low-power-consuming (<100 μW) wireless sensor nodes (WSNs) powered by energy harvesters for different applications [35,36]. The results obtained in this study prove that the proposed BBBH is applicable for powering such WSNs using ultralow-frequency vibration sources as an input with the aid of a proper rectifier circuit to stabilize, rectify, and amplify the voltage generated [9,37–39].

Table 10. Performance comparison of the BBBH with existing designs.

Reference	Descriptions	Types of Piezoelectric Materials	Speed/Frequency	Motion Type	Average Output Power
Li, 2010 [27]	PEH with curved L-shaped proof mass	PZT (T215-H4-103Y)	4 km/h	Walking	49 μW
Fan, 2017 [32]	A shoe mounted PEH	PZT-5H	2 km/h	Walking	30 μW
Izadgoshasb, 2018 [25]	PEH with cantilever beam	MFC (M2814-P2)	6 km/h	Walking	N.A.
Izadgoshasb, 2019 [26]	PEH with a double-pendulum system	MFC (M2814-P2)	1 Hz	Walking	36 μW
Jiang, 2020 [23]	V-shaped bent beam harvester	PZT-5H	1 Hz	Feet stamping	100 μW
Nastro, 2022 [33]	Wearable Ball-Impact PEH	PZT (RS-pro 285–784)	4 Hz	Wrist rotations	9.65 μW
Wang, 2022 [34]	A multi-folded-beam PEH	PZT-5H	9.2 Hz	Running	29 μW
Proposed BBBH	Bent branched beam PEH	MFC (M2814-P2)	2 km/h	Walking	168 μW

4. Conclusions

In this study, a novel bent branched beam PEH (BBBH) was proposed to improve the PEH's energy generation capability from ultralow-frequency ambient vibration sources, such as human motion. A series of experimental and numerical activities were conducted to investigate the performance of the proposed design in terms of operating bandwidth, voltage generation, idle time, and power generation. Initially, the proposed BBBH was numerically studied and geometrically optimized. The BBBH was then fabricated and experimentally tested. Its performance was then compared with the conventional cantilever and a V-shaped beam harvester. Mechanical shaker tests and human motion tests (walking and running motions) were conducted to assess the feasibility of the proposed design in practical applications.

Three close resonance peaks (4.09 Hz, 8.72 Hz, and 10.56 Hz) were achieved by the proposed BBBH within the ultralow-frequency range (1 Hz to 10 Hz), indicating a widening operating bandwidth compared to the CBH and VBH designs. At its first resonance frequency, BBBH could achieve a peak voltage output of 44 V. Moreover, BBBH could perform remarkably at its second natural frequency, which was around 8 Hz (FEA 8.72 Hz), with a voltage output of 62 V. It was noted in the mode shape analysis, that both branches were vibrating in the same direction for the first, second, third, and fifth resonances which could help to generate maximum strain in MFC. The gaining of a maximum strain in MFC leads to the generation of a higher output voltage by enhancing the efficiency of the energy harvester.

Additionally, the BBBH showed considerable voltage and power generation improvement in all tests. The highest average power of 786 μW was obtained at 4 Hz during the shaker test and 813 μW during the human motion test (running). The idle time was also greatly reduced compared to the CBH (only 1/24 in the shaker test at 1 Hz under 1 g acceleration and 1/6 in the case of walking motion). These improvements are significant for a simple harvester design such as the BBBH, which operates without the aid of additional components, including excessive proof masses, magnets, and stoppers. The geometrical parameters of the harvester were selected as a proof of concept, and the harvester can be fabricated as a miniaturized device as required for practical human motion applications. With the aid of a rectifier circuit for power management, the proposed BBBH can potentially be used as a standalone energy provider for low-power-consuming devices such as WSNs and medical implants. For future work, focus can be placed on understanding the influence of multidirectional vibrations on the bent branches of the proposed harvester and developing a theoretical model to comparatively study the experimental results.

Author Contributions: Conceptualization, I.E.P.; data curation, I.E.P. and M.E.; investigation, I.E.P. and M.E.; methodology, I.E.P. and Y.Y.L.; software, I.E.P. and A.M.T.; supervision, Y.Y.L.; validation, A.M.T.; writing—original draft, I.E.P.; writing—review and editing, Y.Y.L., A.M.T., M.U. and C.L. All authors have read and agreed to the published version of the manuscript.

Funding: This research was funded by Southern Cross University, Australia.

Institutional Review Board Statement: Not Applicable.

Informed Consent Statement: Not Applicable.

Data Availability Statement: Data could be provided upon a request to the corresponding author.

Acknowledgments: This research work was supported by Southern Cross University, Australia. The authors gratefully acknowledge the tremendous support and guidance given by Iman Izadgoshasb and Liya Zhao. Furthermore, the authors would like to convey thanks to Erick Wittstock for his outstanding professional assistance in the fabrication process of the prototype.

Conflicts of Interest: The authors declare no conflict of interest.

References

1. Maamer, B.; Boughamoura, A.; El-Bab, A.M.F.; Francis, L.A.; Tounsi, F. A review on design improvements and techniques for mechanical energy harvesting using piezoelectric and electromagnetic schemes. *Energy Convers. Manag.* **2019**, *199*, 111973. [CrossRef]
2. Erturk, A.; Inman, D.J. *Piezoelectric Energy Harvesting*; John Wiley & Sons: Hoboken, NJ, USA, 2011.
3. Kathpalia, B.; Tan, D.; Stern, I.; Erturk, A. An experimentally validated model for geometrically nonlinear plucking-based frequency up-conversion in energy harvesting. *Smart Mater. Struct.* **2017**, *27*, 015024. [CrossRef]
4. Yang, Z.; Zhou, S.; Zu, J.; Inman, D. High-performance piezoelectric energy harvesters and their applications. *Joule* **2018**, *2*, 642–697. [CrossRef]
5. Li, Z.; Yang, Z.; Naguib, H.E. Introducing revolute joints into piezoelectric energy harvesters. *Energy* **2020**, *192*, 116604. [CrossRef]
6. Shan, X.; Tian, H.; Chen, D.; Xie, T. A curved panel energy harvester for aeroelastic vibration. *Appl. Energy* **2019**, *249*, 58–66. [CrossRef]
7. Yang, Z.; Wang, Y.Q.; Zuo, L.; Zu, J. Introducing arc-shaped piezoelectric elements into energy harvesters. *Energy Convers. Manag.* **2017**, *148*, 260–266. [CrossRef]
8. Sun, R.; Li, Q.; Yao, J.; Scarpa, F.; Rossiter, J. Tunable, multi-modal, and multi-directional vibration energy harvester based on three-dimensional architected metastructures. *Appl. Energy* **2020**, *264*, 114615. [CrossRef]
9. Edla, M.; Lim, Y.Y.; Deguchi, M.; Padilla, R.V. A Novel Discontinuous Mode Piezoelectric Energy Harvesting Circuit for Low-Voltage Applications. In Proceedings of the 2021 31st Australasian Universities Power Engineering Conference (AUPEC), Perth, Australia, 21–30 September 2021; pp. 1–5.
10. Abohamer, M.; Awrejcewicz, J.; Amer, T. Modeling of the vibration and stability of a dynamical system coupled with an energy harvesting device. *Alex. Eng. J.* **2023**, *63*, 377–397. [CrossRef]
11. Abohamer, M.K.; Awrejcewicz, J.; Starosta, R.; Amer, T.S.; Bek, M.A. Influence of the motion of a spring pendulum on energy-harvesting devices. *Appl. Sci.* **2021**, *11*, 8658. [CrossRef]
12. Lin, Z.; Al Ba'ba'a, H.; Tol, S. Piezoelectric metastructures for simultaneous broadband energy harvesting and vibration suppression of traveling waves. *Smart Mater. Struct.* **2021**, *30*, 075037. [CrossRef]
13. Kouritem, S.A.; Al-Moghazy, M.A.; Noori, M.; Altabey, W.A. Mass tuning technique for a broadband piezoelectric energy harvester array. *Mech. Syst. Signal Process.* **2022**, *181*, 109500. [CrossRef]
14. Fan, K.; Liu, S.; Liu, H.; Zhu, Y.; Wang, W.; Zhang, D. Scavenging energy from ultra-low frequency mechanical excitations through a bi-directional hybrid energy harvester. *Appl. Energy* **2018**, *216*, 8–20. [CrossRef]
15. Khalid, S.; Raouf, I.; Khan, A.; Kim, N.; Kim, H.S. A review of human-powered energy harvesting for smart electronics: Recent progress and challenges. *Int. J. Precis. Eng. Manuf.-Green Technol.* **2019**, *6*, 821–851. [CrossRef]
16. Izadgoshasb, I.; Lim, Y.Y.; Vasquez Padilla, R.; Sedighi, M.; Novak, J.P. Performance enhancement of a multiresonant piezoelectric energy harvester for low frequency vibrations. *Energies* **2019**, *12*, 2770. [CrossRef]
17. Wu, H.; Tang, L.; Yang, Y.; Soh, C.K. A novel two-degrees-of-freedom piezoelectric energy harvester. *J. Intell. Mater. Syst. Struct.* **2013**, *24*, 357–368. [CrossRef]
18. Zhang, G.; Hu, J. A branched beam-based vibration energy harvester. *J. Electron. Mater.* **2014**, *43*, 3912–3921. [CrossRef]
19. Upadrashta, D.; Yang, Y. Trident-shaped multimodal piezoelectric energy harvester. *J. Aerosp. Eng.* **2018**, *31*, 04018070. [CrossRef]
20. Zhao, Y.; Qin, Y.; Guo, L.; Tang, B. Modeling and experiment of a v-shaped piezoelectric energy harvester. *Shock. Vib.* **2018**, *2018*, 7082724. [CrossRef]
21. Erturk, A.; Renno, J.M.; Inman, D.J. Modeling of piezoelectric energy harvesting from an L-shaped beam-mass structure with an application to UAVs. *J. Intell. Mater. Syst. Struct.* **2009**, *20*, 529–544. [CrossRef]
22. Su, W.-J.; Zu, J.W. Modeling of V-shaped beam-mass piezoelectric energy harvester: Impact of the angle between the beams. In Proceedings of the ASME International Mechanical Engineering Congress and Exposition, Houston, TX, USA, 9–15 November 2012; pp. 573–579.

23. Jiang, W.; Wang, L.; Zhao, L.; Luo, G.; Yang, P.; Ning, S.; Lu, D.; Lin, Q. Modeling and design of V-shaped piezoelectric vibration energy harvester with stopper for low-frequency broadband and shock excitation. *Sens. Actuators A Phys.* **2021**, *317*, 112458. [CrossRef]
24. Li, X.; Upadrashta, D.; Yu, K.; Yang, Y. Analytical modeling and validation of multi-mode piezoelectric energy harvester. *Mech. Syst. Signal Process.* **2019**, *124*, 613–631. [CrossRef]
25. Izadgoshasb, I.; Lim, Y.Y.; Lake, N.; Tang, L.; Padilla, R.V.; Kashiwao, T. Optimizing orientation of piezoelectric cantilever beam for harvesting energy from human walking. *Energy Convers. Manag.* **2018**, *161*, 66–73. [CrossRef]
26. Izadgoshasb, I.; Lim, Y.Y.; Tang, L.; Padilla, R.V.; Tang, Z.S.; Sedighi, M. Improving efficiency of piezoelectric based energy harvesting from human motions using double pendulum system. *Energy Convers. Manag.* **2019**, *184*, 559–570. [CrossRef]
27. Li, W.G.; He, S.; Yu, S. Improving power density of a cantilever piezoelectric power harvester through a curved L-shaped proof mass. *IEEE Trans. Ind. Electron.* **2009**, *57*, 868–876. [CrossRef]
28. Camilloni, E.; DeMaso-Gentile, G.; Scavongelli, C.; Orcioni, S.; Conti, M. Piezoelectric energy harvesting on running shoes. In *Mobile Networks for Biometric Data Analysis*; Springer: Berlin, Germany, 2016; pp. 91–107.
29. Wei, S.; Hu, H.; He, S. Modeling and experimental investigation of an impact-driven piezoelectric energy harvester from human motion. *Smart Mater. Struct.* **2013**, *22*, 105020. [CrossRef]
30. Moro, L.; Benasciutti, D. Harvested power and sensitivity analysis of vibrating shoe-mounted piezoelectric cantilevers. *Smart Mater. Struct.* **2010**, *19*, 115011. [CrossRef]
31. Southgate, D.; Childs, P.; Bull, A. *Sports Innovation, Technology and Research*; World Scientific: Singapore, 2016.
32. Fan, K.; Liu, Z.; Liu, H.; Wang, L.; Zhu, Y.; Yu, B. Scavenging energy from human walking through a shoe-mounted piezoelectric harvester. *Appl. Phys. Lett.* **2017**, *110*, 143902. [CrossRef]
33. Nastro, A.; Pienazza, N.; Baù, M.; Aceti, P.; Rouvala, M.; Ardito, R.; Ferrari, M.; Corigliano, A.; Ferrari, V. Wearable Ball-Impact Piezoelectric Multi-Converters for Low-Frequency Energy Harvesting from Human Motion. *Sensors* **2022**, *22*, 772. [CrossRef]
34. Wang, J.-X.; Li, J.-C.; Su, W.-B.; Zhao, X.; Wang, C.-M. A multi-folded-beam piezoelectric energy harvester for wideband energy harvesting under ultra-low harmonic acceleration. *Energy Rep.* **2022**, *8*, 6521–6529. [CrossRef]
35. Jang, J.; Berdy, D.; Lee, J.; Peroulis, D.; Jung, B. A wireless condition monitoring system powered by a sub-100/spl mu/W vibration energy harvester. *IEEE Trans. Circuits Syst. I Regul. Pap.* **2012**, *60*, 1082–1093. [CrossRef]
36. Zhang, F.; Zhang, Y.; Silver, J.; Shakhsheer, Y.; Nagaraju, M.; Klinefelter, A.; Pandey, J.; Boley, J.; Carlson, E.; Shrivastava, A. A batteryless 19µW MICS/ISM-band energy harvesting body area sensor node SoC. In Proceedings of the 2012 IEEE International Solid-State Circuits Conference, San Francisco, CA, USA, 19–23 February 2012; pp. 298–300.
37. Edla, M.; Lim, Y.Y.; Vasquez Padilla, R.; Deguchi, M. An Improved Rectifier Circuit for Piezoelectric Energy Harvesting from Human Motion. *Appl. Sci.* **2021**, *11*, 2008. [CrossRef]
38. Edla, M.; Lim, Y.Y.; Deguchi, M.; Padilla, R.V.; Izadgoshasb, I. An improved self-Powered H-bridge circuit for voltage rectification of piezoelectric energy harvesting system. *IEEE J. Electron Devices Soc.* **2020**, *8*, 1050–1062. [CrossRef]
39. Edla, M.; Mikio, D.; Izadgoshasb, I.; Mahmud, M.A.P.; Kouzani, A.Z. Self-powered boost-converter for power optimisation and piezo garden lights. *Smart Mater. Struct.* **2022**, *31*, 045021. [CrossRef]

energies

MDPI

Article

A Voltage Doubler Boost Converter Circuit for Piezoelectric Energy Harvesting Systems

Abdul Haseeb [1,*], Mahesh Edla [2], Mustafa Ucgul [1], Fendy Santoso [1] and Mikio Deguchi [3]

1 Faculty of Science and Engineering, Southern Cross University, East Lismore, NSW 2480, Australia
2 Department of Research and Development, Electronics Engineer, MYRA Corporate, Wollongbar Industrial Area, Alstonville, NSW 2477, Australia
3 Department of Electronics and Control Engineering, National Institute of Technology, Niihama College, Niihama 790-8570, Japan
* Correspondence: a.haseeb.10@student.scu.edu.au

Abstract: This paper describes the detailed modelling of a vibration-based miniature piezoelectric device (PD) and the analysis modes of operation and control of a voltage doubler boost converter (VDBC) circuit to find the PD's optimal operating conditions. The proposed VDBC circuit integrates a conventional voltage doubler (VD) circuit with a step-up DC-DC converter circuit in modes 1–4, while a non-linear synchronisation procedure of a conventional boost converter circuit is employed in modes 5–6. This integration acted as the voltage boost circuit without utilising duty cycles and complex auxiliary switching components. In addition, the circuit does not require external trigger signals to turn on the bidirectional switches. This facilitates the operation of VDBC circuit at very low AC voltage ($V_{ac} \geq 0.5$ V). Besides this, the electrical characteristics of VDBC circuit's input (i.e., PD) perfectly concurs with the studied testing scenarios using impedance power sources (mechanical shaker). Firstly, the proposed circuit which can rectify the PD's output was tested at both constant input voltage with varying excitation frequency and constant excitation frequency with varying input voltage. Next, a small-scale solar battery was charged to validate the feasibility of the performance of the proposed VDBC circuit. The proposed circuit achieved a maximum output voltage of 11.7 V_{dc} with an output power of 1.37 mW. In addition, the rectified voltage waveform is stable due to the sminimisation of the ripples. In addition, the performance of VDBC circuit was verified by comparing the achieved results with previously published circuits in the literature. The results show that the proposed VDBC circuit outperformed existing units as described in the literature regarding output voltage and power. The developed rectifier circuit is suitable for various real-life applications such as energy harvesting and battery charging.

Keywords: piezoelectric energy harvesting; voltage doubler; boost converter; self-powered voltage boost converter; ripple reduction; rectifier

1. Introduction

A piezoelectric device [1–3] (PD) transforms the mechanical energy (ME) [1,4], which is available in the environment, into electrical energy (EE) [5,6] in the form of an alternating current (AC). Scientific researchers have had a surge in interest in piezoelectric energy harvesting (PEH) systems during the last decade [1,3,7,8]. However, the PD's are constrained to limited power applications, usually on the scale of milliwatts and microwatts [9–14]. This is mainly due to its internal properties, namely internal capacitance and resistance [1,3,7,15]. They can resist high pressure but are small devices compared to the other electrical devices, which produce high amounts of current and voltage [16]. In addition, another concern is that high-frequency vibrations are required to efficiently exploit these PD's, as they generate very low power and voltage due to their internal characteristics. However, most vibrations are expected to be in the range of 0.1–1 KHz [8], whereas a PD can function at several hundred kilohertz. Despite that, the primary objective of the PEH field is to power miniature electronic devices. However, the PD generates AC in the PEH system due to ambient vibrations in the environment.

Nevertheless, to operate electronic devices, the AC generated from the PD has to be transformed into direct current/voltage (DC) [3,7,15]. Consequently, the PEH entails a power converter circuit (i.e., rectifier circuit) for rectification, facilitating AC-DC conversion. Thus, the rectifier circuit plays a crucial role between the vibrating PD and the storage device in the PEH field.

A conventional full-bridge rectifier (FBR) circuit is the simplest technique for AC to DC conversion [13,14,17–23]. However, the FBR circuit exhibits inefficient conversion efficiency because of forward voltage [24,25] and V_f losses across the diodes [17,26,27], while most power electronic devices require 3.3 V_{dc}. As a result, effective electronic circuits and active materials must be involved to overcome forward voltage issues and maximise the PD's outcomes.

Various single [15,28–31] and two-stage [3,29,32,33] power conversion circuits have been described in the previous literature to overcome this issue. A single-stage circuit converts AC generated by the vibrating PD into DC, whereas a two-stage circuit confines two processes, such as AC to DC and DC to DC. A framework that explains the operation of PEH in both scenarios, termed single and two-stage power conversion circuits, is depicted in Figure 1.

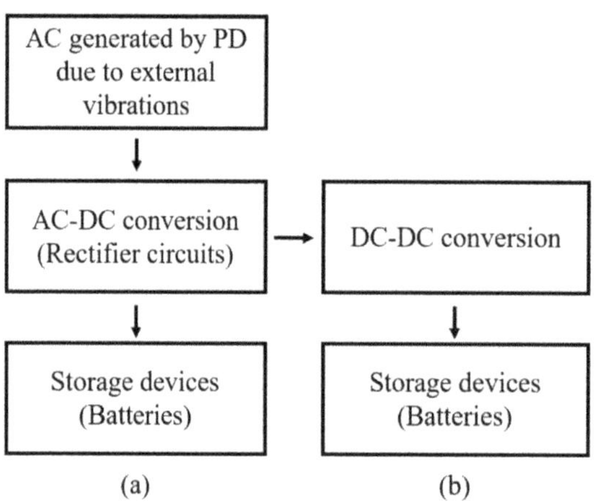

Figure 1. A framework of PEH power-conversion circuits: (**a**) a single stage, (**b**) two-stage.

To attain effective optimisation and achieve maximum power at the output, Edla et al. [1,7,34] attempted to use minimal power electronic components in the H-Bridge rectifier (HBR) circuit. This research involved the congregation of high and low frequencies. Besides, the output voltage and power of the HBR circuit are higher than the conventional FBR circuit. This originates from the fact that the FBR circuit embodies diodes with forward voltage and V_f losses of 0.6 V and 0.15 V. In addition, the rectified voltage waveform from the HBR circuit had fewer ripples. However, the rectified voltage output of the HBR circuit remained low, and the proposed research has not discussed the issue of ripple reduction in the rectified output voltage waveform in detail.

Later, Hsieh et al. [35] proposed a two-stage rectifier circuit to attain maximum output power at load. The proposed circuit comprised a maximum power extraction (MPE) circuit and a regulator for dual-stage circuit operations. To attain maximum power output across the battery, the proposed circuit controlled the switching frequency of the step-down converter circuit by using a controller. Nevertheless, the switching losses were higher than the output power due to the switching frequency.

One other approach to attain maximum power at the output is employing a non-linear system and using external switches, namely, metal-oxide-semiconductor-field-effect-transistor (MOSFET), switching frequency, duty cycles, and passive electronic components, such as inductors and capacitors. A range of non-linear approaches was proposed by [17,22,36–42]. However, the preceding circuits were incorporated with auxiliary circuits, transformers, diodes, and bulky inductors. Thus, the power consumption in the used power electronic components is high, which is unsuitable for PEH applications.

Later, Shareef et al. [26] proposed a power electronic circuit to attain a rectifier-less AC-DC conversion by employing the synchronous process using an inductor. In addition, a cold-startup feature was also included in the proposed circuit to achieve the optimised output from the proposed circuitry. It achieved a gleaning output power of 254 μW at 0.65 mV_{ac} input voltage. However, the proposed methodology was adamant, as it demanded subsidiary components, such as three capacitors, three PDs, a digital logic controller, a polarity detector, and an inductor of milli-range. Thus, the proposed circuit utilised high power and inflexibility in the PEH applications.

In addition, synchronised switching harvesting techniques were proposed by [17,27,43]. In comparison to the FBR circuit, the testing results significantly improved and boosted the input AC into high DC (i.e., output). Nevertheless, the method conceded additional components: an oscillator, polarity detector, controller, diodes, logic gates, and transformers. Consequently, the AC-DC conversion losses and power consumption of the proposed circuit were higher than the power generated by a PD. Additionally, it required high voltage to kickstart the conversion process.

Recently Jeong et al. [44,45] proposed an efficiency-based multi-output charging system to improve the overall conversion efficiency by adaptably dividing the input power across different PV cells based on their power level. This proposed methodology was useful when PV cells are partly shaded. In addition, the proposed system is implemented by adopting CMOS technology.

Edla et al. [1,19] recently conducted an experimental study where they facilitated an HBR circuit to a self-powered dual-stage HBR rectifier (DSHBR) circuit. Numerous high-frequency and high-voltage applications were carried out using the DSHBR circuit. The results verified that the DSHBR circuit successfully enhanced the output voltage and power compared to the HBR circuit. However, the proposed method incorporated an additional DC-DC converter and Zener diode, resulting in increased power losses, complex design, and high cost. In general, the DSHBR circuit generated low output power at load. However, the problems with ripples in the rectified voltage waveform are not fully covered.

Edla et al. [31] recently conducted an experimental study on HBRJT circuit to achieve high voltage gain by employing the Joule thief topology. The proposed HBRJT circuit congregated the HBR and Joule thief circuit for effective AC-DC power conversion with minimal circuit components. The circuit achieved a gleaning output of 13.5 V_O with 0.51 mW output power. However, this topology ignored the issues of ripples in the gained output voltage. Besides, the winding losses in the transformer are not discussed.

Later, Zhao et al. [46] conducted an experimental study on developing an impedance model for a dual stage photovoltaic (PV) inverter. The proposed model was developed to achieve the impedance characteristics of PV as well as provide a solution for the potential stability of an inverter linked to a poor grid. The proposed model comprised of PV arrays, a boost converter and a backend inverter circuit, along with LCL output filter. The proposed study was also validated using PSCAD based simulation results.

Later, Yang et al. [47] proposed a conventional neural network based sub-dictionary predictor to soptimise the conventional atomic decomposition algorithm for power quality disturbance signals. It is verified that the proposed methodologies successfully sminimised the need for additional computation sources while improving the accuracy of sub-dictionary predictor power quality signals.

It is worth emphasising that the vibrating PD generates a very low AC voltage. However, most power electronic devices require a minimum of 3.3 V_{dc} with a stable DC voltage waveform. Nevertheless, the previous researchers [1,7,17,48–50] and in the abovementioned literature focused on size reduction, power optimisation, and minimising conversion losses, while fewer attempts have been made on ripple reduction and AC-DC conversion without employing external operating circuits and complex IC's. Thus, there is a need for a more suitable power-conversion circuit that can minimise the usage of multiple diodes and employ a non-linear switching process that can boost the low AC voltage generated by the vibrating PD to high DC voltage with minimised ripples.

To overcome the abovementioned issues, the authors adopted a traditional voltage doubler (VD) circuit described in the literature [51,52] and extended it with the abovementioned features, such as boosting the low AC voltage to a high DC voltage, self-powered switching process. Therefore, the voltage doubler boost converter (VDBC) circuit is proposed, which is depicted in Figure 2b.

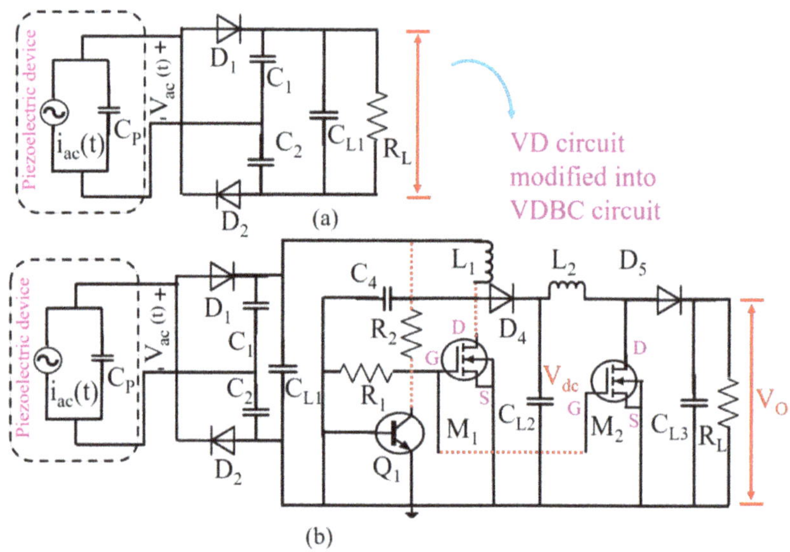

Figure 2. Existing and proposed circuits: (**a**) VD and (**b**) VDBC. (M_1–M_2: MOSFET switches, D: Drain, G: Gate, S: Source).

The proposed VDBC circuit minimises the usage of additional diodes to limit the forward voltage losses. It can increase the PD's low input AC voltage to a high DC output voltage at different vibration frequencies. In addition, the proposed circuit also reduces the ripples in the rectified output voltage waveform and charges the solar battery continuously without employing the additional switching circuits.

The proposed VDBC circuit possesses six modes of operation. Modes 1–4 garnered the combination of a conventional VD circuit for the AC-DC conversion process with a step-up DC-DC converter for effective power conversion, while modes 5–6 include an improved non-linear synchronisation procedure of a conventional boost converter circuit [51,52], to mitigate ripples in the rectified voltage waveform. Besides, the proposed VDBC circuit is a self-powered switching approach, eliminating the need for additional switching components. Consequently, the losses are improved due to high power consumption and stress between the switches and parasitic components.

As this study reports to be an extension of a VD circuit [53] and previously published literature switching circuits [52], the proposed circuit is unique in the way that it implements the VDBC circuit to enhance the power flow towards the load capacitor, C_{L3} with minimum power consumption, less stress across switches and using zero parasitic components during the switching process. Herein, the proposed VDBC circuit is implemented by using the methods described in [22,26,27,54], conventional boost converters in [37,55], and switching rectifier circuits in [1,7,17,56,57].

Despite the proposed circuit employing existing methodologies, there has been no attempt to combine the VD circuit and the standard DC-DC approach with application in charging garden lights. As a general rule of energy harvesting, the VDBC circuit makes no attempt to use external switching circuits to reduce conversion losses and costs. As the VDBC circuit comprises a simple power electronic circuit that operates without needing active logic controllers, it is expected that the VDBC circuit minimises the necessity for external switching circuits while increasing the overall system's performance. Additionally, due to the irregular behaviour of vibration, charging a battery and continually powering LED lights is quite challenging [3,58]. As a result, the proposed VDBC can also resolve this issue when LED bulbs refer to solar battery lights. Other key benefits of the VDBC circuit are highlighted below:

1. The proposed VDBC circuit implemented MOSFET switches. This reduced stress in the switching process, thus achieving higher output voltage (Vdc) and power.
2. As highlighted in the literature, it is developed without employing complex, extraneous circuits and polarity indicators, which leads to lower conversion losses and complexity.

This paper is divided as follows: Section 2 discusses the internal characteristics of PD and PEH circuits, while experimental results and discussion are described in Section 3. The conclusion is described in Section 4.

2. PD Internal Characteristics and PEH Circuits

This section reveals the internal features of PD's circuit and frequently employed rectifier circuits in the PEH field. The VDBC circuit is also illustrated, in detail, in the following paragraphs.

2.1. PD Circuit Model

A typical PD may exhibit both current and voltage sources when it is mechanically exciting [3,26,27,59], as depicted in Figure 3a. However, the internal capacitor, C_P, is parallel when it is considereda current source. While in series, it is regarded as a voltage source.

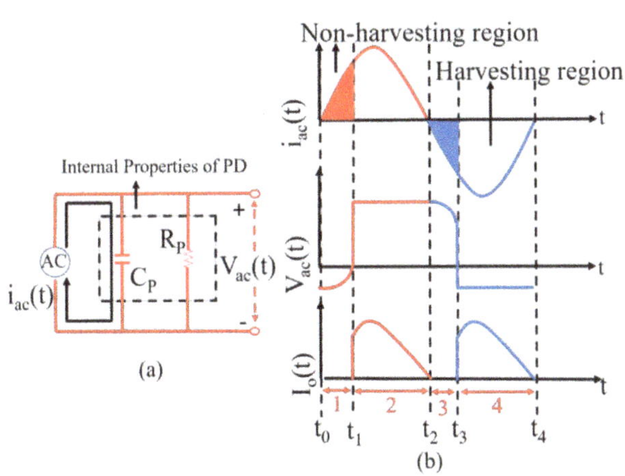

Figure 3. (**a**) PD representation as a current source; (**b**) the current and voltage outputs of a vibrating PD.

The harvested current and voltage deviate from the mechanical vibrations in each half-cycle, as depicted in Figure 3b. Initially, the current generated by the PD due to the mechanical excitation has to charge its internal capacitor, C_P. The charging duration of the internal capacitor is shown in interval 1 in Figure 3b. During this period, the rectifier circuit does not operate and charge the load capacitor, as there is no current flow from the PD. This process endures until the PD's magnitude voltage, V_{ac} (t), and load capacitor becomes identical.

Once the C_P is charged completely, then interval 2 begins. During interval 2, the magnitude of the PD's voltage and load capacitor voltage becomes equal, and the rectification process begins. A similar trend is also predicted in the negative half-cycle [60]. In particular, the internal capacitor charging period occurs in intervals 1 and 3, while the rectification occurs in intervals 2 and 4. In other words, intervals 1 and 3 are referred to as non-harvesting regions, while intervals 2 and 4 are referred to as harvesting regions. During this process, the current harvested by a PD can be written as [1,3,61]:

$$i_{ac}\ (t) = \widehat{I_{ac}}\ \sin(\omega t) \tag{1}$$

where $\widehat{I_{ac}}$ is the magnitude of harvested current, ω is the angular frequency, while t represents the time period.

2.2. VD Circuit

Figure 2a comprises a VD circuit in shunt with the PD, and its corresponding current and voltage waveforms are delineated in Figure 4. As depicted in Figure 2a, the current generated by the PD must charge its internal capacitor, C_P. At this instant, the PD voltage lags behind the total forwarding and rectified voltage (i.e., $V_{PD} \leq V_D + V_{dc}$), where the V_D represents the forward voltage losses across diodes. As a result, the VD circuit does not conduct, and the capacitor C_{L1} remains uncharged. This occurs in both half-cycles of input voltage [3,62]. Intervals 1 and 3 are referred to as non-harvesting regions (i.e., PD charging the internal capacitor, C_P). The VD circuit starts conducting when the PD's magnitude equals the rectified voltage (i.e., $V_{PD} \geq V_D + V_{dc}$). Consequently, the capacitor, C_{L1}, starts charging. Intervals 2 and 4 are referred to as harvesting regions (i.e., PD charges the load capacitor, C_{L1}). The process of energy accumulation across C_{L1} in both half cycles is represented as follows.

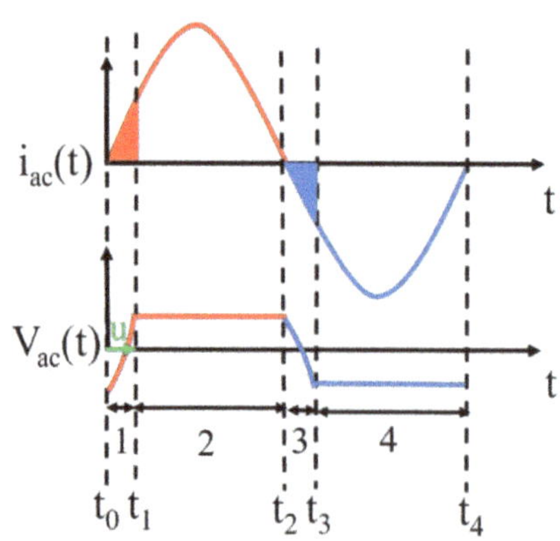

Figure 4. The current and voltage outputs generated by PD. (u: polarisation current used to charge the internal capacitor of PD).

Mode 1: Positive half-cycle

During interval 1 ($t_o < t \leq t_1$):

- C_P = Charging (PD delivers no output to load)

- C_{L1} = Not charging
- D_1–D_2: OFF

During interval 2 ($t_1 < t \leq t_2$):

- C_P = Discharging (PD delivers output to load)
- D_1: ON
- D_2: OFF
- $V_{PD} = V_{dc}$ (C_{L1}: Charged)

Mode 2: Negative half-cycle

During interval 3 ($t_2 < t \leq t_3$):

- C_P = Charging (PD delivers no output to load)
- C_{L1} = Not charging
- D_1-D_2: OFF

During interval 4 ($t_3 < t \leq t_4$):

- C_P = Discharging (PD delivers output to load)
- D_1: OFF
- D_2: ON
- $V_{PD} = V_{dc}$ (C_{L1}: Charged)

The capacitor, C_{L1}, is charged during both half-cycles of the PD. The output power of the VD circuit varies with the changing in rectified voltage [1,27,63], and the high output power turns out, when

$$V_{dc} = \frac{\widehat{I_{ac}}\,(t)}{2\,\omega C_P} \tag{2}$$

In addition, the current output across the capacitor, C_{L1}, can be depicted as [8,63]:

$$\begin{cases} 0, & 0 \leq t \leq u \\ \frac{C_{L1}}{C_{L1} + C_P} \quad \widehat{I_{ac}}|\sin(\omega t) \approx, & u \leq t \leq \pi \end{cases} \tag{3}$$

Given that $C_{L1} \geq C_P$, then most of the current will be output:

$$\frac{C_{L1}}{C_{L1} + C_P}\,\widehat{I_{ac}} \approx i_{ac} \tag{4}$$

The rectified output power, P_O (t), can be stated as a function of the rectified voltage, V_{dc}:

$$\langle P_O\,(t)\rangle = \frac{2\,V_{dc}}{\pi} - (I_{ac} - V_{dc}\omega C_P) \tag{5}$$

2.3. *Voltage Doubler Boost Converter Circuit*

Herein, a VDBC circuit is proposed and studied, one which boosts the low rectified voltage, operates at low AC voltage from PD, and reduces ripples in the rectified voltage output. The proposed VDBC circuit is depicted in Figure 2b. It is designed to work in 6 modes of operation.

Modes 1–2 comprise a VD circuit for AC-DC conversion, while modes 3–4 consist of a step-up DC-DC converter for effective power conversion. Modes 5–6 include an improved non-linear synchronisation procedure of a conventional boost converter circuit. It is worth noting that modes 1–2 of the VDBC circuit are similar to the exact operation of the VD circuit since the proposed VDBC circuit is stated to be an extension of the VD circuit [53] and literature circuits [17,52,55].

When the rectified voltage from the VD circuit is collected across the capacitor, C_{L1}, modes 3–4 (i.e., DC-DC conversion) start conducting. Mode 3 begins when the inductor, L_1, charges, while mode 4 initiates when L_1 discharges. The complete working operation of both modes 3–4 is depicted in Figure 5.

Mode 3: (Charging process of inductor L_1)

The stored voltage in the capacitor, C_{L1}, charges the inductor, L_1. From here, it flows to the base, B of the transistor, Q_1, via a capacitor, C_4. Q_1 behaves as a voltage amplifier that amplifies the low voltage at the base, B, and generates a high output at the collector, C. Thus, the collector output activates the MOSFET switch, M_1, and it starts conducting. Consequently, the inductor starts charging and increases its current. During this time, the voltage accumulated across C_{L1} starts charging the inductor, L_1, slowly from zero to peak, as the inductor prevents a sudden change in current. During this, the diode, D_4, is OFF, and no output flows towards the capacitor, C_{L2}.

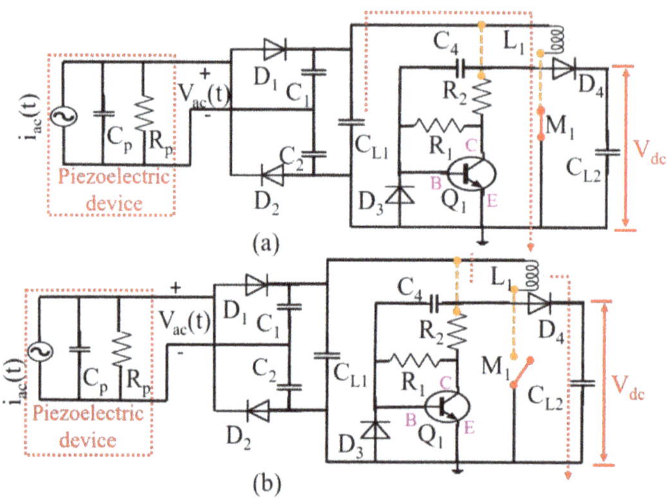

Figure 5. Working operation of VDBC circuit in modes 3 and 4: (**a**) mode 3 (inductor L_1 is charging) and (**b**) mode 4 (inductor L_1 is discharging). (B: Base, C: Collector, E: Emitter).

Mode 4: (Discharging process of Inductor L_1)

During this mode, when L_1 gets saturated, the current flow to the transistor, Q_1 is terminated, and M_1 is turned OFF. Thus, the current stored in L_1 freewheels via a diode, D_4, and charges the capacitor, C_{L2}. As the capacitor prevents a sudden change in voltage, it charges slowly from zero to peak voltage level. A detailed working principle of both modes 3 and 4 is depicted in Figure 5a,b.

Once the rectified voltage fully charges the capacitor C_{L2}, modes 5–6 begin to operate.

In these modes, an improved non-linear synchronisation procedure of a conventional boost converter circuit is implemented to alleviate the ripples in the rectified output voltage stored in C_{L2}. This converter circuit is utilised to sustain a stable and fixed rectified output voltage at the load capacitor, C_{L3}.

Considering the ideal DC-DC converter (modes 5–6) circuit having zero losses, the same rectified current and voltage, which was stored in C_{L2} must be passed through it and stored in the load capacitor, C_{L3}. Consequently, the rectified voltage and current are regarded as constant because they are DC. Therefore, the output power, P_O, at the load capacitor, C_{L3}, can be determined by the product of the voltage (V_O) across the load capacitor, C_{L3}, and the current (I_{CL3}) flowing from the load resistor, R_L.

$$P_O = V_O \times I_{CL3} \tag{6}$$

The DC-DC converter comprises 2 modes, namely modes 5 and 6. Mode 5 starts working when the MOSFET switch, M_2, is turned ON, while mode 6 begins when M_2 is turned OFF. A similar gate signal is employed to turn ON the MOSFET switch, M_2 in mode

5, as used in Modes 3–4. The complete working operation of both modes 5 and 6 is depicted in Figure 6.

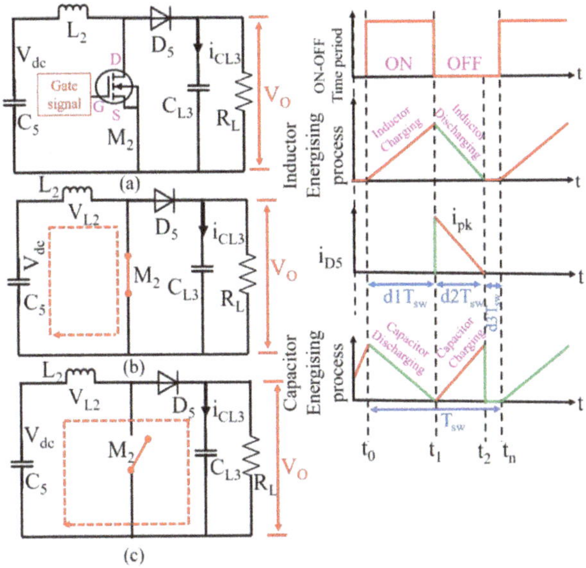

Figure 6. (**a**) Circuit diagram of modes 5–6, (**b**) mode 5 (M_2 switch is on), (**c**) mode 6 (M_2 switch is off) and linked waveforms. (V_{L2}: Inductor current, T_{SW}: time period of the switching process, d_1T_{SW}, d_2T_{SW}: ON and OFF period of the control circuit, i_{D5}: diode.)

Mode 5: (When MOSFET switch M_2 turns ON)

During this period, the MOSFET switch M_2 turns on. Thus, the stored rectified output voltage in the load capacitor, C_{L2}, envelopes through C_{L2}–L_2–M_2–C_{L2} to charge the inductor, L_2. During this period, the diode, D_5, is reverse biased, and no current flows towards the load capacitor, C_{L3}.

Mode 6: (When MOSFET switch M_2 turns OFF)

During this period, the switch M_2 turns off. Thus, the accumulated current across the inductor L_2 during mode 5 is freewheeled via the diode D_5 to charge the load capacitor, C_{L3}.

It should be worth emphasising that the proposed VDBC circuit does not entail an external power source to trigger the MOSFET switches across any of the six modes. Thus, switching pulses, logic gates, and integrated circuits [54] are not required to operate the proposed circuit's MOSFET switches (M_1 and M_2).

This study assumes that the inductor and capacitors are ideal semiconductor devices. The subsequent equations are stated when the MOSFET switches are conducting (Figure 6) [8].

ON Period ($t_o < t < d_1T_W$):

When the switch is in a conduction state:

$$V_{L2}(t)\ V_{dc} \tag{7}$$

$$i_{CL3} = -V_O\ (t)/R \tag{8}$$

Small ripple approximation for $V_O(t)$:

$$V_{L2} = (t) \approx V_{dc} \tag{9}$$

$$i_{CL3} \approx -V_O(t)/R \tag{10}$$

OFF Period ($d_1T_w < t < (d_1 + d_2)T_{sw}$):
When the switch is in a conduction state:

$$V_{L2}(t) = V_{dc} - V_O(t) \tag{11}$$

$$i_{CL3} = i_{dc}(t) - V_O(t)/R \tag{12}$$

Small ripple approximation for $V_O(t)$:

$$V_{L2}(t) \approx V_{dc} - V_O(t) \tag{13}$$

$$i_{CL3} \approx i_{dc}(t) - V_O(t)/R \tag{14}$$

Empowering the load resistances ($(d_1 + d_2)T_{sw} < t < T_{sw}$):

$$VL_2(t) = 0,\ i_{dc}\lim_{\delta\to 0} 0 \tag{15}$$

$$i_{CL3} = -V_O(t)/R \tag{16}$$

Small ripple approximation for $V_O(t)$:

$$V_{L2}(t) = 0 \tag{17}$$

$$i_{CL3} \approx -V_O(t)/R \tag{18}$$

where d represents the duty cycle, Tsw refers to the time period of the switching process, V_{L2} is the inductor current, i_{CL3} is the current flowing through the load resistor, and V_o is the voltage across the load capacitor.

During the switching interval, the MOSFET switch is turned on, while the diode, D_5, is turned off with zero current. Thus, a negligible switching loss can be anticipated. As the result of the above assumptions, diode and inductor waveforms, capacitor charge balance and inductor voltage-sec balance, and the output voltage through modes 5–6 can be stated as follows [56]:

$$\frac{V_O}{V_{dc}} = \frac{1 \pm \sqrt{1 + 4d_1^2/K}}{2a} \tag{19}$$

where V_O represents the generated output voltage of the VDBC circuit, V_{dc} is the rectified output voltage stored in the capacitor, and C_{L2} is the operation of modes 1–4.

3. Experimental Results and Discussions

A conventional piezoelectric cantilever beam consists of an aluminium beam (dimensions: 205 × 20 × 1 mm), with the first end attached to the mechanical vibration shaker (APS—113), and the opposite end carries two permanent magnets. The magnets were utilised as proof mass coupled at the nook of the aluminium beam. A microfiber composite (MFC) patch (M2814-P2, 37 mm × 17 mm × 0.180 mm, C_P = 33.90 nF) was glued at the fixed end of the aluminium beam, facing the maximum strain occurrence. A function generator (Agilent 33210A, The Agilent Technologies, Santa Clara, California, USA) was employed to provide a sinewave signal to a power amplifier (2706, B&K Agilent). This amplifier boosted the sinusoidal signal before triggering the mechanical shaker. This shaker operated on input amplitude and vibrational frequency to produce mechanical vibrations to excite the piezoelectric cantilever beam. An inductive sensor can be applied to track the acceleration of the base excitation beam. Thus, due to these mechanical vibrations, the patched PD generated an AC signal, V_{ac}, which was taken as an input for the proposed VDBC circuit.

To validate the applicability of the VDBC circuit, three different testing scenarios were implemented. In Test 1, the proposed VDBC circuit was tested at constant excitation frequency under varying input voltage. In Test 2, the proposed circuit was analysed at constant input voltage under varying excitation frequencies, as highlighted in Table 1.

However, in Test 3, the proposed circuit was used to charge a solar-powered battery. The generated output voltage, V_O, through the VDBC circuit was stored in C_{L3} in both Test 1 and Test 2. In addition, each load resistor, R_L, was connected in parallel with C_{L3} and outputs were measured using a multimeter.

Table 1. Testing Scenarios.

Sources of Excitation	Test Scenarios	Frequency (Hz)	Input Voltage (V_{ac})	Load Capacitor (μF)	Types of PD	Load Resistors (KΩ)
Mechanical Shaker	Test 1	5, 10, 15, 20	5	10	MFC	R_1 = 100 R_2 = 300 R_3 = 620 R_4 = 910
	Test 2	100	0.5, 0.7, 1.0, 1.25	10		
	Test 3	5	5	(solar battery) 1.2 V_{dc}, 4 mA		100

3.1. Varying Excitation Frequency at Fixed Input Voltage

The output voltage and power attained from Test 1 are depicted in Figure 7a,b, respectively. Here, the output voltage and power obtained from the VDBC circuit are compared at different input frequencies (5–20 Hz) under varying load resistances ranging from (100–910 KΩ).

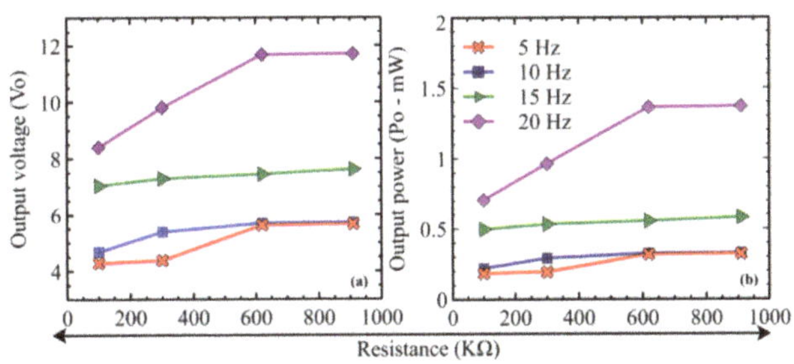

Figure 7. Outcomes of Test 1 at $5V_{ac}$, (**a**) output voltage and (**b**) output power.

According to Figure 7, regardless of applied input voltage and frequency, the output voltage is boosted with the increase in load resistances. It is prominent that the VDBC achieves the maximum output voltage at a higher load resistance of 910 KΩ.

Firstly, a 5 V_{ac} input voltage is given to the VDBC circuit at a 5 Hz excitation frequency. Taking a positive half-cycle, mode 1 started working and conducted the AC-DC conversion process. Consequently, the rectified output voltage is stored in the capacitor, C_{L1} (Figure 7). A similar operation happened in the negative half-cycle. Once C_{L1} is fully charged, it initiates the current flow in modes 3–4 (i.e., DC-DC converter).

The rectified voltage stored in C_{L1}, as the consequence of modes 1–2, stimulates the current flow towards the base of the transistor, Q_1, after passing through the inductor, L_1 and coupling capacitor, C_4. The transistor, Q_1, acted as a common emitter voltage amplifier, amplifying the base input voltage to a higher collector output voltage. The collector output activated the MOSFET switch, M_1's gate and allowed the drain to source current flow, resulting in charging the L_1 (mode 3).

Once L_1 is fully saturated, it blocks the current flow towards Q_1. As a result, the M_1 stopped conducting. Thus, the stored current in L_1 initiated the current flow towards the capacitor, C_{L2}, via forwarding biased diode, D_4 (Mode 4). Once C_{L2} is completely charged, modes 5–6 begin to operate.

In mode 5, the stored rectified voltage in capacitor C_{L2}, V_{dc}, as the result of modes 3–4, causes a magnetic field across the inductor, L_2. It is to be noted that the MOSFET switch,

M_2, is turned on and off during this time duration. Thus, the V_{dc} acted as a short circuit, causing an increased magnetic field across L_2. The prime factor behind this boosted magnetic field across L_2 is that the quicker the magnetic field variates, the higher the potential difference it causes. The magnetic field through L_2 is enveloped as M_1 is switched on. However, when M_2 is switched off, the stored charge across L_2 forward biased the diode, D_5 and charges the load capacitor, C_{L3}. Later, each load resistance is connected in parallel, and output is captured using a multimeter.

Later, the remaining input frequencies were applied under a similar input voltage. The resulting output voltage was measured, as depicted in Figure 7a. Once the output voltage was stored in C_{L3}, the current and voltage across R_L were measured. Using Equation (6), the output power was calculated, depicted in Figure 7b.

Another observation revealed that the excitation frequency considerably impacts the overall performance of the VDBC circuit. Higher output voltage and power values were achieved at higher excitation frequencies. This is because the capacitive inductance of PD drops at higher switching frequencies [2]. Thus, the PD generates a higher AC voltage, V_{ac}, which leads to increased output voltage and power. Figure 8 represents the output voltage and power under constant input voltage and at varying excitation frequencies.

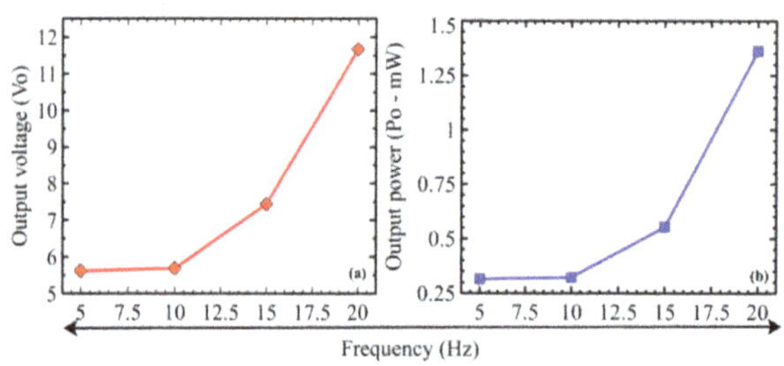

Figure 8. Outcomes of Test 1 at various excitation frequencies (**a**) output voltage and (**b**) output power.

It is worth noting that the main goal of the proposed study is to validate the performance of the VDBC circuit by minimising power conversion, the diode's forward voltage, and ripple losses in the rectified voltage waveform, thus achieving a high output voltage and power. The peak output voltage of 11.7 V_{dc} is achieved at a frequency of 20 Hz. However, the least voltage of 5.66 V_{dc} is accumulated across the load resistor at a 5 Hz input frequency.

3.2. *Varying Input Voltage at Fixed Excitation Frequency*

The performance of the VDBC circuit was tested under a fixed excitation frequency at a varying input voltage. Here the excitation frequency was kept constant at 100 Hz, and the VDBC circuit was subjected to the PD's variable input AC voltage (Table 1). The resulting output voltage and power from Test 2 are depicted in Figure 9a,b.

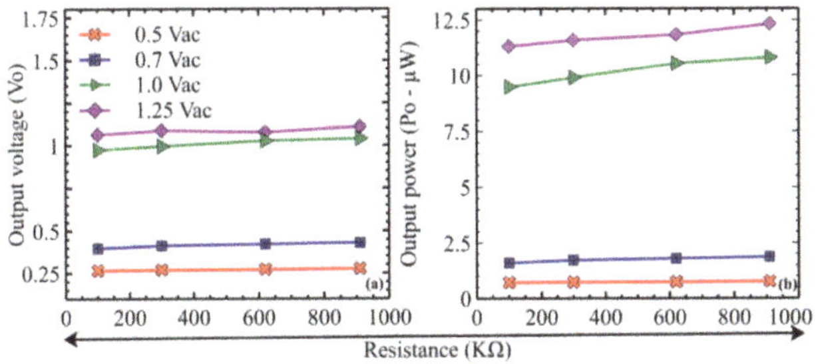

Figure 9. Outcomes of Test 2 at 100 Hz, (**a**). Output voltage and (**b**). Output power.

Initially, 0.5 V_{ac} input voltage was applied across the VDBC circuit at 100 Hz excitation frequency. As the applied input voltage, V_{ac} exceeded the threshold voltage of both MOSFET and transistor switches, the proposed circuit started conducting in modes 1–6, as discussed in Section 3.1. This achieved 0.28 V_{dc}, which was stored across the load capacitor, C_{L3}. Later, the remaining input voltages were applied, and the resulting output voltage and power were calculated and plotted in Figure 9a,b.

To further understand the behaviour of the VDBC circuit, the relation between the output voltage and power with the increasing load resistances was observed (Figure 9). Figure 9 shows that output voltage is enhanced with the increase in the load resistance, regardless of the applied excitation frequency. The highest output voltage was observed at 910 KΩ; however, the least was obtained at 100 KΩ. This behaviour was consistent with all applied input voltages and followed Ohm's law [3,7].

Further examination of Figure 9 reveals that the magnitude of applied input voltage considerably influences the overall performance of the VDBC circuit. The higher-output voltage and power were obtained at a higher input voltage, V_{ac}. This is because, at higher input voltages, V_{ac}, the drain to source resistance path in the MOSFET switch, is reduced. Thus, the output voltage and power are increased. Figure 9 represents the output voltage and power at various input voltages over 100 Hz excitation frequency.

3.3. VDBC Application: Solar Battery Charging

To validate the real-world application of the proposed circuit, the power generated by the VDBC circuit was utilised to charge a solar battery (1.2 V_{dc}, 4 mA battery used as solar cell-powered garden lighting). This battery was powered at a 5 Hz input frequency with an input voltage of 5 V_{ac}, under similar experimental circumstances.

The solar battery was charged up to 1.078 V_O in 32 min at 5 V_{ac}, 5 Hz input parameters. A 1 cm × 1.5 cm solar panel and a DC-DC converter typically charge a similar battery in 4–5 h. It is noted that electronic components of low threshold voltage are chosen for effective power conversion. However, the parameter rating must be set carefully for safe experimental operation. Figure 10 depicts the experimental illustration of the solar battery while turned off and on.

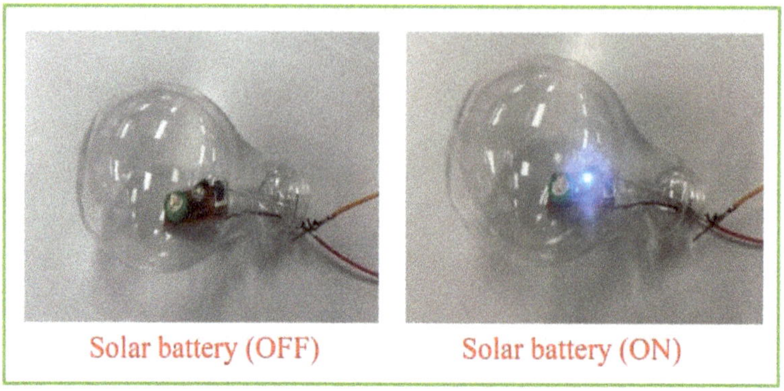

Figure 10. Experimental set-up of solar battery when turned ON and OFF.

3.4. Comparison of VDBC with the Conventional VD Circuit

The output voltage waveform of the VDBC circuit acquired from the oscilloscope at 5 Hz, at 5 V_{ac} input voltage, is summarised in Figure 11. Figure 11 shows that the VDBC circuit eliminated a majority of the ripples at the given frequency.

The rectified voltage waveforms from the VDBC were also compared with those rectified by VD. Both circuits were tested under similar scenarios (5 V_{ac} at 5 Hz frequency). It should be noted that the VD circuit possesses only the rectifier circuit (i.e., AC-DC conversion). The VD circuit charged the capacitor, C_{L1}, which was discharged at the load resistor. Thus, a high number of ripples are observed in the rectified output voltage waveform. These ripples were reduced by the VDBC circuit, which incorporates both AC to DC and DC to DC conversion processes, before the rectified voltage is stored in a load capacitor, C_{L3}.

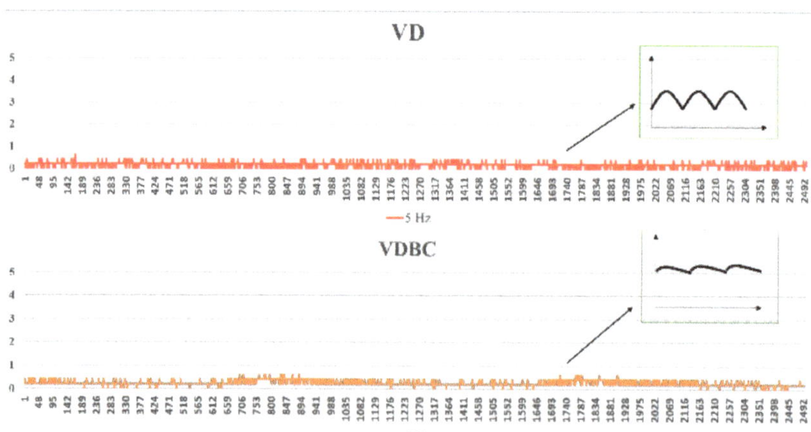

Figure 11. A comparison of rectified voltage waveforms generated by VD and VDBC circuits.

Consequently, the VDBC circuit outperforms the VD circuit regarding the stability of the rectified output voltage waveform. Thus, the secondary objective of minimising ripples in the rectified output voltage waveform was proved in this chapter.

Recalling that the main objective of this study is to rectify and boost the low AC voltage which is generated by PD to a higher DC output voltage and power, this has been achieved by the proposed VDBC circuit at both low and higher excitation frequencies.

Besides, the output voltage waveform generated from the VDBC circuit is stable compared to the circuits described in the existing literature.

In addition, the results demonstrated that the proposed VDBC circuit outperformed the existing literature circuits in terms of output voltage and power. For the sake of comparison, Table 2 lists the generated output power of similar circuits published in the literature.

Table 2. Comparison to previously published literature circuits.

Circuit Sources	Input Voltage (V_{ac})	DC Voltage (V_{dc})	Output Power (μ/mW)	External Power Supply
[27]	8	-	0.8 W	YES
[3]	5	-	22 μW	NO
[64]	0.65	1.8	75 μW	NO
[26]	3.5	-	254 μW	YES
[57]	0.4	3.3	54.4 μW	YES
[65]	0.4	3.3	40 μW	YES
[66]	1.2	-	30 μW	YES
[8]	3	-	364.5 μW	NO
VDBC	5	11.7	1.368 mW	NO

4. Conclusions

This study focuses on extracting higher output voltage by taking the benefits of MOSFETs, which have low threshold voltage, and mitigating ripples in the output voltage waveform of miniature vibrating PEH, which converts ME into EE for charging solar batteries. The proposed VDBC circuit rectified and boosted the low AC voltage, V_{ac}, generated by vibrating PD in the PEH system. The proposed VDBC circuit integrated a previously established VD with a step-up boost converter (modes 1–4) for both AC to DC and DC to DC conversions, while an extended switching circuit using a DC-DC converter in modes 5–6 stabilised the rectified voltage generated from modes 1–4. The feasibility of the proposed VDBC circuit was verified through a series of experimental analyses at both a fixed excitation frequency and a fixed input voltage. In Test 1, the proposed VDBC circuit converted a 5 Vac and 20 Hz input AC voltage to 11.7 Vdc and achieved a output power of 1.368 mW. However, the proposed prototype converted a 1.25 Vac and 100 Hz input AC voltage to 1.11 Vdc with an output power of 12.321 μW. It is noted that proposed circuit is limited to low voltage applications only. For ultra-low voltages, the proposed prototype is not suitable as the input voltage will be dissipated across circuit components and there will be no output at the load. For future work, modification of the VDBC circuit to incorporate the non-linear process is recommended.

Author Contributions: Conceptualization, A.H.; methodology, A.H.; software, A.H.; validation, A.H; formal analysis, A.H.; investigation, A.H.; resources, A.H.; data curation, A.H.; writing—original draft preparation, A.H.; writing—review and editing, A.H., M.E. and M.U.; visualisation, A.H.; supervision, M.E., F.S., M.U. and M.D.; project administration, A.H.; funding acquisition, Not applicable. All authors have read and agreed to the published version of the manuscript.

Funding: This research received no funding.

Data Availability Statement: Data is available on request due to privacy and restrictions.

Acknowledgments: The work is supported by Southern Cross University, Faculty of Science and Engineering, Lismore, NSW, 2480.

Conflicts of Interest: The authors declare no conflict of interests.

Symbols and Abbreviation

Symbol	Explanation
V_f	Diode Forward Voltage
V_{dc}	DC Voltage
V_o	Output Voltage
t	Time Period
V_D	Diode Forward Voltage Losses
C_1, C_2, C_{L1}, C_{L2}	Capacitor
P_o	Rectified Output Power
D_1–D_5	Diodes
M_1–M_2	MOSFET Switches
B	Base
E	Emitter
T_{sw}	Time Period of Switching Process
G	Gate
S	Source
C_{L3}	Load Capacitor
C_P	Internal Capacitor of PD
ω	Angular Frequency
i_{ac}	AC Current
V_{PD}	Voltage Across PD Electrodes
u	Polarization Current used to Charge the PD
V_{ac}	AC Voltage
R_1–R_2	Resistors
L_1–L_2	Inductors
C	Collector
R_L	Load Resistor
d_1–d_2	Duty Cycles
D	Drain
Abbreviation	**Explanation**
PD	Piezoelectric Device
ME	Mechanical Energy
EE	Electrical Energy
AC	Alternating Current
PEH	Piezoelectric Energy Harvesting
FBR	Full Bridge Rectifier
PSCAD	Power System Computer Aided Designs
MOSFET	Metal Oxide Field Effect Transistor
DSHBR	Dual Stage H-Bridge Rectifier Circuit
VDBC	Voltage Doubler Boost Converter
DC	Direct Current
MPE	Maximum Power Extraction
HBR	H-Bridge Rectifier
LCL	Latching Current Limiter

References

1. Edla, M.; Lim, Y.Y.; Deguchi, M.; Padilla, R.V.; Izadgoshasb, I. An improved self-powered H-bridge circuit for voltage rectification of piezoelectric energy harvesting system. *IEEE J. Electron Devices Soc.* **2020**, *8*, 1050–1062. [CrossRef]
2. Edla, M.; Lim, Y.Y.; Mikio, D.; Padilla, R.V. Non-Linear Switching Circuit for Active Voltage Rectification and Ripples Reduction of Piezoelectric Energy Harvesters. *Energies* **2022**, *15*, 709. [CrossRef]
3. Edla, M.; Lim, Y.Y.; Padilla, R.V.; Mikio, D. Design and application of a self-powered dual-stage circuit for piezoelectric energy harvesting systems. *IEEE Access* **2021**, *9*, 86954–86965. [CrossRef]
4. Garbuio, L.; Lallart, M.; Guyomar, D.; Richard, C.; Audigier, D. Mechanical energy harvester with ultralow threshold rectification based on SSHI non-linear technique. *IEEE Trans. Ind. Electron.* **2009**, *56*, 1048–1056. [CrossRef]
5. Dunn, B.; Kamath, H.; Tarascon, J.-M.J.S. Electrical energy storage for the grid: A battery of choices. *Science* **2011**, *334*, 928–935. [CrossRef]

6. Gutowski, T.; Dahmus, J.; Thiriez, A. Electrical energy requirements for manufacturing processes. In Proceedings of the 13th CIRP International Conference on Lfe Cycle Engineering, Leuven, Belgium, 31 May–2 June 2006; pp. 623–638.
7. Edla, M.; Lim, Y.Y.; Padilla, R.V.; Deguchi, M.J.A.S. An improved rectifier circuit for piezoelectric energy harvesting from human motion. *Appl. Sci.* **2021**, *11*, 2008. [CrossRef]
8. Edla, M.; Deguchi, M.; Izadgoshasb, I.; Mahmud, M.A.P.; Kouzani, A.Z. Self-powered boost-converter for power optimisation and piezo garden lights. *Smart Mater. Struct.* **2022**, *31*, 045021. [CrossRef]
9. Beeby, S.P.; Torah, R.N.; Tudor, M.J.; Glynne-Jones, P.; O'Donnell, T.; Saha, C.R.; Roy, S. A micro electromagnetic generator for vibration energy harvesting. *J. Micromech. Microeng.* **2007**, *17*, 1257–1265. [CrossRef]
10. Edla, M.; Lim, Y.Y. An improved piezoelectric energy harvesting circuit for reducing the internal loss. In Proceedings of the 83rd Researchfora International Conference, New Delhi, India, 29–30 January 2020; pp. 4–8.
11. Izadgoshasb, I.; Lim, Y.Y.; Lake, N.; Tang, L.; Padilla, R.V.; Kashiwao, T. Optimising orientation of piezoelectric cantilever beam for harvesting energy from human walking. *Energy Convers. Manag.* **2018**, *161*, 66–73. [CrossRef]
12. Izadgoshasb, I.; Lim, Y.Y.; Tang, L.; Padilla, R.V.; Tang, Z.S.; Sedighi, M. Improving efficiency of piezoelectric based energy harvesting from human motions using double pendulum system. *Energy Convers. Manag.* **2019**, *184*, 559–570. [CrossRef]
13. Rahman, A.; Farrok, O.; Islam, R.; Xu, W. Recent Progress in Electrical Generators for Oceanic Wave Energy Conversion. *IEEE Access* **2020**, *8*, 138595–138615. [CrossRef]
14. Xi, S.; Li, W.; Guo, J.; Liang, J. A self-powered piezoelectric energy harvesting interface circuit based on adaptive SSHI with fully integrated switch control. In Proceedings of the 2020 IEEE International Symposium on Circuits and Systems (ISCAS), Sevilla, Spain, 10–21 October 2020; pp. 1–4.
15. Edla, M.; Lim, Y.Y.; Mikio, D.; Padilla, R.V. A Single-Stage Rectifier-Less Boost Converter Circuit for Piezoelectric Energy Harvesting Systems. *IEEE Trans. Energy Convers.* **2021**, *37*, 505–514. [CrossRef]
16. Erturk, A.; Inman, D.J. An experimentally validated bimorph cantilever model for piezoelectric energy harvesting from base excitations. *Smart Mater. Struct.* **2009**, *18*, 025009. [CrossRef]
17. Du, S.; Jia, Y.; Zhao, C.; Amaratunga, G.A.J.; Seshia, A.A. A Nail-Size Piezoelectric Energy Harvesting System Integrating a MEMS Transducer and a CMOS SSHI Circuit. *IEEE Sens. J.* **2019**, *20*, 277–285. [CrossRef]
18. Du, S.; Jia, Y.; Zhao, C.; Amaratunga, G.A.; Seshia, A.A. A fully integrated split-electrode SSHC rectifier for piezoelectric energy harvesting. *IEEE J. Solid-State Circuits* **2019**, *54*, 1733–1743. [CrossRef]
19. Edla, M.; Lim, Y.Y.; Deguchi, M.; Padilla, R.V. A Novel Discontinuous Mode Piezoelectric Energy Harvesting Circuit for Low-Voltage Applications. In Proceedings of the 2021 31st Australasian Universities Power Engineering Conference (AUPEC), Perth, Australia, 26–30 September 2021; pp. 1–5.
20. Huang, Y.-T.; Li, C.-H.; Chen, Y.-M. A Modified Asymmetrical Half-Bridge Flyback Converter for Step-Down AC–DC Applications. *IEEE Trans. Power Electron.* **2019**, *35*, 4613–4621. [CrossRef]
21. Wang, H.; Zhu, J.; He, T.; Zhang, Z.; Lee, C. Programmed-triboelectric nanogenerators—A multi-switch regulation methodology for energy manipulation. *Nano Energy* **2020**, *78*, 105241. [CrossRef]
22. Xia, H.; Xia, Y.; Shi, G.; Ye, Y.; Wang, X.; Chen, Z.; Jiang, Q. A Self-Powered S-SSHI and SECE Hybrid Rectifier for PE Energy Harvesters: Analysis and Experiment. *IEEE Trans. Power Electron.* **2020**, *36*, 1680–1692. [CrossRef]
23. Zhu, J.; Wang, H.; Zhang, Z.; Ren, Z.; Shi, Q.; Liu, W.; Lee, C. Continuous direct current by charge transportation for next-generation IoT and real-time virtual reality applications. *Nano Energy* **2020**, *73*, 104760. [CrossRef]
24. Blank, V.; Bormashov, V.; Tarelkin, S.; Buga, S.; Kuznetsov, M.; Teteruk, D.; Kornilov, N.; Terentiev, S.; Volkov, A. Power high-voltage and fast response Schottky barrier diamond diodes. *Diam. Relat. Mater.* **2015**, *57*, 32–36. [CrossRef]
25. Teo, J.; Tan, R.H.; Mok, V.; Ramachandaramurthy, V.K.; Tan, C. Impact of bypass diode forward voltage on maximum power of a photovoltaic system under partial shading conditions. *Energy* **2020**, *191*, 116491. [CrossRef]
26. Shareef, A.; Goh, W.L.; Narasimalu, S.; Gao, Y. A Rectifier-Less AC–DC Interface Circuit for Ambient Energy Harvesting from Low-Voltage Piezoelectric Transducer Array. *IEEE Trans. Power Electron.* **2018**, *34*, 1446–1457. [CrossRef]
27. Tabesh, A.; Fréchette, L.G. A low-power stand-alone adaptive circuit for harvesting energy from a piezoelectric micropower generator. *IEEE Trans. Ind. Electron.* **2009**, *57*, 840–849. [CrossRef]
28. Ammar, M.B.; Bouattour, G.; Bouhamed, A.; Sahnoun, S.; Fakhfakh, A.; Kanoun, O. AC-DC Single Phase Rectifiers for Nanocomposite based Flexible Piezoelectric Energy Harvesters. In Proceedings of the 2021 18th International Multi-Conference on Systems, Signals & Devices (SSD), Monastir, Tunisia, 22–25 March 2021; pp. 228–234.
29. Li, D.; Wang, C.; Cui, X.; Chen, D.; Fei, C.; Yang, Y. Recent progress and development of interface integrated circuits for piezoelectric energy harvesting. *Nano Energy* **2022**, *94*, 106938. [CrossRef]
30. Quattrocchi, A.; Montanini, R.; De Caro, S.; Panarello, S.; Scimone, T.; Foti, S.; Testa, A. A New Approach for Impedance Tracking of Piezoelectric Vibration Energy Harvesters Based on a Zeta Converter. *Sensors* **2020**, *20*, 5862. [CrossRef]
31. Srinivasan, R.; Mangalanathan, U.; Gandhi, U.; Karlmarx, L.R. Bridgeless active rectifier for piezoelectric energy harvesting. *IET Circuits Devices Syst.* **2019**, *13*, 1078–1085. [CrossRef]
32. Covaci, C.; Gontean, A. Two-Stage Converter for Piezoelectric Energy Harvesting Using Buck Configuration. In Proceedings of the 2020 IEEE 26th International Symposium for Design and Technology in Electronic Packaging (SIITME), Pitesti, Romania, 21–24 October 2020; pp. 408–411.

33. Sarker, M.R.; Julai, S.; Sabri, M.F.M.; Said, S.M.; Islam, M.; Tahir, M. Review of piezoelectric energy harvesting system and application of soptimisation techniques to enhance the performance of the harvesting system. *Sens. Actuators A Phys.* **2019**, *300*, 111634. [CrossRef]
34. Kashiwao, T.; Izadgoshasb, I.; Lim, Y.Y.; Deguchi, M. Optimization of rectifier circuits for a vibration energy harvesting system using a macro-fiber composite piezoelectric element. *Microelectron. J.* **2016**, *54*, 109–115. [CrossRef]
35. Hsieh, J.-C.; Tsai, T.-H. An AC-DC wind energy harvesting circuit with extended input-voltage range and 95% tracking efficiency. In Proceedings of the 2018 IEEE International Symposium on Circuits and Systems (ISCAS), Florence, Italy, 27–30 May 2018; pp. 1–4.
36. Dini, M.; Romani, A.; Filippi, M.; Tartagni, M. A nanopower synchronous charge extractor IC for low-voltage piezoelectric energy harvesting with residual charge inversion. *IEEE Trans. Power Electron.* **2015**, *31*, 1263–1274. [CrossRef]
37. Eltamaly, A.M.; Addoweesh, K.E. A novel self-power SSHI circuit for piezoelectric energy harvester. *IEEE Trans. Power Electron.* **2016**, *32*, 7663–7673. [CrossRef]
38. Lee, M.; Yang, J.; Park, M.-J.; Jung, S.-Y.; Kim, J. Design and Analysis of Energy-Efficient Single-Pulse Piezoelectric Energy Harvester and Power Management IC for Battery-Free Wireless Remote Switch Applications. *IEEE Trans. Circuits Syst. I Regul. Pap.* **2017**, *65*, 366–379. [CrossRef]
39. Lefeuvre, E.; Badel, A.; Brenes, A.; Seok, S.; Woytasik, M.; Yoo, C.-S. Analysis of piezoelectric energy harvesting system with tunable SECE interface. *Smart Mater. Struct.* **2017**, *26*, 035065. [CrossRef]
40. Lu, S.; Boussaid, F. A highly efficient P-SSHI rectifier for piezoelectric energy harvesting. *IEEE Trans. Power Electron.* **2015**, *30*, 5364–5369. [CrossRef]
41. Pollet, B.; Despesse, G.; Costa, F. A New Non-Isolated Low-Power Inductorless Piezoelectric DC–DC Converter. *IEEE Trans. Power Electron.* **2019**, *34*, 11002–11013. [CrossRef]
42. Xia, H.; Xia, Y.; Ye, Y.; Qian, L.; Shi, G.; Chen, R. Analysis and Simulation of Synchronous Electric Charge Partial Extraction Technique for Efficient Piezoelectric Energy Harvesting. *IEEE Sensors J.* **2018**, *18*, 6235–6244. [CrossRef]
43. Liang, J.; Liao, W.H. Improved design and analysis of self-powered ssynchronised switch interface circuit for piezoelectric energy harvesting systems. *IEEE Trans. Ind. Electron.* **2011**, *59*, 1950–1960. [CrossRef]
44. Jeong, J.; Shim, M.; Maeng, J.; Park, I.; Kim, C. An Efficiency-Aware Cooperative Multicharger System for Photovoltaic Energy Harvesting Achieving 14% Efficiency Improvement. *IEEE Trans. Power Electron.* **2019**, *35*, 2253–2256. [CrossRef]
45. Lian, Q.; Han, P.; Mei, N. A Review of Converter Circuits for Ambient Micro Energy Harvesting. *Micromachines* **2022**, *13*, 2222. [CrossRef] [PubMed]
46. Zhao, E.; Han, Y.; Lin, X.; Yang, P.; Blaabjerg, F.; Zalhaf, A.S. Impedance characteristics investigation and oscillation stability analysis for two-stage PV inverter under weak grid condition. *Electr. Power Syst. Res.* **2022**, *209*, 108053. [CrossRef]
47. Han, Y.; Feng, Y.; Yang, P.; Xu, L.; Zalhaf, A.S. An efficient algorithm for atomic decomposition of power quality disturbance signals using convolutional neural network. *Electr. Power Syst. Res.* **2022**, *206*, 107790. [CrossRef]
48. Do, X.-D.; Nguyen, H.-H.; Han, S.-K.; Lee, S.-G. A rectifier for piezoelectric energy harvesting system with series ssynchronised switch harvesting inductor. In Proceedings of the 2013 IEEE Asian Solid-State Circuits Conference (A-SSCC), Singapore, 11–13 November 2013; pp. 269–272.
49. Guilar, N.J.; Amirtharajah, R.; Hurst, P.J. A Full-Wave Rectifier With Integrated Peak Selection for Multiple Electrode Piezoelectric Energy Harvesters. *IEEE J. Solid-State Circuits* **2008**, *44*, 240–246. [CrossRef]
50. Hamlehdar, M.; Kasaeian, A.; Safaei, M.R. Energy harvesting from fluid flow using piezoelectrics: A critical review. *Renew. Energy* **2019**, *143*, 1826–1838. [CrossRef]
51. Bimbhra, P.; Kaur, S. *Power Electronics*; Khanna publishers: New Delhi, India, 2012; Volume 2.
52. Rashid, M.H. *Power Electronics Handbook*; Butterworth-Heinemann: Oxford, UK, 2017.
53. Kushino, Y.; Koizumi, H. Piezoelectric energy harvesting circuit using full-wave voltage doubler rectifier and switched inductor. In Proceedings of the 2014 IEEE Energy Conversion Congress and Exposition (ECCE), Pittsburgh, PA, USA, 14–18 September 2014; pp. 2310–2315.
54. Lefeuvre, E.; Audigier, D.; Richard, C.; Guyomar, D. Buck-Boost Converter for Sensorless Power Optimization of Piezoelectric Energy Harvester. *IEEE Trans. Power Electron.* **2007**, *22*, 2018–2025. [CrossRef]
55. Erickson, R.W.; Maksimovic, D. *Fundamentals of Power Electronics*; Springer Science & Business Media: Berlin, Germany, 2007.
56. Das, A.; Gao, Y.; Kim, T.T.-H. A 220-mV Power-on-Reset Based Self-Starter With 2-nW Quiescent Power for Thermoelectric Energy Harvesting Systems. *IEEE Trans. Circuits Syst. I Regul. Pap.* **2016**, *64*, 217–226. [CrossRef]
57. Wang, H.; Tang, Y.; Khaligh, A. A Bridgeless Boost Rectifier for Low-Voltage Energy Harvesting Applications. *IEEE Trans. Power Electron.* **2013**, *28*, 5206–5214. [CrossRef]
58. Khan, M.B.; Kim, D.H.; Han, J.H.; Saif, H.; Lee, H.; Lee, Y.; Kim, M.; Jang, E.; Hong, S.K.; Joe, D.J.; et al. Performance improvement of flexible piezoelectric energy harvester for irregular human motion with energy extraction enhancement circuit. *Nano Energy* **2019**, *58*, 211–219. [CrossRef]
59. Du, S.; Amaratunga, G.A.; Seshia, A.A. A cold-startup SSHI rectifier for piezoelectric energy harvesters with increased open-circuit voltage. *IEEE Trans. Power Electron.* **2018**, *34*, 263–274. [CrossRef]
60. Ottman, G.; Hofmann, H.; Lesieutre, G. Optimised piezoelectric energy harvesting circuit using step-down converter in discontinuous conduction mode. *IEEE Trans. Power Electron.* **2003**, *18*, 696–703. [CrossRef]

61. Edla, M.; Deguchi, M.; Lim, Y.Y. A self-powered H-Bridge joule theory circuit for piezoelectric energy harvesting systems. *Power Electron. Devices Compon.* **2022**, *3*, 100015. [CrossRef]
62. Edla, M. Development of High-Efficiency Rectifier Circuits for Piezoelectric Energy Harvesting Systems. Ph.D. Thesis, Southern Cross University, Lismore, Australia, 2021.
63. Ottman, G.; Hofmann, H.; Bhatt, A.; Lesieutre, G. Adaptive piezoelectric energy harvesting circuit for wireless remote power supply. *IEEE Trans. Power Electron.* **2002**, *17*, 669–676. [CrossRef]
64. Hehn, T.; Hagedorn, F.; Maurath, D.; Marinkovic, D.; Kuehne, I.; Frey, A.; Manoli, Y. A fully autonomous integrated interface circuit for piezoelectric harvesters. *IEEE J. Solid-State Circuits* **2012**, *47*, 2185–2198. [CrossRef]
65. Yu, L.; Wang, H.; Khaligh, A. A Discontinuous Conduction Mode Single-Stage Step-Up Rectifier for Low-Voltage Energy Harvesting Applications. *IEEE Trans. Power Electron.* **2016**, *32*, 6161–6169. [CrossRef]
66. Kwon, D.; Rincón-Mora, G.A. A 2-μm BiCMOS Rectifier-Free AC–DC Piezoelectric Energy Harvester-Charger IC. *IEEE Trans. Biomed. Circuits Syst.* **2010**, *4*, 400–409. [CrossRef] [PubMed]

energies

MDPI

Article

Inverter-Less Integration of Roof-Top Solar PV with Grid Connected Industrial Drives

M. Ryyan Khan [1,*,†], Intekhab Alam [2,†] and M. Rezwan Khan [2,†]

1 Department of Electrical and Electronic Engineering, East West University, A/2 Jahurul Islam Avenue, Jahurul Islam City, Aftabnagar, Dhaka 1212, Bangladesh
2 Department of Electrical and Electronic Engineering, United International University, United City, Badda, Dhaka 1212, Bangladesh
* Correspondence: ryyan@ewubd.edu
† Thesis authors contributed equally to this work.

Abstract: Green energy from Solar PV is getting increased attention in the industries due to the falling price of solar panels in the world market. A grid-tied inverter is one of the major components in such a system, where the DC energy from PV is converted to AC and synchronized with the grid to obtain power sharing between the PV and the grid for the industrial drives. In this paper, a DC link has been proposed instead of an AC link for interconnection between the solar PV system and the grid to run those industrial drives. In most modern industrial applications, induction motors are driven by VVVF (Variable Voltage and Variable Frequency) inverters to achieve efficient speed control. The inverters commonly have a rectifier section at the front end that rectifies the input AC to DC and the DC is then used in PWM mode to generate the required voltage and frequency for the induction motor operating under variable speed and load conditions. Such an inverter can use both AC or DC as the input so long the supply voltage has the right value for the inverter to operate. In our proposition, we eliminate the grid-tied inverter and use a DC link, created from the rectified AC and the regular Solar PV, to obtain the power-sharing between the PV output and the grid. Using the DC link output directly to energize the VVVF inverter has an impact on the performance of the inverter. In the proposed system, the solar PV array is designed in such a way that the grid remains as the supplementary power source only to supplement any shortfall in the PV output due to variable sunshine conditions. The control circuit used in this novel technique is inexpensive, efficient, and simple in design when compared to the grid-tied inverters. The proposed system has been implemented at Niagara Textiles in Gazipur, Bangladesh. The experimental/practical results are presented to validate the basic concept. Around a 20% reduction in the cost of energy has been reported in this paper, with a more than 90% efficient system. This will definitely make solar PV energy more competitive with regular energy and attractive to industries for its simplicity.

Keywords: green energy; solar PV; grid-tied inverter; inverter-less

1. Introduction

The ominous threat of global warming has triggered enhanced awareness amongst policymakers, utilities, and consumers alike. Countries in general, irrespective of their geographical location or economic affluence, are considering renewable energy resources as a major thrust area [1–3]. Earlier, the green energy concept was stigmatized as an expensive form of energy meant only for the rich [4,5]. However, the price of the PV modules in the world market has come down to a level such that solar PV-based energy generation has become commercially viable, even for countries with moderate sunshine, so that they can compete with the national grids [6,7]. The most common form of solar PV and grid connection involves grid-tied inverters, where the DC output of solar PV is converted to an AC that is compatible with the grid voltage. These grid-tied inverters should have the following characteristics:

1. The output voltage needs to be sinusoidal;
2. Voltage and frequency of the output must match with those of the grid;
3. The output must have the right phase to connect to the grid.

Additionally, the inverter must draw power from the PV such that maximum power point operation is ensured. Satisfying all the requirements in the grid-tied inverter require complex control circuits, which is reflected in their cost. The design of the grid-tied inverters has improved significantly over the years and present-day inverters can have an efficiency of more than 95% under full load operation. However, under varying load conditions due to varying sunshine, the average efficiency remains more than 90% [8,9].

Despite their high efficiency, grid-tied inverters are still expensive and account for around 15–20% of the PV energy cost [9]. Usually, the cost of the PV energy is within a very narrow margin of profitability and any uncertainty in the PV output such as periodic cleaning of the dirt [10,11] on the PV modules to keep the output to the designed level can eat up the narrow margin of profit. Such factors have a discouraging effect on the industrial uptake of PV systems and any reduction in the energy cost can have a significant positive impact. With the development of power electronics, many of the appliances and drives have electronic controls inside them and are effectively insensitive to the type of input voltage (AC or DC) supply [12]. At the same time, many of the appliances have their own control circuit that enables the appliances to work over a wide range of input voltages. This is particularly true for the industrial motor drives [13–15] and it opens up the opportunity of integrating solar PV to grid-connected systems without using grid-tied inverters. Such a mechanism is expected to be of lower cost compared to the cost of the grid-tied inverters, as already discussed, and can have efficiency as high as or higher than the usual grid-tied inverters. Higher efficiency and lower cost can contribute towards a less expensive PV-based power that can have a positive impact in popularizing PV systems in grid-connected applications. Although such a concept has been proposed in small-scale household cooking systems with cooking appliances using DC loads [16], to the best of our knowledge, a similar system has not been studied with three-phase industrial drives. Here, we have presented the performance of such a system under varying sunshine and load conditions indicating that it can be a very effective low-cost solution for industrial applications.

In this paper, we present the concept and the design considerations for a rooftop inverter-less PV system to be connected with grid-connected industrial drives. An inverter-less system along with rooftop solar PV was implemented in Niagara Textiles, a compliant textile industry in Gazipur, Bangladesh, to verify the performance of the system. Experimental/practical data were collected to observe the power-sharing between the solar panel array and the grid under different loads and sunshine conditions. Results show that load-sharing control was smooth and efficient under varying load and sunshine conditions. After this introduction section, Section 2 describes the issues and proposal of our research. The proposed system model has been explained in Section 3. Section 4 clarifies the system modeling. The findings and results are discussed in Section 5. Lastly, the conclusion has been drawn in Section 6.

2. Inverter-Less Integration of Solar PV with Industrial Drives

The rooftop application of solar PV is gaining momentum, particularly in industries in Bangladesh, as there are a number of economic reasons associated with it. Infrastructure Development Company Limited (IDCOL), Bangladesh [17] has so far financed 11 rooftop solar systems and the average energy cost is approximately USD 0.075/kWh (with 80% loan at a subsidized interest rate of 6%). The cost of captive power generation in Bangladesh is USD 0.055/kWh where a subsidized gas supply is available and the cost of grid power is USD 0.11/kWh. Although IDCOL estimated the cost of inverter based solar PV energy to be lower than the grid energy, it was realized later that most of the industries did not consider the realistic loss of energy due to temperature rise in the solar cells [18–22] and accumulation of dust and dirt on the solar panels [23–28]. When all these losses are taken into account, the actual cost of energy is expected to be higher. So, further reduction in the cost of solar PV-based energy is essential to make solar PV power competitive to the grid power in Bangladesh, when IDCOL funding with subsidized interest rate is exhausted.

The main reasons for solar PV applications in the industrial sector in Bangladesh are:

1. A large number of industries receive natural gas (Methane) from the national gas grid at a price much lower (less than half) than the international price [29]. They use this gas for captive electricity generation and as fuel for boilers for steam generation. As the production of gas from the local gas fields is dwindling, the government has taken up a plan to import gas and supply it to the gas network to sustain the industries. However, the cost of gas is expected to increase by at least 50% within a few months or so [29]. This has instigated considerable concerns for the industries to look for alternative energy resources and due importance is given to solar PV.
2. The falling price of solar PV has made the cost of solar energy competitive with the captive generation. Any increase in gas price will make solar PV even more attractive.
3. For industries not having enough captive generation, solar PV has become a very attractive alternative as grid electricity costs are much higher (almost 40% higher) than captive electricity generation.

However, the main challenges of using PV in industries are as follows:

1. Solar PV output is intermittent in nature with varying output varying with cloud, fog or other weather related uncertainties.
2. Most industries run for long hours every day and the industry must arrange for alternative sources for night hours.
3. One possible option is to have a battery or some other form of energy storage, which is still very expensive and is not economically viable for most industries.
4. Having grid-tied inverters to connect the solar PV to the internal power line is one of the best solutions. However, as already mentioned, in moderate-sized PV systems, less than 500 kWp, the cost of inverters constitutes approximately 15–20% of the energy cost with a complex operation system [30].

Using the proposed system, there is scope to make the PV power less expensive and it is our estimation that the cost of the energy can be reduced by close to 20%. Such a reduction in the energy cost can make them more popular and we expect that more industries will take interest in generating a higher percentage of green energy reducing the overall carbon emission [29,31].

The basic block diagram of the proposed system is shown below in Figure 1. The 3-phase AC supply is rectified to convert it to DC. The PV array is chosen such that it has the maximum power point voltage close to the rectified grid voltage (which automatically means that the open circuit voltage will be approximately 25% higher than the rectified grid voltage) [9]. The −ve terminal of the PV out and the rectified grid voltage is connected to form a common −ve terminal. This common −ve terminal, the +ve terminal of the rectified grid voltage, and the +ve terminal of the PV output are now connected to the three input terminals of the variable voltage variable frequency (VVVF) inverter (DC to AC converter) of the industrial drive via a 3-pole (3-phase) magnetic contactor as shown in Figure 1. The value of the capacitor at the PV terminals is essential for the efficient operation of the system as the VVVF inverter operates in pulse width modulation (PWM) mode.

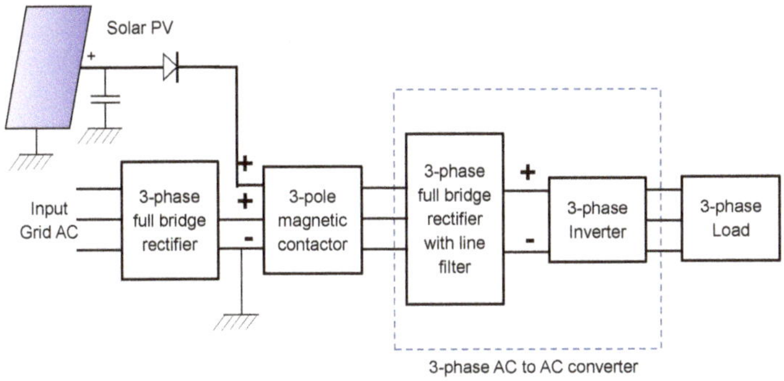

Figure 1. Block diagram of the control circuit to drive the existing motor controller.

3. The Proposed System Model

In the proposed system, we suggest that the number of PV modules connected in series should be such that the DC voltage obtained by the rectification of the grid voltage is close to the maximum power point voltage of the PV array. So, there is no voltage level converter used so far as the PV output is concerned and this will save the cost of the hardware and at the same time reduce the energy loss occurring due to voltage conversion. The VVVF inverter, driving the motor, has a maximum rated input voltage to avoid damage due to over-voltage. While designing the PV array, it is to be taken into consideration that the open circuit voltage of the PV array does not exceed this limit.

Mathematically, these conditions can be expressed as

$$N = int\left(\frac{V_{G_{DC}}}{V_{MP}} + 1\right) \tag{1}$$

where N is the number of panels to be connected in series. $V_{G_{DC}}$ is the rectified 3-phase grid voltage. V_{DC} is the DC operating voltage of the VVVF inverter and V_{MP} is the maximum power point voltage of a single array, V_{OC} is the open circuit voltage of a single PV module and $V_{OC_{Array}}$ is the open circuit voltage of the PV array, V_{DC_m} is the maximum DC voltage that is allowed to apply at the VVVF inverter input terminals. Here, int(x) indicates a mathematical operation that selects an integral value from x by discarding the fractional part. Along with the condition mentioned in Equation (2), it is to be ensured that the open circuit voltage of the PV array does not exceed the allowed voltage limit that can be applied to a PV module, i.e.,

$$N \times V_{OC} < PV_{module\text{-}maximum\text{-}allowed\text{-}voltage} \tag{2}$$

Typically, the rectified grid voltage $V_{G_{DC}}$ ranges within 540–560 V for the VVVF inverters and the maximum allowed voltage for a PV module is less than 1000 V. So, the voltage condition required in Equation (2) is usually within the safe voltage range for an individual PV module.

3.1. Design of the Panel Array

Once the value of N is obtained from Equations (1) and (2), we design an array of PV panels by connecting N number of solar panels in series. To determine the number of parallel branches in the PV array, we need to estimate the required installed capacity of the solar panel array, P_{array}. Given individual panel power rating P_P (Watt peak), we can calculate the number of parallel strings A_i in the PV array as follows:

$$A_i = int\left(\frac{P_{array}}{N \times P_P}\right) \tag{3}$$

It is very likely that $P_{array}/(N \times P_P)$ will be a non-integer, and we need to make a round off of the value of A_i by choosing the lower round off after adding 0.5.

While choosing the value of P_{array}, we need to consider the roof space, the recommended tilt angle at that geographical location, and the amount of green energy the industry intends to consume. It is recommended that we choose P_{array} such that its production does not exceed the power requirement of the connected load under the expected highest sunshine condition at that geographical location, as any excess power from the panel cannot be delivered to the grid in the proposed system.

3.2. The Control Circuit

As discussed earlier, many industrial machines are driven by VVVF inverters. Due to the rectifiers at the first stage of the VVVF circuits, these inherently can run from a DC source with a voltage close to its rated AC peak voltage. For our analysis, a number of AC to AC converters (VVVF inverters) that drive variable speed knitting machines were considered as the load. The AC to AC converter has a 3-phase full bridge rectifier at its input end and a variable frequency inverter at the output end as shown in the schematic diagram in Figure 1. The 3-phase grid supply is first rectified through an externally connected bridge rectifier and the DC output is given as input to the two of the three terminals of the AC to AC converter via a 3-pole magnetic contactor. The negative end of the PV array is connected to the negative end of the external bridge rectifier that converts the grid AC to DC. The positive terminal of the PV array, the positive terminal of the grid DC and the common negative terminal are connected to the three input terminals of the magnetic contactor and the output of the magnetic contactor is then connected to the three inputs of the AC to AC converter. The choice of the terminals has no impact on the operation of the circuit. The circuit arrangement ensures that there is no backflow of power from the panel to the grid or from the grid to the panel. The magnetic coil of the contactor is energized from the grid ensuring that the magnetic contactor disconnects the AC to AC converter from the grid and the solar PV to avoid any accident due to high PV voltage when the grid power is off. In a developing country such as Bangladesh, load shedding is not an uncommon phenomenon and it is important to isolate the load from the PV during load shedding. This is enforced by the magnetic contactor whose magnetic coil is energized by the grid voltage. When there is load shedding the magnetic coil is de-energized and contact with the PV is disconnected. This ensures personnel safety during grid power failure or repair and maintenance of the machine. There is a 1000 μF capacitor connected at the output of the PV array which makes sure that the PV voltage remains steady during the switched mode operation of the inverter. This is important to keep the PV voltage always close to the maximum power point voltage for efficient operation.

3.3. Operating Conditions and Expected Losses

While PV output and the grid are connected via the rectifying circuits (shown in Figure 1), the following two conditions may occur during power sharing between the grid and the panel.

3.3.1. Condition 1: When the Load Power Is Less Than the Maximum Power of the Solar PV

Under this condition, PV output at its maximum power point is higher than the power consumption of the load. This may occur due to strong sunshine or very low power requirement from the load. Under such a condition, the PV terminal voltage increases to a value higher than the grid voltage to adjust to the load power. As per the design of the control circuit, any increase in the PV voltage will reverse bias the rectifier circuit on the grid side (as the grid side voltage is almost constant) and power flow from the grid will stop. This is an unwanted situation as the power flow from the PV panels are less than their capability to deliver and should be avoided.

3.3.2. Condition 2: When the PV Power Is Less Than the Load Power

Under this condition, the rectified grid voltage determines the DC voltage at the input side of the inverter and the current flow from the PV panel will correspond to the DC operating voltage. As per the design, the DC voltage obtained from the rectified grid supply is slightly below the maximum power point (MPP) voltage of the PV.

Therefore, PV supplies the majority of the load power requirement while operating close to MPP. The remaining fraction of the total load power is drawn from the grid. The percentage of power share between the PV and the grid will depend on the sunshine condition. Solar PV maximum power point voltage decreases from the standard testing condition (STC) value under lower sunshine conditions. Therefore, under varying sunshine, solar PV operates at the grid voltage but suffers some degree of power loss due to the deviation of the maximum power point voltage. Our numerically simulated results presented in Section 5.1 show that the loss remains well within 5% for an optimally designed system.

While designing the PV array, we should ensure that the situation described in condition 1 rarely takes place. Unlike the MPPT (maximum power point tracking) inverters, the control circuit is very simple and can handle the rapid fluctuations in the sunshine conditions without having any adverse effect on the efficiency. The I-V characteristics of the solar panel ensure that, under condition 2, PV delivers its power to the load and only the remaining fraction of the load-power requirement is drawn from the grid. During the night hours, there is no PV output and the full power is drawn from the grid.

The array is designed in such a way that the operating voltage of the system is set slightly lower (3–5% lower) than the STC maximum power point voltage of the PV array. This is to make sure that the operating DC voltage of the system (i.e., inverter inside the machine) never goes higher than the MPP voltage of the PV array due to the fluctuation of the grid voltage. The power output of a PV array reduces rapidly if the operating voltage goes beyond the maximum power point. Hence, the array is designed such that the operating DC voltage never exceeds the maximum power point voltage of the PV.

As far as the efficiency of the control circuit is concerned, losses in the rectifiers are the main losses, considering that there is no significant loss in the connecting wires or the magnetic contactor. In the case of the bridge rectifier, two diodes remain in conduction while under operation and the expected voltage drop V_{diode} will be of the order of 2 V ($\sim$1 V for each diode). Considering the operating DC voltage of the inverter to be V_{DC}, any current (I) will deliver load power of $P(= I \times V_{DC})$ and the corresponding power loss in the two diodes will be $P_{Loss}(= I \times V_{diode})$. So, the loss fraction of the load power will be $(I \times V_{diode})/(I \times V_{DC}) = V_{diode}/V_{DC} \approx (2\text{ V})/V_{DC}$. Similarly, in the PV circuit, there is only one diode in the conduction path and the expected loss fraction in the PV portion of the circuit is (1 V)/V_{DC} ($\sim$1 V drop in a single diode). The usual voltage of V_{DC} will be close to 535 V_{DC} for a 3-phase 400 V AC system. This corresponds to 0.37% and 0.18% loss in the grid and PV portion of the control circuit, respectively. This is no doubt a very low loss figure for a circuit connecting PV to the grid. Loss in the magnetic contactor [32] will depend on the type chosen and should not exceed 10 W.

4. System and Model Setup

In the following subsections, we will discuss the experimental system setup and the numerical model considerations.

4.1. System Configuration

We implemented our scheme in a small knitting section of Niagara Textiles, where four knitting machines were energized by the system. Each of the machines was rated at 3.5 kW, and the inverters driving the machines were rated at 4 kW. The actual load of the knitting machines varied with the thickness and type of yarn used and speed of operation. As the total maximum power of the knitting machines was 14 kW, we decided to perform the experiment with a much lower PV size ($\sim$5 kWp) so that the power generated from the solar PV is never higher than the load. This was to ensure the maximum utilization of solar energy. The grid line was a 3-phase 400 V supply, but there was some variation of the line voltage, which varied from 390 V to 410 V depending on the internal load condition of the textile industry. The DC value of the rectified voltage correspondingly varied from 525 to 550 V.

The solar panels that were available in the market had MPP voltages of 18.65 V, 31.34 V, 36.4 V, and higher. While using Equation (1), we found that V_{MPP} of 31.34 V produces the value of N to be 18 when the $V_{G_{DC}}$ is considered 537.5 V (halfway between 525 and 550 V). Therefore, we chose to use 18 panels in series. The rated power of the PV panels available in the market varied from 250–330 W_p and we chose 275 W_p panels for our purpose so that the total watt peak for the PV array of 18 panels is 4.95 kW that is close to the targeted 5 kW. Each panel has 60 cells in series making the MPP voltage of 31.34 V as presented in Table 1. The MPP voltage of the array for N = 18 corresponds to 564.12 V. The average operating voltage of 535 V is about 5% lower than the rated MPP voltage. This is, as discussed earlier, to ensure efficient operation of the PV panels when the grid voltage fluctuates or the sunshine varies. Table 1 summarizes the design factors chosen for the PV array. The PV array was put on the roof of one of the buildings such that there is no shading/partial shading from any other structures nearby. The panels were tilted at an angle of 20° to the south, considering the latitude of the location to be 23° N.

Table 1. Module and array configuration ratings.

W_p of the PV module, W	275
V_{OC}, V	38.28
I_{sc}, A	9.29
V_m, V	31.34
I_m, A	8.77
Number of PV modules	18
Rating of the array W_p, kW	4.95

Temperature coefficients of the short circuit current, open circuit voltage, and maximum power are 0.058%/°C, −0.33%/°C, and −0.4%/°C, respectively.

4.2. Numerical Model

Typically, the equivalent circuit of a solar cell is represented by a current source connected in parallel with a diode and a resistance R_{sh}. A series resistance R_s takes into account the resistance inside the solar cell while the load current flows through the semiconductor and contact of the p-n junction. Figure 2 shows the widely used equivalent circuit.

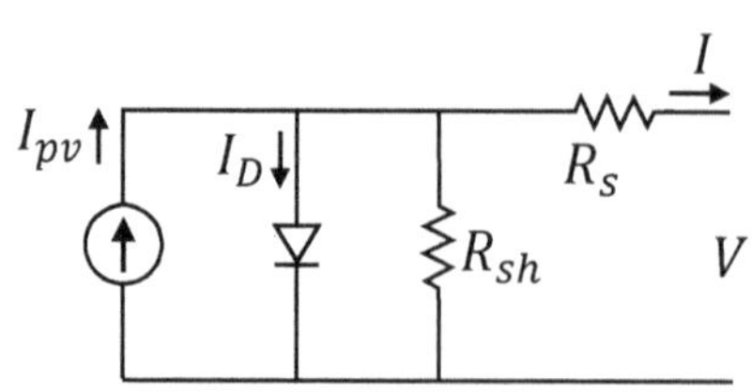

Figure 2. Circuit model of a solar cell.

The load current I_L [33] is given by

$$I_L = I_{sc} - I_0\left(\exp\frac{(V + I_L R_s)}{kT/e} - 1\right) - \frac{(V + I_L R_{sh})}{R_{sh}} \tag{4}$$

where e is the electronic charge, k is the Boltzmann constant and T is the temperature in K. I_{sc} is the short circuit current for a given input radiation, V is the output voltage of a PV cell. The solar module used has 60 cells in series and we have used 18 panels in series. Therefore, the voltage $V = V_L/(60 \times 18)$, where V_L is the PV array terminal voltage. During the open circuit condition, I_L is zero and the diode current is much larger than the current through the shunt resistance R_{sh}. Therefore, we have $I_0 = 1.44 \times 10^{-10}$A, under the standard testing condition listed in Table 1. We numerically modeled the solar module I–V given the I_0 value—to match the voltage and the current at the MPP to the ratings, we found R_s and R_{sh} values of 4.47 mΩ and 10 Ω, respectively.

5. Results and Discussions

5.1. Numerical Analysis

The grid AC voltage in Bangladesh can fluctuate from 390 to 410 V depending upon the load condition of the textile mill. As there is no maximum power point tracking in the proposed control circuit, the PV array usually operates at a voltage that corresponds to the DC voltage obtained by the rectification of the grid AC. Hence, there is a likelihood that the array will operate at a voltage off the MPP and will not extract the maximum possible power.

If the operating voltage is too much off the MPP voltage, we may lose a significant amount of PV power. So, to get an idea about the loss of PV power due to deviation from MPP voltage, we simulated the maximum power point voltage for the PV array under different sunshine conditions with the parameters given in Table 1. The simulation result is presented in Figure 3.

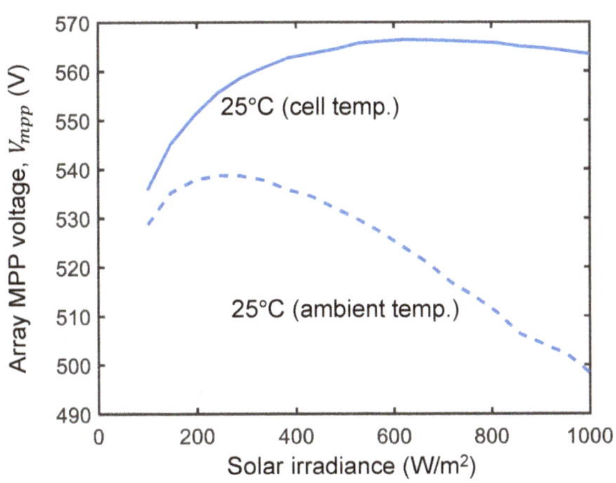

Figure 3. Variation of MPP voltage for variation of sunshine.

The figure shows that the maximum power point voltage at cell temperature of 25 °C varies from near 565 V at sunshine of 1000 W/m^2 to a lower voltage of 555 V for sunshine of 250 W/m^2. The average sunshine in Bangladesh is close to 600 W/m^2 [34] for the effective sunshine hours of 8 h and the corresponding MPP voltage is 565 V. However, the MPP voltage drops significantly if we take into consideration the actual cell temperature under different sunshine conditions. We used the Skoplaki model [35] to calculate the cell temperatures based on the experimentally obtained ambient temperature and plane-of-array irradiance. Considering the ambient temperature at 25 °C, the MPP voltage drops with higher incident radiation and is shown by the dotted line in Figure 3. Under the 1-sun condition, our calculations show that the cell temperature becomes close to 55 °C and MPP voltage reaches as low as 500 V. This forms the basis for our choice of the PV array voltage. On observation in the Niagara Textile Mill, we found that the DC voltage at the load mostly remains close to 535 V with the occasional dip to 530 V. We simulated the performance of the solar panel under different DC voltages for varying sunshine conditions and the results are presented in Figure 4. Within an operating DC voltage of 520–540 V, we may lose some power from the PV panel array due to the deviation from the MPP of the panels. If the cell temperature is kept constant at 25 °C, then the loss never exceeds ~4% for a wide range of sunshine varying from 250–1000 W/m^2, as shown in Figure 4a. On the other hand, if we simulate the losses considering the actual cell temperature under different sunshine conditions, then the losses remain below 5% except for the high radiation intensity of 1 sun when the loss becomes ~7.5% for an operating voltage of 540 V DC, see Figure 4b. However, the sunshine intensity in Bangladesh has an average value close to 650 W/m^2 and seldom reaches 1000 W/m^2 and the operating voltage mostly remains around 535 V. So, it is quite reasonable to assume that the losses due to deviation from the MPP voltage remain within 5%.

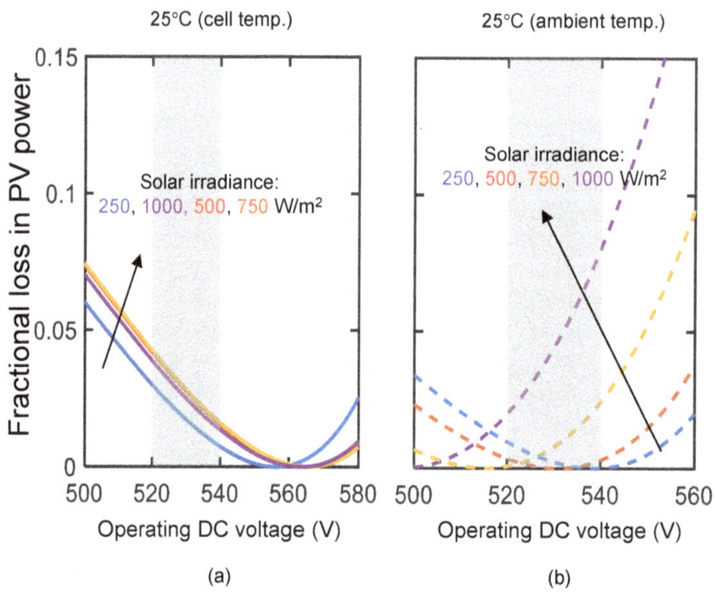

Figure 4. Loss of solar PV power due to deviation in the MPP voltage under varying sunshine conditions for different operating DC voltage are shown assuming (**a**) STC conditions, i.e., cell temperature at 25 °C, and (**b**) practical conditions, i.e., ambient temperature of 25 °C. Load power is greater than 5 kW.

As explained in Section 3.3, the low power consumption by the load connected to the inverter may result in losses due to higher PV generation. Considering the factors mentioned above, it is important to estimate the power extraction efficiency of our system under different load and sunshine conditions. If P_{MP} is the power of the array at MPP under a given sunshine condition and P_{PV} is the PV power received at the load then we can define a power extraction efficiency η_{ex}, where

$$\eta_{ex} = \frac{P_{PV}}{P_{MP}} \tag{5}$$

The net load power is

$$P_L = P_{PV} + P_G, \tag{6}$$

where P_G is the load power consumed from the grid.

The simulated results in Figure 5 show the simulated extraction efficiency of the PV array for two different conditions. The solid line is for a fixed cell temperature of 25 °C (i.e.,

STC). The dashed line is for an ambient temperature of 25 °C incorporating the variation in cell temperature (using the Skoplaki model, as mentioned earlier) and the variation of MPP power of the panels due to varying sunshine for a given load power of P_L.

The PV-array was designed for $P_{MP} \approx 5$ kW (rated under 1000 W/m^2 irradiance)—we can expect the output to proportionately (approximately) reduce with decreasing intensity. As long as the panel MPP is lower than the load requirement $P_{MP} < P_L$, the load draws the maximum possible power from the panel and the rest from the grid. Of course, we will always have $P_{MP} < P_L$ at lower sunshine, resulting in high extraction efficiencies η_{ex}, as seen in Figure 5.

For example, for a load power of 3 kW, the PV power P_{MP} remains lower than the load power P_L when the sunshine remains lower than 650 W/m^2. As the sunshine increases further, the PV power becomes higher than the load power and the extraction efficiency starts falling rapidly, as can be seen in Figure 5. This is an expected outcome as the PV system's excess power cannot be supplied to the grid.

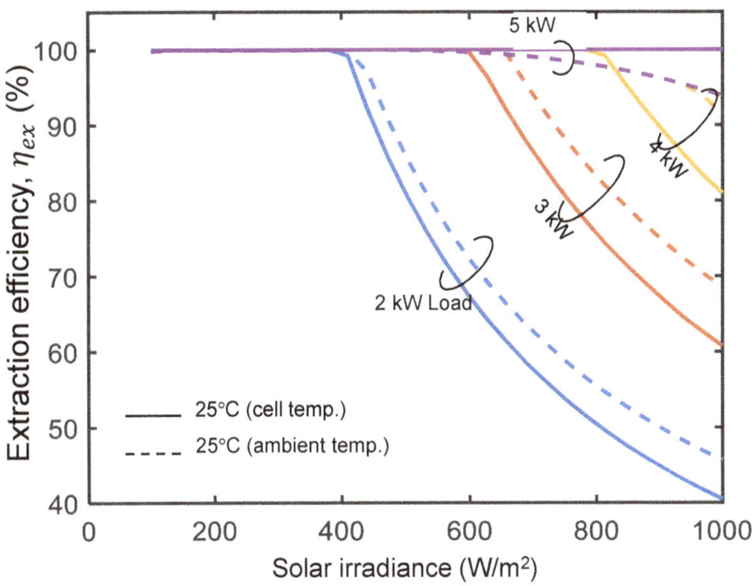

Figure 5. Solar energy extraction efficiency of the system for different load power under varying sunshine conditions. Load voltage is assumed to be 535 V_{DC}.

It may be intuitively expected that η_{ex} remains close to 100% when we have a matching load of 5 kW. However, for sunshine higher than 700 W/m^2, the cell temperature becomes quite high and the extraction efficiency becomes lower, even for $P_L > P_{MP}$, due to lowering of the MPP voltage from the grid voltage (the simulation results consider an operating DC voltage of 535 V). As already discussed earlier, the extraction efficiency for $P_L > P_{MP}$ remains above 95% (meaning a loss of 5%) for a wide range of sunshine (<950 W/m^2), which should be an acceptable figure for the sunshine conditions in Bangladesh.

The above simulations provide us with the following insights:

1. Unlike the grid-tied inverters, the DC link presented here is unable to transmit power to the grid. So, the design should ensure that the connected load must not be less than the PV generation, as higher PV generation will result in low energy extraction efficiency from the PV array.
2. The presented control circuit does not track the maximum power point of the PV array but instead supplies power at a voltage generated by rectification of the grid

supply. The PV array is designed such that the maximum power point is close to the operating DC voltage to limit the power extraction loss $(1-\eta_{ex})$. So, it is important to estimate the possible loss of energy when the grid voltage fluctuates.

3. The operating point of the load voltage should remain within a range such that the actual extraction of power from the grid remains high under varying load conditions, grid voltage fluctuations, and variations in the sunshine.

5.2. Experimental Results

Figure 6 shows the picture of the PV array installed on the roof of the Niagara textiles. Four knitting machines, each having a maximum power of 3.5 kW, were connected to the DC grid as described earlier and the total estimated load was expected to vary from 6 kW to 10 kW, depending on the size of the cotton yarn used and the speed of the machines determined by the production process. The knitting machines driven by the VVVF inverters are rated at a maximum input voltage of 570 V_{AC} (corresponding peak value of 800 V). The PV system of 18 panels connected in series has an open circuit voltage of 690 V, which is well within the maximum voltage rating of the inverter. The 3-phase full wave rectifier module had a maximum reverse voltage of 1000 V and a maximum current of 60 A which is much higher than the required current capacity. The choice of the rectifier module was based on its ready availability in the local market and the price was USD 4.50. The magnetic contactor used was a 3-pole 700 V_{AC} with a current rating of 40 A. The power-sharing between the solar PV and the grid under different running conditions, as measured, is given in Table 2.

Figure 6. Picture of the solar PV array installed at the roof of Niagara Textiles.

The measurements were taken at different times of the day to have different levels of sunshine and also under different load conditions. The column P_{MP} was numerically calculated (as discussed in Section 5.1) for the measured sunshine data given in the table at 25 °C ambient temperature. As can be seen from Table 2, the lowest grid power delivered was 0.6 kW, which indicates that the PV power was always less than the load power. If we compare the η_{ex} presented in Table 2, we find that the value remains below 95%, less than predicted by the simulation results. This is due to the fact that the measured power includes other losses such as line loss, loss in the rectifier network, etc. The extraction efficiency (η_{ex}) value seems to reduce for lower sunshine conditions. It may be mentioned here that the lower sunshine values correspond to the morning or the afternoon hours when the incidence angle of the sunshine was higher causing more reflection from the panel surface. We also cannot eliminate the possibility of 1–2% measurement errors. However, it is important that the overall extraction efficiency (including all the losses) remains higher than 90% which is reasonably high.

Table 2. Estimation of extraction efficiency in our experiments.

Irradiance, $\frac{W}{m^2}$	Load Voltage, V_{dc}	Measured PV Power, P_{PV}, kW	Power from Grid, kW	P_{MP}, kW (Calculated)	% η_{ex}
266.7	538	1.07	5.5	1.25	85.6
444.4	536	1.9	4.8	2.08	91.3
705.9	540	3.01	0.6	3.22	93.5
740.7	540	3.19	0.9	3.36	94.9

6. Conclusions

Modern-day industrial drives are inverter driven and the input AC from the grid is rectified right away to feed the internal DC bus of the inverter. In this research, we fed DC directly to the inverter terminals to get rid of the grid-tied inverters for lower cost and higher efficiency. The proposed system shows a seamless power-sharing mechanism that gives priority to solar power drawing only the shortfall from the grid. Although no power point tracking is used for the solar panels, the array of the solar panels is designed in such a way that the operating DC voltage is close to the maximum power point voltage of the PV array. Simulations show that the loss of power due to deviation in the maximum power point is less than 5% even under higher cell temperatures. The loss in the control circuit is estimated to be less than 1% and we achieved more than 90% power extraction from the panels over a wide range of sunshine and load conditions except for very low sunshine. The elimination of grid-tied inverters is expected to reduce the cost of energy from such rooftop solar by around 20%, which may prove to be very attractive for industries. The limitation of the proposed system is its inability to export PV power to the grid when the industry is not in operation. So, it is important to do further research to find an effective way to utilize solar PV output during the holidays when the industry is closed.

Author Contributions: Conceptualization, I.A. and M.R.K. (M. Rezwan Khan); Methodology, I.A. and M.R.K. (M. Rezwan Khan); Software, M.R.K. (M. Ryyan Khan); Validation, M.R.K. (M. Ryyan Khan); Formal analysis, M.R.K. (M. Ryyan Khan) and M.R.K. (M. Rezwan Khan); Investigation, M.R.K. (M. Ryyan Khan), I.A. and M.R.K. (M. Rezwan Khan); Data curation, M.R.K. (M. Ryyan Khan); Writing—original draft, I.A.; Writing—review & editing, M.R.K. (M. Ryyan Khan); Supervision, M.R.K. (M. Rezwan Khan); Project administration, M.R.K. (M. Rezwan Khan); Funding acquisition, M.R.K. (M. Rezwan Khan). All authors have read and agreed to the published version of the manuscript.

Funding: Funded by Institute for Advanced Research, Publication Grant of United International University, Ref. No.: IAR-2022-Pub-041 (https://iar.uiu.ac.bd/, accessed on 20 January 2023).

Data Availability Statement: The data can be made available upon reasonable request to the authors via email.

Acknowledgments: The authors would like to express their sincere gratitude to the Niagara Textiles for their support in implementing the project.

Conflicts of Interest: The authors declare no conflict of interest.

Abbreviations

The following abbreviations are used in this manuscript:

V_L	Load Voltage
V_{OC}	Open circuit voltage
$V_{G_{DC}}$	Rectified grid voltage
V_{MP}	Maximum power point voltage of a single array
P_P	Individual Panel Power in Watt Peak W_p
V_{PV_m}	Maximum voltage of the PV array
I_{sc}	Short circuit current
I_m	Panel maximum current

References

1. Roy, D.; Rout, U.K.; Jonalagadda, S.; Behera, P. A Brief Analysis on Solar Status of India. In Proceedings of the International Conference on Innovative Mechanisms for Industry Applications, Bengaluru, India, 21–23 February 2017; Volume 1, pp. 451–456. [CrossRef]
2. de Sousa Stilpen, D.V.; Cheng, V. Solar Photovoltaics in Brazil: A Promising Renewable Energy Market. In Proceedings of the International Renewable and Sustainable Energy Conference (IRSEC), Marrakech, Morocco, 10–13 December 2015; Volume 3. [CrossRef]
3. Sultan, H.M.; Kuznetsov, O.N.; Diab, A.A.Z. Site selection of large-scale grid-connected solar PV system in Egypt. In Proceedings of the IEEE NW Russia Young Researchers in Electrical and Electronic Engineering Conference, Moscow and St. Petersburg, Russia, 29 January–1 February 2018; pp. 813–818. [CrossRef]
4. Solar Pulse. Are Solar Panels Just for Rich People? *Priceconomics Data Studio* 21 June 2016. Available online: https://priceonomics.com/are-solar-panels-just-for-rich-people/ (accessed on 20 January 2023).
5. Hinckley, S. Coastal California Has a Reputation as a Leader in Green Energy—But Only the Rich are Benefiting. Available online: www.businessinsider.com/only-the-rich-are-benefiting-from-green-energy-in-california-2018-2 (accessed on 12 February 2018).
6. Calma, J. Solar Energy Reaches Historically Low Costs: Solar is 'The New King'. *The Verge* 13 October 2020. Available online: https://www.theverge.com/2020/10/13/21514902/solar-energy-cost-historic-low-energy-agency-outlook-2020 (accessed on 20 January 2023).
7. Niranjan, A. Falling Solar Panel Prices Spell Sunny Future for Clean Energy. *DW.com* 25 May 2020. Available online: https://www.dw.com/en/cheap-solar-energy-prices-explained/a-53590607 (accessed on 20 January 2023).
8. Nipon Ketjoy, W.C.A.; Mensin, P. Analysis of factors affecting efficiency of inverters: Case study grid-connected PV systems in lower northern region of Thailand. *Energy Rep.* **2021**, *7*, 3857. [CrossRef]
9. Rashid, M.H. *Alternative Energy in Power Electronics*; Elsevier: Amsterdam, The Netherlands, 2015.
10. Gostein, M.; Caron, J.R.; Littmann, B. Measuring soiling losses at utility-scale PV power plants. In Proceedings of the 2014 IEEE 40th Photovoltaic Specialist Conference (PVSC), Denver, CO, USA, 8–13 June 2014; pp. 0885–0890. [CrossRef]
11. Zhao, W.; Lv, Y.; Wei, Z.; Yan, W.; Zhou, Q. Review on dust deposition and cleaning methods for solar PV modules. *J. Renew. Sustain. Energy* **2021**, *13*, 032701. [CrossRef]
12. Rezwan Khan, M.; Lutful Kabir, S.M.; Choudhury, M.A. Resurrection of DC: An Exposition for Future Power System. In Proceedings of the 2018 10th International Conference on Electrical and Computer Engineering (ICECE), Dhaka, Bangladesh, 20–22 December 2018. [CrossRef]
13. Bose, B.K. *Power Electronics and Motor Drives: Advances and Trends*; Academic Press: Cambridge, MA, USA, 2006. [CrossRef]
14. Eisenbrown, R.E. AC Drives, Historical and Future Perspective of Innovation and Growth. In Proceedings of the 25th Anniversary of The Wisconsin Electric Machines and Power Electronics Consortium (WEMPEC), University of Wisconsin, Madison, WI, USA, 18 May 2008; pp. 6–10.
15. Narendra, A.; Venkataramana Naik, N.; Kumar Panda, A.; Tiwary, N. A Comprehensive Review of PV Driven Electrical Motors. *Sol. Energy* **2020**, *195*, 278–303. [CrossRef]
16. Khan, M.R.; Alam, I. A Solar PV-Based Inverter-Less Grid-Integrated Cooking Solution for Low-Cost Clean Cooking. *Energies* **2020**, *13*, 5507. [CrossRef]
17. Infrastructure Development Co., Ltd. Available online: http://idcol.org/home/rooftopsolar (accessed on 15 October 2020).
18. Amelia, A.R.;Irwan, Y.M.; Leow, W.Z.; Irwanto, M.; Safwati, I.; Zhafarina, M. Investigation of the Effect Temperature on Photovoltaic (PV) Panel Output Performance. *Int. J. Adv. Sci. Eng. Inf. Technol.* **2017**, *6*, 682–688.
19. Thong, L.W.; Murugan, S.; Ng, P.K.; Sun, C.C. Analysis of Photovoltaic Panel Temperature Effects on its Efficiency. In Proceedings of the 2nd International Conference on Electrical Engineering and Electronics Communication System 2016, Ho Chi Minh, Vietnam, 18–19 November 2016.
20. Fesharaki, V.J.; Dehghani, M.; Fesharaki, J.J.; Tavasoli, H. The Effect of Temperature on Photovoltaic Cell Efficiency. In Proceedings of the 1st International Conference Emerging Trends in Energy Conservation, Bordeaux, France, 6–10 November 2011; pp. 20–21.

21. Mohamed, A.; Khatibi, T. Correlation for estimating solar cell temperature based on a tropical field operation of a photovoltaic system. In Proceedings of the 2014 IEEE/PES Transmission & Distribution Conference & Exposition (T&D), Chicago, IL, USA, 14–17 April 2014; pp. 1–5.
22. Manasrah, A.; Masoud, M.; Jaradat, Y.; Bevilacqua, P. Investigation of a Real-Time Dynamic Model for a PV Cooling System. *Energies* **2022**, *15*, 1836. [CrossRef]
23. Mithhu, M.M.H.; Rima, T.A.; Khan, M.R. Global analysis of optimal cleaning cycle and profit of soiling affected solar panels. *Appl. Energy* **2021**, *285*, 116436. [CrossRef]
24. Jahangir, J.B.; Al-Mahmud, M.; Shakir, M.S.S.; Mithhu, M.M.H.; Rima, T.A.; Sajjad, R.N.; Khan, M.R. Prediction of Yield, Soiling Loss, and Cleaning Cycle: A Case Study in South Asian Highly Construction-Active Urban Zone. In Proceedings of the 2020 47th IEEE Photovoltaic Specialists Conference (PVSC), Calgary, AB, Canada, 15 June–21 August 2020; pp. 1371–1374. [CrossRef]
25. Ilse, K.K.; Figgis, B.W.; Naumann, V.; Hagendorf, C.; Bagdahn, J. Fundamentals of soiling processes on photovoltaic modules. *Renew. Sustain. Energy Rev.* **2018**, *98*, 239–254. [CrossRef]
26. Maghami, M.R.; Hizam, H.; Gomes, C.; Radzi, M.A.; Rezadad, M.I.; Hajighorbani, S. Power loss due to soiling on solar panel: A review. *Renew. Sustain. Energy Rev.* **2016**, *59*, 1307–1316. [CrossRef]
27. Jones, R.K.; Baras, A.; Al Saeeri, A.; Al Qahtani, A.; Al Amoudi, A.O.; Al Shaya, Y.; Alodan, M.; Al-Hsaien, S.A. Optimized Cleaning Cost and Schedule Based on Observed Soiling Conditions for Photovoltaic Plants in Central Saudi Arabia. *IEEE J. Photovoltaics* **2016**, *6*, 730–738. [CrossRef]
28. Jones, R.K. Solving the Soiling Problem for Solar Power Systems. *Joule* **2019**, *3*, 2298–2300. . [CrossRef]
29. Bangladesh Energy Regulatory Commission. Available online: http://www.berc.org.bd/site/view/notices (accessed on 12 February 2023).
30. Schultis, D.-L. Effective Volt/var Control for Low Voltage Grids with Bulk Loads. *Energies* **2022**, *15*, 1950. [CrossRef]
31. Chowdhury, S.A.; Khan, M.Z.R. The Net Metering Guideline of Bangladesh-Potential and Way Forward. In Proceedings of the 2020 11th International Conference on Electrical and Computer Engineering (ICECE), Dhaka, Bangladesh, 17–19 December 2020; pp. 435–438 . [CrossRef]
32. Lin, S.; Huang, X. Power Loss Analysis of AC Contactor at Steady Closed State with Electromagnetic-Thermal Coupling Method. *J. Inf. Hiding Multimed. Signal Process.* **2017**, *8*, 290–299.
33. Dongaonkar, S.; Karthik, Y.; Wang, D.; Frei, M.; Mahapatra, S.; Alam, M.A. On the nature of shunt leakage in amorphous silicon p-i-n solar cells. *IEEE Electron Device Lett.* **2010**, *31*, 1266–1268. [CrossRef]
34. Obaidullah, M.; Sarkar, M.A.R.; Martinac, I. Performance and Life Cycle Cost Analysis of a Stand Alone PV Lighting System. *J. Energy Environ.* **2004**, *3*, 25–32.
35. Skoplaki, E.; Palyvos, J.A. Operating temperature of photovoltaic modules: A survey of pertinent correlations. *Renew. Energy* **2009**, *34*, 23–29. [CrossRef]

energies

MDPI

Article

Oscillating-Foil Turbine Performance Improvement by the Addition of Double Gurney Flaps and Kinematics Optimization

Benoît Genest and Guy Dumas *

CFD Laboratory LMFN, Department of Mechanical Engineering, Laval University, 1065 Avenue de la Médecine, Québec, QC G1V 0A6, Canada
* Correspondence: gdumas@gmc.ulaval.ca

Abstract: Refinement of the performance of a fully constrained oscillating-foil turbine is carried out via the addition of passive double Gurney flaps. Flaps ranging from $h_{GF} = 0.005c$ to $0.075c$ are added at the trailing edge of the NACA 0015 blade of turbines operating in high-efficiency regimes without leading-edge vortex shedding (LEVS). Performance improvements are determined using 2D numerical simulations with an unsteady Reynolds-averaged Navier–Stokes (URANS) approach. Based on a recent study of the double Gurney flaps on stationary foils, instantaneous power-extraction coefficients are analyzed and modifications of the foil's kinematics are tested in order to fully benefit from the Gurney flaps' performance improvements. Modifications to the pivot point location of the foil, to the pitch-heave phase, and to the pitching amplitude of the turbine are considered. Improvements are found for all turbine cases studied, including some of the previously optimal cases reported in the literature. The double Gurney flaps, being a simple and passive device, offer great practical application potential. They represent an efficient refinement to already robust and high-performance oscillating-foil turbines operating without the perceived benefit of leading-edge vortex shedding, an essential characteristic for actual, finite-span applications.

Keywords: oscillating-foil turbine; Gurney flap; CFD; URANS

1. Introduction

Amidst the current global energy crisis and growing concerns for the energy independence of various countries, hydrokinetic turbines offer huge untapped potential and are more relevant than ever. With an estimated 3000 GW of available tidal power [1], different concepts of hydrokinetic turbines have emerged over the last decades. One interesting concept, the oscillating-foil turbine (OFT), offers a performance potential on par with classical horizontal-axis and cross-flow turbines, as well as a rectangular power-extraction plane particularly well suited to riverbed applications, and a simple blade geometry that makes them an attractive manufacturing proposition.

OFTs have been the subject of research since the early 1980s [2] and research has intensified since the mid-2000s up to today [3–29]. Advancements in the field have tackled two main challenges : (1) finding the best means of extracting power from the turbine's motion and (2) improving the hydrodynamic performance of the turbine, either through kinematic or geometric modifications. Improving the overall energy-harvesting performance of the OFT remains vital for technology adoption and will be investigated here via the addition of double Gurney flaps at the trailing edge of the OFT's blade.

1.1. Power-Extraction

Extracting the available hydrokinetic power from the oscillating-foil turbine's blade has been a central challenge since the inception of the first prototypes, and the subject of many subsequent developments. The first turbine prototype developed by McKinney and DeLaurier [2] employed fully constrained kinematics by imposing motion and synchronization between the two degrees of freedom of the foil (pitch and heave) through mechanical means, a novel idea at the time, effectively creating a single degree of freedom through which power can be extracted by connecting an electric generator.

Various means of synchronizing the two motions of the OFT, such as four-bar linkages [3] or Scotch-yoke mechanisms [2] have been tested. Other mechanical advancements include the use of a lever-arm producing a quasi-heave trajectory via an arc motion instead of the typical straight-line vertical heave, thus reducing mechanical complexity [3–5]. However, mechanical challenges remain for constrained-motion OFTs. As such, experimental prototypes have succeeded to extract power [3], but large ventures, such as the Engineering Business Limited "Stingray" [30], did not prove the commercial potential at the time. Currently, there are no known commercial applications of the OFT.

With the more recent development of fully passive [4,6–8] and semi passive [9–11] OFT concepts, taking advantage of the fluid–structure interaction to eliminate the challenging mechanical coupling of the motions, some mechanical limitations have been partially overcome.

Peng and Zhu [6] were among the first to explore numerically fully passive turbines, reaching efficiencies up to 0.25. Veilleux and Dumas [7] investigated the fully passive aeroelastic problem at $Re = 5 \times 10^5$, reaching 2D efficiencies up to 0.34. Their findings were later supported by the experimental work of Boudreau et al. [8] who, using structural parameters based on Veilleux and Dumas's optimal case, observed self-sustained fully passive motions with an harvesting efficiency of 0.31 at a lower $Re = 2.1 \times 10^4$.

Semi passive turbines were first researched by Shimizu et al. [9] and Zhu and Peng [10] in a concept where pitch was imposed and heave obtained from free-motions. A novel concept where heave is imposed and pitch motion left passive was introduced by Boudreau et al. [11]. This concept offers the most interesting application, because the heave controller can also extract and convert the harvested power. Efficiencies up to 0.454 were reached for a reduced frequency $f^* = 0.20$, pitch amplitude $\theta_0 = 84.7°$, and phase lag of $\phi = 95.4°$ (deduced from free motions). Interestingly, the authors' methodology for finding initial structural parameters employed a reverse-solving approach, based on a fully constrained foil motion. This highlights the relevance, still, of studying imposed-motion OFTs, despite the interesting real-world applicability of semi and fully passive OFTs. Imposed-motion OFTs offer the advantage of complete control over operating parameters, an advantage that will become apparent further in this paper in the study of hydrodynamic improvements.

1.2. Hydrodynamic Performance Improvements

Early research on the OFT began as an extension from flapping and oscillating wings used in propulsion. Exceeding the limit for propulsion, given by the feathering parameter χ [31], researchers managed to reach the power-extraction regime and to obtain noteworthy performances [32]. Motion kinematics and geometric characteristics—such as foil shape—were then studied and led to 2D performances now reaching up to 0.49 [12].

The pioneering study of McKinney and DeLaurier [2] experimentally achieved efficiencies up to 0.28 after observing the effect of varying three motion parameters: reduced frequency, phase angle, and pitching amplitude. Heave was fixed, frequency was obtained rather than imposed, and an effective maximum angle of attack could be inferred.

Davids [33] conducted a thorough investigation of the OFT, both numerically and experimentally. Using a potential-flow panel-code, he was able to observe the effect of five parameters on the power extraction and performance of the turbine. Ideal pivot-point location was found around $x_p = 0.3c$ for a motion phase around $\phi = 90°$. A best case was determined manually and reached up to 0.30 efficiency. Davids concluded that heaving amplitude and operating frequency were the main governing parameters of performance, whilst other parameters, though connected to each other, could be adjusted to provide good performance. His experimental conclusions highlighted the difficulties associated with high operating reduced frequencies and the important effects of dynamic stall. Maximum effective angles of attack were kept low to avoid flow separation, not considered by the potential-flow solver.

Dumas and Kinsey [13] conducted one of the first Navier–Stokes, high-resolution, unsteady computational studies of the OFT at $Re = 1100$, finding an efficiency of $\eta = 0.34$ for $f^* = 0.15$ and $\theta_0 = 75°$. Extending their work, Kinsey and Dumas [14], conducted a thorough parametric study into the effect of motion parameters f^*, H_0 and θ_0, of geometric parameters such as foil thickness and pivot-point location (x_p), and of viscosity from $Re = 500$ to 10^4. They determined that motion parameters govern the predicted performances, and that the leading-edge vortex shedding mechanism appeared of crucial importance to performance, especially its synchronization.

Kinsey and Dumas [15] later found efficiencies up to $\eta = 0.43$ for an imposed-motion OFT at $Re = 5 \times 10^5$. They also developed one of the first approaches to predicting optimal parameters, drawing from their conclusions that optimal performances are found around an effective angle of attack of $\alpha_0 = 33°$ and $\dot{\alpha}_{max} = 0.55$. They predicted an optimal case for $H_0 = c$ using $f^* = 0.172$ and $\theta_0 = 80.27°$ (untested). Their actual optimal case ($f^* = 0.16$, $\theta_0 = 85°$ and $H_0 = 1.5c$) however showed no LEVS and had an effective maximum angle of attack of $\alpha_0 = 28.55°$.

Picard-Deland et al. [12] also studied imposed-motion OFTs at $Re = 5 \times 10^5$ but employing a sinusoidal effective angle of attack profile instead of the pitching function usually employed. They were able to obtain high efficiencies up to 0.44, over a wide range of heaving amplitudes from $H_0 = c$ up to $15c$. Further improvements were obtained by modifying the angle of attack function to maintain a high AOA over a large portion of the operating cycle, yielding performances up to 0.49. For all heaving amplitudes studied, they were able to obtain the best cases without the presence of LEVS, a remarkable and important conclusion in light of the challenges associated with vortex synchronization in real-world, finite-span applications where 3D effects occur.

In fact, the presence of leading-edge vortex shedding (LEVS) and their proper synchronisation with the blade motion has often been thought to be a crucial aspect driving the OFT performance [14,19]. However, many recent studies have demonstrated that it is not the case, and that high performance and efficiency—often higher than with LEVS—can be achieved without LEVS [12,15]. Further, many 3D investigations have highlighted how difficult LEVS timing is to maintain once 3D effects are factored in [16,20].

The three-dimensional application of the OFT has also been studied by Kinsey and Dumas [16] who found that 3D performance can be about 90% of the 2D-predicted performance for an aspect-ratio greater than 10 if endplates are used. The spanwise vortex uncorrelation was found to be the most dramatic effect of a finite-span blade to impact the OFT. Drofelnik and Campobasso [20] also confirmed the spanwise uncorrelation of LEVS and found an efficiency lowered by 12% for a 3D blade with an aspect-ratio of 10 and endplates, in line with Kinsey and Dumas.

1.3. Geometric Modifications

Geometric modifications or additions have been suggested for the OFT from its inception. McKinney and Delaurier [2] even concluded their seminal paper, noting that an articulated active flap was to be the subject of a future study.

Different foil shapes, thin and thick conventional foils [14], flat plates, ellipses, and c-shaped foils [4] have all been studied but have not surpassed the performances of classical OFTs with symmetrical NACA airfoils. Chordwise (2D) flexibility was added to a thin-plate OFT with imposed-motions by Jeanmonod and Olivier [21], with limited success for already high-performance cases. Active camber was tested by others, either through chord-wise flexibility [22,23] or through an articulated trailing-edge flap [24,25], bringing performance improvements at the heavy cost of complexity.

The Gurney flap (GF) consists of a small flat plate perpendicular to the chord-line or to the foil's surface at its location in the vicinity or precisely at the trailing-edge. Single GFs are a known way to improve the lift coefficient and sometimes performance (L/D) of airfoils, provided their height h_{GF} is appropriately scaled with the local boundary layer thickness [34]. Double Gurney flaps (see Figure 1 for an illustration) are a lesser known alternative that offers a rare advantage of improving airfoil characteristics whilst maintaining symmetry and producing the same aerodynamics at positive and negative angles of attack. The current authors have recently published an investigation into the double Gurney flap applied to a NACA 0015 airfoil [35] providing scaling rules and physical explanations for their workings. At fixed angles of attack and for $Re = 5 \times 10^5$, double Gurney flaps were shown to improve the maximum lift by over 30% for $h_{GF} = 0.02c$, and the maximum performance of an airfoil (taken to be l/d) by over 5% at large C_L values for $h_{GF} = 0.005c$. The lift improvements found, along with symmetrical aerodynamic characteristics, should prove extremely valuable in OFT applications.

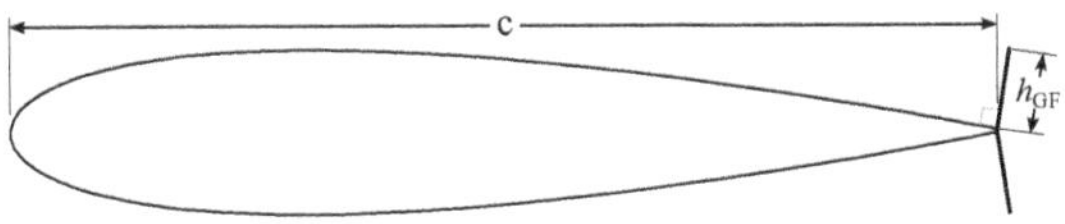

Figure 1. Double Gurney flap geometry and definition.

Double Gurney flaps have already been added to OFTs in numerical simulations by Xie et al. [26]. At a Reynolds number of 10,000, the authors found that double Gurney flaps up to $h_{GF} = 0.03c$ were beneficial to the power coefficient of the turbine and could enhance it up to 22%. Their low Reynolds number study focused on two small amplitudes for the angle of attack of $\alpha_0 = 10°$ and 15°, far from the optimal turbine performance found by Kinsey and Dumas [15] and Picard-Deland et al. [12], allowing for more performance-increase potential.

Active geometric improvements have also been studied in the context of GFs. Zhu et al. [27] numerically tested passive double Gurney flaps and compared them with an active Gurney flap, always positioned on the pressure side of the foil. They found good improvement over the base cases for fixed double Gurney flaps, with $h_{GF} = 0.015c$ yielding the best increases, but found much higher performances for the active flap, increasing η above 0.45 with $h_{GF} = 0.075c$. However, their active flap consumed no power and was instantaneously repositioned through mesh boundaries switching.

Sun et al. [28] also studied the movable Gurney flap, finding conclusions similar to Zhu et al. [27] in that a single movable GF of $h_{GF} = 0.07c$ yielded the largest performance increases, this time when considering the power consumed to move the GF on either side of the blade.

Overall, active geometric improvements to the turbine always require specific activation timing and phase with respect to the foil's motion, and add notable mechanical complexity in real-world applications. Their applicability will be subject to a delicate act of balancing between performance and simplicity, this dilemma often favoring simplicity, thus reliability, especially in early adoption stages.

The reader interested in a more in-depth overview of the OFT should refer to the comprehensive review of Wu et al. [29].

1.4. Scope

The present study aims at determining the effect on performance of double Gurney flaps added to existing high-performance imposed-motion OFT cases.

In the spirit of providing insightful conclusions, aimed towards commercial adoption, the current work will focus on the pre-existing best-performing cases that display no leading-edge vortex shedding (LEVS) behavior. Indeed, the 2D study of those cases is suitable to 3D extension, more so than cases relying on LEVS, yet they provide excellent performance as shown in the literature. Thus, we select cases based on canonical motion parameters and on a maximum angle of attack of the blade $\alpha_0 = 29°$, following the proposed optimal method by Kinsey and Dumas [15].

A passive geometric improvement is selected; namely, the addition of a double Gurney flap on the OFT's blade. This further promotes a simple turbine on which the effect of various motion and geometric parameters is studied, and their effect on performance is determined. Optimal double GF height will be determined for improving performance, the effect of various motion parameters will be considered, and the physical effects of fitting GFs will be explained. Conclusions with respect to the application of the double GF are reached and subsequent works are suggested.

2. Methodology

2.1. Equations of Motion

The oscillating-foil turbine's kinematics are composed of motion in two degrees of freedom—pitch and heave—prescribed through Equations (1) and (2) following the same convention used by Kinsey and Dumas in [15]. By imposing a kinematic phase between pitch and heave through ϕ, the two degrees of freedom effectively become one and we obtain a periodic motion of the blade.

$$h(t) = H_0 \, sin(2\pi f t + \phi) \tag{1}$$

$$\theta(t) = \theta_0 \, sin(2\pi f t) \tag{2}$$

The heaving amplitude H_0 is an important motion parameter of the turbine and was taken equal to c for the base cases studied. Phase shift between both motions ϕ is usually set to 90° [29], but was varied here from 80° to 100°.

Although a sinusoidal pitching function (Equation (2)) is employed, the pitching amplitude θ_0 of the cases studied was selected based on maximum effective angle of attack α_0. The effective angle of attack function is expressed by Equation (3), again following the same convention as in [15]:

$$\alpha(t) = arctan(\frac{-V_y}{U_\infty}) - \theta(t), \tag{3}$$

where U_∞ is the free-stream velocity, $V_y(t)$ is the heaving velocity (dh/dt). Similarly $\dot{\theta}$ will later be obtained by time-derivation of $\theta(t)$ ($d\theta/dt$).

By selecting a maximum effective angle of attack α_0—defined as the $max(\alpha(t))$—of 29° for the base case, we impose all other parameters of the OFT, except θ_0, replace $\theta(t)$ in Equation (3) by its expression (Equation (2)), and solve implicitly for θ_0. It should be noted that the approximation $\alpha_0 = |\alpha_{T/4}|$ is often used in other studies, but is only valid for $\phi = 90°$ and thus should be avoided here, as ϕ will be varied.

The dimensional frequency f is used in Equations (1) and (2), yet the reduced frequency is a more indicative parameter of an OFT's motion, as defined in Equation (4). This parameter was varied between 0.12 and 0.20 in the current study, thus investigating the parametric space around the optimum predicted by Kinsey and Dumas [15] at $f^* = 0.172$ when $H_0 = c$.

$$f^* = \frac{fc}{U_\infty} \tag{4}$$

The turbine kinematics is prescribed about the blade's pitching axis (x_p), defined from the foil's leading-edge to the pivot-point, and was varied in the current study between 0.25c and 0.5c. Here, a method is developed to reproduce quasi-similar foil motion whilst using different pivot locations. This is detailed below, with the help of Figure 2 supporting the explanation.

Imposing the maximum amplitude for a new pivot location $x_{p,2}$ determines the new heaving amplitude $H_{0,2}$. This amplitude replaces $H_0 = c$, the latter only valid for base cases at $x_p = c/3$. Movement of any point on the foil's chord-line still follows a sinusoidal motion, only with a different phase. The new phase ϕ_2 is therefore determined by comparing the instants in the cycle where maximum heaving amplitude is reached, for both the original and the new pivot location. The relative phase shift thus obtained modifies the existing motion phase ϕ, and becomes the new ϕ_2.

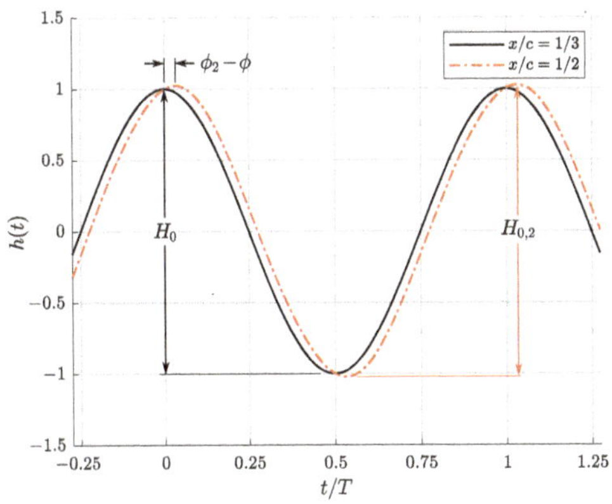

Figure 2. Heaving motion function $h(t)$ of two locations along the chord (x/c) for a pivot location $x_p = c/3$, showing how the new heaving amplitude and motion phase are determined.

Adjusting these two parameters, ϕ and H_0, cannot achieve true kinematic similarity, because the foil undergoes a fore-and-aft motion through an oscillation cycle. This motion depends on the distance of any point on the chord-line from the pivot axis and on the pitching amplitude θ_0. Since the OFT has no degree of freedom in the x-axis, no compensation is possible. This effect is of the order of $\pm 0.1c$ at the leading edge for pivot locations between $0.25c$ and $0.5c$.

2.2. Geometry and Physical Parameters

We select a NACA 0015 for the OFT's blade, a symmetrical airfoil widely used in OFT studies [7,8,11,14,15] and thus offering a good basis for comparison. The NACA 0015 foil is relatively thick and helps in delaying leading-edge flow separation, as found in [14], whilst still offering good performances. The turbine operates at a chord-based Reynolds number of 5×10^5, the same as other OFT studies [7,12,15,18] and as the double Gurney flap study of Genest and Dumas [35].

Double Gurney flaps of individual heights ranging from $h_{GF} = 0.005c$ to $0.075c$ are fitted to the foil's trailing edge, perpendicular to the foil's surface, as illustrated in Figure 1. The flaps are modeled by zero-thickness walls and have the exact same form-factor as in [35].

2.3. Power Extraction and Efficiency

The power generated by the turbine, by unit-length span, is derived from the instantaneous forces applied on the foil in the directions of motions. Namely, $Y(t)$ is defined as the force applied in the heaving axis (y-axis), and $M(t)$ is the moment about the pitching center x_p. These instantaneous signals can be normalized to yield C_Y (Equation (5)) and C_M (Equation (6)):

$$C_Y(t) = \frac{Y(t)}{\frac{1}{2}\rho U_\infty^2 bc}, \quad (5)$$

$$C_M(t) = \frac{M(t)}{\frac{1}{2}\rho U_\infty^2 bc^2}. \quad (6)$$

Those forces, when acting in the direction of motion, do work and thus produce power (Equation (7)), which can be normalized and cycle-averaged to yield a mean power coefficient $\overline{C_P}$ in Equation (8). The same normalization can be applied to each individual motion contributing to C_P to yield $C_{P,Y}$ and $C_{P,\theta}$.

$$P(t) = Y(t)V_y(t) + M(t)\dot{\theta}(t) \tag{7}$$

$$\overline{C_P} = \frac{\overline{P}}{\frac{1}{2}\rho U_\infty^3 bc} \tag{8}$$

The most usual metric to express turbine efficiency is to compare the power extracted by the turbine against the full kinetic energy flux through the extraction plane. In the case of a free-stream turbine, efficiency is subject to the limit of Betz [36] which is 0.59 in no-confinement condition. In its most rigorous definition, the extraction plane is taken as the window swept by any point on the blade at any given time through a cycle. The extraction plane is therefore the span, b (here a unit-length span), multiplied by d, the swept area, as illustrated in Figure 3. Efficiency η is thus defined by Equation (9).

$$\eta = \frac{\overline{P}}{\frac{1}{2}\rho U_\infty^3 bd} \tag{9}$$

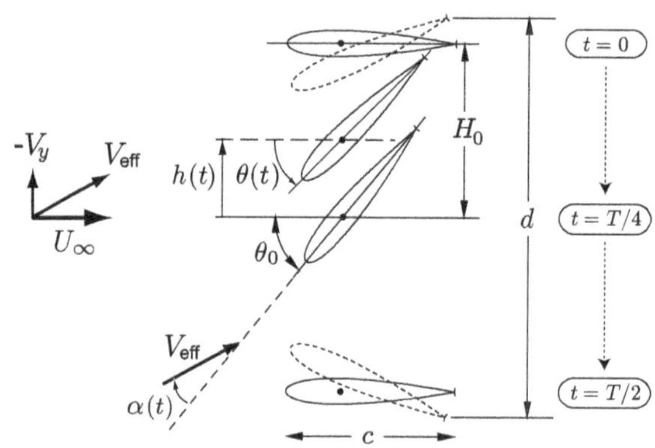

Figure 3. Representation of the kinematic parameters of the OFT equipped with a double GF (adapted with permission from Kinsey and Dumas [15]).

A thorough discussion on the different metrics of efficiency was given by Kinsey and Dumas [15] and should be referred to for further detail.

2.4. Numerics and Meshing

Numerical simulations are performed in Star-CCM+ 14.02.012, a finite-volumes code, solving for the unsteady Reynolds-averaged Navier–Stokes (URANS) equations in two dimensions. Foil motion is imposed through user-defined functions in Star-CCM+, through a combination of a moving near-foil mesh, representing the foil and its immediate surrounding region, and a larger, fixed, flow domain representing the full computed domain and its boundary conditions. The moving mesh is overlaid using the overset mesh method (chimera grid) in StarCCM+ over the larger static mesh, the latter having been progressively refined in a zone covering the entire moving mesh motion to provide a high-quality interface. This combination is represented in Figure 4.

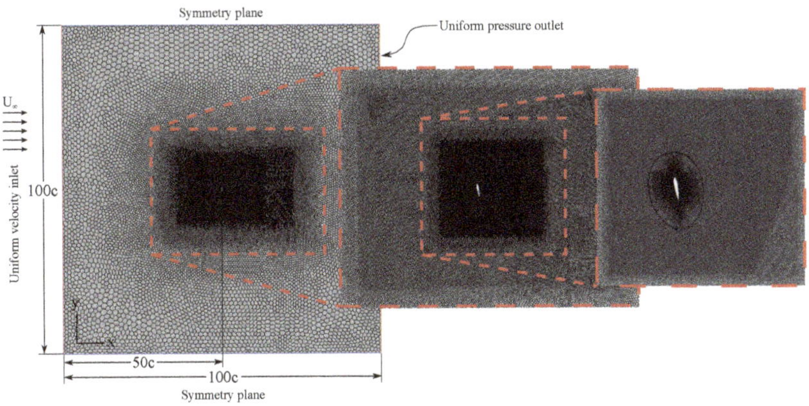

Figure 4. Mesh and domain specifications, including boundary conditions.

The moving near-foil meshes used in the author's previous work on Gurney flaps applied to static NACA airfoils [35] were used again here. They consist of specific meshes for every double Gurney flap's heights, generated around the same architecture consisting of 730 nodes surrounding the airfoil. The nodes expand in the normal direction outwards from the surface of the foil, particularly around the trailing edge where the Gurney flaps are located, in order to insert a no-thickness wall of varying height representing all the studied flaps. Necessary refinements were carried out near the wall for every mesh representing different flap heights to ensure adequate flow resolution.

Simulations are run at a time step corresponding with 2000 data points for every oscillation cycle, as found to be adequate by Kinsey and Dumas [18] and Picard-Deland et al. [12] for very similar simulation parameters. An implicit second order time-differentiation scheme is employed. The resulting time-step is also checked to remain below what is prescribed by Genest and Dumas [35], to capture all the physics of the Gurney flap. The number of iterations per time-step is limited to 135 or by an asymptotic limit on drag, heaving-force, and pitching moment coefficients residue set to 7×10^{-5} over the last 24 inner iterations.

Turbulence modeling is carried out with the one-equation Spalart–Allmaras model with curvature correction, as used again by Kinsey and Dumas [18] and which was found to be appropriate for single-foil OFT simulation at $Re = 5 \times 10^5$. It is also the same model used by Genest and Dumas [35] for Gurney flap simulation.

Simulations were run long enough to provide cycle-to-cycle converged results with a variation of the moving average of performance (η) over the last 5 cycles smaller than 0.001, depending on the type of convergence observed in the simulations (see Kinsey and Dumas [15] for further detail).

2.5. Validation and Independence of Numerical Parameters

Validation consisted of reproducing a specific case of interest for OFTs (without GFs) that was thoroughly documented by Kinsey and Dumas [18] and subsequently reproduced by Picard-Deland et al. [12], with the following kinematic parameters : $f^* = 0.16$, $\theta_0 = 75°$, $H_0 = c$ at $\phi = 90°$ and $x_p = c/3$. Those kinematic parameters yield operating conditions with a maximum effective angle of attack of 29.85° and no LEVS, which are very close to the ones selected for the current study ($\alpha_0 = 29°$ and no LEVS), and thus provide a solid basis for validation.

The power coefficient C_P and the heaving force coefficient C_Y signals are plotted in Figure 5 for three different mesh resolutions, along with the same signals obtained from Kinsey and Dumas [18]. The sensitivity study is also carried out for the time-wise resolution with 1000, 2000 and 4000 time-steps per cycle (TS/cycle), and the validation data is presented in Table 1 alongside the mesh-resolution sensitivity data.

Table 1. Results for numerical parameters sensitivity studies and comparison with Kinsey and Dumas [18]. (**Bold** indicates the case with numerical parameters retained for the study).

Case	Time Steps/ Cycle	Cells (Near-Foil)	$\hat{C}_Y$	$\hat{C}_M$	$\overline{C_P}$
Reference [18]	2000	–	3.168 (+2.64%)	0.613 (−0.75%)	1.023 (+2.02%)
Coarse mesh	2000	35,000	3.016 (−2.28%)	0.618 (+0.06%)	0.988 (−1.51%)
Standard (Medium mesh)	**2000**	**66,600**	**3.087**	**0.618**	**1.003**
Fine mesh	2000	145,000	3.099 (+0.41%)	0.618 (+0.04%)	1.006 (+0.30%)
Coarse TS/cycle	1000	66,600	3.126 (+1.29%)	0.616 (−0.20%)	1.002 (−0.07%)
Fine TS/cycle	4000	66,600	3.069 (−0.58%)	0.619 (+0.18%)	1.003 (−0.01%)

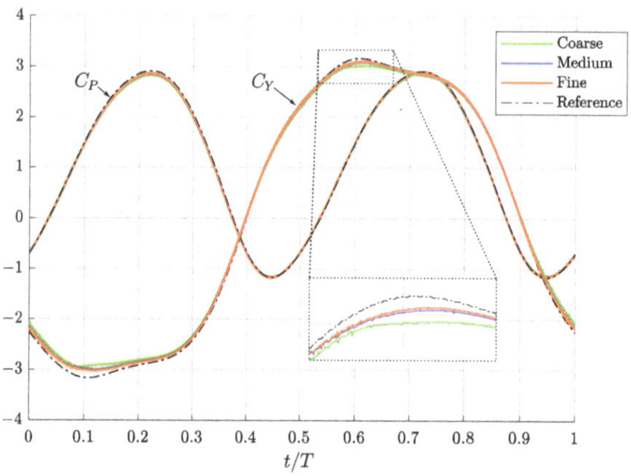

Figure 5. Comparison of instantaneous C_P and C_Y for 3 mesh resolutions and with Kinsey and Dumas [18] reference, for case $f^* = 0.16$, $\theta_0 = 75°$, $H_0 = c$, $\phi = 90°$, $x_p = c/3$ (without GF).

A close match in signals shape and amplitude is obtained for the two finest mesh resolutions in Figure 5. The peak amplitude of the instantaneous signals for heaving-force coefficient $\hat{C}_Y$ and pitching-moment coefficient $\hat{C}_M$ of all cases used in the sensitivity study are compared in Table 1 with data from Kinsey and Dumas [18]. Our base case "Standard" shows a 2.02% variation from the reference on power extraction coefficient.

Our "Standard" mesh resolution was deemed satisfactory, as further refinement brought no change in signal shape and a smaller than 0.5% change in both signals amplitudes and on $\overline{C_P}$. The selected resolution of 2000 time-steps per cycle yielded a maximum deviation of −0.58% compared with the finer-resolved case amplitudes, and near-undetectable variation over the averaged quantity $\overline{C_P}$.

A mesh sensitivity study for the various GF heights used was already carried out by the authors in a previous study [35], for the exact meshes employed in the current study, and should thus be referred to.

3. Results and Discussion

3.1. Impact of Double Gurney Flaps

The effect of the double GF on the OFT is first studied for a base case $f^* = 0.18$ yielding the best performance ($\eta = 0.401$). Gurney flaps of heights ranging from $h_{\mathrm{GF}} = 0.005c$ to $0.075c$ are added and the effect on performance η is presented in Figure 6.

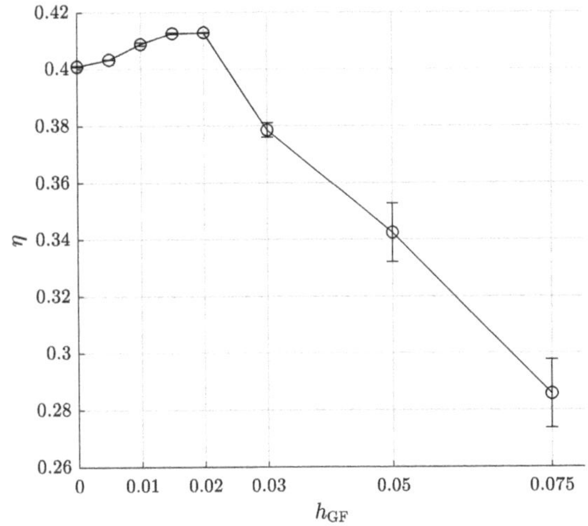

Figure 6. Effect of Gurney flaps on the performance of an OFT operating at $f^* = 0.18$, $H_0 = c$, $\alpha_0 = 29°$, $\phi = 90°$ fitted with various heights of double Gurney flaps (error bars showing variation of η over last 5 cycles).

Performance is improved by the addition of double GFs, for heights up to $h_{\text{GF}} = 0.02c$. Maximum performance of 0.413 is achieved for a relative improvement of +3%. Taller Gurney flaps have an adverse effect on performance, with $h_{\text{GF}} = 0.03c$ decreasing performance below that of the base case and successively taller flaps having more adverse effects and yielding a more chaotic cycle-to-cycle variation, as shown by the error bars plotted.

Figure 7 shows the reversal portion of the oscillating cycle between $t/T = 0.375$ and 0.625, that is, the moment in the cycle when the blade undergoes rapid pitching motion combined with a small to null heaving velocity, through which the heaving motion direction is reversed. This part of the cycle is where LEVS normally occur when the effective angle of attack is larger. We see that fitting a moderate double GF of $h_{\text{GF}} = 0.02c$ has a qualitatively small effect on the vorticity field, whereas tall GFs of 0.05c modify the flow field dramatically, introducing trailing-edge vortex shedding originating from the GFs and interacting with the OFT blade in a non-synchronized fashion, leading to the large cycle-to-cycle variations observed in Figure 6.

Going further and looking at the instantaneous power coefficients for the heaving motion and the pitching motion in Figure 8a, we observe that the double GF $h_{\text{GF}} = 0.02c$ has increased the maximum instantaneous heaving power extraction around $t/T = 0.625$ to 0.75. This is due to a larger vertical force (by means of increased lift) acting in the same direction as the heaving motion. The $h_{\text{GF}} = 0.05c$ flaps have generated smaller maximum $C_{P,Y}$ increases, whilst adding heavy shedding that is not synchronized with the heaving cycle and thus generates important cycle-to-cycle variations. The negative contribution of the pitching motion is also observed through $C_{P,\theta}$. Both cases with flaps show increased power consumption during the reversal portion of the cycle. The combined contribution of both motions is shown in Figure 8b, and displays larger peaks for all cases equipped with flaps, both in power extraction ($C_P > 0$) and in power consumption ($C_P < 0$). This signal, once cycle-averaged ($\overline{C_P}$), was shown in Figure 6 to increase for $h_{\text{GF}} = 0.02c$ but to decrease for $h_{\text{GF}} = 0.05c$. In short, $h_{\text{GF}} = 0.02c$ provides a net positive balance between increased heaving force and increased power consumption during reversal, whilst $h_{\text{GF}} = 0.05c$ increases the power consumption of the reversal cycle but provides limited increases in heaving force.

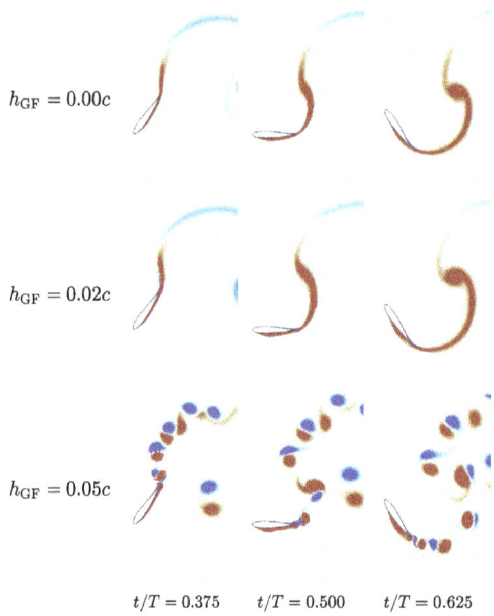

Figure 7. Normalized z-vorticity field during reversal for three double GF heights at $f^* = 0.18$. Positive vorticity in red, negative vorticity in blue.

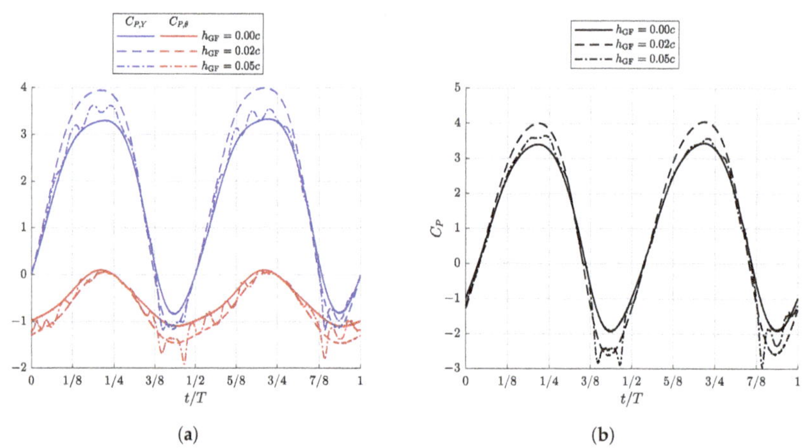

Figure 8. Instantaneous signals for three GF cases at $f^* = 0.18$ and $\alpha_0 = 29°$. (**a**) Power coefficients of heave and pitch. (**b**) Total power extraction coefficient.

3.2. Operating Frequency

The effect of the double GF is now studied over five reduced operating frequencies ranging from $f^* = 0.12$ to 0.20. Gurney flaps of various heights are fitted and the performance is plotted along with the base case (no flaps) in Figure 9. All f^* cases are improved by the addition of GFs, with maxima reached between $h_{GF} = 0.015c$ and $0.02c$ in all instances. For $f^* = 0.12$, 0.14 and 0.20, noteworthy performance improvements can be achieved up to $h_{GF} = 0.03c$, but for the other operating cases, performance decreases past $h_{GF} = 0.02c$. Furthermore, higher operating frequencies ($f^* = 0.16$ to 0.20) display a sharp decrease in performance past $h_{GF} = 0.03c$, whereas lower reduced frequency cases have a more progressive decrease in performance.

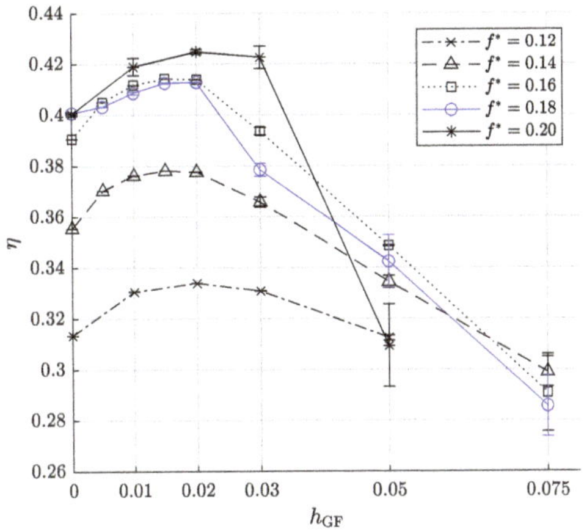

Figure 9. Effect of Gurney flaps on the performance of OFTs operating at reduced frequencies from $f^* = 0.12$ to 0.20, fitted with double GFs of $h_{\mathrm{GF}} = 0.005c$ to $0.075c$.

We now concentrate on the effect of small GFs in relation to reduced operating frequency. The same f^* between 0.12 and 0.20 are selected, and flaps of $h_{\mathrm{GF}} = 0.01c$ and $0.02c$ are considered. The performance and its relative gain are obtained in Figure 10.

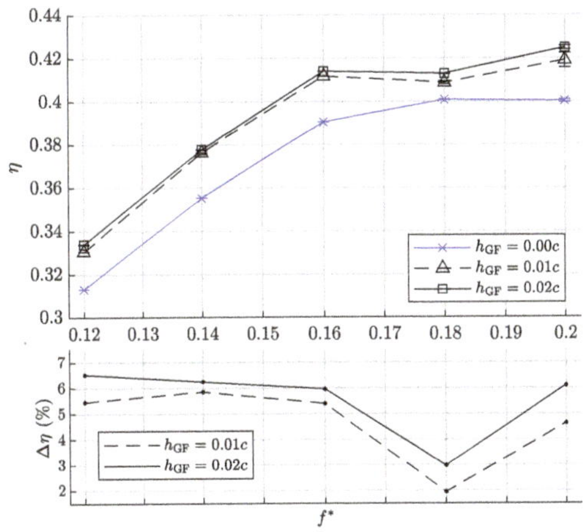

Figure 10. Effect on performance (**top**) and relative performance improvement (**bottom**) of OFTs operating at different reduced frequencies (f^*) with $\alpha_0 = 29°$ when fitted with double GFs of $h_{\mathrm{GF}} = 0.01c$ and $0.02c$.

We can see that, aside from our selected base case $f^* = 0.18$, performance with the addition of double GFs can be improved by more than +6.0% when using a $h_{\mathrm{GF}} = 0.02c$. The base case experiences a lower performance improvement of +3.0%. This can be analyzed by looking at Figure 11, which displays the vorticity field during reversal for the $f^* = 0.16$ case. Comparing the flow field with that of $f^* = 0.18$ in Figure 7, we observe that the latter has introduced a weaker vorticity region, fore of the GF located on the lower-side of the blade, which is not present on the case at $f^* = 0.16$. This suggests that the $f^* = 0.18$ case is near a limit for fully attached cases, hence the lower performance improvements noted.

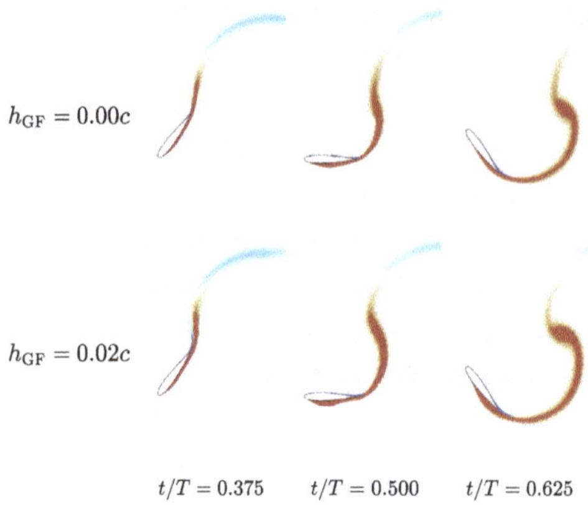

Figure 11. Normalized z-vorticity field during reversal for two double GF heights at $f^* = 0.16$. Positive vorticity in red, negative vorticity in blue.

3.3. Pivot-Point Location

Pivot-point location was varied while keeping the kinematics quasi-similar and the maximum effective angle of attack constant, following the method described earlier in Section 2.1. The effect of pivot location x_p is presented in Figure 12.

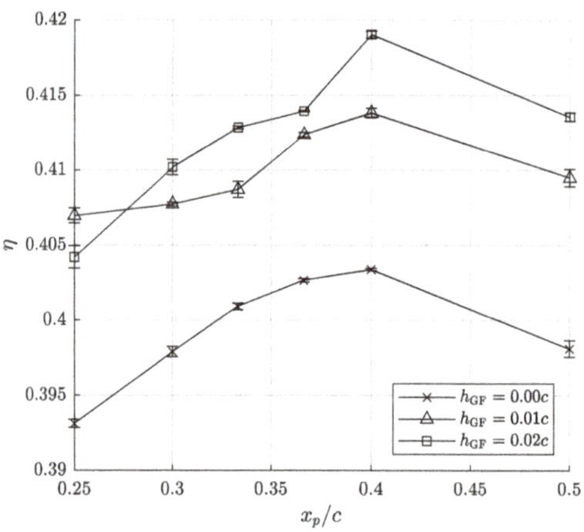

Figure 12. Effect on performance of fitting double GFs on OFTs with different pivot-point locations (x_p) at $f^* = 0.18$ and $\alpha_0 = 29°$.

We see that moving the pivot-point whilst keeping α_0 constant has a definite impact on performance. An optimum for the base case (no GF) can be found around $x_p = 0.4c$ with $\eta = 0.403$. Fitting double GFs can improve the performance of all cases, with a relative impact similar at most pivot-point locations. Maximum performance of $\eta = 0.419$ is reached for $x_p = 0.4c$ when a double GF of $h_{GF} = 0.02c$ is fitted (relative improvement of +3.9% over no-flap case at $x_p = 0.4c$ and of +4.5% over the base case at $x_p = c/3$).

Varying pivot-point location without reproducing quasi-similar motions through the method described earlier imparts kinematics changes which impact vortex generation during the reversal motion of the foil. For instance, to keep α_0 constant, the pitching amplitude must be adjusted as well as heaving amplitude, which impact the maximum angular acceleration of the blade.

3.4. Motion Phase

The effect of motion phase on turbine performance and double GF effectiveness is presented in Figure 13. Double Gurney flaps are fitted to OFTs operating at a fixed $f^* = 0.18$, and the phase between pitch and heave motions is varied from $\phi = 80°$ to $100°$, whilst keeping $\alpha_0 = 29°$ and $H_0 = c$. The effect on η is noticeable on both the base case ($\phi = 90°$) and on cases with double GFs fitted, and is of greater magnitude than simply fitting GFs.

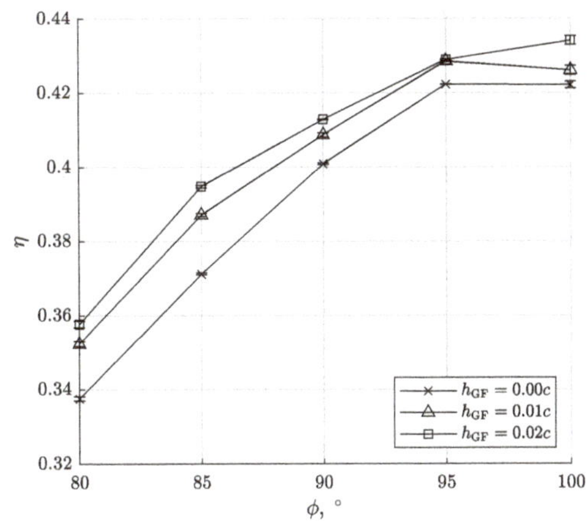

Figure 13. Effect on performance of fitting double GFs on OFTs with different motion phases at $f^* = 0.18$ and $\alpha_0 = 29°$.

Observing the impact of motion phase on the flow field in Figure 14 shows that the motion phase dramatically impacts the reversal portion of the oscillation cycle. A large leading-edge vortex is now apparent at $\phi = 100°$, although it is not strictly shed, whereas $\phi = 80°$ shows thinner vorticity region on the lower-side compared with the base case at $\phi = 90°$ presented in Figure 7. This indicates that the criterion selected to avoid LEVS of $\alpha_0 = 29°$ is not strictly sufficient if $\phi \neq 90°$.

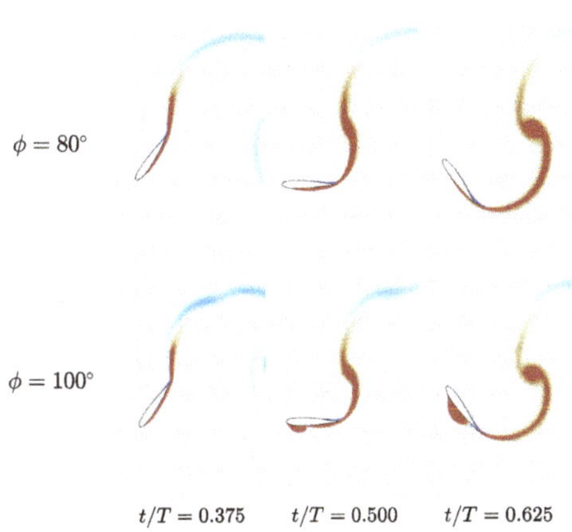

Figure 14. Normalized z-vorticity field during reversal for two motion phases at $f^* = 0.18$. Positive vorticity in red, negative vorticity in blue.

3.5. Maximum Effective Angle of Attack

A maximum effective angle of attack of 29° was selected as a starting point in the current study in order to have efficient cases without LEVS. The performance of OFTs operating between $f^* = 0.12$ and 0.20 was presented earlier in Figure 10, and the relative performance improvements when fitted with double Gurney flaps was shown to remain below +7%.

Performance and its relative improvement is now compared in Figure 15 for $f^* = 0.14$ to 0.18 between OFTs operating at $\alpha_0 = 29°$ and at $\alpha_0 = 25°$, before and after fitting double GFs of $h_{GF} = 0.02c$.

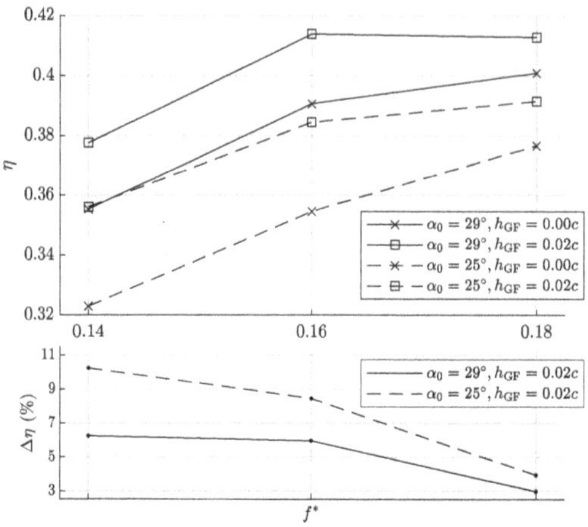

Figure 15. Effect on performance (**top**) and on relative performance improvement (**bottom**) of a $h_{GF} = 0.02c$ double GF fitted to turbines operating at $\alpha_0 = 29°$ and $\alpha_0 = 25°$.

Performance is still improved in all turbines, with relative improvements ranging from ≈+3% to more than +10%. The fact that the same GFs ($h_{GF} = 0.02c$) on different turbines produce different results is now evidenced. The highest relative performance improvement (+10.2%) is obtained for the lowest-performing no-flap case ($f^* = 0.14$ and $\alpha_0 = 25°$). Further, turbines operating at $\alpha_0 = 25°$ all produce a lower base performance, but see greater improvements by the addition of double GFs, compared with $\alpha_0 = 29°$. Relative improvements are greater for lower reduced-frequency cases ($f^* = 0.14$), which also display the smallest performance in the first place.

Nevertheless, maximum performance is still obtained with the higher α_0 case and the highest reduced frequency, as found in Section 3.2. This highlights the importance of selecting already high-performance turbine cases when testing geometric improvements, and the misleading message of reporting large relative improvements based on tests conducted on initially poor-performance turbines.

3.6. Discussion

The optimal double Gurney flap height for the base turbine case studied is located between $h_{GF} = 0.015c$ and $0.02c$. For stationary airfoils, the optimal double GF that improved performance—i.e., L/D ratio—was found by Genest and Dumas [35] to be $h_{GF} = 0.005c$ at $Re = 5 \times 10^5$ on a NACA 0015 foil, but the lift coefficient was found to be increased by GFs up to $h_{GF} = 0.02c$. OFTs can therefore benefit from much taller flaps, compared to foils at fixed angles of attack. This is in part due to the fact that the OFT's performance is derived from power-extraction, which is directly related to lift and moment on the blade, but to a lesser degree to drag.

In fact, the best double GFs used here ($h_{GF} = 0.015c$ and $0.02c$) were found by Genest and Dumas [35] to have an increased coefficient of drag, a greater lift slope. and a more negative and negatively sloped moment coefficient compared with a clean airfoil. This led to a reduced airfoil performance (L/D) when compared with smaller flaps or no-flap airfoils. When applied to turbines, it is expected that heaving power extraction would be increased by lift, and that pitching power extraction would be diminished. This is indeed what we observed in Figure 8a.

From those observations, we can infer that the main beneficial characteristic of an airfoil fitted with a double GF and used as an OFT blade is to have a greater lift slope. The more negative pitching moment associated with the double GF, a device fitted at the trailing-edge that shifts the pressure distribution aftwards [35], is detrimental to power-extraction. This negative effect could have been reduced by changing pivot-point location or motion phase, yet similar performance improvements were found when tested in Sections 3.3 and 3.4.

Finally, we should note that the conclusions derived from this study of the double GF may only be applicable to a single turbine, and not necessarily to turbine arrays, where the increased turbine drag will impact wake-recovery and could be detrimental to the array's power extraction.

4. Conclusions

The numerical investigation of double Gurney flaps applied to OFT turbines operating with a maximum effective angle of attack of $\alpha_0 = 29^\circ$ without the presence of LEVS has demonstrated the interest of the double GF to improve the power-extraction performance of turbines already operating with high extraction performance. Performance improvements from +3.0% to +10.2% were found over all cases observed.

The conclusions of this paper can be summarized as follows:

- Overall, the effect of adding double GFs remains modest, but has been shown to always be beneficial if its height is kept within the prescribed limits found (max $h_{GF} = 0.02c$). This limit suggests that the optimal double GFs for OFT performance are the ones yielding the highest lift coefficient on static airfoils, regardless of airfoil performance (L/D).
- Selecting a maximum effective angle of attack α_0 is not a sufficient criterion to avoid LEVS. Indeed, changing motion phase has shown that leading edge vortices can start to appear, yet remain attached and are not being shed.
- Changing pivot-point location also has an important effect on performance, yet the impact of double GFs still remains qualitatively the same.
- Adding double GFs is a passive modification, applicable even to the highest-performing OFTs, that offers a good potential for making the OFT an even more attractive concept in the development of hydrokinetic turbines.

Future works could take advantage of the increased lift slope of the double GF by using a proportionally longer power extraction period. For instance, this could be performed by imposing large heaving amplitudes and a modified pitching function that locates the reversal motion only at the limits of heaving, such as that explored by Picard-Deland et al. [12] for cases of very large heaving amplitudes.

Additionally, the effect of Reynolds number on the optimal Gurney flap height should now be studied, stemming from the conclusions of Giguère et al. [34] and Genest and Dumas [35] regarding the scaling relative to the boundary layer thickness. A boundary layer thickness scaling in the context of OFTs would be difficult to quantify, and as shown by the taller GFs employed, might not be fully applicable. Hence, the effect of Reynolds number on the GFs' performance improvements would provide more insight into their proper scaling.

Finally, studying the effect of perturbed flow conditions at the inlet could provide insight into the effect of double GFs on the boundary layer robustness and characteristics.

Author Contributions: Conceptualization, B.G. and G.D.; Methodology, B.G.; Validation, B.G.; Formal analysis, B.G. and G.D.; Investigation, B.G.; Resources, G.D.; Data curation, B.G.; Writing—original draft, B.G.; Writing—review & editing, G.D.; Visualization, B.G.; Supervision, G.D.; Funding acquisition, G.D. All authors have read and agreed to the published version of the manuscript.

Funding: This research was made possible through access to the high performance computing resources provided by Calcul Québec (calculquebec.ca) and the Digital Research Alliance of Canada (alliancecan.ca) (RAPI qcy-255), and was funded by the Natural Sciences and Engineering Research Council of Canada (NSERC Discovery Grant RGPIN-2018-04527 and NSERC CGS M scholarship).

Data Availability Statement: Not applicable.

Acknowledgments: The authors would like to acknowledge all members of the CFD Laboratory LMFN, and in particular Kevin Gunther, for their everlasting support and many insightful discussions. We also wish to thank Thomas Kinsey for graciously providing the original data used in Ref. [18].

Conflicts of Interest: The authors declare no conflict of interest, and the funders had no role in the design of the study; in the collection, analyses, or interpretation of data; in the writing of the manuscript; or in the decision to publish the results.

Abbreviations

The following abbreviations are used in this manuscript:

GF	Gurney flap
LEVS	Leading-edge vortex shedding
OFT	Oscillating-foil turbine
TS	time-step
URANS	Unsteady Reynolds-averaged Navier–Stokes

References

1. Lemonis, G.; Cutler, J. Wave and Tidal Energy Conversion. In *Encyclopedia of Energy*; Elsevier: Amsterdam, The Netherlands, 2004; pp. 385–396. [CrossRef]
2. McKinney, W.; DeLaurier, J. The Wingmill : An Oscillating-Wing Windmill. *J. Energy* **1981**, *5*, 109–115. [CrossRef]
3. Kinsey, T.; Dumas, G.; Lalande, G.; Ruel, J.; Méhut, A.; Viarouge, P.; Lemay, J.; Jean, Y. Prototype testing of a hydrokinetic turbine based on oscillating hydrofoils. *Renew. Energy* **2011**, *36*, 1710–1718. [CrossRef]
4. Jiang, W.; Zhang, D.; Xie, Y. Numerical investigation into the effects of arm motion and camber on a self-induced oscillating hydrofoil. *Energy* **2016**, *115*, 1010–1021. [CrossRef]
5. Sitorus, P.E.; Ko, J.H. Power extraction performance of three types of flapping hydrofoils at a Reynolds number of 1.7E6. *Renew. Energy* **2019**, *132*, 106–118. [CrossRef]
6. Peng, Z.; Zhu, Q. Energy harvesting through flow-induced oscillations of a foil. *Phys. Fluids* **2009**, *21*, 123602. [CrossRef]
7. Veilleux, J.C.; Dumas, G. Numerical optimization of a fully-passive flapping-airfoil turbine. *J. Fluids Struct.* **2017**, *70*, 102–130. [CrossRef]
8. Boudreau, M.; Dumas, G.; Rahimpour, M.; Oshkai, P. Experimental investigation of the energy extraction by a fully-passive flapping-foil hydrokinetic turbine prototype. *J. Fluids Struct.* **2018**, *82*, 446–472. [CrossRef]
9. Shimizu, E.; Isogai, K.; Obayashi, S. Multiobjective Design Study of a Flapping Wing Power Generator. *J. Fluids Eng.* **2008**, *130*, 021104. [CrossRef]
10. Zhu, Q.; Peng, Z. Mode coupling and flow energy harvesting by a flapping foil. *Phys. Fluids* **2009**, *21*, 033601. [CrossRef]
11. Boudreau, M.; Gunther, K.; Dumas, G. Investigation of the energy-extraction regime of a novel semi-passive flapping-foil turbine concept with a prescribed heave motion and a passive pitch motion. *J. Fluids Struct.* **2019**, *84*, 368–390. [CrossRef]
12. Picard-Deland, M.; Olivier, M.; Dumas, G.; Kinsey, T. Oscillating-Foil Turbine Operating at Large Heaving Amplitudes. *AIAA J.* **2019**, *57*, 5104–5113. [CrossRef]
13. Dumas, G.; Kinsey, T. Eulerian simulations of oscillating airfoils in power extraction regime. *Adv. Fluid Mech. VI* **2006**, *52*, 10.
14. Kinsey, T.; Dumas, G. Parametric Study of an Oscillating Airfoil in a Power-Extraction Regime. *AIAA J.* **2008**, *46*, 1318–1330. [CrossRef]
15. Kinsey, T.; Dumas, G. Optimal operating parameters for an oscillating foil turbine at Reynolds number 500,000. *AIAA J.* **2014**, *52*, 1885–1895. [CrossRef]
16. Kinsey, T.; Dumas, G. Three-Dimensional Effects on an Oscillating-Foil Hydrokinetic Turbine. *J. Fluids Eng.* **2012**, *134*, 071105. [CrossRef]
17. Kinsey, T.; Dumas, G. Optimal Tandem Configuration for Oscillating-Foils Hydrokinetic Turbine. *J. Fluids Eng.* **2012**, *134*, 031103. [CrossRef]
18. Kinsey, T.; Dumas, G. Computational Fluid Dynamics Analysis of a Hydrokinetic Turbine Based on Oscillating Hydrofoils. *J. Fluids Eng.* **2012**, *134*. 021104. [CrossRef]
19. Young, J.; Ashraf, M.A.; Lai, J.C.S.; Platzer, M.F. Numerical Simulation of Fully Passive Flapping Foil Power Generation. *AIAA J.* **2013**, *51*, 2727–2739. [CrossRef]

20. Drofelnik, J.; Campobasso, M.S. Comparative turbulent three-dimensional Navier–Stokes hydrodynamic analysis and performance assessment of oscillating wings for renewable energy applications. *Int. J. Mar. Energy* **2016**, *16*, 100–115. [CrossRef]
21. Jeanmonod, G.; Olivier, M. Effects of chordwise flexibility on 2D flapping foils used as an energy extraction device. *J. Fluids Struct.* **2017**, *70*, 327–345. [CrossRef]
22. Liu, W.; Xiao, Q.; Zhu, Q. Passive Flexibility Effect on Oscillating Foil Energy Harvester. *AIAA J.* **2016**, *54*, 1172–1187. [CrossRef]
23. Dang, H.N.L.; Jeong, D.; Ko, J.H. Study of the power performance of a variable-camber hydrofoil used in a flapping tidal stream turbine. *J. Mar. Sci. Technol.* **2022**, *27*, 1148–1162. [CrossRef]
24. Liu, G.; Tian, C.; Wu, L.; Liu, X. Energy extraction of oscillating foil with an active deflecting trailing-edge flap. *J. Renew. Sustain. Energy* **2021**, *13*, 064501. [CrossRef]
25. Sun, G.; Wang, Y.; Xie, Y.; Ma, P.; Zhang, Y. Hydrodynamic and energy extraction properties of oscillating hydrofoils with a trailing edge flap. *Appl. Ocean Res.* **2021**, *110*, 102530. [CrossRef]
26. Xie, Y.; Jiang, W.; Lu, K.; Zhang, D. Numerical investigation into energy extraction of flapping airfoil with Gurney flaps. *Energy* **2016**, *109*, 694–702. [CrossRef]
27. Zhu, B.; Huang, Y.; Zhang, Y. Energy harvesting properties of a flapping wing with an adaptive Gurney flap. *Energy* **2018**, *152*, 119–128. [CrossRef]
28. Sun, G.; Wang, Y.; Xie, Y.; Lv, K.; Sheng, R. Research on the effect of a movable gurney flap on energy extraction of oscillating hydrofoil. *Energy* **2021**, *225*, 120206. [CrossRef]
29. Wu, X.; Zhang, X.; Tian, X.; Li, X.; Lu, W. A review on fluid dynamics of flapping foils. *Ocean Eng.* **2020**, *195*, 106712. [CrossRef]
30. The Engineering Business Limited. *Stingray Tidal Energy Device—Phase 3*; Tech. Rep.; Crown Copyright; The Engineering Business Limited: Stocksfield, UK, 2005.
31. Anderson, J.M.; Streitlien, K.; Barrett, D.S.; Triantafyllou, M.S. Oscillating foils of high propulsive efficiency. *J. Fluid Mech.* **1998**, *360*, 41–72. [CrossRef]
32. Jones, K.; Platzer, M., Numerical computation of flapping-wing propulsion and power extraction. In Proceedings of the 35th Aerospace Sciences Meeting and Exhibit, Reno, NV, USA, 6–9 January 1997.
33. Davids, S.T. A Computational and Experimental Investigation of a Flutter Generator. Master's Thesis, Naval Postgraduate School, Monterey, CA, USA, 1999.
34. Giguère, P.; Dumas, G.; Lemay, J. Gurney Flap Scaling for Optimum Lift-to-Drag Ratio. *AIAA J.* **1997**, *35*, 1888–1890. [CrossRef]
35. Genest, B.; Dumas, G. Numerical Investigation into Single and Double Gurney Flaps for Improving Airfoils Performance. *J. Aircr.* 2023, *in press*.
36. Betz, A. Das Maximum der theoretisch möglichen Ausnützung des Windes durch Windmotoren. *Z. Gesamte Turbinenwesen* **1920**, *26*, 307–309.

energies

MDPI

Article

Complex Positioning System for the Control and Visualization of Photovoltaic Systems

Filip Žemla *, Ján Cigánek, Danica Rosinová, Erik Kučera and Oto Haffner

Faculty of Electrical Engineering and Information Technology, Slovak University of Technology in Bratislava, 841 04 Bratislava, Slovakia; jan.ciganek@stuba.sk (J.C.)
* Correspondence: filip.zemla@stuba.sk

Abstract: This paper presents a proposal of a complex mechatronic system that enhances the effectivity of obtaining energy from renewable resources. The main focus is on the photovoltaic energy system, which obtains electricity from the conversion of solar radiation through photovoltaic crystalline silicon-based panels. The design of the complex mechatronic system consists of several steps. The structural design of the photovoltaic panel positioning unit in the form of a three-dimensional model is made in the selected modelling programming environment. Subsequently, a propulsion system is proposed for the designed structure, the functionality of which is verified in the programming environment Automated Dynamic Analysis of Mechanical Systems. The control system design using a programmable logical controller is also presented. The corresponding control algorithm is designed in the programming environment Step7 and covers the optimal positioning of photovoltaic panels. The developed application in the WinCC environment provides a visualization of the positioning control process. The conclusion is devoted to the assessment of the obtained results for the proposed complex mechatronic system for photovoltaic panel positioning in comparison with photovoltaic panels in fixed installation. The presented results were obtained by simulations.

Keywords: mechatronic systems; visualization; control systems; energy systems; photovoltaic panels; programmable logical controller (PLC); supervisory control and data acquisition (SCADA) systems

1. Introduction

Today, few people can imagine everyday life without electricity, but few realize the significant environmental impact that its production has on the scope of our needs. Energy demands are constantly increasing, as evidenced by the projected increase in the current electricity consumption by 30–35% every decade. Considering the fact that approximately 2/3 of the world's production comes from the thermal conversion of non-renewable sources such as coal and natural gas, which are significant producers of greenhouse gases and CO_2 emissions, the negative impact on the environment is widely known. For several decades, only around 20% of the world's electricity production was obtained from renewable sources due to the geographical or climatic requirements, or due to the high production cost [1]. Over the past 10 years, this share has increased to almost 30% and can be expected to continue to grow. Therefore, it is extremely important to develop and improve the efficiency of these forms of obtaining electrical energy, which motivates the presented research involving the design of a complex mechatronic system for the control and monitoring of the production of electrical energy using a photovoltaic system. In the design, a monocrystalline panel is considered, because it achieves higher efficiency with optimal placement of the photovoltaic panel. Our solution focuses on the design of the positioning system and is not limited to this type of photovoltaic panels.

This paper deals with the positioning of photovoltaic panels (PV) in order to track the trajectory of the Sun's movement in the sky so that its rays fall on the surface of the panels at the best possible angle. Thus, higher efficiency in collecting solar energy is achieved.

At the same time, we have to design such a control system that its energy consumption is significantly lower than the benefit of collecting solar energy.

In work [2], several methods are available for tracking the Sun for the purpose of positioning photovoltaic panels. These methods are based on soft computing approaches and for this reason, they are significantly more computationally demanding than our proposed solution.

Several projects have been published that deal with more complex solutions for the PV placement and energy harvesting. Nevertheless, the positioning system can be further improved, which would increase generating electricity from solar power. In Table 1, several existing systems are summarized. From the overview of the projects, it can be observed that the efficiency of positioning systems for PV has been continuously improving over the past few decades.

Table 1. General results [3,4].

Algorithm	References	Error	Gain in Energy Production Compared with a Non-tracking System
Closed-loop Control	Maish (1990) [5]	1°	-
	Enslin (1992) [6]	-	10–15%
	Brown et al. (1993) [7]	<0.01°	-
	Kalogirou (1996) [8]	0.05–0.2°	-
	Khalifa et al. (1998) [9]	-	75%
	Falbel et al. (2002) [10]	0.05°	-
	Al-Mohamad (2004) [11]	-	20%
	Abdallah (2004) [12]	-	15–44%
	Aiuchi et al. (2004) [13]	0.1°	-
Open-loop Control	Blanco-Muriel et al. (2001) [14]	0.08°	-
	Abdallah et al. (2004) [15]	-	41%
	Reda et al. (2004) [16]	0.0003°	-
	Chen F. et al. (2006) [17]	0.02°	-
	Chen F. et al. (2007) [18]	0.2°	-
	Grena (2008) [19]	0.0027°	-
	Chong et al. (2009) [20]	-	-

Compared to [21], we offer a much simpler calculation of the Sun's trajectory during the day. The presented solution is comprehensive and includes the entire positioning system design methodology, from structural design, through control design, to visualization of the positioning process and statistical evaluation of energy collection.

In [22], the authors offered a solution without a positioning system, but achieved higher efficiency in energy collection when using additional mirrors compared to a fixed panel without mirrors by 53%. In a related work [23], the authors report an increase in efficiency by 6.8% using a positioning system. The positioning algorithm was supplemented with the values of added sensors such as a GPS module, magnetometer, RTC module, gyroscope and temperature and humidity sensor. The work does not mention a specific control system or its own consumption for positioning.

Positioning systems are available on the market, see e.g., [24,25]. Their price is around EUR 900 and they offer a 35–40% increase in electricity productivity. In [26], a review of recent results regarding the effectiveness of different solutions for the placement of PV systems is available. Most of the compared dual-axis systems show efficiencies in the range of 20–45% compared to fixed PV panels. Our solution aims to achieve higher efficiency at a comparable price of the construction, including the positioning system and its control.

This paper includes an analysis of the photovoltaic energy system, where the current state as well as trends in this area are considered. A large emphasis is put on the factors that affect the efficiency of energy conversion, especially for silicon panels. A complex mechatronic system is subsequently designed for these panels, ensuring the positioning of the photovoltaic panel directly at the Sun during the day, to maximize the efficiency of converting solar radiation into electrical energy. The work consists of the design of the mechanical structure (console) as well as the software solution for the PLC and the visualization of the proposed positioning. The conclusion of the work is devoted to the evaluation of the profit of the positioning of the photovoltaic panel compared to fixed oriented panels.

The main contribution of this paper can be summarized as the design and control of the photovoltaic panel positioning system, so that we can significantly increase the generated electrical energy when compared to a static photovoltaic system. The control considers the minimization of the energy consumption required for positioning the panel using a stepper motor. The development of a complex mechatronic system is based on co-simulation, as follows: Automated Dynamic Analysis of Mechanical Systems (MSC Adams)–physical model of construction, PV and stepper motors; MATLAB Simulink–simulation of shaft positioning. A partial task is the design of a specific stepper motor and its parameters. The verification of the proposed design is achieved by a simulation experiment comparing the net energy gained from the designed photovoltaic system and from the fixed photovoltaic system.

This contribution is focused on the design of the simple construction that meets the specified requirements for robustness, stability, accuracy and resistance to external influences, as well on the design of the control system for positioning the photovoltaic panel with minimum consumption to maximize the gain of the generated electrical energy.

2. Problem Formulation and Preliminaries

Since the work deals with the photovoltaic energy system, in this section, we recall the corresponding physical background and take a closer look at photovoltaic cells and positioning systems.

2.1. *Photovoltaics—Theoretical Background*

Mankind has been interested in photovoltaics, or the conversion of solar radiation into electrical energy, since an important discovery was made in 1839 by the French physicist Alexandre Edmond Becquerel, who observed the interaction of solar radiation with chemical processes generating an electric current in an electric circuit consisting of liquids and metals [27]. Czochralski's method made it possible to create a semiconductor PN junction, based on which the first photovoltaic silicon cell with an efficiency of 4.5% was produced in 1954 [28]. In the following years, production technologies began to improve as well as experiments with new materials in order to increase the low efficiency of converting solar energy into electricity. The developments reached photovoltaic cells with an efficiency of 44% in laboratory conditions [29].

Photovoltaics, with its long history, is still experiencing its greatest boom today thanks to modern technologies that have allowed for a reduction in the production price of photovoltaic panels and mainly due to the growing pressure to protect the environment. The intense development in this area brings a constant supply of innovations with increasing conversion efficiency, which makes photovoltaics increasingly lucrative in the field of renewable energy sources. This is evidenced by the worldwide annual increase of 30% in the installed capacity of power plants that use solar radiation to produce electricity [30].

Solar energy spreads from the Sun continuously in all directions through space, decreasing with the distance that solar energy has to travel; therefore, the energy falling on the Earth's atmosphere is only a small fraction of the generated energy. The intensity falling on the Earth's atmosphere can be expressed by (1), where the total solar radiation H_{sun} is obtained from the Stefan–Boltzmann law on the thermal radiation of bodies (2), which is the product of the Stefan–Boltzmann constant and the bodies' absolute temperature. When calculating the radiation intensity, we multiply the total solar radiation by the area of the

Sun and divide this by the area over which the solar radiation extends at the distance of our globe. The calculation is illustrated in Figure 1 [31,32].

$$H_0 = \frac{4\pi R_{sun}^2}{4\pi D^2} H_{sun} = \frac{R_{sun}^2}{D^2} H_{sun}, \quad (1)$$

$$\text{where } H_{sun} = \sigma T^4, \quad (2)$$

in which: σ—Stefan–Boltzman constant $[\frac{W}{m^2K^4}]$; T—absolute temperature [K]; H_{sun}—intensity of radiation on the surface of the Sun $[\frac{W}{m^2}]$; R_{sun}—radius of the Sun [m]; D—Earth distance from the Sun [m]; H_0—radiation intensity at a given distance $[\frac{W}{m^2}]$.

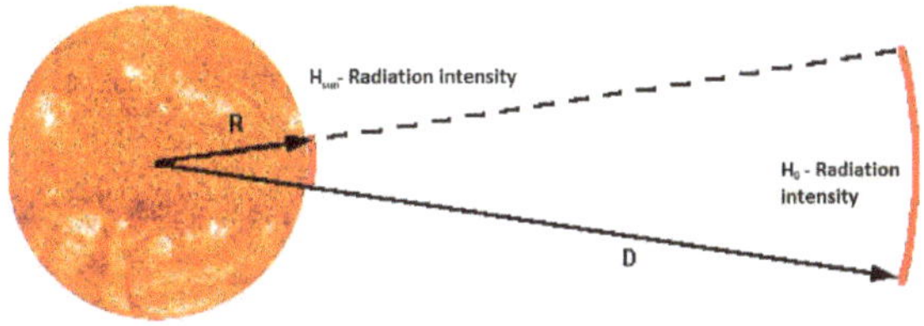

Figure 1. Spreading the intensity of solar radiation [33].

Substituting the distance between the Earth and the Sun into Equation (1), we can obtain the value of the solar constant H_{const} = 1353 kW/m^2, which indicates the intensity of solar radiation falling on the Earth's atmosphere [32]. The solar constant is generally used for photovoltaic calculations, but the actual value of the solar intensity falling on the surface of the Earth's atmosphere is variable during the year, since the trajectory of the Earth's orbit around the Sun is an ellipse. The elliptical trajectory causes the value of the actual radiation intensity to vary in the range of 3.4% of the solar constant, while it reaches its maximum in January, and the value can be expressed using (3). Based on this equation, we can calculate the value of the incident radiation intensity on the upper layer of the atmosphere for a given day of the year n [34,35].

$$H = H_{const}\left(1 + 0.033\cos\left(\frac{360(n-2)}{365}\right)\right), \quad (3)$$

in which: H—radiation intensity $[\frac{W}{m^2}]$; H_{const}—solar constant $[\frac{W}{m^2}]$; n—day of the year.

Solar radiation falls on the Earth's surface after radiation from the solar core, traveling through space and through the Earth's atmosphere in the following three forms: direct, diffuse and reflected solar radiation. The reflected component is very limited in nature, as far as the Earth's surface is concerned, but this value can be slightly higher in the case of water bodies, and in the case of populated areas where reflections occur from, for example, buildings. Nevertheless, the use of this component for the benefit of a photovoltaic power plant is quite problematic [36].

Photovoltaic power plants on the Earth's surface mainly use direct sunlight and diffuse radiation. The main production component is direct sunlight, but the diffuse component makes up a significant part of the production and in some cases, it even exceeds the gain from direct sunlight. For example, in Central Europe, with fixed-oriented panels, production from the diffuse component with a 90% share in global radiation prevails in the winter period. In annual averages, it ranges between 50 and 70% of global radiation, while the term global radiation refers to the sum of direct and diffuse solar radiation [36,37].

However, it should be remembered that the density of the energy flow on the Sun's mantle is around 63 MW/m^2. At the boundary of the Earth's atmosphere, it reaches a value of around 1366 kW/m^2, and after passing through the atmosphere to the Earth's surface under ideal conditions, solar radiation reaches an energy density of up to 1 kW/m^2 [34]. However, if we look at this on a global scale, 18×10^{16} W of solar energy falls on the

Earth under favourable conditions [38,39]. Since the Sun is about halfway through its life cycle, which lasts 10 billion years, we can consider this resource as an inexhaustible, highly prospective and, above all, clean source of energy.

On average, around 3500 Wh/m^2 falls on the territory of the Slovak Republic per day and the difference in this value between northern and southern Slovakia varies slightly, up to 15% of global radiation and the maximum value is reached in southern Slovakia. In northern Slovakia, however, there are areas with local extremes in areas with high altitudes. In the annual summary, the global radiation falling on the territory of the Slovak Republic amounts to 1200–1350 kWh/m^2 per year (Figure 2). The values are determined based on measurements of meteorological stations in Slovakia over 15 years [40].

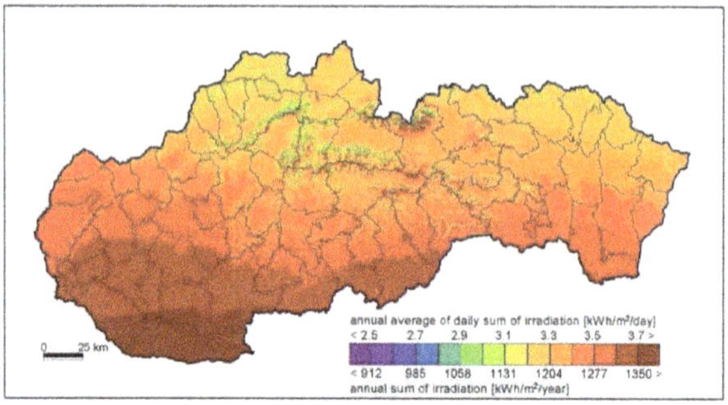

Figure 2. Solar radiation in the atmosphere on the territory of Slovak Republic [40].

For analysis in photovoltaics, however, knowledge of the measured average values of incident solar energy for a given period is not always sufficient, and it is necessary to calculate the influence of the atmosphere on the amount of incident solar energy. As we have discussed above, solar radiation when passing through the atmosphere is affected by various particles in the atmosphere. Their mutual interaction can be manifested either by absorption, reflection, or their subsequent emission. The calculations are based on the knowledge of the solar constant and the 30% fraction of absorption and scattering of solar radiation during the shortest path through the atmosphere, i.e., when the Sun is at its zenith [41,42].

The relative position of the Sun to the observer is variable in time, and thus the effective amount of atmosphere through which solar radiation passes also changes. The influence of this atmospheric mass is expressed by the *AM* coefficient, calculated on the basis of (4) [41]. Therefore, the angle from the zenith θ determines the angle at which the Sun is located from the zenith. The resulting atmospheric mass coefficient is, therefore, $AM = 1$ and if the angle from the zenith is $\theta = 0°$ at the same time as the angle increases, the *AM* coefficient also increases [41].

$$AM = \frac{1}{\cos(\theta)}, \tag{4}$$

in which: θ—angle from the zenith [°].

However, for our needs in photovoltaic system design, it is necessary to know the intensity value of either direct or global radiation. This relationship is derived based on the knowledge of the solar constant $H_{const} = 1.353\ kW/m^2$, the amount transferred by the atmosphere in the proportion of 70% and the empirical constant obtained by observations, which describes the inhomogeneity in the individual layers of the atmosphere and is equal to 0.678. Based on these data, using the coefficient of the atmospheric layer, we obtain an equation for calculating the intensity of direct solar radiation I_D (5) [34,42].

$$I_D = H_{const} 0.678^{\left(AM^{0.678}\right)}, \tag{5}$$

in which: *AM*—atmospheric mass coefficient.

It is obvious that with increasing altitude, the influence of the atmosphere will decrease, and the value of the solar intensity will increase. The effect of altitude is not significant in common ground photovoltaic systems and is, therefore, neglected in most calculations. However, if the investigated area is located at significantly higher altitudes, the real value of the solar radiation intensity could differ significantly. An empirical dimensionless constant obtained by measurements at a height of several kilometres above sea level (α = 0.14) is, therefore, added to the relationship; thus, the resulting (6) determines the value of the solar radiation intensity I_D with respect to the altitude h [43].

$$I_D = H_{const}\left[(1-\alpha h)0.678^{\left(AM^{0.678}\right)} + \alpha h\right], \tag{6}$$

in which: α—empirical constant equal to 0.14, h—altitude [km].

To determine the intensity of global radiation, it is also necessary to know the share of the diffuse radiation component and this value is usually equal to a value close to 10% of direct solar radiation. Calculation of global radiation intensity is, therefore, possible based on (7), assuming a cloudless day and impact on a vertically inclined plane [44].

$$I_G = 1.1 I_D, \tag{7}$$

in which: I_D—intensity of direct sunlight $[\frac{\text{kW}}{\text{m}^2}]$.

The value of the intensity of the diffuse component of solar radiation can be determined more precisely based on more exact calculations that consider more parameters, but for the needs of this work, the value obtained from (7) is sufficient. Figure 3 shows the spectral power density of global radiation, i.e., its intensity for selected weights of atmospheric coefficients *AM* and the angle of incoming rays from the Sun to the Earth surface h [45,46].

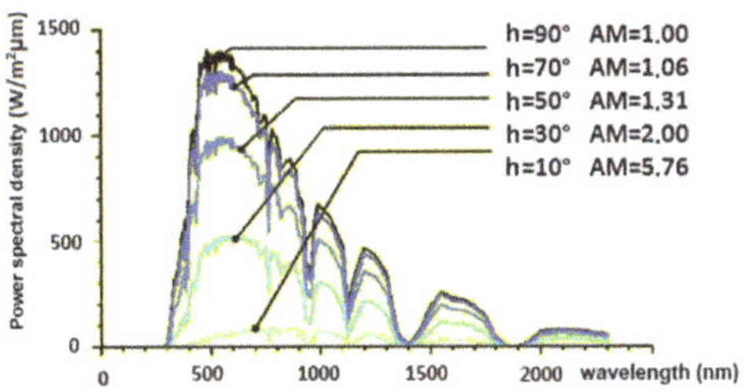

Figure 3. Solar radiation in the atmosphere on the territory of the Slovak Republic [40].

The incident electromagnetic radiation can be represented by a given wavelength and intensity or by the number of photons with energy W. A photon interprets a quantum of electromagnetic radiation, which represents a certain discrete amount of energy. So, electromagnetic radiation is a flow of photons carrying a finite amount of energy. Energy W can be expressed based on Planck's law [47].

$$W = hv = \frac{hc}{\lambda}. \tag{8}$$

in which: h—Planck's constant [J s]; v—radiation frequency $[\frac{1}{\text{s}}]$; c—speed of light $[\frac{\text{m}}{\text{s}}]$; λ—wavelength [μm].

Based on the width of the band gap W_G of the semiconductor material, the so-called absorption edge is defined, which determines the maximum wavelength of the incident radiation that the given semiconductor is able to absorb. For higher values of wavelengths, the energy of photons is no longer sufficient for absorption events and the given substance behaves as transparent. The range of wavelengths absorbed by a semiconductor can be expressed by (9). For silicon, the absorption edge is approximately λ_{MAX} = 1.1 μm [48].

$$\lambda \leq \frac{1.24}{W_G}, \tag{9}$$

in which: W_G—the width of the forbidden band [eV].

Based on the above relations, electromagnetic radiation will be absorbed in a semiconductor if its energy is higher than the band gap $W \geq W_G$. In addition to absorption, events such as spontaneous emission or stimulated emission can occur in semiconductors when interacting with electromagnetic radiation, as illustrated in Figure 4. Below, we consider only absorption for generating electrical energy based on the conversion from electromagnetic radiation. Spontaneous emission can, however, under certain circumstances occur in photovoltaic cells in the form of a non-radiative transition, when the energy released by the recombination of an electron from the conduction band with a hole in the valence band is sold to the crystal lattice in the form of thermal energy. However, this phenomenon is highly suppressed by the asymmetry of the PN junction. Stimulated emission is not essential for the needs of this work, so it will not be analyzed [49].

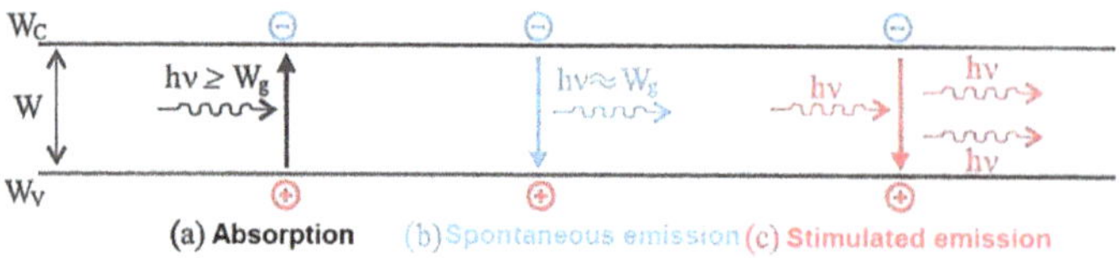

Figure 4. Interactive events [50].

The size of the energy barrier in the PN junction is given by the size of the diffusion voltage U_D. The size of the diffusion voltage is determined by the temperature and the concentration of acceptors and donors determined by the doping of the silicon structure with impurities. The resulting relationship for calculating the size of the diffusion stress U_D is given by (10) [51].

$$U_D = \frac{kT}{e}\ln\left(\frac{N_D N_A}{n_i^2}\right), \tag{10}$$

in which: k—Boltzmann's constant $[\frac{J}{K}]$; N_D—concentration of donors $[\frac{1}{\mathrm{m}^3}]$; N_A—concentration of acceptors $[\frac{1}{\mathrm{m}^3}]$; n_i—intrinsic concentration $[\frac{1}{\mathrm{m}^3}]$.

This description generally applies to semiconductor structures without external influences, i.e., without a connected external source or circuit, and also for our needs of photovoltaics in an unlit state. Through the interaction of the PN junction with the radiant flux and its influence on the energy levels, photons interact with the crystal structure, which is manifested by the generation of electron and hole pairs. Due to the energy barrier and electric potential at the space charge interface, these pairs are separated, and electrons can freely pass from the P layer to the N layer. However, if a pair is generated and subsequently splits in the N layer, this allows only an extremely small number of electrons to pass into the P layer. So, the consequence of illumination is an increase in the concentration of minor charge carriers. The difference between the Fermi levels during illumination is given by the potential difference, which corresponds to the photovoltaic voltage U_P. However, the size of the photovoltaic voltage cannot be greater than the difference of the original curvature of the energy levels, i.e., it cannot be greater than the diffusion voltage U_D. In practice, this means for silicon photovoltaic cells a value of around 0.6 V.

2.2. Photovoltaic Cells

The converters of solar energy into electrical one are based on physical principles. Photovoltaic cells of the first generation are based on crystalline silicon structures and a

P-N junction. Silicon in this generation is processed as monocrystalline or multi-crystalline. However, the absorption coefficient of silicon α determines the thickness of the semiconductor at least 100 µm for the given intensity of solar radiation to absorb the maximum incident energy. The calculation of the absorption thickness of the semiconductor is based on (11). From this fact, it follows that a relatively large amount of material in the form of crystalline silicon is required for the production of such photovoltaic cells, which is the reason why they are also called bulk photovoltaic cells [49–53].

$$E_{opt} = E_{0opt}\exp(-\alpha y), \quad (11)$$

in which: E_{opt}—intensity of radiation in a semiconductor $[\frac{W}{m^2}]$; E_{0opt}—ntensity of incident radiation $[\frac{W}{m^2}]$; α—absorption coefficient $[\frac{1}{m}]$; y—depth in semiconductor [m].

The structure of the photovoltaic cell is determined by the technological processing of silicon, while the technology of monocrystalline or multicrystalline processing is diametrically different. The production of a multicrystalline structure is significantly simpler and cheaper, which, however, is reflected in the slightly lower efficiency of the photovoltaic cell, but this difference is in the order of percent. The processing of monocrystalline silicon is significantly more demanding for technological equipment than the energy intensity of the production process. For both structures, as in semiconductor technology in general, a high purity of the material is necessary, which again makes the given process difficult and demanding, both financially as well as energetically, but research in this area has progressed and more innovative chemical processes for cleaning silicon are under development [54].

The most important step in the preparation of silicon structures is the creation of a P-N junction, which ensures the entire process of converting sunlight into electrical energy. The creation of the P-N junction was described earlier. The N-type semiconductor layer is created by the diffusion of phosphorus (P) and the diffusion is applied to the top layer of the silicon structures. The N-type layer created by diffusion is disproportionately smaller in width compared to the P-type layer, thus creating an important asymmetry of the P-N junction (Figure 5).

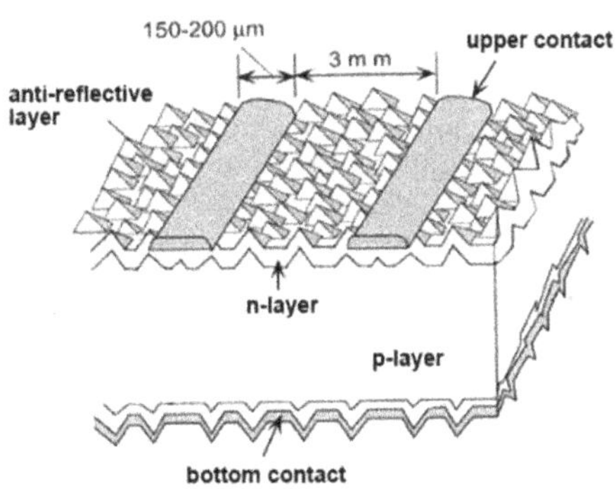

Figure 5. The structure of the first-generation cells [54].

The result of most research in the field of photovoltaics includes a functional prototype of a photovoltaic cell. To be able to compare photovoltaic cells with each other, it was necessary to introduce parameters that describe their quality. Most of these parameters are given by measurements of a specific product, whether they are DC or AC measurements. A very important indicator of cell functionality based on measured values is the volt-ampere characteristic (VACH). However, to enable comparison of these results on a global scale, standard test conditions have been introduced that define parameters with a great influence on the resulting values that describe the quality of the photovoltaic cell [55,56].

Standard test conditions:

- Lighting intensity 1000 W/m^2,

- Radiation spectrum AM 1.5,
- Temperature 25 °C.

To achieve standard test conditions, especially for the required spectrum of radiation as well as its intensity, a solar simulator can be used. With the help of several optical elements and systems, this device is able to imitate the required solar spectrum even with the prescribed intensity [56].

Photovoltaic cells are designed for a number of applications, be it energy or electronics, and their purpose remains the same, namely the generation of electricity. In the field of electronics, they find application mainly in devices that require a kind of energy independence, which can make a significant impact on the photovoltaic market. In this paper, the application in the energy industry is considered, where we require larger energy capacities from photovoltaics. In both cases, however, the efficiency of the energy conversion of solar radiation into electricity and the price of the photovoltaic cell are key. Research has dealt with the optimization of these important parameters for several decades [57,58].

2.3. Positioning of Photovoltaic Panels

Currently, positioning devices for photovoltaics are available in several designs and the main division is in the number of axes around which the photovoltaic panel, or field, can be rotated. Based on this number, we know one-axis and two-axis positioning devices. One-axis configuration assumes that one axis is fixed, and the other axis is positioned during the day. In the two-axis configuration, both axes are positioned, while their positioning direction consists of rotation around the vertical axis and a change in tilt with respect to the horizontal plane or positioning around the east–west axis and the north–south axis. Optimal pointing based on the position of the Sun is possible after sensing optical parameters or based on a model of the position of the Sun [58,59]. The complex positioning system proposed in this work considers the two-axis configuration and is based on model control; therefore, it must calculate the position of the Sun.

The position of the Sun relative to the observer on Earth is given by the Earth's orbit around the Sun, the rotation of the globe around its own axis and the position of the observer on Earth (Figure 6). However, the corresponding value is variable over time and the exact position can be calculated for a specific day and time. There are several calculation models, differing in their accuracy and calculation complexity. For our needs of positioning silicon-based photovoltaic cells, even less accurate and simpler calculation models are sufficient. The lower accuracy of the used model has practically zero impact when used in practice. The differences are minimal and are almost not reflected in the angular dependence of the generated power. The position of the Sun relative to the observer is expressed based on the following two angles: the azimuth and elevation angle. However, to calculate them, it is necessary to know other values explained below.

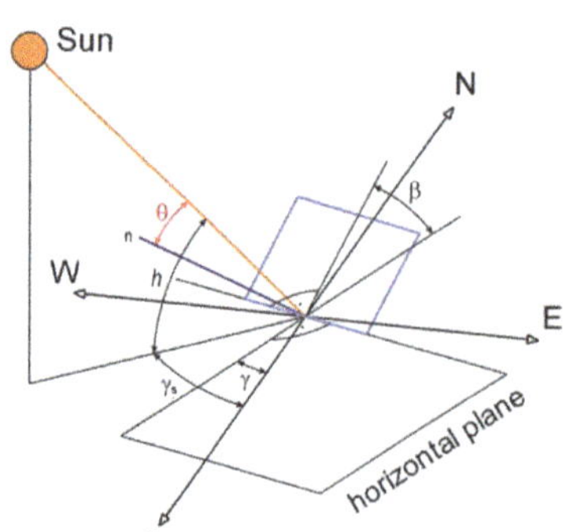

Figure 6. Geometric movement of the Sun relative to the receiving surface characteristics [46].

The angle of declination δ is the basis of practically every calculation for the position of the Sun. It expresses the angle between the connecting line of the centres of the Earth and the Sun relative to the plane created by the Earth's equator. It ranges from $-23.45°$ to $23.45°$. Only the day of the year is needed for the calculation, which can be observed in the following equation [60–62].

$$\delta = 23.45\sin\left(\frac{360}{365}(d-81)\right), \tag{12}$$

in which: d—day of the year.

The hour angle ω expresses the angle between the parallel solar radiation and the straight line given by the centre of the Earth and the position of the observer. It is positive in the morning hours, equals zero at lunchtime and gradually decreases with time. Since we need this value that is dependent on local time, its calculation must be divided into several steps and adjusted to the appropriate form. To calculate the local time, it is necessary to know the time zone ΔG_{TM} and the longitude D. The first step is the calculation of the time Equation (13) from which we obtain the EoT parameter. This is followed by the calculation of the time correction factor (14) TC and local solar time (15) LST, on the basis of which it is possible to determine the hour angle ω (16) [60–63].

$$EoT = 9.87\sin(2B) - 7.53\sin(B) - 1.5\sin(B), B = \frac{360}{365}(d-81), \tag{13}$$

$$TC = 4(D - 15\Delta G_{TM}) + EoT, \tag{14}$$

$$LST = LT + \frac{TC}{60}, \tag{15}$$

$$\omega = 15(LST - 12), \tag{16}$$

in which: D—longitude [°]; ΔG_{TM}—time zone; LT—investigated local time.

The elevation angle α expresses the angle between the plane of the observer and the straight line formed by the observer and the Sun, thus expressing how high the Sun is on the horizon (17). It can be determined by knowing the declination angle δ, the hour angle ω and the latitude ϕ [58–61].

$$\alpha = \cos^{-1}(\sin(\Phi)\sin(\delta) - \cos(\delta)\cos(\omega)\cos(\Phi)), \tag{17}$$

Azimuth A_Z is an angle that expresses the compass position of the Sun with respect to the observer, while it is calculated from the north pole in a clockwise direction, i.e., at solar noon $A_Z = 180°$. Another representation is that at solar noon, $A_Z = 0°$ and towards the east, it acquires negative values. Azimuth can be expressed based on the above angles (18) [60–63].

$$A_Z = \cos^{-1}\left(\frac{\sin(\delta)\cos(\Phi) - \cos(\delta)\sin(\Phi)\cos(\omega)}{\cos(\alpha)}\right), \tag{18}$$

Based on given relationships, we can determine the position of the Sun relative to the observer for a specific position on Earth and a specific time using the angle of elevation and azimuth. For this reason, we do not need to have images of the position of the Sun available, and when rotating the right photovoltaic panel, we will only need the current rotation in both planes.

3. Design and Implementation of the Complex Mechatronic Positioning System

In this section, we describe the design and implementation of the mechatronic positioning device and the implementation of the control system. Finally, we present a software application for energy system visualization. The design of a positioning system aims at enhancing efficiency of obtaining energy from a renewable source. In our case, as has already been mentioned, it is a positioning mechatronic system on two axes that ensures the optimal tilting of the photovoltaic panel perpendicular to the incident sunlight. Photovoltaic panels, for which this positioning system is intended, are the most widespread on the market and their technology is the most researched, so they are panels based on crystalline silicon. As mentioned in the analysis of the energy system, the angle under which the solar radiation falls on the plane created by the surface of the panel has a fundamental influence on the amount of incident energy and its subsequent transformation into electrical energy. Thus, the positioning device increases profit in generated power during the day in comparison with a fixed panel, but it consumes a certain amount of the obtained energy for positioning. The amount of energy input to the positioning must be minimized; therefore, the design of the mechanical as well as the drive system is very important.

3.1. Hardware Design and Implementation of Mechatronic Positioning Device

Hardware design basically consists of the following three main parts: the photovoltaic panel, console, and drive unit. The design for the optimal photovoltaic panel is also determined by the geographical location in which the given positioning system will be installed. For different geographical locations, optimal construction designs and their control can be different. In this work, we consider the geographical location of Bratislava; therefore, the design of the mechatronic system will be primarily focused on the corresponding latitude [63]. In this subsection, the requirements for the design of a mechatronic positioning device for photovoltaic panels for roof applications are defined. The following requirements for the positioning system are subsequently analyzed in order to fulfil them:

1. Movement variability, which enables us to orient the photovoltaic panel perpendicularly to the direction of the Sun's rays for each position of the Sun for a given geographic location, while the maximum permissible deviation is 15° from the perpendicular incidence of the rays on the panel at the extreme of the Sun's positions.
2. The positioning accuracy of the normal vector of the photovoltaic panel to the desired point determined by the azimuth and elevation angle relative to the positioning device with an accuracy of $\pm 2.5°$. Positioning accuracy must be ensured over the full range of motion of the device.
3. The robustness of sensing the angle of rotation of individual tilting axes of the photovoltaic panel.
4. The design of the structure and mechanical parts of the mechanism must meet all the attributes for the minimum energy requirements for the propulsion system.
5. Ensure a fixed position over the full range of the positioning, which will prevent rotation around both axes caused by external influences.
6. The structure must be installable on the flat roof of the building, while it must be able to withstand weather effects of wind, rain, temperature and sunlight. Even under these influences, the positioning system must achieve a sufficient service life as well as minimal maintenance during the operation of the device.
7. A variable console that allows fixing available panels on the market, as the dimensions of the panels are not standardized and may differ slightly from different manufacturers. The bracket must be able to carry the load due to the influence of the panel as well as other influences. The permissible load capacity of the installed panel is at least 25 kg and the necessity of the possibility of height variability, which allows us to find an equilibrium position for different panels.
8. The design shall allow multiple positioning devices to be connected to one common drive system. The design shall allow the connection of individual positioning devices for both tilting axes.
9. A complex mechatronic positioning system of photovoltaic cells should have as small weight as possible, under all prerequisites for mechanical resistance and durability. The weight is important because of the installations on the roof of the building.
10. The system must be manufacturable on the basis of available materials and technologies in common practice, with the lowest possible production price. The system

must be capable of being assembled and allow for the possible replacement of any component without significant intervention.

The range of mobility of the mechatronic positioning system for the purpose of increasing energy efficiency, and therefore increasing the generated power of photovoltaic cells, depends on the position of the Sun. The position of the Sun changes during the day and the trajectory of the Sun across the sky is dependent on the latitude of the given energy system.

We can express the position of the Sun using two angles, namely the elevation angle and the azimuth angle relative to the system (for more details, see Section 2.3), that is, to the geographic location of the system installation [33,63]. For applications of this system, we start from the geographical location of Bratislava, where the latitude is S 48°1′ and longitude is E 17°11′ [63]. The annual course of the elevation angle at solar noon in Bratislava is shown in Figure 7. It is obvious that azimuth changes during the light phase of the day depending on the current season, as it is shown in Figure 8.

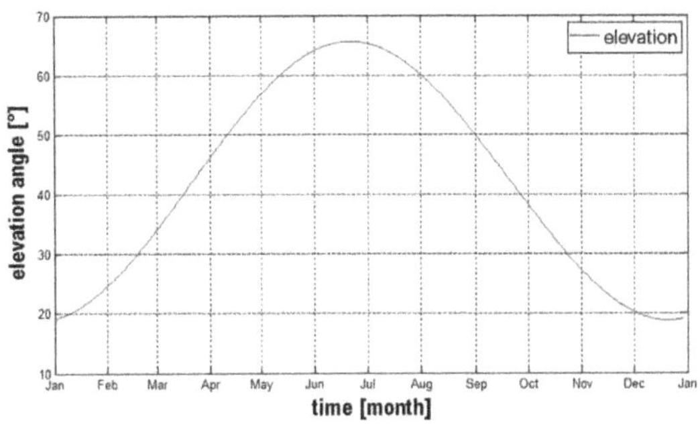

Figure 7. The annual course of the elevation angle at solar noon.

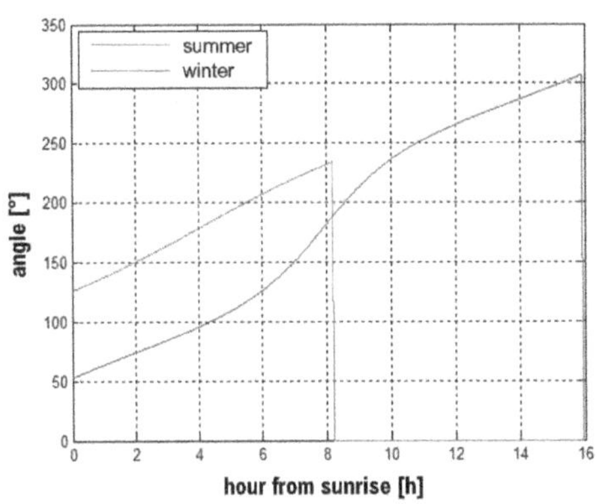

Figure 8. Azimuth curves for the solstices in the light phase of the day.

The variability of the console is based on requirement 7 and is supposed to provide several essential functions. In addition to the need for a sufficient load-bearing capacity, it must also fulfil other essential parameters related to the energy requirement of positioning as well as the universality of the mechatronic system for positioning photovoltaic cells, which can be of different sizes from different manufacturers. Note that the knowledge of these parameters at the stage of design preparation is essential and it is very important to analyze them before the design process.

Structural design takes place in the CATIA program environment [64]. This development tool enables modelling of 3-D objects in a virtual reality environment. It contains many modules that provide various functions or analyses. For the needs of our work, the

PART DESIGN module, which enables 3-D modelling of individual bodies, is perfectly suited and the ASSEMBLY DESIGN module allows the individual modelled bodies to be interconnected into complex mechanical systems. Both modules fall under the category called MECHANICAL DESIGN.

The modelled mechanical systems using the mentioned modules are composed of several bodies that are fixed to each other by bindings. Bindings can be different, but because it is a mechanical system, it also contains bindings that introduce a certain number of degrees of freedom between individual bodies into the system. It must also be noted that the dynamics of movements, and thus also the action of forces, will be solved through the MSC ADAMS program, which is much more suitable for these tasks.

At the beginning of the design and modelling process is the photovoltaic panel, from which the subsequent design of the structural solution is based. The photovoltaic panel, i.e., its dimensions and weight, represents the input and an uninfluenceable parameter of the design. It is, therefore, obvious that the design of other components of the mechatronic positioning device will be based on those values. Its basic parameters are shown in Figure 9.

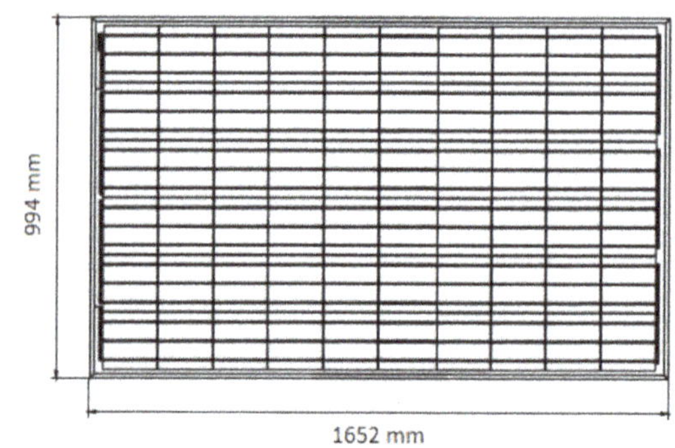

Figure 9. Photovoltaic panel drawing.

Basic parameters of the photovoltaic panel:

- Technology: polycrystalline;
- Nominal power: 250 Wp;
- Efficiency: 15.2%;
- Dimensions: 1652 × 994 × 40 (mm)
- Weight: 20 kg (without frame);
- Composition: frame, glass, module, EVA and back cover;
- Glass thickness: 3.2 mm.

A view of the frame model can be observed in Figure 10a and a cross-section of the internal structure with single layers of the panel is shown in Figure 10b. The final model of the photovoltaic panel is shown in Figure 11a with the used materials such as aluminium, glass and the photovoltaic module.

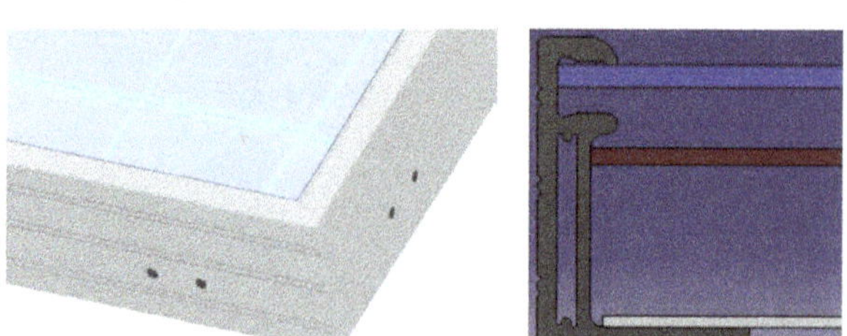

(**a**) (**b**)

Figure 10. Frame detail (**a**); cut through the panel (**b**).

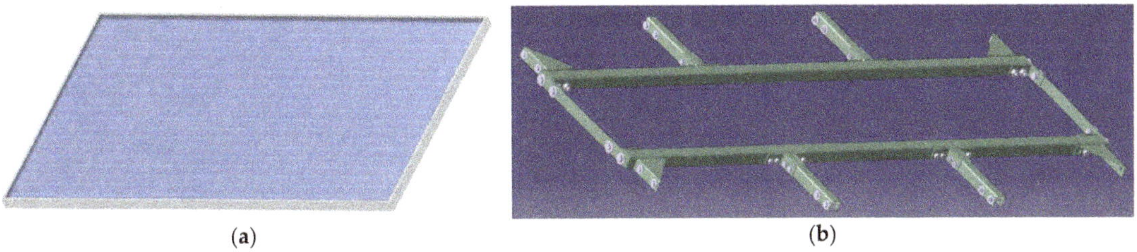

Figure 11. Photovoltaic panel model (**a**); bracket without fasteners (**b**).

The design of the console is based on the established requirements and parameters of the photovoltaic panel. The main conditions for the proposal are sufficient load capacity, the possibility of the secure attachment of the panel, the mentioned variability of fastening elements and the possibility of height adjustment in relation to other components of the positioning system. Great emphasis is also placed on assembly ability, as well as the use of real and available structural elements. The base of the console construction is shown in Figure 11b.

The entire system consists of the base of the positioning system, drive, and shaft position sensor. In Figure 12, we can observe a two-axis complex positioning system of photovoltaic cells. Based on positioning using individual actuators and the collision detection function, the extreme positions of the system were found. With the help of these extreme positions, it is possible to determine the range of system mobility expressed in the rotation angles of the individual shafts of the rotation axes (Table 2). These angles do not directly express the direction of the normal vector of the panel.

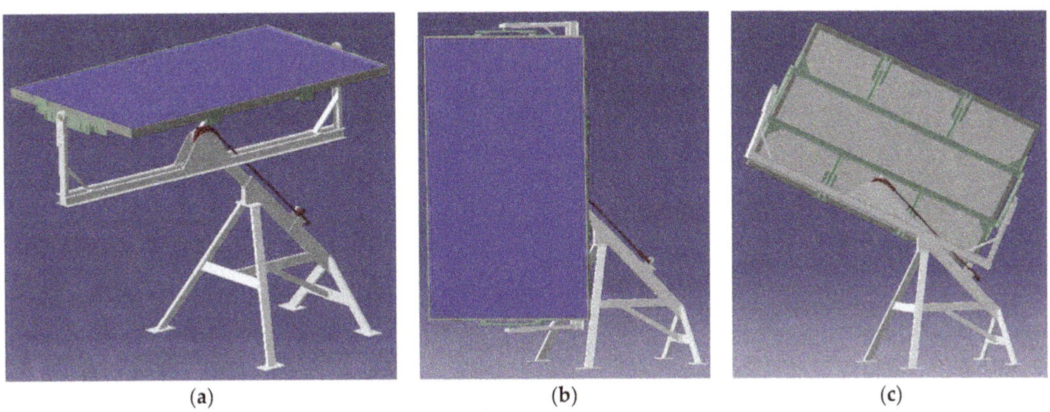

Figure 12. Complex positioning system (**a**); extreme positions of the mechatronic positioning system (**b**,**c**).

Table 2. The mobility range of the designed photovoltaic panel positioning system.

	North–South Axis	East–West Axis
From	−20°	82°
To	−75°	75°

The aim is to design a system that ensures optimal conditions for maximizing the efficiency of the energy system. The evaluation of the energy demand of positioning can be determined if the electrical parameters of the motor with which the given photovoltaic system is positioned are known. In order to correctly choose the drives, i.e., the positioning motors, it is necessary to know the static and dynamic moments in individual positions and movements.

We use a stepper motor as an actuator for the positioning of the photovoltaic panel. To specify the parameters of a stepper motor, a mechanical analysis of the positioning system was performed. Thus, the maximum torques arising during the positioning of individual axes were obtained. The analysis was carried out in the positions that correspond to the maximum torques; therefore, it has a sufficient informative value for the stepper motor specification.

The choice of the stepper motors for each axis was, therefore, based on the size of the step, sufficient torque as well as the minimal energy requirements. The choice of the gear ratio was also very important, because a larger ratio will ensure a greater torque, but at the expense of the number of steps the drive must perform. However, with the number of steps, the time of its operation and thus its energy consumption also increases; therefore, it is important to find the right compromise for the choice of this parameter. For the needs of using a stepper motor, a mechanical analysis of the positioning system was performed, based on which the maximum torques that arise during the positioning of individual axes were given. The analysis was performed in the positions of the largest moments (Table 3); therefore, it has a sufficient informative value for the dimensioning of the stepper motor.

Table 3. Results of the mechanical analysis of the positioning system.

Results of Static Analysis of the Positioning System	
Maximum static moment of the VZ axis	0.6 Nmm
Maximum static moment of the SJ axis	28.5 Nmm
Results of the Dynamic Analysis of the Positioning System	
Maximum dynamic moment of the VZ axis	0.52 Nmm
Maximum dynamic moment of the SJ axis	0.97 Nmm

The parameters of the selected stepper motor that correspond to all the requirements are summarized in Table 4.

Table 4. Parameters of the selected stepper motor.

Parameter	Value
Nominal torque	1.89 Nm
Nominal voltage	3.2 V
Nominal current	2.8 A
Number of windings	4
Winding resistance	1.13 Ω
Inductance	3.6 mH
Rotor inertia	480 g·cm^2
Holding torque	0.068 Nm

3.2. *Simulation Software of Mechatronic Positioning Device*

After designing the structure solution, we know the working range of mobility, based on which we can calculate the energy gain compared to fixed photovoltaic panels. The proposed complex mechatronic positioning system can be verified at this stage thanks to the capabilities of advanced computer technology and software equipment. In the MSC ADAMS program environment [65], the designed model of the positioning device can be controlled using co-simulation with the MATLAB computing environment [66]. The cooperation of these two programs is based on exporting the model from MSC ADAMS to MATLAB. This export is based on the definitions that determine which parameters should be available for the output model, whether they are input or output variables. However, for the needs of functionality verification, it is important to verify the functionality of the stepper motor for individual tilt axes, for the moments arising in the system, as well as the control frequency of the stepper motor. Therefore, the simulation scheme (Figure 13) contains blocks that describe the behavior of individual components of a complex mechatronic system. The basis is a block that represents the designed model of the photovoltaic panel (orange block in Figure 13). The inputs to this model are the tilting moments on individual positioning axes. The outputs are the moments that were analyzed and the tilt position of the individual axes.

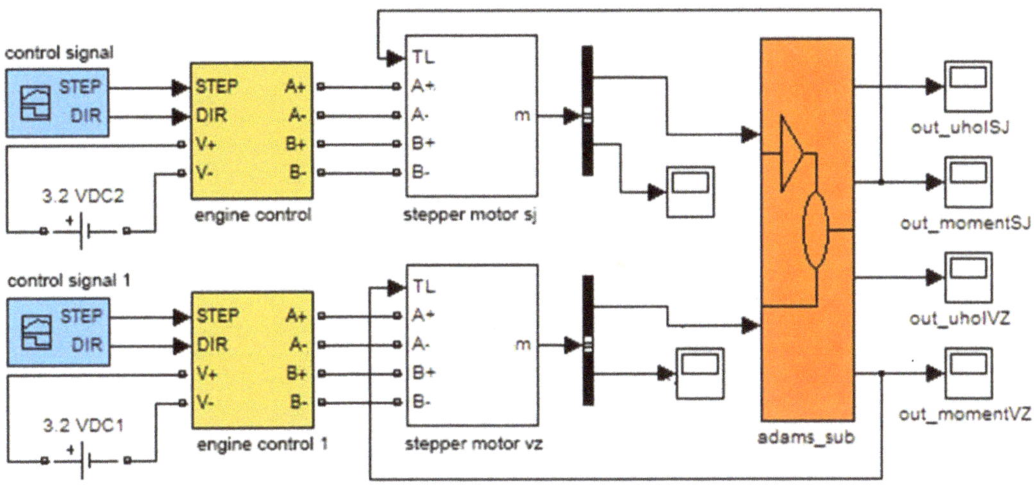

Figure 13. Simulation scheme of a complex positioning system.

Other blocks in the simulation scheme are the stepper motors (white blocks) that position the panel in individual axes. The outputs of these blocks are the torques generated by the stepper motors based on the input signals from the engine control blocks (yellow blocks). The input to the stepper motor block is the power supply of the individual windings, as well as the load torque. The control of the stepper motor (yellow blocks) ensures the switching of the voltages on the individual windings and ensures the change in the shaft position through pulse-width modulation. The input signal to the engine control block is a control signal that determines the operation as well as its direction. The second input is the supply voltage, which is transferred to the motor windings. The control frequency is 10 Hz and the block that describes the stepper motor has the parameters corresponding to a stepper motor designed at the end of Section 3.1.

The task of this simulation is to verify the correct operation of the motors under the influence of the load arising during the positioning of the individual axes of the system. If the stepper motor was loaded with a large torque, the step could be lost; thus, the positioning of the system could malfunction. In simulation results, we can monitor this phenomenon, whether the motor step is not lost due to the load.

The resulting responses (Figure 14) show the electromagnetic moment of the stepper Wmotor, i.e., the moment by which we position the device and especially the position of the motor shaft, and oscillations indicate the mentioned malfunction. The response is shown at a higher load and the results show the moment generated by the stepper motor, as well as a stable step change in a given interval with a change in size corresponding to one step, thus verifying the functionality of the photovoltaic panel positioning system designed in this way.

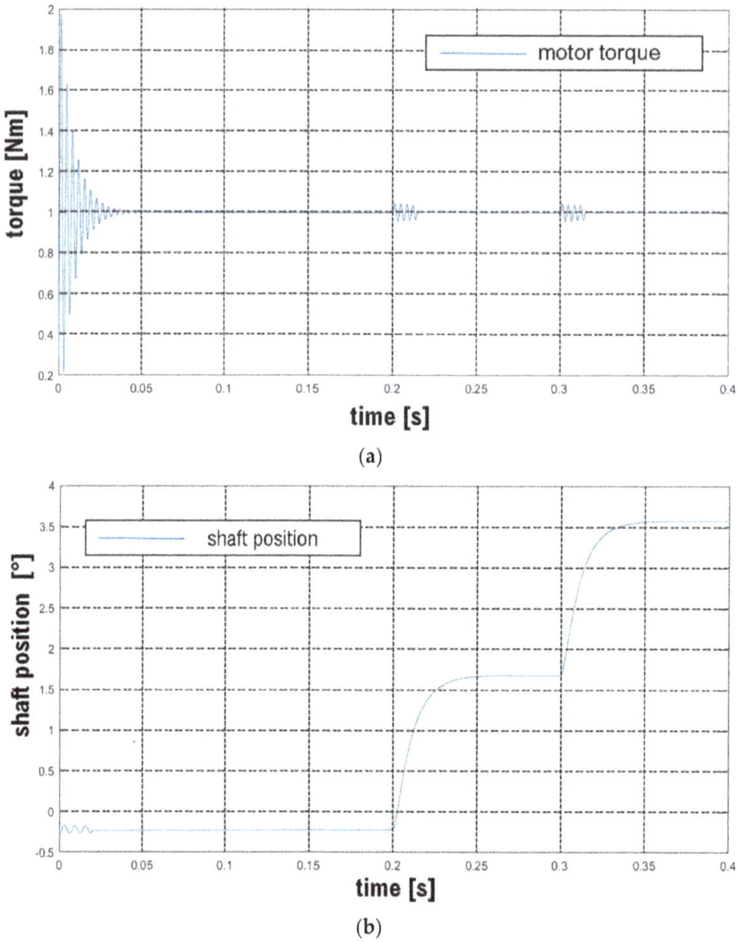

Figure 14. Simulation responses for motor torque (**a**); shaft position (**b**).

It can be observed in Figure 14 that the co-simulation of the position control for the east–west axis of the photovoltaic system in the MATLAB Simulink environment took place as expected, i.e., with each request to change the position from the control signal block (see Figure 13). This change will be reflected by vibrations for the response of the stepper motor torque and the following correct change in position of the entire photovoltaic panel.

The x-axis shows the time in seconds, and the y-axis shows the motor torque in Nm in the upper part of the picture, and the turning angle in ° in the lower part.

3.3. Program Implementation of the PLC Control System

A control element that ensures the operation of the system must also be part of the complex mechatronic system for positioning photovoltaic cells. Its function primarily consists of the control of stepper motors in order to efficiently position the photovoltaic cells in relation to the energy source, i.e., towards the Sun. Optimal positioning ensures optimal efficiency of obtaining electrical energy from the photovoltaic energy system. In addition to the control of the stepper motors, the controller must also calculate the optimal pointing based on the position model of the Sun, since this is a design without any optical sensing of the Sun's position. The positioning system must be able to determine when it is optimal to position the photovoltaic panel, that is, the moment when the deviation of the normal vector of the panel and the position of the Sun is greater than optimal. However, the optimality as well as the possibility of positioning the photovoltaic panel towards the Sun are also determined by weather conditions, which the control system must take into account.

The program solution can only be verified in practice by running the program on a real PLC control system. In such cases, another part of the STEP7 software [67] is used to verify the functionality of the program, which is the PLCSIMULATOR, that enables the tuning of the given program. The PLCSIMULATOR is, therefore, an essential part of program implementation in the case when a real PLC is not available.

In practice, the use of PLC devices would depend on the scale of the photovoltaic panel system; in the case of one photovoltaic panel, a Siemens Logo PLC device would be sufficient, the current price of which is approximately EUR 150. In the case of several devices with up to 10 photovoltaic panels, a PLC device S7-1200 or equivalent would be needed, the price of which is about EUR 500. In addition, in the case of several devices over 10 photovoltaic panels, it would be necessary to use a PLC device S7-1500 or equivalent with input/output modules, where the price would start from EUR 2000 and more, depending on the number of panels.

The control algorithm, as well as the entire positioning system, are designed to minimize their own consumption. The control program solution has a fundamental influence on the running of the drives, while it is necessary to avoid unnecessary steps or that the drives run due to the incorrect positioning of the system. This is the priority of the design of the entire control algorithm, which must be ensured if the energy input to the positioning system of the photovoltaic panels is to be minimized.

The software solution itself is designed to minimize the consumption of the control system, that is, it performs only the immediate necessary steps. For the positioning of the photovoltaic panels, a frequent change is not necessary; therefore, it is not necessary to perform all the calculations in each program cycle. The duration of the execution of one program cycle is determined by the length and complexity of the program, but it varies in units of a maximum of tens of ms. Thus, it is sufficient that the deviation of the Sun's position is compared to the normal vector of the photovoltaic panel, which is preceded by the calculation of the Sun's position, as well as the direction of the vector, only once every minute using a cyclic PLC interrupt. Therefore, as part of the execution of each program cycle, only an evaluation of the possible fault conditions or operator interventions is realized.

The calculation of the position of the Sun is composed of several functions that ensure the calculation of all necessary parameters. After determining the position of the Sun, the direction of the normal vector of the photovoltaic panel is calculated. The calculation of the normal vector is given by the structural solution of the positioning system and is based on the value obtained by the position sensors on the individual positioning axes. After calculating these two values, the control system can evaluate the deviation and, based on the determined value, declare whether it is necessary to change the current position or not. When requesting a change in location, a subsequent evaluation of weather effects is necessary, which, if favourable, is followed by a positioning algorithm that ensures the optimal positioning of the photovoltaic panel with the aim of maximizing energy gain. The positioning algorithm determines the optimal direction of the normal vector of the panel; based on this direction, it determines the required rotation of the individual positioning axes of the system. After determining the desired value as well as the direction of the desired change, it starts the individual drives until they acquire the desired value. The control algorithm is clearly summarized in the flowchart of the positioning software solution in Figure 15.

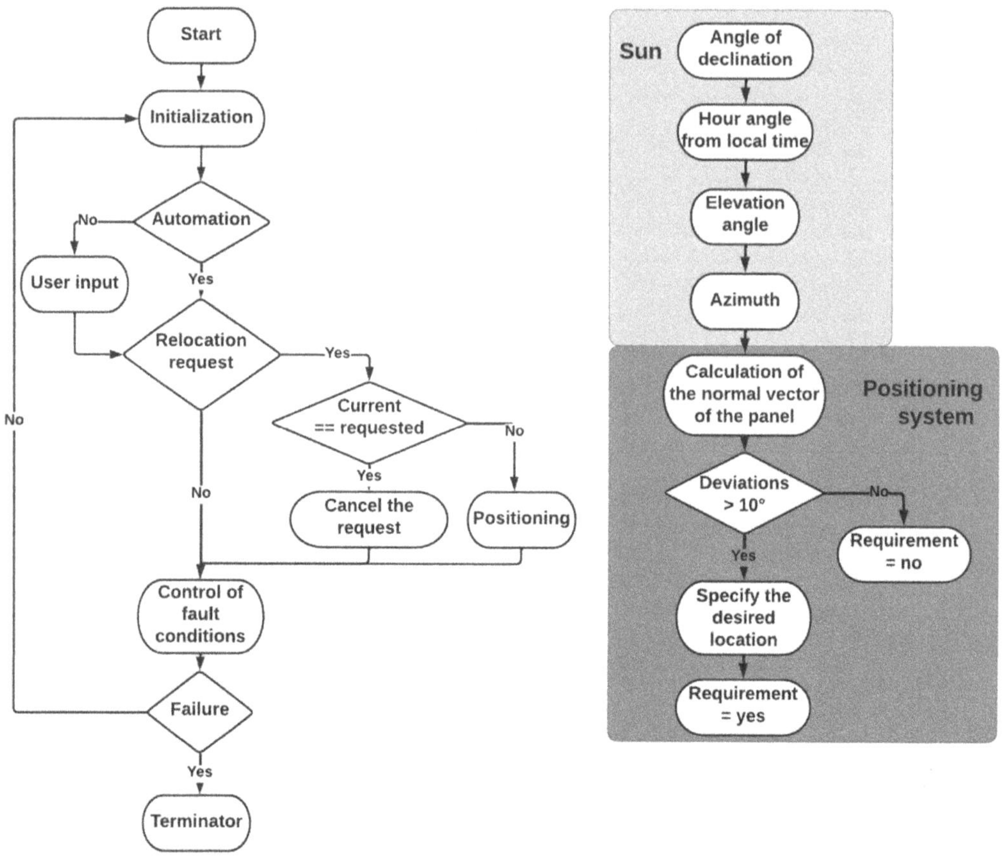

Figure 15. Development diagram of the control algorithm.

The development of the photovoltaic panel positioning control system was based on the following requirements, and we can conclude that the proposed algorithm fulfilled them.

(1) Sufficiently accurate calculation of the position of the Sun with respect to the positioning system of photovoltaic panels.
(2) Identification of unsuitable conditions for panel positioning in terms of energy demand in relation to gain, i.e., in cloudy periods when direct sunlight does not pass through. It is considered essential to use a solar radiation intensity sensor for the real system. In case the measured value is below the specified minimum, the panel will be moved into the default position.
(3) Weather effects in the form of high wind speed must be considered, which is important especially for safety reasons. In this condition, it is necessary to minimize the stress of the structure and to prevent an increased energy demand, caused by the positioning of the panel under strong wind. In case of wind speed above the critical value, the control system must ensure the position in which the system is least affected by this factor.
(4) The positioning algorithm must ensure rotation to the optimum position. The optimum position is given by the current position of the Sun, and the control system must be able to ensure that the normal vector of the panel is positioned to this point. The position must be determined accurately, and the positioning must be performed

with a minimum number of steps, without searching for the optimal position by positioning the panel.

(5) The control system must be robust under various unfavourable conditions, for example in the event of a power failure, where the system must remain fixed, and after restarting, it must be able to identify its position and optimize it by a potential correction with a minimum number of steps.

(6) The control system must be able to identify potential malfunctions and take adequate measures to correct them, and if this is not possible, ensure that the operator is informed of the malfunction.

(7) Positioning to any direction of the normal vector of the positioning device panel, specified by the operator.

3.4. Creation of a Software Application for Visualization of the Photovoltaic System

It is necessary to create a program environment that enables a human–machine and machine–human interface with the given controlled process of positioning photovoltaic panels. For our photovoltaic energy process equipped with a positioning photovoltaic panel system, it is important to have an overview of the positioning equipment as well as the position of the Sun to verify the functionality of the positioning process. The visualization application should give the operator an overview of the theoretical value of the incident power, but especially of the electrical power generated by the power system based on conversion from solar radiation. The visualization must be implemented in conjunction with the PLC control system of the positioning device. The WINCC FLEXIBLE program environment enables the creation of HMI interfaces for these control systems [68]. The main parts of the proposed visualization system are the screen header, main screen, overview of alarms, graphic evaluation, and manual operation control panel. These parts are described in detail in the subsections below.

The design of the visualization environment is based on the requirements and needs of the given positioning process. Therefore, the structure of the visualization environment is directed to provide a clear view of the process with all the important parameters displayed. These are divided into sections according to their affiliation, allowing the operator quick access to the necessary data as well as the possibility to intervene in positioning the photovoltaic panels. This is performed in the form of an arbitrary intervention and can trigger any state of the positioning device. The structure of the visualization environment enables the operator to have an overview of the progress of the process variables using a clear display on the screen with a graphical design. Another important screen of the visualization environment is the list of alarms, which allows an overview of the operation of the device from the point of view of faults and warning messages for the selected period.

When designing a visualization environment, the availability of key information and the ability to actively intervene at any time, i.e., on every screen, is important. Therefore, the operator must be able to select the manual or automatic mode of operation of the positioning system or have the possibility to switch between the different screens that offer the visualization of the positioning of the photovoltaic panels.

3.4.1. Screen Header

The header of the visualization screen consists of two parts, the part equipped with navigation and control buttons and the part that displays the current values that correspond to the controlled process. The part containing the buttons is displayed in each screen. The representation of the header of the visualization environment is shown in Figure 16, where both mentioned parts of the header are displayed.

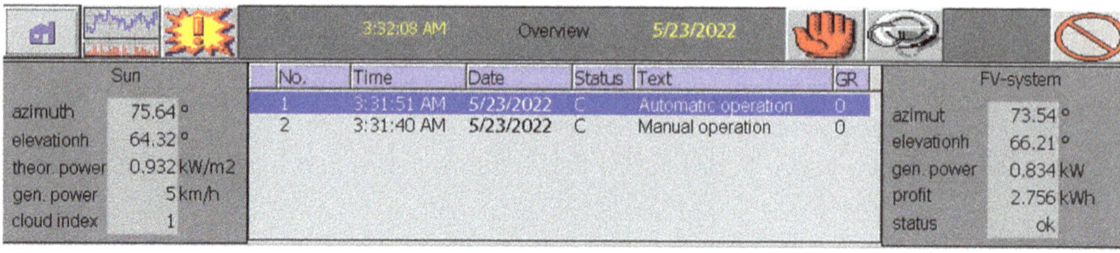

Figure 16. Header of the visualization environment.

The navigation part of the header of the visualization environment is located at the very top of the header, and therefore also at the top of all visualization screens. Its implementation is solved using a template, which ensures its application to all screens. The navigation section contains the following elements in order from the left to the right:

- Main screen panel button,
- Graphical evaluation panel button,
- Alarm overview panel button,
- Middle part with current time data and the panel just opened,
- Button for activating manual operation,
- Button for activating automatic operation,
- Ending the visualization.

The overview part is located right below the navigation panel of the header of the visualization environment. This overview section offers a clear display of the most important information about the process, as well as active error reports. The overview section is divided into three segments, the left segment is an overview of information about the Sun, followed by an overview of active alarm and fault reports, and the last segment is dedicated to important information about the positioning system.

3.4.2. Main Screen

The main screen of the designed visualization environment offers an overview of all available information about the process. The screen consists of the mentioned header and other elements offering a clear display of the given values. The main part of the screen is again divided into several segments that display values related to and belonging to specific elements of the designed complex mechatronic system for positioning photovoltaic panels.

The main part of the screen of the visualization environment is made up of three sections that belong to individual parts of the screen. The structure of individual sections is clearly located in individual parts of the given screen. In the left part of the screen, there are sections of the generated power; in the central part, there is a section of the visual display of the positioning device together with the outputs of the measured values, through which it is possible to display the rotation angles around the individual axes. In the right part, there is a section dedicated to the meteorological station. The resulting design of the main screen is illustrated in Figure 17.

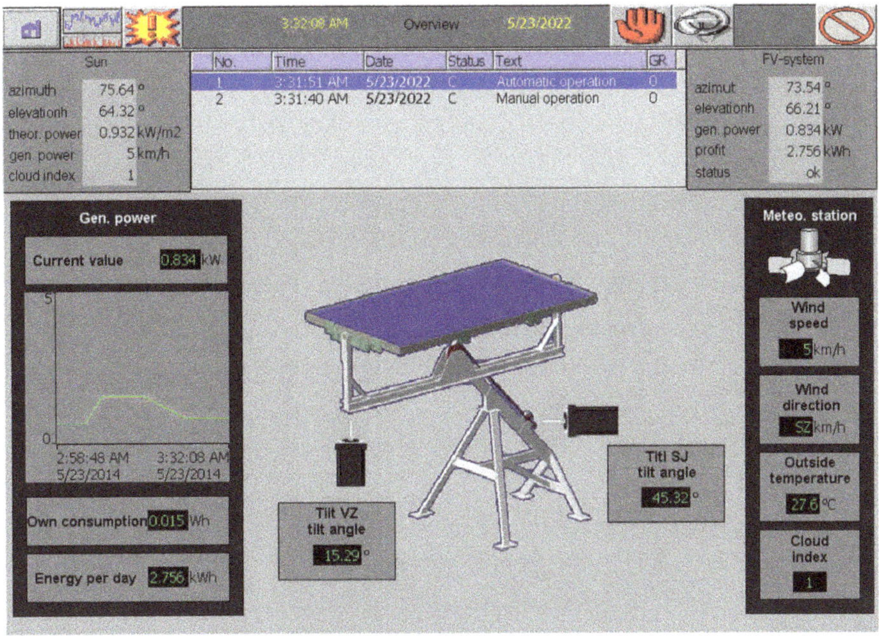

Figure 17. The main screen of the visualization environment.

3.4.3. Overview of Alarms

The operator of the complex mechatronic system for positioning photovoltaic panels must supervise the correct operation of the device. After evaluating any fault condition or warning message, the operator is informed about this fact in the form of a warning sign by writing a fault message in the corresponding light in the overview part of the screen. However, as stated above, only active fault or warning messages are displayed in this screen (Figure 18). After they have passed, the message from this window disappears; therefore, it is important to provide the operator with a retrospective evaluation of faults.

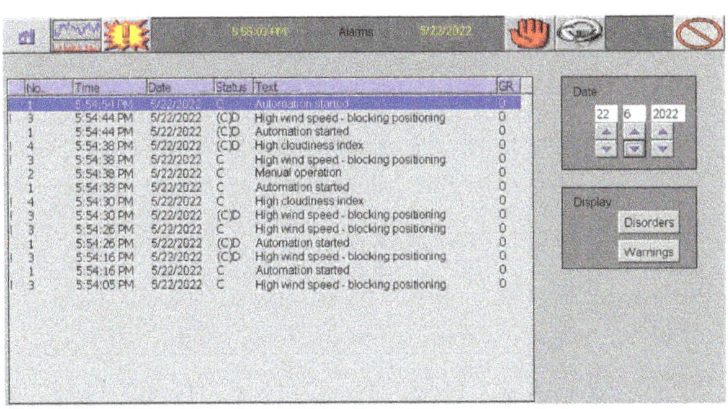

Figure 18. Alarm messages screen.

For the retroactive evaluation of fault conditions as well as warning messages, it is necessary to create a screen within the visualization environment that provides an overview of fault messages and their history. Within this screen, the operator must also have the option to display these fault and warning messages for a specified day in the past. The size of the volume of data that can be retrospectively evaluated is given by the database system, which, however, is not the included in this work.

3.4.4. Graphic Evaluation

The design of the visualization environment allows the operator, in addition to observing the current values, the possibility to observe the behavior of the determined variables for the selected period. The screen dedicated to the graphical evaluation of the process variables allows the operator to observe the selected variables using the progress in the graph. The operator has the option of looking back in time through the mentioned graph. However, to ensure the desired selection, the screen must also contain control elements that enable the operator to perform the given action (Figure 19).

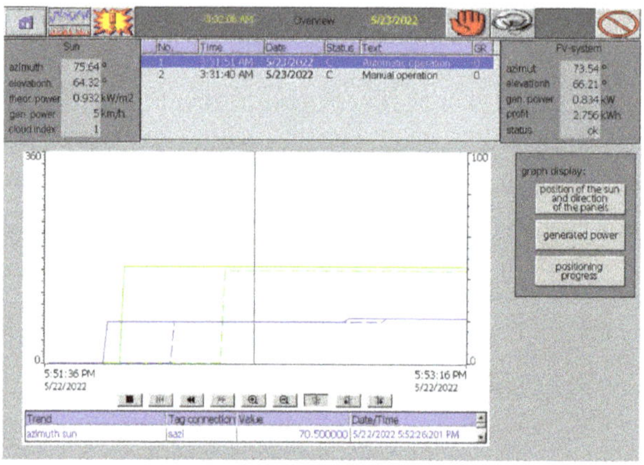

Figure 19. Graphical evaluation screen of process variables.

3.4.5. Manual Operation Control Panel

The device operator has the possibility to influence the positioning device based on a specific request. In practice, this means that the operator has the option to choose the manual operating mode of the control system, during which the automatic positioning of the photovoltaic panels to the optimal direction is not carried out. During this selected mode of operation, the operator has the possibility to perform any positioning of the panel (Figure 20). The need for this option is because maintenance and service operations may occur during operation, when the operator must be able to ensure the positioning of the system in the selected position to perform these operations.

The proposed solution of the visualization environment enables manual operation to be invoked using the header, which contains a button for activating the manual operation of the positioning control system. After pressing the button, manual operation is immediately selected, and the manual operation controller window is displayed. Starting manual control is possible from any screen. The user has the option to hide one of the four predefined positions, which can be achieved by pressing the corresponding button. These predefined positions make it easy for the operator to recall the most frequently used positions.

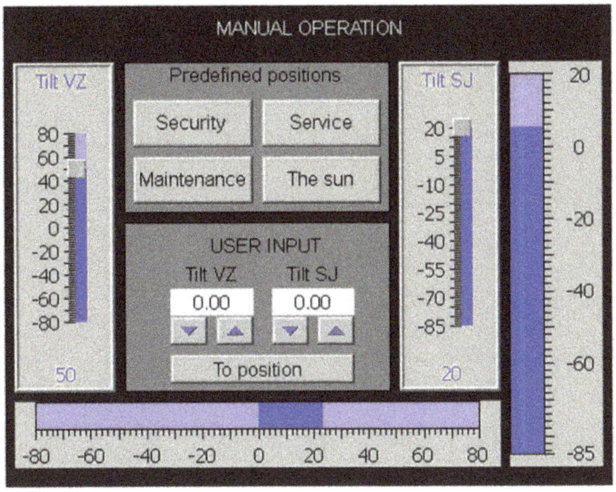

Figure 20. Manual operation controller.

4. Results

This section presents the results of the simulation experiments with the simulation model developed in the previous section and its evaluation. The evaluation process as well as the simulation calculation are performed in the MATLAB program environment, which assists with the calculations, further graphical evaluation of the positioning process and the evaluation of the energy balance. For the overall evaluation of the results of the complex mechatronic system for positioning photo-voltaic panels in energy applications, it was necessary to perform several partial analyses.

The first analysis included in the corresponding calculations is the analysis of the movement of the Sun during the year, while simultaneously evaluating the incident solar energy on the optimally oriented plane. The plane is optimally oriented when its normal vector points exactly to the Sun, the incident rays are perpendicular and there is maximum efficiency in converting solar energy into electrical energy. The value given in this analysis tells us the amount of incident energy per year in a given area (Figure 21). The calculations start from the geographical position and a specific time, which determines the position of the Sun. Based on this position, it is possible to determine the atmospheric effect coefficient for specific input parameters. The size of this atmospheric effect coefficient determines the amount of solar energy passing through the atmosphere without absorption to the Earth's surface (Figure 22). Its size has a fundamental influence on the solar energy falling on the optimally oriented plane on the surface of the Earth.

Another part of the analysis concerns the design of the positioning algorithm for the photovoltaic panel, which means that in the computing environment, the course of the positioning process is simulated throughout the year based on the trajectories of the Sun. Then, the amount of energy incident on such an inclined plane can be evaluated.

When simulating the positioning algorithm of a mechatronic system designed for power systems, it is necessary to also simultaneously evaluate the power consumption required for this positioning with the panel positioning. The energy expended in positioning the photovoltaic panel must be as small as possible if the energy gain is to be ensured compared to the fixed panels.

The evaluation of the achieved results is based on a simulation, which will ensure the recalculation of values throughout the year by minutes based on the position of the Sun, thus determining the light parts of the day. Calculations of incident energy as well as a positioning algorithm are subsequently applied to the data that determine the position of the Sun during the day. Simultaneously with the evaluation of these factors, the energy incident on the fixed-oriented panel is also evaluated, based on which we have to determine the energy gain.

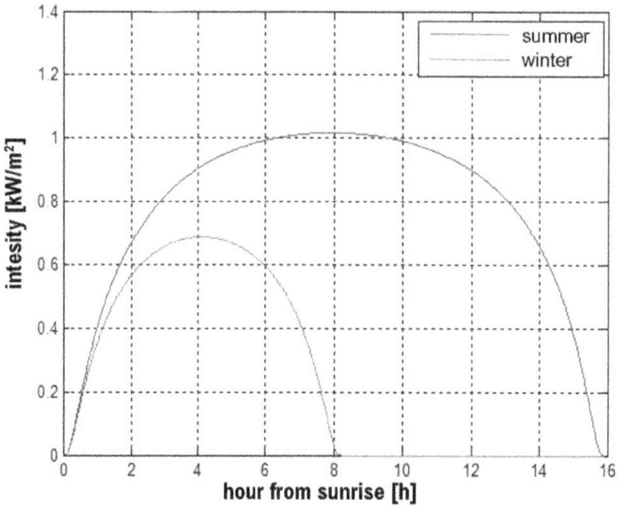

Figure 21. The daily course of solar radiation intensity during the days of the solstice.

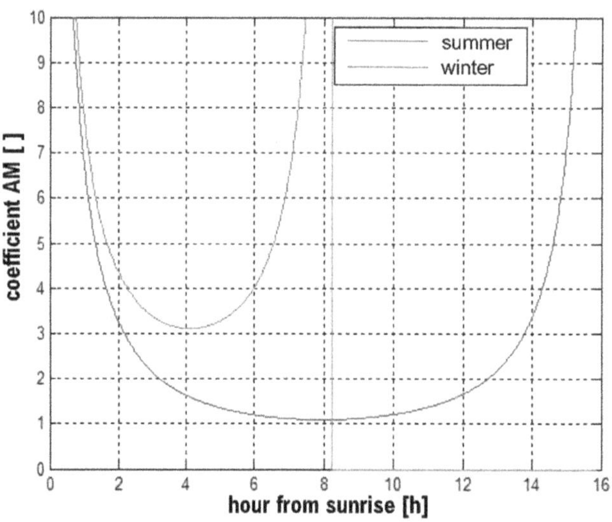

Figure 22. Daily course of the atmospheric effect coefficient during the days of the solstice.

When sunlight hits the inclined surface of the panel at a certain angle, reflection occurs, while the intensity of the incident radiation on the deviated plane is reduced by the corresponding values. The intensity of the radiation incident during the year on a panel positioned by the system or fixed can be calculated based on the above-mentioned influence. The effect of this decrease in intensity is given by the size of the deviation angles in individual directions that express the position of the Sun. The calculation used to evaluate the intensity incident on the I_P panel (19) is based on the intensity of the radiation incident on the optimally inclined plane and the relationship describing the angle difference (Figures 23 and 24).

$$I_P = I_G(\cos(h)\sin(\beta)\cos(\gamma - \gamma_s) + \sin(h)\cos(\beta)), \quad (19)$$

in which: I_G—intensity of radiation in a semiconductor $[\frac{kW}{m^2}]$; h—the elevation angle of the Sun [°]; β—panel tilt [θ]; γ—azimuth of the panel [θ]; γ_s—azimuth of the Sun [θ].

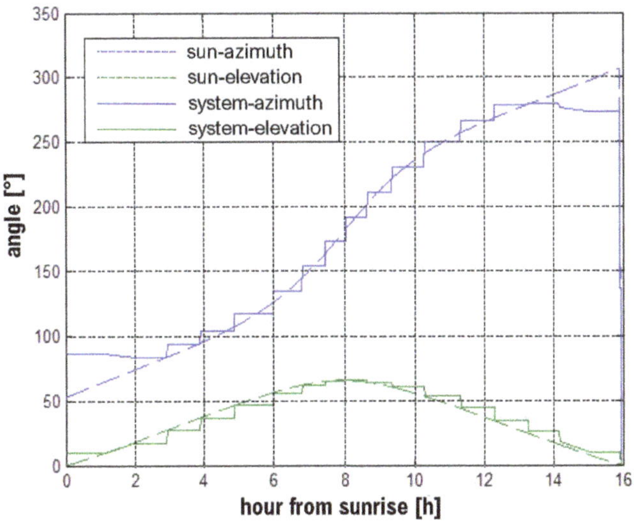

Figure 23. Daily course of system positioning—summer solstice.

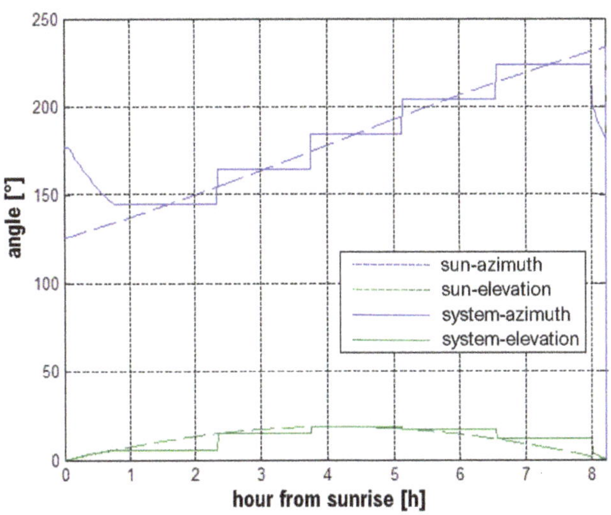

Figure 24. Daily course of system positioning—winter solstice.

By simulating individual daily trends, we can determine the gains for individual days of the year. However, here we are still talking about the potential generated power under the mentioned conditions and without considering the energy input to the positioning system. The annual course of the generated energy of the photovoltaic energy system in positioning and fixed implementation is shown in Figure 25.

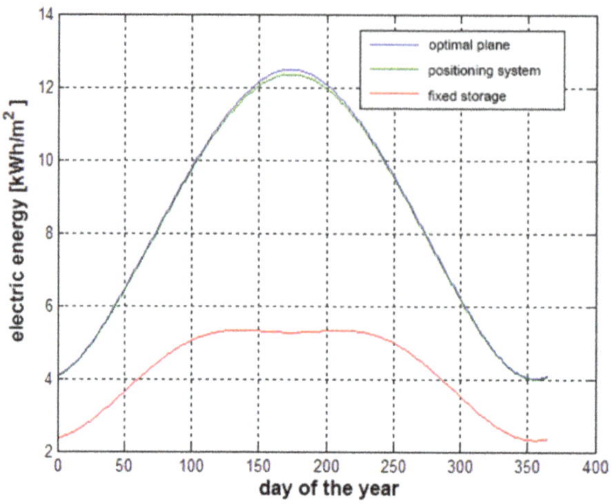

Figure 25. Potentially generated energy per day.

The potentially generated energy depicted in Figure 25 shows the benefit of the positioning device over fixed storage, with the positioning system ensuring the maximization of the potential amount of energy that can be produced. A significant increase can be observed in the summer season, due to the mentioned effects of the azimuth at sunrise and sunset, which is behind the plane of the fixed panel and there are significant deviations between the angles of the position of the Sun and the normal vector of the panel during the day. This problem is completely solved by the designed complex mechatronic system for positioning the photovoltaic panel.

However, to ensure the optimal position that maximizes the generated energy, the photovoltaic panel positioning system must spend a certain amount of energy on position changes during the year. We can express the electrical energy spent on the positioning of the system based on the knowledge of the drive system, as well as its operation time throughout the year. We can express the operating time based on the control frequency of stepper motors with a given nominal power, as well as the motors that operate the brake release mechanism on both axes. The parameters of the propulsion system are in Table 5. Based on these parameters, the results of the annual engine run analysis, as well as the calculated total energy input, are determined and summarized in Table 6.

Table 5. Parameters of the propulsion system.

Rated Engine Power	Rated Brake Power
8.96 W	15 W

Table 6. Evaluation of the energy deposit.

Number of Motor Steps	Drive Operation Time	Energy Input
837,219 steps/year	23.3 h/year	0.9 kWh/year

In the summary evaluation, the listed values are obtained from the analysis of the results and provide an illustrative example of the potential of using the positioning device in applications with one panel on the roofs of buildings. It should be noted that it is valid for the entire range of latitude as well as its surroundings, given the geographical location of Bratislava. Evaluation of the work results is summarized in Tables 7 and 8 with the consumption of both stepper motors. The visualization is not constantly displayed; it is only an information panel that starts when the user intervenes. It is similar a system add-on; therefore, its consumption is not included in the evaluation.

Table 7. Comprehensive results for a real photovoltaic panel (conversion efficiency: 17%, active area: 1.6 m^2).

Energy Falling on an Optimally Inclined Plane	Potential of Generated Energy—Positioning System	Potential of Generated Energy—Fixed System
3083 kWh/m^2/year	3057 kWh/m^2/year	1544 kWh/m^2/year

Table 8. General results.

Theoretical Generated Energy—Positioning System	Theoretical Generated Energy	Energy Deposit—Positioning System
832 kWh/m^2/year	420 kWh/m^2/year	0.9 kWh/m^2/year

The energy gain of the positioning system compared to the fixed system can be expressed as follows:

$$832 - 420 - 0.9 = 411.1 \text{ kWh/m}^2 \text{ per year (98\%)}.$$

The results show the theoretical value of the energy gain of the energy system using a positioning device compared to a fixed device. It must be noted, however, that it is given on the assumption of cloudless days and without the influence of diffuse radiation, which would reduce the total stated energy gain, because its influence is not included in the calculations. The achieved results indicate a significantly higher potential gain of generated energy compared to the commercially available solutions, e.g., [24,25].

All the requirements from Section 3.1 can be considered as fulfilled. Requirements 1 and 2 are fulfilled by the appropriate selection of the stepper motor. With the structural design solution, we ensured requirements 3, 4, 5, 6, 7, 9 and 10. By equipping the system with a tilt sensor, we also contributed to the verification of robustness (requirement 3). By choosing a PLC, the variability of the connection of several systems was ensured (requirement 8).

5. Conclusions

The result of the work is the design of a complex positioning device for photovoltaic panels that ensures the optimal inclination of the panel to achieve the maximum energy gain of the system. The proposal consists of a functional 3-D model of the structural solution in the CATIA program environment, a specific drive system in the form of real stepper motors that ensure optimal positioning based on the PLC control system, which is intended for the given proposal. This PLC did not have any influence on the energy gain calculations in the previous section.

The control algorithm that delivers the function of the positioning device is solved by a program written in the STEP7 environment, which ensures the operation of the entire mechatronic design, i.e., mainly the positioning of the photovoltaic panels to the optimal position. The course of the overall control, as well as the energy process, can be monitored and influenced through the visualization of the given process.

The visualization is solved using the WINCC program and provides the corresponding functions to the operator of the energy process. The conclusion of the work is the evaluation of the results achieved with the designed positioning system, ensuring the optimal gain of electrical energy. Based on the analyses, the results of the energy gain of the designed solution are determined against the energy gain from the fixed installation system, while the theoretical value of the gain of the designed positioning device is 98% compared to the fixed solution. However, the value would be smaller in practice, because the analysis does not include the influence of diffuse radiation. The presented complex design of the positioning PV system is believed to compete with the existing designs, for example see [26].

The next step in the future work would be to build a physical device, which we would test in real conditions and compare with the theoretical design.

Author Contributions: Conceptualization, F.Ž. and J.C.; methodology, J.C.; software, F.Ž.; validation, F.Ž., O.H. and E.K.; formal analysis, E.K. and D.R.; investigation, J.C.; resources, J.C., F.Ž. and D.R.; data curation, J.C. and E.K.; writing—original draft preparation, F.Ž. and J.C.; writing—review and editing, J.C. and D.R.; visualization, F.Ž.; supervision, J.C. and D.R.; project administration, J.C.; funding acquisition, J.C. All authors have read and agreed to the published version of the manuscript.

Funding: This research received no external funding.

Data Availability Statement: Data sharing is not applicable to this article.

Acknowledgments: The work has been supported by the Slovak Grant Agency VEGA 1/0107/22, KEGA 039STU-4/2021, Scientific Grant APVV-21-0125 and by the Operational Program Integrated Infrastructure for the project: "Research and development of the applicability of autonomous flying vehicles in the fight against the pandemic caused by COVID-19", Project no. 313011ATR9, co-financed by the European Regional Development Fund and by. Special thanks go to our student Ing. Miroslav Šimončič for his cooperation in the fulfilment of software tasks.

Conflicts of Interest: The authors declare no conflict of interest.

References

1. Our World in Data (2022)—Our World in Data based on BP Statistical Review of World Energy. Available online: https://ourworldindata.org/grapher/electricity-prod-source-stacked (accessed on 20 February 2023).
2. Saravanan, S.; Ramesh Babu, N. Maximum power point tracking algorithms for photovoltaic system—A review. *Renew. Sustain. Energy Rev.* **2016**, *57*, 192–204. [CrossRef]
3. Alboteanu, L.; Manolea, G.; Ravigan, F. Positioning systems for solar panels placed in isolated areas. *Ann. Univ. Craiova Electr. Eng.* **2006**. Available online: https://www.researchgate.net/publication/228860210_Positioning_systems_for_solar_panels_placed_in_isolated_areas/citations#fullTextFileContent (accessed on 20 February 2023).
4. Rhif, A. A Position Control Review for a Photovoltaic System: Dual Axis Sun Tracker. *IETE Tech. Rev.* **2011**, *28*, 479. [CrossRef]
5. Maish, A.B. Performance of a self-aligning solar array tracking controller. In Proceedings of the IEEE Photovoltaic Specialists Conference, Kissimimee, FL, USA, 21–25 May 1990.
6. Enslin, J.H.R. Maximum power point tracking: A cost saving necessity in solar systems. *Renew. Energy* **1992**, *2*, 543–549. [CrossRef]
7. Brown, D.G.; Stone, K.W. High accuracy/low cost tracking system for solar concentrators using a neural network. In Proceedings of the 28th Intersociety Energy Conversion Engineering Conference, Atlanta, GA, USA, 8–13 August 1993.
8. Kalogirou, S.A. Design and construction of a one-axis sun-tracking system. *Sol. Energy* **1996**, *57*, 465–469. [CrossRef]
9. Khalifa, A.N.; Al-Mutawalli, S.S. Effect of two-axis sun tracking on the performance of compound parabolic concentrators. *Energ. Convers. Manag.* **1998**, *39*, 1073–1079. [CrossRef]
10. Falbel, G.; Puig-Suari, J.; Peczalski, A. Sun oriented and powered, 3 axis and spin stabilized cubesats. In Proceedings of the IEEE Aerospace Conference, Big Sky, MT, USA, 9–16 March 2002.
11. Al-Mohamad, A. Efficiency improvements of photo-voltaic panels using a sun-tracking system. *Appl. Energy* **2004**, *79*, 345–354. [CrossRef]
12. Abdallah, S. The effect of using sun tracking systems on the voltage-current characteristics and power generation of flat plate photovoltaics. *Energy Convers. Manag.* **2004**, *45*, 1671–1679. [CrossRef]
13. Aiuchi, K.; Yoshida, K.; Katayama, Y.; Nakamura, M.; Nakamura, K. Sun tracking photo-sensor for solar thermal concentrating system. In Proceedings of the International Solar Energy Conference, Portland, OR, USA, 11–14 July 2004.
14. Blanco-Muriel, M.; Alarcón-Padilla, D.C.; López-Moratalla, T.; Lara-Coira, M. Computing the solar vector. *Sol. Energy* **2001**, *70*, 431–441. [CrossRef]
15. Abdallah, S.; Nijmeh, S. Two axes sun tracking system with PLC control. *Energy Convers. Manag.* **2004**, *45*, 1931–1939. [CrossRef]
16. Reda, I.; Andreas, A. Solar position algorithm for solar radiation applications. *Sol. Energy* **2004**, *76*, 577–589. [CrossRef]
17. Chen, F.; Feng, J.; Hong, Z. Digital sun sensor based on the optical vernier measuring principle. *Meas. Sci. Technol.* **2006**, *17*, 2494–2498. [CrossRef]
18. Chen, F.; Feng, J. Analogue sun sensor based on the optical nonlinear compensation measuring principle. *Meas. Sci. Technol.* **2007**, *18*, 2111–2115. [CrossRef]

19. Grena, R. An algorithm for the computation of the solar position. *Sol. Energy* **2008**, *82*, 462–470. [CrossRef]
20. Chong, K.K.; Wong, C.W. General formula for on-axis sun tracking system and its application in improving tracking accuracy of solar collector. *Sol. Energy* **2009**, *83*, 298–305. [CrossRef]
21. Gasparac, I.; Vrazic, M.; Petrovic, I. Expected and measured values of luminance sensors in dual-axis photovoltaic positioning system. In Proceedings of the 2nd IMEKO TC 11 International Symposium Metrological Infrastructure, Dubrovnik Riviera, Croatia, 15–17 June 2011.
22. Bahaidarah, H.M.; Tanweer, B.; Gandhidasan, P.; Rehman, S. A Combined Optical, Thermal and Electrical Performance Study of a V-Trough PV System—Experimental and Analytical Investigations. *Energies* **2015**, *8*, 2803–2827. [CrossRef]
23. Plachta, K. Autonomous Tracking Controller for Photovoltaic Systems Using Global Positioning System. In Proceedings of the 2018 IEEE International Conference on Environment and Electrical Engineering and 2018 IEEE Industrial and Commercial Power Systems Europe (EEEIC/I&CPS Europe), Palermo, Italy, 12–15 June 2018; pp. 1–5. [CrossRef]
24. Etop Alternative Energy, Active Construction for Photovoltaic Panels Sunflower-SF44V2P. Available online: https://etopalternativeenergy.sk/produkt/aktivna-konstrukcia-na-fotovoltaicke-panely-sunflower-sf44v2p/ (accessed on 12 February 2023).
25. Amazon, ECO-WORTHY Solar Panel Dual Axis Tracking System (Increase 40% Power) with Tracker Controller, Complete Solar Tracker Kit, Ideal for Different Solar Panels, for Yard/Farm/Field. Available online: https://www.amazon.com/ECO-WORTHY-Dual-Solar-Tracking-System/dp/B00JYAIS9W?th=1 (accessed on 12 February 2023).
26. Seme, S.; Štumberger, B.; Hadžiselimović, M.; Sredenšek, K. Solar Photovoltaic Tracking Systems for Electricity Generation: A Review. *Energies* **2020**, *13*, 4224. [CrossRef]
27. Lincot, D. The new paradigm of photovoltaics: From powering satellites to powering humanity. *Comptes Rendus Phys.* **2017**, *18*, 381–390. Available online: https://www.sciencedirect.com/science/article/pii/S1631070517300555 (accessed on 12 February 2023). [CrossRef]
28. Goetzberger, A.; Hebling, C.; Schock, H.-W. Photovoltaic materials, history, status and outlook. *Mater. Sci. Eng. R Rep.* **2003**, *40*, 381–390. Available online: https://www.sciencedirect.com/science/article/pii/S0927796X0200092X?casa_token=cbw-ZIYcs8wAAAAA:_g-vYcOOjCnDRxDySeJzYzBa-n7Ws8l_qbdq_1WwZokkG_dsIRxcP1wZ_v9DlBmPknP66T2ETw (accessed on 15 February 2023). [CrossRef]
29. Schaller, R.D.; Sykora, M.; Pietryga, J.M.; Klimov, V.I. Seven Excitons at a Cost of One: Redefining the Limits for Conversion Efficiency of Photons into Charge Carriers. *Nano Lett.* **2006**, *6*, 424–429. [CrossRef] [PubMed]
30. Alkasrawi, M.; Abdelsalam, E.; Alnawafah, H.; Almomani, F.; Tawalbeh, M.; Mousa, A. Integration of Solar Chimney Power Plant with Photovoltaic for Co-Cooling, Power Production, and Water Desalination. *Processes* **2021**, *9*, 2155. [CrossRef]
31. Welons, M. The Stefan-Boltzmann Law, College of Wooster Physics Department, Ohio, USA. 2007. Available online: https://physics.wooster.edu/wp-content/uploads/2021/08/Junior-IS-Thesis-Web_2007_Wellons.pdf (accessed on 15 February 2023).
32. Ramírez-Moreno, M.A.; González-Hernández, S.; Angulo-Brown, F. The role of the Stefan–Boltzmann law in the thermodynamic optimization of an n-Müser engine. *Phys. A Stat. Mech. Its Appl.* **2016**, *444*, 914–921. Available online: https://www.sciencedirect.com/science/article/pii/S037843711500953X (accessed on 15 February 2023). [CrossRef]
33. Honsberg, C.B.; Bowden, S.G. Photovoltaics Education Website. 2019. Available online: www.pveducation.org (accessed on 15 February 2023).
34. Wang, W. On the utilization and development of solar energy in Tibet. *E3S Web Conf.* **2021**, *329*. Available online: https://www.e3s-conferences.org/articles/e3sconf/abs/2021/105/e3sconf_gesd2021_01082/e3sconf_gesd2021_01082.html (accessed on 15 February 2023). [CrossRef]
35. Jenkins, J.H.; Fischbach, E.; Buncher, J.B.; Gruenwald, J.T.; Krause, D.E.; Mattes, J.J. Evidence of correlations between nuclear decay rates and Earth–Sun distance. *Astropart. Phys.* **2009**, *32*, 42–46. Available online: https://www.sciencedirect.com/science/article/pii/S092765050900084X?casa_token=V9abI0tmyM0AAAAA:nAFM-xdapi1Ry1AyPsQ_matymOchIklzvX4NT7JrTdJkra9CLvhvc-souZSA41pA45c7ocTONw (accessed on 18 February 2023). [CrossRef]
36. Murphy, D.M. Effect of Stratospheric Aerosols on Direct Sunlight and Implications for Concentrating Solar Power. *Environ. Sci. Technol.* **2009**, *43*, 2784–2786. [CrossRef] [PubMed]
37. Attivissimo, F.; Lanzolla, A.M.L.; Passaghe, D.; Paul, M.; Gregory, D.; Knox, A. Photovoltaic-thermoelectric modules: A feasibility study. In Proceedings of the 2014 IEEE International Instrumentation and Measurement Technology Conference (I2MTC) Proceedings, Montevideo, Uruguay, 12–15 May 2014; pp. 659–664.
38. ASC Chemistry for Life, Energy from the Sun. Available online: https://www.acs.org/content/acs/en/climatescience/energybalance/energyfromsun.html (accessed on 18 February 2023).
39. Wang, Q.; Li, S.; Jiang, F. Uncovering the impact of the COVID-19 pandemic on energy consumption: New insight from difference between pandemic-free scenario and actual electricity consumption in China. *J. Clean. Prod.* **2021**, *313*, 127897. [CrossRef]
40. Hofierka, J.; Cebecauer, T. Spatially distributed assessment of solar resources for energy applications in Slovakia. *Acta Fac. Studiorum Humanit. Nat. Univ. Prešoviensis. Prírodné Vedy Folia Geogr.* **2008**, *12*, 97–114.
41. Nijegorodov, N.; Luhanga, P.V.C. Air mass: Analytical and empirical treatment; an improved formula for air mass. *Renew. Energy* **1996**, *7*, 57–65. [CrossRef]
42. Li, Y.; He, C.; Zuo, L.; Zhao, F.; Zhan, L.; Li, X.; Xia, R.; Yip, H.; Li, C.; Liu, X.; et al. High-Performance Semi-Transparent Organic Photovoltaic Devices via Improving Absorbing Selectivity. *Adv. Energy Mater.* **2021**, *11*, 2003408. [CrossRef]

43. Landis, G.A.; Vo, T. Photovoltaic performance in the Venus environment. In Proceedings of the 2009 34th IEEE Photovoltaic Specialists Conference (PVSC), Philadelphia, PA, USA, 7–12 June 2009.
44. Karakaya, E. Finite Element Method for forecasting the diffusion of photovoltaic systems: Why and how? *Appl. Energy* **2016**, *163*, 464–475. [CrossRef]
45. Nayak, P.; Garcia-Belmonte, G.; Kahn, A.; Bisquert, J.; Cahen, D. Photovoltaic efficiency limits and material disorder. *Energy Environ. Sci.* **2012**, *5*, 6022–6039. [CrossRef]
46. Matuška, T. Solar thermal systems (MP 1.6.11). 2012. Available online: https://profesis.ckait.cz/dokumenty-ckait/mp-1-6/mp-1-6-11/ (accessed on 18 February 2023).
47. Reinders, A.; Verlinden, P.; van Sark, W. *Photovoltaic Solar Energy: From Fundamentals to Applications*; John Wiley & Sons: Hoboken, NJ, USA, 2017; p. 720.
48. Keevers, M.J.; Green, M.A. Absorption edge of silicon from solar cell spectral response measurements. *Appl. Phys. Lett.* **1995**, *66*, 174–176. [CrossRef]
49. Kira, M.; Jahnke, F.; Hoyer, W.; Koch, S.W. Quantum theory of spontaneous emission and coherent effects in semiconductor microstructures. *Prog. Quantum Electron.* **1999**, *23*, 189–279. [CrossRef]
50. University of Žilina, Faculty of Management Science and Informatics, Optoelectronic Components. Available online: https://www.kis.fri.uniza.sk/~ludo/e-Publikacia/elektronika/kap9/elektronika.html (accessed on 18 February 2023).
51. dos Santos, L.L.C.; Canha, L.N.; Bernardon, D.P. Projection of the diffusion of photovoltaic systems in residential low voltage consumers. *Renew. Energy* **2018**, *116*, 384–401. [CrossRef]
52. Mills, A.; Le Hunte, S. An overview of semiconductor photocatalysis. *J. Photochem. Photobiol. A Chem.* **1997**, *108*, 1–35. [CrossRef]
53. Pleskov, Y. *Semiconductor Photoelectrochemistry*; Springer: New York, NY, USA, 1986; ISBN 978-1-4684-9080-0.
54. Benda, V.; Cerna, L. A Note on Limits and Trends in PV Cells and Modules. *Appl. Sci.* **2022**, *12*, 3363. [CrossRef]
55. Obukhov, S.G.; Plotnikov, I.A.; Sheryazov, S.K. Methods of effective use of solar power system. In Proceedings of the 2016 2nd International Conference on Industrial Engineering, Applications and Manufacturing (ICIEAM), Chelyabinsk, Russia, 19–20 May 2016.
56. Emery, K.A. Solar simulators and I–V measurement methods. *Sol. Cells* **1986**, *18*, 251–260. [CrossRef]
57. Kang, R.; Jeong, T.; Lee, B. Lead-Free Perovskite and Improved Processes and Techniques for Creating Future Photovoltaic Cell to Aid Green Mobility. 2022. Available online: https://www.intechopen.com/online-first/84068 (accessed on 18 February 2023).
58. Jäger-Waldau, A. *Overview of the Global PV Industry, Reference Module in Earth Systems and Environmental Sciences*; Elsevier: Amsterdam, The Netherlands, 2013; ISBN 9780124095489.
59. Alternative Energy Tutorials, Solar Panel Orientation—PV Solar Panel Orientation and Positioning. 2015. Available online: https://www.alternative-energy-tutorials.com/solar-power/solar-panel-orientation.html (accessed on 18 February 2023).
60. Walraven, R. Calculating the position of the Sun. *Sol. Energy* **1978**, *20*, 393–397. [CrossRef]
61. Energy Education, Solar Panel Orientation. 2018. Available online: https://energyeducation.ca/wiki/index.php?title=Solar_panel_orientation&oldid=6109 (accessed on 21 February 2023).
62. Marsh, J. What's the Best Direction and Angle for My Solar Panels? Available online: https://news.energysage.com/solar-panel-performance-orientation-angle/ (accessed on 21 February 2023).
63. Božiková, M.; Bilčík, M.; Madola, V.; Szabóová, T.; Kubik, L'.; Lendelová, J.; Cviklovič, V. The Effect of Azimuth and Tilt Angle Changes on the Energy Balance of Photovoltaic System Installed in the Southern Slovakia Region. *Appl. Sci.* **2021**, *11*, 8998. [CrossRef]
64. Catia. Available online: https://www.3ds.com/products-services/catia/ (accessed on 21 February 2023).
65. MSC Adams. Available online: https://hexagon.com/products/product-groups/computer-aided-engineering-software/adams (accessed on 23 February 2023).
66. Matlab. Available online: https://www.mathworks.com/products/matlab.html (accessed on 21 February 2023).
67. Siemens Simatic Step7. Available online: https://new.siemens.com/global/en/products/automation/industry-software/automation-software/tia-portal/software/step7-tia-portal.html (accessed on 23 February 2023).
68. Siemens WinCC. Available online: https://new.siemens.com/global/en/products/automation/industry-software/automation-software/scada/simatic-wincc-v7.html (accessed on 23 February 2023).

sensors

MDPI

Communication

Improving the Output Efficiency of Triboelectric Nanogenerator by a Power Regulation Circuit

Wenbo Li [1,2], Baichuan Leng [1], Shengyu Hu [1] and Xiaojun Cheng [1,2,*]

[1] Beijing Institute of Nanoenergy and Nanosystems, Chinese Academy of Sciences, Beijing 101400, China; liwenbo@binn.cas.cn (W.L.); lengbaichuan@binn.cas.cn (B.L.); hushengyu@binn.cas.cn (S.H.)

[2] School of Nanoscience and Technology, University of Chinese Academy of Sciences, Beijing 100049, China

* Correspondence: chengxiaojun@binn.cas.cn

Abstract: Triboelectric nanogenerator (TENG) is a promising technology for harvesting energy from various sources, such as human motion, wind and vibration. At the same time, a matching backend management circuit is essential to improve the energy utilization efficiency of TENG. Therefore, this work proposes a power regulation circuit (PRC) suitable for TENG, which is composed of a valley-filling circuit and a switching step-down circuit. The experimental results indicate that after incorporating a PRC, the conduction time of each cycle of the rectifier circuit doubles, increasing the number of current pulses in the TENG output and resulting in an output charge that is 1.6 fold that of the original circuit. Compared with the initial output signal, the charging rate of the output capacitor increased significantly by 75% with a PRC at a rotational speed of 120 rpm, significantly improving the utilization efficiency of the TENG's output energy. At the same time, when the TENG powers LEDs, the flickering frequency of LEDs is reduced after adding a PRC, and the light emission is more stable, which further verifies the test results. The PRC proposed in this study can enable the energy harvested by the TENG to be utilized more efficiently, which has a certain promoting effect on the development and application of TENG technology.

Keywords: triboelectric nanogenerator; power regulation circuit; performance improvement; conduction time; two-stage output

1. Introduction

Energy is an indispensable factor for social development. With the increasing severity of the energy crisis and the increasing requirements for environmental protection, the demand for sustainable and efficient energy conversion and storage technologies is becoming more and more urgent [1–3]. In recent years, the innovation of new energy and the utilization of environmental energy have been research hotspots, which are necessary for energy storage and optimization of related renewable energy [4,5]. As a promising energy harvesting technology, triboelectric nanogenerator (TENG) has attracted considerable attention due to its sustainability and low cost [6,7], as it has been proven capable of harvesting energy from various sources, including human motion, wind, vibration and water currents, which makes it highly versatile in practical applications [8–12]. At the same time, the application directions of TENG are also very wide, covering smart phones, wearable devices, wireless sensor networks, fitness equipment and other fields [13]. It also has broad application prospects in environmental monitoring, intelligent transportation, smart home, medical and health fields [14,15].

Despite its well-established theoretical basis and significant advantages, the TENG faces several challenges that hinder its efficiency [16,17]. Furthermore, stability remains a significant problem in practical applications [18–20]. Therefore, in-depth research on the performance and mechanism of TENG is essential for promoting its practical application [21,22]. Scholars have explored and conducted experimental research, including theoretical analysis and derivation, material selection and preparation, structure design and optimization, as well as performance testing and analysis, to address the challenges related to the TENG [23–27]. In addition to studying TENG's characteristics, developing a power management circuit that matches the output of TENG is crucial to improving its efficiency [28,29]. Hu et al. reported a simple and adjustable automatic spark switch circuit, which achieved energy accumulation and rapid release [30]. Cheng et al. proposed a dual-loop circuit, reducing the energy loss in the diode to 47.1% [31]. Zhang et al. demonstrated a universal power management circuit that achieved an efficiency of 80% in matching the impedance of a TENG [32]. A power management circuit can convert the TENG's alternating current (AC) output into a direct current (DC) output, which can power electronic devices or store energy in batteries or capacitors [33,34]. Maximizing harvested energy by increasing energy conversion efficiency is a significant challenge in TENG technology [35]. A suitable power management circuit can effectively enhance the TENG's output performance, playing a vital role in promoting its practical application.

In this paper, a power regulation circuit (PRC) suitable for TENG is proposed, which is composed of a first-stage valley-filling circuit and a second-stage switching step-down circuit, and its electrical performance is tested by a typical freestanding-mode TENG. Unlike common power management circuits that use switch circuits to reduce the duty cycle of the TENG, a PRC increases the conduction time of the TENG to increase the output charge and improve the energy utilization efficiency of the TENG. The experimental results show that adding the PRC increases the power supply time of the TENG in each cycle, thereby increasing the pulse number of the TENG output signal and the output charge is 1.6-fold higher than the initial value. Compared with the initial output signal, the charging rate of the output capacitor increases by 75%, improving the TENG's performance significantly. The stability of the PRC is further proved by frequency conversion charging experiments. At the same time, when the TENG supplies power to LEDs with a motion frequency of 2 Hz, the LEDs using the PRC flicker less than LEDs powered by direct rectification, the light is also more stable, further verifying the test results. The proposed PRC can efficiently utilize the output energy of the TENG, providing new insights into the storage and utilization of the TENG's energy.

2. Structure and Mechanism

Figure 1 presents a schematic diagram of a TENG and a PRC system, where Figure 1a illustrates the specific structure of the TENG. As the power supply segment of the PRC, the TENG is used to verify the applicability of the PRC for the TENG's power management and compared with the signal output after through the PRC. The TENG adopts a rotating freestanding mode, comprising a rotor, a shell and a shaft. The shell structure is made of an acrylic tube with a diameter of 94 mm, with six pairs of copper electrodes attached to its inner wall. The rotor is made of polylactide (PLA) material and uniformly inlaid with six pieces of fluorinated ethylene propylene (FEP) films, forming the power generation unit of the TENG. The TENG is driven by a rotating motor, and the output signals of the system are measured by a programmable electrometer and data acquisition system, which are processed and stored by LabVIEW software and computer. Figure 1b shows the schematic diagram of the PRC system, which consists of two-stage circuits. The first stage comprises a rectifier and a valley-filling circuit to increase the conduction time of the rectifier bridge during the TENG's unit motion cycle and enhance the initial output energy. The second-stage circuit steps down and stores the previous stage circuit to better match the output load. A PMOS transistor is used as a control switch, automatically turned on and off based on the magnitude of the first-stage output signal. To facilitate the experiment's debugging and component parameter determination, a PRC system was built on a breadboard, acting as a passive power management circuit directly connected to the TENG's output.

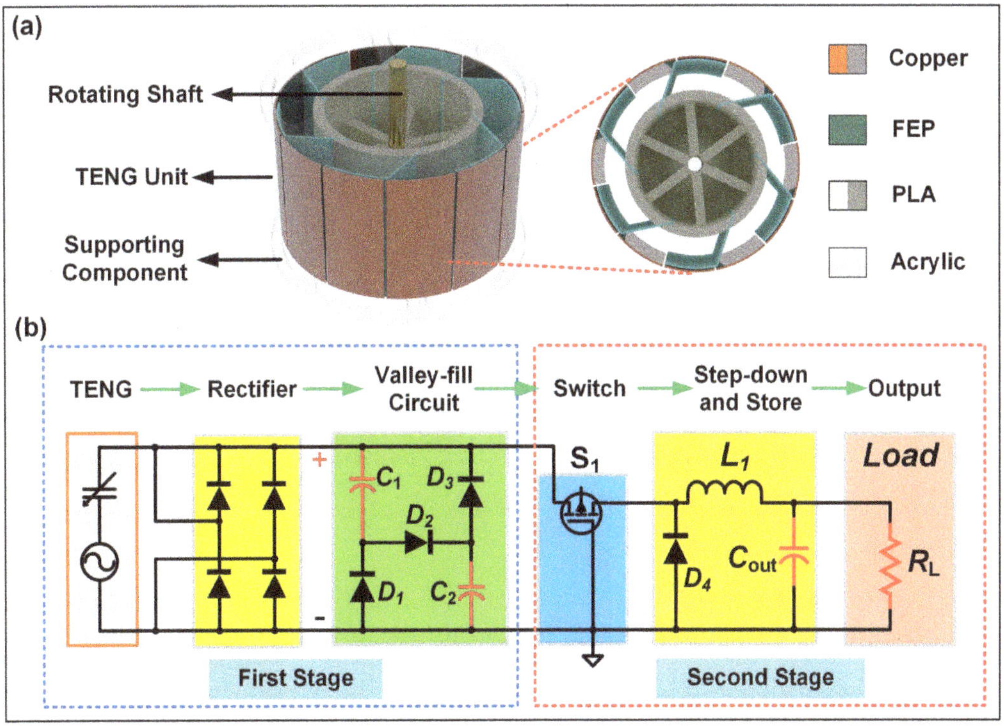

Figure 1. Design schematic diagram of the TENG and the PRC. (**a**) System structure of the TENG. (**b**) The circuit diagram of the PRC.

Figure 2 depicts the specific working principle of the TENG, which mainly comprises four processes. As presented in Figure 2a, according to the triboelectric sequence table, the FEP film surface easily gains electrons, resulting in a negative surface potential. Conversely, the copper electrode surface readily loses electrons, leading to a positive surface potential. In the initial state (Figure 2a(i)), the rotor remains stationary, and the FEP film is in full contact with copper electrode I. Consequently, contact electrification causes charge transfer, and both the copper electrode and the FEP film acquire equal amounts of positive and negative charges. When the FEP film is rotated to the state illustrated in Figure 2a(ii), it contacts both copper electrode I and copper electrode II, inducing electrostatic induction that drives electrons on electrode II towards electrode I, resulting in a secondary electrode under external load, and the current from electrode I to electrode II. As the rotor continues to move to the state displayed in Figure 2a(iii), the FEP film comes into complete contact with copper electrode II, and electrostatic equilibrium is reached, with no charge flowing. Subsequently, the rotor rotates to state iv (Figure 2a(iv)), under external load conditions, and a potential difference drives the generation of current in the opposite direction to the previous half-cycle, eventually returning to the initial state, initiating a new cycle. Therefore, continuous rotation of the TENG produces a continuous AC signal. Additionally, COMSOL was employed to simulate the electrostatic field, exhibiting the power generation mechanism of the TENG under different motion states (Figure 2b).

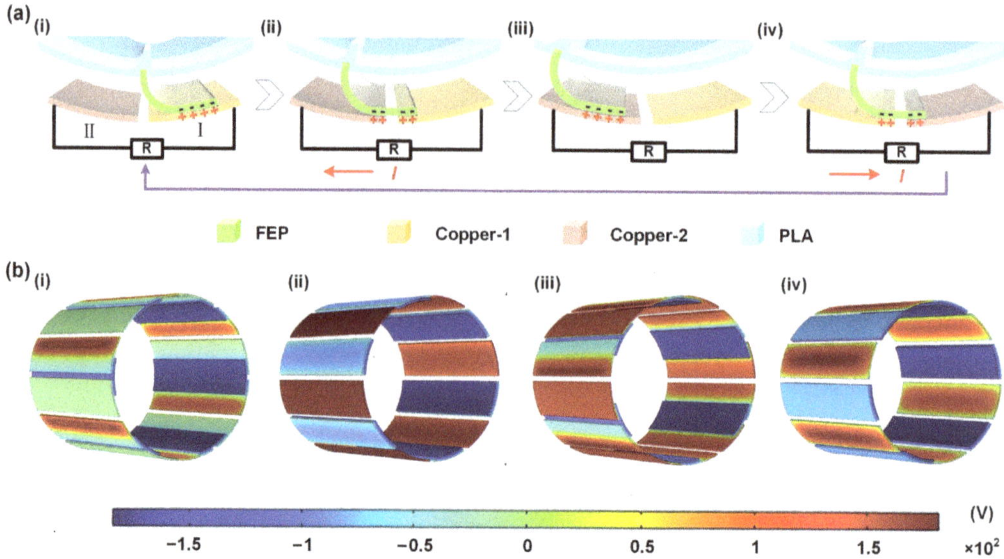

Figure 2. Principle analysis of the TENG: (**a**) The specific working principle of the TENG, (**i–iv**) the charge transfer relationship between the FEP film and the copper electrode when the TENG rotates; (**b**) (**i–iv**) simulation diagram of charge distribution in different states of the TENG.

Figure 3 illustrates the specific working process of the PRC. Figure 3a represents the first-stage circuit, which comprises a rectifier and a passive valley-filling circuit. To illustrate the functioning of the first-stage circuit, Figure 3b presents its detailed working principle. When the voltage of the TENG is high, the output current signal follows the path illustrated in Figure 3b(i). The current charges C_1 and C_2, and D_2 conducts while D_1 and D_3 are cut off. In contrast, when the voltage is low, the capacitors C_1 and C_2 are discharged through D_1 and D_3 while D_2 is cut off, as shown in Figure 3b(ii). Meanwhile, when selecting a capacitor, it must be ensured that the values of C_1 and C_2 are equal, and in order to match the internal resistance of the TENG, the selection of the capacitor should not be too large. The selected capacitor value is 0.1 μF. Furthermore, neglecting the voltage drop across the diodes, when the TENG output voltage attains V_m, the voltage drops across C_1 and C_2 are both 1/2 V_m. Subsequently, when the output voltage of the TENG drops to 1/2 Vm, the rectifier bridge diode is cut off. At this point, D_1 and D_3 conduct, and the two capacitors are connected in parallel to power the load. Compared to the basic rectifier circuit, the power supply after rectification needs to be higher than V_m to provide power to the external load. Therefore, as shown in Figure 3c, compared to the case of directly rectifying the output (Figure 3c(i)), the first-stage circuit increases the power supply time of the external load within one cycle, and due to the longer conduction time, only a small peak current is required to fulfill the power requirement within one cycle (Figure 3c(ii)). The second-stage output circuit, as illustrated in Figure 3d, comprises a PMOS switch and a step-down storage section. Figure 3e demonstrates the specific working process. Initially, when the output voltage Vi of the first stage reaches the threshold voltage of the switch S_1, that is, when the source voltage of the PMOS transistor is 2 to 3 V higher than the gate voltage, the energy is transferred from the first-stage circuit to the second-stage circuit, and S_1 is passed through the change in the voltage is automatically turned on and off, and assists the second-stage circuit to realize the voltage reduction, realizing a regulation process of the passive switch. The output current powers the capacitor C_{out} and the load via the inductor L_1, as depicted in Figure 3e(i). Conversely, when the voltage falls below V_i, the switch S_1 is tuned off, causing the energy stored in the inductor to be released to C_{out} and the load. At this point, D_4 is turned on, creating a closed loop, as shown in Figure 3e(ii). Ultimately, when the energy stored in C_{out} reaches a certain threshold, it becomes capable of supplying stable power to the output load, as displayed in Figure 3e(iii). By employing the aforementioned two-stage circuits, the PRC system can accomplish the entire working process.

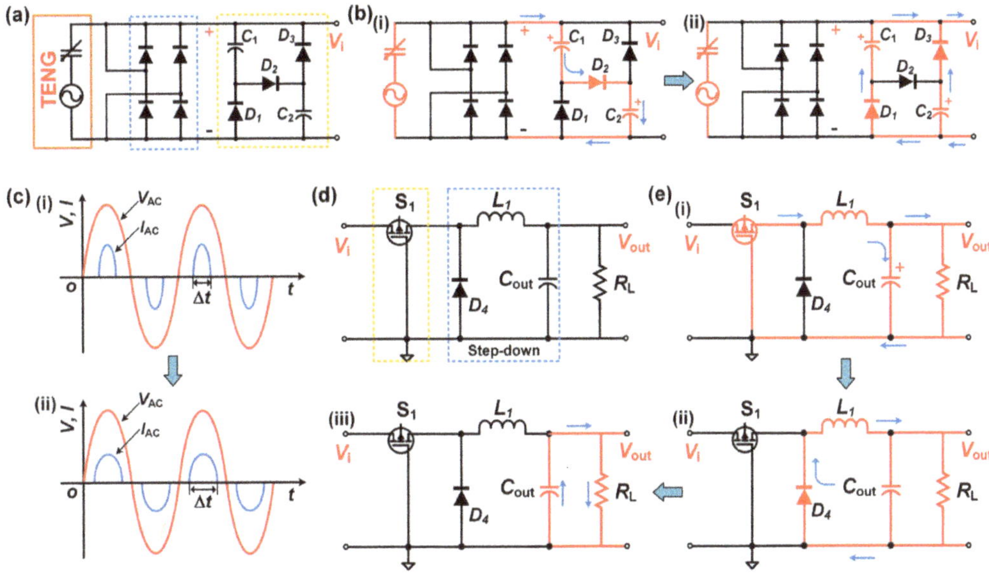

Figure 3. PRC-specific workflow: (**a**) the circuit diagram of the first stage in the PRC system; (**b**) the workflow of the first-stage circuit in the PRC, (**i**) series charging and (**ii**) parallel discharging of capacitors.; (c) the output mechanism diagram of the first-stage circuit in the PRC; (**d**) the circuit diagram of the second stage in the PRC system; (**e**) (**i–iii**) the workflow of the second-stage circuit within one cycle in the PRC.

3. Results and Discussion

3.1. Basic Output Performance

Figure 4 illustrates the fundamental performance tests of the TENG and the PRC. Specifically, Figure 4a(i–iii) exhibit the open-circuit voltage, short-circuit current, and transferred charge output signals of the TENG at different rotational speeds, respectively. Notably, the peak values of the open-circuit voltage and transferred charge remain nearly constant with increasing rotational frequency, whereas the short-circuit current gradually increases, having established the essential output performance and stable operating state of the TENG. Figure 4b(i,ii), respectively, demonstrate the short-circuit currents outputted by the TENG rectification and the TENG integrated with a first-stage circuit, at a consistent rotational speed of 120 rpm (unless stated otherwise, all subsequent experimental conditions were conducted at this rotational speed). The current pulse signals in Figure 4b(ii) appear to be more closely packed. To further elucidate the effect of the first-stage circuit in the PRC on the TENG output signal, Figure 4b(iii) compares the two sets of current signals in Figure 4b(i,ii) with amplification. It can be concluded that after the first-stage circuit, the number of pulse currents output by the TENG per unit time doubled, and the transferred charge amount increased by 60%. Although the peak current value slightly decreases, the accumulated energy output during the entire cycle shows a significant increase, which indicates that the first-stage circuit effectively increases the conduction time of the rectifier bridge diode of the TENG, leading to improved output power. It is worth noting that in the PRC system, the inductance L_1 plays a critical role in determining the circuit performance, particularly when the frequency of TENG motion is relatively high and the switch is rapidly turned on and off. To investigate this effect, experiments were carried out using different inductance values, as shown in Figure 4c(i–iii). C_{out} with values of 10 μF, 22 μF, and 47 μF was used to test the charging time. Since the motion frequency of the TENG itself is not high, μH level inductance values are chosen. The capacitance charging curves show that the charging rate of C_{out} increases with an increasing inductance value. However, when the value of L_1 exceeds 330 μH, the delay caused by inductance becomes more significant, leading to a reduction in the charging rate. Based on our results, the optimal value of L_1 is between 150 μH and 330 μH.

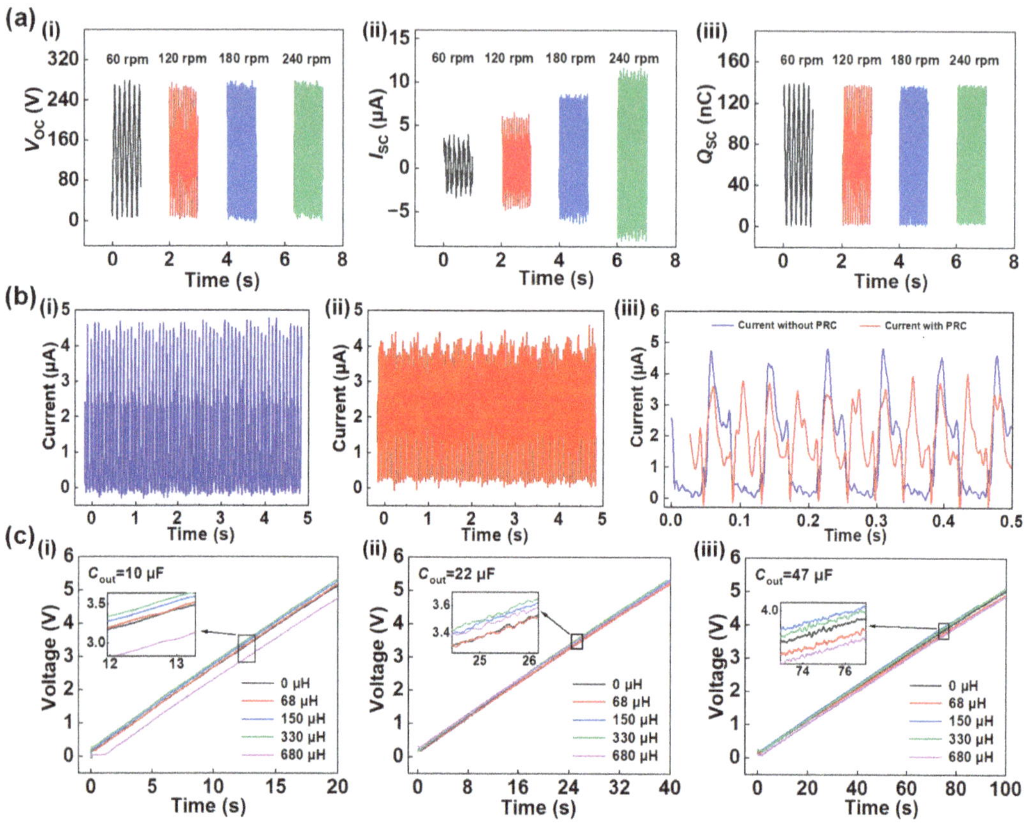

Figure 4. Basic output characteristics of the TENG and the PRC. (**a**) (**i–iii**) The original signals of voltage, current and charge of the TENG at different frequencies; (**b**) (**i**) short-circuit current output by direct rectification and (**ii**) by the PRC of the TENG, (**iii**) amplification of the current signal for comparison.; (**c**) (**i–iii**) influence of different inductance values on charging characteristics of output capacitor.

3.2. Demonstration of the PRC

Once the basic principles and parameters of the PRC system are determined, Figure 5 presents the final output performance analysis of the PRC system, along with comparative experiments and application demonstrations. Figure 5a shows the load charging characteristic curves for different C_{out} of 10 μF, 22 μF, and 47 μF, indicating that the internal resistance of the TENG should be considered when choosing the load. If the C_{out} is insufficiently large, the load voltage will fluctuate, leading to poor load-bearing capacity. Moreover, when the capacitance of C_{out} remains constant, the charging speed of C_{out} will be faster with a larger load. Comparative experiments were conducted as shown in Figure 5b(i,ii), where capacitors were charged under two different output conditions, one with the PRC system and the other without it. The results demonstrate that the charging rate of capacitors increased by 75% when the capacitor is charged to 5 V with the PRC system included. In addition, the charging time is significantly reduced with the inclusion of the PRC system. In order to verify the stability of the PRC system, Figure 5c presents a bar chart of the charging time of three groups of output capacitors charged to 5 V under different TENG motion frequencies. The charging time ratios of the three groups of capacitors are also tested at different frequencies using 3 Hz as the reference frequency, as shown in the curve in Figure 5c, the three curves almost coincide and can be fitted into one curve, indicating that the charging time varies proportionally with the motion frequency, and the entire circuit system is very stable. Finally, Figure 5d demonstrates the actual application capability of the PRC circuit through an experimental demonstration. Figure 5d(i) shows the testing system diagram, and the power supply of LEDs under two output modes, direct rectification and PRC output, is tested at a TENG speed of 120 rpm. The experimental findings indicate that when the LEDs are connected to the PRC (as illustrated in Figure 5d(i) and Video S1 in the Supplementary Materials), the flashing phenomenon is reduced and the overall lighting process becomes more stable. These demonstration results are consistent with the test data, thus confirming that the PRC can enhance the output efficiency of the TENG and improve the output stability.

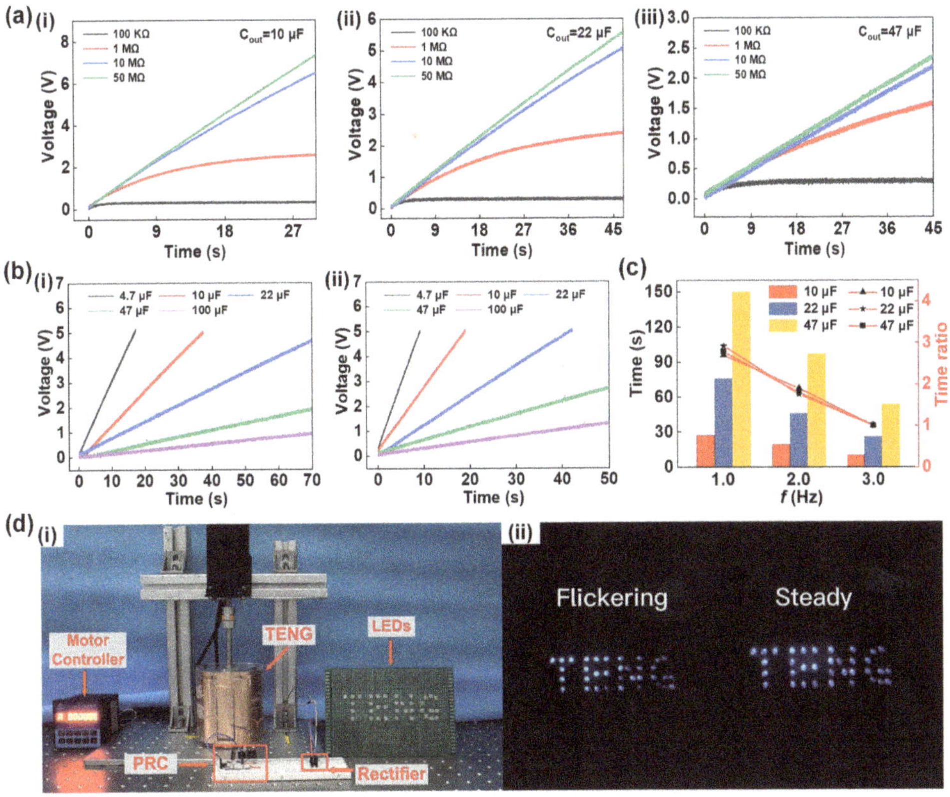

Figure 5. The output performance test and application demonstration of the PRC. (**a**) (**i–iii**) Curves of the charging voltage and the charging rate of C_{out} under different loads; (**b**) comparison experiment of output capacitor charging (**i**) with PRC and (**ii**) without the PRC; (**c**) the relationship between frequency and charging time; (**d**) (**i**) experimental test physical diagram and (**ii**) application demonstration of the TENG powering LEDs in two output modes.

4. Conclusions

In summary, the PRC system increases the power supply time of the TENG in each cycle, thereby increasing the number of pulses in the TENG output current signal. As a result, the total output energy per unit cycle increases, and compared to the initial output signal, the charging rate of the output capacitor increases by 75% after the PRC is added, improving the output efficiency of the TENG significantly. The PRC consists of a first-stage valley-filling circuit and a second-stage switching step-down circuit. The circuit performance was tested by a rotation-type freestanding-mode TENG. The frequency conversion charging experiment further demonstrated the stability of the PRC. At the same time, when the TENG supplies power to the LEDs at 120 rpm, the LEDs using the PRC flicker less and emit more stable light than the LEDs supplied by direct rectification, further validating the test results. The PRC proposed in this study can more effectively harvest and utilize the TENG's energy, which has a certain promoting effect on the storage and utilization of the output energy of the TENG.

Supplementary Materials: The following are available online at https://www.mdpi.com/article/10.3390/s23104912/s1, Video S1: Demonstration of lighting LEDs by the TENG with and without a PRC.

Author Contributions: Conceptualization, W.L and X.C.; software, W.L.; investigation, W.L., B.L., S.H. and X.C.; resources, X.C.; writing—original draft preparation, W.L., B.L., S.H. and X.C.; writing—review and editing, X.C.; supervision, X.C. All authors have read and agreed to the published version of the manuscript.

Funding: This research received no external funding.

Institutional Review Board Statement: Not applicable.

Informed Consent Statement: Informed consent was obtained from all subjects involved in this study.

Data Availability Statement: Not applicable.

Conflicts of Interest: The authors declare no conflict of interest.

Nomenclature

TENG	Triboelectric Nanogenerator
PRC	Power Regulation Circuit
PLA	Polylactide
FEP	Fluorinated Ethylene Propylene
PMOS	Positive channel Metal Oxide Semiconductor
C_1, C_1	Capacitance of the first-stage circuit
D_1, D_2, D_3, D_4	Diode
L_1	Inductance
C_{out}	Output Capacitance
R_L	Load
V_i	Input Voltage
V_{out}	Output Voltage

References

1. Le, C.D.; Vo, C.P.; Vu, D.L.; Nguyen, T.H.; Ahn, K.K. Water electrification based triboelectric nanogenerator integrated harmonic oscillator for waste mechanical energy harvesting. *Energy Convers. Manag.* **2022**, *251*, 115014. [CrossRef]
2. Niu, S.M.; Wang, X.F.; Yi, F.; Zhou, Y.S.; Wang, Z.L. A universal self-charging system driven by random biomechanical energy for sustainable operation of mobile electronics. *Nat. Commun.* **2015**, *6*, 8975. [CrossRef] [PubMed]
3. Elsanadidy, E.; Mosa, I.M.; Luo, D.; Xiao, X.; Chen, J.; Wang, Z.L.; Rusling, J.F. Advances in Triboelectric Nanogenerators for Self-powered Neuromodulation. *Adv. Funct. Mater.* **2023**, *33*, 2211177. [CrossRef]
4. Curto, D.; Favuzza, S.; Franzitta, V.; Guercio, A.; Amparo Navarro Navia, M.; Telaretti, E.; Zizzo, G. Grid Stability Improvement Using Synthetic Inertia by Battery Energy Storage Systems in Small Islands. *Energy* **2022**, *254*, 124456. [CrossRef]
5. Harada, K.; Yabe, K.; Takami, H.; Goto, A.; Sato, Y.; Hayashi, Y. Two-step approach for quasi-optimization of energy storage and transportation at renewable energy site. *Renew. Energy* **2023**, *211*, 846–858. [CrossRef]

6. Zhang, X.S.; Gao, Q.; Gao, Q.; Yu, X.; Cheng, T.H.; Wang, Z.L. Triboelectric Rotary Motion Sensor for Industrial-Grade Speed and Angle Monitoring. *Sensors* **2021**, *21*, 1713. [CrossRef]
7. Wang, Z.L. Triboelectric nanogenerators as new energy technology and self-powered sensors—Principles, problems and perspectives. *Faraday Discuss.* **2014**, *176*, 447–458. [CrossRef]
8. Kim, W.G.; Kim, D.W.; Tcho, I.W.; Kim, J.K.; Kim, M.S.; Choi, Y.K. Triboelectric Nanogenerator: Structure, Mechanism, and Applications. *ACS Nano* **2021**, *15*, 258–287. [CrossRef]
9. Wang, Z.L.; Jiang, T.; Xu, L. Toward the blue energy dream by triboelectric nanogenerator networks. *Nano Energy* **2017**, *39*, 9–23. [CrossRef]
10. Pathak, M.; Kumar, R. Synchronous Pre-biasing of Triboelectric Nanogenerator for Enhanced Energy Extraction. *IEEE Trans. Power Electron.* **2022**, *37*, 11552–11566. [CrossRef]
11. Zhang, B.S.; Li, W.B.; Ge, J.W.; Chen, C.G.; Yu, X.; Wang, Z.L.; Cheng, T.H. Single-material-substrated triboelectric-electromagnetic hybrid generator for self-powered multifunctional sensing in intelligent greenhouse. *Nano Res.* **2023**, *16*, 3149–3155. [CrossRef]
12. Zhao, B.; Zhao, K.; Wang, X.C.; Liang, J.R.; Chen, Z.Y. Series Synchronized Triple Bias-Flip Circuit: Maximizing the Usage of a Single Storage Capacitor for Piezoelectric Energy Harvesting Enhancement. *IEEE Trans. Power Electron.* **2021**, *36*, 6787–6796. [CrossRef]
13. Fang, L.; Zheng, Q.W.; Hou, W.C.; Gu, J.Y.; Zheng, L. A self-powered tilt angle sensor for tall buildings based on the coupling of multiple triboelectric nanogenerator units. *Sens. Actuators A Phys.* **2023**, *349*, 114015. [CrossRef]
14. Zhang, B.S.; Zhang, S.; Li, W.B.; Gao, Q.; Zhao, D.; Wang, Z.L.; Cheng, T.H. Self-Powered Sensing for Smart Agriculture by Electromagnetic–Triboelectric Hybrid Generator. *ACS Nano* **2021**, *15*, 20278–20286. [CrossRef] [PubMed]
15. Wang, Z.L. Self-Powered Nanotech. *Sci. Am.* **2008**, *298*, 82–87. [CrossRef] [PubMed]
16. Li, Y.H.; Yu, J.R.; Wei, Y.C.; Wang, Y.F.; Feng, Z.Y.; Cheng, L.Q.; Huo, Z.W.; Lei, Y.Q.; Sun, Q.J. Recent Progress in Self-Powered Wireless Sensors and Systems Based on TENG. *Sensors* **2023**, *23*, 1329. [CrossRef] [PubMed]
17. Duan, Q.S.; Peng, W.Q.; He, J.X.; Zhang, Z.J.; Wu, Z.C.; Zhang, Y.; Wang, S.F.; Nie, S.X. Rational Design of Advanced Triboelectric Materials for Energy Harvesting and Emerging Applications. *Small Methods* **2023**, *7*, 2201251. [CrossRef]
18. Wang, H.M.; Xu, L.; Bai, Y.; Wang, Z.L. Pumping up the charge density of a triboelectric nanogenerator by charge-shuttling. *Nat. Commun.* **2020**, *11*, 4203. [CrossRef]
19. Cheng, G.; Lin, Z.H.; Lin, L.; Du, Z.l.; Wang, Z.L. Pulsed Nanogenerator with Huge Instantaneous Output Power Density. *ACS Nano* **2013**, *7*, 7383–7391. [CrossRef]
20. Wang, Z.L.; Song, J.H. Piezoelectric Nanogenerators Based on Zinc Oxide Nanowire Arrays. *Science* **2006**, *312*, 242–246. [CrossRef]
21. Xi, F.; Pang, Y.; Liu, G.; Wang, S.; Li, W.; Zhang, C.; Wang, Z.L. Self-powered intelligent buoy system by water wave energy for sustainable and autonomous wireless sensing and data transmission. *Nano Energy* **2019**, *61*, 1–9. [CrossRef]
22. Zargari, S.; Daie Koozehkanani, Z.; Veladi, H.; Sobhi, J.; Rezania, A. A new Mylar-based triboelectric energy harvester with an innovative design for mechanical energy harvesting applications. *Energy Convers. Manag.* **2021**, *244*, 114489. [CrossRef]
23. Cao, X.L.; Xiong, Y.; Sun, J.; Xie, X.Y.; Sun, Q.J.; Wang, Z.L. Multidiscipline Applications of Triboelectric Nanogenerators for the Intelligent Era of Internet of Things. *Nano-Micro Lett.* **2022**, *15*, 14. [CrossRef] [PubMed]
24. He, W.C.; Shan, C.C.; Fu, S.K.; Wu, H.Y.; Wang, J.; Mu, Q.J.; Li, G.; Hu, C.G. Large Harvested Energy by Self-Excited Liquid Suspension Triboelectric Nanogenerator with Optimized Charge Transportation Behavior. *Adv. Mater.* **2023**, *35*, 2209657. [CrossRef] [PubMed]
25. Rasel, M.S.; Maharjan, P.; Park, J.Y. Hand clapping inspired integrated multilayer hybrid nanogenerator as a wearable and universal power source for portable electronics. *Nano Energy* **2019**, *63*, 103816. [CrossRef]
26. Wang, Z.; Tang, Q.; Shan, C.C.; Du, Y.; He, W.C.; Fu, S.K.; Li, G.; Liu, A.P.; Liu, W.L.; Hu, C.G. Giant performance improvement of triboelectric nanogenerator systems achieved by matched inductor design. *Energy Environ. Sci.* **2021**, *14*, 6627–6637. [CrossRef]
27. Xu, L.Q.; Xuan, W.P.; Chen, J.K.; Zhang, C.; Tang, Y.Z.; Huang, X.W.; Li, W.J.; Jin, H.; Dong, S.R.; Yin, W.L.; et al. Fully self-powered instantaneous wireless humidity sensing system based on triboelectric nanogenerator. *Nano Energy* **2021**, *83*, 105814. [CrossRef]
28. Fan, F.-R.; Tian, Z.-Q.; Wang, Z.L. Flexible triboelectric generator. *Nano Energy* **2012**, *1*, 328–334. [CrossRef]
29. Zhang, H.M.; Marty, F.; Xia, X.; Zi, Y.L.; Bourouina, T.; Galayko, D.; Basset, P. Employing a MEMS plasma switch for conditioning high-voltage kinetic energy harvesters. *Nat. Commun.* **2020**, *11*, 3221. [CrossRef]
30. Wang, Z.; Liu, W.L.; He, W.C.; Guo, H.Y.; Long, L.; Xi, Y.; Wang, X.; Liu, A.P.; Hu, C.G. Ultrahigh Electricity Generation from Low-Frequency Mechanical Energy by Efficient Energy Management. *Joule* **2021**, *5*, 441–455. [CrossRef]
31. Zhang, Z.H.; Gu, G.Q.; Zhang, W.H.; Gu, G.X.; Shang, W.Y.; Liu, Y.; Cheng, G.; Du, Z. Double loops power management circuit of pulsed triboelectric nanogenerator with enhanced efficiency at low operating voltage and its application in self-powered flue gas monitoring system. *Nano Energy* **2023**, *110*, 108360. [CrossRef]
32. Xi, F.; Pang, Y.; Li, W.; Jiang, T.; Zhang, L.; Guo, T.; Liu, G.; Zhang, C.; Wang, Z.L. Universal power management strategy for triboelectric nanogenerator. *Nano Energy* **2017**, *37*, 168–176. [CrossRef]
33. Jiang, D.W.; Lian, M.Y.; Xu, M.J.; Sun, Q.; Xu, B.B.; Thabet, H.K.; El-Bahy, S.M.; Ibrahim, M.M.; Huang, M.N.; Guo, Z.H. Advances in triboelectric nanogenerator technology—Applications in self-powered sensors, Internet of things, biomedicine, and blue energy. *Adv. Compos. Hybrid Mater.* **2023**, *6*, 57. [CrossRef]

34. Wu, H.; Wang, S.; Wang, Z.K.; Zi, Y.L. Achieving ultrahigh instantaneous power density of 10 MW/m^2 by leveraging the opposite-charge-enhanced transistor-like triboelectric nanogenerator (OCT-TENG). *Nat. Commun.* **2021**, *12*, 5470. [CrossRef]
35. Zhu, P.C.; Zhang, B.S.; Wang, H.Y.; Wu, Y.H.; Cao, H.J.; He, L.B.; Li, C.Y.; Luo, X.P.; Li, X.; Mao, Y.C. 3D printed triboelectric nanogenerator as self-powered human-machine interactive sensor for breathing-based language expression. *Nano Res.* **2022**, *15*, 5098–5104. [CrossRef]

applied sciences

MDPI

Article

Vibration Energy Harvester Based on Bilateral Periodic One-Dimensional Acoustic Black Hole

Zihao Zhang [1,2], Hai Wang [1,2,*], Chunlai Yang [1,2,*], Hang Sun [1,2] and Yikai Yuan [1,2]

[1] School of Mechanical Engineering, Anhui Polytechnic University, Wuhu 241000, China; zihao_0216@163.com (Z.Z.); hang961203@163.com (H.S.); yyikai666@gmail.com (Y.Y.)
[2] Anhui Key Laboratory of Advanced Numerical Control & Servo Technology, Wuhu 241000, China
* Correspondence: wanghai@ahpu.edu.cn (H.W.); ycl@ahpu.edu.cn (C.Y.)

Abstract: An acoustic black hole (ABH) has been applied in the regulation of structural performance to form the aggregation effect of elastic waves in the local area of the structure, which has been used in energy harvesting in recent years. The piezoelectric vibration energy harvester (VEH) integrated with the beam of a bilateral periodic 1D ABH is proposed in this study. The theoretical model of the proposed VEH is established and analyzed based on the transfer matrix method. The performance of the VEHs is numerically simulated by COMSOL Multiphysics. The simulation results show that the performance of the bilateral ABH beam is higher than its traditional counterpart. Finally, the performance of the proposed VEH is validated in an experimental system. The experimental results show that the peak output voltage of the VEH Model 3 can reach 169.16 V, which is 1.9 times that of the traditional one. In the optimal impedance matching, the output power of the third bilateral VEH is 2.7 times that of the traditional ABH, which can reach 91.52 mW.

Keywords: acoustic black hole; vibration energy harvester; Gaussian pulse; output voltage

1. Introduction

Vibrations can be seen anywhere in industrial production and the surrounding circumstances, especially in structures such as beams or rods in civil engineering and mechanical engineering [1–3]. This kind of redundant vibration is usually harmful and severely affects reliability and accuracy in various mechanical and precision equipment [4,5]. However, the health status of the mechanical equipment and engineering structures can be monitored by wireless sensor network (WSN) nodes in real time. Traditional WSN [6] nodes are powered by batteries, which have a limited lifetime and are hard to maintain, especially in harsh environments and remote areas. Therefore, it is extremely important to develop a novel power supply method that can replace batteries with a long lifetime and is easy to maintain.

Harvesting energy from the vibrations in the surrounding environment where the WSN nodes are deployed, which can be converted to electrical energy, should be a possible solution. The device that can harvest vibration energy and convert it to electrical energy is a vibration energy harvester (VEH). VEHs can be divided into piezoelectric, electromagnetic and electrostatic types according to the operation principle. The piezoelectric VEH has wide employment through its simple structure and high electromechanical transfer coefficient [7]. A cantilever is used as the primary structure of a traditional piezoelectric VEH with a narrow operation bandwidth and low harvesting efficiency.

The nonlinear effect can be used to improve the performance of traditional VEHs. Nonlinear structures are widely used due to their simple structure, which can be integrated into the cantilever beam in a VEH. As a nonlinear structure, the ABHs have wide employment in the field of vibrational reduction, noise reduction and energy harvesting due to the unique energy focus effect. ABHs have structures with geometrical or material parameters gradually decreasing t in thin-walled components such as beams and benches, which can focus the elastic wave [8]. Gao et al. [9–11] proposed several kinds of one-dimensional double-leaf ABH beams with nested damping units, unit periodic arrays and spatial V-shaped bending. The low-frequency band gap and vibration characteristics of the structure were analyzed. The application of an ABH structure in energy harvesting has gained a lot of attention in recent years. Deng [12] uses the Lagrange equation to derive the combined equation of the bending vibration of the ABH beam with the piezoelectric layer. The influence of the parameters of the ABH and piezoelectric layer on energy harvesting efficiency is analyzed. Zhao et al. [13] established an electromechanical coupling model of an ABH cone structure with a surface-mounted piezoelectric transducer and analyzed the dynamic response with numerical simulation. Simulation results show that the response of the system is better than the traditional structure. Ji et al. [14] proposed an energy harvester integrated with a high-intensity ABH and validated it through experimental tests.

In addition, the characteristic that acoustic black holes can gather energy is also applicable to vibration suppression. Du et al. [15] proposed a new solution to embed 2D ABHs on the support plate to suppress the transmission of compressor vibration to the refrigerator body. Liang et al. [16] focus on the mid- and low-frequency performance of plates embedded with the array of an ABH for energy focalization and vibration and noise suppression. Lyu et al. [17] proposed a type of metamaterial plate enabling in-plane ultra-wide vibration isolation in engineering equipment development and experiments show that the effective attenuation ability is compared with the traditional hexagonal lattice.

A piezoelectric VEH with a bilateral periodic 1D ABH is proposed in this paper. The vibration energy can be accumulated in the center of the ABH due to the focus effect and converted into electrical energy by the piezoelectric plates attached to the ABH surface. The theoretical model of the energy harvester with a bilateral periodic 1D ABH structure is established, and the performance of the above VEH is discussed through numerical simulation in COMSOL and verified by experimental tests.

2. VEH Integrated with Bilateral Periodic 1D ABH

2.1. Structure of VEH

The traditional VEH with an ABH is a cantilever beam with a unilateral ABH embedded. As shown in Figure 1, a VEH with a bilateral 1D ABH structure is proposed, where the piezoelectric plates attached to the ABH are ignored. Three different types of VEHs with an ABH are designed and analyzed.

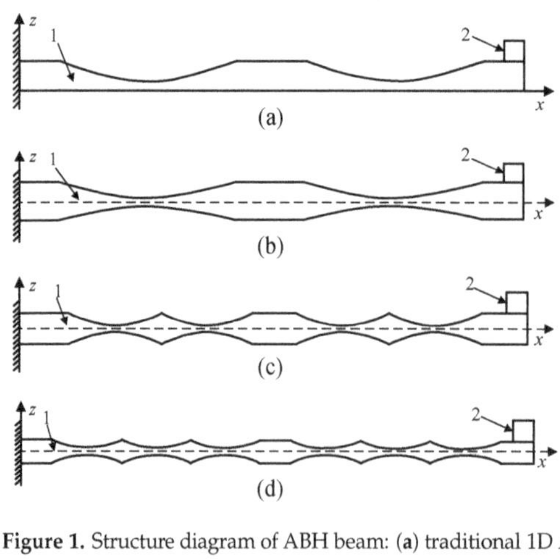

Figure 1. Structure diagram of ABH beam: (**a**) traditional 1D ABH beam, (**b**) bilateral 1D ABH Model 1, (**c**) bilateral 1D ABH Model 2, (**d**) bilateral 1D ABH Model 3 (1: major structure; 2: quality mass).

The collection of vibration energy is realized by the ABHs embedded in the cantilever of the VEH. The operation bandwidth can be adjusted by tuning the parameters of ABHs and the attached mass. The mass added to the edge of the beam greatly reduces the natural frequency of the VEH (The VEHs discussed next all contain mass blocks).

2.2. Operation Principle of VEH

As shown in Figure 2, the thickness of the ABH fits the formula $h(x) = \varepsilon x^m$, where $h(x)$ and represents the thickness variation of the beam; ε is the undetermined coefficient and m is a power exponent. The structural stiffness is ensured by extending the ABH portion to a truncated platform at a minimum thickness, which is determined by the fabrication process. The operation principle of the proposed VEH is as follows: Vibration is generated in the cantilever beam when an external excitation signal is applied on the constrained end. The elastic waves according to the vibration are transmitted in the cantilever beam and concentrated in the center of the ABH zone due to the focus effect. The concentrated vibration energy is converted into electrical energy by the attached piezoelectric plate in the center of the ABHs.

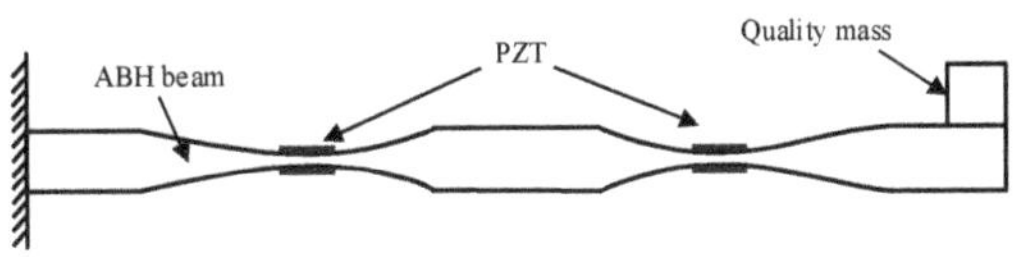

Figure 2. Structure diagram of piezoelectric energy harvester.

2.3. Dynamic Modeling of ABH Beam

As shown in Figure 3, the left side of the beam is fix constrained and the right side is left free. It is assumed that the external excitation $Y(x,t) = Y_0 \sin Pt$ is applied at $0.1\ x_L$ from the left end, where Y_0 is the amplitude of the external excitation, and P is its frequency. The thickness of the constant thickness part of the beam is h_m, x_A is the length of a single ABH unit, h_t is the truncated thickness of the ABH and x_t is the truncated extended platform length. The adverse effects of the truncated platform can be offset to some extent by properly adding damping material to the surface of the curved portion of the ABH. x_E is the length of the periodic joint in the middle and x_L is the total length of the beam.

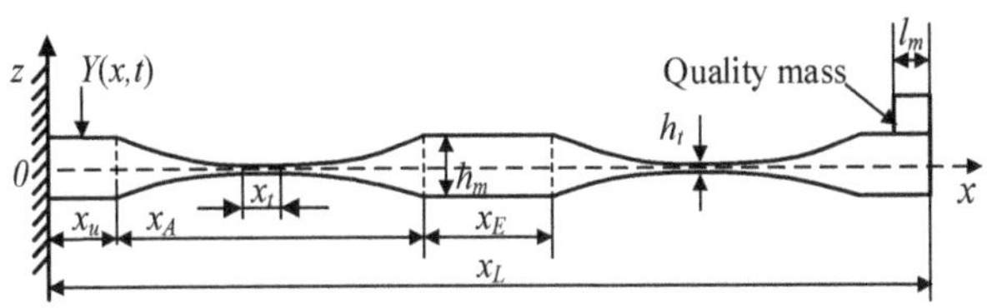

Figure 3. Bilateral one-dimensional periodic ABH beam Model 1.

The shear deformation of the cantilever beam is neglected because the length of the ABH beam is much larger than its thickness, and the transverse vibration is dominant in practice. The undamped transverse vibration equation of the beam based on the Euler–Bernoulli beam theory is [18]:

$$\rho T^B(x)\frac{\partial^2 w(x,t)}{\partial t^2} + \frac{\partial^2 M^T(x,t)}{\partial x^2} = Y(x,t), \tag{1}$$

where ρ is the material density of the ABH beam, $T^B(x)$ is the cross-section area of the beam and $M^T(x,t)$ is the bending moment of the beam.

The ABH beam is a variable cross-section beam, which can be approximately equivalent to a linear combination of N equal cross-section beams as shown in Figure 4. When the N is large enough, the ABH beam can be approximated by the equivalent segmented beam, and then the vibration of the equivalent beam is analyzed by the transfer matrix method. The transfer matrix method is that any uniform beam in the segmented beam is analyzed separately to determine its corresponding transfer matrix, and then the whole beam is combined and analyzed according to the continuity between the segmented beams to obtain the whole vibration response of the ABH beam.

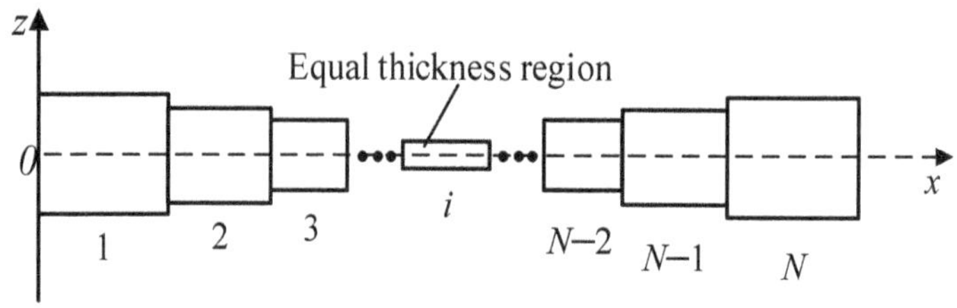

Figure 4. Sectional diagram of ABH beam.

For the single-section uniform beam after discretization, if the damping layer is neglected, the transverse free vibration equation of the number i uniform beam can be expressed as [19]:

$$\rho T_i^B(x)\frac{\partial^2 w(x,t)}{\partial t^2}+\frac{\partial^2 M_i^T(x,t)}{\partial x^2}=Y(x,t), x_i \leq x < x_{i+1}, i=1,2,\ldots N. \tag{2}$$

Let $w(x,t)$ and $j(x,t)$ be the components of the beam in the x-axis and z-axis directions, respectively, as shown in Figure 5. The shear and rotation of the beam are ignored since the thickness of the beam is much smaller than its length, where r and θ are the beam's axial strain and rotation angle after flexural deformation. Then, the length of the deformed element can be expressed as [20]:

$$dxw' = \sqrt{\left(\frac{dw}{dx}\right)^2+\left(1+\frac{dj}{dx}\right)^2}dx, \tag{3}$$

and the axial strain of the beam is:

$$r=\frac{dx'-dx}{dx}, \tag{4}$$

the axial strain of the beam can be neglected when the excitation of the ABH beam is mainly transverse vibration, which means $r = 0$. At this time, the following can be obtained:

$$j' = \sqrt{1-w'^2}-1, \tag{5}$$

$$\tan\theta = \frac{w'}{1+j'}, \tag{6}$$

by substituting Equation (5) into Equation (6), the rotation angle θ can be expressed as:

$$\theta = \arctan\left(\frac{w'}{\sqrt{1-w'^2}}\right), \tag{7}$$

The Taylor expansion of Equation (7) can be obtained:

$$\theta = w' + \frac{1}{6}w'^3 + o(w'^5), \tag{8}$$

the curvature λ_θ of the beam element after deformation can be expressed as:

$$\lambda_\theta = \theta' = w'' + \frac{1}{2}w'^2 w'', \tag{9}$$

it can be obtained that the nonlinear strain ε_B generated by the ABH beam during vibration is:

$$\varepsilon_B = r - z\lambda_\theta = -z(w'' + \frac{1}{2}w'^2 w''). \tag{10}$$

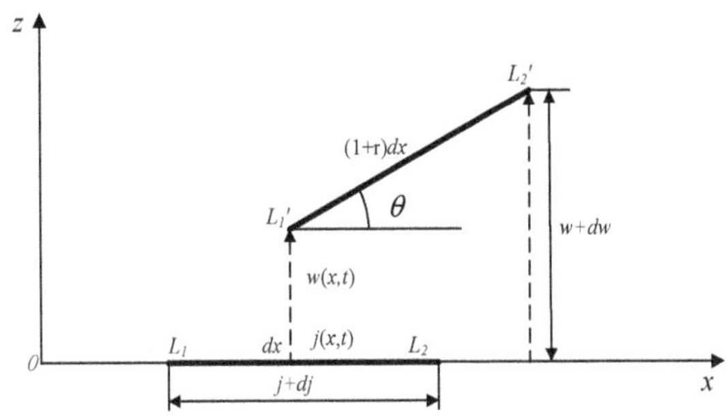

Figure 5. Geometric relationship of the deformed element.

Without considering the damping layer, the bending moment of the uniform beam in the first section number i can be expressed as:

$$M_i^T(x,t) = -\int_{-\frac{h(x)}{2}}^{\frac{h(x)}{2}} zb\sigma_B dz, \tag{11}$$

where σ_B is the inner stress of the ABH beam, b is the width of the ABH beam, substituting the stress–strain constitutive relation $\sigma_B = E_B\varepsilon_B$ and Equation (10) into Equation (11), the bending moment of the uniform beam number i can be expressed as:

$$M_i^T(x,t) = E_B I_i(x)\left[\frac{\partial^2 w}{\partial x^2} + \frac{1}{2}\left(\frac{\partial w}{\partial x}\right)^2 \frac{\partial^2 w}{\partial x^2}\right], \tag{12}$$

where E_B is the elastic modulus of the beam, and $I_i(x)$ is the section moment of inertia of the number i uniform beam, which is expressed as:

$$I_i(x) = \frac{bh^3(x_i)}{12}, \tag{13}$$

substituting Equation (12) into Equation (2), the vibration equation of the number i uniform beam is written as:

$$\begin{aligned}&\rho T_i^B(x)\tfrac{\partial^2 w}{\partial t^2} + E_B I_i^B(x)\left[\tfrac{\partial^4 w}{\partial x^4} + \tfrac{1}{2}\left(\tfrac{\partial w}{\partial x}\right)^2\tfrac{\partial^4 w}{\partial x^4} + 3\tfrac{\partial w}{\partial x}\tfrac{\partial^2 w}{\partial x^2}\tfrac{\partial^3 w}{\partial x^3} + \left(\tfrac{\partial^2 w}{\partial x^2}\right)^3\right] \\ &= Y(x,t), \quad x_i \le x \le x_{i+1}, \quad i = 1,2,\ldots N.\end{aligned} \tag{14}$$

The entire ABH beam is composed of multiple uniform beams after discretization, and the cross-section area and cross-section moment of the inertia of each uniform beam are different. The composite parameters of each node are introduced as follows:

$$T^B(x) = \begin{cases} T_1^B, & 0 = x_1 \le x \le x_2, \\ \vdots & \vdots \\ T_i^B, & x_i \le x \le x_{i+1}, \\ \vdots & \vdots \\ T_K^B, & x_K \le x \le x_{K+1} = x_L \end{cases} \quad I^B(x) = \begin{cases} I_1^B, & 0 = x_1 \le x \le x_2, \\ \vdots & \vdots \\ I_i^B, & x_i \le x \le x_{i+1}, \\ \vdots & \vdots \\ I_K^B, & x_K \le x \le x_{K+1} = x_L \end{cases}, \tag{15}$$

in summary, the transfer matrix method mechanical model of the ABH beam considering the nonlinear strain can be expressed as:

$$\rho T^B(x)\frac{\partial^2 w}{\partial t^2} + E_B I^B(x)\left[\frac{\partial^4 w}{\partial x^4} + \frac{1}{2}\left(\frac{\partial w}{\partial x}\right)^2\frac{\partial^4 w}{\partial x^4} + 3\frac{\partial w}{\partial x}\frac{\partial^2 w}{\partial x^2}\frac{\partial^3 w}{\partial x^3} + \left(\frac{\partial^2 w}{\partial x^2}\right)^3\right] \\ = Y(x,t), \quad 0 \le x \le x_L. \tag{16}$$

3. Simulation and Analysis

3.1. Structure and Material

The response of the VEH with a bilateral 1D ABH is simulated and compared with a traditional VEH with a traditional 1D ABH. The cutting length of each ABH region is set to 90 mm for simplicity. The geometry parameters of these structures are listed in Table 1.

Table 1. Parameters of ABH beams.

Parameter	Traditional ABH	Model 1	Model 2	Model 3
x_L		290 mm		
h_m		10 mm		
b		20 mm		
ε	0.00546	0.00273	0.01525	0.05125
m		2		
x_t		12 mm		
h_t		1.7 mm		
l_m		8 mm		
x_A		90 mm		
x_E		40 mm		
x_u		30 mm		

3.2. Simulation of VEH with ABHs

3.2.1. Natural Frequency of ABH Beams

The grid division in the FEM model of the ABH beam (take bilateral 1D ABH Model 1 as an example) is shown in Figure 6. As shown in Figure 7, the left end of the VEH is fix-constrained (the red arrow indicates the initial stress distribution under the external force at the free end), the right end is attached with a mass block and set free and the natural frequencies are acquired, as shown in Figure 8. The natural frequency of the ABH beams decreases with the increment of the ABH arrays. The specific frequency values in the figure are shown in Table A1 in Appendix A.

Figure 6. Grid division of beam in COMSOL.

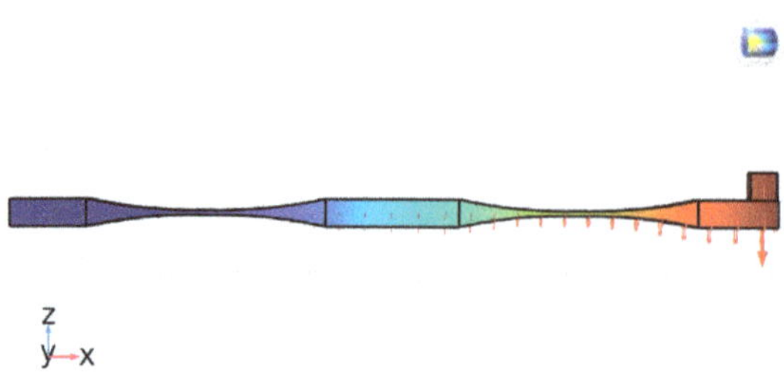

Figure 7. Simulation diagram of beam in COMSOL.

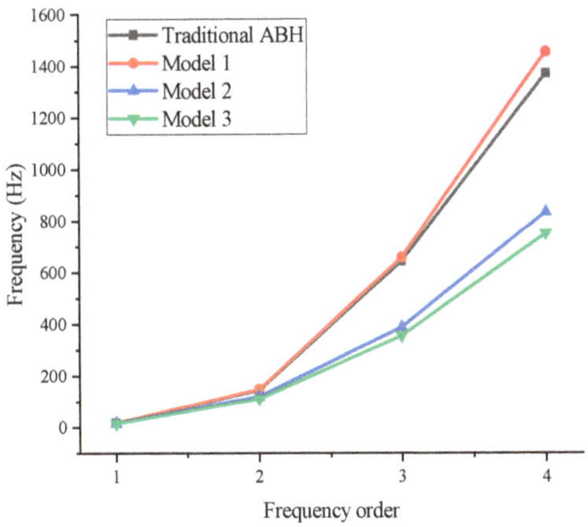

Figure 8. First four-order natural frequency.

3.2.2. Transient Response under Gaussian Pulse

The characteristic of a Gaussian pulse is energy concentration, which is beneficial to compress the signal spectrum and reduce the adjacent channel interference. The excitation in the actual environment can be better simulated by Gaussian pulses. The broadband Gaussian pulse is used as the excitation signal during the transient simulation of the VEH with ABHs. The formula of the broadband Gaussian pulse is as below:

$$\cos\left[6000\pi\left(t-5\times10^{-4}\right)\right]e^{-\frac{t-5\times10^{-4}}{3\times10^{-8}}} \tag{17}$$

The Gaussian pulse has a wide frequency band and short excitation time, and its center frequency is set to 3000 Hz. Time domain and frequency domain images of Gaussian pulses are shown in Figures 9 and 10, respectively. In the simulation, a given displacement of 5×10^{-5} m is loaded in the rightmost boundary of the beam to obtain an ideal excitation effect. The energy focus effect here of the ABH is reflected by the energy density of the ABH region, which can be validated by calculating the focused energy of the ABH zone. The Gaussian impulse response in the time domain is shown in Figure 11. Simulation results show that the energy focus effect of the ABH increased with the number of ABH arrays.

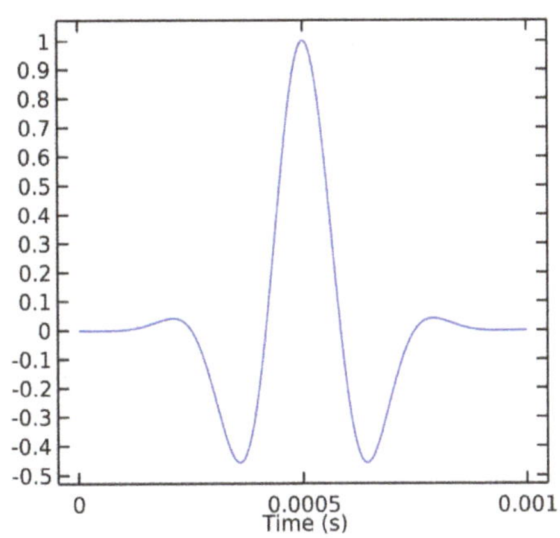

Figure 9. Normalized time domain for Gaussian pulse.

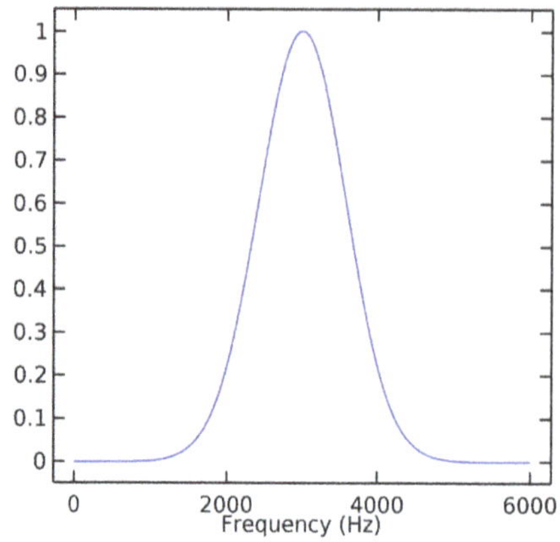

Figure 10. Normalized frequency domain for Gaussian pulse.

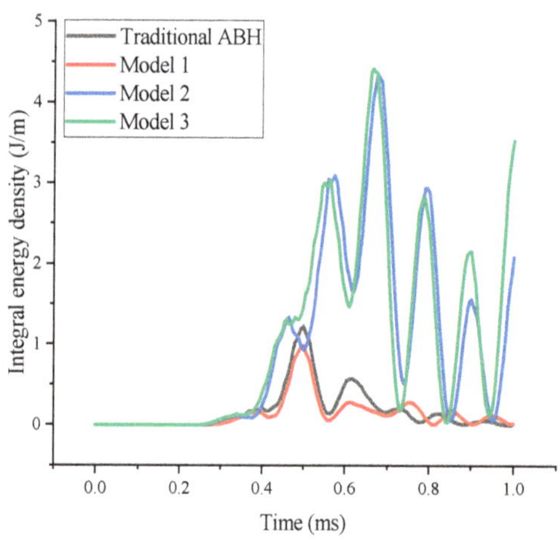

Figure 11. Integral energy density of ABH region under Gaussian pulse.

The kinetic energy density and strain of ABH Model 1 are shown in Figure 12. It can be seen that the energy is mainly concentrated in the acoustic black hole. The kinetic energy density nephograms of the other three structures are shown in Appendix A.

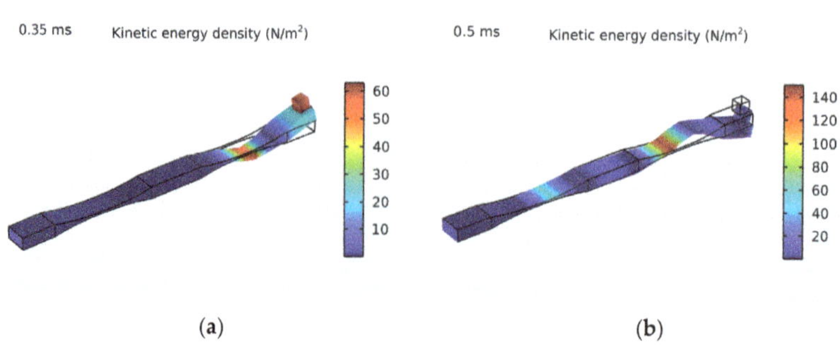

Figure 12. *Cont.*

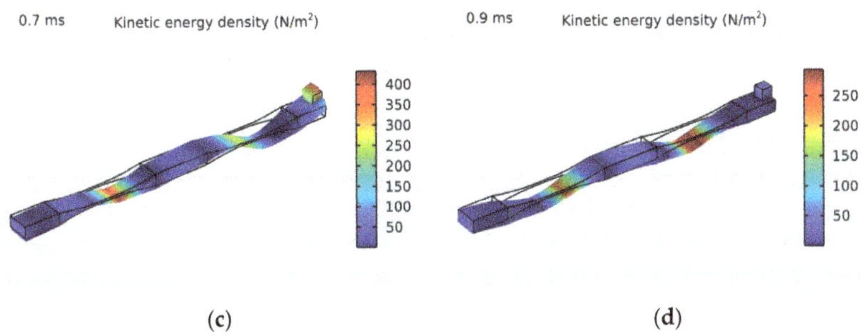

Figure 12. Kinetic energy density and strain of ABH Model 1: (**a**) ABH in 0.35 ms, (**b**) ABH in 0.5 ms, (**c**) ABH in 0.7 ms, (**d**) ABH in 0.9 ms.

The response of the VEH with ABHs under Gaussian pulse is transformed by FFT, as shown in Figure 13. The VEH with more ABH periodic arrays can store more vibration energy in a wide frequency range of around 3000 Hz.

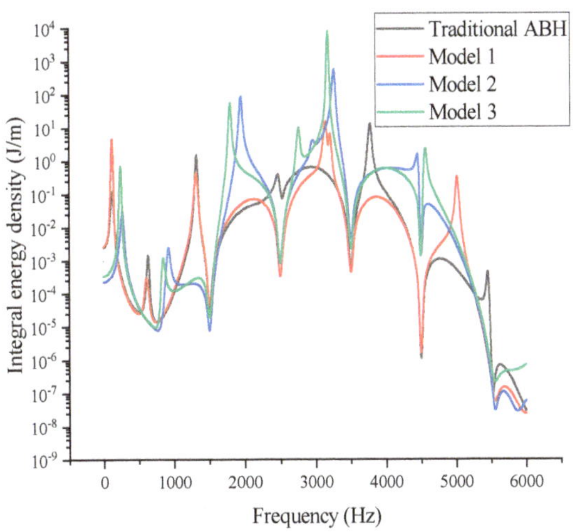

Figure 13. Integral energy density of ABH region under Gaussian pulse (FFT).

3.2.3. Analysis in Frequency Domain

The analysis of these structures in a frequency domain is implemented under the condition that the rightmost end is applied with a 5 N force. The frequency band is divided into low frequency (0–2000 Hz), intermediate frequency (2000–4000 Hz) and high frequency (4000–6000 Hz). The response of the four VEHs with the ABH is compared with the velocity of the surfaces of the ABH central platform, as shown in Figure 14. Then, the root mean square (RMS) processing is performed on the simulation results, so the overall response of each structure in different frequency bands is observed, as shown in Figure 15.

As shown in Figure 14, the VEH with a bilateral ABH can obtain higher vibration energy mainly concentrated in the low-frequency range. Figure 15 shows that the RMS vibration velocity of the bilateral ABH structure is higher than the traditional counterpart, and the vibration velocity of Model 3 is about 3.84 times that of the traditional ABH in the low-frequency range.

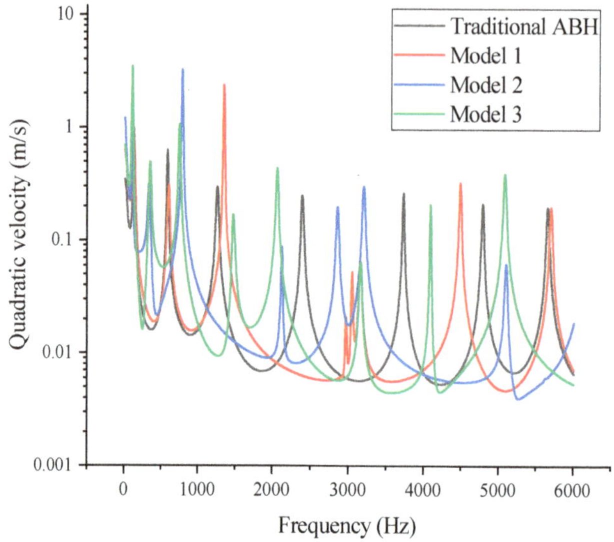

Figure 14. The quadratic velocity of ABH region.

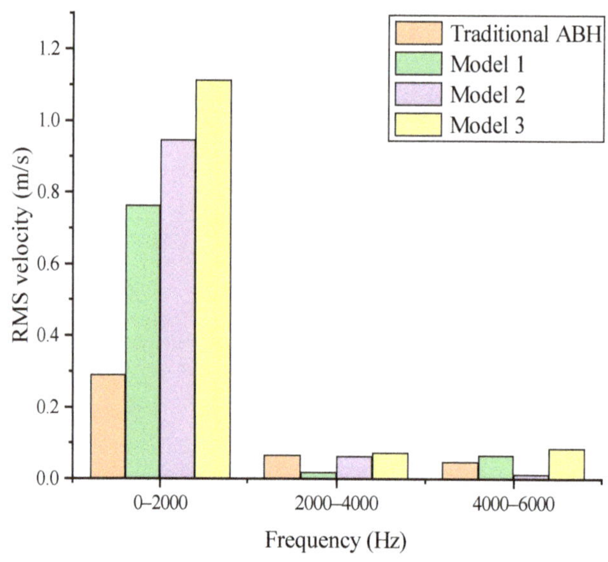

Figure 15. The RMS velocity of ABH region.

4. Experimental Test

4.1. Experimental System

The performance of VEHs with ABHs is tested in an experimental system. The main components of the experimental platform are shown in Figure 16.

The working process of the whole experimental system is as follows: the signal generated by the signal generator will be amplified by the power amplifier to drive the vibrator to vibrate, and the electrical signal generated by VEH will be collected by the data recorder and transmitted to the host computer for display.

The prototype of four different models of VEHs is shown in Figure 17. These beams were fabricated by 3D printing and the piezoelectric plates with a size of 20 × 12 × 0.3 mm were pasted on the ABH part. The specific parameters of each structure are consistent with Table 1.

These beams are made of AlSi10Mg, the attached mass block is iron and the attached piezoelectric plate is made of PZT-5H with a 0.3 mm thickness truncated platform. The parameters of the material are listed in Table 2.

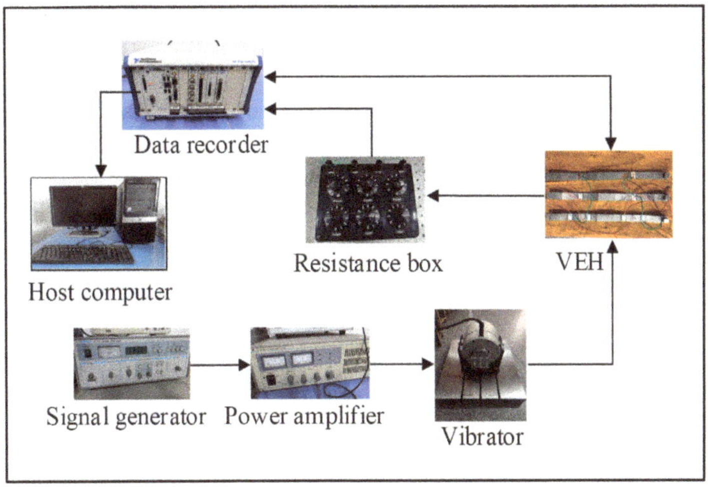

Figure 16. Experimental system for energy harvesting performance.

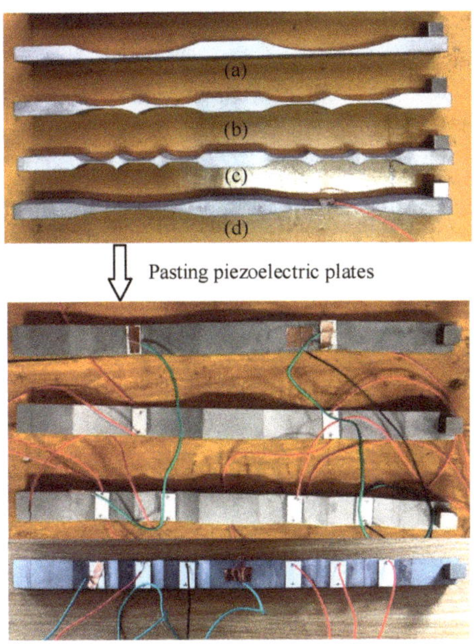

Figure 17. Energy harvester structure: (**a**) traditional 1D ABH beam, (**b**) bilateral 1D ABH Model 2, (**c**) bilateral 1D ABH Model 3, (**d**) bilateral 1D ABH Model 1.

Table 2. Parameters of size and material of energy harvester.

Parameter	AlSi10Mg	Iron	PZT-5H
Young's modulus (GPa)	74.01	200	60.61
Density (kg/m^3)	2790	7870	7500
Poisson's ratio	0.33	0.29	0.31
Damping loss factor	0.02	0.03	0.31

4.2. Results and Discussion

4.2.1. Output Voltage of VEH with ABH

The output voltage of VEHs with an ABH in the frequency domain is shown in Figure 18, where the external load resistance is 10 kΩ. The peak output voltage of the VEH with a traditional ABH is 8.5 V at 153 Hz, and the peak output voltage of the VEH with the bilateral ABH Model 1 is 10.5 V at 146 Hz separately. The peak output voltage of the VEH with ABH Model 2 is 22.1 V at 140 Hz, and the peak output voltage of the VEH with ABH Model 3 is 29.2 V at 130 Hz. The peak output voltage of the VEH with the ABH increased with the number of ABH arrays. The peak output voltage of VEH Model 3 is 3.4 times that of the traditional counterpart. The specific natural frequency and output voltage are shown in Table A2 in Appendix A.

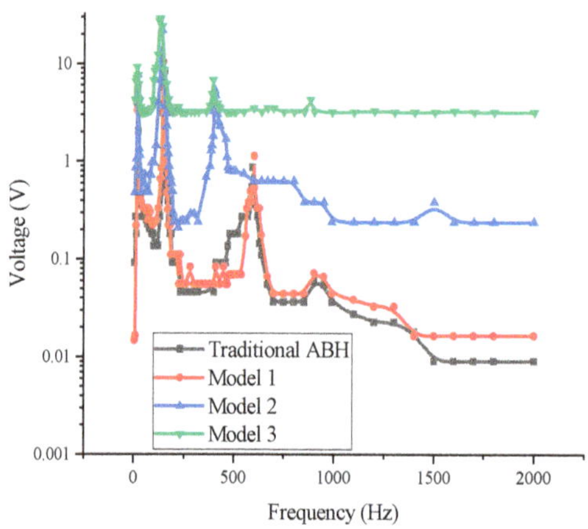

Figure 18. Voltage frequency response of several beams.

A comparison of the experimental and simulated natural frequencies of the VEHs is shown in Table 3. The deviation between the simulation and experiment is mainly due to the insufficient precision of actual machining and the vibration interference of the vibrator itself cannot be completely eliminated.

Table 3. Comparison of natural frequency.

Structure/Natural Frequency (Hz)	First			Second		
	Simulation	Test	Deviation	Simulation	Test	Deviation
Traditional ABH	18.1	20	9.5%	144.9	153	5.3%
Model 1	18.2	20	9.0%	149.6	146	2.5%
Model 2	16.4	18	8.9%	120.7	140	13.8%
Model 3	15.2	17	10.6%	110.5	130	13.5%

4.2.2. Optimal Impedance Matching of VEH with ABH

The second value of the natural frequency of each structure is used as the frequency of the external excitation, and the impedance matching is performed under the sinusoidal excitation with an amplitude of 3 N. The output voltage and power of VEHs with an ABH in the optimal impedance matching are shown in Figures 19 and 20.

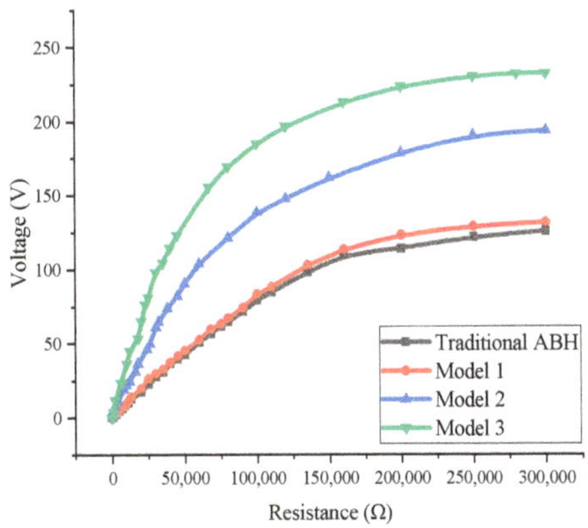

Figure 19. Output voltage of VEHs with different load resistances.

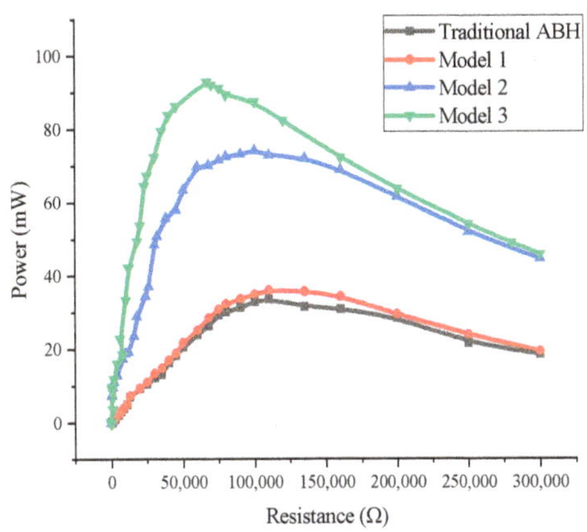

Figure 20. Output power of VEHs with different load resistances.

The optimal impedance value of the traditional VEH with an ABH is 115 kΩ, the output peak voltage is 84.78 V and the output power reaches 33.63 mW. The optimal value of VEH Model 1 is 110 kΩ, the output peak voltage is 88.36 V and the output power reaches 35.62 mW. The optimal value of VEH Model 2 is 100 kΩ, the output peak voltage is 137.58 V, and the output power reaches 75.01 mW. The optimal value of VEH Model 3 is 67 kΩ, the output peak voltage is 169.16 V, and the output power is 91.52 mW. The peak output voltage of VEH Model 3 is 1.9 times that of the VEH with the traditional ABH, and the peak output power of VEH Model 3 is 2.7 times that of the VEH with a traditional ABH. The harvesting efficiency of the VEH with an ABH can be improved significantly with the number of ABH arrays.

4.2.3. Transient Voltage Response Test of VEH with ABH

Each VEH with an ABH is excited with a sinusoidal signal in its second natural frequency, which is connected with optimal impedance matching. The transient voltage response of each VEH is measured under sinusoidal excitation with an amplitude of 3 N.

Figure 21 shows the time domain response of the transient output voltage of VEHs at optimal impedance. The average output voltage of each VEHs is shown in Figure 22. The average output voltage increases with the number of ABH arrays.

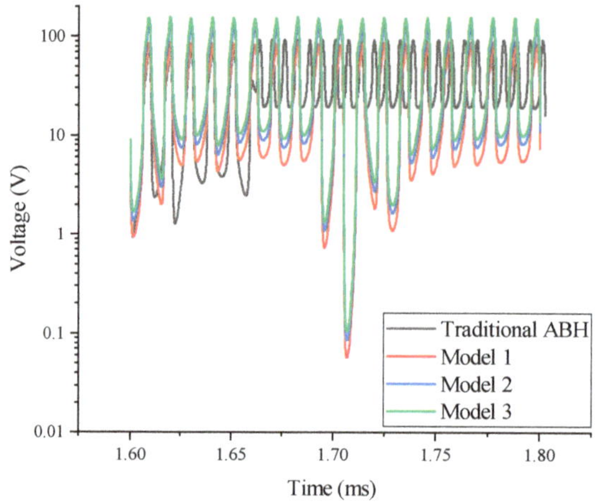

Figure 21. Transient output voltage of VEHs in optimal impedance.

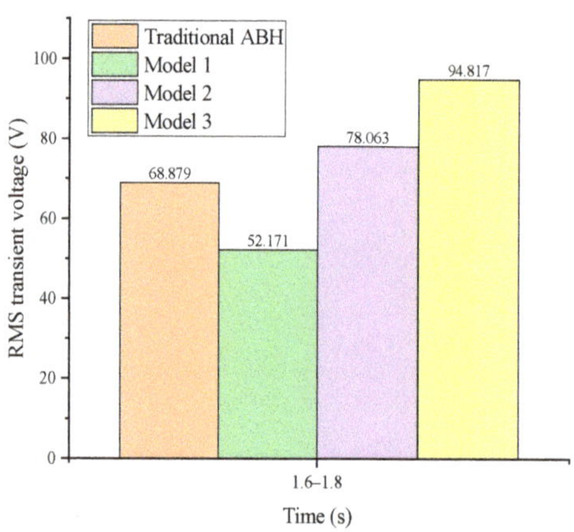

Figure 22. Transient average output voltage of VEHs with optimal impedance.

5. Conclusions

A piezoelectric VEH with a bilateral periodic one-dimensional ABH is proposed in this paper. First, the energy focus effect of an ABH in the VEH is simulated by COMSOL under a Gaussian pulse excitation. The simulation results show that the beam with a bilateral ABH is easier to generate more energy with than the traditional one. The performance of the VEHs with an ABH is tested in an experiment. The experimental results show that the output voltage and power of the VEH with a bilateral periodic 1D ABH is higher compared with the VEH with a traditional one-dimensional ABH and increased with the number of ABH arrays. The main research results are as follows:

(1) After the best impedance matching is completed: the output power of the traditional VEH reaches 33.63 mW, that of VEH Model 1 can reach 35.62 mW, that of VEH Model 2 can reach 75.01 mW and that of VEH Model 3 can reach 91.52 mW.

(2) Under the same external conditions: the output voltage of the traditional VEH reaches 84.78 V, that of VEH Model 1 can reach 88.36 V, that of VEH Model 2 can reach 137.58 V and that of VEH Model 3 can reach 169.16 V.

The energy harvesting performance of VEH Model 3 is the best. The results show that the energy collection efficiency will increase with the increase in the ABH arrays. It is expected that the proposed VEH with an ABH may provide an effective energy supply option for WSN sensor nodes, help to monitor the life of some equipment and, finally, promote green and sustainable development.

Author Contributions: Conceptualization, Z.Z. and H.W.; methodology, Z.Z. and C.Y.; software, Z.Z.; validation, H.W. and H.S.; formal analysis, Z.Z. and C.Y.; investigation, H.W. and H.S.; resources, C.Y.; data curation, H.W.; writing—original draft preparation, Z.Z. and Y.Y.; writing—review and editing, H.W. and Y.Y.; visualization, Z.Z. and H.W.; supervision, C.Y.; project administration, H.W. and C.Y.; funding acquisition, H.W. All authors have read and agreed to the published version of the manuscript.

Funding: Support was received from the overseas study visit and training program for outstanding young backbone talents of Anhui Province (gxgwfx2021035); the innovation team of Anhui Polytechnic University; the Special Display and Imaging Technology Anhui Technology Innovation Center open project support (2020AJ06001), Anhui Polytechnic University—Jiujiang District Industrial Collaborative Innovation Special Fund Project (2022cyxtb4); the Anhui Provincial Natural Science Foundation (No. 1908085QA13); the National Natural Science Foundation of China (No. 11902001) and the Natural Science Research Project of Institutions of Higher Education in Anhui Province (No. KJ2017A114).

Institutional Review Board Statement: Not applicable.

Informed Consent Statement: Not applicable.

Data Availability Statement: Not applicable.

Conflicts of Interest: The authors declare no conflict of interest.

Appendix A

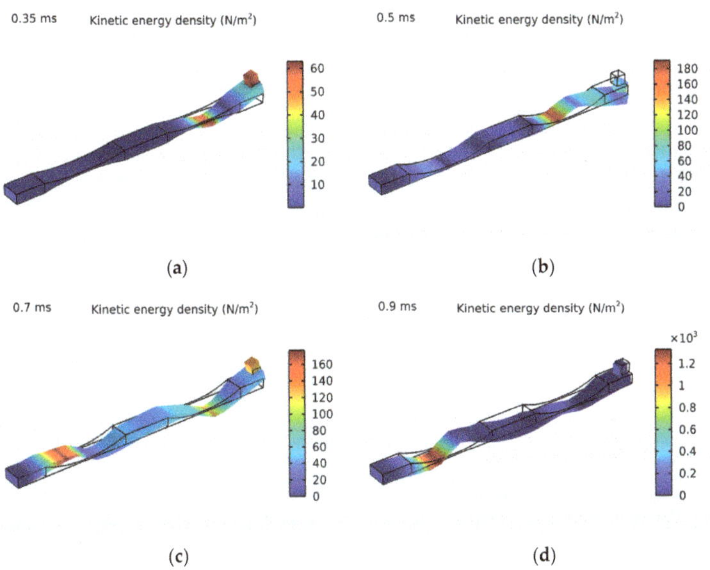

Figure A1. Kinetic energy density and strain of traditional ABH: (**a**) ABH in 0.35 ms, (**b**) ABH in 0.5 ms, (**c**) ABH in 0.7 ms, (**d**) ABH in 0.9 ms.

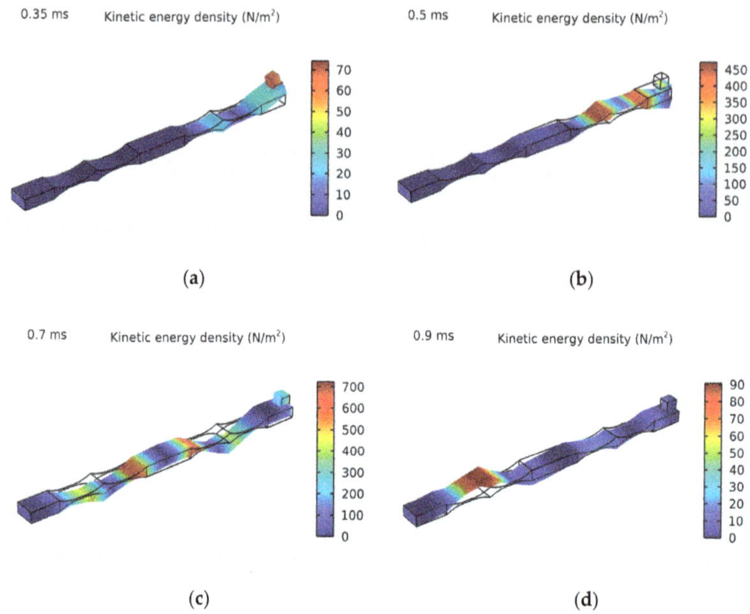

Figure A2. Kinetic energy density and strain of ABH model 2: (**a**) ABH in 0.35 ms, (**b**) ABH in 0.5 ms, (**c**) ABH in 0.7 ms, (**d**) ABH in 0.9 ms.

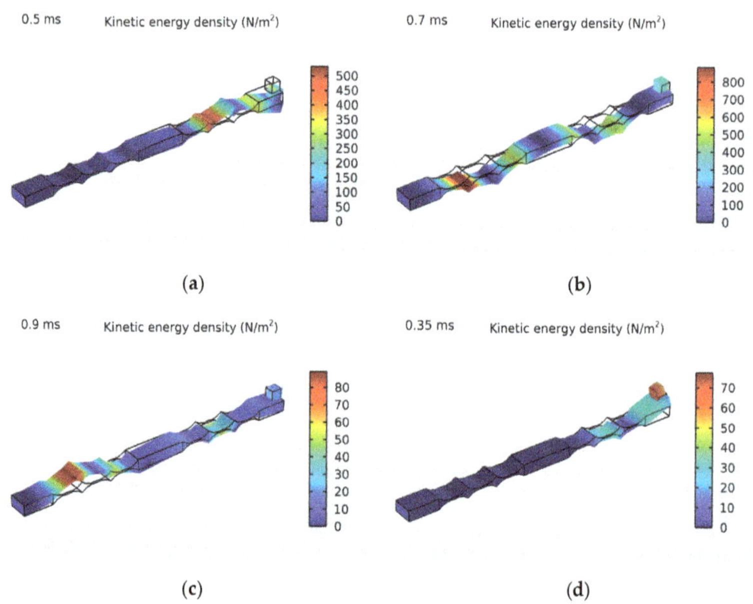

Figure A3. Kinetic energy density and strain of ABH Model 3: (**a**) ABH in 0.35 ms, (**b**) ABH in 0.5 ms, (**c**) ABH in 0.7 ms, (**d**) ABH in 0.9 ms.

Table A1. First four-order natural frequency in simulation.

Order Number/Structure	Traditional	Model 1	Model 2	Model 3
1 (Hz)	18.143	18.183	16.446	15.222
2 (Hz)	144.95	149.64	120.74	110.51
3 (Hz)	643.7	657.18	388.83	353.96
4 (Hz)	1368.3	1452.7	833.02	750.32

Table A2. First four-order natural frequency and output voltage in experiment.

Order Number/Structure	Traditional	Model 1	Model 2	Model 3
1 (Hz)	20	20	18	17
Output voltage (V)	4.08	5.83	7.82	9.36
2 (Hz)	153	146	140	130
Output voltage (V)	8.48	10.52	22.08	29.21
3 (Hz)	595	603	403	395
Output voltage (V)	0.86	1.14	5.34	6.89
4 (Hz)	900	900	800	780
Output voltage (V)	0.06	0.07	0.63	4.3

References

1. Bosso, N.; Magelli, M.; Zampieri, N. Application of low-power energy harvesting solutions in the railway field: A review. *Veh. Syst. Dyn.* **2021**, *59*, 841–871. [CrossRef]
2. Chen, F.; Yang, C.; Guo, Z.; Wang, Y.; Ma, X. A Magnetically Controlled Current Transformer for Stable Energy Harvesting. *IEEE Trans. Power Deliv.* **2022**, *38*, 212–221. [CrossRef]
3. Li, Z.; Yan, Z.; Luo, J.; Yang, Z. Performance comparison of electromagnetic energy harvesters based on magnet arrays of alternating polarity and configuration. *Energy Convers. Manag.* **2019**, *179*, 132–140. [CrossRef]
4. Li, Y. Design and Implementation of Electrostatic Energy Harvesters with Green Nanomaterials. Ph.D. Thesis, University of Guelph, Guelph, ON, Canada, 2018.
5. Yang, Y.; Wang, S.; Stein, P.; Xu, B.X.; Yang, T. Vibration-based energy harvesting with a clamped piezoelectric circular diaphragm: Analysis and identification of optimal structural parameters. *Smart Mater. Struct.* **2017**, *26*, 045011. [CrossRef]
6. Qi, L.; Pan, H.; Pan, Y.; Luo, D.; Yan, J.; Zhang, Z. A review of vibration energy harvesting in rail transportation field. *Iscience* **2022**, *25*, 103849. [CrossRef] [PubMed]
7. Li, H.; Doaré, O.; Touzé, C.; Pelat, A.; Gautier, F. Energy harvesting efficiency of unimorph piezoelectric acoustic black hole cantilever shunted by resistive and inductive circuits. *Int. J. Solids Struct.* **2022**, *238*, 111409. [CrossRef]
8. Krylov, V.V. New type of vibration dampers utilising the effect of acoustic 'black holes'. *Acta Acust. United Acust.* **2004**, *90*, 830–837.
9. Gao, N.; Wang, B.; Lu, K.; Hou, H. Complex band structure and evanescent Bloch wave propagation of periodic nested acoustic black hole phononic structure. *Appl. Acoust.* **2021**, *177*, 107906. [CrossRef]
10. Gao, N.; Wei, Z.; Hou, H.; Krushynska, A.O. Design and experimental investigation of V-folded beams with acoustic black hole indentations. *J. Acoust. Soc. Am.* **2019**, *145*, EL79–EL83. [CrossRef] [PubMed]
11. Gao, N.; Wei, Z.; Zhang, R.; Hou, H. Low-frequency elastic wave attenuation in a composite acoustic black hole beam. *Appl. Acoust.* **2019**, *154*, 68–76. [CrossRef]
12. Deng, J.; Guasch, O.; Zheng, L.; Song, T.; Cao, Y. Semi-analytical model of an acoustic black hole piezoelectric bimorph cantilever for energy harvesting. *J. Sound Vib.* **2021**, *494*, 115790. [CrossRef]
13. Zhao, L.; Conlon, S.C.; Semperlotti, F. Broadband energy harvesting using acoustic black hole structural tailoring. *Smart Mater. Struct.* **2014**, *23*, 065021. [CrossRef]
14. Ji, H.; Liang, Y.; Qiu, J.; Cheng, L.; Wu, Y. Enhancement of vibration based energy harvesting using compound acoustic black holes. *Mech. Syst. Signal Process.* **2019**, *132*, 441–456. [CrossRef]
15. Du, X.; Fu, Q.; Zhang, J.; Zong, C. Numerical and Experimental Study on Suppression Effect of Acoustic Black Hole on Vibration Transmission of Refrigerator Compressor. *Appl. Sci.* **2021**, *11*, 8622. [CrossRef]
16. Liang, H.; Liu, X.; Yuan, J.; Bao, Y.; Shan, Y.; He, T. Influence of Acoustic Black Hole Array Embedded in a Plate on Its Energy Propagation and Sound Radiation. *Appl. Sci.* **2022**, *12*, 1325. [CrossRef]
17. Lyu, X.; Ding, Q.; Ma, Z.; Yang, T. Ultra-Wide Bandgap in Two-Dimensional Metamaterial Embedded with Acoustic Black Hole Structures. *Appl. Sci.* **2021**, *11*, 11788. [CrossRef]
18. Sun, J.Q. Vibration and sound radiation of non-uniform beams. *J. Sound Vib.* **1995**, *185*, 827–843. [CrossRef]

19. Wagg, D.; Neild, S. *Nonlinear Vibration with Control*; Springer: New York, NY, USA, 2010.
20. Nayfeh, A.H.; Mook, D.T. *Nonlinear Oscillations*; Wiley: New York, NY, USA, 1995.

energies

Article

Design of Flexible FeCoSiB/ZnO Thin-Film Multiferroic Module for Low-Frequency Energy Harvesting

Yan Guo [1], Chen Yang [2] and Bin Huang [2,*]

[1] College of Science & Technology, Ningbo University, Ningbo 315300, China; guoyan@nbu.edu.cn
[2] Piezoelectric Device Laboratory, Faculty of Mechanical Engineering & Mechanics, Ningbo University, Ningbo 315211, China; yc0467@126.com
* Correspondence: huangbin@nbu.edu.cn

Abstract: Multiphase magnetoelectric (ME) composites deposited on flexible substrates have been widely studied, which can respond to ambient mechanical, magnetic, and electric field excitations. This paper reports an investigation of flexible FeCoSiB/ZnO thin-film generators for low-frequency energy harvesting based on three substrates. Both hard substrate Si and flexible substrates (Polyethylene terephthalate (PET) and Polyimide (PI)) are adopted to make a comparison of energy conversion efficiency. For the single ME laminate, a PET-based flexible ME generator presents the best ME coupling performance with an average coupling voltage output of ~0.643 mV and power output of ~41.3 nW under the alternating magnetic field of 40 Oe and 20 Hz. The corresponding ME coupling coefficient reaches the value of 321.5 mV/(cm·Oe) for this micrometer scale harvester. Flexible ME modules with double cantilevered ME generators are further designed and fabricated. When two PET-based generators are connected in series, the average voltage output and power are ~0.067 mV and ~0.447 nW, respectively. Although the energy harvested by ME thin-film generators is much smaller than bulk multiferroic materials, it proves the feasibility of using flexible FeCoSiB/ZnO generators for harvesting ambient magnetic energy and supplying sustainable electronic devices in the future.

Keywords: magnetoelectric coupling; flexible; thin-film; energy harvesting; low frequency

1. Introduction

With the rapid development of electronic technology, the development of intelligent devices towards miniaturization, integration, multi-function, and flexibility has led to the emergence of flexible electronic products, breaking the application bottleneck and service scope of traditional electronic products [1–4]. The flexibility and ductility of flexible substrates enable Micro-Electro-Mechanical System (MEMS) devices based on flexible substrates to achieve integrated display, sensing, signal acquisition, and other functions to meet different demands [5–9]. At present, one of the main problems faced by flexible electronic devices is that the battery is insufficient for continuous power supply or inconvenient to replace the battery [10–12]. In addition to improving the energy density of the battery, energy from the external environment can be harvested and converted into electrical energy in order to solve the problem of power supply. The harvested energy can be stored in a battery or directly used to drive the electronic components as part of the energy supply element. To this end, researchers have developed a variety of energy-harvesting devices to power electronic products, including triboelectric nanogenerators, piezoelectric nanogenerators, photovoltaic devices, thermoelectric devices, and magnetoelectric composite devices [13–17]. The integration of acquisition devices on flexible substrates can give full play to the flexible and ductile properties of flexible electronic devices. In recent years, flexible materials based on polymers have received extensive attention, for example, polyimide (PI), polyterephthalic acid (PET), polydimethylsiloxane (PDMS), and other polymers used in flexible sensors, generators, and resonators [18–21].

Magnetoelectric (ME) thin-film materials are multifunctional materials with both magnetostrictive and piezoelectric properties to realize the electric polarization of magnetic field-controlled materials or electric field-induced magnetic ordering [22–27]. ME thin-film materials have excellent application prospects in ME sensors, memories, and transducers [28–31]. They can also be used for non-contact wireless energy harvesting systems, which have many advantages and can be applied to the power supply of flexible wearable and implantable electronic devices. For energy-harvesting devices, improving the energy conversion efficiency and expanding the harvesting range of ME structures are effective ways to increase the harvested energy. Chen et al. [32] proposed a method that can harvest the surrounding magnetic energy in a wide frequency range, not simply at the resonant frequency. Shashank et al. [33] proposed an anisotropic self-biased dual-phase (magnetostrictive phase and piezoelectric phase) low-frequency magneto-mechano-electric energy harvester with giant energy density. They compared the output voltage under single-source excitation and dual-phase superimposed excitation and found that the output voltage driven by the dual-phase acquisition mechanism was significantly strengthened compared with the single-phase excitation. Subsequently, Ryu et al. [34] proposed a PMN-PZT/Ni/Pi multilayer flexible ME energy harvester with an output voltage of ~9.5 V and an output power density of ~46 mW/(cm^3/Oe^2) under a low-frequency driving magnetic field of ~500 μT and 60 Hz. It can light up 35 LED lamps in parallel with the turn on/off frequency of ~1 Hz and can also drive a wireless sensor network module under the magnetic field of ~700 μT. Wu [35] proposed a multiferroic module to harvest biomechanical energy to supply mobile electronic devices. The reported open-circuit voltage and the short-circuit current of the developed block ME composite-based energy-harvesting device were ~17 V and ~7.2 μA, respectively, which can be used for temperature/humidity sensors or night running indicators. Zou et al. [36] fabricated an array of microbridge-structured ME thin-film sensors with 1 μm PZT film and FeCoSiB layer deposited on a Si substrate. They reported that the ME coupling coefficient could be ~16.87 V/(cm·Oe) in resonance and the sensitivity was enhanced greatly when three identical sensors were connected in series. Pham et al. [37] studied the effect of magnetic layer thickness on the inverse ME effect of ME nanocomposite materials under a low-biasing magnetic field and found that the induced magnetization under an electric field reached 245%. Nan et al. [38] conducted a quantitative analysis of the ME behavior of ultra-thin multiferroic heterostructures, which were co-mediated by strain and charge.

The above literature is mainly about ME energy harvesting either for block ME composites or for ME composites on hard substrates. There are also studies of micro/nano size ME thin films based on flexible substrates for energy harvesting and sensing devices. The fabrication method of flexible ME thin films can be referred to in the following two works. He et al. [39] deposited ZnO films on flexible PI and PET films to fabricate dual-mode flexible surface acoustic wave-based ultraviolet light sensors and investigated the sensitivities of the two resonant modes. Bai et al. [40] fabricated a ME surface acoustic wave resonator based on a flexible PI substrate and found that it had a high electromechanical coupling coefficient when the thickness of the piezoelectric layer and magnetostrictive layer were both 1.1 μm. However, the reported voltage and power of ME thin-film harvesters are usually in microvolt and nanowatt levels for energy harvesting and sensing, which are much smaller compared with the block ME devices [41]. Thus, more effort has to be made to develop more sophisticated power supply devices based on flexible ME composites. Furthermore, material composition, device fabrication, and potential application scenarios have to be also more deeply investigated.

In this work, we present FeCoSiB/ZnO ME thin-film generators fabricated on flexible substrates, which are capable of converting magnetic energy into electrical energy under alternating magnetic fields. The present generators are designed to harvest low-frequency magnetic energy and can be applied to various scenarios, such as walking, running, vehicle vibration, and so on. For the present devices, the piezoelectric layer and magnetostrictive layer are deposited on three different substrates to study the ME coupling performance of

ME thin films. The cantilevered structure is adopted to investigate the coupling voltage under alternating magnetic fields with low frequencies. The results will show that the PET-based ME thin-film cantilevered structure has the highest conversion efficiency with an average voltage output of ~0.643 mV and power output of ~41.3 nW under the alternating magnetic field of 40 Oe and 20 Hz. The corresponding ME coupling coefficient will also be calculated. Moreover, a flexible multiferroic ME module is further developed and fabricated with a 3D-printed frame, a movable permanent magnet, and two cantilevered flexible ME thin-film generators clamped at the top of the frame. This module can simulate the low-frequency alternating magnetic field by the movable magnet. With a series connection of two generators, the average voltage output and power output of PET-based flexible ME module are ~0.067 mV and ~0.447 nW, respectively, which are much larger than the output of Si-based module under low-frequency excitation signals. Although the harvested energy is not as significant as other bulk ME devices, the results prove that the proposed multiferroic film module can harvest magnetic energy and realize the ME coupling under low-frequency excitations. This work demonstrates the feasibility of flexible ME thin-film generators used for low-frequency energy harvesting and they can be served as a power supply source for directly supplying various sensors in the future.

2. Materials and Methods

2.1. Fabrication of FeCoSiB/ZnO Laminates on Substrates

Iron cobalt silicon boron (FeCoSiB) and zinc oxide (ZnO) [42,43] with considerable magnetostrictive coefficients and piezoelectric coefficients were used as the magnetostrictive phase and piezoelectric phase in our work. FeCoSiB is a kind of magnetostrictive material widely used as magnetic sensors that have good mechanical and magnetic performance such as large permeability, large magnetostriction coefficient, low resistivity, and low Curie temperature. ZnO is also suitable for fabricating flexible devices compared with other piezoelectric ceramics. Both magnetron sputtering targets are cuboid-shaped and were deposited on three substrates (Si, PI, and PET) by magnetron sputtering (Discovery-635 sputtering system), shown in Figure 1a. The sputtered size of the two phases was 20 mm × 12 mm in length and width, and the substrate was cut into the size of 30 mm × 12 mm × 200 μm. The substrate is 10 mm longer than the sputtered functional layers for clamping and much thicker than the functional layers. The substrate was cleaned with acetone solution, ethanol solution, and hydrofluoric acid solution to prevent the adhesion of organic matter and stains from affecting the experimental results before deposition. It should be noted that hydrofluoric acid solution is only used to clean the silica on the silicon surface instead of all substrates. The chromium (Cr) layer was first deposited between the electrode layer and the substrate to prevent the film from debonding and curling deformation. The copper (Cu) layer was deposited on the substrate as a lower electrode, and ZnO and FeCoSiB layers were deposited in the next steps. Finally, Cu was deposited on the ME layer again as the upper electrode layer. After sputtering, the deposited substrate was placed in the chamber for 2 h to prevent the layered structure from cracking due to thermal expansion and contraction of the film.

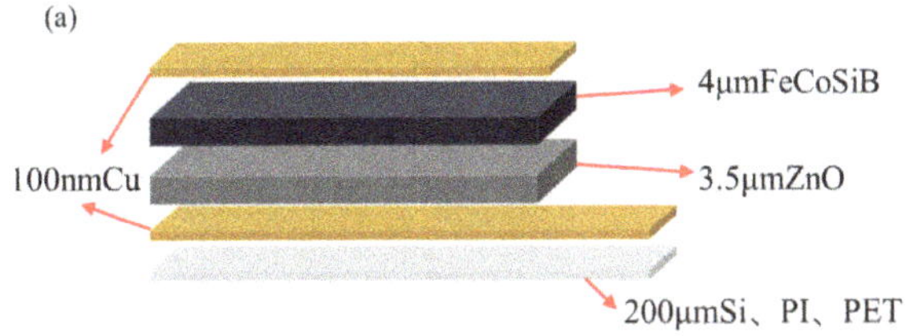

Figure 1. *Cont.*

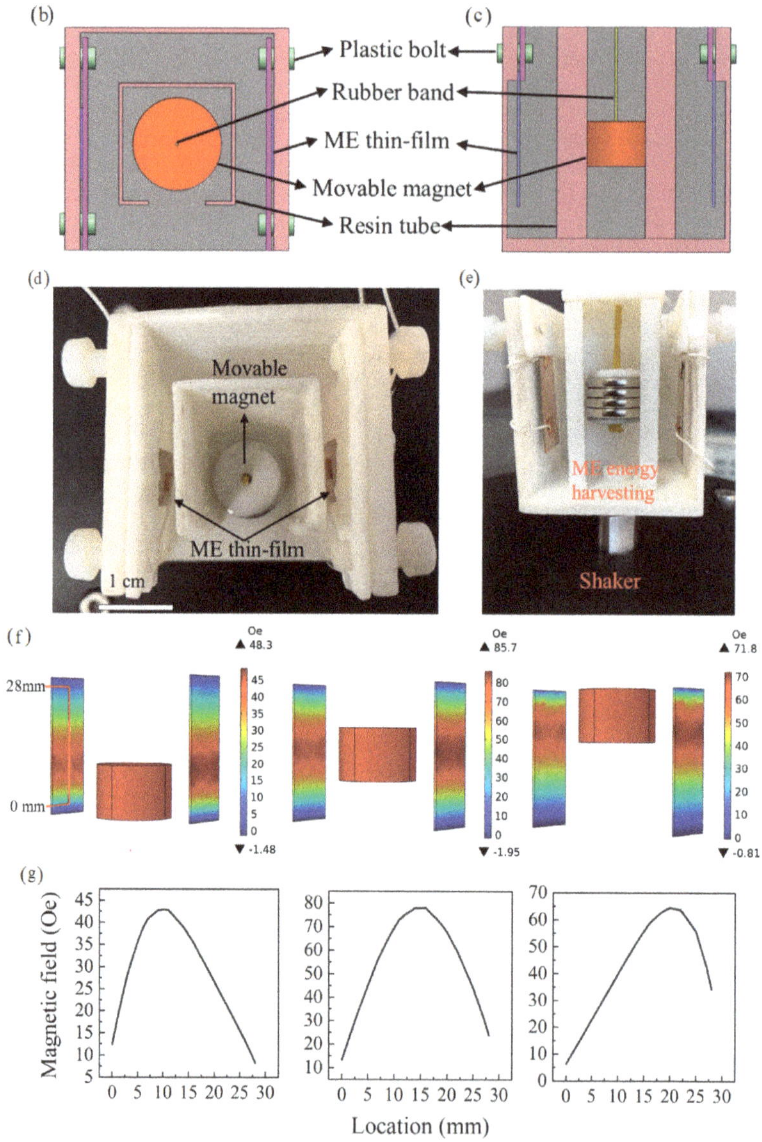

Figure 1. Framework of ME thin-film energy harvester: (**a**) initial design of ME thin-film composition. (**b**) top view of functional structure. (**c**) orthographic view. (**d**) top view of double cantilevered ME module. (**e**) orthographic view of double cantilevered ME module. (**f**) COMSOL model for magnetic field simulation with three locations of the movable magnet. (**g**) Magnetic field changes along the laminate with three locations of the movable magnet.

2.2. Design of Double Cantilevered Flexible ME Thin-Film Module

The double cantilevered ME thin-film module is designed and fabricated with flexible substrate and Si substrate. The frame of the ME module was fabricated by 3D printing using resin with a bottom area of 45 × 45 mm^2 and a height of 50 mm. A pair of ME thin-film laminates were clamped with a plastic bolt at the top of the frame on two sides of the resin tube, shown in Figure 1b,c. The NdFeB permanent magnet with the size of

Φ18 mm × 3 mm was hung by a rubber band in the resin tube to keep it moving vertically and provide an alternative magnetic field to the ME devices under mechanical vibration excitation by the shaker. The fabricated ME module is shown in Figure 1d,e. The permanent magnet is vertically movable, and the magnetic field to ME devices changes during the vertical motion of the magnet. The permanent magnet NdFeB N35 is provided by Shanxi Dajinhua Magnetic Co., Ltd., and the residual flux density Br is 1.25T. Due to the symmetric configuration, two ME devices are expected to have similar ME coupling performance. For the purpose of improving the efficiency of energy harvesting, quadrangular, hexagonal, and octagonal shapes of the 3D-printed frame can be designed in the future, which means four, six, and eight ME thin-film laminates can be fastened at the top. However, we only used two laminates to present the feasibility, which is enough to prove efficiency. We use COMSOL package to simulate the magnetic field applied to the ME laminate as shown in Figure 1f,g. For the magnetic field, no current interface is used with the magnetic flux conservation condition. The simulation is carried out by the steady-state analysis with a physics-controlled mesh of normal size. The results show that at three different vertical locations of the magnet, the magnitudes of the magnetic field applied to the laminate are different, with the middle location of the magnet presenting the maximum field and the lower limit position displaying the minimum field to the ME laminate. The peak values for the three cases are about 43 Oe, 78 Oe, and 65 Oe, respectively.

2.3. Characterization of ME Thin-Film

In this paper, iron cobalt silicon boron (FeCoSiB) and zinc oxide (ZnO) were sputtered on three substrates that are placed in the magnetron sputtering chamber simultaneously so as to control the same fabrication condition. The fabricated specimens are shown in Figure 2a, and they are supposed to have the same thickness as the functional layers. Tapes are attached at the top and bottom corners of the substrates before sputtering so that we can distinguish them by the colors of the substrates in the tape-attached area. Since the sputtering parameters and temperature have a great influence on the thickness of the film deposited, the sputtering rate measured by the step meter may be influenced and different from the expected rate with the increase in deposition time and temperature. Therefore, we measured the thickness of the specimens by using scanning electron microscopy (SEM). Before scanning the cross-sectional area of the specimen, we sputtered platinum (Pt) to the specimens to increase the conductivity. Since the flexible substrate is prone to collapse under high-voltage scanning, only the functional layers deposited on the Si substrate are presented to demonstrate the cross-sectional thickness. From the SEM figure, as shown in Figure 2b for the Si substrate, we can clearly see the lamination from this specimen, and we further use the energy dispersive spectrometer (EDS) to analyze the elements of each layer as shown in Figure 2c–e, which represent the bottom Cu layer, ZnO layer, and FeCoSiB layer, respectively. Figure 2c shows the elements in the bottom electrode layer, which has a large fraction of Cu. However, Si, Fe, and Zn elements can also be found since these elements can be easily penetrated the lower layers. The Zn element is shown in Figure 2d but not presented in Figure 2e, implying that Figure 2d is the ZnO layer and Figure 2e is the FeCoSiB layer. From the XRD figure in Figure 2f, we can see that the peak of ZnO film is shown near 34.14° and the peak of FeCo is shown near 43.06°. The crystal orientations are (002) and (110) for ZnO and FeCo. For the specimens, ZnO film has better grain orientation on Si and PET substrates than PI substrate based on the heights of peaks. The thickness of the ZnO layer is about 3.5 μm, and the thickness of the FeCoSiB layer is about 4 μm, which can help calculate the coupling coefficient in the following sections. The magnetic hysteresis loop of FeCoSiB film on the Si substrate is shown in Figure 3. It is found that the hysteresis loops along the long and short axes are not significantly different, indicating that the films are isotropic in plane.

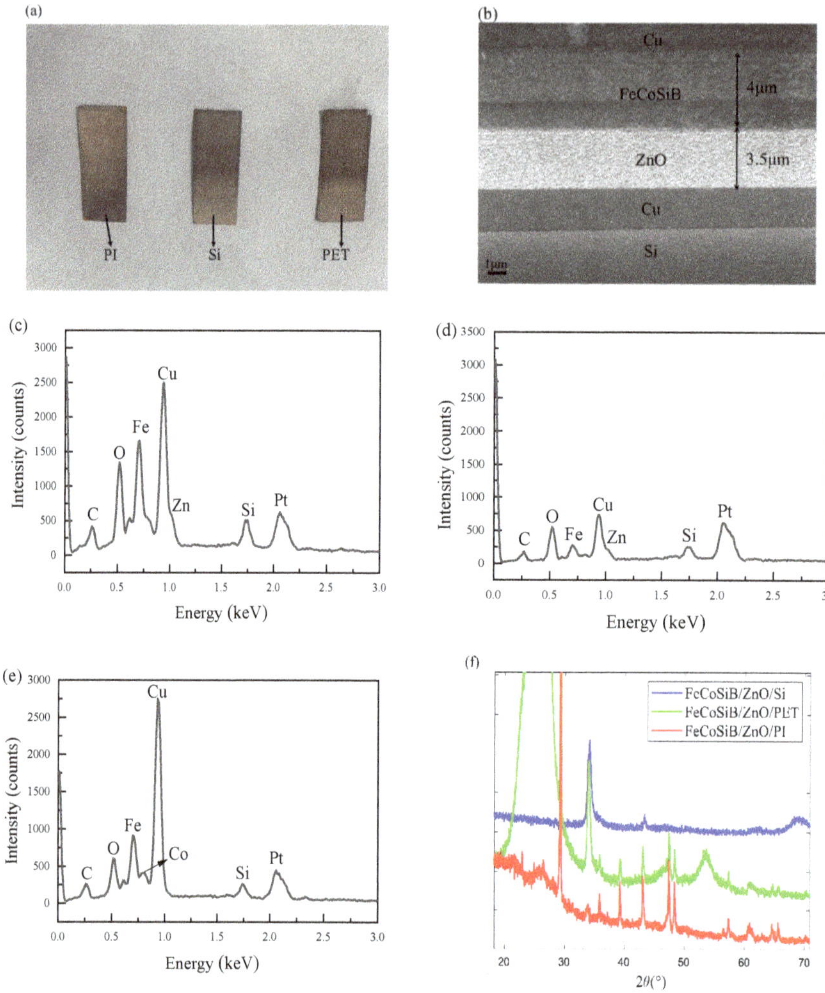

Figure 2. Characterization of ME thin-film: (**a**) fabricated specimens. (**b**) SEM figure of the fabricated specimen on Si substrate. EDS results of (**c**) Cu layer, (**d**) ZnO layer, (**e**) FeCoSiB layer, and (**f**) XRD result.

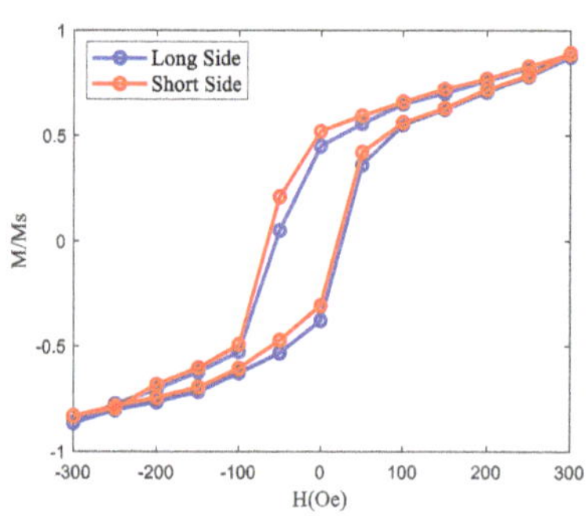

Figure 3. The hysteresis loop FeCoSiB film on Si substrate.

3. Results

3.1. ME Performance of Single Cantilevered FeCoSiB/ZnO Thin-Film Generator

An experiment test platform was built for the performance test of FeCoSiB/ZnO thin-film generators as shown in Figure 4. The AC magnetic field was generated by the Helmholtz coil driven by the signal generator (Tektronix AFG 1022) that is connected to the power amplifier (Aigtek ATA-3080). The ME thin-film generator was clamped and placed in the middle of the coil as so to guarantee a uniform magnetic field applied to the ME device, and the probe of the Gauss meter was fixed as close as possible to the ME thin-film generator to measure the magnetic field accurately. To measure the voltage and power output of the ME device, a multimeter (Keithley DMM-7510) was used and connected to a laptop to demonstrate the time history of output signals. The internal resistance of the multimeter is 10 MΩ when measuring the output voltage, which is much larger than the resistance of the device. With this testing system, we measured the voltage output, current output, and power output with different amplitudes and frequencies of the magnetic field. The ME performance of magnetoelectric composites can be evaluated by the magnetoelectric coupling coefficient as $\alpha_{me} = V/(dH)$ [44], where V is the voltage difference between the upper and lower surface of the piezoelectric layer measured by the multimeter, d is the thickness of the piezoelectric layer, and H is the magnitude of the applied magnetic field.

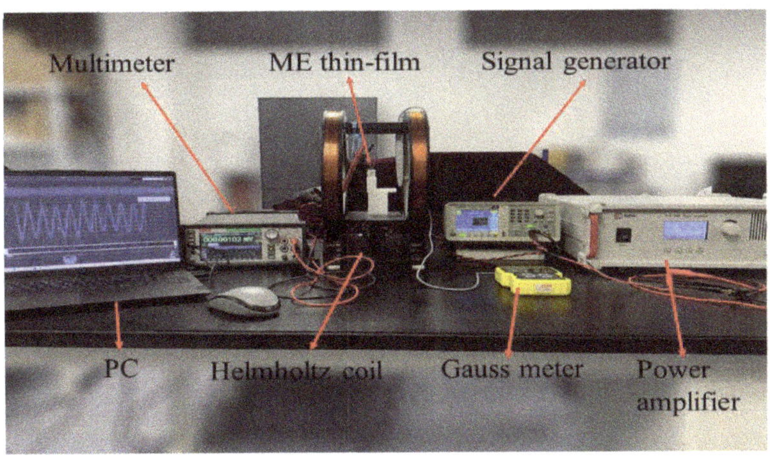

Figure 4. The ME testing system for thin-film energy harvester.

The single cantilevered ME laminates based on three substrates under various low-frequency magnetic fields are investigated for the voltage output. Four driving alternating magnetic fields (2 Hz, 4 Hz, 10 Hz, and 20 Hz) with amplitudes of 10 Oe, 20 Oe, 30 Oe, and 40 Oe, respectively, are applied to three ME laminates. It is expected that the flexible substrates and hard substrates will influence the coupling differently due to the material properties. These results are illustrated in Figures 5–7 for PI substrate, PET substrate, and Si substrate, respectively. For the PI substrate, as shown in Figure 4, it is first found that by increasing the amplitude of the driving magnetic field, the magnitude of the voltage output increases significantly. When the frequency of the driving magnetic field increases from 2 Hz to 20 Hz, the maximum magnitude of the voltage output increases from 0.179 mV to 0.875 mV for the case of 40 Oe. This result shows that at a low-frequency range, with increasing the frequency of the driving magnetic field, the coupling voltage output increases significantly. For the PET substrate, the results are shown in Figure 6 for four driving frequencies and four amplitudes of the magnetic field. Correspondingly, it is found that the maximum amplitudes of the output voltage are 0.214 mV, 0.418 mV, 0.989 mV, and 1.050 mV for the case of 40 Oe, respectively. Thus, with PET substrate, the

ME laminate presents better coupling performance due to the smaller Young's modules of PET material. Figure 7 shows the voltage output of the ME laminate on the Si substrate. As expected, Si is much stiffer than the polymer substrates so that the voltage output is the smallest among the three cases. From the graph, it is found that the maximum amplitudes of the output voltage are 0.143 mV, 0.275 mV, 0.492 mV, and 0.582 mV for the case of 40 Oe, respectively. Consequently, we plot the effect of the magnitude of the driving magnetic field on the coupling voltage output in Figure 8 for three substrates. We can easily find the linear relationship between the magnitude of the magnetic field and voltage output for three cases. In the present analysis range, the voltage output increases linearly with the increase in the magnetic field. The increase in frequency also increases the voltage output. However, the increment from 10 Hz to 20 Hz is not as significant as the increment from 2 Hz to 4 Hz or 4 Hz to 10 Hz. This implies that the ME performance may not increase significantly when the frequency reaches some value at the low-frequency range. In a word, the coupling of ME thin-film with PET substrate is the strongest among the three cases that are mainly determined by the smallest elastic modulus of PET substrate.

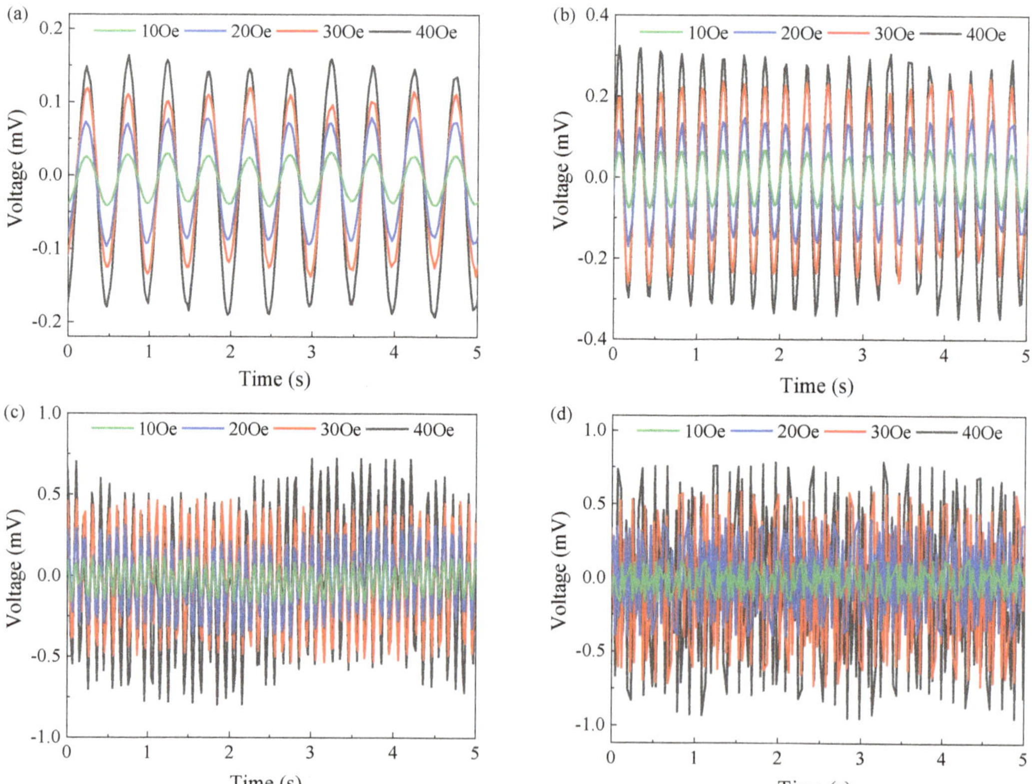

Figure 5. The effect of magnitude and frequency of the driving magnetic field on the voltage output of ME laminate on PI substrate: (**a**) 2 Hz, (**b**) 4 Hz, (**c**) 10 Hz, and (**d**) 20 Hz.

To further investigate the ME-coupling performance, we calculate the average voltage output and power for the above results. Since the current is supposed to be only several picoamperes by using the multimeter, the average power is calculated by using the internal resistance of the multimeter 10 MΩ and the average voltage output instead of measuring the current for power calculation. The results are shown in Figure 9 for the three substrates. The average voltage and power reflect that the harvested energy intensity is better than the peak value. The maximum average voltage outputs under 40 Oe magnetic field are 0.532 mV, 0.643 mV, and 0.355 mV for Pi, PET, and Si, respectively. In addition, the power outputs are 28.3 nW, 41.3 nW, and 12.6 nW for Pi, PET, and Si, respectively. The tendency of the curve changes with respect to the magnitude of the magnetic field and alternating frequency is similar to the curves of maximum value shown in Figure 8. However, the average output power increases dramatically when the frequency increases from 4 Hz to 10 Hz. We further calculate the ME-coupling coefficients by using the average voltage output, which is 266 mV/(cm·Oe), 321.5 mV/(cm·Oe), and 177.5 mV/(cm·Oe) for Pi, PET, and Si, respectively. The current outputs are also measured with the multimeter internal resistance of 1 kΩ, which can be regarded as the load resistance. It should be noted that the resistance of the multimeter is automatically adjusted based on the measurement range. The current outputs under four frequencies and four amplitudes of the magnetic field are given in Figures 10–12 for the three substrates. For the PI substrate, the maximum current outputs are 149.1 pA, 207.5 pA, 234.0 pA, and 276.5 pA when the exciting frequencies are 2 Hz, 4 Hz, 10 Hz, and 20 Hz, respectively. The average current outputs are shown in Figure 13, which demonstrates that the current output performance of PET is also better than those of PI and Si substrates. Thus, these results demonstrate that the single cantilevered ME laminate on PET substrate presents the best ME-coupling performance under low-frequency magnetic field excitation.

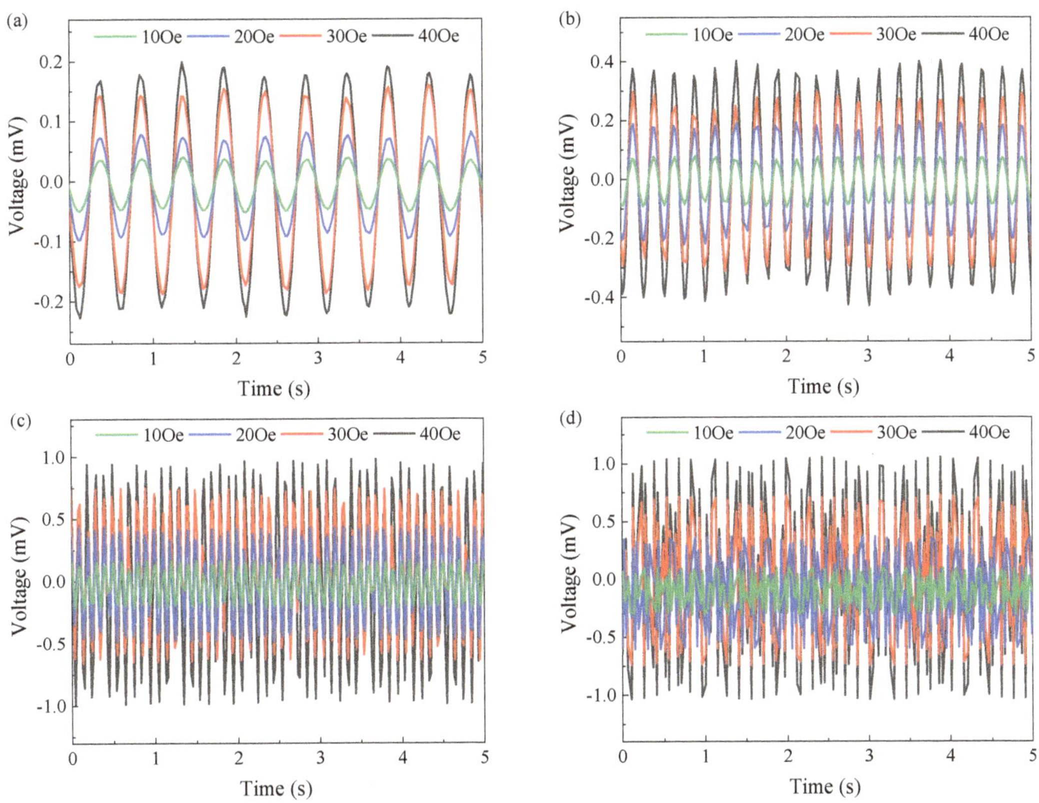

Figure 6. The effect of magnitude and frequency of the driving magnetic field on the voltage output of ME laminate on PET substrate: (**a**) 2 Hz, (**b**) 4 Hz, (**c**) 10 Hz, and (**d**) 20 Hz.

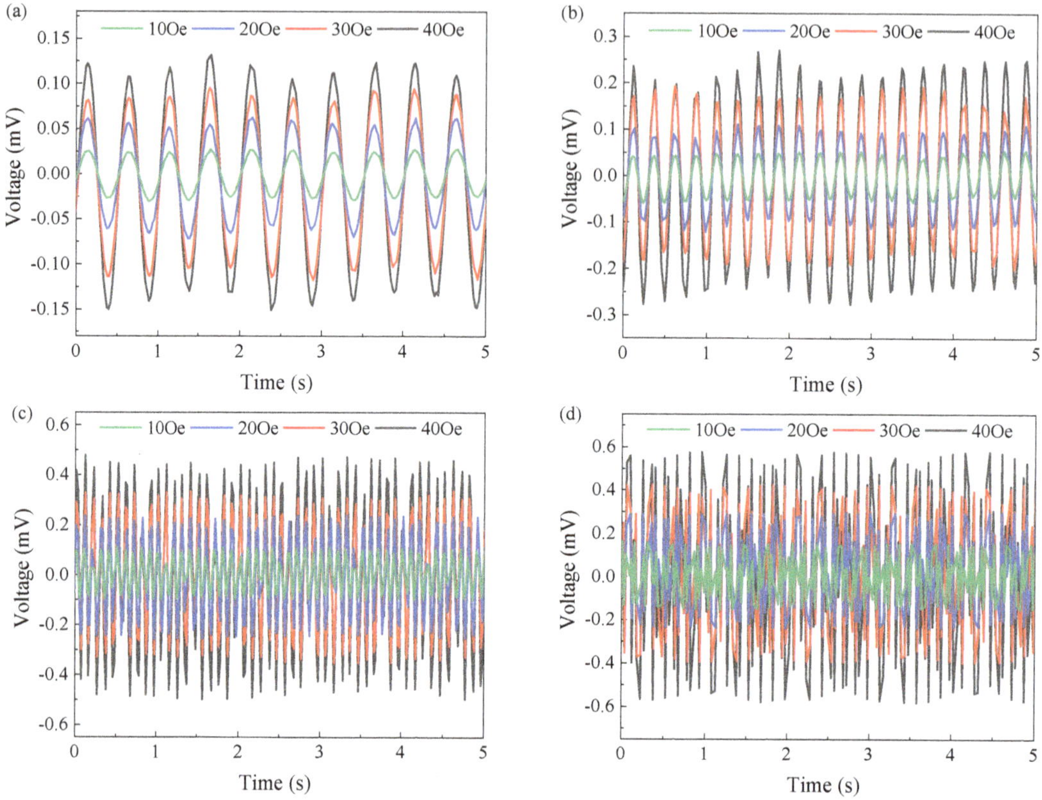

Figure 7. The effect of magnitude and frequency of the driving magnetic field on the voltage output of ME laminate on Si substrate: (**a**) 2 Hz, (**b**) 4 Hz, (**c**) 10 Hz, and (**d**) 20 Hz.

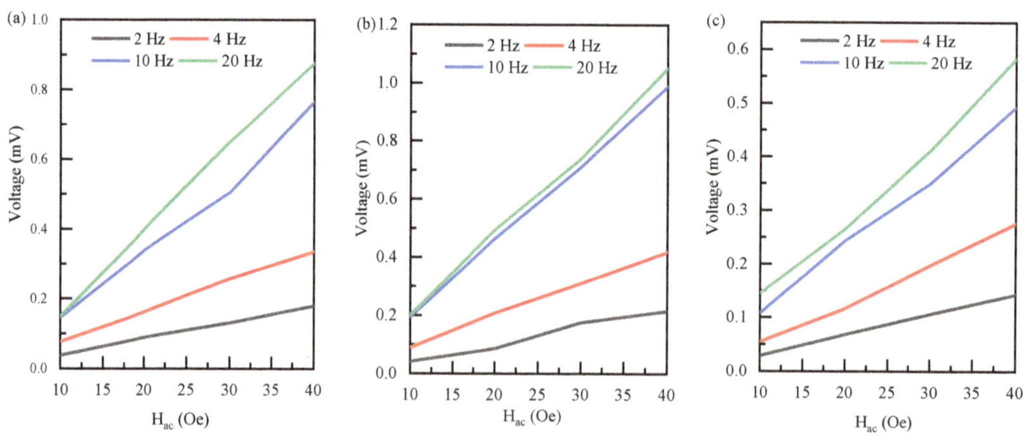

Figure 8. Effect of the magnitude of driving magnetic field on the coupling voltage output with different frequencies: (**a**) PI, (**b**) PET, and (**c**) Si.

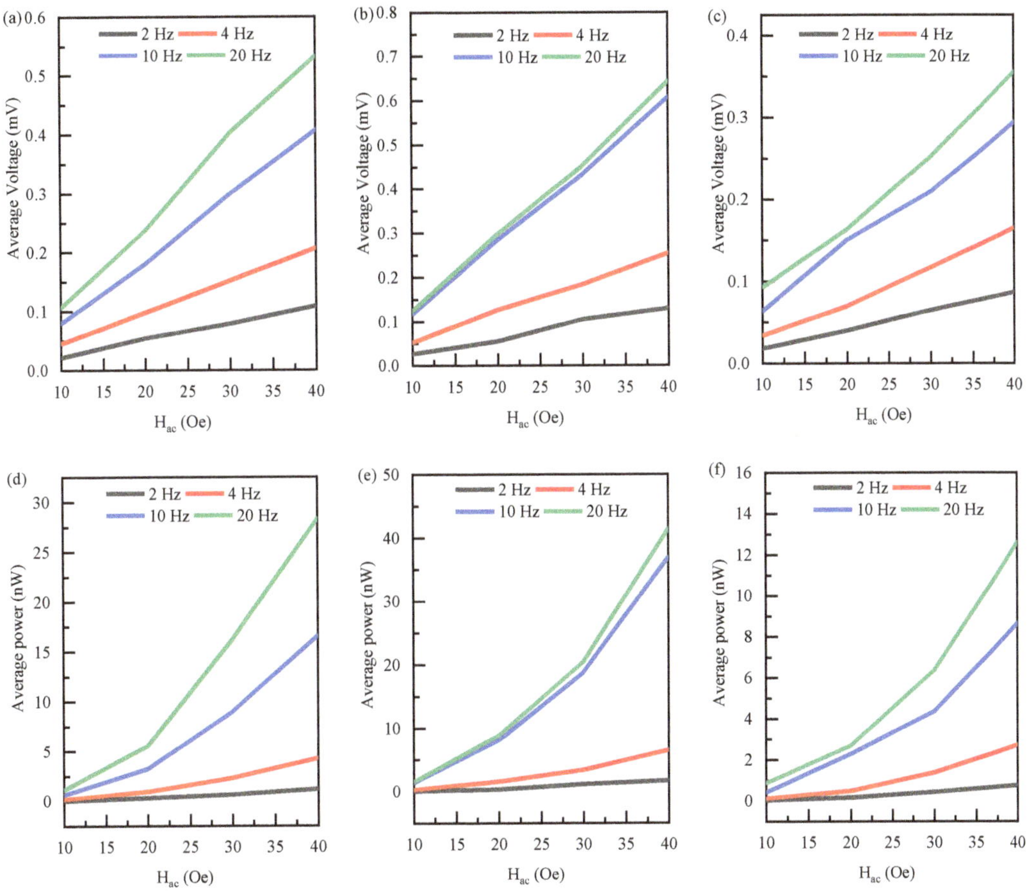

Figure 9. Output of average voltage and power for different substrates: (**a**) and (**d**) PI, (**b**) and (**e**) PET, (**c**) and (**f**) Si.

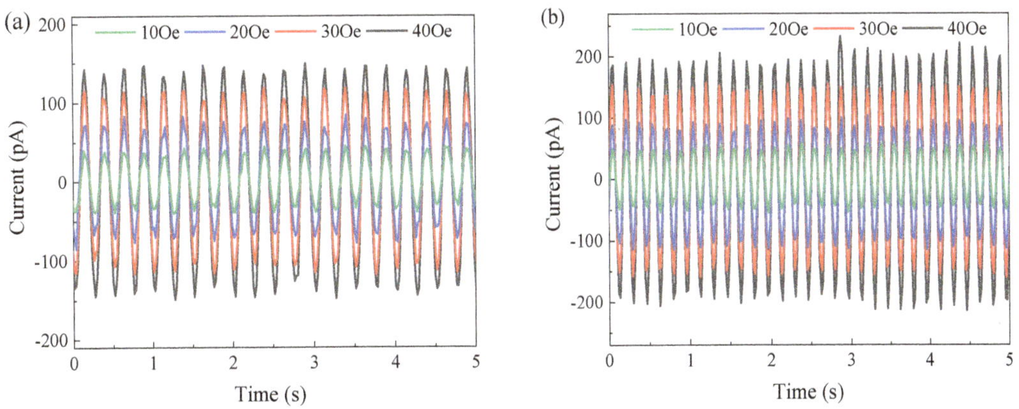

Figure 10. *Cont.*

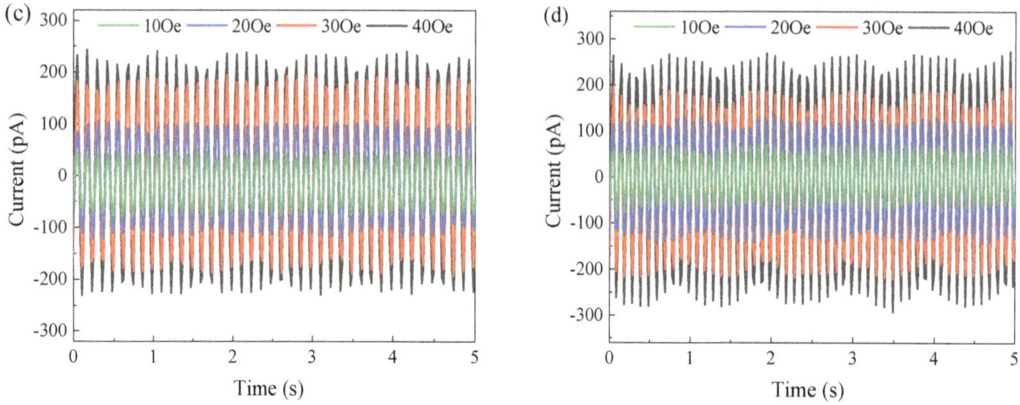

Figure 10. The effect of magnitude and frequency of the driving magnetic field on the current output of ME laminate on PI substrate: (**a**) 2 Hz, (**b**) 4 Hz, (**c**) 10 Hz, and (**d**) 20 Hz.

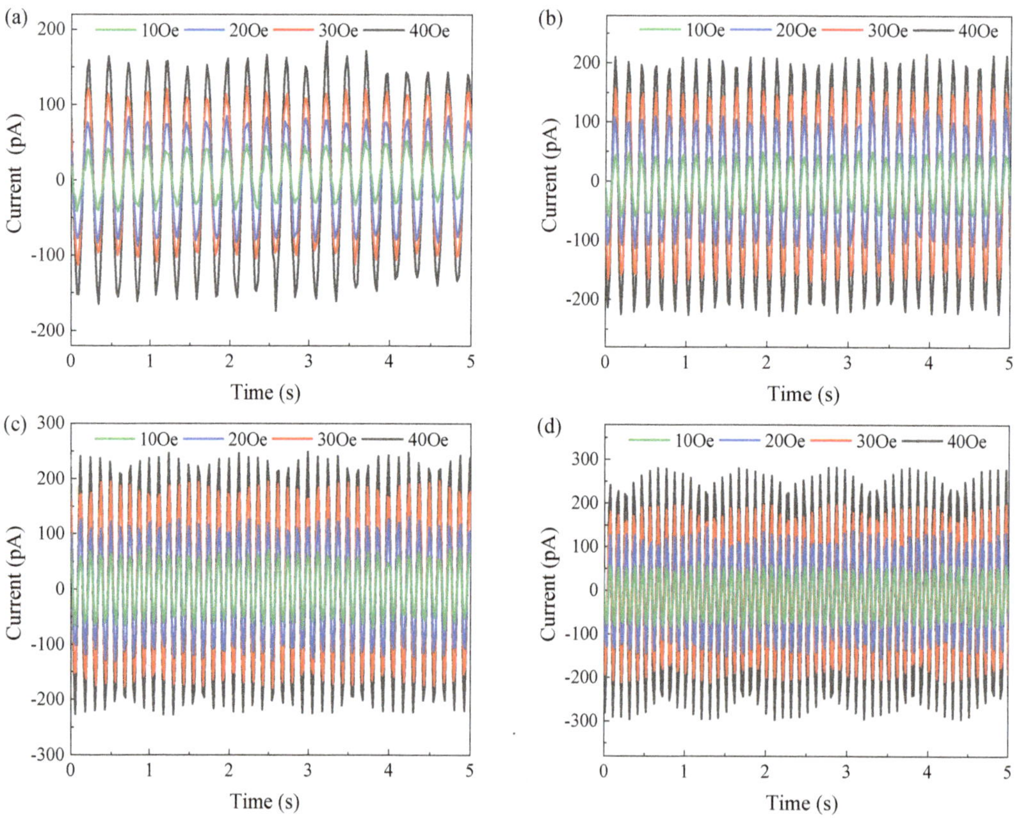

Figure 11. The effect of magnitude and frequency of the driving magnetic field on the current output of ME laminate on PET substrate: (**a**) 2 Hz, (**b**) 4 Hz, (**c**) 10 Hz, and (**d**) 20 Hz.

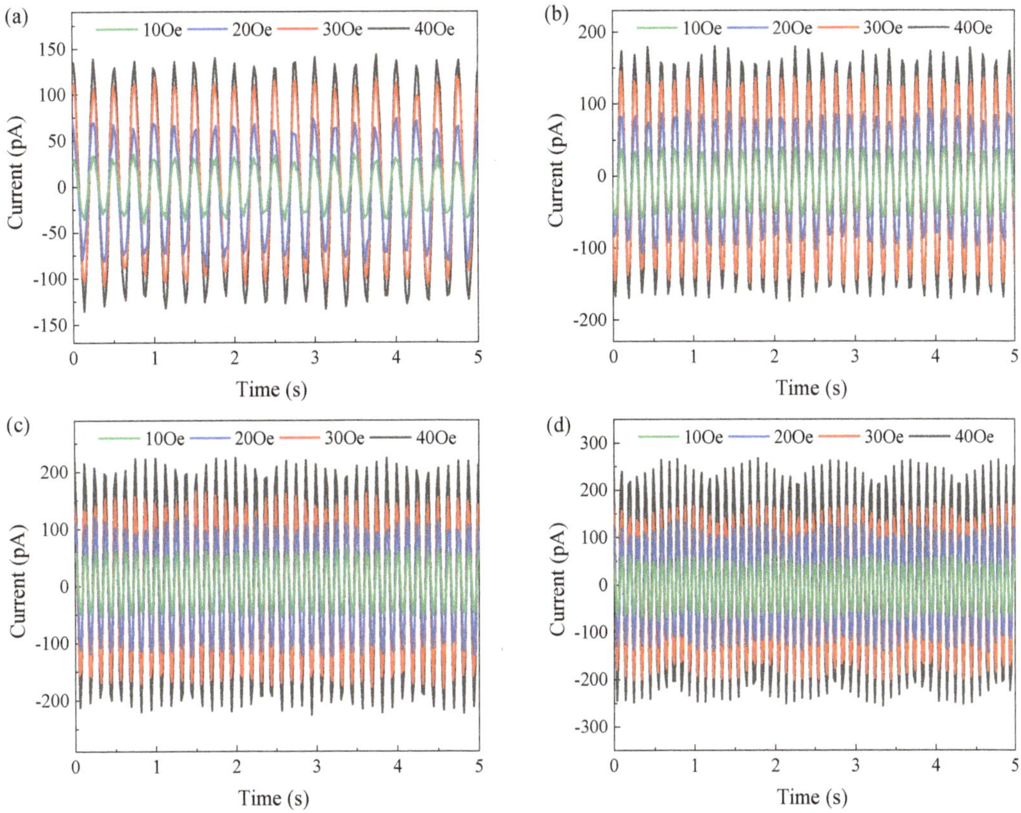

Figure 12. The effect of magnitude and frequency of the driving magnetic field on the current output of ME laminate on Si substrate: (**a**) 2 Hz, (**b**) 4 Hz, (**c**) 10 Hz, and (**d**) 20 Hz.

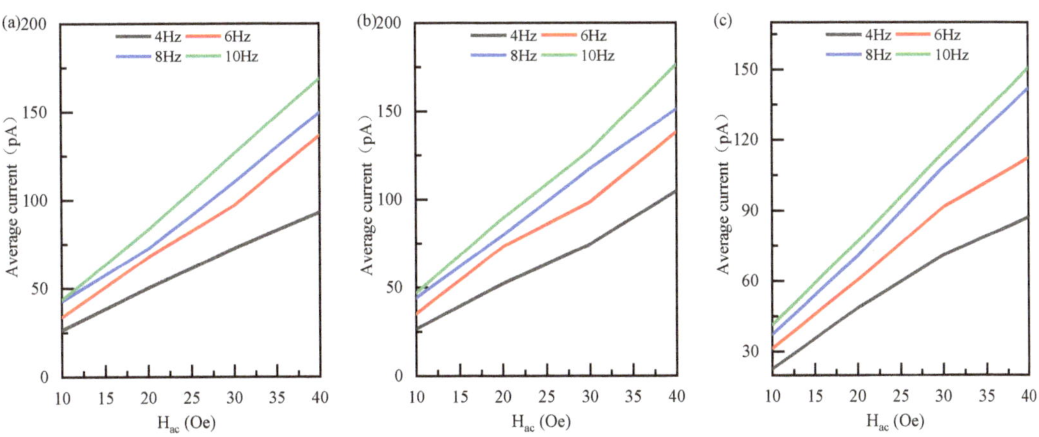

Figure 13. Output of average current for different substrates: (**a**) PI, (**b**) PET, (**c**) Si.

3.2. Demonstration of Double Cantilevered Flexible ME Module for Low-Frequency Energy Harvesting

The demonstration of a double cantilevered ME module for low-frequency energy harvesting was conducted with a mechanical excitation by a shaker, as shown in Figure 1e. The shaker is connected to a signal generator and a power amplifier to generate harmonic and vertical motion. The vertical motion of the shaker excites the frame and causes the permanent magnet to move up and down resulting in the magnetic field change applied to the ME devices. Thus, this module with the external shaker can simulate energy harvesting from an external low-frequency magnetic field, and the whole process can be regarded as a mechano-magneto-electric energy conversion system. According to the previous investigation, single laminate on PET substrate presents better ME performance than PI substrate. Only PET and Si substrates are adopted for the ME module under low-frequency (2 Hz, 3 Hz, and 4 Hz), and the results are shown in Figures 14 and 15, respectively. Both voltage output and their power spectral density (PSD) are presented in these figures. We measure the voltage output of sample #1, sample #2, and two samples connected in series as well. In Figure 14, it is found that sample #1 and sample #2 have similar outputs since they are fabricated by the same process, where the coupling voltage outputs of the two samples are 0.022 mV and 0.022 mV, 0.053 mV and 0.050 mV, and 0.058 mV and 0.054 mV at 2 Hz, 3 Hz, and 4 Hz, respectively. In addition, the output in series is almost the sum of the two individual outputs, where the coupling voltage outputs are 0.041 mV, 0.095 mV, and 0.105 mV, respectively. It is also found that as the frequency increases, the voltage output of the ME module also increases. The results in Figure 14(a-2,b-2,c-2) are obtained by calculating the PSD of the voltage for the conditions of 2 Hz, 3 Hz, and 4 Hz, respectively. The resonance peaks in these figures can be observed, which represent high energy density under specific frequencies. For example, these peaks in Figure 14(a-2) are at the frequencies of 2.4 Hz, 4.8 Hz, 7.3H z, and 9.6H z for the load condition of 2 Hz. Similarly, for the load condition of 3 Hz and 4 Hz, there are also several peaks presented, but with different magnitudes and locations. However, most of them are within 10 Hz implying that the excitation at the low-frequency peaks can induce large energy density for the proposed device.

Figure 15 shows the ME voltage output of the module based on the Si substrate as well as the PSD results. The same mechanical excitation is applied to the module in order to make a comparison with the PET substrate. In this case, the experimental results show that the two samples output the voltages of 0.017 mV and 0.015 mV, 0.040 mV and 0.036 mV, and 0.045 mV and 0.038 mV at 2 Hz, 3 Hz, and 4 Hz, respectively, while for the series-connected double cantilevered ME module, the coupling voltage outputs are 0.028 mV, 0.071 mV, and 0.078 mV, respectively. These values are much smaller than those of the PET substrate. The PSD results are also presented to demonstrate the frequency characteristics, where the peaks can be easily found at different locations and with different magnitudes. These peaks are also found to be within 10–15 Hz, which proves that it is possible for low-frequency energy harvesting. Thus, by comparing the coupling voltage output of the double cantilevered ME module with PET substrate and Si substrate, the device with PET substrate exhibits much more excellent performance due to its flexibility and ductility, which makes the double cantilevered flexible ME module more efficient in energy harvesting at low frequencies. Similarly, Table 1 gives the average voltage output and power output of the double cantilevered flexible ME energy harvester to reveal the energy harvesting efficiency more properly. The values given in Table 1 are much smaller than those in Section 3.1 since the effective magnetic field change to the ME devices is much smaller than those generated by the Helmholtz coil driven by a signal generator and power amplifier. As shown in Figure 1g, although the change in a maximum magnetic field can be about 30 Oe when the magnet moves from the lowest position to the highest position, its effective change to the whole laminate is much smaller. This results in a much smaller output for the double-cantilevered ME module in low-frequency energy harvesting. The hanging spring also affects the movement of the permanent magnet from the top to the bottom. However, the results prove the efficiency and feasibility of using a flexible ME module for low-frequency energy harvesting. As we discussed earlier, to harvest more energy from the change of low-frequency magnetic field, more cantilevered ME laminates can be fixed in the present structure due to the superposition of voltage output for the ME module connected in series. That will increase the voltage and power output considerably.

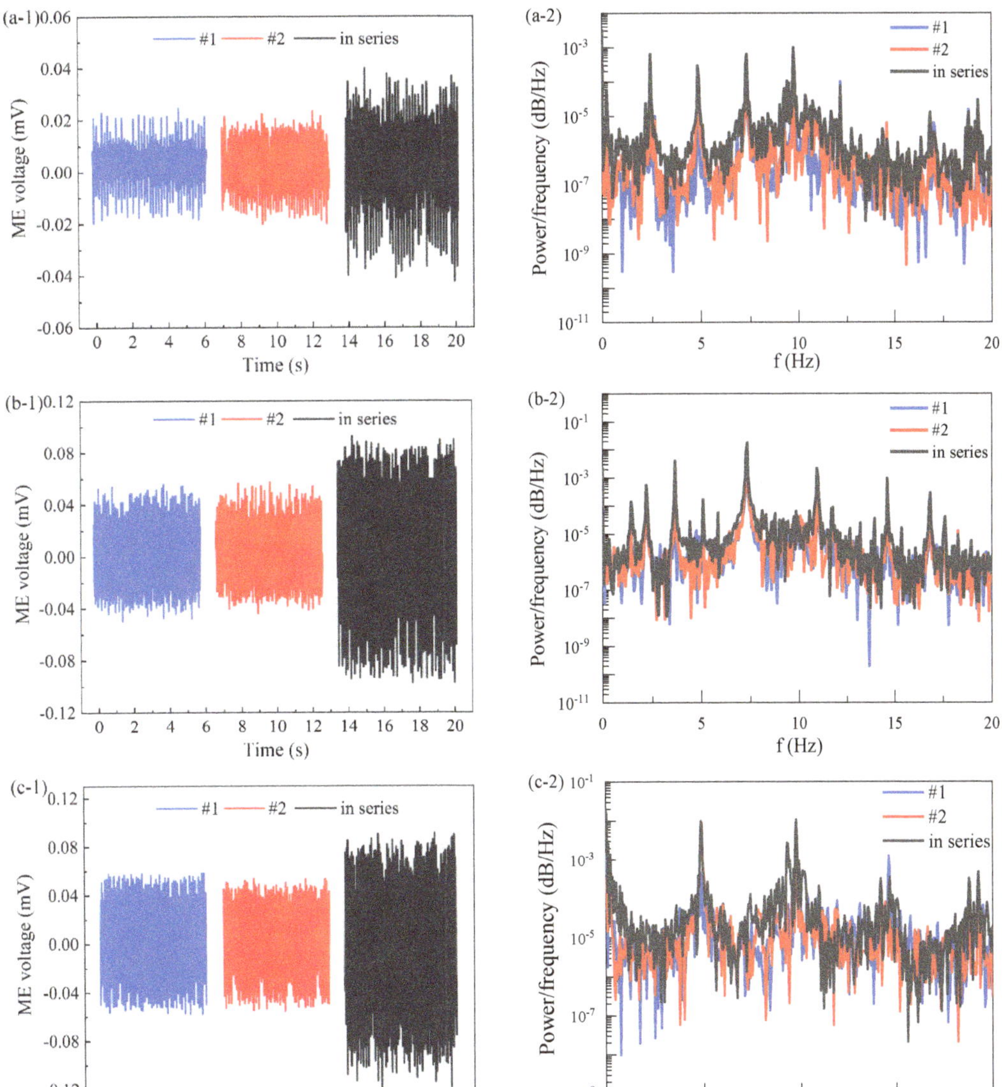

Figure 14. Voltage output and PSD of the voltage based on PET substrate for laminate 1 and 2 separately, and connected in series: (**a-1**) voltage and (**a-2**) its PSD under 2 Hz magnetic field, (**b-1**) voltage and (**b-2**) its PSD under 3 Hz magnetic field, and (**c-1**) voltage and (**c-2**) its PSD under 4 Hz magnetic field.

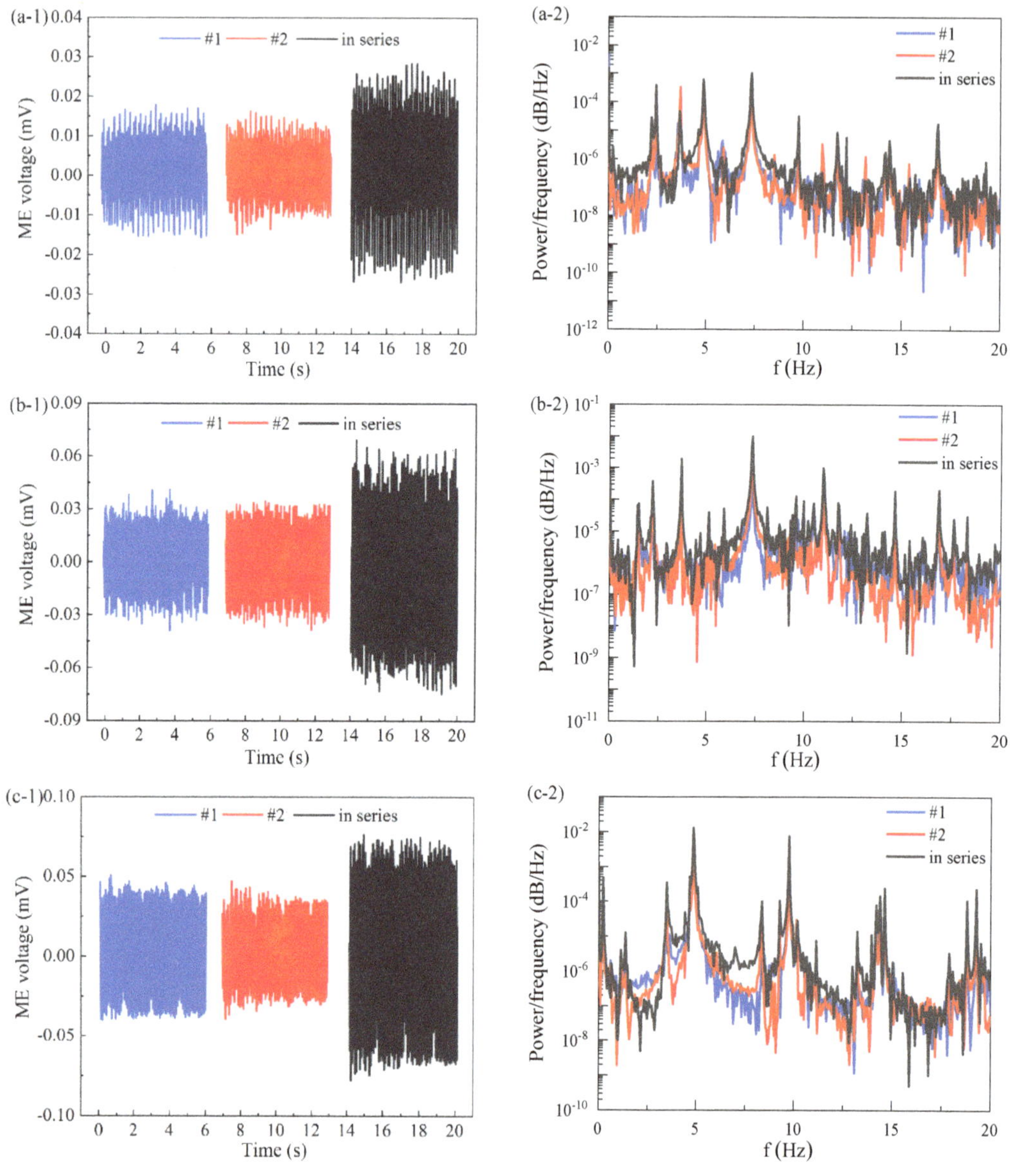

Figure 15. Voltage output and PSD of the voltage based on Si substrate for laminate 1 and 2 separately, and connected in series: (**a-1**) voltage and (**a-2**) its PSD under 2 Hz magnetic field, (**b-1**) voltage and (**b-2**) its PSD under 3 Hz magnetic field, and (**c-1**) voltage and (**c-2**) its PSD under 4 Hz magnetic field.

Table 1. Average voltage output and power of the double cantilevered flexible ME module.

Substrate	Hz \ Laminate	2	3	4
PET	#1	0.014 mV	0.034 mV	0.037 mV
		0.020 nW	0.114 nW	0.137 nW
	#2	0.014 mV	0.032 mV	0.034 mV
		0.020 nW	0.101 nW	0.118 nW
	In series	0.026 mV	0.061 mV	0.067 mV
		0.068 nW	0.366 nW	0.447 nW
Si	#1	0.011 mV	0.025 mV	0.029 mV
		0.012 nW	0.065 nW	0.082 nW
	#2	0.010 mV	0.023 mV	0.024 mV
		0.009 nW	0.053 nW	0.059 nW
	In series	0.018 mV	0.045 mV	0.050 mV
		0.032 nW	0.205 nW	0.247 nW

4. Conclusions

In this work, we demonstrated low-frequency energy harvesting using flexible ME generators deposited on three substrates. Due to the ME-coupling effect, the fabricated micrometer scale ME laminates can successfully convert magnetic energy into electrical energy under a low-frequency magnetic field. The effects of the amplitude and frequency of the magnetic field on the coupling voltage output are investigated in the present work. It is found that single cantilevered ME laminate based on PET substrate possesses the best ME coupling performance. The average voltage output is ~0.643 mV, the output power is ~41.3 nW under the alternating magnetic field of 40 Oe and 20 Hz, and the corresponding ME coupling coefficient is 321.5 mV/(cm·Oe). We further designed the double cantilevered ME module for harvesting low-frequency magnetic energy generated by a moving permanent magnet. The results of the harvester with a pair of flexible PET based ME laminates connected in series show that the average voltage output and power are ~0.067 mV and ~0.447 nW, respectively. These results demonstrate the feasibility of using flexible substrates for ME harvesters, which show better performance than hard substrates. The PSD results demonstrate the frequency characteristics of the measurements and prove that the device can harvest a low-frequency magnetic field. Since the ME thin films are at the micrometer scale, the harvested voltage and power are at the millivolt and nanowatt levels that are much smaller than bulk ME harvesters and still not enough to directly supply to electronic devices currently. However, it is enough for the energy consumption of microelectronics and sensing devices. There are ways to improve energy harvesting efficiency, such as increasing film thickness, using more ME laminates, depositing films with large coupling coefficients, and improving the structural design. The present design of a flexible multiferroic ME module can harvest low-frequency magnetic energy from many scenarios and could be used as sensors or supplied to wearable and portable electronic devices in the near future.

Author Contributions: Conceptualization, B.H.; methodology, Y.G. and C.Y.; formal analysis, B.H.; investigation, C.Y.; data curation, Y.G. and C.Y.; writing—original draft preparation, Y.G. and C.Y.; writing—review and editing, B.H.; visualization, Y.G.; supervision, B.H.; project administration, B.H.; funding acquisition, B.H. All authors have read and agreed to the published version of the manuscript.

Funding: This work was supported by the National Natural Science Foundation of China (Grant no. 11702150), Natural Science Foundation of Zhejiang Province (Grant no. LY21A020003), Natural Science Foundation of Ningbo (Grant no. 202003N4015).

Data Availability Statement: The data presented in this study are available in the article.

Conflicts of Interest: The authors declare no conflict of interest.

References

1. Jiang, D.; Shi, B.; Ouyang, H.; Fan, Y.; Wang, Z.L.; Chen, Z.-M.; Li, Z. A 25-year bibliometric study of implantable energy harvesters and self-powered implantable medical electronics researches. *Mater. Today Energy* **2020**, *16*, 100386. [CrossRef]
2. Kim, J.; Son, D.; Lee, M.; Song, C.; Song, J.-K.; Koo, J.H.; Lee, D.J.; Shim, H.J.; Kim, J.H. A wearable multiplexed silicon nonvolatile memory array using nanocrystal charge confinement. *Sci. Adv.* **2016**, *2*, e1501101. [CrossRef]
3. Pang, C.; Lee, C.; Suh, K.-Y. Recent advances in flexible sensors for wearable and implantable devices. *J. Appl. Polym. Sci.* **2013**, *130*, 1429–1441. [CrossRef]
4. Lü, C.; Zhang, Y.; Zhang, H.; Zhang, Z.; Shen, M.; Chen, Y. Generalized optimization method for energy conversion and storage efficiency of nanoscale flexible piezoelectric energy harvesters. *Energy Convers. Manag.* **2019**, *182*, 34–40. [CrossRef]
5. Forrest, S.R. The path to ubiquitous and low-cost organic electronic appliances on plastic. *Nature* **2004**, *428*, 911–918. [CrossRef] [PubMed]
6. Xue, H.; Yang, Q.; Wang, D.; Luo, W.; Wang, W.; Lin, M.; Liang, D.; Luo, Q. A wearable pyroelectric nanogenerator and self-powered breathing sensor. *Nano Energy* **2017**, *38*, 147–154. [CrossRef]
7. Sekitani, T.; Yokota, T.; Zschieschang, U.; Klauk, H.; Bauer, S.; Takeuchi, K.; Takamiya, M.; Sakurai, T.; Someya, T. Organic nonvolatile memory transistors for flexible sensor arrays. *Science* **2009**, *326*, 1516–1519. [CrossRef] [PubMed]
8. O'Handley, R.C.; Huang, J.K.; Bono, D.C.; Simon, J. Improved wireless, transcutaneous power transmission for in vivo applications. *IEEE Sens. J.* **2008**, *8*, 57–62. [CrossRef]
9. Hwang, G.T.; Byun, M.; Jeong, C.K.; Lee, K.J. Flexible piezoelectric thin-film energy harvesters and nanosensors for biomedical applications. *Adv. Healthc. Mater.* **2015**, *4*, 646–658. [CrossRef]
10. Goodenough, J.B.; Park, K.S. The Li-ion rechargeable battery: A perspective. *J. Am. Chem. Soc.* **2013**, *135*, 1167–1176. [CrossRef]
11. Niu, S.; Wang, X.; Yi, F.; Zhou, Y.S.; Wang, Z.L. A universal self-charging system driven by random biomechanical energy for sustainable operation of mobile electronics. *Nat. Commun.* **2015**, *6*, 8975. [CrossRef]
12. Zheng, Q.; Tang, Q.; Wang, Z.L.; Li, Z. Self-powered cardiovascular electronic devices and systems. *Nat. Rev. Cardiol.* **2021**, *18*, 7–21. [CrossRef] [PubMed]
13. Fan, F.-R.; Tian, Z.-Q.; Lin Wang, Z. Flexible triboelectric generator. *Nano Energy* **2012**, *1*, 328–334. [CrossRef]
14. Wang, X.; Song, J.; Liu, J.; Wang, Z.L. Direct-current nanogenerator driven by ultrasonic waves. *Science* **2007**, *316*, 102–105. [CrossRef]
15. Zhang, P.; Li, W.; Li, S.; Wang, Y.; Xiao, W. Reliability assessment of photovoltaic power systems: Review of current status and future perspectives. *Appl. Energy* **2013**, *104*, 822–833. [CrossRef]
16. Morozovska, A.N.; Eliseev, E.A.; Svechnikov, G.S.; Kalinin, S.V. Nanoscale electromechanics of paraelectric materials with mobile charges: Size effects and nonlinearity of electromechanical response of SrTiO3films. *Phys. Rev. B* **2011**, *84*, 045402. [CrossRef]
17. Onuta, T.-D.; Wang, Y.; Long, C.J.; Takeuchi, I. Energy harvesting properties of all-thin-film multiferroic cantilevers. *Appl. Phys. Lett.* **2011**, *99*, 203506. [CrossRef]
18. Li, B.; Kavaldzhiev, M.N.; Kosel, J. Flexible magnetoimpedance sensor. *J. Magn. Magn. Mater.* **2015**, *378*, 499–505. [CrossRef]
19. Gaspar, J.; Fonseca, H.; Paz, E.; Martins, M.; Valadeiro, J.; Cardoso, S.; Ferreira, R.; Freitas, P.P. Flexible magnetoresistive sensors designed for conformal integration. *IEEE Trans. Magn.* **2017**, *53*, 1–4. [CrossRef]
20. Pan, C.T.; Liu, Z.H.; Chen, Y.C. Study of broad bandwidth vibrational energy harvesting system with optimum thickness of PET substrate. *Curr. Appl. Phys.* **2012**, *12*, 684–696. [CrossRef]
21. Yang, Y.; Jung, J.H.; Yun, B.K.; Zhang, F.; Pradel, K.C.; Guo, W.; Wang, Z.L. Flexible pyroelectric nanogenerators using a composite structure of lead-free KNbO(3) nanowires. *Adv. Mater.* **2012**, *24*, 5357–5362. [CrossRef] [PubMed]
22. Wan, J.G.; Wang, X.W.; Wu, Y.J.; Zeng, M.; Wang, Y.; Jiang, H.; Zhou, W.Q.; Wang, G.H.; Liu, J.M. Magnetoelectric $CoFe_2O_4$–$Pb(Zr,Ti)O_3$ composite thin films derived by a sol-gel process. *Appl. Phys. Lett.* **2005**, *86*, 122501. [CrossRef]
23. Ryu, H.; Murugavel, P.; Lee, J.H.; Chae, S.C.; Noh, T.W.; Oh, Y.S.; Kim, H.J.; Kim, K.H.; Jang, J.H. Magnetoelectric effects of nanoparticulate Pb(Zr0.52Ti0.48)O_3–$NiFe_2O_4$ composite films. *Appl. Phys. Lett.* **2006**, *89*, 102907. [CrossRef]
24. Park, J.H.; Jang, H.M.; Kim, H.S.; Park, C.G.; Lee, S.G. Strain-mediated magnetoelectric coupling in $BaTiO_3$-Co nanocomposite thin films. *Appl. Phys. Lett.* **2008**, *92*, 062908. [CrossRef]
25. Zavaliche, F.; Zheng, H.; Mohaddes-Ardabili, L.; Yang, S.Y.; Zhan, Q.; Shafer, P.; Reilly, E.; Chopdekar, R.; Jia, Y. Electric field-induced magnetization switching in epitaxial columnar nanostructures. *Nano Lett.* **2005**, *5*, 1793. [CrossRef] [PubMed]
26. Dörr, K.; Thiele, C.; Bilani, O.; Herklotz, A.; Schultz, L. Dynamic strain in magnetic films on piezoelectric crystals. *J. Magn. Magn. Mater.* **2007**, *310*, 1182–1184. [CrossRef]
27. Yang, J.J.; Zhao, Y.G.; Tian, H.F.; Luo, L.B.; Zhang, H.Y.; He, Y.J.; Luo, H.S. Electric field manipulation of magnetization at room temperature in multiferroic $CoFe_2O_4/Pb(Mg_{1/3}Nb_{2/3})_{0.7}Ti_{0.3}O_3$ heterostructures. *Appl. Phys. Lett.* **2009**, *94*, 212504. [CrossRef]
28. Nan, T.; Hui, Y.; Rinaldi, M.; Sun, N.X. Self-biased 215 MHz magnetoelectric NEMS resonator for ultra-sensitive DC magnetic field detection. *Sci. Rep.* **2013**, *3*, 1985. [CrossRef]

29. Yao, Y.P.; Liu, Y.K.; Dong, S.N.; Yin, Y.W.; Yang, S.W.; Li, X.G. Multi-state resistive switching memory with secure information storage in $Au/BiFe_{0.95}Mn_{0.05}O_3/La_{5/8}Ca_{3/8}MnO_3$ heterostructure. *Appl. Phys. Lett.* **2012**, *100*, 193504. [CrossRef]
30. Li, Z.; Wang, J.; Lin, Y.; Nan, C.W. A magnetoelectric memory cell with coercivity state as writing data bit. *Appl. Phys. Lett.* **2010**, *96*, 162505. [CrossRef]
31. Lu, Y.; Chen, J.; Cheng, Z.; Zhang, S. The PZT/Ni unimorph magnetoelectric energy harvester for wireless sensing applications. *Energy Convers. Manag.* **2019**, *200*, 112084. [CrossRef]
32. Zhang, C.L.; Chen, W.Q. A wideband magnetic energy harvester. *Appl. Phys. Lett.* **2010**, *96*, 123507. [CrossRef]
33. Patil, D.R.; Zhou, Y.; Kang, J.-E.; Sharpes, N.; Jeong, D.-Y.; Kim, Y.-D.; Kim, K.H.; Priya, S.; Ryu, J. Anisotropic self-biased dual-phase low frequency magneto-mechano-electric energy harvesters with giant power densities. *APL Mater.* **2014**, *2*, 046102. [CrossRef]
34. Ryu, J.; Kang, J.-E.; Zhou, Y.; Choi, S.-Y.; Yoon, W.-H.; Park, D.-S.; Choi, J.-J.; Hahn, B.-D.; Ahn, C.-W. Ubiquitous magneto-mechano-electric generator. *Energy Environ. Sci.* **2015**, *8*, 2402–2408. [CrossRef]
35. Wu, H.; Tatarenko, A.; Bichurin, M.I.; Wang, Y. A multiferroic module for biomechanical energy harvesting. *Nano Energy* **2021**, *83*, 105777. [CrossRef]
36. Li, H.; Zou, Z.; Yang, Y.; Shi, P.; Wu, X.; Ou-Yang, J.; Yang, X.; Zhang, Y.; Zhu, B. Microbridge-structured magnetoelectric sensor array based on pzt/fecosib thin films. *IEEE Trans. Magn.* **2020**, *56*, 1–4. [CrossRef]
37. Hong, N.T.M.; Duc, N.H.; Thang, P.D. Converse magnetoelectric effect in PZT/NiFe/CoFe nanocomposites. *Int. J. Nanotechnol.* **2013**, *10*, 206–213. [CrossRef]
38. Nan, T.; Zhou, Z.; Liu, M.; Yang, X.; Gao, Y.; Assaf, B.A.; Lin, H.; Velu, S.; Wang, X. Quantification of strain and charge co-mediated magnetoelectric coupling on ultra-thin Permalloy/PMN-PT interface. *Sci. Rep.* **2014**, *4*, 3688. [CrossRef]
39. He, X.L.; Zhou, J.; Wang, W.B.; Xuan, W.P.; Yang, X.; Jin, H.; Luo, J.K. High performance dual-wave mode flexible surface acoustic wave resonators for UV light sensing. *J. Micromech. Microeng.* **2014**, *24*, 055014. [CrossRef]
40. Liu, W.; Lu, Y.W.; Hu, W.B.; Bai, F.M. Magnetic surface acoustic wave resonator based on flexible substrate. *Piezoelectr. Acoustoopt.* **2020**, *42*, 439–443.
41. Zhao, Y.; Gao, S.; Huang, X.; Huo, W.; Xu, K. Fully flexible electromagnetic vibration sensors with annular field confinement origami magnetic membranes. *Adv. Funct. Mater.* **2020**, *30*, 2001553. [CrossRef]
42. Chen, S.; Yang, X.; Ouyang, J.; Lin, G.; Jin, F.; Tong, B. Fabrication and characterization of shape anisotropy AlN/FeCoSiB magnetoelectric composite films. *Ceram. Int.* **2014**, *40*, 3419–3423. [CrossRef]
43. Zayer, N.K.; Greef, R.; Rogers, K.; Grellier, A.; Pannell, C.N. In situ monitoring of sputtered zinc oxide films for piezoelectric transducers. *Thin Solid Films* **1999**, *352*, 179–184. [CrossRef]
44. Zhou, Y.; Li, C.-J.; Pan, Y.-R. Magnetoelectric effect analysis of magnetostrictive/piezoelectric laminated composites. *Acta Phys. Sin.-Chin. Ed.* **2018**, *67*, 077702. [CrossRef]

micromachines

MDPI

Article

Nano Groove and Prism-Structured Triboelectric Nanogenerators

Resul Saritas [1,*], Majed Al-Ghamdi [2], Taylan Memik Das [3,4], Omar Rasheed [5], Samed Kocer [1], Ahmet Gulsaran [4], Asif Abdullah Khan [6], Md Masud Rana [6], Mahmoud Khater [7], Muhammed Kayaharman [6], Dayan Ban [6], Mustafa Yavuz [4] and Eihab Abdel-Rahman [1]

1 Department of Systems Design Engineering, University of Waterloo, Waterloo, ON N2L 3G1, Canada; skocer@uwaterloo.ca (S.K.); eihab@uwaterloo.ca (E.A.-R.)
2 National Security Program King Abdulaziz City for Science and Technology, KACST, Riyadh 12354, Saudi Arabia
3 Department of Mechanical Engineering, University of Kirikkale, Kirikkale 71450, Turkey; taylandas@gmail.com
4 Department of Mechanical and Mechatronics Engineering, University of Waterloo, Waterloo, ON N2L 3G1, Canada; agulsaran@uwaterloo.ca (A.G.); myavuz@uwaterloo.ca (M.Y.)
5 Department of Mechanical Engineering, University of Toronto, Toronto, ON M5S 1A1, Canada; rasho1959@gmail.com
6 Department of Electrical and Computer Engineering, University of Waterloo, Waterloo, ON N2L 3G1, Canada; aakhan@uwaterloo.ca (A.A.K.); mm4rana@uwaterloo.ca (M.M.R.); dban@uwaterloo.ca (D.B.)
7 Department of Mechanical Engineering, King Fahd University of Petroleum and Minerals, Dhahran 31261, Saudi Arabia; mkhater@uwaterloo.ca
* Correspondence: resulsaritas@gmail.com

Abstract: Enhancing the output power of triboelectric nanogenerators (TENGs) requires the creation of micro or nano-features on polymeric triboelectric surfaces to increase the TENGs' effective contact area and, therefore, output power. We deploy a novel bench-top fabrication method called dynamic Scanning Probe Lithography (d-SPL) to fabricate massive arrays of uniform 1 cm long and 2.5 µm wide nano-features comprising a 600 nm deep groove (NG) and a 600 nm high triangular prism (NTP). The method creates both features simultaneously in the polymeric surface, thereby doubling the structured surface area. Six thousand pairs of NGs and NTPs were patterned on a $6 \times 5\,cm^2$ PMMA substrate. It was then used as a mold to structure the surface of a 200 µm thick Polydimethylsiloxane (PDMS) layer. We show that the output power of the nano-structured TENG is significantly more than that of a TENG using flat PDMS films, at 12.2 mW compared to 2.2 mW, under the same operating conditions (a base acceleration amplitude of 0.8 g).

Keywords: triboelectric; nano features; nano groove and prism; PDMS

1. Introduction

Demand for distributed electrical power has seen a significant growth as Internet of Things applications come to market creating demand for dense sensor networks made of many millions of nodes [1,2]. Localized power supply is an attractive approach to power those networks, which is driving a search for new power supply modalities. Waste mechanical vibrations is a promising distributed power source readily available in many environments [3,4].

Waste vibrations have been converted into electrical energy using piezoelectric [5], electromagnetic [6], and electrostatic nanogenerators [7]. All of these methods are potentially clean, highly efficient, and environmentally friendly [8]. A new type of electrostatic nanogenerators, triboelectric nanogenerators (TENGs), was introduced in 2012 [9]. TENGs convert mechanical vibrations into electrical energy by exploiting the triboelectric effect to provide the voltage source necessary for electrostatic induction [10–12]. Upon separation of a metallic and a dielectric film, the metal loses electrons, becoming positively charged while the dielectric gains electrons, becoming negatively charged [13,14]. The two charged films serve as a voltage source within a variable capacitor [15]. In turn, the capacitor drives a current between its plates as they move with respect to each other under the influence of mechanical vibrations [16].

TENGs can provide a power density in the range of mW/cm^2 sufficient to power many autonomous sensor nodes and other self-powered portable electronic devices [17,18]. A variety of materials with opposite polarity have been implemented in TENGs to improve their efficiency, particularly polytetrafluoroethylene (PTFE) and poly-dimethylsiloxane (PDMS), which are frequently chosen as negative triboelectric material due to their high tendency to receive electrons [19–21].

The output power of TENGs has also been enhanced by creating nano structured triboelectric surfaces, thereby increasing the contact area between the two triboelectric films. This allows more freely moving electrons to leave the metallic film to the dielectric film resulting in a higher voltage drop between the two films and increasing the output power of the TENG [22]. Two classes of fabrication methods have been adopted to achieve this goal. The first creates those structures using oxides, by the etching of TiO_2 nanowires and nano-sheets into Ti films [23], chemical synthesis of SiO_2 nanoparticles on SiO_2 films [24], and growing of ZnO nanorods on indium tin oxide (ITO) films [25], or metals, by etching of gold flowers into gold films [26]. The second creates nano-features, such as PDMS nanopillars [27], PDMS triangular pyramids, and PDMS triangular ridges [26,28,29], by molding polymers using standard imprint lithography (SIL).

Currently, the creation of those nano-features requires the use of cleanroom facilities, for electrochemcial deposition [30] or standard soft lithography [31], thereby imposing cost and access barriers on TENG technology. Alternatively available low-cost and rapid fabrication methods such as the CO_2 laser ablation method are not able to produce features less than 50 µm, which limits the maximum surface area in turn power generation. In this study, we employ a novel and low-cost bench-top method to fabricate micro- and nano-features. The method allows us to introduce massive arrays of one centimeter-long nano triangular prisms (NTPs) and nano grooves (NGs) on a Poly(methyl methacrylate) (PMMA) substrate. It employs a micro-resolution 3D stage equipped with a micro-sized tip via a spring-damper system. In total, twelve thousand NGs and NTPs are patterned in a $6 \times 5\,cm^2$ PMMA substrate, which is used as a mold to transfer those features into PDMS substrates using spin coating. The fabricated PDMS films serve as the negative triboelectric layer while aluminum (Al) foil is used as the positive triboelectric layer. The output power of the fabricated TENG is compared to that of an identical TENG made using a flat surface PDMS layer.

2. The TENG

2.1. Fabrication of Nano-Structured Films

The fabrication method employs a micro-resolution needle mounted to a micro-resolution 3D stage via a spring-damper mechanism. As the needle traverses the polymeric substrate surface, non-uniformity in the morphology of the substrate surface generates time-variation in the magnitude and direction of the contact force, which in turn leads to needle vibrations. The spring-damper mechanism acts to dissipate those vibrations and stabilizes the contact force, thereby generating uniform features in the substrate. An optical camera equipped with a 6× objective lens and 2× optical magnifier is attached to the stage to monitor the patterning process. Figure 1a shows the fabrication setup used to pattern the PMMA substrate.

The desired pattern and a constant contact force are defined in an interface software SonoGuide, Version 2.42 [32] which translates it into a 3D trajectory written in machine language. The controller of the micro-stage follows that trajectory as it traverses the needle tip over the substrate surface while tracking the process using the optical camera. Upon first contact, the of Z-axis actuator commands a displacement z(t) resulting in the tip penetrating the substrate surface with an average displacement of x(t) and a spring compression force of

$$F(t) = k(z(t) - x(t)) \quad (1)$$

where k is the spring stiffness given as 100 N/m, acting on the tip.

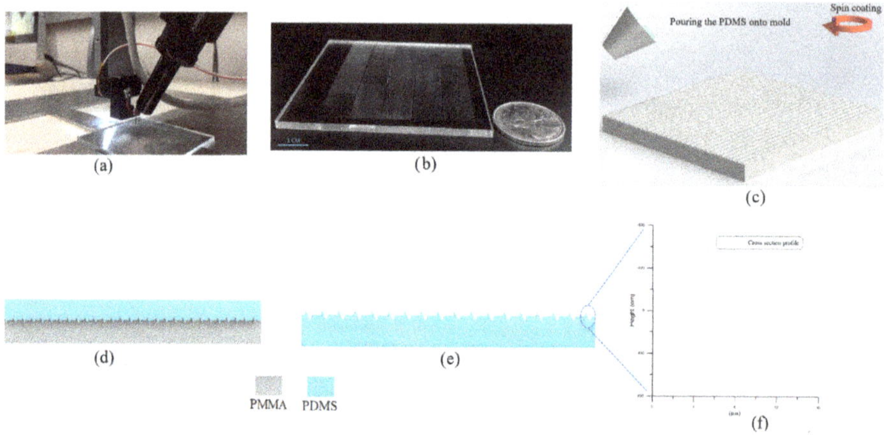

Figure 1. (**a**) The fabrication setup. (**b**) An array of 6000 NGs and NTPs on a $6 \times 5\,\text{cm}^2$ area of a PMMA substrate. (**c**) Spin coating of PDMS onto the PMMA mold. (**d**) Curing of PDMS. (**e**) Peel off of the PDMS film from the mold. (**f**) A cross-sectional profile of the film obtained using AFM microscopy. The profile shows the NGs and NTPs.

The desired pattern is made of a massive array of uniform 1 cm long, 2.5 µm wide 600 nm deep NGs and 600 nm high NTPs were patterned. The feature dimensions, width and height, are controlled by varying the substrate-tip contact force while the length is commanded by the micro-stage to the desired length. The sharpness of the needle tip, and the hardness of the substrate determine whether patterning results in plastic deformations or chip removal is required. In this case, the contact force and the needle tip orientation with respect to the trajectory were calibrated to plastically deform the PMMA substrate. As the deformation created NGs in the surface, the material flowing out of them created NTPs on their edges, resulting in the simultaneous creation of two nano features in a single patterning event. The dual nano features increase the effective surface area, thus enhancing triboelectricity. The patterned PMMA substrate, shown in Figure 1b exhibiting an array of 6000 cm long NGs and NTPs, is then used as a mold to cast PDMS films.

The PDMS base and its curing agent are mixed at a weight ratio of 5:1 and spin-coated on the PMMA mold, Figure 1c. The mold and cast PDMS, Figure 1d, are placed for 5 min on a hot-plate for curing at 80 °C. Multiple spin-coating at 500 RPM and curing cycles are undertaken to reach a thickness of 200 µm. The PDMS film is then peeled off from the mold resulting in the surface pattern shown schematically in Figure 1e. The surface topography of the fabricated film was measured using an atomic force microscope (Bruker AFM, Camarillo, CA, USA). Figure 1f shows the cross-sectional profile of the NGs and NTPs on the fabricated PDMS film. The bases of the NGs and NTPs were measured to be 2.5 µm wide and their apexes were measured to be 600 nm high.

2.2. Fabrication of the TENG

Figure 2a shows optical microscope (Nikon Inverted Microscope, Tokyo, Japan) images of uniformly patterned 2.5 µm wide and 600 nm deep and 1 cm long pairs of NGs and NTPs.The area shown, $4 \times 3\,\text{mm}^2$, exhibits 85 pairs of NGs and NTPs. A close-up view of the patterns in a $0.75 \times 0.565\,\text{mm}^2$ area is shown in Figure 2b and an AFM image of the plastically deformed NGs and flown NTPs in the area of $15 \times 15\,\mu\text{m}^2$ is shown Figure 2c.

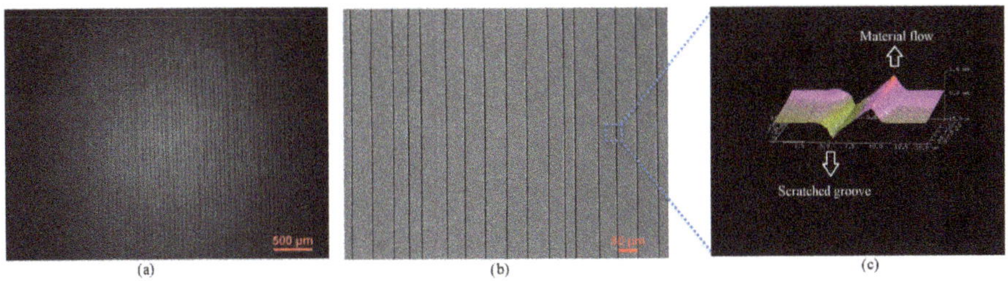

Figure 2. (**a**) A microscopic image of an array made of 85 pairs of NGs and NTPs in a $4 \times 3\,\text{mm}^2$ area. (**b**) A close-up microscopic image of a $750 \times 565\,\mu\text{m}^2$ area. (**c**) AFM image of the NGs and NTPs in a $0.75 \times 0.565\,\text{mm}^2$ area illustrating plastic deformation and material flow.

The TENG consists of structural support elements and energy triboelectric harvesting elements. The former include two backbone PMMA plates, $8 \times 8\,\text{cm}^2$ in area and 0.87 cm in thickness, to provide structural support and an inertial energy harvesting mass. Four linear guides, 60 mm long and 3.2 mm in diameter steel rods, are placed at the four corners of the plates to maintain alignment between the plates as they oscillate against each other under host vibrations. Four concentric steel springs, 24 mm long with an outer diameter of 4.8 mm, inner diameter of 4.5 mm, and spring stiffness of 90 N/m, are housed around the linear guides to maintain separation between the plates. A $6 \times 5\,\text{cm}^2$ PMMA stage is attached to the bottom backbone plate using double-sided tape to control the capacitive gap. The harvesting elements are the PDMS film and two Al foil films, all with identical planar dimensions to the stage.

Using double-sided tape, an Al film is attached to the top backbone plate and another to the stage to serve as the positive triboelectric film and negative electrode, respectively. The flat bottom surface of the PDMS film is pressed onto the negative electrode while avoiding any air gaps at the interface.

Holes with a diameter of 3.2 mm were laser cut into the corners of the backbone plates within their inactive margins. The linear guides were inserted into the bottom backbone plate holes and fixed to it using super glue. Next, the springs were placed around the linear guides. Finally, the guides are inserted through the top backbone plate holes. The top plate was allowed to move freely up and down along the linear guides.

2.3. Working Principle

Figure 3 shows a schematic of the harvesting elements and their interaction throughout the harvesting cycle. The TENGs generate an alternating current as the external (host) vibrations cycle the harvesting elements through contact and separation states and the TENG undergoes the corresponding triboelectrification and electrostatic induction stages. Initially, the two electrodes are neutral and separate from each other. Under external vibrations, the top Al electrode and PDMS film come into contact. At this point, triboelectric charge generation begins as electrons are injected from the Al to the PDMS. As separation begins, the top plate moves upward under the influence of the springs' restoring force acting against the internal electric field between the charges distributed on the two triboelectric surfaces. As a result, electrons flow from the bottom (negative) electrode to the top (positive) electrode under the influence of electrostatic induction between the PDMS film on the bottom electrode. This process terminates when the top plate reaches maximum separation distance. As it reverses direction and approaches the bottom plate, a current flows in the reverse direction as electrons flow from the top electrode under the influence of the electrostatic induction of the PDMS film on it to the bottom electrode. The cycle terminates with the triboelectric films coming back into contact.

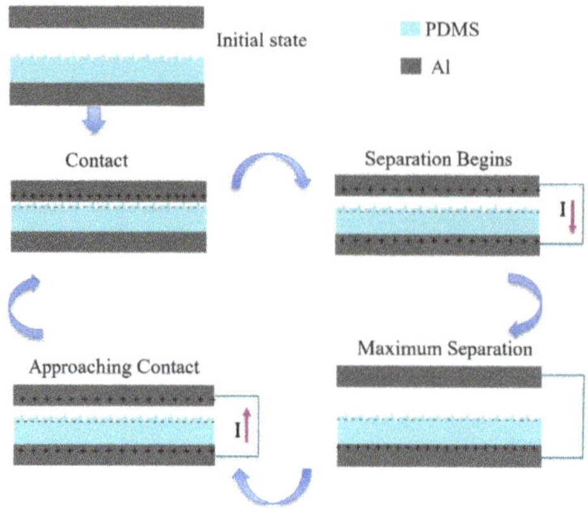

Figure 3. Triboelectric cyclic harvesting of mechanical energy.

3. Results and Discussion

Figure 4 shows the experimental setup of the triboelectric energy harvester. An electromagnetic shaker is driven by an amplifier (LABWORKS pa-138, Labworks Inc., Costo Mesa, CA, USA) and close loop controlled by a vibration controller (VR9500 Revolution, Michigan, CA, USA) using acceleration feedback. The output voltage and current of the TENGs are measured using an Oscilloscope (Tektronix-TDS2004C, Beaverton, OR, USA) via a high-input impedance (70 MΩ) probe. A low-noise current amplifier (SRS Stanford Research Systems-SR570 Preamplifier, Sunnyvale, CA, USA) is employed to measure the short-circuit current.

The flat and nano-structured PMDS films shown in Figure 5 serve as the negative triboelectric material. After peeling off the nano-structured PDMS film from the PMMA mold, its surface features are the reverse of those in the mold, namely six thousand NGs and NTPs each 1 cm long, 2.5 µm wide, and 600 nm deep and high, respectively.

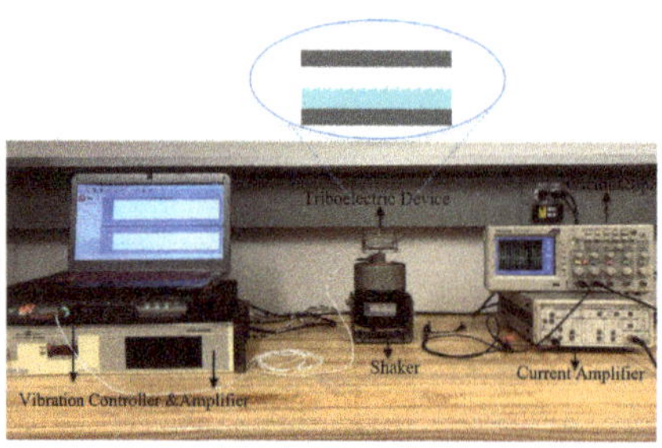

Figure 4. The experimental setup mounts the TENG on electromagnetic shaker and uses a current amplifier and an oscilloscope to measure the output signal.

The shaker was used to deliver base accelerations of the form:

$$a(t) = A\cos(2\pi f t) \quad (2)$$

with an amplitude of $A = 0.8\,\text{g}$ and a frequency of $f = 8\,\text{Hz}$ to the TENGs. The controller was used to maintain a constant acceleration amplitude as the resistive load was varied. To overcome the high impedance of the open-circuit TENG, the output voltage was measured through a 70 MΩ input impedance probe. Figure 5c compares the open-circuit voltage of the flat PMDS TENG to the nano-structured PMDS. It can be seen that the nano-structured film allowed the TENG to increase the peak-to-peak voltage from the range of V_{pp} = 300–400 V to the range V_{pp} = 1100–1400 V.

Figure 5d compares the short-circuit current through a 50 Ω resistor of the flat PMDS TENG to that of the nano-structured PMDS TENG. The peak-to-peak current generated by the nano-structured TENG was measured at $I_{pp} = 200\,\mu A$ compared to $I_{pp} = 40\,\mu A$ for the flat TENG. These results show that the nano-structured film increases the output voltage and current of the TENG four-fold compared to the flat film.

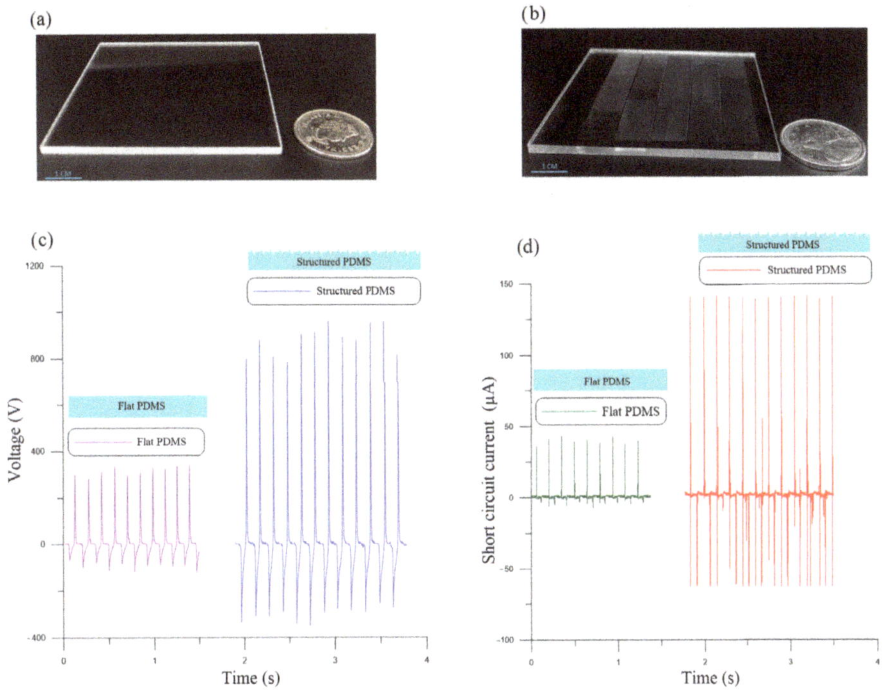

Figure 5. (**a**) The flat and (**b**) nano-structured molds used to cast the PDMS films. (**c**) Open-circuit voltage and (**d**) short-circuit current of the flat and nano-structured TENGs.

Nano-structuring the PDMS film increases its contact area with the Al films, thereby improving its ability to create a voltage difference across the TENG plates and, therefore, the output power. Specifically, the PDMS NTPs bend during contact with the Al film introducing sliding and friction between the two surfaces which allows thee PDMS to break the bonds between electrons on the Al surface more efficiently. Cycle-on-cycle variability in the impact, sliding, and friction processes result in the variability of open-circuit voltage observed in Figure 5c.

We compared the output power of the flat PDMS TENG to that of the nano-structured PDMS TENG by measuring the current and voltage they generated across a variable resistive load R connected between their terminals as per the schematic in Figure 6a. Parallel resistance imposed by the probe in evaluating the load resistance was accounted for the equivalent load. The current dropped and the voltage increased as the load resistance was increased from $R = 2\,\Omega$ to $R = 120\,M\Omega$. For the nano-structured TENG, the measured peak-to-peak current, the red line in Figure 6c, dropped from $I_{pp} = 245\,\mu A$ to 8.8 μA, and the voltage, the blue line in Figure 6c, increased from $V_{pp} = 0.5$ V to 1328 V.

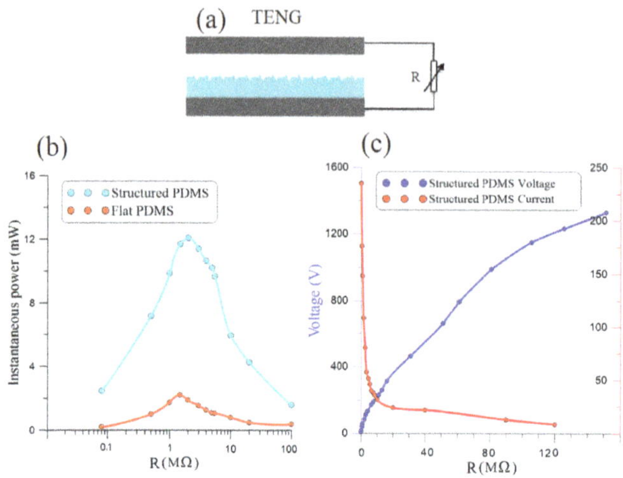

Figure 6. (**a**) A schematic illustrating measurement of the voltage and current generated by the TENG across a resistive load R. (**b**) Comparison of the output power of the nano-structured to the flat TENGs across R. (**c**) The measured voltage and current generated by the nano-structured TENG across R.

The measured voltage and current were used to evaluate the instantaneous output power across the resisitive load as:

$$P(t) = V(t)I(t) \tag{3}$$

Figure 6b compares the maximum instantaneous power $P_{\max}$ of the flat TENG (red line) to that of the nano-structured TNEG (cyan line). We found that the optimal resistance of the flat TENG was $R_{opt} \approx 1.5\,\text{M}\Omega$ while that of the nano-structured TENG was $R_{opt} \approx 2.0\,\text{M}\Omega$. We also found that the nano-structured TENG consistently outperformed the flat TENG irrespective of the load reaching an optimal maximum instantaneous output power of $P_{\max} = 12.2\,\text{mW}$ compared to $P_{\max} = 2.2\,\text{mW}$ for the flat TENG. This shows that introducing nano-structures into the surface of the PDMS film increased the TENG's power generation capacity by more than five-fold.

We also examined the ability of the TENG to store energy by connecting it to a storage capacitor using the circuit layout shown in Figure 7a. A bridge rectifier, consisting of four Schottky diodes, was used to convert the alternating (AC) output current of the TENG to direct current (DC) to charge the capacitor. We measured the voltage across the capacitor as the harvester charged it by harvesting kinetic energy from the vibrations waveform described above.

Figure 7b shows the evolution over time of the voltage $V(t)$ across five storage capacitors ranging in capacitance from $C = 0.1\,\mu\text{F}$ to $4.7\,\mu\text{F}$. We found that the energy stored in the capacitor, evaluated as

$$E = \frac{1}{2}CV^2 \tag{4}$$

ranged from $E = 5.2\,\mu\text{J}$ in 4s for the $C = 0.1\mu\text{F}$ capacitor to $E = 18\,\mu\text{J}$ in 16 s for the $C = 3.3\,\mu\text{F}$ capacitor. While neither the on-resistance losses of the diodes or the capacitor size were optimized, the results show that stored energy was delivered at a rate better than $1\,\mu\text{W}$.

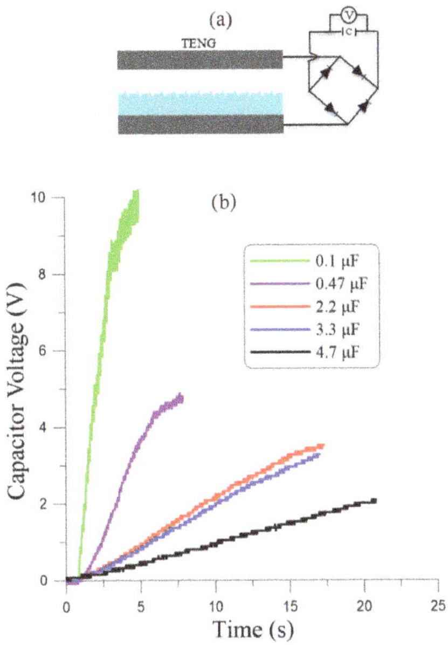

Figure 7. (**a**) A schematic of the TENG and its charging circuit comprising a capacitor and a bridge rectifier. (**b**) Measured capacitor voltage as a function of time for five capacitance values.

Finally, a schematic of the TENG demonstration with LEDs is given in Figure 8a, and we showed the comparative functionality of the TENGs by deploying them to directly power arrays of series-connected red, white, yellow, and blue LEDs, illustrated in Figure 8b. The array of LEDs was inserted into the energy harvesting circuit described above in place of the storage capacitor. The nano-structured PDMS TENG was able to light up a bank of 180 red, white, yellow, and blue LEDs shown in Figure 8c, whereas the maximum array size the flat PDMS TENG was able to light up was 67 LEDs.

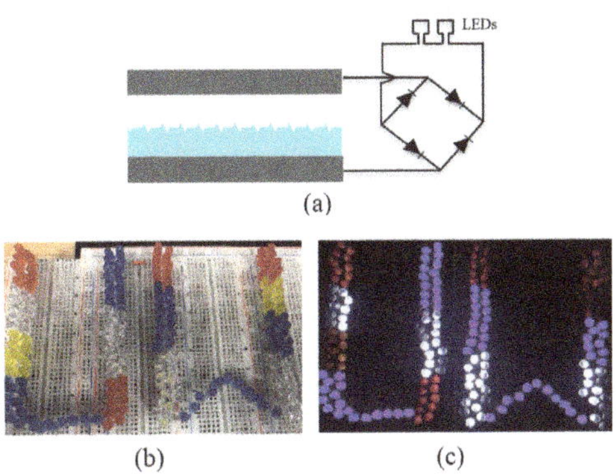

Figure 8. (**a**) A schematic of the TENG demonstration where they were deployed to power an array of series-connected red, white, yellow, and blue LEDs via a bridge rectifier. Images of an array of 180 LEDs while their power source, a nano-structured TENG, was (**b**) off and (**c**) on.

4. Conclusions

We introduced a novel, rapid, uniform, and low-cost process for high-speed (1 mm/s) fabrication of long (1 cm) nano-structures (NGs and NTP) on the surface of PMMA substrates without a need for cleanroom facilities. We used this process to fabricate PMMA molds which, in turn, were deployed to produce nano-structured thin PDMS films. Those films were then employed as the negative triboelectric material in TENGs. Surface nano-structuring was found to increase the output power from 2.2 mW for flat PDMS TENGs to 12.2 mW for nano-structured PDMS TENGs. Ultimately, nano-structured TENGs proved to be a more effective power source able to light an array of 180 LEDs when the flat TENGs were only able to light 67 LEDs.

Author Contributions: Conceptualization, R.S.; Methodology, M.A.-G., T.M.D., A.G. and A.A.K.; Validation, S.K. and M.K. (Mahmoud Khater); Investigation, O.R.; Resources, M.Y.; Data curation, M.M.R., M.K. (Muhammed Kayaharman), D.B. and M.Y.; Writing—original draft, E.A.-R.; Writing—review & editing, R.S. All authors have read and agreed to the published version of the manuscript.

Funding: This research received no external funding.

Data Availability Statement: Not applicable.

Acknowledgments: The authors would like to thank "Performance Micro Tool" for providing us with carbide sharp tips. We would like to also acknowledge KACST for their support.

Conflicts of Interest: The authors declare no conflict of interest.

References

1. Bai, P.; Zhu, G.; Lin, Z.H.; Jing, Q.; Chen, J.; Zhang, G.; Ma, J.; Wang, Z.L. Integrated multilayered triboelectric nanogenerator for harvesting biomechanical energy from human motions. *ACS Nano* **2013**, *7*, 3713–3719. [CrossRef] [PubMed]
2. Wang, Z.L.; Chen, J.; Lin, L. Progress in triboelectric nanogenerators as a new energy technology and self-powered sensors. *Energy Environ. Sci.* **2015**, *8*, 2250–2282. [CrossRef]
3. Nie, J.; Ren, Z.; Shao, J.; Deng, C.; Xu, L.; Chen, X.; Li, M.; Wang, Z.L. Self-Powered Microfluidic Transport System Based on Triboelectric Nanogenerator and Electrowetting Technique. *ACS Nano* **2018**, *12*, 1491–1499. [CrossRef] [PubMed]
4. Chen, J.; Zhu, G.; Yang, W.; Jing, Q.; Bai, P.; Yang, Y.; Hou, T.C.; Wang, Z.L. Harmonic-resonator-based triboelectric nanogenerator as a sustainable power source and a self-powered active vibration sensor. *Adv. Mater.* **2013**, *25*, 6094–6099. [CrossRef]
5. Kim, S.W.; Lee, T.G.; Kim, D.H.; Lee, K.T.; Jung, I.; Kang, C.Y.; Han, S.H.; Kang, H.W.; Nahm, S. Determination of the appropriate piezoelectric materials for various types of piezoelectric energy harvesters with high output power. *Nano Energy* **2019**, *57*, 581–591. [CrossRef]
6. Bendame, M.; Abdel-Rahman, E.; Soliman, M. Wideband, low-frequency springless vibration energy harvesters: Part I. *J. Micromech. Microeng.* **2016**, *26*, 115021. [CrossRef]
7. Niu, S.; Liu, Y.; Wang, S.; Lin, L.; Zhou, Y.S.; Hu, Y.; Wang, Z.L. Theory of sliding-mode triboelectric nanogenerators. *Adv. Mater.* **2013**, *25*, 6184–6193. [CrossRef]
8. Xu, W.; Huang, L.B.; Wong, M.C.; Chen, L.; Bai, G.; Hao, J. Environmentally Friendly Hydrogel-Based Triboelectric Nanogenerators for Versatile Energy Harvesting and Self-Powered Sensors. *Adv. Energy Mater.* **2017**, *7*, 1–8. [CrossRef]
9. Fan, F.R.; Tian, Z.Q.; Lin Wang, Z. Flexible triboelectric generator. *Nano Energy* **2012**, *1*, 328–334. [CrossRef]
10. Chun, J.; Ye, B.U.; Lee, J.W.; Choi, D.; Kang, C.Y.; Kim, S.W.; Wang, Z.L.; Baik, J.M. Boosted output performance of triboelectric nanogenerator via electric double layer effect. *Nat. Commun.* **2016**, *7*, 1–9. [CrossRef]
11. Dharmasena, R.D.; Jayawardena, K.D.; Mills, C.A.; Deane, J.H.; Anguita, J.V.; Dorey, R.A.; Silva, S.R. Triboelectric nanogenerators: Providing a fundamental framework. *Energy Environ. Sci.* **2017**, *10*, 1801–1811. [CrossRef]
12. Jin, C.; Kia, D.S.; Jones, M.; Towfighian, S. On the contact behavior of micro-/nano-structured interface used in vertical-contact-mode triboelectric nanogenerators. *Nano Energy* **2016**, *27*, 68–77. [CrossRef]
13. Niu, S.; Wang, S.; Lin, L.; Liu, Y.; Zhou, Y.S.; Hu, Y.; Wang, Z.L. Theoretical study of contact-mode triboelectric nanogenerators as an effective power source. *Energy Environ. Sci.* **2013**, *6*, 3576–3583. [CrossRef]
14. Pan, C.T.; Chen, Y.J.; Liu, Z.H.; Huang, C.H. Design and fabrication of LTCC electro-magnetic energy harvester for low rotary speed. *Sens. Actuators A Phys.* **2013**, *191*, 51–60. [CrossRef]
15. Zi, Y.; Wang, J.; Wang, S.; Li, S.; Wen, Z.; Guo, H.; Wang, Z.L. Effective energy storage from a triboelectric nanogenerator. *Nat. Commun.* **2016**, *7*, 1–8. [CrossRef]
16. Wang, Y.; Yang, Y.; Wang, Z.L. Triboelectric nanogenerators as flexible power sources. *NPJ Flex. Electron.* **2017**, *1*, 1–9. [CrossRef]
17. Wang, S.; Lin, L.; Wang, Z.L. Nanoscale triboelectric-effect-enabled energy conversion for sustainably powering portable electronics. *Nano Lett.* **2012**, *12*, 6339–6346. [CrossRef]
18. Askari, H.; Khajepour, A.; Khamesee, M.B.; Saadatnia, Z.; Wang, Z.L. Piezoelectric and triboelectric nanogenerators: Trends and impacts. *Nano Today* **2018**, *22*, 10–13. [CrossRef]
19. Seol, M.L.; Han, J.W.; Moon, D.I.; Meyyappan, M. Triboelectric nanogenerator for Mars environment. *Nano Energy* **2017**, *39*, 238–244. [CrossRef]

20. Kim, S.; Gupta, M.K.; Lee, K.Y.; Sohn, A.; Kim, T.Y.; Shin, K.S.; Kim, D.; Kim, S.K.; Lee, K.H.; Shin, H.J.; et al. Transparent flexible graphene triboelectric nanogenerators. *Adv. Mater.* **2014**, *26*, 3918–3925. [CrossRef]
21. Wang, S.; Lin, L.; Wang, Z.L. Triboelectric nanogenerators as self-powered active sensors. *Nano Energy* **2015**, *11*, 436–462. [CrossRef]
22. Zhu, G.; Lin, Z.H.; Jing, Q.; Bai, P.; Pan, C.; Yang, Y.; Zhou, Y.; Wang, Z.L. Toward large-scale energy harvesting by a nanoparticle-enhanced triboelectric nanogenerator. *Nano Lett.* **2013**, *13*, 847–853. [CrossRef] [PubMed]
23. Lin, Z.H.; Xie, Y.; Yang, Y.; Wang, S.; Zhu, G.; Wang, Z.L. Enhanced triboelectric nanogenerators and triboelectric nanosensor using chemically modified TiO2 nanomaterials. *ACS Nano* **2013**, *7*, 4554–4560. [CrossRef] [PubMed]
24. Lin, Z.H.; Cheng, G.; Wu, W.; Pradel, K.C.; Wang, Z.L. Dual-mode triboelectric nanogenerator for harvesting water energy and as a self-powered ethanol nanosensor. *ACS Nano* **2014**, *8*, 6440–6448. [CrossRef]
25. Ko, Y.H.; Nagaraju, G.; Lee, S.H.; Yu, J.S. PDMS-based triboelectric and transparent nanogenerators with ZnO nanorod arrays. *ACS Appl. Mater. Interfaces* **2014**, *6*, 6631–6637. [CrossRef]
26. Park, S.J.; Seol, M.L.; Jeon, S.B.; Kim, D.; Lee, D.; Choi, Y.K. Surface Engineering of Triboelectric Nanogenerator with an Electrodeposited Gold Nanoflower Structure. *Sci. Rep.* **2015**, *5*, 1–7. [CrossRef]
27. Dudem, B.; Huynh, N.D.; Kim, W.; Kim, D.H.; Hwang, H.J.; Choi, D.; Yu, J.S. Nanopillar-array architectured PDMS-based triboelectric nanogenerator integrated with a windmill model for effective wind energy harvesting. *Nano Energy* **2017**, *42*, 269–281. [CrossRef]
28. Zhang, J.H.; Li, Y.; Du, J.; Hao, X.; Huang, H. A high-power wearable triboelectric nanogenerator prepared from self-assembled electrospun poly (vinylidene fluoride) fibers with a heart-like structure. *J. Mater. Chem. A* **2019**, *7*, 11724–11733. [CrossRef]
29. Zhang, J.H.; Li, Z.; Xu, J.; Li, J.; Yan, K.; Cheng, W.; Xin, M.; Zhu, T.; Du, J.; Chen, S.; et al. Versatile self-assembled electrospun micropyramid arrays for high-performance on-skin devices with minimal sensory interference. *Nat. Commun.* **2022**, *13*, 5839. [CrossRef]
30. Zhu, G.; Pan, C.; Guo, W.; Chen, C.Y.; Zhou, Y.; Yu, R.; Wang, Z.L. Triboelectric-generator-driven pulse electrodeposition for micropatterning. *Nano Lett.* **2012**, *12*, 4960–4965. [CrossRef]
31. He, X.; Mu, X.; Wen, Q.; Wen, Z.; Yang, J.; Hu, C.; Shi, H. Flexible and transparent triboelectric nanogenerator based on high performance well-ordered porous PDMS dielectric film. *Nano Res.* **2016**, *9*, 3714–3724. [CrossRef]
32. SonoPlot SonoGuide Software. Available online: https://www.sonoplot.com/ (accessed on 7 September 2021).

applied sciences

MDPI

Article

Potential of Variable Geometry Radial Inflow Turbines as Expansion Machines in Organic Rankine Cycles Integrated with Heavy-Duty Diesel Engines

Fuhaid Alshammari [1,*], Abdullah Alghafis [2], Ibrahim Alatawi [1], Ahmed S. Alshammari [3], Ahmed Alzamil [3] and Abdullah Alrashidi [4]

1 Mechanical Engineering Department, Engineering College, University of Hail, Hail 2440, Saudi Arabia; ib.alatawi@uoh.edu.sa
2 Mechanical Engineering Department, College of Engineering, Qassim University, Buraydah 52571, Saudi Arabia; a.alghafis@qu.edu.sa
3 Electrical Engineering Department, Engineering College, University of Hail, Hail 2440, Saudi Arabia; ahm.alshammari@uoh.edu.sa (A.S.A.); aa.alzamil@uoh.edu.sa (A.A.)
4 Engineering College, Northern Border University, Arar 91431, Saudi Arabia; abdullah.alrashidi@nbu.edu.sa
* Correspondence: fu.alshammari@uoh.edu.sa; Tel.: +966-501938208

Abstract: This work evaluates the feasibility of utilizing an organic Rankine cycle (ORC) for waste heat recovery in internal combustion engines to meet the stringent regulations for reducing emissions resulting from the combustion of fossil fuels. The turbine is the most crucial component of the ORC cycle since it is responsible for power production. In this study, a variable geometry radial inflow turbine is designed to cope with variable exhaust conditions. A variable geometry turbine is simply a radial turbine with different throat openings: 30, 60, and 100%. The exhaust gases of a heavy-duty diesel engine are utilized as a heat source for the ORC system. Different engine operating points are explored, in which each point has a different exhaust temperature and mass flow rate. The results showed that the maximum improvements in engine power and brake specific fuel consumption (BSFC) were 5.5% and 5.3% when coupled to the ORC system with a variable geometry turbine. Moreover, the variable geometry turbine increased the thermal efficiency of the cycle by at least 20% compared to the system with a fixed geometry turbine. Therefore, variable geometry turbines are considered a promising technology in the field and should be further investigated by scholars.

Keywords: variable geometry; radial inflow turbine; organic Rankine cycle; waste heat recovery

1. Introduction

Manufacturers are currently under pressure to improve the efficiency of internal combustion engines in terms of fuel consumption and exhaust emissions [1]. In fact, the combustion of fossil fuels is a primary source of harmful gases, such as CO_2 and NOx [2]. Diesel engines, in fact, are among the most significant polluters due to their widespread applications. Most of the heat wasted in internal combustion engines comes from exhaust gases [3]. In fact, only 30 to 40% of fuel energy is converted into useful work [4]. To harness this wasted energy, waste heat recovery technologies can be integrated.

Recently, organic Rankine cycle (ORC) as a waste heat recovery technology has received great attention among scholars and in the industry sector due to the availability of components, its simplicity, and reliability [5,6]. Moreover, ORC systems exhibit great thermal efficiency in comparison to other WHR technologies under a variety of operating situations [7]. The organic Rankine cycle is simply a Rankine cycle operating with organic fluids instead of steam. Due to the low boiling point of some organic fluids, they can be utilized in low to medium heat source applications [8,9]. In addition, the densities of organic fluids are relatively high, which results in smaller systems compared to steam;

hence, there are lower costs [10]. Typical Rankine cycles consist of four main components, namely, an evaporator, turbine, condenser, and pump. The turbine is one of the most important components in the system since this is the medium where power production takes place [11].

Variable geometry turbines are of valuable interest especially in internal combustion engines in order to cope with the transient operating conditions in the engine. Conventional radial turbines (fixed geometry turbines) suffer from low efficiencies at part-load conditions [12]. A variable geometry radial turbine is simply a radial turbine that has movable stator vanes which provide a quick response to the change in engine load/speed [13]. In addition, by utilizing variable geometry vanes, it is possible to adjust the mass flow rate and to balance the volume of the working fluid [13]. The concept of variable geometry turbines is not new and has been extensively studied for turbocharging. However, applying such a technology in ORC systems is an area in which little available literature exists. Therefore, the investigated literature is limited to applications of the ORC with variable geometry turbines. The application of ORCs in internal combustion engines, covering different types of expansion machines, was extensively reviewed in the authors' study [14]. The authors also experimentally tested the potential of ORCs for waste heat recovery using fixed geometry turbines in several studies [15].

Thaddaeus et al. [16] simulated a 1.25 L Zetec-SE DOHC engine model coupled with an organic Rankine cycle and a variable geometry turbine. The results showed that the variable geometry turbine can improve the thermal efficiency and net power by 5.59% and 3 kW, respectively. Manente et al. [17] evaluated the performance of an ORC considering geothermal resources as the heat sources. Although a variable geometry turbine was considered, the authors did not point out the potential of the turbine. They concluded that the net power output is higher at low ambient temperatures than at warm ones. Lebedevas and Čepaitis [18] examined the decarbonization of marine transport using an ORC and a variable geometry turbine. The results showed that the highest system efficiency of 15% was achieved by utilizing the variable geometry turbine. Schuster et al. [19] investigated the thermodynamic optimization of organic Rankine cycles considering both fixed and variable geometry radial inflow turbines. The results proved the efficacy of variable geometry turbines when operating at off-design conditions. Other scholars integrated fixed geometry radial inflow turbines or other types of turbines, such as axial ones. Guillaume et al. [20] tested the potential of improving heavy-duty diesel engine performance using an ORC and a fixed geometry radial turbine. The results showed that the radial turbine presented a maximum power of 2.8 kW. Wang et al. [21] investigated the possibility of capturing the wasted heat of engine coolants and exhaust gases using a supercritical ORC. An improvement of 8% in fuel efficiency was obtained. Furukawaet et al. [22] concluded from their experiment that the engine BSFC could be decreased by 7.5% when adapting an ORC to capture the wasted heat in the engine coolant. Recently, Chen et al. [5] analyzed the performance of a preheated ORC as a waste heat recovery system in a ship two-stroke low-speed main engine. The authors utilized the wasted heat in both exhaust gases and coolants. The maximum cycle power and efficiency were 491.3 kW and 12.84%, respectively.

It is obvious that studies that investigate the potential of variable geometry turbines in engine–ORC systems are extremely limited. In fact, although [16,17,19] evaluated the benefits of variable geometry turbines, they did not cover the design process of the turbine. Moreover, the authors in studies [16–19] did not study the effects of the variable geometry turbines on the engines' power and BSFC. Therefore, the aim of this study is to shed more light on the design process of the turbine and explore its efficacy in improving heavy-duty diesel engines in terms of power and BSFC. It is worth mentioning that the authors have published several studies on the design and optimization of fixed geometry radial inflow turbines [23].

2. System Description

The proposed system consists of a 10.3 L heavy-duty diesel engine and the ORC system as shown in Figure 1. Stages 1 to 5 represent the turbine stage as shown in Figures 2 and 3.

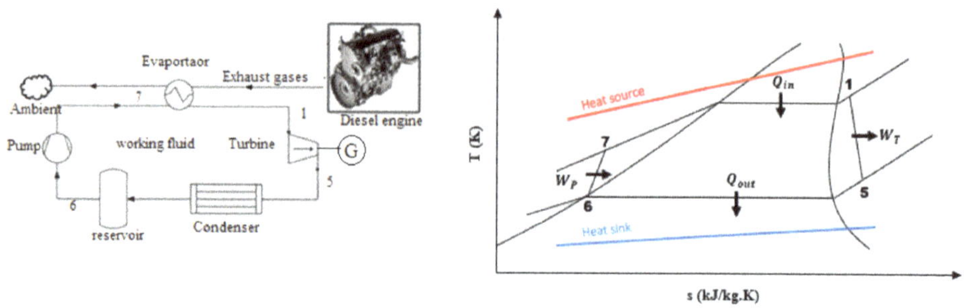

Figure 1. The proposed system along with T-s diagram of the thermodynamic cycle.

2.1. Engine Modeling

The data of the heavy-duty diesel engines were provided by the Saudi Electricity Company in Hail, Saudi Arabia. To explore the potential of the radial inflow turbine on the engine power and BSFC, three different engine operating points were explored as shown in Table 1.

Table 1. Engine properties at different engine operating conditions.

Engine Point	P1	P2	P3
Speed [rev/min]	750	1200	2200
Torque [Nm]	234	1388	1500
BSFC [g/kWh]	277	188	221
Mass flow rate of exhaust gases [kg/s]	0.310	0.380	0.500
Temperature of exhaust gases [K]	688	679	810

The powertrain power (engine + ORC) can be obtained using Equation (1) and the engine BSFC can be calculated using Equation (2).

$$P_{powertrain} = W_{net} + P_{engine} \tag{1}$$

$$BSFC = \frac{engine\ fuel\ consumption}{P_{powertrain}} \tag{2}$$

2.2. Organic Rankine Cycle Modeling

A typical ORC is composed of four components, namely, an evaporator, condenser, turbine, and pump. The liquid organic fluid is firstly heated to a superheated state by absorbing the wasted heat in the exhaust gas of the engine via the evaporator. The superheated fluid then drives the turbine to generate power. For the cycle to repeat, the superheated fluid is condensed to the liquid phase and pumped again to the evaporator. The mathematical model of the ORC system is detailed below.

The energy balance through the evaporator is presented in Equation (3).

$$Q_{Evap} = \dot{m}_{exh}(h_{exh,in} - h_{exh,out}) = \dot{m}_{wf}\left(h_{wf,out} - h_{wf,in}\right) \tag{3}$$

Similarly, the energy rejected through the condenser can be obtained using Equation (4).

$$Q_{cond} = \dot{m}_{wf}\left(h_{wf,in} - h_{wf,out}\right) \tag{4}$$

To ensure a liquid state at the outlet of the condenser (pump inlet), a sub-cooling of 2 °C is assumed. A 65% pump efficiency η_{pump} is assumed. Therefore, the power consumed by the pump can be obtained from Equation (5).

$$P_{pump} = \frac{\dot{m}_{wf}\left(h_{s,wf,out} - h_{wf,in}\right)}{\eta_{pump}} \tag{5}$$

The cycle thermal efficiency can be obtained using the following (6). The numerator of Equation (6) is simply the cycle network.

$$\eta_{thermal} = \frac{W_{output} - W_{Pump}}{Q_{Evap}} \tag{6}$$

It is worth mentioning that NOVEC 649 is used as the working fluid in the cycle because it is non-flammable and has a near-zero GWP. Moreover, NOVEC649 produces lower back pressure at the evaporator outlet and lower turbine rotational speed.

2.3. *Modeling of the Radial Inflow Turbine*

Volute, stator vanes, and rotor blades are the three primary parts of the radial inflow turbine, as depicted in Figure 2. Figure 3 presents a meridional view of the turbine stage. The flow is firstly collected in the volute, accelerated, and further expanded through the stator vanes and directed to the rotor blades at the proper angle. Finally, the fluid rotates the rotor blades, converting the kinetic energy into mechanical power. The turbine is coupled to an electrical generator to convert the mechanical power into electrical power. The design process is depicted in Figure 4 and detailed in the following subsections. Figure 5 presents an enthalpy–entropy diagram of the turbine stage.

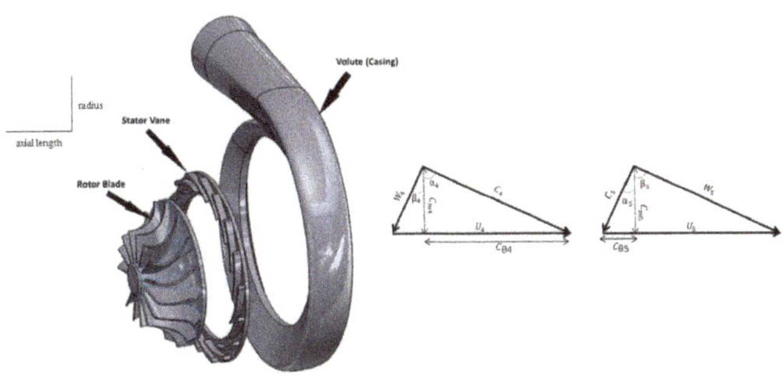

Figure 2. Turbine architecture (**left**) and velocity triangles of the rotor (**right**).

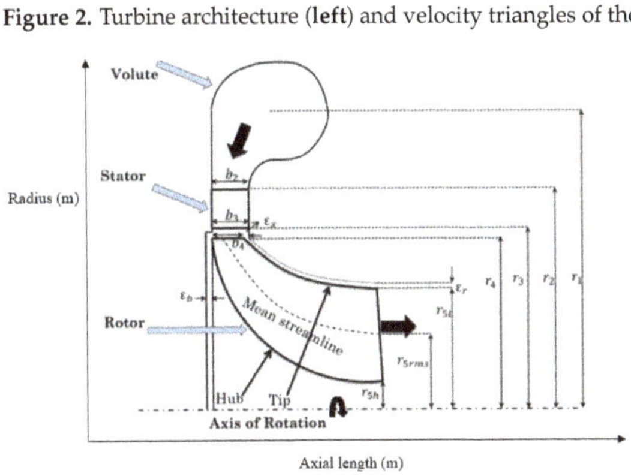

Figure 3. Meridional view of the turbine stage.

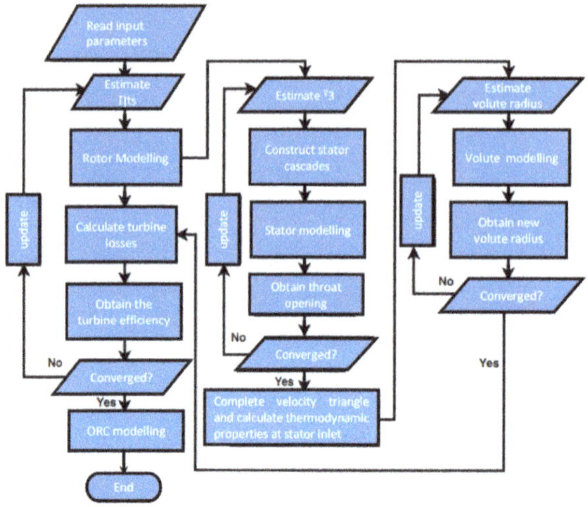

Figure 4. Flow chart of design process of the radial inflow turbine.

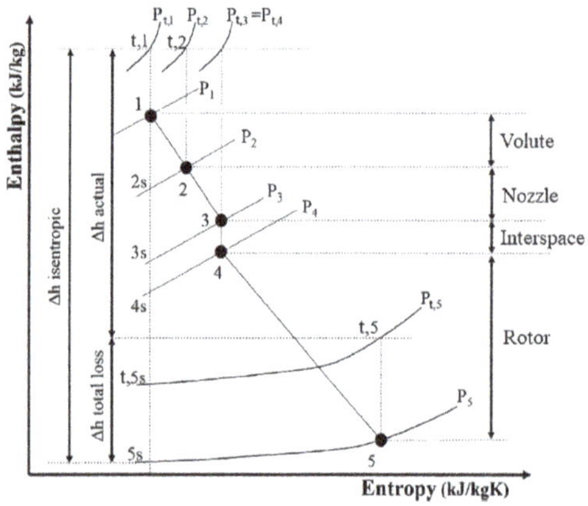

Figure 5. An enthalpy–entropy diagram of the turbine stage.

2.3.1. Rotor Modeling

The rotor is simulated using two common non-dimensional parameters, namely, loading coefficient Ψ and flow coefficient φ. These parameters are shown in Equations (7) and (8) [24].

$$\varphi = \frac{C_{m5}}{U_4} \tag{7}$$

$$\Psi = \frac{\Delta h_{act}}{U_4^2} \tag{8}$$

Knowing that loading and flow coefficients are given, the blade speed U4 and the meridional velocity C_{m5} can be obtained as per Equations (7) and (8). The rotor radius r_4 and blade height b_4 can be obtained utilizing Equations (9) and (10), respectively. After that, the blockage factor K_{B4} and blade thickness t_{b4} are calculated using Equations (11) and (12), respectively. The rotor number of blades Z_r can be calculated using the Glassman correlation [25], expressed in Equation (13).

$$r_4 = \frac{U_4}{\omega} \tag{9}$$

$$\dot{m}_{calc} = 2(1 - K_{B4})\pi r_4 \rho_4 b_4 C_{m4} \tag{10}$$

$$K_{B4} = 1 - \frac{Z_r t_{b4}}{2\pi r_4 sin\beta_4} \tag{11}$$

$$t_{b4} = 0.04\, r_4 \tag{12}$$

$$Z_r = \frac{\pi}{30}(110 - \alpha_4)\tan\alpha_4 \tag{13}$$

The thermodynamic properties of the working fluid can be calculated as per Equations (14)–(17), completing the inlet geometry of the rotor. Refprop, version 9 [26] software was utilized to obtain the thermodynamic properties of the fluid.

$$h_4 = h_{o4} + \frac{1}{2}C_4^2 \tag{14}$$

$$P_{04} = P_{01} - \frac{\rho_{01}\Delta h_{is}(1 - \eta_{ts})}{4} \tag{15}$$

$$\{h_{04}, P_{04}\} = EoS(\rho_{04},\ T_{04},\ a_{04},\ s_{05},\ fluid) \tag{16}$$

$$\{h_4, S_4\} = EoS(\rho_4,\ T_4,\ a_4,\ P_4, fluid) \tag{17}$$

Turbine power can be obtained using the Euler equation, Equation (18). As proposed by Aungier [27], the meridional speed C_{m5} can be calculated using Equation (19).

$$W_{output} = \dot{m}(U_4 C_{\theta 4} - U_5 C_{\theta 5}) \tag{18}$$

$$C_{m5} = C_{m4}\left(1 + 5\left(\frac{b_4}{r_4}\right)^2\right) \tag{19}$$

The smallest distance between two adjacent blades is the rotor throat o_5. It can be obtained considering the behavior of the exit Mach number M_{5rel}, as expressed in Equation (20)

$$\begin{array}{ll} o_5 = \frac{s_5 C_{m5}}{W_5} & for \quad M_{5rel} < 1 \\ o_5 = \frac{s_5 C_{m5}\rho_5}{\rho_* W_*} & for \quad M_{5rel} \geq 1 \end{array} \tag{20}$$

2.3.2. Stator Modeling

Three different stator openings are investigated, namely, 30%, 60%, and 100%, as shown in Figure 6. The stator opening is controlled by varying the stator width, o_3, as shown in Equation (21). The stator throat width is shown in Figure 6, where s_v is the nozzle pitch, as shown in Figure 6. ρ_* and a_* are the sonic condition density and the speed of sound, respectively.

$$\begin{cases} o_3 = s_v sin\alpha_3 & for \quad M_3 < 1 \\ o_3 = \frac{C_{m3}\rho_3}{\rho_* a_*} & for \quad M_3 \geq 1 \end{cases} \tag{21}$$

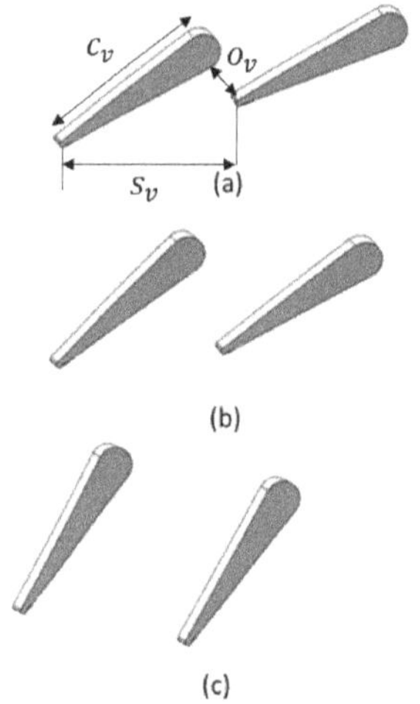

Figure 6. Different stator openings: (**a**) 30%, (**b**) 60%, and (**c**) 100%.

The design method of the turbine nozzle is conducted in an iterative manner, as depicted in the flowchart in Figure 4. The tangential velocity at the stator outlet $C_{\theta 3}$ is obtained using Equation (22). As shown in the flowchart of the design process, the nozzle exit absolute velocity C_3 is found in an iterative manner. The static enthalpy h_3 is then found using Equation (23). The remaining thermodynamic properties of the working fluid are calculated, as shown in Equation (24). The mass flow rate $(\dot{m}_{calc})$ is recalculated using Equation (25) until convergence is satisfied.

$$C_{\theta 3} = \frac{C_{\theta 4} \cdot r_4}{r_3} \tag{22}$$

$$h_3 = h_{o3} + \frac{1}{2} C_3^2 \tag{23}$$

$$\{S_3 = S_4, h_3\} = EoS(\rho_3,\ T_3,\ a_3,\ P_3, fluid) \tag{24}$$

$$\dot{m}_{calc} = 2\pi r_3 \rho_3 b_3 C_{m3} \tag{25}$$

2.3.3. Volute Modeling

For simplicity, a circular section is assumed and Equations (26)–(29) are solved in an iterative manner. SC is the swirl coefficient, which is given a value of 0.95 as suggested by [24].

$$C_1 = \frac{r_2 C\theta_2}{r_1 SC} \tag{26}$$

$$h_1 = h_{01} - \frac{C_1^2}{2} \tag{27}$$

$$\{h_1, s_1\} = EoS(\rho_1,\ T_1,\ a_1,\ P_1, fluid) \tag{28}$$

$$A_1 = \frac{\dot{m}}{\rho_1 C_1} \tag{29}$$

2.3.4. Losses Modeling

To verify the original assumption of the total to static efficiency η_T, the rotor losses model is used. The turbine rotor is responsible for the majority of the losses. The most common rotor losses as outlined in [24,28] are incidence loss, passage loss, tip clearance loss, windage loss, and exit energy; these are included in the present model and summarized in Table 2. In a subsequent step, the rotor losses are summed together using Equation (30).

$$\Delta h_{\text{loss,rotor}} = \Delta h_{\text{incidence}} + \Delta h_{\text{passage}} + \Delta h_{\text{tip}} + \Delta h_{\text{windage}} + \Delta h_{\text{exit}} \quad (30)$$

Table 2. Losses model across the radial turbine stage.

Loss	Equation
Incidence Loss	$\Delta h_{incidence} = \frac{1}{2}\left[W_4 sin(\beta_4 - \beta_{4,opt})\right]^n$
Passage Loss	$\Delta h_{\text{passage}} = \frac{1}{2}\left(2f_t \frac{L_h}{D_h}\overline{W}^2 + \frac{r_4 C_4^2}{r_c Z_r}\right)$
Tip Clearance Loss	$\Delta h_{\text{tip}} = \frac{U_4^3 Z_r}{8\pi}\left(K_a \varepsilon_a C_a + K_r \varepsilon_r C_r + K_{a,r}\sqrt{(\varepsilon_a \varepsilon_r C_a C_r)}\right)$
Windage Loss	$\Delta h_{\text{windage}} = k_f \frac{\overline{\rho} U_4^3 r_4^2}{2\dot{m}_{wf}}$
Exit Energy Loss	$\Delta h_{\text{exit}} = \frac{1}{2} C_5^2$

3. Results and Discussions

This section presents the effects of the variable geometry turbine on the overall system performance at different engine operating points. The results are also compared with those obtained using a fixed geometry turbine.

3.1. Effects of Stator Openings on Cycle and Turbine Performances

First, the effects of stator openings on turbine efficiency are investigated as shown in Figure 7. All engine operating points present a similar trend, where turbine efficiency increases, reaching a maximum value at P3, then decreases at P3 as shown in Figure 7a. At P1, the 60% opening presents substantially higher turbine efficiency with 57.5%, while P3 presents the lowest turbine efficiency with a value of 42.4%. The 30% opening presents the lowest turbine efficiency because of the excessive heat for the low mass flow rates of the exhaust gas as shown in Table 1. In other words, the mass flow rate of the working fluid decreases at low openings. Therefore, the excessive heat results in an extremely superheated fluid at the turbine inlet. Moreover, radial turbine losses (tip clearance and incidence, as shown in Table 2) increase at higher turbine inlet temperatures, which deteriorates the turbine efficiency as per Equation (30). Regarding stator openings, the 100% opening is not always favorable due to the increased fluid capacity that requires more heat to be fully vaporized.

Figure 7c depicts the behavior of turbine power with different stator openings and various engine operating points. It is clear that the 100% opening is superior for all the investigated engine points. This is expected, since turbine power is directly related to the mass flow rate and the enthalpy drop through the turbine. P3 with a 100% stator opening presents the highest turbine power with 18.5 kW. The 30% and 60% stator openings present maximum values of 12.60 and 16 kW, respectively. Again, as the mass flow rate decreases, the turbine power decreases according to Equation (18). Figure 7c shows that the 30 and 60% throat openings present comparable turbine power at engine point P2. This is due to the comparable exhaust temperatures as per Table 1.

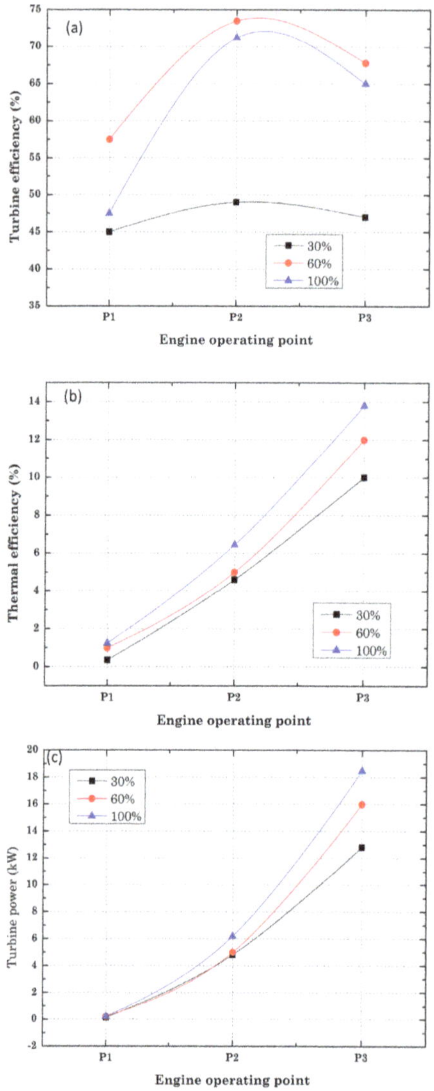

Figure 7. Effects of stator opening on (**a**) turbine efficiency, (**b**) turbine power, and (**c**) cycle thermal efficiency at different engine operating points.

The effects of stator openings on the thermal efficiency are depicted in Figure 7b. The cycle thermal efficiency is directly related to turbine power and the absorbed heat in the evaporator. Thermal efficiency follows the trend of turbine power qualitatively. As per Equation (6), the cycle thermal efficiency is a function of turbine power. As the turbine power increases, the cycle thermal efficiency increases. As already depicted in Figure 7c, the turbine power increases and, hence, the cycle efficiency increases. The 100% opening presents the highest thermal efficiency for all the engine operating points, followed by the 60% opening. The maximum thermal efficiency of the system is 13.8% and obtained at P3, with a 100% opening. The lowest thermal efficiency is 0.4% and presented by engine operating point P1 with a 30% stator opening.

3.2. Comparison between Fixed and Variable Geometry Turbines

The potential of utilizing a variable geometry turbine instead of a fixed one is clearly depicted in Figure 8. It can be seen from Figure 8a that turbine efficiency for variable geometry is always higher than with a fixed geometry turbine for the various engine operating points. For a fixed geometry turbine, the mass flow rate is fixed. Therefore, and as explained earlier, the evaporator produces excessive heat which results in extremely superheated fluid. A variable geometry turbine, on the other hand, has the ability to adapt the wasted heat in the engine, which substantially improves the turbine performance. The minimum and maximum improvements in the turbine efficiency with variable geometry in comparison to that with fixed geometry are 20% and 70%, respectively. Similarly, turbine power is significantly higher when integrated with a variable geometry turbine, as shown in Figure 8b. This is because of the variable mass flow rate that directly influences the turbine power as per Equation (18). At P3, for instance, the variable geometry turbine presents 18 kW turbine power, while the fixed geometry turbine presents 14.7 kW. At P1, in fact, the turbine power could be doubled when utilizing a variable geometry turbine compared to a fixed one.

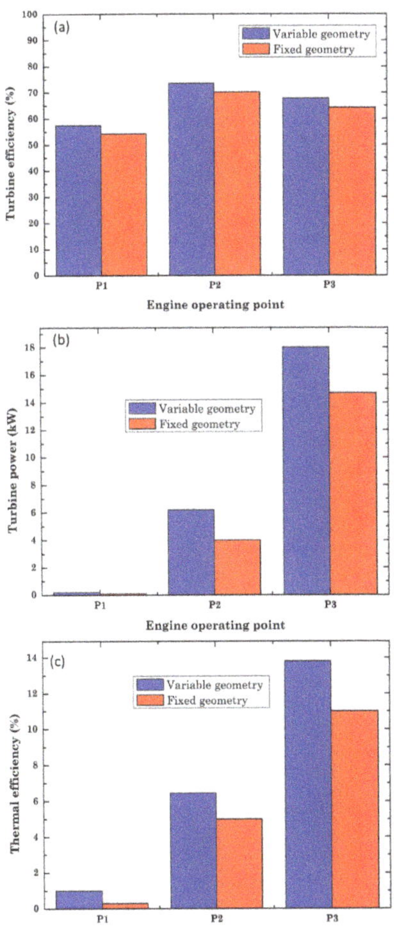

Figure 8. Comparison between fixed and variable geometry turbines for (**a**) turbine efficiency, (**b**) turbine power, and (**c**) cycle thermal efficiency at different engine operating points.

Since cycle thermal efficiency is directly related to turbine performance, it presents similar trends, as shown in Figure 8c. A minimum improvement of 4.48% can be achieved when a variable geometry turbine is utilized. At P3, the advantage of utilizing a variable geometry turbine over a fixed one is clear. The cycle efficiency of the variable geometry turbine is 25.45% higher than that of the fixed geometry turbine.

The main aim of coupling the ORC system to the engine was to explore the feasibility of such a system in improving engine performance. Figure 9 shows that the engine performance is enhanced in terms of power and BSFC after utilizing the ORC system. This enhancement is expected since the ORC power output is added to the engine power as per Equation (1). Similarly, and as per Equation (2), the engine BSFC improves by improving the engine power. As presented in Figure 9a, the engine power could improve by at least 1% when integrated with an ORC system with a variable geometry turbine. At P3, in fact, the engine power may increase by 5.5% when bottomed with an ORC system. Engine BSFC presents a similar trend, as shown in Figure 9b. Figure 9 also depicts the benefits of utilizing a variable geometry turbine instead of a fixed geometry turbine. It is clear that, a variable geometry turbine is favorable over different engine operating points.

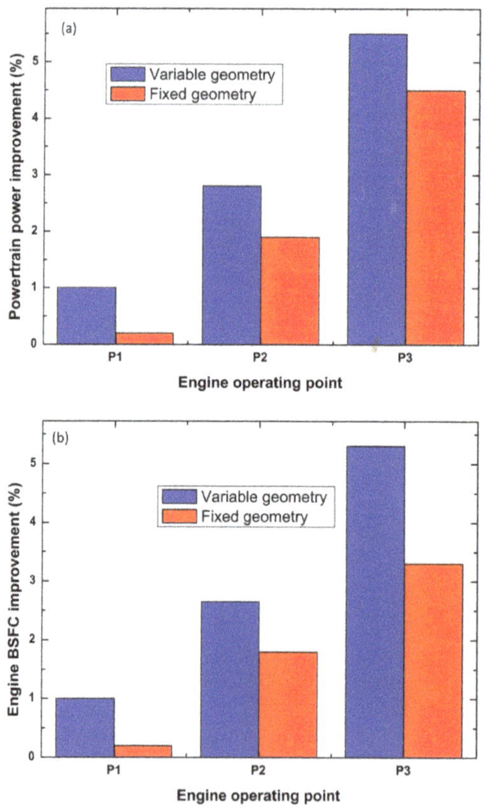

Figure 9. Comparison between fixed and variable geometry turbines for (**a**) powertrain power and (**b**) engine BSFC.

4. Conclusions

The current study explored the potential of integrating a variable geometry radial turbine instead of a fixed geometry turbine for an engine–ORC system. The study was carried out using different engine operating conditions, where different exhaust temperatures and mass flow rates were investigated. Moreover, different stator openings were explored: 30, 60, and 100%. The results confirmed the potential of utilizing variable geometry turbines in improving the overall system performance. This is because variable geometry turbines can adapt to the different operating conditions of the exhaust gases by changing the stator opening accordingly. The conclusions can be summarized as follows:

- The total to static efficiency of a variable geometry turbine is 25.45% higher than that of a fixed geometry turbine.
- Due to the increased mass capacity with a variable geometry radial turbine, the powertrain power increased substantially when integrated with an ORC as a bottom cycle. This increase could be as high as 5.5%.
- The ORC performance benefited significantly from the utilization of the variable geometry turbine. The ORC efficiency could be improved by at least 20%.
- The results also showed that even at low engine torque and power, P1, the overall system performance improved substantially by integrating an ORC system with a variable geometry turbine. However, this improvement is trivial (1%) and cannot justify the added complexity and cost.
- It is worth noting that the ORC system has a large weight, which will significantly affect its benefits. In the future, this phenomenon will be investigated in addition to cost analysis.

Author Contributions: Conceptualization, F.A. and I.A.; Data curation, F.A., A.A. (Abdullah Alghafis), I.A., A.A. (Ahmed Alzamil) and A.A. (Abdullah Alrashidi); Formal analysis, F.A. and A.A. (Abdullah Alghafis); Funding acquisition, F.A.; Investigation, F.A., A.S.A. and A.A. (Abdullah Alrashidi); Methodology, F.A., I.A., A.S.A. and A.A. (Ahmed Alzamil); Project administration, F.A.; Resources, A.S.A.; Software, F.A. and A.A. (Abdullah Alrashidi); Supervision, F.A.; Validation, F.A., I.A., A.S.A. and A.A. (Ahmed Alzamil); Visualization, F.A.; Writing—original draft, F.A.; Writing—review and editing, A.A. (Abdullah Alghafis), I.A., A.S.A., A.A. (Ahmed Alzamil) and A.A. (Abdullah Alrashidi). All authors have read and agreed to the published version of the manuscript.

Funding: This research was funded by the Scientific Research Deanship at the University of Ha'il—Saudi Arabia through project number "RD-21 003".

Data Availability Statement: Data are contained within the article.

Conflicts of Interest: The authors declare no conflict of interest.

Nomenclature

1–5	Stations through turbine	Subscript	
a	Speed of sound [m/s]	h	Hub, hydraulic
A	Area	r	Radial, rotor
b	Blade height [m]	s	Insentropic, stator
K	Blockage factor [-]	t	Tip, total
C	Absolute velocity [m/s]	Greek Symbols	
C_m	Meridional velocity [m/s]	η	Efficiency [-]
C_θ	Tangential velocity [m/s]	β	Relative angle [deg]
Ca	Axial coefficient [-]	ε	Clearance [m]
Cr	Radial coefficient [-]	ρ	Density [kg/m^3]
d	Diameter [m]	α	Absolute flow angle [deg]
h	Enthalpy [kJ/kg]	γ	Stator setting angle [deg]
K_a	Discharge coefficient of the axial component [-]	Abbreviations	
K_r	Discharge coefficient of the radial component [-]	BSFC	Brake specific fuel consumption
$K_{a,r}$	Cross coupling coefficient of the axial and radial components [-]	CFD	Computational fluid dynamics

M	Mach number [-]	GWP	Global warming potential
$\dot{m}$	Mass flow rate [kg/s]	ICE	Internal combustion engine
o	Throat opening [m]	NOx	Nitrogen oxide
P	Pressure [kPa]	ORC	Organic Ranke cycle
s	Entropy [kJ/kg.k]	WHR	Waste heat recovery
T	Temperature [K]		
U	Tip speed [m/s]		
w	Relative velocity [m/s]		
W	Work [kW]		
z	Axial length [m]		

References

1. Seher, D.; Lengenfelder, T.; Gerhardt, J.; Eisenmenger, N.; Hackner, M.; Krinn, I. Waste Heat Recovery for Commercial Vehicles with a Rankine Process. In Proceedings of the 21st Aachen Colloquium Automobile and Engine Technology, Aachen, Germany, 8–10 October 2012.
2. Liu, L.; Wang, K.; Wang, S.; Zhang, R.; Tang, X. Assessing energy consumption, CO_2 and pollutant emissions and health benefits from China's transport sector through 2050. *Energy Policy* **2018**, *116*, 382–396. [CrossRef]
3. Galindo, J.; Ruiz, S.; Dolz, V.; Royo-Pascual, L.; Haller, R.; Nicolas, B.; Glavatskaya, Y. Experimental and thermodynamic analysis of a bottoming Organic Rankine Cycle (ORC) of gasoline engine using swash-plate expander. *Energy Convers. Manag.* **2015**, *103*, 519–532. [CrossRef]
4. Zhao, R.; Li, W.; Zhuge, W.; Zhang, Y.; Yin, Y. Numerical study on steam injection in a turbocompound diesel engine for waste heat recovery. *Appl. Energy* **2017**, *185*, 506–518. [CrossRef]
5. Chen, W.; Fu, B.; Zeng, J.; Luo, W. Research on the Operational Performance of Organic Rankine Cycle System for Waste Heat Recovery from Large Ship Main Engine. *Appl. Sci.* **2023**, *13*, 8543. [CrossRef]
6. Badescu, V.; Aboaltabooq, M.H.K.; Pop, H.; Apostol, V.; Prisecaru, M.; Prisecaru, T. Avoiding malfunction of ORC-based systems for heat recovery from internal combustion engines under multiple operation conditions. *Appl. Therm. Eng.* **2019**, *150*, 977–986. [CrossRef]
7. Berchiche, A.; Guenoune, M.; Belaadi, S.; Léonard, G. Optimal Energy Integration and Off-Design Analysis of an Amine-Based Natural Gas Sweetening Unit. *Appl. Sci.* **2023**, *13*, 6559. [CrossRef]
8. Liu, P.; Shu, G.; Tian, H.; Feng, W.; Shi, L.; Xu, Z. Preliminary experimental comparison and feasibility analysis of CO2/R134a mixture in Organic Rankine Cycle for waste heat recovery from diesel engines. *Energy Convers. Manag.* **2019**, *198*, 111776. [CrossRef]
9. Lu, Y.; Wang, Y.; Wang, L.; Yuan, Y.; Liu, Z.; Roskilly, A.P. Experimental Investigation of a Scroll Expander for Power Generation Part of a Resorption Cogeneration. *Energy Procedia* **2015**, *75*, 1027–1032. [CrossRef]
10. Yang, J.; Li, Y.; Tan, H. Integrated hydrogen liquefaction process with a dual-pressure organic Rankine cycle-assisted LNG regasification system: Design, comparison, and analysis. *Appl. Energy* **2023**, *347*, 121372. [CrossRef]
11. He, Y.; Wu, H.; Zhang, Y.; Jin, T.; Xing, Z. Experimental Research on the Effects of Suction Ports on Twin Screw Expander Performance. *Appl. Sci.* **2022**, *12*, 3979. [CrossRef]
12. Rijpkema, J.; Andersson, S.B.; Munch, K. Experimental study of an organic Rankine cycle with R1233zd(E) for waste heat recovery from the coolant of a heavy-duty truck engine. *Energy Convers. Manag.* **2021**, *244*, 114500. [CrossRef]
13. Saidur, R.; Rezaei, M.; Muzammil, W.K.; Hassan, M.H.; Paria, S.; Hasanuzzaman, M. Technologies to recover exhaust heat from internal combustion engines. *Renew. Sustain. Energy Rev.* **2012**, *16*, 5649–5659. [CrossRef]
14. Alshammari, F.; Karvountzis-Kontakiotis, A.; Pesyridis, A.; Usman, M. Expander Technologies for Automotive Engine Organic Rankine Cycle Applications. *Energies* **2018**, *11*, 1905. [CrossRef]
15. Alshammari, F.; Pesyridis, A.; Alshammari, A.S.; Alghafis, A.; Alatawi, I.; Alzamil, A. Potential of capturing transportation wasted heat for better fuel economy and electricity generation: Comprehensive testing. *Energy Convers. Manag.* **2022**, *267*, 115939. [CrossRef]
16. Thaddaeus, J.; Pesiridis, A.; Karvountzis-kontakiotis, A. Design of Variable Geometry Waste Heat Recovery Turbine for High Efficiency Internal Combustion Engine. *Int. J. Sci. Eng. Res.* **2016**, *7*, 1001–1017.
17. Manente, G.; Toffolo, A.; Lazzaretto, A.; Paci, M. An Organic Rankine Cycle off-design model for the search of the optimal control strategy. *Energy* **2013**, *58*, 97–106. [CrossRef]
18. Lebedevas, S.; Čepaitis, T. Research of organic rankine cycle energy characteristics at operating modes of marine diesel engine. *J. Mar. Sci. Eng.* **2021**, *9*, 1049. [CrossRef]
19. Schuster, S.; Markides, C.N.; White, A.J. Design and off-design optimisation of an organic Rankine cycle (ORC) system with an integrated radial turbine model. *Appl. Therm. Eng.* **2020**, *174*, 115192. [CrossRef]
20. Guillaume, L.; Legros, A.; Desideri, A.; Lemort, V. Performance of a Radial-Inflow Turbine Integrated in an ORC System and Designed for a WHR on Truck Application: An Experimental Comparison between R245fa and R1233zd. *Appl. Energy* **2017**, *186*, 408–422. [CrossRef]

21. Wang, E.; Yu, Z.; Zhang, H.; Yang, F. A regenerative supercritical-subcritical dual-loop organic Rankine cycle system for energy recovery from the waste heat of internal combustion engines. *Appl. Energy* **2017**, *190*, 574–590. [CrossRef]
22. Furukawa, T.; Nakamura, M.; Machida, K.; Shimokawa, K. A Study of the Rankine Cycle Generating System for Heavy Duty HV Trucks. *SAE Tech. Pap.* **2014**, *1*, 7. [CrossRef]
23. Alshammari, F.; Ben Khedher, N.; Ben Said, L. Development of an automated design and off-design radial turbine model for solar organic Rankine cycle coupled to a parabolic trough solar collector. *Appl. Therm. Eng.* **2023**, *230*, 120677. [CrossRef]
24. Moustapha, H.; Zelesky, M.F.; Baines, N.C.; Japikse, D. *Axial and Radial Turbines*, 1st ed.; Concepts NREC: White River Junction, VT, USA, 2003.
25. Glassman, A. *Computer Program for Design Analysis of Radial-Inflow Turbines*; NASA: Washington, DC, USA, 1976.
26. Lemmon, E.; Huber, M.; McLinden, M. *NIST Standard Reference Database 23: Reference Fluid Thermodynamic and Transport Properties-REFPROP, Version 9.0*; National Institute of Standards and Technology, Standard Reference Data Program: Gaithersburg, MD, USA, 2010.
27. Aungier, R.H. *Turbine Aerodynamics: Axial-Flow and Radial-Inflow Turbine Design and Analysis*, 1st ed.; American Society of Mechanical Engineers: New York, NY, USA, 2006.
28. Rahbar, K.; Mahmoud, S.; Al-Dadah, R.K.; Moazami, N. Modelling and optimization of organic Rankine cycle based on a small-scale radial inflow turbine. *Energy Convers. Manag.* **2015**, *91*, 186–198. [CrossRef]

applied sciences

MDPI

Article

Design of an Energy Harvester Based on a Rubber Bearing Floating Slab Track

Shuo Yu [1], Zheng Li [2], Hao Jin [3,*], Donghao Yin [3] and Jiajia Yan [1]

[1] PowerChina Huadong Engineering Corporation Limited, Hangzhou 311122, China; yu_s@hdec.com (S.Y.); yan_jj@hdec.com (J.Y.)
[2] Transport Planning and Research Institute, Ministry of Transport, Beijing 100028, China; lizheng@tpri.org.cn
[3] School of Transportation, Southeast University, Nanjing 211189, China; 220215103@seu.edu.cn
* Correspondence: jinhao@seu.edu.cn

Abstract: It is known that a large amount of vibration mechanical energy will be generated during train operation. If the mechanical energy can be obtained from the track structure, it can greatly optimize the energy configuration of the metro. Currently, most sensors are limited to disk or cymbal structures and are not used in the track bed; different from existing structures, this paper designs a spherical energy harvester based on a rubber bearing floating slab track, wherein the size range of the spherical energy harvesting structure was approximately determined based on the geometric spatial relationships of the actual track bed internal structure. Compared to the traditional disk and cymbal energy harvesters, the mechanical and electrical properties of the spherical energy harvesting structure was studied by a numerical simulation method, and the optimal size of the spherical energy was determined by calculation. The main conclusions are as follows: (1) Compared with the traditional disk harvester and cymbal harvester, the spherical harvester had better mechanical and electrical properties. (2) By calculating the output power of energy harvesters under load matching, we found that the output power of the spherical harvester was two orders of magnitude higher than that of the disk harvester and 53% higher than that of the cymbal harvester. (3) The optimum size of the spherical harvester was when the thickness of piezoelectric layer was 2 mm, the radius was 16 mm, the axial ratio of the spherical shell was 0.4, and the thickness of the spherical shell was 4 mm.

Keywords: rail transit; sensor; floating slab track; energy harvester; piezoelectric layer dimension

1. Introduction

In recent years, urban rail transit has been widely accepted as a large-capacity and low-carbon transportation mode. With the increasing mileage of rail transit operation, its operational safety has been paid more and more attention [1]. To ensure the safety of rail transit during the operation period, it is necessary to carry out health monitoring on its engineering structure. Traditional monitoring sensor networks are mainly connected to a mains power supply or battery, which has some problems such as high maintenance cost and environmental pollution. Since a large amount of mechanical energy will be generated by train operation in urban rail transit system [2], it is of great significance to design a kind of energy harvester to capture electric energy from the track structure.

At present, many scholars have designed energy harvesters for rail transit infrastructure in order to capture energy from different structures and positions, mainly including rails, fasteners, and sleepers. For the rail, Li et al. [3] designed a broadband cantilever energy harvester, which installed a broadband piezoelectric array on one side of the rail to capture the track vibration mechanical energy during train operation at various frequencies. Due to the train vibration frequency deviating from its working frequency band in real environment, it did not achieve the expected effect. Zhang et al. [4] developed a portable and efficient electromagnetic energy harvesting system with supercapacitors, which was installed at the bottom of rails to convert the energy of track vibration into electric energy and provide stable power output for external loads. For the fastener, Yuan et al. [5] designed a piezoelectric energy harvester installed under the fastener backing plate to capture the energy generated by train vibration, and the power obtained was sufficient to power the wireless sensor. For the sleeper, Wang et al. [6] designed an energy harvester for the upper part of the sleeper based on the electromagnetic energy collection technology. It featured a gear structure that converted the irregular pulse bidirectional linear vibration of the track into adjustable unidirectional rotational motion to improve the output power and reliability. Lin et al. [7] also designed an enabling structure for the upper part of sleeper, using a spring pretensioning and resetting device for quick installation without anchoring it into the railroad foundation. In addition to the design of energy harvesters starting from common track infrastructure components, scholars have carried out corresponding energy capture research on floating slab track. Hou et al. [8] proposed a small but high bearing capacity piezoelectric energy harvester for steel spring floating slab track, which was mainly installed between track slab and steel spring supporting structure to improve the energy harvesting efficiency. At present, the design of energy harvesters for a floating slab track cannot cover all types, so it still needs further study.

The piezoelectric energy harvesting structure utilizes the positive piezoelectric effect of piezoelectric materials to convert mechanical energy into electrical energy [9], thus achieving energy recovery and reuse. The materials involved in piezoelectric energy harvesting mainly include piezoelectric ceramics (PZT), polyvinylidene fluoride (PVDF), and piezoelectric fibers. Piezoelectric materials, especially PZT, are generally not directly exposed to the working environment in the process of energy conversion due to their own brittleness, weak bearing capacity, small displacement, and other characteristic limitations, but packaging structures are added outside them to form piezoelectric energy harvesters [10]. At present, there are many studies on piezoelectric energy harvesters. The common energy harvesters are cantilever beam type [11–15], cymbal type [16–18], multilayer stack type [19,20], bridge type [21,22], arch type [23–25], etc., and there are new energy harvesters [26–29] 2,201,2424 made of composite materials, such as piezoelectric fiberboard (MFC), polyvinylidene fluoride (PVDF), and piezoelectric film.

The cymbal energy harvester is the earliest energy harvester designed and studied, and it is also recognized as energy harvester with comprehensive performance. It consists of metal caps at the upper and lower ends and piezoelectric layers in the middle, so it can capture radial and axial vibration energy at the same time and have high conversion efficiency. Kim et al. [30] tested the ability of a cymbal energy harvester to obtain electrical energy in a dynamic vibration environment, finding that metal end caps can increase ceramic endurance so that the whole structure can bear a higher load. Zhao et at. [31] used finite element software to analyze the geometrical parameters of cymbal energy harvesters applied to asphalt pavements, and the results showed that the energy harvesting efficiency was related to many factors, such as cymbal spacing, end cap thickness, and cavity height.

The improvement and application of the cymbal energy harvester structure has also been concerned by many researchers. Based on the traditional cymbal energy harvester, G. Yesner et al. [32] designed a bridge structure energy harvester with metal end caps and a new type of electrode. Compared with the traditional cymbal piezoelectric energy harvester, the energy obtained was increased, and the failure period was obtained by cyclic loading of the prepared structure. It was found that the inconsistent thickness of the bonding layer was the cause of the structure failure. Liu et al. [33] designed another radial layered cymbal piezoelectric energy harvester. To match the vibration frequency under traffic load, the metal end cap was calibrated by theoretical calculation and experimental test. The output power of the structure was not high, but a scheme to optimize the cymbal energy harvester structure was provided. The disadvantage of a cymbal piezoelectric energy harvester is that the metal end caps will cause stress concentration at the inner edge of ceramic plate, resulting in damage or fracture of the ceramic plate. In addition, the manufacturing process of a cymbal energy harvester is complex, and there are limitations in practical application.

The sensors mentioned in the current article are limited to disk or cymbal structure and not used in the track bed; different from existing structures, this paper designed a spherical energy harvester based on a rubber bearing floating slab track, and the size range of the spherical energy harvesting structure was approximately determined based on the geometric spatial relationships of the actual track bed internal structure. Compared to the traditional disk and cymbal energy harvester, the mechanical and electrical properties of the spherical energy harvesting structure were studied by the numerical simulation method. Meanwhile, in order to improve the energy harvesting efficiency of the spherical energy harvester, the geometric parameters such as piezoelectric layer thickness, piezoelectric layer radius, spherical shell axial ratio, and spherical shell thickness were analyzed to obtain the optimal geometric size.

The architecture of the paper is organized as follows: Section 2 introduces the spatial location where spherical sensors are suitable for harvesting energy in the track bed. Section 3 proposes the dimensions and parameters of spherical energy harvesting structure. Section 4 introduces the performance differences between the disk, cymbal, and spherical energy harvesting structure. Section 5 introduces the optimization process of geometric parameters of the spherical energy harvesting structure.

2. Energy Harvesting Position of Rubber Bearing Floating Slab Track

When the energy harvester is laid in urban rail transit, the existing research mainly focuses on the position close to the vibration source, such as rail and fastener, but it needs to occupy some bounded space, so its application scope will be limited in the rail transit operation environment with strict limit requirements. At the same time, there is a possibility of loosening and dropping of the installation under the long-term reciprocating train operation environment, which has certain potential safety hazards. Therefore, the designed energy harvester in this paper is considered to be arranged inside the floating slab track, thus reducing the impact on the safety of train operation. To make full use of the mechanical energy generated during train operation and realize large-scale power generation, energy harvesters are mainly in the form of an array. An array of energy harvesters is formed under each rubber support, and its position and arrangement are shown in Figure 1.

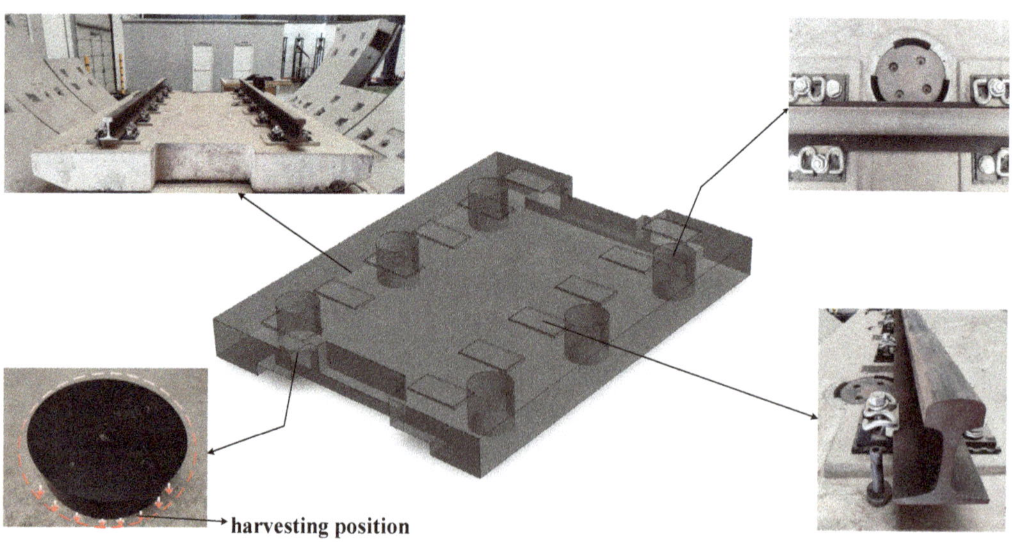

Figure 1. Energy harvesting position.

3. Proposal of the Spherical Energy Harvester

The disk energy harvester is the simplest type, being mainly composed of the piezoelectric layer in the middle and force transmission structures at both ends. The piezoelectric layer is made of piezoelectric ceramic (PZT-5H), while the force transmission structure is made of steel, and they are both wafer-shaped. The geometric dimensions of the disk energy harvester are shown in Figure 2.

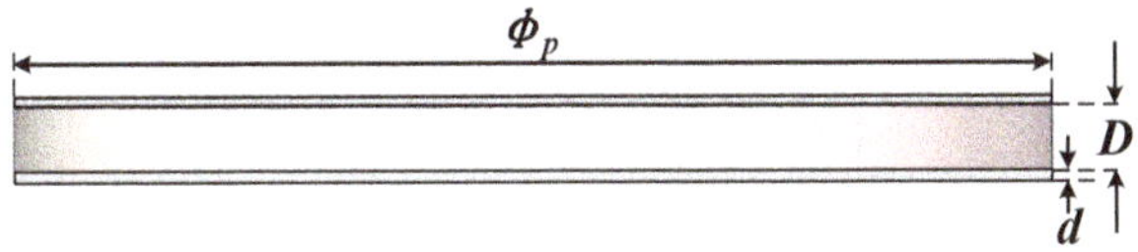

Figure 2. Dimensions of the disk energy harvester.

Here, $d = 0.3$ mm denotes the thickness of force transmission structures at both ends; $D = 2$ mm indicates the thickness of the piezoelectric layer; and $\phi_p = 32$ mm shows the diameter of the disk energy harvester.

The cymbal energy harvester is one kind of energy harvester with better performance, mainly consisting of the piezoelectric element and metal caps on both sides, which are generally bonded by epoxy resin or anchored by bolts. The material used is consistent with that of the disk energy harvester, that is, PZT-5H is used for piezoelectric elements, and steel is used for metal caps.

Due to the special structure of the metal cap, the axial load applied on the metal cap is converted and amplified into two components along the axial and radial directions of the piezoelectric layer, resulting in a larger voltage than that of the general disk energy harvester. The geometric dimensions of the cymbal energy harvester are shown in Figure 3 [31].

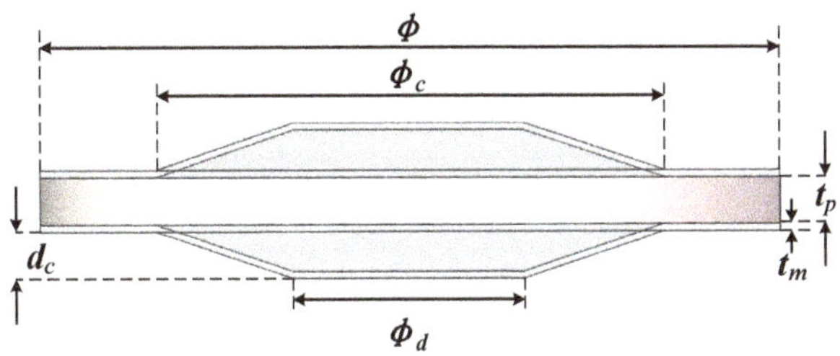

Figure 3. Dimensions of the cymbal energy harvester.

Here, $t_m = 0.3$ mm denotes the thickness of the metal cap; $t_p = 2$ mm indicates the thickness of the piezoelectric layer; $\phi_C = 22$ mm is the inner cavity bottom diameter; $\phi_d = 10$ mm is the inner cavity top diameter; $d_c = 2$ mm indicates the height of the inner cavity; and $\phi = 32$ mm denotes the diameter of cymbal energy harvester.

The spherical energy harvester studied in this paper is composed of the piezoelectric layer and spherical shells at both ends. The piezoelectric layer is a circular sheet structure with a thickness of 2 mm and a diameter of 32 mm. The outer diameter of spherical shells at both ends is the same as that of piezoelectric layer, that is, 32 mm. The whole structure is shown in Figure 4. Similarly, the piezoelectric layer is made of PZT-5H, and the spherical shell is made of steel [34].

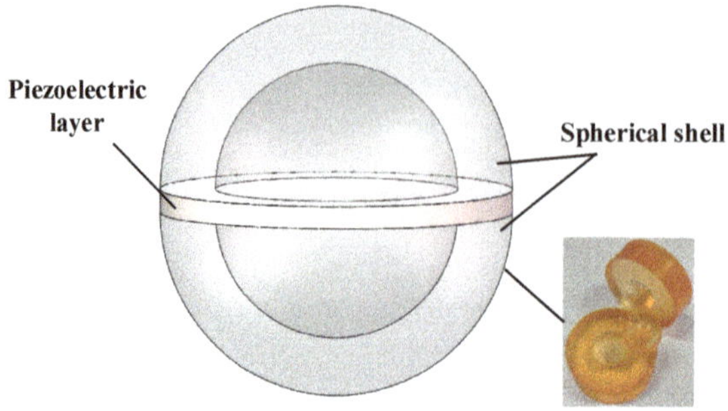

Figure 4. Spherical harvester energy.

The material parameters of PZT-5H and steel used for each energy harvester are shown in Tables 1 and 2 [8].

Table 1. Material parameters of PZT-5H.

Parameter	Value	Unit	Description
S_{33}^{E}	20.7×10^{-12}	m^2/N	Flexibility coefficient
d_{33}	593×10^{-12}	C/N	Piezoelectric strain constant
ε_{r33}^{T}	3400	/	Relative dielectric constant
C_{11}^{E}	1.27×10^{11}	N/m^2	Elastic coefficients in direction 1
C_{12}^{E}	8.02×10^{10}	N/m^2	Elastic coefficients in direction 2
C_{13}^{E}	8.47×10^{10}	N/m^2	Elastic coefficients in direction 3
C_{33}^{E}	1.17×10^{11}	N/m^2	Elastic coefficients in direction 4
C_{44}^{E}	2.30×10^{10}	N/m^2	Elastic coefficients in direction 5
C_{55}^{E}	2.30×10^{10}	N/m^2	Elastic coefficients in direction 6
C_{66}^{E}	2.35×10^{10}	N/m^2	Elastic coefficients in direction 7
e_{31}	17.03	C/m^2	Piezoelectric stress constants at X-axis
e_{33}	23.24	C/m^2	Piezoelectric stress constants at Y-axis
e_{15}	−6.62	C/m^2	Piezoelectric stress constants at Z-axis
ε_{r33}^{S}	1433.6	/	Relative dielectric constant
ε_{r11}^{S}	1704.4	/	Stress-charge constant
ρ	7500	kg/m^3	Density

Table 2. Material parameters of steel.

Parameter	Elastic Modulus/Pa	Density/(kg/m^3)	Poisson's Ratio
Steel	2×10^{9}	7850	0.3

4. Performance Comparison of Energy Harvesters

4.1. Numerical Model

In the COMSOL Multiphysics software, the output voltage of each energy harvester was simulated by using the coupled physical field of solid mechanics and electrostatic. In order to simulate the voltage output characteristics of energy harvesters under the same stress state, energy harvesters of different structural types were embedded in the cylindrical concrete structure with a height of 50 mm and a diameter of 32 mm, where the concrete strength grade is C35. The overall model is shown in Figure 5.

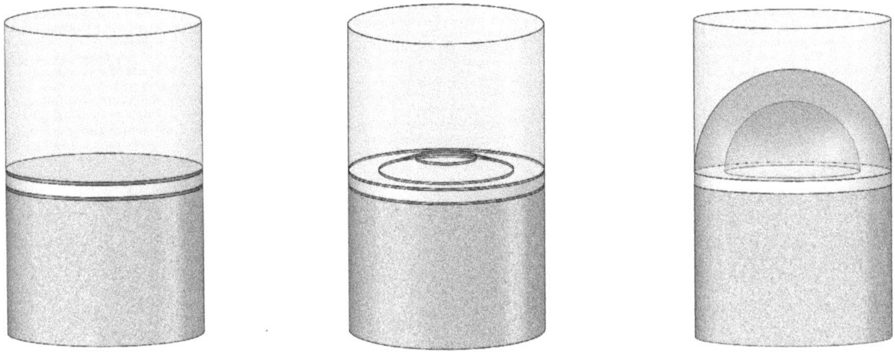

(a) Disk energy harvester (b) Cymbal energy harvester (c) Spherical energy harvester

Figure 5. Numerical model of energy harvesters.

After assigning each part to corresponding materials, a base vector coordinate system was defined so that the piezoelectric material was polarized along the direction of the load. A uniform load of 0.3 MPa was set at the top of the concrete column with the direction perpendicular to its upper surface, and a uniform circumferential pressure of 0.2 MPa was set. At the same time, fixed constraints were set at the bottom of concrete columns. The model set the ground boundary as the bottom of the piezoelectric layer, the terminal boundary as the top of the piezoelectric layer, and the terminal type as electric charge. The initial potential of the piezoelectric layer was 0, and the zero-charge boundary was set in the circumferential direction.

4.2. Mechanical and Electrical Performance

Since the electrical signal output from the energy harvester varies with excitation, an energy harvesting circuit is required to capture or store the electric energy generated by the structure [35,36]. While the energy harvesting circuit also has certain requirements for the output signal of the energy harvester, it generally requires the minimum output voltage of the energy harvester to be around 0.5~1.4 V. If the output voltage of the energy harvester is less than the minimum turn-on voltage, the circuit will be in an open circuit and will not be able to transmit or store electric energy. Therefore, the output voltage of the energy harvester is one important electrical characteristic.

As shown in Figure 6a, the stress of the disk energy harvester was concentrated at the force transmission plate, and the peak stress was 0.45 MPa. As shown in Figure 6b, the output voltage of the disk energy harvester had an obvious laminar distribution along the axial direction, and the maximum value was 0.62 V. The total electricity generated was calculated to be 1.09×10^{-9} J.

As shown in Figure 7a, the stresses in the cymbal energy harvester were concentrated at the top and bottom edges of metal caps with a peak stress of 3.62 MPa. As shown in Figure 7b, the output voltage of the cymbal energy harvester was obviously distributed in layers along the axial direction, and the maximum value was 6.5 V. The total electricity generated by this energy harvester was calculated to be 9.37×10^{-8} J.

Compared with the disk energy harvester, the output voltage and total output energy of the cymbal energy harvester were greatly increased. The output voltage was about 10 times higher than that of the disk energy harvester, and the total output energy was an order of magnitude higher than that of the disk energy harvester. However, the cymbal energy harvester showed a significant stress concentration at the change of geometric sizes, and the peak stress was about eight times higher than that of the disk energy harvester.

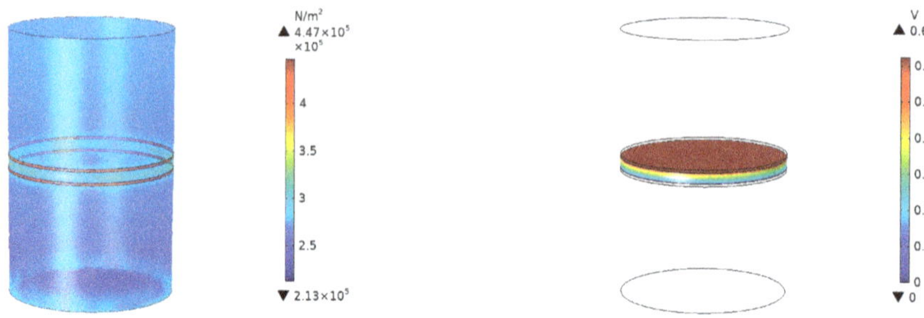

(**a**) Stress distribution (**b**) Output voltage

Figure 6. Mechanical and electrical properties of disk energy harvester.

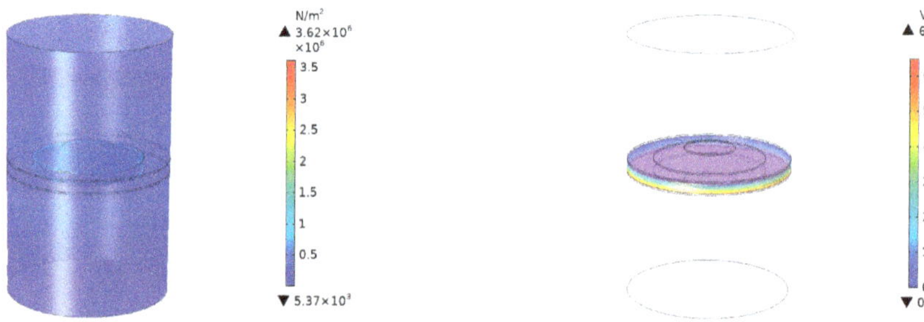

(**a**) Stress distribution (**b**) Output voltage

Figure 7. Mechanical and electrical properties of the cymbal energy harvester.

As shown in Figure 8a, the stress of the spherical energy harvester was concentrated at the bottom edge of the spherical shell, but there was an obvious uniform gradual change of stress along the surface of the spherical shell, thus effectively reducing the peak stress, which was 1.25 MPa. As shown in Figure 8b, the output voltage of the spherical energy harvester also showed obvious laminar distribution along the axial direction, and the maximum value was 6.7 V. The total electricity generated by the spherical energy harvester was calculated to be 1.17×10^{-7} J.

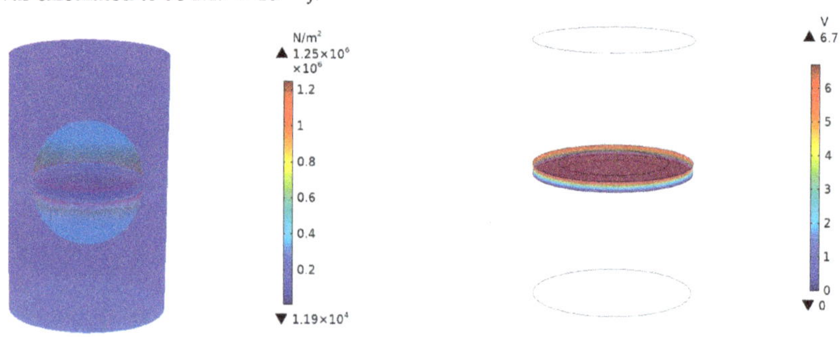

(**a**) Stress distribution (**b**) Output voltage

Figure 8. Mechanical and electrical properties of the spherical energy harvester.

Compared with the disk energy harvester, the total output energy of the spherical energy harvester was greatly improved, being two orders of magnitude higher than that of the disk energy harvester. And compared with the cymbal energy harvester, although the spherical energy harvester also showed some stress concentration at the bottom of the spherical shell, the peak stress was reduced by 65% compared with the cymbal one due to the good stress homogeneity of the spherical shell structure. The spherical energy harvester had both good mechanical properties as well as superior electrical properties. Its output voltage was 11 times more than that of the disk energy harvester, and the total output electric energy was 20% more than that of the cymbal energy harvester.

4.3. Load Match

In general, the piezoelectric energy harvester can be equivalent to an AC power supply with internal resistance, and part of the generated electric energy was consumed by internal resistance, while the other part was used for the load resistor to do work. Therefore, for the piezoelectric energy harvester, load resistance matching is particularly important, and a suitable external load is beneficial to improve the output performance of the energy harvester. For the piezoelectric energy harvester, it is generally regarded as a sinusoidal alternating voltage source with constant amplitude, while for the piezoelectric material itself, it has capacitive and resistive properties. Figure 9 shows the equivalent circuit model of the piezoelectric energy harvester. Here, V is the voltage source; R_0 is the equivalent internal resistor of the energy harvester; C_0 is the equivalent capacitance; and R_L is the load resistor.

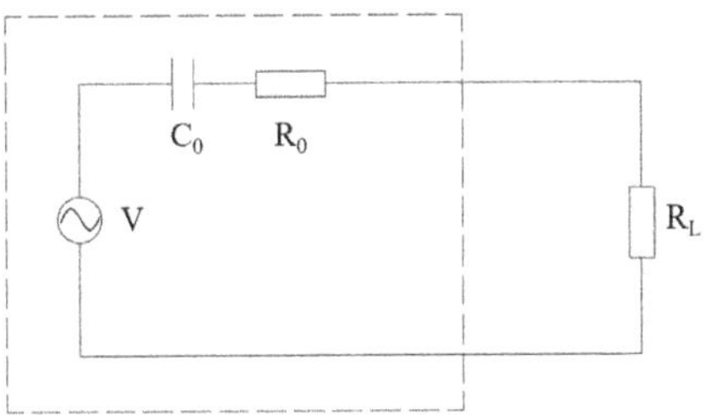

Figure 9. Equivalent circuit of the energy harvester.

For the above equivalent circuit, the capacitance C_0 is calculated as Equation (1):

$$C_0 = \frac{\bar{\varepsilon}_{33}A}{H} \tag{1}$$

where $\bar{\varepsilon}_{33} = \varepsilon_{r33}^T \times \varepsilon_0$ ($\varepsilon_0 = 8.85 \times 10^{-12}$ F/m) represents the vacuum dielectric constant; ε_{r33}^T denotes the relative dielectric constant; $\bar{\varepsilon}_{33}$ indicates the dielectric constant in F/m; A is the piezoelectric layer area; and H is the piezoelectric layer thickness.

The capacitive impedance $R_{C_0} = \frac{1}{2\pi f C_0}$, resistance $R_0 = \frac{tan\delta}{2\pi f C_0}$, and the ratio of impedance to capacitive reactance is $tan\delta = \frac{R_0}{R_{C_0}}$. PZTs are capacitive elements, and the capacitive reactance is far greater than the impedance, so the impedance of PZT can generally be ignored and regarded as a pure capacitive element. Therefore, the general value of $tan\delta$ is very small and negligible, and the load voltage is

$$V_L = V\left|\frac{R_L}{R_L + R_0 + R_{C_0}}\right| \approx V\left|\frac{R_L}{R_L + R_{C_0}}\right| \tag{2}$$

Since the voltage source is a sinusoidal AC voltage, the load power is

$$P_L = \frac{V_L^2}{2R_L} \approx V^2\left|\frac{R_L}{2\left(R_L + R_{C_0}\right)^2}\right| \tag{3}$$

From Equation (3), the output power is maximum when the external load resistance R_L is equal to the internal impedance R_{C_0}. The value of the load impedance at this time is called the matching impedance, that is,

$$R_{match} = |R_{C_0} + R_0| \approx |R_{C_0}| = \frac{1}{\omega C_0} = \frac{1}{2\pi f \cdot C_0} \tag{4}$$

Therefore, the matching impedance is not only related to the capacitance value of the piezoelectric energy harvester, but also related to the excitation frequency. The output power when the load is matched is

$$R_{match} \approx \frac{V^2}{8R_L} \tag{5}$$

The load match model of disk energy harvester is established in COMSOL. In the global definition, the load resistance "R_{load}" is defined, and the resistor R_L is added with the resistance value "R_{load}", that is, the load resistor is set between the output end of the piezoelectric energy harvester and the ground node of the circuit. After setting up, frequency domain analysis can be used to study the effect of external load resistance on the output power of piezoelectric energy harvester, and the output excitation frequency used was 200 Hz. The voltage on the load resistor varied with the resistance, as shown in Figure 10a, and the power varied with the resistor, as shown in Figure 10b.

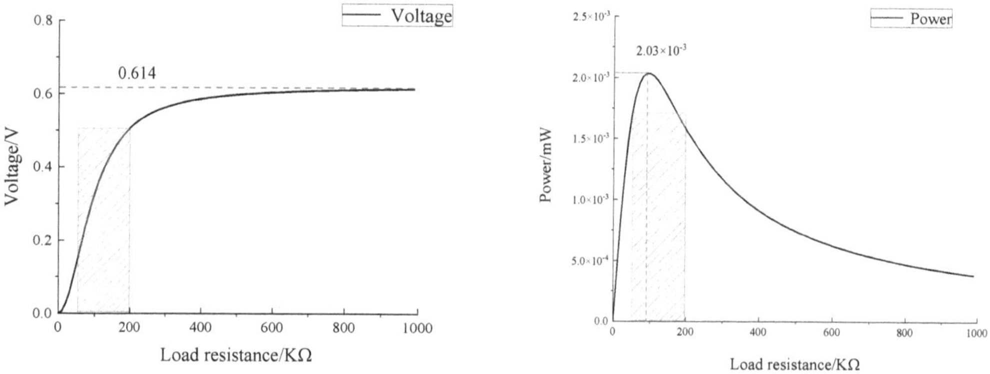

(**a**) Voltage (**b**) Power

Figure 10. Influence of load resistance on the electrical output characteristics of the disk energy harvester.

From the results, it can be seen that as the circuit load resistance increased, the output voltage increased monotonically and the rate of increase (i.e., the slope of the curve) decreased. When the load resistance increased to a near open circuit, the output voltage was 0.61 V, which approached the open circuit voltage of 0.62 V, and the curve tended to be flat. While the output power was low when the load resistance was very small or very large, it first increased and then decreased with the increase in the load resistance, and there was a maximum value. In addition, as can be seen from the shaded part under the curve, when the load resistance R_L was in a certain range (50~200 KΩ), the output power of the disk energy harvester was about 1.7×10^{-3} mW, which was at a low level. And the output voltage was less than 0.5 V in this resistance range, making it difficult to achieve the minimum input requirements of the energy capture circuit, so the electrical output performance of the disk energy harvester was found to be relatively ordinary. In this range, the load resistance that made the output power optimal was the "matching impedance". Therefore, when $R_L = 90$ KΩ, the maximum output power was 2.03×10^{-3} mW, and the internal resistance of the piezoelectric energy harvester was calculated to be 90 KΩ.

The load match model of the cymbal energy harvester was established, and the effect of external load resistance on the output power of the cymbal energy harvester was studied by frequency domain analysis. The results of voltage variation with resistance on the load resistor are shown in Figure 11a, and the results of power variation with resistance are shown in Figure 11b.

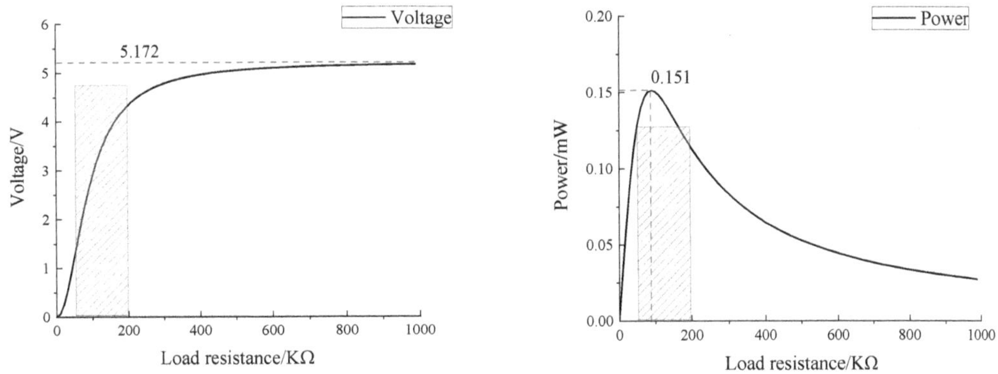

(**a**) Voltage (**b**) Power

Figure 11. Influence of load resistance on the electrical output characteristics of the cymbal energy harvester.

According to the results, as the circuit load resistance increased, the voltage across the resistor increased monotonically and at a decreasing rate. When the load resistance increased to a near open circuit, the output voltage was 5.2 V, which approached the open circuit voltage of 6.5 V, and the curve tended to flatten out. While the output power was low when the load resistance was very small or very large, it first increased and then decreased with the increase in the load resistance, and there was a maximum value in the process. When the load resistance R_L was within a certain range (50~200 KΩ), the output voltage of the cymbal energy harvester was able to reach the input requirement of the energy harvesting circuit, which was greater than 1.4 V. And the output power reached a high level, up to 0.13 mW. Meanwhile, when $R_L = 90\text{K}\Omega$, the maximum output power was 0.15 mW, and the internal resistance of the cymbal piezoelectric energy harvester was calculated to be 90 KΩ.

Compared with the disk energy harvester, the cymbal energy harvester had the same matching impedance of 90 KΩ. But the output power of the cymbal energy harvester was two orders of magnitude higher than that of the disk type at a better load resistance, and it was up to the microwatt level.

The load match model of the spherical energy harvester was established, and the effect of external load resistance on the output power of the energy harvester was studied through frequency domain analysis. The voltage on the load resistor varied with the resistance, as shown in Figure 12a, and the power varied with the resistance, as shown in Figure 12b.

As can be seen from Figure 12, with the increase in circuit load resistance, the voltage across the resistor increased monotonously, and the growth rate decreased. When the load resistance increased to a near open circuit, the output voltage was 6.53 V, which approached the open circuit voltage of 6.7 V, and the curve tended to level off at this time. The output power was lower when the load resistance was very small or very large, and the law was consistent with other energy harvesters, that is, the power first increased and then decreased with the increase in load resistance, and there was a maximum value in the process.

In addition, it can be seen from the shaded part under the curve that when the load resistance R_L was within a certain range (50~200 KΩ), the output voltage of the energy harvester was greater than 1.4 V, being able to meet the input requirements of the energy harvesting circuit. And the output power was able to reach a high level, up to 0.2 mW.

When $R_L = 100$ KΩ, the maximum output power was 0.23 mW, and the internal resistance of the spherical piezoelectric energy harvester was calculated to be 100 KΩ.

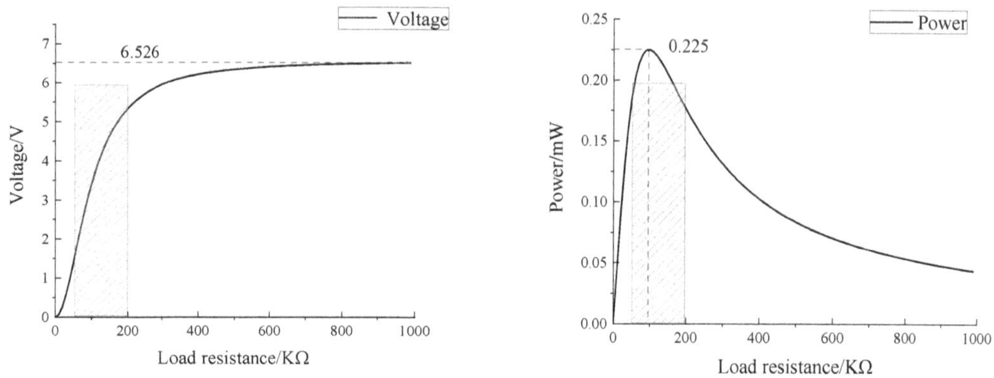

(**a**) Voltage (**b**) Power

Figure 12. Influence of load resistance on the electrical output characteristics of the spherical energy harvester.

Compared with the disk and cymbal energy harvester, the matching impedance of the spherical energy harvester was slightly larger than the other two, at 100 KΩ. At the better load resistance, the output power of the spherical energy harvester was two orders of magnitude higher than that of the disk one and 53% higher than that of the cymbal energy harvester, which was also able to reach the microwatt level.

The output electrical properties of the three energy harvesting structures was as shown in Table 3.

Table 3. Comparison of electrical properties.

Variable	Disk Structure	Cymbal Structure	Spherical Structure
Output voltage/V	0.62	6.5	6.7
Electrical output power/mW	2.03×10^{-3}	0.15	0.23

The same load of 0.3 MPa was applied to the disk, cymbal, and spherical energy harvesting structures, and the power output efficiencies were 2.03×10^{-3} mW, 0.15 mW, and 0.23 mW, respectively.

In terms of the three types of energy harvesting structures, the disk energy harvester is simple in structure, easy to make, and long in service, but its energy harvesting efficiency is low. The energy harvesting efficiency of the cymbal type is higher than that of the disk type, but due to its complex structure, it can easily produce stress concentration, which is not conducive to long-term service. Compared with the disk energy harvesting structure, the output of the spherical energy harvesting structure is greatly improved. Compared with the cymbal energy harvesting structure, although the spherical energy harvesting structure also has a certain stress concentration phenomenon at the bottom of the spherical shell, the stress concentration phenomenon is weakened due to the good stress equalization of the spherical shell structure. Therefore, the spherical energy harvesting structure not only has good mechanical properties, but also has better electrical properties.

5. Parameter Optimization Design

5.1. Parameter Analysis

To investigate the optimal size of the spherical energy harvester, the parameter analysis of its geometry was carried out to obtain the optimal energy harvesting effect. The parametric geometry is shown in Figure 13.

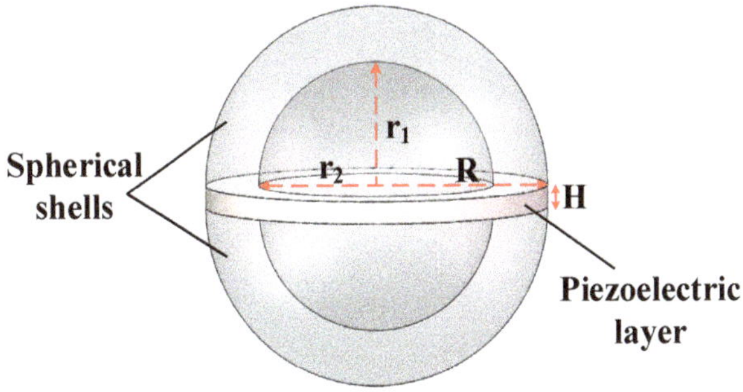

Figure 13. Geometric model of the spherical energy harvester.

Here, H represents the thickness of the piezoelectric layer, R represents the radius of the piezoelectric layer, r_1 represents the vertical axis radius of the inner cavity of the spherical shell, and r_2 represents the horizontal axis radius of the inner cavity of the spherical shell. The outer radius of the spherical shell is aligned with the radius of the piezoelectric layer. The axis ratio I is defined as $I = r_1/r_2$, which is used to characterize the ellipticity of the hollow shell; the thickness of the spherical shell is $T = R\text{-}r_2$. The parameters of the initial spherical energy harvester are shown in Table 4.

Table 4. Initial size parameters.

Parameter	Value	Unit	Description
H	2	mm	Piezoelectric layer thickness
R	16	mm	Piezoelectric layer radius
r_1	11	mm	Vertical axis radius of the inner cavity of the spherical shell
r_2	11	mm	Horizontal axis radius of the inner cavity of the spherical shell
I	1	1	Axis ratio
T	5	mm	Spherical shell thickness

To obtain the optimal size of the spherical harvester, the dimensions of the piezoelectric layer and hollow shell were parametrically analyzed, and the parameters mainly included the piezoelectric layer thickness H, the piezoelectric layer radius R, the axis ratio I, and the spherical shell thickness T. The material properties of each working condition were consistent during the analysis, and PZT-5H was used for the piezoelectric layer.

5.2. Piezoelectric Layer Thickness

The parametric numerical model of the spherical energy harvester was established according to the initial size, and the parametric scanning was defined. Thickness H scanning range was 0.5~8 mm, defined H = range (0.5, 0.5, 8). The peak stress, peak vertical displacement, peak voltage, and total electricity of the spherical energy harvester with different piezoelectric layer thicknesses were obtained.

The peak stress under different piezoelectric layer thicknesses is shown in Figure 14. The peak stress increased with the thickness of the piezoelectric layer, and the overall curve showed an increasing trend, although there were small fluctuations at some thickness values. With the increase in piezoelectric layer thickness H from 0.5 mm to 8 mm, the peak stress of the energy harvester increased from 1.3 MPa to 1.56 MPa, and the whole increase was 20%. Figure 15 shows the peak vertical displacement under different piezoelectric layer thicknesses. The peak vertical displacement decreased with the increase in piezoelectric layer thickness, and the decreasing trend was basically linear. When the thickness of the piezoelectric layer increased, the peak value of vertical displacement decreased from 2.47×10^{-4} mm to 2.29×10^{-4} mm, with a decrease of 7.3%.

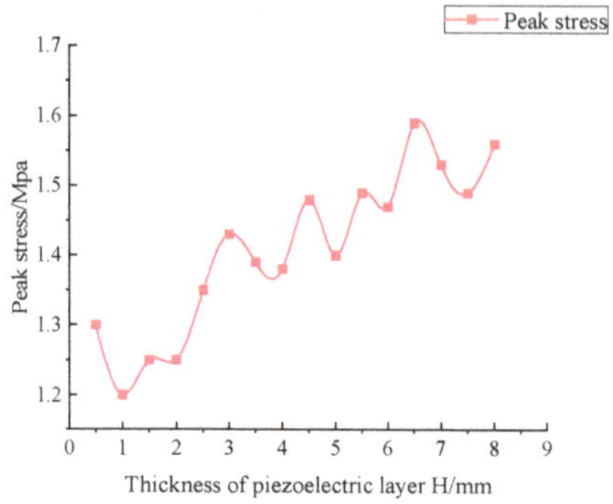

Figure 14. Peak stress under different piezoelectric layer thicknesses.

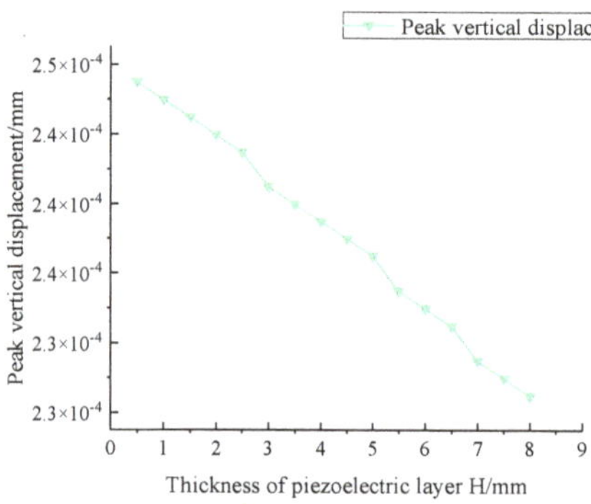

Figure 15. Peak vertical displacement under different piezoelectric layer thicknesses.

As can be seen from Figure 16, the peak voltage increased with the increase in piezoelectric layer thickness, being almost linear as a whole. The peak voltage of the energy harvester increased from 1.6 V to 20.8 V, with a total rise of 13 times. Figure 17 shows the change of total electricity with the increase in piezoelectric layer thickness. It can be seen that the total electricity increased with the increase in piezoelectric layer thickness, and it increased from 2.66×10^{-8} J to 3.10×10^{-7} J, with an increase of 11.7 times.

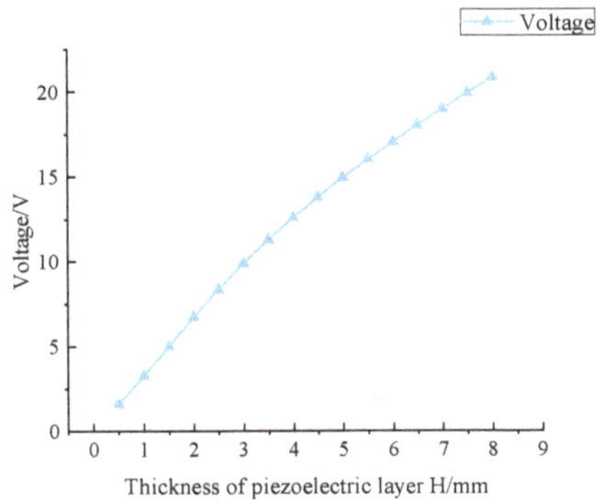

Figure 16. Peak voltage under different piezoelectric layer thicknesses.

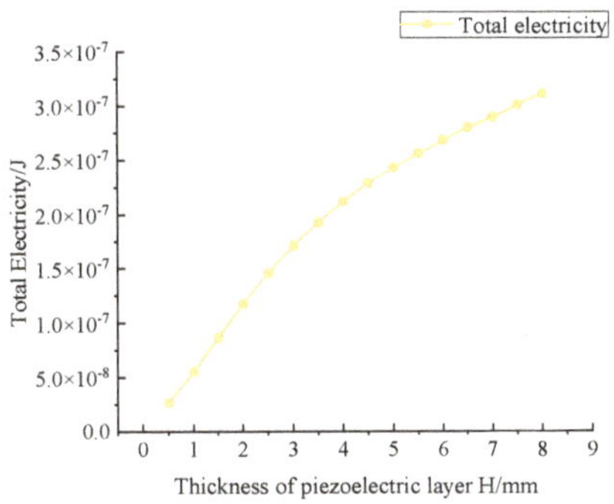

Figure 17. Total electricity under different piezoelectric layer thicknesses.

Considering that the thickness of the piezoelectric layer exceeded 7 mm, the processing cost increased sharply and the material stability became poor. As can be obtained from Equation (1), with the increase in thickness, the capacitance C_0 decreased, which was not conducive to matching with the load capacitance C_1, resulting in low output efficiency when containing an external circuit. Therefore, the thickness of the piezoelectric layer should not be too thick, and if it is too thin, the total electric output will be too small. The minimum total electrical energy output was taken as a 1×10^{-7} J order of magnitude, and the minimum thickness of the piezoelectric layer under the electric energy output of this order of magnitude was 2 mm, so the optimal thickness of the piezoelectric layer was selected as 2 mm comprehensively.

5.3. Piezoelectric Layer Radius

In the study of the effect of piezoelectric layer radius on the energy-harvesting structure, parametric scanning was defined for the numerical model. The radius R scan range was from 6 to 23 mm, and R = range (6, 1, 23) was defined in the parameter value list. The peak stress, peak vertical displacement, peak voltage, and total electricity of the spherical energy harvester under different piezoelectric layer radii were also obtained.

The peak stress under different piezoelectric layer radii is shown in Figure 18, and the peak stress increased with the increase in the piezoelectric layer radius. The radius R of the piezoelectric layer increased from 6 mm to 23 mm, and the peak stress of the energy harvester increased from 1.02 MPa to 1.57 MPa, with an increase of 53.9%. Figure 19 shows the variation of the peak vertical displacement, which decreased and then increased with the radius of the piezoelectric layer. The minimum displacement value existed when the radius was between 18 and 21 mm, which was 2.41×10^{-4} mm. With the change of the piezoelectric layer radius, the peak vertical displacement decreased from 2.87×10^{-4} mm to 2.43×10^{-4} mm in general, which was a decrease of 15.5%.

As can be seen from Figure 20, the peak voltage increased with the increase in the piezoelectric layer radius and then decreased, and the maximum value was 6.7 V when the piezoelectric layer radius R = 16 mm. With the increase in radius, the peak voltage of the energy harvester increased from 5.6 V to 6.5 V, which was 16% higher. The change of total electricity is shown in Figure 21. It can be seen that the total electricity increased with the increase in the piezoelectric layer radius, and it increased from 1.13×10^{-8} J to 2.3×10^{-7} J with the increase in the piezoelectric layer radius R from 6 mm to 23 mm, with a total increase of 20.4 times.

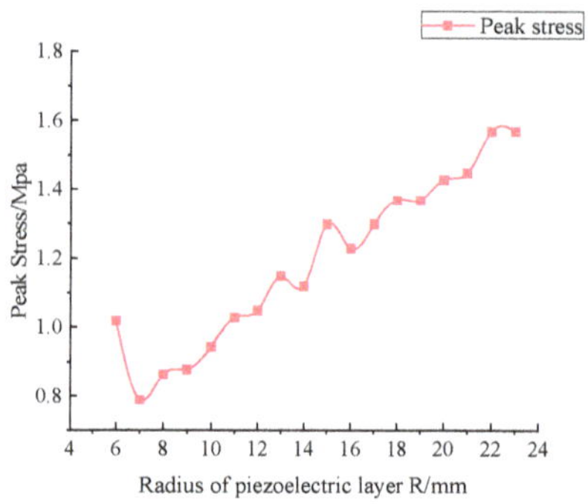

Figure 18. Peak stress under different piezoelectric layer radii.

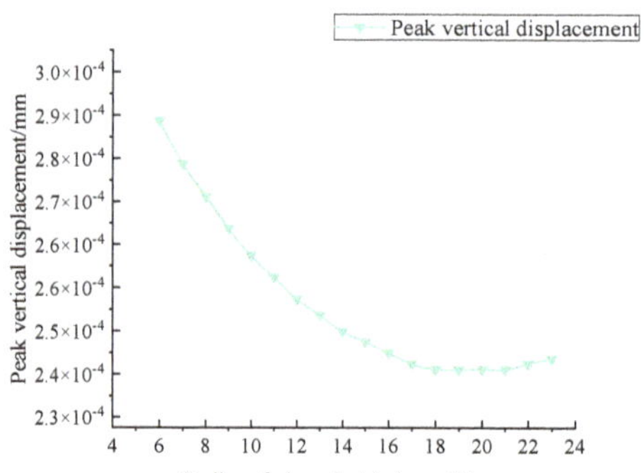

Figure 19. Peak vertical displacement under different piezoelectric layer radii.

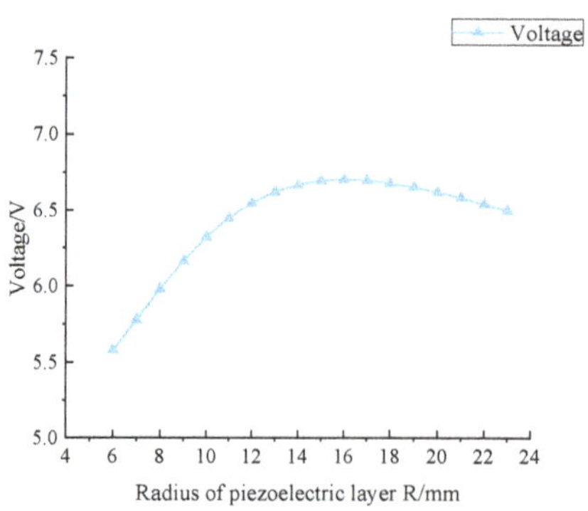

Figure 20. Voltage under different piezoelectric layer radii.

When the radius of piezoelectric layer varied from 6 mm to 23 mm, the peak stress and vertical displacement were within the safety range, so the electrical properties were given priority to determine the radius size. When the radius of the piezoelectric layer R was 16 mm, the peak voltage of the energy harvester can be taken to the maximum value of 6.7 V, and the minimum output voltage requirement (0.4~1.5 V) can be met. At this time, the total electricity was 1.17×10^{-7} J, so the optimal radius of piezoelectric layer was 16 mm.

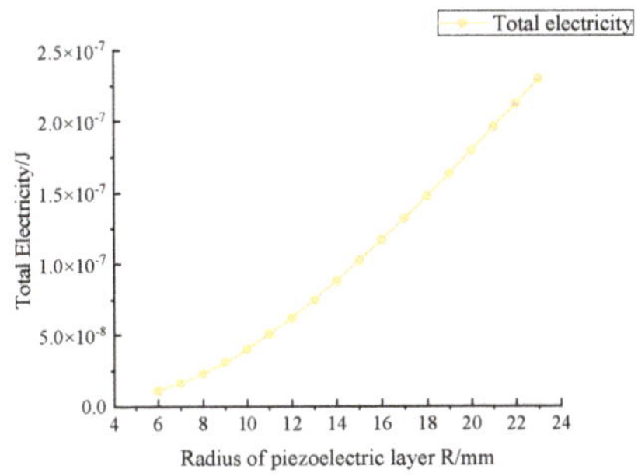

Figure 21. Total electricity under different piezoelectric layer radii.

5.4. Axial Ratio of Spherical Shell

When studying the effect of the axial ratio of the spherical shell on the energy harvester, the vertical axis radius of the inner cavity of the spherical shell should be kept consistent, at 11 mm. Because $I = r_1/r_2$, the axial ratio can be controlled by changing the radius r_2. In the numerical model, R_2 was defined as a parametric scan with a range of 5~30 mm, and R2 = range (5, 1, 30) was defined in the parameter value list. A simple calculation shows that the corresponding axial ratio range was 0.37~2.20. As the axis ratio increased, the geometry of the spherical shell slowly changed from a flat ellipsoid to a hanging ellipsoid. The stress nephogram of the spherical energy harvester with different axial ratios obtained by the COMSOL simulation is shown in Figure 22.

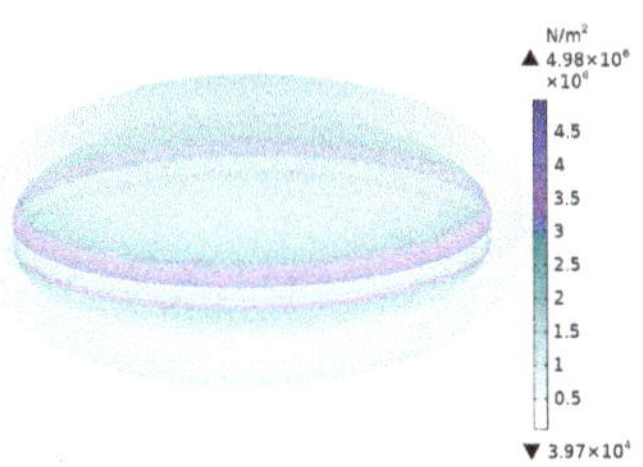

(**a**) I = 0.37

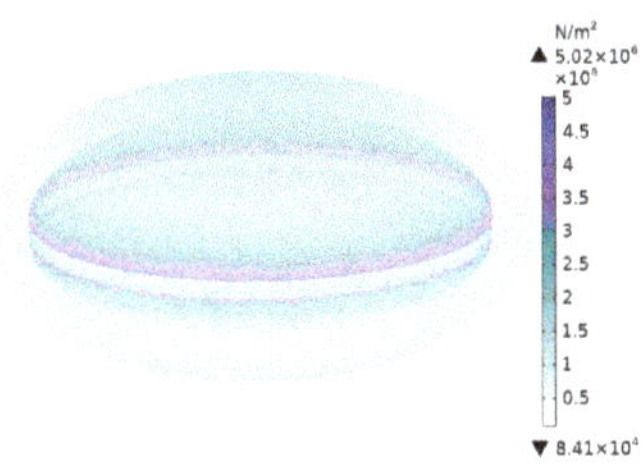

(**b**) I = 0.38

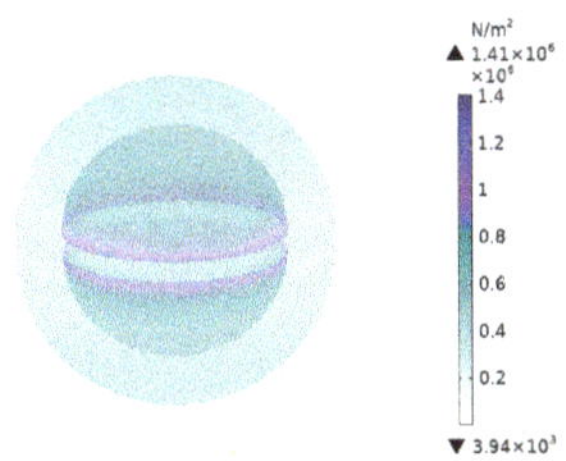

(**c**) I = 0.92

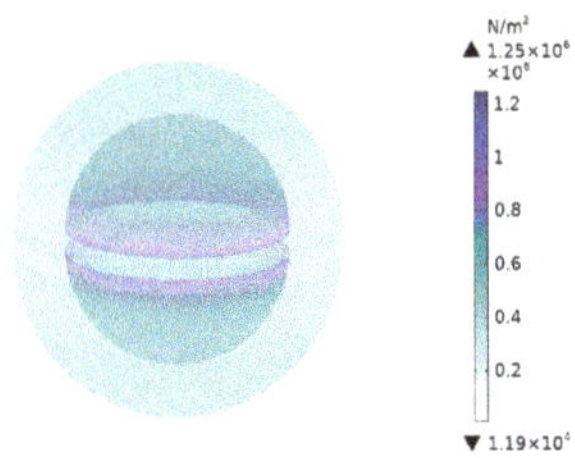

(**d**) I = 1.00

Figure 22. *Cont.*

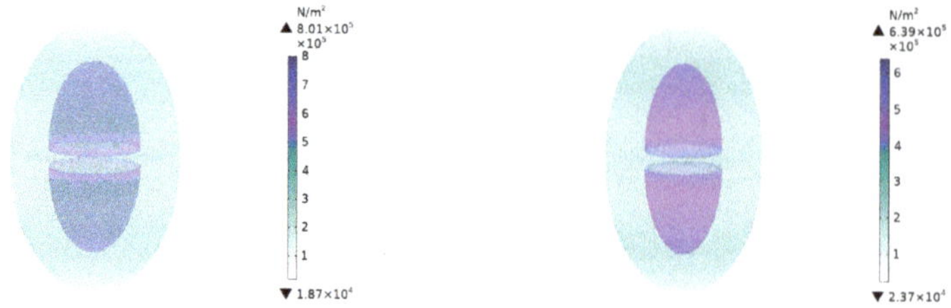

(**e**) I = 1.83 (**f**) I = 2.20

Figure 22. Stress nephogram of the spherical energy harvester with different axial ratios.

Meanwhile, the peak stress, peak vertical displacement, peak voltage, and total electricity of the spherical energy harvester under different axial ratios of the spherical shell were obtained, as shown in Figures 23–26.

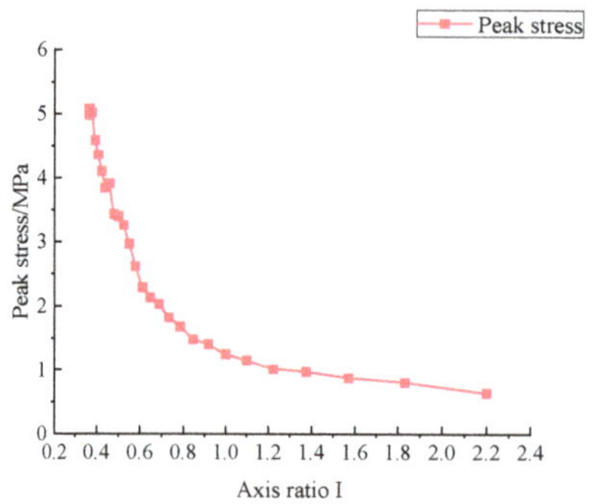

Figure 23. Peak stress under different axial ratios of the spherical shell.

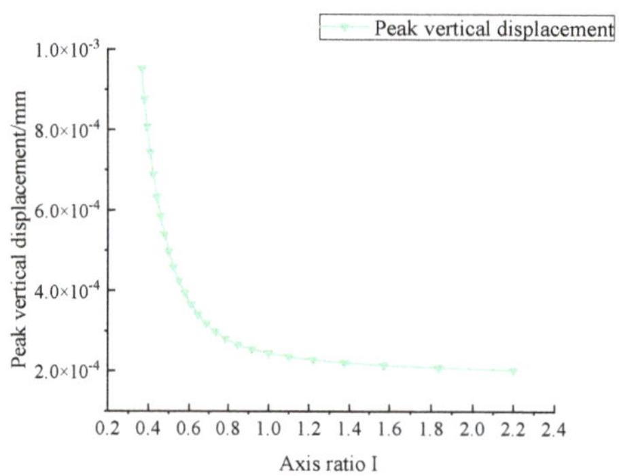

Figure 24. Peak vertical displacement under different axial ratios of the spherical shell.

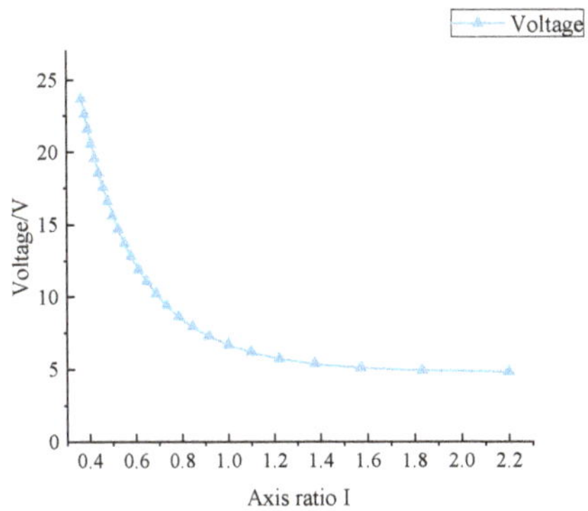

Figure 25. Peak voltage under different axial ratios of the spherical shell.

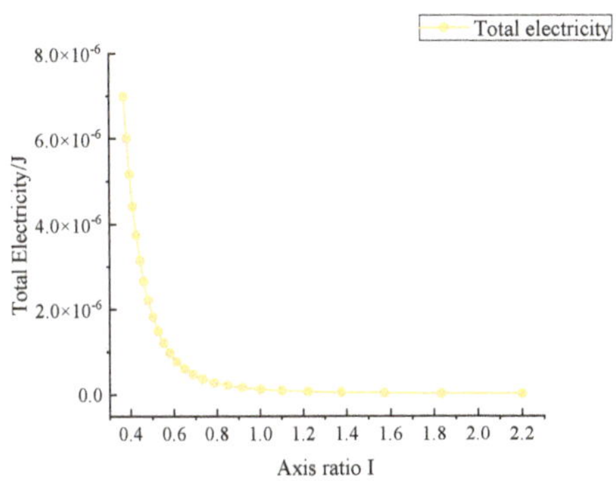

Figure 26. Total electricity under different axial ratios of the spherical shell.

As shown in Figure 23, the peak stress decreased with the increase in the axial ratio of the spherical shell, and the decreasing rate became slower and slower. When the axial ratio of the spherical shell increased from 0.37 to 2.2, the peak stress of the energy harvester decreased from 5.08 MPa to 0.64 MPa, with a decrease of 87.4%. Figure 24 shows the variation of the peak vertical displacement, which decreased with the increase in the axial ratio of the spherical shell, and the decreasing rate also became slower and slower. With the increase in the axial ratio of the spherical shell, the peak vertical displacement decreased from 9.51×10^{-4} mm to 2.02×10^{-4} mm, a decrease by 78.8%.

As can be seen from Figure 25, the peak voltage of the energy harvester decreased with the increase in the axial ratio, and the decreasing rate slowed down continuously. The overall peak voltage of the energy harvester decreased from 23.7 V to 4.8 V, with the decrease of 79.8%. As shown in Figure 26, the change of total electricity decreased with the increase in the axis ratio, from 6.98×10^{-6} J to 2.35×10^{-8} J, which was two orders of magnitude lower. The variation of the peak voltage with the axial ratio of the spherical shell was analyzed, and it was concluded that the axial ratio of the spherical shell had a great influence on the electrical properties of the spherical energy harvester.

When the axial ratio of the spherical shell varied from 0.37 to 2.2, the peak stress and vertical displacement of the energy harvester were within the safe range. And the electrical properties such as peak voltage and total electricity changed obviously with the axial ratio, so the electrical properties were given priority to determine the axial ratio of the spherical shell. When the axial ratio of the spherical shell was small, the voltage and electric energy output of the energy harvester became better. That is, when the geometric design of the energy harvester was more inclined to the flat ellipsoid, it had better captive performance. When the axial ratio I was 0.4, the peak voltage was 20.6 V, which met the minimum output voltage requirement (0.4~1.5 V). And the total power was 4.40×10^{-6} J, which was at a better level. Therefore, the optimal axial ratio of the spherical shell was 0.4.

5.5. Spherical Shell Thickness

When studying the influence of the spherical shell thickness on the energy harvester, the horizontal axis radius of the inner cavity of the spherical shell was kept constant, at 11 mm. The axial ratio was controlled by changing the outer radius of spherical shell, i.e., the piezoelectric layer radius R. In the parameter value list, R = range (12, 1, 23) was defined, and the spherical shell thickness T = R-r_2 ranged from 1 mm to 12 mm. The peak stress, peak vertical displacement, peak voltage, and total electricity of the spherical energy harvester with different spherical shell thicknesses were obtained.

As shown in Figure 27, the peak stress decreased with increasing spherical shell thickness. The spherical shell thickness T increased from 1 mm to 12 mm, and the peak stress of the energy harvester decreased from 4.28 MPa to 0.81 MPa, with a decrease in 81.1%. Figure 28 shows the variation of the peak vertical displacement, which decreased with increasing spherical shell thickness from 4.22×10^{-4} mm to 1.65×10^{-4} mm, with a decrease of 60.9%.

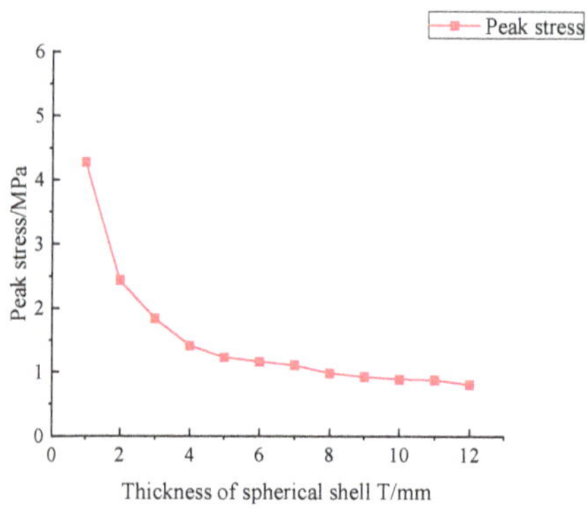

Figure 27. Peak stress under different spherical shell thicknesses.

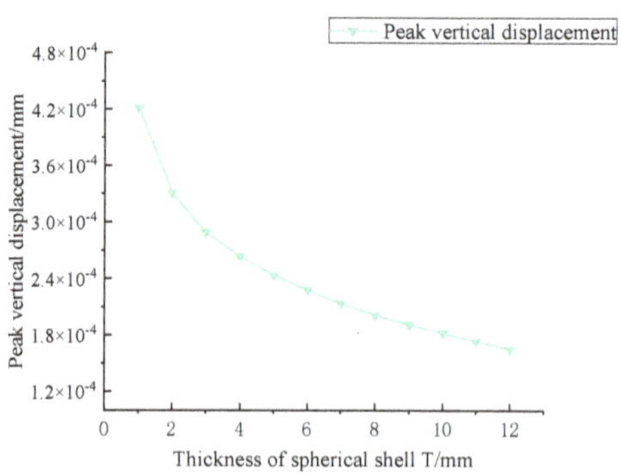

Figure 28. Peak vertical displacement under different spherical shell thicknesses.

As can be seen from Figure 29, the peak voltage first increased and then decreased when the spherical shell thickness was continuously increased, and the peak voltage was 6.7 V when the thickness T = 4 mm. As the thickness increased from 1 mm to 12 mm, the peak voltage of the energy harvester increased from 3.6 V to 5.9 V, with an increase in 63.8%. As shown in Figure 30, the change of total electricity increased with the increase in spherical shell thickness, showing a linear trend. With the increase in spherical shell thickness, the total electricity of energy harvester increased from 3.91×10^{-8} J to 1.86×10^{-7} J, with an increase of 4.8 times.

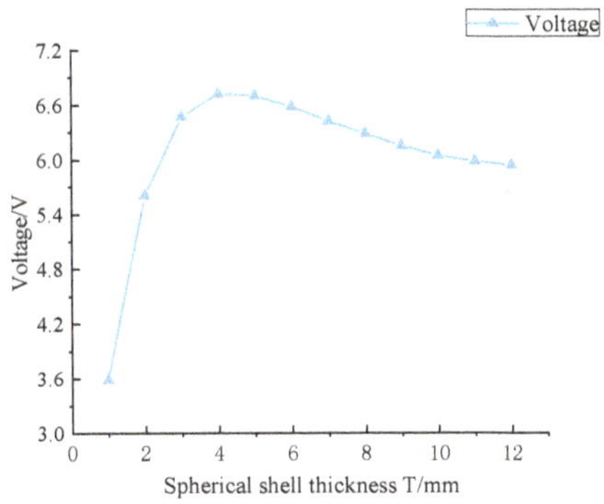

Figure 29. Peak voltage under different spherical shell thicknesses.

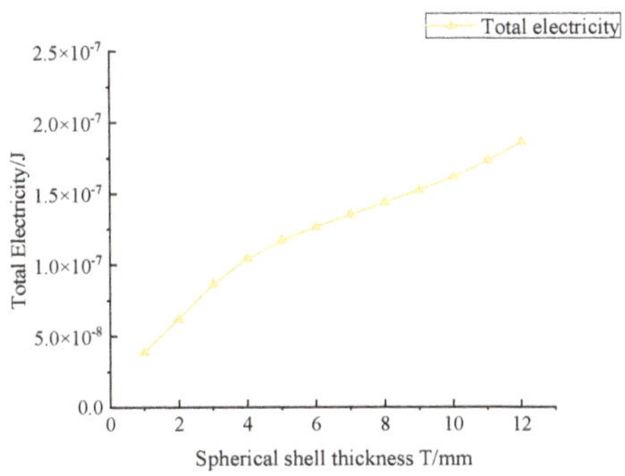

Figure 30. Total electricity under different spherical shell thicknesses.

When the spherical shell thickness varied in the range of 1~12 mm, the peak stress and vertical displacement of the energy harvester were within the safety range, so the electrical properties were also given priority to determine the thickness of the spherical shell. When the thickness of the spherical shell was 4 mm, the peak voltage of the energy harvester can be taken to the maximum value of 6.7 V, which meets the minimum output voltage requirement (0.4~1.5 V). At this time, the total electricity was 1.05×10^{-7} J, and the optimal spherical shell thickness was 4 mm.

Through numerical simulation of the dimension parameters H, R, I, and T of the spherical energy capture structure, it was found that the peak voltage and total electricity of the spherical energy structure increased with the increase in piezoelectric layer thickness H. The peak voltage first increased and then decreased when increasing the piezoelectric layer radius R, and the total electricity increased with the increase in piezoelectric layer radius R. The peak voltage and total electricity decreased with the increase in spherical shell axial ratio I. The peak voltage first increased and then decreased with the spherical shell thickness T getting larger, and the total electricity increased with the thickness T of the spherical shell. Considering obtaining better electrical performance and combining it with the actual situation, the optimal size of the spherical energy harvester was obtained as shown in Table 5, and the energy harvester cell under the optimal size is shown in Figure 31.

Table 5. Optimal size parameters.

Parameter	Value	Unit	Description
H	2	mm	Piezoelectric layer thickness
R	16	mm	Piezoelectric layer radius
r_1	5	mm	Vertical axis radius of the inner cavity of the spherical shell
r_2	12	mm	Horizontal axis radius of the inner cavity of the spherical shell
I	0.4	1	Axis ratio
T	4	mm	Spherical shell thickness

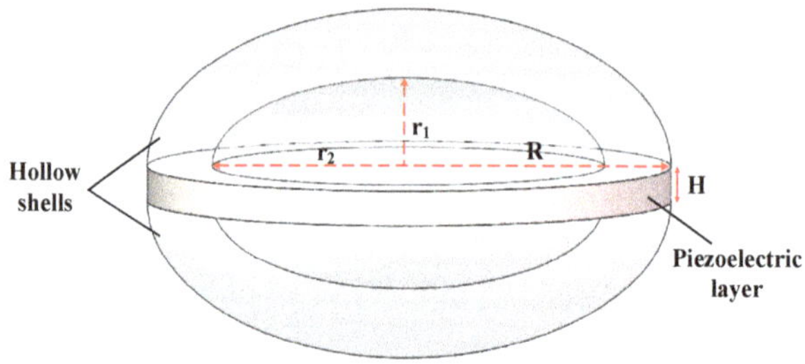

Figure 31. Optimal size model of the spherical energy harvester.

6. Conclusions

For the rubber bearing floating slab track, a spherical energy harvester was designed, and the mechanical and electrical properties of the disk, cymbal, and spherical energy harvesters were compared by numerical simulation. Furthermore, the geometric parameters of the spherical energy harvester were optimized. The conclusions are as follows:

(1) Compared with the disk and cymbal energy harvester, the spherical energy harvester had better electrical and mechanical properties and showed less stress concentration. The output voltage of the spherical energy harvester was 11 times that of the disk one, and the total electricity output was 20% more than that of the cymbal energy harvester.
(2) By calculating the output power of the energy harvesters under load match, we found that the output power of the spherical energy harvester was two orders of magnitude higher than that of the disk one and 53% higher than that of the cymbal one, being able to reach the microwatt level under impedance match.
(3) The optimal dimensions of the spherical energy harvester were piezoelectric layer thickness h = 2 mm, piezoelectric layer radius r = 16 mm, spherical shell axis ratio I = 0.4, and spherical shell thickness t = 4 mm.
(4) The spherical energy harvesters had better mechanical and electrical properties than the traditional energy harvesting structure, so it has broad prospects in the field of energy harvesting of rail transit and other energy harvesting fields.
(5) This paper creatively proposed a new spherical energy harvesting structure and applied it to energy harvesting in the field of rail transit. In the future, it can also be applied to the field of energy harvesting under long-term heavy load such as road surface.

The summary of abbreviations and symbols is shown in Appendix A.

Author Contributions: Conceptualization, S.Y.; methodology, H.J.; software, Z.L.; validation, J.Y.; formal analysis, D.Y.; investigation, S.Y.; resources, Z.L.; data curation, D.Y.; writing—original draft preparation, Z.L.; writing—review and editing, S.Y.; visualization, H.J.; supervision, J.Y.; project administration, H.J.; funding acquisition, H.J. All authors have read and agreed to the published version of the manuscript.

Funding: This research was funded by [the Natural Science Foundation of Jiangsu Province] grant number [BK20211173], [China Postdoctoral Science Foundation] grant number [2022M722958], and [Key R&D Program of Zhejiang] grant number [2023C03182].

Institutional Review Board Statement: Not applicable.

Informed Consent Statement: Not applicable.

Data Availability Statement: The data presented in this study are available on request from the corresponding author. The data are not publicly available due to copyright restrictions.

Conflicts of Interest: Author Shuo Yu and Jiajia Yan were employed by the company PowerChina HuaDong Engineering Corporation Limited. The remaining authors declare that the research was conducted in the absence of any commercial or financial relationships that could be construed as a potential conflict of interest.

Appendix A

d	the thickness of force transmission structures
D	the thickness of the disk piezoelectric layer
ϕ_p	the diameter of the disk energy harvester
t_m	the thickness of the metal cap
t_p	the thickness of the cymbal piezoelectric layer
ϕ_C	the inner cavity bottom diameter
ϕ_d	the inner cavity top diameter
d_c	the height of the inner cavity
ϕ	the diameter of cymbal energy harvester
d_{33}	piezoelectric strain constant
$\mathrm{e}_{31}, \mathrm{e}_{33}, \mathrm{e}_{15}$	the piezoelectric stress constant
ε^S_{r33} and ε^S_{r11}	relative dielectric constants
ε^T_{r33}	relative dielectric constant
$C^E_{11}, C^E_{12}, C^E_{13}, C^E_{33}, C^E_{44}, C^E_{55}, C^E_{66}$	the elastic coefficients
ρ	density
H	piezoelectric layer thickness
R	piezoelectric layer radius
r_1	vertical axis radius of the inner cavity of the spherical shell
r_2	horizontal axis radius of the inner cavity of the spherical shell
I	axis ratio
T	spherical shell thickness
R_0	equivalent internal resistor of the energy harvester
C_0	equivalent capacitance
R_L	load resistor
$\bar{\varepsilon}_{33}$	vacuum dielectric constant
A	piezoelectric layer area
R_{C_0}	capacitive impedance
R_L	external load resistance
R_{match}	matching impedance
V	voltage source

References

1. Xu, L.; Ma, M. Dynamic response of the multilayered half-space medium due to the spatially periodic harmonic moving load. *Soil Dyn. Earthq. Eng.* **2022**, *157*, 107246. [CrossRef]
2. Jin, H.; Tang, S.; Zhao, C.; Jiang, B. Study on vibration propagation characteristics caused by segments joints in shield tunnel. *Int. J. Struct. Stab. Dyn.* **2023**, *23*, 2350156. [CrossRef]
3. Li, J.; Jang, S.; Tang, J. Implementation of a piezoelectric energy harvester in railway health monitoring. *Proc. SPIE* **2014**, *9061*, 90612Q. [CrossRef]
4. Zhang, X.; Zhang, Z.; Pan, H.; Salman, W.; Yuan, Y.; Liu, Y. A portable high-efficiency electromagnetic energy harvesting system using supercapacitors for renewable energy applications in railroads. *Energy Convers. Manag.* **2016**, *118*, 287–294. [CrossRef]
5. Yuan, T.; Yang, J.; Song, R.; Liu, X. Vibration energy harvesting system for railroad safety based on running vehicles. *Smart Mater. Struct.* **2014**, *23*, 125046. [CrossRef]
6. Wang, J.; Penamalli, G.; Zuo, L. Electromagnetic energy harvesting from train induced railway track vibrations. In Proceedings of the 2012 IEEE/ASME 8th IEEE/ASME International Conference on Mechatronic and Embedded Systems and Applications, Suzhou, China, 8–10 July 2012; pp. 29–34. [CrossRef]

7. Lin, T.; Pan, Y.; Chen, S.; Zuo, L. Modeling and field testing of an electromagnetic energy harvester for rail tracks with anchorless mounting. *Appl. Energ.* **2018**, *213*, 219–226. [CrossRef]
8. Hou, W.; Zheng, Y.; Guo, W.; Guo, P. Piezoelectric vibration energy harvesting for rail transit bridge with steel-spring floating slab track system. *J. Clean. Prod.* **2021**, *291*, 125283. [CrossRef]
9. Mohamed, M.; Wu, W.; Moniri, M. Power harvesting for smart sensor networks in monitoring water distribution system. In Proceedings of the 2011 International Conference on Networking, Sensing and Control, Delft, The Netherlands, 11–13 April 2011; pp. 393–398. [CrossRef]
10. Lee, B.; Lin, S.; Wu, W.; Wang, X.; Chang, P.; Lee, C. Piezoelectric mems generators fabricated with an aerosol deposition pzt thin film. *J. Micromech. Microeng.* **2009**, *19*, 065014. [CrossRef]
11. Fang, H.; Liu, J.; Xu, Z.; Dong, L.; Wang, L.; Chen, D.; Cai, B.; Liu, Y. Fabrication and performance of mems-based piezoelectric power generator for vibration energy harvesting. *Microelectron. J.* **2006**, *37*, 1280–1284. [CrossRef]
12. Shen, D.; Park, J.; Ajitsaria, J.; Choe, S.; Wikle, H.; Kim, D. The design, fabrication and evaluation of a mems pzt cantilever with an integrated si proof mass for vibration energy harvesting. *J. Micromech. Microeng.* **2008**, *18*, 055017. [CrossRef]
13. Dagdeviren, C.; Hwang, S.; Su, Y.; Kim, S.; Cheng, H.; Gur, O.; Haney, R.; Omenetto, F.; Huang, Y.; Rogers, J. Transient, biocompatible electronics and energy harvesters based on zno. *Small* **2013**, *9*, 3398–3404. [CrossRef] [PubMed]
14. Cho, K.; Park, H.; Heo, J.; Priya, S. Structure-performance relationships for cantilever-type piezoelectric energy harvesters. *J. Appl. Phys.* **2014**, *115*, 204108. [CrossRef]
15. Cui, Y.; Zhang, Q.; Yao, M.; Dong, W.; Gao, S. Vibration piezoelectric energy harvester with multi-beam. *AIP Adv.* **2015**, *5*, 041332. [CrossRef]
16. Sahi, Y.; Chandan, S. Finite element analysis of lead-free ceramic based cymbal transducer for energy harvesting. *Integr. Ferroelectr.* **2020**, *212*, 81–88. [CrossRef]
17. Chen, Y.; Gao, M.; Yu, W.; Zhang, L. On energy harvesting characteristics of cymbal transducer. *Appl. Mech. Mater.* **2013**, *365–366*, 1318–1323. [CrossRef]
18. Palosaari, J.; Leinonen, M.; Hannu, J.; Juuti, J.; Jantunen, H. Energy harvesting with a cymbal type piezoelectric transducer from low frequency compression. *J. Electroceramics* **2012**, *28*, 214–219. [CrossRef]
19. Kamata, Y.; Yoon, D.; Sasaki, T.; Nozaki, Y.; Yamaura, S.; Sekiguchi, T.; Nakajima, T.; Shoji, S. Stacked piezoelectric energy harvesting device by printing process. *Micro Nano Lett.* **2016**, *11*, 650–653. [CrossRef]
20. Wang, C.; Wang, S.; Li, Q.; Wang, X.; Gao, Z.; Zhang, L. Fabrication and performance of a power generation device based on stacked piezoelectric energy-harvesting units for pavements. *Energy Convers. Manag.* **2018**, *163*, 196–207. [CrossRef]
21. Zhao, H.; Qin, L.; Ling, J. Test and analysis of bridge transducers for harvesting energy from asphalt pavement. *Int. J. Transp. Sci. Technol.* **2015**, *4*, 17–28. [CrossRef]
22. Yao, L.; Zhao, H.; Dong, Z.; Sun, Y.; Gao, Y. Laboratory testing of piezoelectric bridge transducers for asphalt pavement energy harvesting. *Key Eng. Mater.* **2011**, *492*, 172–175. [CrossRef]
23. Wang, F.; Wu, W.; Lozowski, A.; Alizadehyazdi, V.; Amin, A. Energy harvesting with a piezoelectric thunder. In Proceedings of the ASME 2015 International Mechanical Engineering Congress and Exposition, Houston, TX, USA, 13–19 November 2015.
24. Wang, F.; Wang, Z.; Soroush, M.; Abedini, A. Energy harvesting efficiency optimization via varying the radius of curvature of a piezoelectric thunder. *Smart Mater. Struct.* **2016**, *25*, 095044. [CrossRef]
25. Wise, S. Displacement properties of rainbow and thunder piezoelectric actuators. *Sens. Actuators A Phys.* **1998**, *69*, 33–38. [CrossRef]
26. Peddigari, M.; Kim, G.; Park, C.; Min, Y.; Kim, J.; Ahn, C.; Choi, J.; Hahn, B.; Choi, J.; Park, D.; et al. A comparison study of fatigue behavior of hard and soft piezoelectric single crystal macro-fiber composites for vibration energy harvesting. *Sensors* **2019**, *19*, 2196. [CrossRef] [PubMed]
27. Lin, X.; Zhou, K.; Zhang, X.; Zhang, D. Development, modeling and application of piezoelectric fiber composites. *Trans. Nonferrous Met. Soc. China* **2013**, *23*, 98–107. [CrossRef]
28. Mallouli, M.; Chouchane, M. Piezoelectric energy harvesting using macro fiber composite patches. *Proc. Inst. Mech. Eng. Part C J. Mech. Eng. Sci.* **2020**, *234*, 4331–4349. [CrossRef]
29. Panigrahi, B.; Sitikantha, D.; Bhuyan, A.; Panda, H.; Mohanta, K. Dielectric and ferroelectric properties of pvdf thin film for biomechanical energy harvesting. *Mater. Today Proc.* **2021**, *41*, 335–339. [CrossRef]
30. Kim, H.; Batra, A.; Priya, S.; Uchino, K.; Markley, D.; Newnham, R.; Hofmann, H. Energy harvesting using a piezoelectric "cymbal" transducer in dynamic environment. *Jpn. J. Appl. Phys.* **2004**, *43*, 6178–6183. [CrossRef]
31. Zhao, H.; Yu, J.; Ling, J. Finite element analysis of cymbal piezoelectric transducers for harvesting energy from asphalt pavement. *J. Ceram. Soc. Jpn.* **2010**, *118*, 909–915. [CrossRef]
32. Yesner, G.; Jasim, A.; Wang, H.; Basily, B.; Maher, A.; Safari, A. Energy harvesting and evaluation of a novel piezoelectric bridge transducer. *Sens. Actuators A Phys.* **2019**, *285*, 348–354. [CrossRef]
33. Liu, X.; Wang, J. Performance exploration of a radially layered cymbal piezoelectric energy harvester under road traffic induced low frequency vibration. *Iop Conf. Ser. Mater. Sci. Eng.* **2019**, *542*, 12075. [CrossRef]
34. Jin, H.; Li, Z.; Wang, Z.; Tang, S. Vibration energy harvesting from tunnel invert-filling using rubberized concrete. *Environ. Sci. Pollut. Res.* **2023**, *30*, 30167–30182. [CrossRef] [PubMed]

35. Citroni, R.; Di Paolo, F.; Livreri, P. Evaluation of an optical energy harvester for SHM application. *AEU Int. J. Electron. Commun.* **2019**, *111*, 152918. [CrossRef]
36. Sezer, N.; Koc, M. A comprehensive review on the state-of-the-art of piezoelectric energy harvesting. *Nano Energy* **2021**, *80*, 105567. [CrossRef]

sensors

MDPI

Article

Research on Pneumatic Control of a Pressurized Self-Elevating Mat for an Offshore Wind Power Installation Platform

Junguo Cui [1,2,*], Qi Shi [1,2], Yunfei Lin [1,2], Haibin Shi [3], Simin Yuan [3] and Wensheng Xiao [1,2]

1 College of Mechanical and Electrical Engineering, China University of Petroleum, Qingdao 266580, China
2 National Engineering Research Center of Marine Geophysical Prospecting and Exploration and Development Equipment, Qingdao 266580, China
3 Shanghai Zhenhua Heavy Industry Co., Ltd., Shanghai 200125, China
* Correspondence: cuijunguo@upc.edu.cn

Abstract: Efficient deep-water offshore wind power installation platforms with a pressurized self-elevating mat are a new type of equipment used for installing offshore wind turbines. However, the unstable internal pressure of the pressurized self-elevating mat can cause serious harm to the platform. This paper studies the pneumatic control system of the self-elevating mat to improve the precision of its pressure control. According to the pneumatic control system structure of the self-elevating mat, the pneumatic model of the self-elevating mat is established, and a conventional PID controller and fuzzy PID controller are designed and established. It can be seen via Simulink simulation that the fuzzy PID controller has a smaller adjustment time and overshoot, but its anti-interference ability is relatively weak. The membership degree and fuzzy rules of the fuzzy PID controller are optimized using a neural network algorithm, and a fuzzy neural network PID controller based on BP neural network optimization is proposed. The simulation results show that the overshoot of the optimized controller is reduced by 9.71% and the stability time is reduced by 68.9% compared with the fuzzy PID. Finally, the experiment verifies that the fuzzy neural network PID controller has a faster response speed and smaller overshoot, which improves the pressure control accuracy and robustness of the self-elevating mat and provides a scientific basis for the engineering applications of the self-elevating mat.

Keywords: wind power installation platform; pressurized self-elevating mat; pneumatic control system; fuzzy neural network PID

1. Introduction

The offshore wind power installation platform is a necessary piece of equipment for the installation of offshore wind turbines, and the suitability of the platform directly affects the construction efficiency of offshore wind farms, as some existing wind power installation platforms cannot adapt to deep-water operation [1,2]. Based on a project involving the construction of a 2000-ton offshore wind power installation platform, the risk characteristics of offshore platforms during the design stage have been systematically studied [3], and a significant stabilization of these platforms in terms of pitching angle and nacelle acceleration due to the installation of a gyro has been achieved [4]. Previous studies have analyzed the dynamic characteristics of offshore platforms [5,6], summarized the key technologies of offshore platforms [7], proposed that one of the most relevant cost items is the floating substructure [8], and presented the fuzzy control method as an alternative solution for the installation platform [9]. Takagi proposed an innovative jack-up seawall as an effective countermeasure to protect the port and its hinterland from tsunamis [10]. Feng designed a control strategy and algorithm according to the characteristics that the lifting platform needs to display in order to carry the test body, and realized safe and reliable remote control based on the specific motion conditions [11]. According to the working environment and system characteristics of an underwater winch, Hu et al. determined that the underwater winch was driven by a hydraulic servo, and the overall design of the hardware and software architecture of the control system was carried out [12]. Their study involved analyzing risk characteristics, stabilization with a gyro, key technologies, cost considerations, and innovative solutions for offshore wind power installation platforms.

Pneumatic control initially adopts the active pressure regulation method, and through the reasonable design of a pneumatic device, one can achieve high reliability. Wang et al. [13] first proposed the use of a programmable controller (PLC) in pneumatic control systems. Compared with relay control, PLC control has low delay and high reliability. A PLC supports multiple communication protocols and can accurately control and monitor the actuators in the process of establishing a vacuum environment [14]. Li et al. [15] designed a new multi-channel and scalable pressure controller based on STM32, which can output both positive and negative pressure. Yang et al. found that a high-performance vacuum servo system can further enhance the competitiveness of vacuum servo technology [16]. The fuzzy-PID control algorithm provides an improvement to overshoot [17], but can also adapt to the dynamic changes of the system to enhance the anti-interference performance and adaptive ability of the system, and also has a faster response speed compared with the traditional PID algorithm [18,19]. Pneumatic control has evolved from active pressure regulation to PLC-based systems, with advancements in pressure controllers and fuzzy-PID algorithms for improved precision and reliability in vacuum environments.

A robust predictive tracking control (RPTC) approach has been developed to deal with a class of nonlinear SISO systems. To improve the control performance, the RPTC architecture mainly consists of a robust fuzzy PID (RFPID)-based control module and a robust PI grey model (RPIGM)-based prediction module [20], and a self-organizing intelligent controller (SOIC) was proposed for a class of nonlinear systems [21]. An improved genetic algorithm optimization fuzzy controller has been applied to the throttle valve, which exhibits great advantages in terms of speed, stability, and robustness [22]. A simple and robust unknown dynamics estimator (UDE) was also introduced, leading to reduced computational costs and simpler parameter tuning [23]. Zhang et al. focused on two control strategies, namely fuzzy time sampled-data control and fuzzy time-space sampled-data control, and some simulations were presented to verify their effectiveness and superiority [24]. After the introduction of fuzzy rules and a neural network, the system can adjust parameters adaptively and various performances are improved [25,26]. By combining fuzzy control and artificial intelligence, the control performance of the controller can be improved [27]. Using the genetic algorithm to optimize the PID controller based on a fuzzy neural network and the genetic algorithm can improve the self-adaptive ability of the system [28]. Huang et al. proposed a new reachability analysis approach based on Bernstein polynomials which can verify that a neural-network-controlled system has a more general form of activation functions [29]. A PID-neuro-fuzzy model based on a structured artificial neural network and fuzzy rules was proposed in [30], and a multilayer structure was provided to improve the learning ability and flexibility of a fuzzy neural network in [31]. These papers introduced methods such as robust fuzzy PID and PI grey model modules, self-organizing intelligent controllers, genetic algorithm optimization, fuzzy rules, neural networks, and reachability analysis techniques to enhance control adaptability and effectiveness.

In this paper, a kind of offshore wind power installation platform with a pressurized self-elevating mat is proposed. There are relatively few studies on pneumatic control systems for self-elevating mats for wind power installation platforms. A fuzzy PID control method, optimized based on a neural network algorithm, is proposed for a self-elevating mat pressure system in this study. By combining the conventional PID algorithm with fuzzy theory, a fuzzy PID controller is designed which has the response speed of conventional PID and the self-adjusting ability of fuzzy control. The BP neural network algorithm with adaptive learning abilities is used to optimize the fuzzy PID, and the BP neural network is used to calculate the membership rule of the fuzzy control, so as to adjust the PID control parameters in real time and improve the pressure control accuracy of the self-elevating mat. This paper first innovatively designs the pressurized self-elevating mat in conjunction with the wind power installation platform, and establishes the working principle and model of the pressure control system of the pressurized self-elevating mat, and then designs the pneumatic control system and fuzzy controller. On this basis, the fuzzy neural network PID controller is designed, through the simulation of several controllers and experimental verification. To judge the control effect of each controller, we chose the best controller.

2. System Working Principle and Model Establishment

2.1. Working Principle of the Self-Elevating Mat

The proposed offshore wind power installation platform with a pressurized self-elevating mat is shown in Figure 1. When working, the self-elevating mat is filled with air and sits on the bottom of the seabed. However, the internal pressure system of the self-elevating mat on the platform is non-linear and delayed, and if the pneumatic control is not performed in a timely manner, it will cause potential safety hazards for the installation platform. Therefore, research on the pneumatic control technology of self-elevating mats is of great significance to avoiding damage to the self-elevating mat structure and improving the safety of the platform. The concept of a pressurized self-elevating mat is proposed for the first time in this study. Because the self-elevating mat for a wind power installation platform is large, complex, has many actions, and is difficult to control, it is a key innovation point to design the inlet pressure control system of pressurized self-elevating mats by using a fuzzy neural network PID controller.

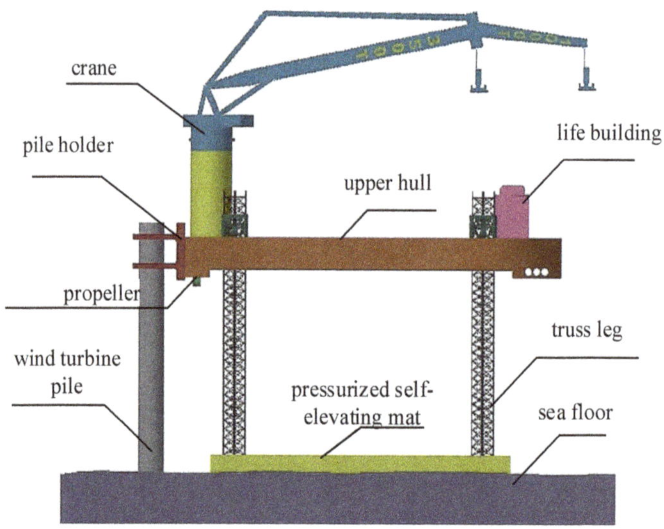

Figure 1. Structure diagram of wind power installation platform.

The self-elevating mat is a closed area, and its surface is connected with gas pipelines and other components. There is also a set of pinion and rack mechanism system installed on the truss legs. During the operation of the wind power installation platform, the pressurized self-elevating mat is filled with seawater and sent to the seabed via the pinion and rack mechanism. Then, high-pressure gas is pumped into the mat for drainage. At this time, the mat filled with air obtains buoyancy, which is transmitted to the upper hull through the truss leg to balance the platform's own weight, and carries out the operation of lifting the upper hull combined with the seabed reaction.

The offshore wind power installation platform is mainly composed of the upper hull, the pressurized self-elevating mat, and four truss legs. The upper hull is equipped with a crane, pile holder, propeller, and other facilities.

The pneumatic control system of the pressurized self-elevating mat is shown in Figure 2, and is mainly composed of the air compressor, gas storage tank, and pipe winding machine on the upper platform, and the solenoid valve, pressure relief valve, and check valve on the self-elevating mat. When the wind power installation platform is in operation, the pressurized self-elevating mat filled with water is required to sit on the seabed. High-pressure gas is generated by the external pressurized equipment and then sent to the self-elevating mat through the gas transmission pipeline. The seawater in the self-elevating mat is discharged and buoyancy is generated, meaning that the upper platform is lifted to realize wind power installation.

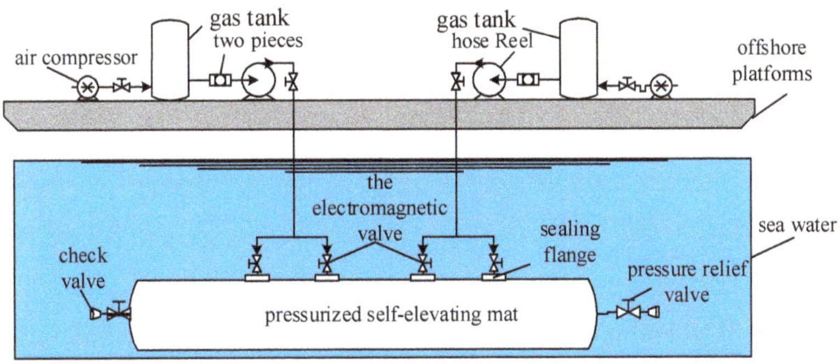

Figure 2. Air supply flow chart of the air source device.

There are three main stages in the work process of the wind power installation platform:

Sinking and drainage process. When the wind power installation platform reaches the work site, the self-elevating mat is sunk through the platform pile leg rack lifting mechanism. In this process, the valves around the self-elevating mat are opened to connect the seawater inside and outside the self-elevating mat box, and there is no pressure difference. When the self-elevating mat is completely seated on the seabed, the pneumatic device is used to fill the inside of the self-elevating mat with high-pressure gas and the seawater in the box is discharged. The buoyancy is generated to lift the upper platform through the spud legs.

Washing out pile and lifting process. The self-elevating mat needs to be returned to the upper platform after the operation finishes. Due to the large adsorption force between the self-elevating mat and the seabed ballast, the pile punching device is required to drive the medium between the self-elevating mat and the seabed surface in order to reduce the adsorption force. At the end of the punching pile, the self-elevating mat is lifted into the upper hull through the pile leg. In order to reduce the force acting on the pile leg, the self-elevating mat is lifted up in a hollow state.

Pressure retaining process. During the self-elevating mat's seating and rising, in order to prevent the structure being crushed by seawater, it is necessary to ensure that the box is under positive pressure (the internal pressure is greater than the external pressure), and the internal and external pressure difference is about one atmosphere. At the same time, it is necessary to monitor the internal pressure and compensate the pressure in order to avoid changes in the internal pressure of the self-elevating mat due to sealing problems or a local leakage.

2.2. Modeling of the Pressure System

(1) Intake model

The throttling element of the pneumatic control system utilizes an electromagnetic proportional valve with a fast-opening characteristic. The relationship between the opening degree and the flow coefficient can be expressed as:

$$K_v = -1.0741e^{-\frac{x}{3874.6}} + 1.0816 \tag{1}$$

This function is fitted by origin using the polynomial method, where x is the opening size, 0~100%, and K_v is the flow coefficient when the opening is x, 0~100%.

The mass flow rate of the solenoid valve is determined by the valve flow coefficient and the inlet–outlet pressure difference, so the mass flow rate of the solenoid proportional valve into the self-elevating mat is [32]:

$$M_m = C_p K_v x P_1 \sqrt{\frac{2}{RT}} \left(\frac{2}{K+1}\right)^{\frac{1}{K-1}} \sqrt{\frac{K}{K+1}} \tag{2}$$

where M_m is the mass flow rate of the valve, kg/s; C_p is the leakage coefficient, which is 0.6~1.0; P_1 is the inlet pressure of the solenoid valve, Pa; T is the temperature, K; K is the adiabatic index, which is 1.4; and R is the gas constant, 8.314 J/(mol·K).

(2) Exhaust model

The self-elevating mat in this study is a closed area, and its surface is connected with the gas transmission pipeline and other components. In actual engineering, the greater the airtightness of the container is, the higher the construction difficulty and cost are. Therefore, a small amount of leakage is allowed in the construction process of the self-elevating mat, but the leakage process must meet the requirements of establishing a pressure-maintaining control system for the self-elevating mat. That is, gas discharge of the self-elevating mat mainly consists of two parts: leakage and pressure relief valve exhaust. Because the gas velocity during leakage is subsonic, the formula for calculating the leakage mass M_{T1} is as follows [33]:

$$M_{T1} = C_p A \rho_i \sqrt{\frac{2K}{K-1} RT \left[\left(\frac{P_2}{P_i}\right)^{\frac{2}{K}} - \left(\frac{P_2}{P_i}\right)^{\frac{K+1}{K}} \right]} \tag{3}$$

where A is the leakage area, m^2; ρ_i is the air density at time t of the box, kg/m^3; and P_i is the pneumatic pressure at box t, Pa.

The pressure relief flow characteristics of the pressure relief valve are largely similar to the leakage process, but the pressure relief area is controlled by the current, so the mass flow rate M_{T2} flowing through the pressure relief valve is calculated as follows:

$$M_{T2} = C_p K_v x A \rho_i \sqrt{\frac{2K}{K-1} RT \left[\left(\frac{P_2}{P_i}\right)^{\frac{2}{K}} - \left(\frac{P_2}{P_i}\right)^{\frac{K+1}{K}} \right]} \tag{4}$$

(3) Transfer function establishment

The intake and exhaust flow of the self-elevating mat are obtained, and so the complete flow equation of the gas mass M_c in the box is as follows:

$$M_c = M_m - M_{T1} - M_{T2} \tag{5}$$

According to the continuity equation, it can be seen that the inflow and outflow of gas inside the self-elevating mat are equal to the change rate of gas mass, and the power source for discharging gas mainly comes from the internal and external pressure difference. This means that the discharged mass can be linearly converted, and Equation (5) can be expressed as:

$$\frac{V}{KRT} \cdot \frac{dP}{dt} = k_m x - k_1 p - k_2 p \tag{6}$$

$$k_{\mathrm{m}} = C_p K_v x \sqrt{\frac{2}{RT}} \left(\frac{2}{K+1}\right)^{\frac{1}{K-1}} \sqrt{\frac{K}{K+1}} \tag{7}$$

where the k_m in (6) is calculated in (7), where k_1 is the flow pressure coefficient at the leak and k_2 is the flow pressure coefficient of the pressure relief valve.

The pneumatic control system of the self-elevating mat controls the pressure change inside the self-elevating mat by controlling the current. The actuator is a solenoid valve, and the controlled object is the internal pressure of the self-elevating mat. The input of the solenoid valve is the control current, and the output is the valve opening. This process can be regarded as a second-order link, and the transfer function can be expressed as [34]:

$$\frac{x(s)}{i(s)} = \frac{\frac{k_v}{K_v}\omega_v^2}{s^2 + 2\xi_v \omega_v s + \omega_v^2} \tag{8}$$

where k_v is the gain coefficient of the solenoid valve, m·A^{-1}; ω_v is the natural frequency of the solenoid valve, rad·s^{-1}; and ξ_v is the damping ratio of the solenoid valve.

Under zero initial conditions, the Laplace transform of Equation (6) is carried out, and the transfer function between the pressure inside the self-elevating mat and the valve opening is as follows:

$$\frac{P(s)}{x(s)} = \frac{k_m}{\frac{V}{KRT} \cdot s + k_1 + k_2} \tag{9}$$

The solenoid valve opening x is taken as the input in the self-elevating mat link, and the internal pressure of the self-elevating mat is taken as the output. Considering the delay of the solenoid valve, the transfer function of the self-elevating mat pneumatic control system can be obtained by incorporating the relevant parameters:

$$G(s) = \frac{P(s)}{x(s)} \cdot \frac{x(s)}{i(s)} = \frac{k_{\mathrm{m}} k_v \omega_v^2}{K_v \left(\frac{V}{KRT} \cdot s + k_1 + k_2\right)(s^2 + 2\xi_v \omega_v s + \omega_v^2)} e^{-\tau s} \tag{10}$$

3. Design of a Fuzzy PID Controller for a Pneumatic Control System

When regulating the internal pressure of the self-elevating mat, the PLC first collects the value of internal pressure and the status of each valve, calculates the deviation between the set pressure and the actual pressure, controls the opening of each valve using the controller to control the gas flow, and realizes the adjustment of the internal pressure of the self-elevating mat. The self-elevating mat of the wind power installation platform has a large volume, a long transmission pipe, and high pressure requirements. Additionally, the self-elevating mat has a large pressure change range and a long pressure control time under different operating conditions. Therefore, it is necessary to meet the gas demand of the self-elevating mat under different operating conditions and ensure that the pressure in the chamber stays positive and quickly reaches the set value, which becomes a difficult problem in the design of pneumatic control systems. In this paper, the fuzzy control theory is used to design the controller, and the neural network is used to optimize the fuzzy controller.

3.1. Fuzzy Control Algorithm

Fuzzy control, using expert experience to make fuzzy rules, can quickly adjust the system control parameters in order to ensure that the system maintains the best control level. By combining the conventional PID algorithm with fuzzy theory, a fuzzy self-tuning PID controller is designed which not only has the high precision and fast response speed of conventional PID control, but also has good robustness and a parameter-adaptive function, so as to achieve the goal of high-quality control and energy savings.

Fuzzy PID can achieve a better control effect according to fuzzy rules, aiming to address the nonlinear and complex model problem of solenoid valve opening flow in the pneumatic control system for a pressurized self-elevating mat. The core of fuzzy PID is the design of the fuzzy controller [35]. The principle of fuzzy PID control is shown in Figure 3. By fuzzifying the input and applying fuzzy reasoning and defuzzification, the correction value of PID parameters is determined, and the parameters of the PID controller are adjusted in real time to improve the control accuracy.

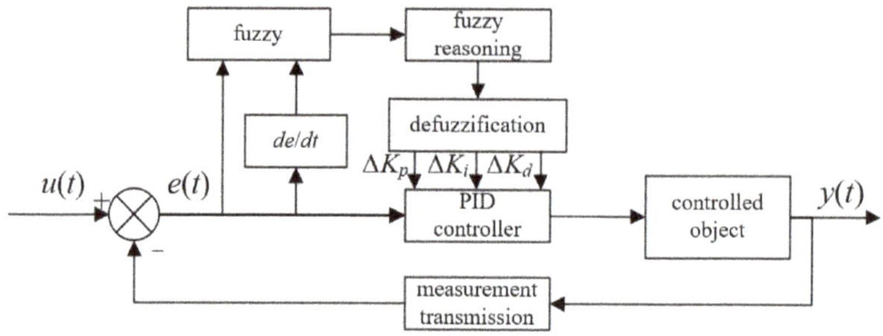

Figure 3. Schematic diagram of fuzzy PID control.

(1) Fuzzifying

In order to obtain a better membership function, the selection should be combined with the characteristics of the system. In this paper, the input adopts a linear trigonometric function to describe the membership function, while the relationship between the current and the opening of the controlled valve is non-linear, and so the output is described by a Gaussian function.

In Figure 3, the input of the fuzzy controller is the pressure deviation e, the rate of change ec, and the output is the correction values of PID parameters ΔK_p, ΔK_i, and ΔK_d. In this paper, the fuzzy domain is divided into subsets (NB, NM, NS, ZO, PS, PM, and PB), and the fuzzy subset corresponds to the uniformly quantized domain $[-3, 3]$. Combined with the characteristics of the pneumatic control system, the input is a kind of membership function described by a linear trigonometric function. Meanwhile, the current of the controlled valve has a nonlinear relationship with the opening, and so the output is described by a Gaussian function. The membership function images are shown in Figure 4.

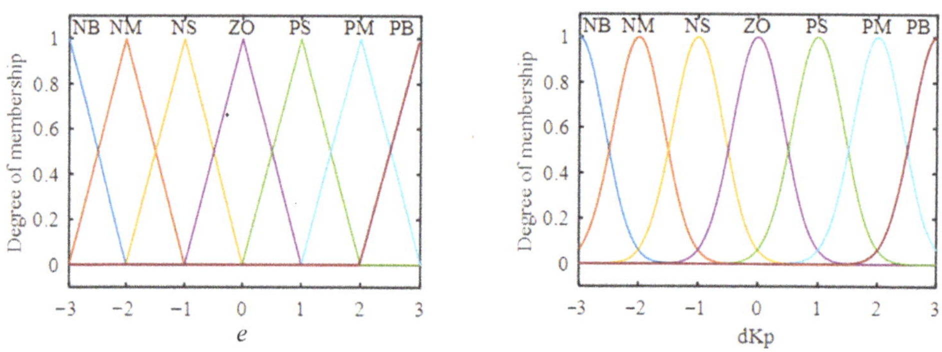

(**a**) Input membership function graph (**b**) Output membership function graph

Figure 4. The membership function of e and ec.

(2) Fuzzy reasoning

The fuzzy reasoning method adopts the Mamdani model to establish the fuzzy PID, and combines expert experience to formulate the relationship between input state quantity $[e, ec]$ and control variable $[\Delta K_p, \Delta K_i, \Delta K_d]$. A total of 49 inference rules are shown in Tables 1–3.

Table 1. ΔK_p fuzzy inference table.

e	ec						
	NB	**NM**	**NS**	**ZO**	**PS**	**PM**	**PB**
NB	PB	PB	PM	PM	PS	ZO	ZO
NM	PB	PB	PM	PS	PS	ZO	NS
NS	PM	PM	PM	PS	ZO	NS	NS
ZO	PM	PM	PS	ZO	NS	NM	NM
PS	PS	PS	ZO	NS	NS	NM	NM
PM	PS	ZO	NS	NM	NM	NM	NB
PB	ZO	ZO	NM	NM	NM	NB	NB

Table 2. ΔK_i fuzzy inference table.

e	ec						
	NB	**NM**	**NS**	**ZO**	**PS**	**PM**	**PB**
NB	NB	NB	NM	NM	NS	ZO	ZO
NM	NB	NB	NM	NS	NS	ZO	ZO
NS	NB	NM	NS	NS	ZO	PS	PS
ZO	NM	NM	NS	ZO	PS	PM	PM
PS	NM	NS	ZO	PS	PS	PM	PB
PM	ZO	ZO	PS	PS	PM	PB	PB
PB	ZO	ZO	PS	PM	PM	PB	PB

Table 3. ΔK_d fuzzy inference table.

e	ec						
	NB	**NM**	**NS**	**ZO**	**PS**	**PM**	**PB**
NB	PS	NS	NB	NB	NB	NM	NM
NM	PS	NS	NB	NM	NM	NS	ZO
NS	ZO	NS	NM	NM	NS	NS	ZO
ZO	ZO	NS	NS	NS	NS	NS	ZO
PS	ZO	ZO	ZO	ZO	ZO	ZO	ZO
PM	PB	NS	PS	PS	PS	PS	PB
PB	PB	PM	PM	PM	PS	PS	PB

(3) Defuzzification

Defuzzification is the process of converting the abstract value result obtained through fuzzy reasoning into a clear value that can be recognized by the actuator through various methods [36]. There are many ways to de-blur, and the calculation method should be reasonably selected according to the control requirements of the pneumatic control system. Commonly used methods include the maximum degree of membership method, the weighted average method, and the center of gravity method. For the center of gravity method, the system output is the barycenter of the area surrounded by the membership function curve and the coordinate axis. The center of gravity method can respond to the input signal even when the input signal does not change much, and the value is smooth and has high precision. Therefore, the center of gravity method is used for defuzzification. The calculation formulas are shown in (11).

$$z_0 = \frac{\int_a^b z\mu_c(z)dz}{\int_a^b \mu_c(z)dz} \tag{11}$$

After the clear ΔK_p, ΔK_i, and ΔK_d are obtained, they should also be adjusted through combination with the actual PID parameters, as shown in Equation (12):

$$K_x = K_{x0} + k_x \cdot \Delta K_x \tag{12}$$

where K_{x0} is the initial value of each PID parameter; k_x is the quantization factor; and ΔK_x is the correction value after sharpening.

3.2. Simulation of Control Algorithm

In the previous section, the self-adjusting fuzzy PID controller was analyzed and designed for the pneumatic control system. In this section, the controller is built using Simulink and simulation experiments are carried out. The fuzzy PID simulation model is shown in Figure 5, and the internal structure of the fuzzy controller is shown in Figure 6.

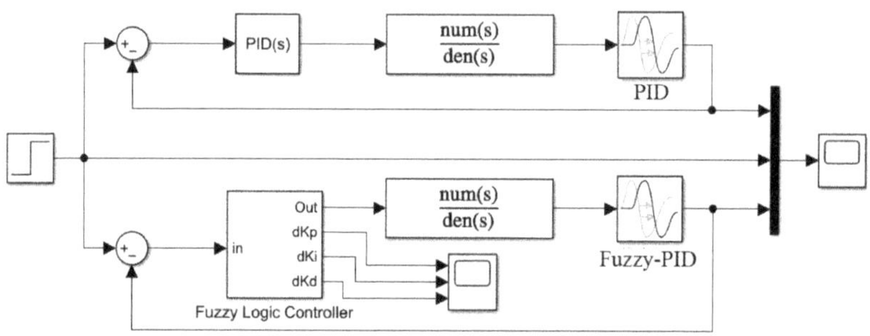

Figure 5. Simulation structure of the fuzzy PID controller.

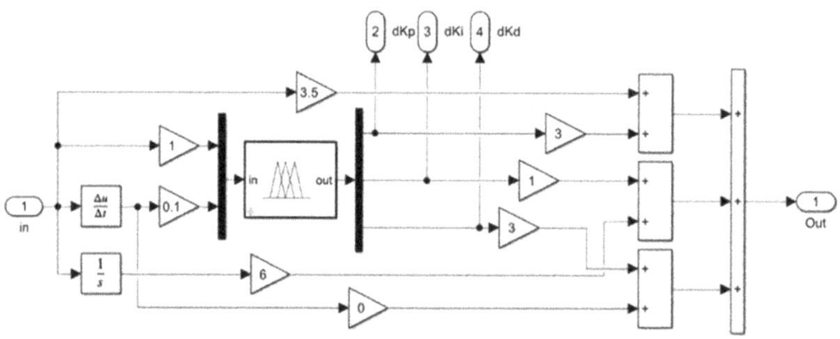

Figure 6. Internal structure of the fuzzy controller.

The inputs of the system are set as the step signal and rectangular wave step signal, respectively, with an amplitude of 10, the simulation times are 10 s and 30 s, respectively, and the three initial parameters of PID are K_p = 3.5, K_i = 6 and K_d = 0, respectively. With the same PID parameter setting, conventional PID control and fuzzy PID control were used to simulate and analyze the pneumatic control system according to the structure in Figure 5, and the response curves are shown in Figure 7a,b.

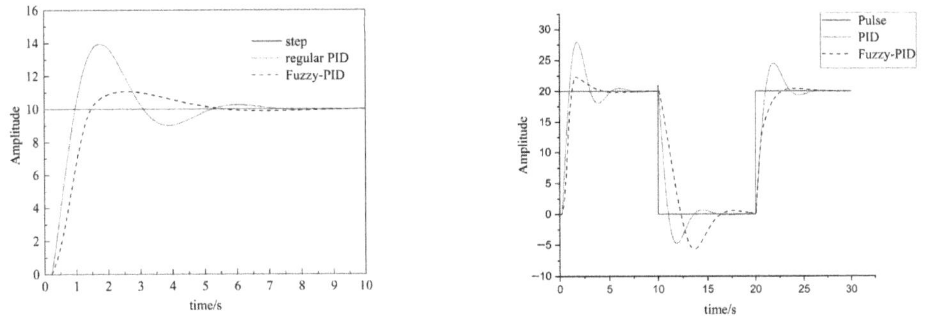

(**a**) response curves with step signal (**b**) response curves with rectangular wave step signal

Figure 7. Fuzzy PID simulation.

It can be seen that, when the input is set as step signal, the rise time of the traditional PID is 0.875 s, the peak value is 13.96, and the stability time is 5.085 s. Meanwhile, the step response rise time of the fuzzy controller is 1.271 s, the peak value is 11.07, and the stability time is 4.865 s. The comparison shows that the fuzzy control has a smaller overdrive and a faster stabilization time. When the input is set as rectangular wave step signal, it can be seen that the fuzzy control has a smaller overdrive but the stabilization times are very close to each other.

The output parameters ΔK_p, ΔK_i, and ΔK_d of the fuzzy controller are shown in Figure 8a and b, respectively.

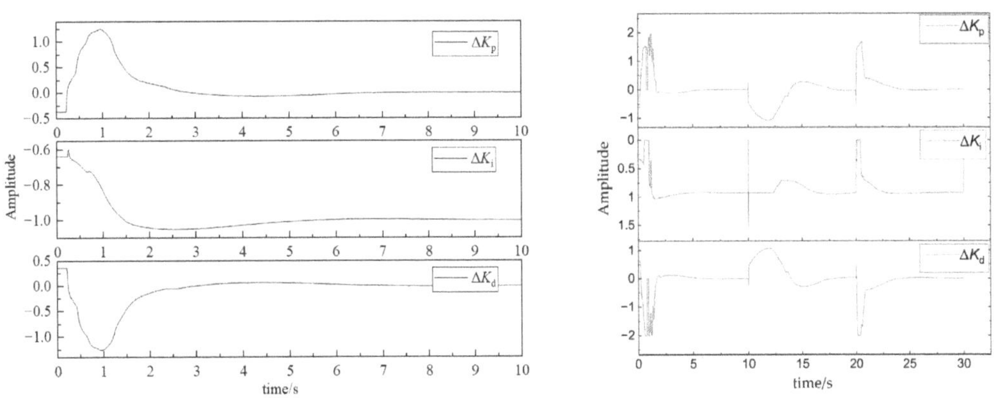

(**a**) the input is step signal (**b**) the input is rectangular wave step signal

Figure 8. Fuzzy control parameter results.

In the pneumatic control system, the feedback of the system or other signals will change due to various accidents, which will cause interference. It is necessary to analyze the anti-interference ability of the system. During the simulation, at 10 s, an interference with an amplitude of 1 is introduced, and the overall simulation time is set as 20 s. The simulation results are shown in Figure 9.

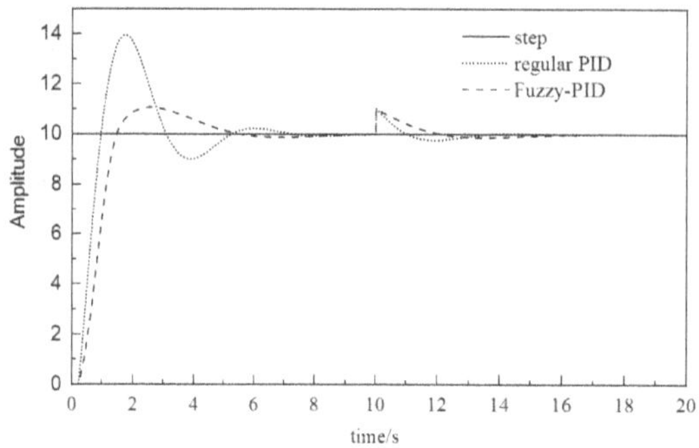

Figure 9. Simulation analysis of interference introduction.

It can be seen from the simulation results that although the response of fuzzy control is more stable than that of conventional PID, the response speed is slower when dealing with interference. In addition, the response speed of fuzzy PID is slow in the initial stage, and the adjustment ability in case of interference is poor, so it is necessary to introduce an artificial intelligence control method to improve fuzzy control.

4. Design of a Fuzzy Neural Network PID Controller for a Pneumatic System

4.1. Fuzzy Neural Network Algorithm

The fuzzy rules and membership function of the fuzzy controller are generally obtained according to expert experience. However, for a controlled object with interference, better fuzzy rules and membership functions can no longer be obtained by means of expert experience, and there are problems such as self-learning. Therefore, the BP neural network is introduced to optimize the fuzzy controller. By combining a neural network and fuzzy PID control, a PID controller based on a fuzzy neural network is formed, and PID parameters are adjusted using the adaptive fuzzy neural network to increase system stability.

The input of the fuzzy neural network PID (FNN-PID) controller is the pressure error e and rate of change ec, and the membership rules are adjusted in real time through the learning algorithm, while the three parameters of PID are the output. The control principle is shown in Figure 10.

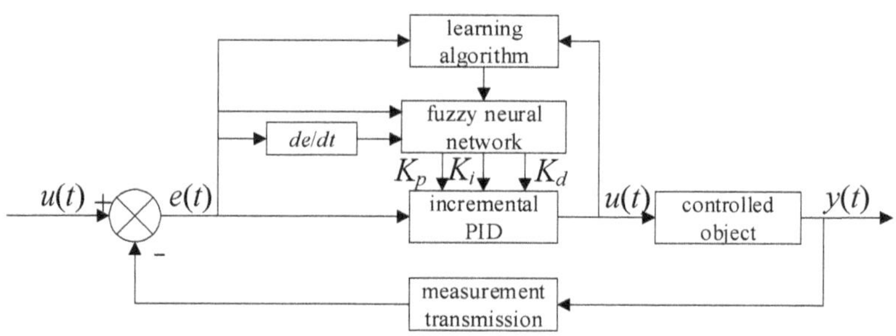

Figure 10. Schematic diagram of FNN-PID control.

(1) FNN-PID controller design

In this paper, the FNN-PID controller is designed based on the Mamdani model, which is a five-layer feedforward approximation network with two inputs and three outputs. The inputs in the input layer are the deviation e and the rate of change ec. After the membership degree fuzzification layer and fuzzy inference layer are normalized in the fuzzy neural model, the output layer performs clear calculations based on the previous layer to obtain the corresponding PID control parameters K_p, K_i, and K_d. The fuzzy neural network structure is shown in Figure 11.

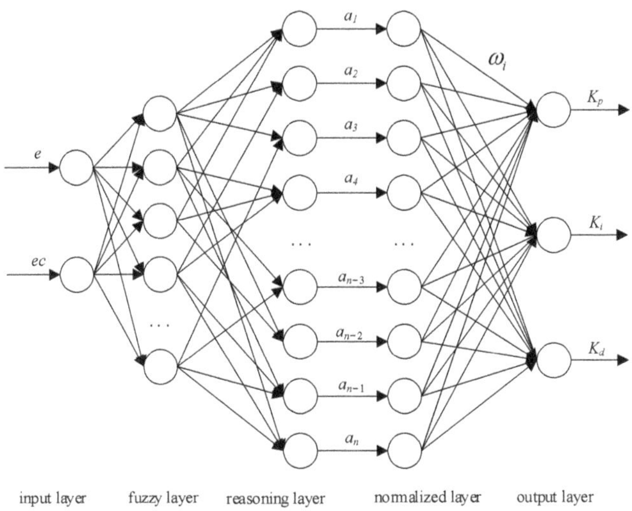

Figure 11. Fuzzy neural network structure.

The first layer is the input layer, whose input vector $x_i = [e, ec]^{\mathrm{T}}$, and this layer has two nodes. The input and output of this layer can be expressed as:

$$\begin{cases} I_i^{(1)} = x_i \\ O_{ij}^{(1)} = I_i^{(1)} \end{cases} \tag{13}$$

where I is the input of the network layer; O indicates network layer input; (1) is the first layer of the network; x_i is the input of the ith node, $I = 1, 2$; and j is the output node of the layer network, $j = 1, 2, \ldots, 7$.

The second layer is the fuzzifying layer. The input is fuzzified and the membership function is determined using a Gaussian function. The fuzziness domain of each input vector is divided into NB, NM, NS, ZO, PS, PM, and PB, which are −3, −2, −1, 0, 1, 2, and 3, respectively. Each node represents one of the fuzzy subsets, and there are 14 nodes in total. The input and output of this layer can be expressed as:

$$\begin{cases} I_i^{(2)} = O_{ij}^{(3)} \\ O_{ij}^{(2)} = u_{ij}(x_i) = \exp\left[-\frac{(x_i - c_{ij})^2}{b_{ij}^2}\right] \end{cases} \tag{14}$$

where c_{ij} is the central value of the membership degree function and b_{ij} is the width of the membership function.

The third layer is the fuzzy reasoning layer. All nodes in the hidden layer have corresponding fuzzy rules. Based on the matching of fuzzy rules, the fitness of each $u_{ij}(x_i)$ value is calculated. The input and output of this layer can be expressed as:

$$\begin{cases} I_i^{(3)} = O_{ij}^{(2)} \\ O_l^{(3)} = \prod_{i=1}^{2} \prod_{j=1}^{7} u_{ij}(x_i) \end{cases} \tag{15}$$

where $l = 1, 2, \ldots, 49$.

The fourth layer is the normalization layer, which limits the results to a certain range and speeds up the convergence rate of the fuzzy neural net. The input and output of this layer can be expressed as:

$$\begin{cases} I_l^{(4)} = O_l^{(3)} \\ O_l^{(4)} = \frac{I_l^{(4)}}{\sum_{l=1}^{49} I_l^{(4)}} \end{cases} \tag{16}$$

The fifth layer is the output layer, which has three nodes. Through the clear calculation of the input of the fourth layer, the final output of the controller and the three control parameters of PID are obtained. The input and output of this layer can be expressed as:

$$\begin{cases} I_l^{(5)} = O_l^{(4)} \\ O_h^{(5)} = \sum_{h=1}^{3} \sum_{l=1}^{49} \omega_{hl} I_l^{(5)} \end{cases} \tag{17}$$

where h is the output number, which is 1,2, or 3, and ω_{hl} is the connection weight between the fourth and fifth layers.

(2) BP learning algorithm

After establishing the fuzzy neural network structure model, it is necessary to modify the central value of the membership function c_{ij}, width b_{ij}, and the connection weight ω_{hl} between the fourth and fifth layers. The performance index function $E(k)$ of the learning algorithm is:

$$E(k) = \frac{1}{2}e(k)^2 = \frac{1}{2}[u(k) - y(k)]^2 \tag{18}$$

where $u(k)$ is the set value of the pressure control system at the KTH iteration; $y(k)$ is the output value; and $e(k)$ is the error.

In order to reduce the control error of the system, the minimum value of fuzzy neural network parameters is searched for through the algorithm gradient. The iterative formulas of each network parameter are as follows:

$$\omega_{ij}(k+1) = \omega_{ij}(k) - \eta\frac{\partial E}{\partial \omega_{ij}} + \alpha(\omega_{ij}(k-1) - \omega_{ij}(k-2)) \tag{19}$$

$$c_{ij}(k+1) = c_{ij}(k) - \eta\frac{\partial E}{\partial c_{ij}} + \alpha(c_{ij}(k-1) - c_{ij}(k-2)) \tag{20}$$

$$b_{ij}(k+1) = b_{ij}(k) - \eta\frac{\partial E}{\partial b_{ij}} + \alpha(b_{ij}(k-1) - b_{ij}(k-2)) \tag{21}$$

where η is the learning rate and α is the momentum factor.

(3) Establishment of FNN-PID controller

After the mathematical model and the learning algorithm of the fuzzy neural network are determined, the incremental PID is introduced into the fuzzy neural network PID controller. Incremental PID only outputs the latest three iterations results. If the system sampling problem occurs, the output of the system will not be affected in the short term. The stability of the system can be increased.

The rule of incremental PID control is shown in Equations (22) and (23):

$$\Delta u(k) = K_p[e(k) - e(k-1)] + K_i e(k) + K_d[e(k) - 2e(k-1) + e(k-2)] \tag{22}$$

$$u(k) = u(k-1) + \Delta u(k) \tag{23}$$

where $u(k)$ and $e(k)$ are the controller output and control error of the system during k iterations, respectively, and K_p, K_i, and K_d are the outputs of the fuzzy neural network model.

4.2. Simulation and Result Analysis

According to the design of the fuzzy network controller in the previous section, a fuzzy neural network PID controller model is established. The S-function is adopted to realize the function of the fuzzy neural network controller. The input of the controller includes set values, deviation, and the output pressure value of the pneumatic system, in which the output is the PID output optimized via the fuzzy neural network. The simulation model is shown in Figure 12.

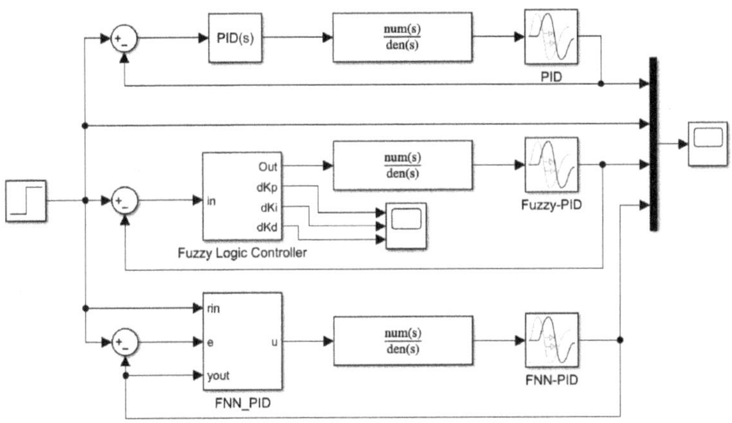

Figure 12. FNN-PID controller simulation structure.

The internal structure of the FNN-PID controller is shown in Figure 13. It can be seen that the controller inputs the deviation obtained by the delayed sampling module, the relevant parameters of the membership function and other values at different sampling times into the fuzzy neural network PID control system established by the S-function. Then, the function of the fuzzy neural network PID is carried out.

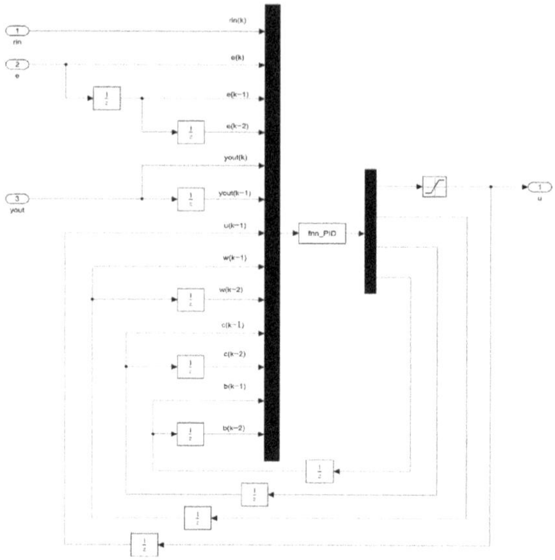

Figure 13. FNN-PID controller.

Using the same simulation time and input value, and introducing interference sources, the simulations of the pneumatic system controlled by conventional PID, fuzzy PID, and fuzzy neural network PID controller are performed, respectively. The simulation results are shown in Figure 14.

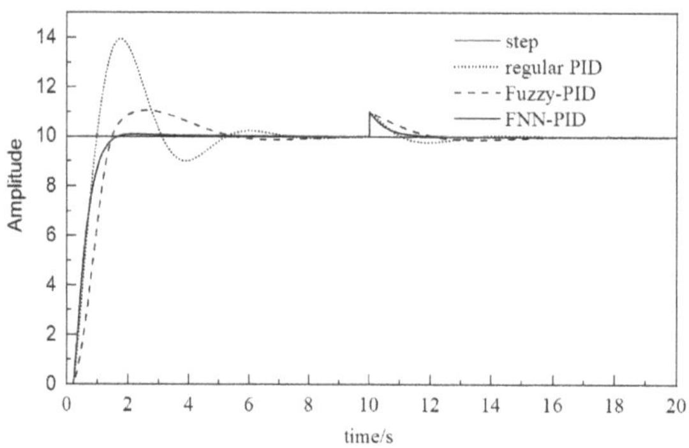

Figure 14. Simulation results under interference.

According to the simulation results in Figure 14, the system performance indexes of the three different controllers are summarized in Table 4.

Table 4. System performance specifications.

Controller	Rise Time/s	Adjustment Ime/s	The Peak	Overshoot	Anti-Interference Adjustment Time/s	Disturbance Overshoot
PID	0.875	5.085	13.96	39.6%	2.88	23%
Fuzzy-PID	1.271	4.865	11.07	10.7%	4.25	13%
FNN-PID	1.050	1.514	10.09	0.9%	1.19	0.03%

It can be seen from Table 4 that the conventional PID has the largest overshoot, the longest adjustment time, and large fluctuations. Compared with conventional PID, the fuzzy PID has a smaller overshoot and a shorter adjustment time, but the rise time is longer and the fluctuation is smaller. Compared with the previous two methods, the fuzzy neural network PID controller has a faster response, minimum overshoot, the shortest adjustment time, and almost no fluctuation. When disturbed, the overshoot of the conventional PID controller is within 23%, and the system can reach a stable state within 2.88 s. The conventional PID control has a long adjustment time, poor robustness, and weak anti-interference ability. The overshoot of the fuzzy PID controller is within 13% and the fluctuation is not large, but the stability time is as long as 4.25 s and the anti-interference ability is poor. The overshoot of the fuzzy neural network PID controller is within 0.03% and the system can reach a stable state within 1.18 s, which indicates that fuzzy neural network PID controller has a quick response, short adjustment time, small fluctuations, and the strongest anti-interference ability.

5. Experimental Verification

Table 5 lists the parameters of the control system based on the system selection, component characteristics, and working environment of the floating body.

Table 5. Control system parameters.

Parameter Name	Parameter Symbol	Parameter Value	Parameter Unit
Parameter Unit pressure loss coefficient	ξ	1.15	—
Pipe density	ρ_0	8.18	$kg{\cdot}m^{-3}$
Gas flow rate in the pipeline	μ	10	$m{\cdot}s^{-1}$
Total length of straight pipe	L_a	280	m
Lost length of pipeline	L_b	50	m
Total area of intake valves	A_1	0.5	m^2
Total area of pressure relief valve	A_2	0.2	m^2
Solenoid valve shape factor	C_L	0.8	—
Adiabatic coefficient	K	1.4	—
Operating temperature	T	295	K
Volume of floating body	V	20,000	m^3
Gas constant	R	8.314	$J{\cdot}(kg{\cdot}K)^{-1}$
Air source pressure	P_0	709,275	Pa
Proportional valve amplification factor	K_a	0.8	$A{\cdot}V^{-1}$
Proportional valve gain	k_v	1.6×10^{-3}	$m{\cdot}A^{-1}$
Solenoid valve damping ratio	ξ_v	0.6	—
Natural frequency of solenoid valve	ω_v	120	$rad{\cdot}s^{-1}$
Pressure sensor coefficient	K_f	1	—

On the basis of the controller design for the self-elevating mat, a self-elevating mat experimental prototype was manufactured. The experimental platform was built to verify the effect of the fuzzy neural network PID in the actual pressure control of a self-elevating mat. Considering the size of the self-elevating mat box and that the focus of this paper was the pneumatic control system, a similar model was adopted for verification. The experimental principle is shown in Figure 15.

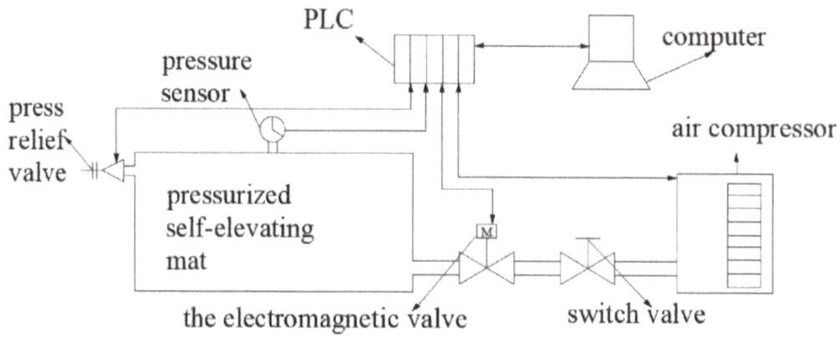

Figure 15. Schematic diagram of pneumatic experiment.

The experimental platform mainly included an air compressor, a PLC, a computer, a self-elevating mat, valves, and pipeline components. During the experiment, the required air volume was provided by the air compressor, which was transported into the self-elevating mat through the pipeline valve. The PLC collected the pneumatic values measured by pressure sensors and calculated the opening of the solenoid valve through different controllers for output, so as to realize constant internal pressure in the self-elevating mat. In this experiment, a leakage was set at the pressure relief valve on the self-elevating mat box, and, by using different control logics, we explored the internal pressure conditions during the charging and holding stages of the self-elevating mat when the charging reached 100 kPa. The quality of each control logic was analyzed by collecting data analysis. The experimental platform is shown in Figure 16.

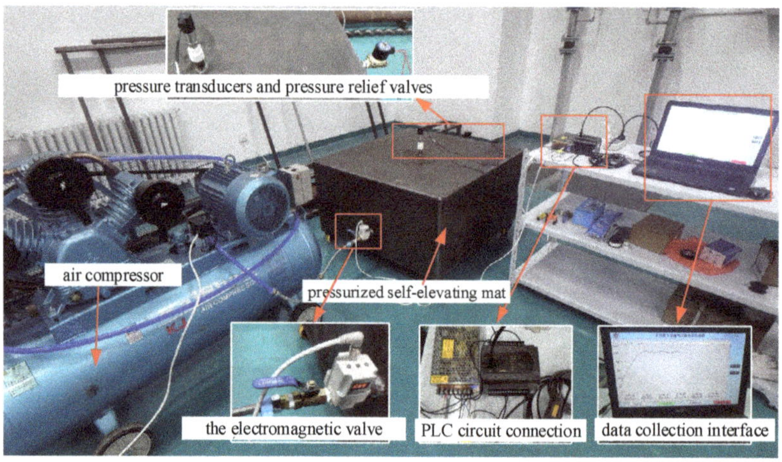

Figure 16. Experimental platform.

Experimental steps: (1) Firstly, we adjusted the controller logic to the PID controller, opened the pressure relief valve to introduce interference, and started the control system; (2) we started pressurizing and maintaining pressure according to the set pressure, and observed the internal pressure changes on the HMI interface and recorded real-time pressure data; and (3) thirdly, the control logic was adjusted to the fuzzy PID and fuzzy neural PID controller, respectively, and operations (1) and (2) were repeated to record the internal pressure change data under different controllers. (4) Finally, we analyzed and compared the results.

According to the pressure experiment procedure, the pressure relief valve was opened to introduce interference, the pressure of the box was pressurized to 100 kPa, and then the pressure was held for a period of time. The pressure data of the self-elevating mat under different controllers were recorded. The experimental curves are shown in Figure 17.

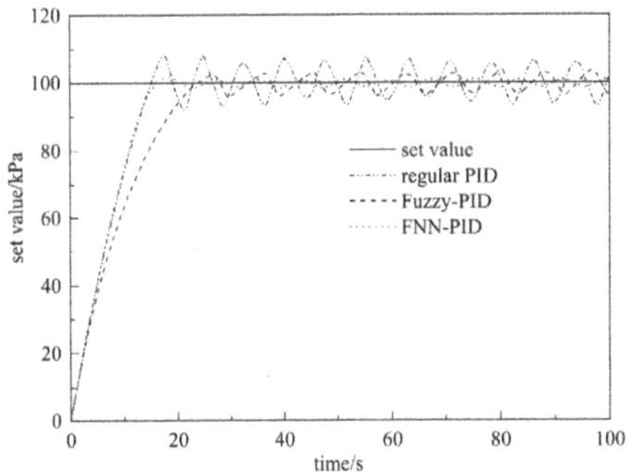

Figure 17. Experimental curve of pressure boost and pressure retention.

According to the experimental results, it can be seen that the pressure system's overshoot under the conventional PID controller is the largest, and the fluctuation is relatively severe. The overshoot of the fuzzy PID controller is smaller than that of the conventional PID, but its response speed is too slow. The optimized fuzzy neural network PID controller has excellent performance in reducing the overshoot, speeding up the response, reducing the adjustment time, and enhancing system stability. Due to the structure of the experiment, and certain interference and instability factors present during the experiment process, such as the instability of the output air source pressure of the plunger-type air compressor, the pressure value fluctuated in the pressure holding stage, resulting in a certain deviation between the experimental results and the simulation results.

6. Discussion

In this paper, aimed at addressing the problem of insufficient pressure control accuracy of self-elevating mats, the scheme of a self-elevating mat pressure control system is designed, the overall structure of the wind power installation platform is determined, the selection of related equipment in the pressure system is completed, the transfer function of the self-elevating mat pressure control system is established, a fuzzy neural network is used to optimize the fuzzy controller, and the operation monitoring system is designed for the control of the self-elevating mat. Finally, the superiority of the fuzzy neural network PID controller is verified by experiments. Research on the self-elevating mat system's structure scheme, the pressure model, the control strategy, simulation analysis, hardware and software implementation, etc., provides a theoretical basis and technical guidance for the practical application of the project, but certain aspects still need to be further developed and improved: the air consumption of the self-elevating mat punching pile system has been simplified in the design, the self-elevating mat leakage model and pressure relief valve model are linearized, and the accuracy can be improved. If the design of the self-elevating mat's pressure control system allows, the experimental verification of the self-elevating mat's pressure operation monitoring system can be completed on site.

7. Conclusions

The pressurized self-elevating mat in this paper is a new type of equipment, and the pneumatic control system for the self-elevating mat is a kind of controlled object that is nonlinear and has a large delay, which is difficult to control. In this paper, the conventional PID controller and fuzzy PID controller are designed, and the PID strategy and fuzzy PID strategy are simulated using Simulink. Comparing the results, it can be seen that the adjustment time and overshoot of the fuzzy PID controller are smaller, but the anti-interference ability is weaker. Therefore, the learning characteristics of the BP neural network are used to optimize the fuzzy PID, and the fuzzy neural network is designed to establish and calculate the membership rule of PID. The simulation results show that the controller optimized using the BP neural network has the best performance with regard to overshoot control, stability time, and anti-interference, which effectively improves the control precision and robustness of the pneumatic control system for the self-elevating mat. Finally, the effectiveness of the simulation results is verified via experiments.

Author Contributions: Conceptualization, J.C., H.S. and W.X.; methodology, Y.L.; software, Q.S.; validation, Q.S., Y.L. and S.Y.; formal analysis, Q.S.; investigation, J.C.; resources, W.X.; data curation, H.S.; writing—original draft preparation, Y.L. and Q.S.; writing—review and editing, J.C.; visualization, Q.S.; supervision, W.X.; project administration, S.Y.; funding acquisition, H.S. All authors have read and agreed to the published version of the manuscript.

Funding: This research was funded by the project of Ministry of Industry and Information Technology of the People's Republic of China (grant no. CJ11N20), the Strategic Research and Consulting project of Chinese Academy of Engineering (grant no. 2022-DFZD-36), and the Fundamental Research Funds for the Central Universities (grant no. 22CX01003A-1).

Institutional Review Board Statement: Not applicable.

Informed Consent Statement: Not applicable.

Data Availability Statement: The data presented in this study are available upon request from the corresponding author.

Conflicts of Interest: Author H.S. and S.Y. were employed by the company Shanghai Zhenhua Heavy Industry Co., Ltd. The remaining authors declare that the research was conducted in the absence of any commercial or financial re-lationships that could be construed as a potential conflict of interest. The authors declare that this study received funding from Ministry of Industry and Information Technology of the People's Republic of China, Chinese Academy of Engineering and the Fundamental Research Funds for the Central Universities. The funder was not involved in the study design, col-lection, analysis, interpretation of data, the writing of this article or the decision to submit it for publication.

References

1. Breton, S.P.; Moe, G. Status, plans and technologies for offshore wind turbines in europe and north america. *Renew. Energ.* **2009**, *34*, 646–654. [CrossRef]
2. Feng, Y.; Lin, H.; Ho, S.L.; Yanc, J.; Dong, J.; Fang, S. Overview of wind power generation in china: Status and development. *Renew. Sust. Energ. Rev.* **2015**, *50*, 847–858. [CrossRef]
3. He, B.; Wu, J.; Xiao, J. Product safety risk assessment approach to sustainable design. *Int. J. Precis. Eng. Manuf.-GT* **2022**, *10*, 783–788. [CrossRef]
4. Fenu, B.; Attanasio, V.; Casalone, P.; Novo, R.; Mattiazzo, G. Analysis of a gyroscopic-stabilized floating offshore hybrid wind-wave platform. *J. Mar. Sci. Eng.* **2020**, *8*, 439. [CrossRef]
5. Dymarski, P.; Dymarski, C.; Ciba, E. Stability analysis of the floating offshore wind turbine support structure of cell spar type during its installation. *Pol. Marit. Res.* **2019**, *26*, 109–116. [CrossRef]
6. Li, J.L.; Xie, Y.H.; Wu, W.G.; Zhang, C. Analysis of the dynamic response of offshore floating wind power platforms in waves. *Pol. Marit. Res.* **2020**, *27*, 17–25.
7. Zhou, B.; Zhang, Z.; Li, G.; Yang, D.; Santos, M. Review of key technologies for offshore floating wind power generation. *Energies* **2023**, *16*, 710. [CrossRef]
8. Ghigo, A.; Cottura, L.; Caradonna, R.; Bracco, G.; Mattiazzo, G. Platform optimization and cost analysis in a floating offshore wind farm. *J. Mar. Sci. Eng.* **2020**, *8*, 835. [CrossRef]
9. Zambrana, P.; Fernandez-Quijano, J.; Fernandez-Lozano, J.J.; Rubio, P.M.M.; Garcia-Cerezo, A.J. Improving the performance of controllers for wind turbines on semi-submersible offshore platforms: Fuzzy supervisor control. *Energies* **2021**, *14*, 6222. [CrossRef]
10. Takagi, H.; Tomiyasu, R.; Oyake, T.; Araki, T.; Takata, Y. Tsunami intrusion through port breakwaters enclosed with self-elevating seawalls. *Ocean Eng.* **2020**, *199*, 107028. [CrossRef]
11. Feng, L.L. Control of Underwater Cable-Driven Suspension Lift Platform. Master's Thesis, Huazhong University of Science and Technology, Wuhan, China, 2021.
12. Hu, H.G.; Chen, Y.H.; Yu, F.; Liu, Q.K.; Wang, B. Design of quasi-real-time communication underwater winch buoy control system. *Mar. Sci.* **2022**, *46*, 131–142.
13. Wang, X.L.; Zhu, W.; Zhou, Y.A. Research on selecting the automatic control modes for the vacuum control. *Vacuum* **1993**, *3*, 1–4.
14. Jin, C.; Zhang, Y.; Ling, W.; Yang, X.; Cheng, J.; Feng, L.; Wan, J.; Wang, H.; Lu, Y.; Li, L.; et al. Design of the vacuum-centralized control system for the space plasma environment research facility based on the experimental physics and industrial control system. *J. Vac. Sci. Technol. B* **2022**, *40*, 1–5. [CrossRef]

15. Li, X.M.; Huang, H.M.; Wen, S.H.; Yuan, Y. Design of a new multi-channel scalable air pressure controller. *J. Yanshan Univ.* **2022**, *46*, 327–335.
16. Yang, G.; Jiang, P.; Lei, L.; Wu, Y.; Li, B. Adaptive backstepping control of vacuum servo system using high-speed on-off valves. *IEEE Access* **2020**, *8*, 129799–129812. [CrossRef]
17. Esleman, E.A.; Önal, G.; Kalyoncu, M. Optimal pid and fuzzy logic based position controller design of an overhead crane using the bees algorithm. *SN Appl. Sci.* **2021**, *3*, 1–13. [CrossRef]
18. Feng, Y.; Wu, M.; Chen, X.; Chen, L.; Du, S. A fuzzy pid controller with nonlinear compensation term for mold level of continuous casting process. *Inf. Sci.* **2020**, *539*, 487–503. [CrossRef]
19. Zaheeruddinsingh, K. Intelligent frequency control in microgrid: Fractional order fuzzy pid controller. *J. Intell. Fuzzy Syst.* **2020**, *39*, 195–212. [CrossRef]
20. Dinh, T.Q.; Marco, J.; Yoon, J.I.; Ahn, K.K. Robust predictive tracking control for a class of nonlinear systems. *Mechatronics* **2018**, *52*, 135–149. [CrossRef]
21. Han, H.G.; Wu, X.L.; Liu, Z.; Qiao, J.F. Design of self-organizing intelligent controller using fuzzy neural network. *IEEE Trans. Fuzzy Syst.* **2018**, *26*, 3037–3111. [CrossRef]
22. Liang, H.B.; Zou, J.L.; Zuo, K.; Muhammad, J.K. An improved genetic algorithm optimization fuzzy controller applied to the wellhead back pressure control system. *Mech. Syst. Signal Process.* **2020**, *142*, 106708. [CrossRef]
23. Na, J.; Li, Y.; Huang, Y.; Gao, G.; Chen, Q. Output feedback control of uncertain hydraulic servo systems. *IEEE Trans. Ind. Electron.* **2019**, *67*, 490–500. [CrossRef]
24. Zhang, R.M.; Zeng, D.Q.; Ju, H.P.; Lam, H.K.; Xie, X.P. Fuzzy sampled-data control for synchronization of t-s fuzzy reaction-diffusion neural networks with additive time-varying delays. *IEEE Trans. Cybern.* **2021**, *51*, 2384–2397. [CrossRef] [PubMed]
25. Shi, Q.; Lam, H.K.; Xuan, C.B.; Chen, M. Adaptive neuro-fuzzy pid controller based on twin delayed deep deterministic policy gradient algorithm. *Neurocomputing* **2020**, *402*, 183–194. [CrossRef]
26. Fang, Y.K.; Cheng, C.Y.; Dong, Z.; Min, H.G.; Zhao, X.M. A fault diagnosis framework for autonomous vehicles based on hybrid data analysis methods combined with fuzzy PID control. In Proceedings of the International Conference on Unmanned Systems, Harbin, China, 27–28 November 2020.
27. Liu, J.W.; Tao, X.; Ma, X.Y.; Feng, K.G.; Chen, J.M. Fuzzy controllers with neural network predictor for second-order linear systems with time delay. *IEEE Access* **2020**, *8*, 206049–206062. [CrossRef]
28. Lei, Y.; Wu, Y.Q.; Yuan, Y.Q.; Chen, C. Intelligent neural network controller optimization and simulation using GA. In Proceedings of the International Conference on Intelligent Transportation, Big Data & Smart City (ICITBS), Changsha, China, 12–13 January 2019.
29. Huang, C.; Fan, J.M.; Li, W.C.; Chen, X.; Zhu, Q. ReachNN: Reachability analysis of neural-network controlled systems. In Proceedings of the 18th International Symposium on Automated Technology for Verification and Analysis (ATVA'20), Virtual Conference, 19–23 October 2020.
30. Moura, J.P.; Fonseca Neto, J.V.; Rego, P.H.M. A neuro-fuzzy model for online optimal tuning of pid controllers in industrial system applications to the mining sector. *IEEE Trans. Fuzzy Syst.* **2019**, *28*, 1864–1877. [CrossRef]
31. Park, D.; Le, T.L.; Quynh, N.V.; Long, N.K.; Hong, S.K. Online tuning of pid controller using a multilayer fuzzy neural network design for quadcopter attitude tracking control. *Front. Neurorobotics* **2021**, *14*, 619350. [CrossRef]
32. Huang, Y. Research on Piezoeiectric Pilot Pneumatic Valve Position Proportional Control System. Master's Thesis, Shanghai Jiao Tong University, Shanghai, China, 2017.
33. Wang, D.Q.; Zhang, P. Study on unsteady gas leakage model of pressure vessel. *Chin. J. Saf. Sci.* **2012**, *22*, 154–158.
34. Chen, Y.C. Research on Liquid Pressure System of Low Pressure Cantilever Casting Machine. Master's Thesis, Inner Mongolia University of Technology, Inner Mongolia, China, 2019.
35. Nguyen, H.T.; Nguyen, M.T.; Nguyen, V.D.H.; Doan, T.T.; Vo, C.P. Designing PID-Fuzzy controller for pendubot system. *Robot. Manag.* **2017**, *22*, 21–27.
36. Muravyova, E.A.; Stolpovskaya, Y.V. Process control system based on a fuzzy controller. *J. Phys. Conf. Ser.* **2020**, *1515*, 1–7. [CrossRef]

MDPI AG
Grosspeteranlage 5
4052 Basel
Switzerland
Tel.: +41 61 683 77 34

MDPI Books Editorial Office
E-mail: books@mdpi.com
www.mdpi.com/books

www.ingramcontent.com/pod-product-compliance
Lightning Source LLC
Chambersburg PA
CBHW041935260926
51850CB00018B/193